Numerical Polynomial Algebra

Hans J. Stetter

Institute for Applied and Numerical Mathematics
Vienna University of Technology
Vienna, Austria

Society for Industrial and Applied Mathematics
Philadelphia

Library of Congress Cataloging-in-Publication Data

Stetter, Hans J., 1930-
    Numerical polynomial algebra / Hans J. Stetter.
        p. cm.
    Includes bibliographical references and index.
    ISBN 0-89871-557-1 (pbk.)
      1. Polynomials. 2. Numerical analysis. I. Title.

QA161.P59S74 2004
  512.9'422—dc22
                                            2004041691

About the cover: The cover art shows the discretized image of the variety of a pseudofactorizable polynomial in three variables; cf. Example 7.13 and Figure 7.5 for the varieties of the pseudofactors.

Contents

Preface

"Numerical Polynomial Algebra" is not a standard designation of a mathematical discipline; therefore, I should start by explaining the title of this book. Historically, in the growth of computational mathematics, which occurred in parallel with the breath-taking explosion in the performance of computational machinery, all areas of mathematics which play a role in the modelling and analysis of real world phenomena developed their branch of Numerical Analysis: Linear Algebra, Differential Equations, Approximation, Optimization, etc. The collective term Numerical *Analysis* turned out to be appropriate: The fact that data and relations from the real world inevitably have a *limited accuracy* make it necessary to embed the computational tasks into metric spaces: Few parts of computational scientific computing can proceed without approximations and without the analytic background (like norms, for example) to deal with the inherent indeterminations. Numerical Linear Algebra is the best-known example: It originated from an embedding of the constructive parts of classical linear algebra into linear functional analysis, and its growth into one of the supporting pillars of scientific computing was driven by the use of *analytic tools* like mappings, norms, convergent iteration, etc. Empirical data could easily be fitted into this conceptual frame so that the approximate solution of approximate linear problems with approximate data could be conceived and implemented.

One area of mathematics did not follow that trend: classical nonlinear algebra. It had undergone a remarkable algorithmic development in the late 19th century; then the axiomatic age had turned it into an abstract discipline. When the symbol manipulation capabilities of electronic computers became evident, a faction of algebraists remembered the algorithmic aspects of their field and developed them into "Computer Algebra," as a computational tool for the solution of constructive problems in *pure mathematics*. They have designed and implemented algorithms which delight the algebraic community; but at the same time, this enterprise has somehow prevented the growth of a numerical nonlinear algebra. The inadequacy of this mathematically interesting project for realistic problems is exposed when the solution of a system of linear equations with numerical coefficients is obtained in the form of fractions of integers with hundreds of digits.

But nonlinear algebraic tasks do exist in scientific computing: Multivariate polynomials are a natural modelling tool. This creates multivariate systems of polynomial equations, multivariate interpolation problems, decomposition problems (factorization) etc.; the modelling of nontrivial geometric constellations alone generates a multitude of nonlinear algebraic problems. These computational tasks from the real world possess (some) data with limited accuracy and there are no exact solutions; thus, they are generally not accessible by the sophisticated exact tools which Computer Algebra has provided. At the same time, they often require a *global structural* analysis of the situation and cannot satisfactorily be solved with general-purpose tools of Numerical Analysis. (The computation of the zeros of *one* univariate polynomial became an exception: Here, algebra and numerical analysis joined ranks to develop efficient and reliable black-box software for the—necessarily approximate—solution of this task.)

Thus, in the late 20th century, a no man's land between computer algebra and numerical analysis had remained on the landscape of scientific computing which invited discovery and cultivation for general usage. But—most surprisingly—this challenge of pioneering a "numerical nonlinear algebra" remained practically unnoticed by the many young mathematicians hungry for success, even by those working in the immediate neighborhood of the glaring white spot. When I accepted that challenge more than 10 years ago and tried to recruit help for my expeditions, my soliciting was met with little resonance. On these expeditions, I have met stimulating mathematical adventures all along the way and interesting unsolved problems wherever I proceeded. Many of these problems are still waiting for their efficient solution.

From the beginning, in stepping into this virgin territory, I found it more important to set up directions and road posts than to investigate and plot small areas meticulously. I believe that I have now gained an overview of large parts of that territory and I wish to communicate my findings in printed form, beyond my many lectures at conferences and seminars over the past years. This has been the motive for writing this book. The more restrictive title "Numerical *Polynomial* Algebra" (instead of the original "Numerical Nonlinear Algebra") expresses the fact that there remain interesting and computationally important areas in nonlinear algebra which I have not even touched.

A number of principles have guided the composition of this text:

The most prominent one is *continuity*: Throughout, all data are from $\mathbb{C}$ or $\mathbb{R}$ so that all quantities and relations are automatically embedded into analysis, as in Numerical Linear Algebra. Derivatives of maps are widely used, not just formally but also quantitatively. This permits an analysis of the sensitivity of results to small changes in the data of a problem ("condition"). Continuity is the indispensable basis for the use of floating-point computation—or any other approximate computation. Concepts which are inherently discontinuous (like g.c.d., radical, etc.) must be reinterpreted or abandoned.

Continuitiy is also a prerequisite for the consideration of *data with limited accuracy* which we systematically assume throughout the text, with a concept of families of neighborhoods as a formal basis. Correctness of a result is replaced by its *validity*, conceived as a continuous property represented by a numerical value not as a discrete property (yes-no): A result is valid if it is the exact result of nearby data, which is established by a backward error analysis. For multi-component quantities, we use weighted maximum norms throughout; but a weighted 2-norm would do just as well.

The interpretation of algebraic relations as continuous maps permits the systematic use of *iterative refinement* as an algorithmic tool. Crude initial results may be refined into sufficiently valid ones by the use of *local linearization*, a standard tool throughout analysis.

Within polynomial algebra proper, I have tried to employ the *quotient ring aspect* of ideals wherever possible. The vector space structure of quotient rings and the linear mapping structure of multiplication permit an ample use of concepts and algorithms from (numerical) linear algebra. The determination of all zeros of a polynomial system from the eigenvectors of the multiplication matrices of the associated quotient ring is the most prominent example.

I have widely used standard *linear algebra notations*. The systematic use of row vectors for coefficients and of column vectors for bases has proved very helpful; within monomial basis vectors, components are always arranged by increasing degree or term order. These conventions may lead to linear systems $b^T A = c^T$ for row vectors and elimination from right to left, which is somewhat nonstandard, but the internal consistency of this notational principle has been an ample reward.

Another guiding principle has been to write a *textbook* rather than a monograph. For a novel area of practical importance—which numerical polynomial algebra is in many ways—it is crucial that students are given the opportunity to absorb its principles. I hope that this book may be used as a text for relevant courses in Mathematics and Computer Science and to help students get acquainted with the numerical solution of quantitative problems in commutative algebra. I have included "Exercises" with all sections of the book; as usual, they are meant to challenge the reader's understanding by confronting him/her with numerical and theoretical problems. Also, most of the numerical examples in the text are not only demonstrations for the relevance of formal results but an invitation for a replication of the indicated computation. The textbook approach has also kept me from including references to technical papers within the text. Instead, I have added "Historical and Bibliographical Notes" at the end of each chapter which put the material into perspective and point to contributors of its development.

The dual nature of the subject area as a part of numerical analysis as well as of polynomial algebra requires that the text be attractive and readable for students and scientists from both fields. As I know from my own experience, a standard numerical analyst knows few concepts and results from commutative algebra, and a standard algebraist has a natural aversion to approximate data and approximate computation which appear as foreign elements in his/her world. Therefore, I have seen it necessary to include low level introductory sections on matters of numerical analysis as well as of polynomial algebra, and I have tried to refrain from highly technical language in either subject area. Thus, a reader well versed in one of the areas must find some passages trivial or naive, but I consider this less harmful than assuming a technical knowledge which part of the intended readership does not possess.

Beyond students and colleagues from numerical analysis and computer algebra, the intended readership comprises experts from various areas in scientific computing. Polynomial algebra provides specialized and effective tools for many of their tasks, in particular for tasks with strong geometric aspects. They may be interested to see how many nontrivial algebraic problems can be solved efficiently in a meaningful way for data with limited accuracy. Altogether, I hope that this book may arouse general interest in a neglected area of computational mathematics where—for a while at least—interesting research projects abound and publishable results lurk behind every corner. This should make the area particularly attractive for scientists in the beginning phases of their careers.

For me personally, my encounter with numerical polynomial algebra has become a crucial event in my scientific life. It happened at a time when, with my advancing age, my interest in mathematical research had begun to decrease. In particular, I had lost interest in highly technical investigations as they are indispensable in any advanced scientific field. At that point, through some coincidences, I became aware of the fact that many fundamental aspects of the numerical treatment of nonlinear algebraic problems had hardly been touched. In my 60s, I began to learn the basics of commutative algebra and to apply my lifelong experience in numerical analysis to it—and my fascination grew with every new insight. This late love affair of my scientific life has gained me 10 or more years of intense intellectual activity for which I can only be grateful. The result is this book, and—like a late lover—I must ask forgiveness for some foolish ideas in it which may irritate my younger and more meticulous colleagues.

This book also marks the end of my active scientific research. I have decided that I will devote my few or many remaining years to other activities which I have delayed long enough. If my mind should, for a short while, continue to tempt me with mathematical ideas and problems, I will simply put them on my homepage for others to exploit. I also owe it to my dear wife Christine who has so often patiently acknowledged the priority of science in our 44 years of married life that this state does not continue to the very end. Without her continuing love and support, this last mark of my scientific life would not have come into existence.

Vienna, July 2003 Hans J. Stetter

Acknowledgments

Fifteen years ago, when I began to get interested in the numerical solving of polynomial systems, my knowledge of commutative algebra was nil. I could not have gained even modest insight into polynomial algebra represented in this book without the advice and help of many colleagues much more knowledgeable in the area; implicitly or explicitly, they have contributed a great deal to my work. I wish to express my gratitude to

my early road companion H. M. Moeller;

B. Buchberger, H. Hong, J. Schicho, F. Winkler at RISC;

my friends at ORCCA, R. Corless, K. Geddes, M. Giesbrecht, D. Jeffrey, I. Kotsireas, G. Labahn, G. Reid, S. Watt, and a number of people at Waterloo Maple Inc.;

my friends and collaborators in China, Huang Y. Zh., Wu W. D., Wu W. Ts., Zhi L. H.;

in the USA, B. Caviness, G. Collins, D. Cox, G. Hoffmann, E. Kaltofen, Y. N. Lakshman, T. Y. Li, D. Manocha, V. Pan, S. Steinberg, M. Sweedler, B. Trager, J. Verschelde;

in Japan, H. Kobayashi, M. T. Noda, T. Sasaki, K. Shirayanagi;

in France, J.-Ch. Faugère, I. Emiris (now back in Greece), D. Lazard, B. Mourrain, M.-F. Roy;

in Italy, D. Bini, P. M. Gianni, M. G. Marinari, T. Mora, L. Robbiano, C. Traverso;

in Spain, L. Gonzáles-Vega, T. Recio;

in Germany, J. Apel, J. Calmet, K. Gatermann, J.v.z. Gathen, T. Sauer, F. Schwarz, W. Seiler, V. Weispfenning;

in Russia, V. Gerdt;

my students J. Haunschmied, V. Hribernig, A. Kondratyev, G. Thallinger;

and the numerous other colleagues all over the world who have discussed matters of polynomial algebra with me on various occasions.

Part I

Polynomials and Numerical Analysis

Chapter 1

Polynomials

In their use as modelling tools in Scientific Computing, polynomials appear, at first, simply as a *special class of functions* from the $\mathbb{C}^s$ to $\mathbb{C}$ (or $\mathbb{R}^s$ to $\mathbb{R}$). Such polynomials are automatically objects of univariate ($s = 1$) or multivariate ($s > 1$) *analysis* over the complex or real numbers. For *linear* polynomials, this fact has played virtually no role in classical linear algebra; but it has become a fundamental aspect of today's *numerical linear algebra* where concepts from analysis (norms, neighborhoods, convergence, etc.) and related results are widely used in the design and analysis of computational algorithms. In an analogous manner, the consideration of polynomial algebra as a part of analysis plays a fundamental role in *numerical polynomial algebra*; it will be widely used throughout this book. In particular, this embedding of algebra into analysis permits the extension of algebraic algorithms to polynomials with coefficients of limited accuracy; cf. Chapter 3.

On the other hand, certain sets of polynomials have special algebraic structures: they may be linear spaces, rings, ideals, etc. Algebraic properties related to these structures may play a crucial role in solving computational tasks involving polynomials, e.g., for finding zeros of polynomial systems; cf. Chapter 2.

In this introductory chapter, we consider various aspects of polynomials which will play a fundamental role in our later investigations.

The following notations will generally be used (but without strict adherence):

scalars and coefficients $\in \mathbb{C}$ or $\mathbb{R}$:	lower-case Greek letters $\alpha, \beta, \gamma, \ldots$
elements (points) $\in \mathbb{C}^s$ or $\mathbb{R}^s$:	lower-case Greek letters $\xi, \eta, \zeta, \ldots$
vectors of coefficients etc. :	lower-case Latin letters $a, b, c, \ldots$
s-dim. variables (indeterminates) :	lower-case Latin letters $x, y, z, \ldots$
polynomials :	lower-case Latin letters $p, q, \ldots$
systems of polynomials :	upper-case Latin letters $P, Q, \ldots$

1.1   Linear Spaces of Polynomials

Definition 1.1. A *monomial* in the s variables $x_1, \ldots, x_s$ is the power product

$$x^j := x_1^{j_1} \ldots x_s^{j_s}, \quad \text{with} \quad j = (j_1, \ldots, j_s) \in \mathbb{N}_0^s; \tag{1.1}$$

j is the *exponent* and $|j| := \sum_{\sigma=1}^s j_\sigma$ the *degree* of the monomial x^j. The set of all monomials in s variables will be denoted by T^s, independently of the notation for the variables. $T_d^s \subset T^s$ is the set of monomials in s variables of degree $\leq d$. $\square$

For example, $x^2 y^3 z$ is a monomial of degree 6 in T^3 and hence contained in T_d^3 for $d \geq 6$. Note that each monomial set T_d^s contains the monomial $1 = x^0 = x^{(0,\ldots,0)}$.

Proposition 1.1. T_d^s contains $\binom{d+s}{d}$ monomials; $\binom{d+s-1}{d}$ of these have exact degree d.

Proof: The proposition follows from fundamental formulas in combinatorics. $\square$

Obviously, the number of different monomials grows rapidly with the number of variables s and the degree d. For example, there are 126 monomials of degree ≤ 5 in 4 variables, and 3003 monomials of degree ≤ 8 in 6 variables. This rapid growth is a major reason for the high computational complexity of many polynomial algorithms.

Definition 1.2. A *complex (real) polynomial*[1] in s variables is a finite linear combination of monomials from T^s with coefficients from $\mathbb{C}$ or $\mathbb{R}$, resp.:

$$p(x) = p(x_1, \ldots, x_s) = \sum_{(j_1, \ldots, j_s) \in J} \alpha_{j_1 \ldots j_s} x_1^{j_1} \ldots x_s^{j_s} = \sum_{j \in J} \alpha_j x^j. \tag{1.2}$$

The set $J \subset \mathbb{N}_0^s$ which contains the exponents of those monomials which are present in the polynomial p (i.e. which have a nonvanishing coefficient) is the *support* of p; $\deg(p) := \max_{j \in J} |j|$ is the *(total) degree* of p. The summands of a polynomial are called *terms*. The exponent and the degree of a term are those of the associated monomial. $\square$

Definition 1.3. A polynomial p with a support J such that $|j| = \deg(p)$ for each $j \in J$ is called *homogeneous*. $\square$

The following is a polynomial of total degree 4 in the 3 variables x, y, z:

$$4 x^2 y^2 - 7 x z^3 + 2 z^4 + 3.5 x^3 - y^2 z - 8.5 x z - 10.$$

The terms of degree 4 form a homogeneous polynomial of total degree 4:

$$4 x^2 y^2 - 7 x z^3 + 2 z^4.$$

Definition 1.4. The set of all complex (real) polynomials in s variables will be denoted by $\mathcal{P}_{\mathbb{C}}^s$ of $\mathcal{P}_{\mathbb{R}}^s$ resp., independently of the notation for the variables. When the coefficient domain is evident, the notation $\mathcal{P}^s$ will be used. $\mathcal{P}_d^s \subset \mathcal{P}^s$ will denote the set of polynomials in s variables of total degree $\leq d$. $\square$

Obviously, $\mathcal{P}_d^s$ is a *linear space (vector space)* over $\mathbb{C}$ or $\mathbb{R}$, resp., of dimension $\binom{d+s}{d}$ (cf. Proposition 1.1); addition and multiplication by a scalar are defined in the natural way. A

---

[1] Throughout this book, only such polynomials are considered; cf. the preface.

generic basis in $\mathcal{P}_d^s$ is furnished by the monomials of $\mathcal{T}_d^s$ arranged in some linear order. With respect to such a basis, the coefficients of a polynomial p are the *components* of p as an element of the linear space, i.e. p is represented by the vector of its coefficients (ordered appropriately). The zero element 0 is the *zero polynomial* with $\alpha_j = 0$ for all j. Vector space computations in $\mathcal{P}_d^s$ (i.e. addition and multiplication by a scalar) are thus reduced to the analogous computations with the coefficient vectors, as in any linear space.

However, polynomials may also be *multiplied*—and multiplication generally results in a polynomial of higher degree which is outside the linear space of the factor polynomials; cf. section 1.3. Therefore, the vector space notation for polynomials can only be used within specified contexts. On the other hand, because of its simplicity, it *should* be used in computations with polynomials wherever it is feasible.

Another reason for a potential inadequacy of vector space notation in dealing with polynomials is the fact that the cardinality $|J|$ of the support J of a polynomial may be very small relative to the magnitude of the associated basis $\mathcal{T}_d^s$ so that almost all components are 0. Such polynomials are called *sparse* in analogy to the use of this word in linear algebra. Multivariate polynomials which appear in scientific computing are generally sparse.

Fortunately, we will often have to deal with linear spaces $\mathcal{R}$ of polynomials from some $\mathcal{P}^s$ with a fixed *uniform support* J so that a fixed monomial basis $\{x^j, j \in J\}$ can be used. Moreover, in these spaces $\mathcal{R}$, multiplication of the element polynomials is defined in a way that it does not lead out of $\mathcal{R}$; hence in spite of their fixed dimensions $|J|$, they are commutative rings, so-called *quotient rings*. We will formally introduce and discuss these objects in section 2.2 and later use them a great deal.

In numerical polynomial algebra, a good deal of the algorithmic manipulations of polynomials are *linear operations*; in this context, we will widely employ the standard notations of numerical linear algebra. To facilitate this practice, we will generally collect the *coefficients* of a polynomial into a *row vector* $a^T = (\ldots \alpha_j \ldots)$ and its *monomials* into a *column vector* $\mathbf{x} = (\ldots x^j \ldots)^T$. Then

$$p(x) = \sum_{j \in J} \alpha_j x^j = (\ldots\ \alpha_j\ \ldots) \begin{pmatrix} \vdots \\ x^j \\ \vdots \end{pmatrix} =: a^T \mathbf{x}. \tag{1.3}$$

The use of *row* vectors for coefficients agrees with the common notation $a^T \mathbf{x}$ for linear polynomials in linear algebra; therefore it is the natural choice. It implies, however, that a linear system for the computation of a coefficient vector a^T appears in the form $a^T A = b^T$. For the sake of a systematic notation (which greatly assists the human intuitive and associative powers), we will not transpose such systems during formal manipulations.

Example 1.1: The monomial vector for polynomials from $\mathcal{P}_d^1$ (univariate polynomials of maximal degree d) is $\mathbf{x} := (1, x, \ldots, x^d)^T$. A shift of the origin to $\xi \in \mathbb{R}$ requires a rewriting to the

reflection at the y-axis, reflection at the origin, etc. For $\mathcal{N} = \{1, x, y, y^2\}$, we have

$$s_{\mathcal{N}}(z_1, z_2, z_3, z_4) \;=\; \det \begin{pmatrix} 1 & 1 & 1 & 1 \\ \sqrt{\tfrac{3}{2}} & -\sqrt{\tfrac{3}{2}} & -\sqrt{\tfrac{3}{2}} & \sqrt{\tfrac{3}{2}} \\ \sqrt{\tfrac{1}{2}} & \sqrt{\tfrac{1}{2}} & -\sqrt{\tfrac{1}{2}} & -\sqrt{\tfrac{1}{2}} \\ \tfrac{1}{2} & \tfrac{1}{2} & \tfrac{1}{2} & \tfrac{1}{2} \end{pmatrix} \;=\; 0,$$

while $\mathcal{N} = \{1, x, y, xy\}$ yields

$$s_{\mathcal{N}}(z_1, z_2, z_3, z_4) \;=\; \det \begin{pmatrix} 1 & 1 & 1 & 1 \\ \sqrt{\tfrac{3}{2}} & -\sqrt{\tfrac{3}{2}} & -\sqrt{\tfrac{3}{2}} & \sqrt{\tfrac{3}{2}} \\ \sqrt{\tfrac{1}{2}} & \sqrt{\tfrac{1}{2}} & -\sqrt{\tfrac{1}{2}} & -\sqrt{\tfrac{1}{2}} \\ \tfrac{\sqrt{3}}{2} & -\tfrac{\sqrt{3}}{2} & \tfrac{\sqrt{3}}{2} & -\tfrac{\sqrt{3}}{2} \end{pmatrix} \;=\; -12.$$

The zeros of the system P above form a square and also have many symmetries; but, in terms of x, y, they are concealed by the rotation of the principal axes of the two conics. $\square$

For ideals with multiple zeros, the missing elements $\partial_{0\ldots0}[z_\mu]$ in the basis $\mathbf{c}_0^T$ of $\mathcal{D}[\mathcal{I}]$ may be replaced with appropriate derivative evaluations at the multiple zeros, cf. sections 2.3.2 and 2.4. We will see in sections 6.3 and 9.3 how m_μ-fold zeros with a specified derivative structure may be interpreted as limiting constellations of m_μ individual zeros. In this sense, the multiple zeros fill the gaps which would otherwise remain in the $\mathbb{C}^{ms}$ of the components of sets of m disjoint zeros.

Each $p \in \mathcal{P}^s$ belongs to a unique residue class $[p]_\mathcal{I}$ mod $\mathcal{I}$ and each residue class has a unique representative in the quotient ring $\mathcal{R}[\mathcal{I}]$; cf. section 2.2.1.

Definition 2.20. For $p \in \mathcal{P}^s$, the *normal form* $NF_\mathcal{I}[p]$ is the unique polynomial in $\mathcal{R}[\mathcal{I}]$ which satisfies

$$p - \mathrm{NF}_\mathcal{I}[p] \in \mathcal{I}. \tag{2.55}$$

With a specified basis vector $\mathbf{b}$ of $\mathcal{R}[\mathcal{I}]$ and the conjugate basis $\mathbf{c}^T$ of $\mathcal{D}[\mathcal{I}]$, Proposition 2.13 implies

$$\mathrm{NF}_\mathcal{I}[p] \;=\; \mathbf{c}^T(p)\,\mathbf{b}(x) \;=\; \sum_{\mu=1}^{m} c_\mu(p)\,b_\mu(x). \tag{2.56}$$

2.5.2   Border Bases of Polynomial Ideals

In order to complete step (i) of the procedure described below Theorem 2.27, we must also find the multiplication matrices A_σ of $\mathcal{R}[\mathcal{I}]$ with respect to the basis $\mathbf{b}$.

Definition 2.21. For a specified closed set $\mathcal{N} \subset \mathcal{T}^s$, we define the following sets in $\mathcal{T}^s$:

(i) The *corner set*

$$C[\mathcal{N}] := \{ x^j \in \mathcal{T}^s \; : \; x^j \notin \mathcal{N}, \text{ but } \textit{each} \text{ negative neighbor of } x^j \in \mathcal{N} \}; \tag{2.57}$$

(ii) the *border set*

$$B[\mathcal{N}] := \begin{cases} \{ x^j \in \mathcal{T}^s \; : \; x^j \notin \mathcal{N}, \text{ but } \textit{some} \text{ negative neighbor of } x^j \in \mathcal{N} \}, \\ \{ x^j \in \mathcal{T}^s \; : \; x^j \notin \mathcal{N}, \; x^j \text{ a positive neighbor of some } x^{j'} \in \mathcal{N} \}; \end{cases} \tag{2.58}$$

motionless. One will automatically assume that several measurements of the water level, at the same location and within a short time interval, will produce deviating values; most of these may lie between 5.15 and 5.17, but one will not be surprised also to obtain a value like 5.19. In this case, the water level would be adequately described by the empirical quantity (5.16 m, 1 cm). □

When we consider the indetermination introduced into an algebraic problem by the indetermination of its empirical data, the *intrinsic* data are considered as *fixed*. Therefore, it will generally be convenient to consider their specification as a *part of the specification of the algebraic problem* and reserve the term "data" to the empirical data of the problem.

Formally, with an empirical data quantity we associate a *family of neighborhoods*, parametrized by a positive real parameter δ in the following way:

Definition 3.2. The empirical quantity $(\bar{\alpha}, \varepsilon)$, with the *specified value* $\bar{\alpha} \in \mathbb{C}$ or $\mathbb{R}$ and the *tolerance* $\varepsilon > 0$, defines the *family of neighborhoods*

$$N_\delta(\bar{\alpha}, \varepsilon) := \{\tilde{\alpha} : |\tilde{\alpha} - \bar{\alpha}| \le \delta\,\varepsilon\}, \quad \delta \in \mathbb{R}_0; \tag{3.2}$$

for *real* $\bar{\alpha}$, it must be specified (or clear from the context) whether the $\tilde{\alpha}$ in $N_\delta(\bar{\alpha}, \varepsilon)$ may be in $\mathbb{C}$ or whether they are restricted to the real line. □

Naturally, one could replace the word "neighborhood" in the above by the word "interval." However, we believe that the term neighborhood conveys more of the *vagueness* which is inherent in empirical data: Beyond a certain accuracy, their value is simply not defined, and the boundary is not sharp (as it is in an interval) but *blurred*; this is why one must consider a *family* of such neighborhoods rather than one interval. Perhaps the notion of a "data cloud" is best suited to describe that situation.

Obviously, our concept of empirical data is closely related to the concept of "fuzzy data" which plays a useful role in various engineering applications; cf. the related literature. In a simplified view, the introduction of a probability distribution for δ would turn our empirical data items into fuzzy data items. We have abstained from using this approach for two reasons:

- probabilities are foreign to the algebraic world and they would have introduced an awkward formalism;

- in most of the real-world applications, the choice of a probability distribution for an empirical quantity cannot be based on substantial information.

Rather than dealing with formal probabilities, we associate with the parameter δ in (3.2) a *validity scale*: The values $\tilde{\alpha} \in N_\delta(\bar{\alpha}, \varepsilon)$, with $\delta = O(1)$, are considered as *valid instances* of the empirical quantity $(\bar{\alpha}, \varepsilon)$. (Here and throughout the book, we abuse the symbol $O(1)$ to denote real numbers "of the order 1," without any relation to a limit process!) Generally, one may visualize the relation between values of the parameter δ and an intuitive concept of validity by the scale

$$\delta : \quad 0 \quad \dots \quad 1 \quad \dots \quad 3 \quad \dots \quad 10 \quad \dots \quad 30 \quad \dots$$

$$\text{valid} \qquad \begin{array}{c}\text{probably}\\\text{valid}\end{array} \qquad \begin{array}{c}\text{possibly}\\\text{valid}\end{array} \qquad \begin{array}{c}\text{probably}\\\text{invalid}\end{array} \qquad \text{invalid} \tag{3.3}$$

In a particular context, a more precise adjustment of the validity scale to values of δ may be possible for the experts who have modelled the situation. In any case, while the validity of $\tilde{a}$

basis vector

$$\mathbf{x} - \xi \;=\; \begin{pmatrix} 1 \\ x - \xi \\ (x-\xi)^2 \\ \cdots \\ (x-\xi)^d \end{pmatrix} \;=\; \begin{pmatrix} 1 & & & & 0 \\ -\xi & 1 & & & \\ \xi^2 & -2\xi & 1 & & \\ \vdots & & & \ddots & \ddots & \\ \pm\xi^d & \cdots & & \cdots & -d\xi & 1 \end{pmatrix} \begin{pmatrix} 1 \\ x \\ x^2 \\ \vdots \\ x^d \end{pmatrix} \;=:\; \Xi\,\mathbf{x}\,;$$

the corresponding rewriting of a polynomial p is simply achieved:

$$p(x) \;=\; a^T \mathbf{x} \;=\; a^T \Xi^{-1}\, \Xi\, \mathbf{x} \;=\; \bar{a}^T (\mathbf{x} - \xi) \;=\; \sum_{j=0}^{d} \bar{\alpha}_j \, (x - \xi)^j. \quad \square$$

Naturally, we may freely use bases other than monomial for linear spaces of polynomials if it is advantageous for the understanding of a situation or for the design and analysis of computational algorithms. For example, we may wish to have a basis which is orthogonal w.r.t. some special scalar product, or which has other desirable properties.

Polynomials, in particular multivariate ones, often occur in sets or *systems*. We denote systems of polynomials by capital letters; e.g.,

$$P(x) \;=\; \{p_\nu(x), \; \nu = 1(1)n\}\,.$$

Notationally and operationally, such systems will often be treated as *vectors* of polynomials:

$$P(x) \;=\; \begin{pmatrix} p_1(x) \\ \vdots \\ p_n(x) \end{pmatrix}. \tag{1.4}$$

It is true that this notation implies an order of the polynomials in the system which has originally not been there. But such an (arbitrary) order is also generated by the assignment of subscripts and generally without harm.

Exercises

1. (a) Consider the linear space $\mathcal{P}_4^2$ (cf. Definition 1.4). What is its dimension? Introduce a monomial basis; consider reasons for choosing various orders for the basis monomials x^j in the basis vector $\mathbf{x}$.

(b) Differentiation w.r.t. x_1 and x_2, resp., are linear operations in $\mathcal{P}_4^2$. For a fixed basis vector $\mathbf{x}$, which matrices D_1, D_2 represent differentiation so that $\frac{\partial}{\partial x_i} \mathbf{x} = D_i\, \mathbf{x}$, $i = 1, 2$. Which matrix represents $\frac{\partial^2}{\partial x_1 \partial x_2}$? How can you tell from the D_i that all derivatives of an order greater than 4 vanish for $p \in \mathcal{P}_4^2$?

(c) With $p(x) = a^T \mathbf{x}$, show that the coefficient vector of $\frac{\partial}{\partial x_i} p(x)$ is $a^T D_i$. Check that $D_1 D_2 = D_2 D_1$. Explain why the commutatitivity is necessary and sufficient to make the notation $q(D_1, D_2)$, with $q \in \mathcal{P}^2$, meaningful. What is the coefficient vector of $q(\frac{\partial}{\partial x_1}, \frac{\partial}{\partial x_2})\, p(x)$?

2. (a) According to Example 1.1, the coefficient vector $\bar{a}^T$ of $p(x) = a^T \mathbf{x} \in \mathcal{P}_d^1$ w.r.t. the basis $(\ldots (x - \xi)^j \ldots)^T$ is $\bar{a}^T = a^T \, \Xi^{-1}$. Derive the explicit form of Ξ^{-1}.

(b) By Taylor's Theorem, the components $\bar{\alpha}_j$ of $\bar{a}^T$ are also given by $\bar{\alpha}_j = \frac{1}{j!} \frac{\partial^j}{\partial x^j} p \, (\xi)$. Show that this leads to the same matrices in the linear transformation between a^T and $\bar{a}^T$.

3. (a) The classical Chebyshev polynomials $T_\nu \in \mathcal{P}_\nu^1$ are defined by

$$T_0(x) := 1, \quad T_1(x) := x, \quad T_{\nu+1}(x) := 2 x \, T_\nu(x) - T_{\nu-1}(x), \quad \nu = 2, 3, \ldots .$$

Show that $T_\nu(1) = 1$, $T_\nu(-1) = (-1)^\nu$, $\forall \nu$; $T_\nu(0) = 0$ for ν odd and $= (-1)^{\nu/2}$ for ν even. Derive the same relations from the identity

$$T_\nu(\cos \varphi) \; = \; \cos \nu \varphi, \quad \varphi \in [0, \pi]. \tag{1.5}$$

(b) Consider the representations $p(x) = a^T \mathbf{x} = b^T \, (T_0(x), \ldots, T_d(x))^T$ for $p \in \mathcal{P}_d^1$. Which matrices M and M^{-1} represent the transformations $b^T = a^T M^{-1}$ and $a^T = b^T M$.

(c) The T_ν satisfy $\max_{x \in [-1,1]} |T_\nu(x)| = 1$, as is well-known and also follows, e.g., from (1.5). For $p(x) = b^T \, (T_0(x), \ldots, T_d(x))^T$, this implies $\max_{x \in [-1,1]} |p(x)| \leq \sum_{\nu=0}^{d} |\beta_\nu|$ (why?). Which bound for $|p|$ in terms of the monomial coefficients a^T follows from b)?

1.2 Polynomials as Functions

In this section, we recall some analytic aspects of polynomials regarded as functions. While the linear polynomials of linear algebra constitute a particular simple class of functions whose analytic aspects are trivial or straightforward, this is no longer the case for polynomials of a total degree $d > 1$. For example, the trivial polynomial $p(x) = x^2 + a$ maps the two disjoint real points ξ and $-\xi$, $\xi \neq 0$, to the same point $\xi^2 + a \in \mathbb{R}$ so that p is not bijective. Furthermore, the image of $\mathbb{R}$ is only the interval $[a, \infty)$.

The fact that $\mathbb{R}$ is not an algebraically closed field causes well-known complications when *real* polynomials are regarded as functions between domains in real space only. Therefore, throughout this book, we will mainly consider polynomials as functions between *complex* domains; note that real polynomials may also be considered as having a complex domain and range. Except if stated otherwise, individual polynomials in s variables or systems of n such polynomials will be regarded as mappings

$$p \, : \; \mathbb{C}^s \, \to \, \mathbb{C} \qquad \text{or} \quad P \, : \; \mathbb{C}^s \, \to \, \mathbb{C}^n .$$

However, we must clarify the notation $\mathbb{C}$: In our algebraic context, it will *always* denote the *open* complex "plane" *without* the point ∞. For us, "$|x|$ very large" is a near-singular situation, as in most other areas of numerical analysis. This is particularly important for our use of the multidimensional complex spaces $\mathbb{C}^s$, with their analytically intricate structure at ∞. In any case, our *intuition* for the $\mathbb{C}^s$, $s > 1$, is extremely restricted so that we may often have the $\mathbb{R}^s$ in mind when we are formally dealing with the $\mathbb{C}^s$. Compare also Proposition 1.4 and the remark following it.

Multivariate *differential operators* will play an important role in some parts of this book; we use the following notation for them:

Definition 1.5. For $j \in \mathbb{N}_0^s$,

$$\partial_j := \frac{1}{j_1! \ldots j_s!} \frac{\partial^{|j|}}{\partial x_1^{j_1} \ldots \partial x_s^{j_s}} \tag{1.6}$$

is a differentiation operator *of order* $|j|$. A polynomial $q(x) = \sum_{j \in J} b_j x^j \in \mathcal{P}_d^s$ defines the differential operator

$$q(\partial) := \sum_{j \in J} b_j \, \partial_j \, . \qquad \square \tag{1.7}$$

In examples, we may also use shorthand notations like p_{x_1} for $\partial_1 p = \frac{\partial}{\partial x_1} p$. The *factors* in (1.6) simplify a number of expressions; cf., e.g., the expansion (1.9) below. In particular, the well-known Leibniz rule for the differentiation of products takes the simple form

$$\partial_j (p \cdot q) = \sum_{k \leq j} \partial_{j-k} \, p \cdot \partial_k \, q \, , \tag{1.8}$$

where $\leq$ is the generic partial ordering in $\mathbb{N}_0^s$.

Proposition 1.2. For $p \in \mathcal{P}_d^s$, we have $\partial_j \, p \in \mathcal{P}_{d-|j|}^s$, and all derivatives of an order $> d$ vanish identically.

Proof: $\quad \partial_\sigma x^j = \begin{cases} j_\sigma \, x^j / x_\sigma & \text{if } x_\sigma \text{ divides } x^j, \\[2mm] 0 & \text{if } x_\sigma \text{ does not divide } x^j \end{cases} . \qquad \square$

Proposition 1.3. For $p \in \mathcal{P}_d^s$, $\xi \in \mathbb{C}^s$, the expansion of p in powers of $\bar{x} = x - \xi$ (the Taylor expansion about ξ) is

$$p(x) = p(\xi + \bar{x}) = \sum_{\delta=0}^{d} \sum_{|j|=\delta} (\partial_j \, p)(\xi) \, \bar{x}^j =: \overline{p}(\bar{x}; \xi) \, . \tag{1.9}$$

Proof: $\quad$ The proof follows from the binomial theorem and (1.6). $\quad \square$

Example 1.2: For $x = (y, z)$, $\bar{x} = (\bar{y}, \bar{z}) \in \mathbb{C}^2$, $p(y, z) = 5 \, y^3 z - 2 \, y^2 z^2 + z^4$, $\xi = (\eta, \zeta) = (2, -1)$:

$$\begin{aligned}
\overline{p}(\bar{y}, \bar{z}) = {}& p(\eta, \zeta) + p_y(\eta, \zeta) \, \bar{y} + p_z(\eta, \zeta) \, \bar{z} \\
& + \tfrac{1}{2} p_{yy}(\eta, \zeta) \, \bar{y}^2 + p_{yz}(\eta, \zeta) \, \bar{y}\bar{z} + \tfrac{1}{2} p_{zz}(\eta, \zeta) \, \bar{z}^2 \\
& + \tfrac{1}{6} p_{yyy}(\eta, \zeta) \, \bar{y}^3 + \tfrac{1}{2} p_{yyz}(\eta, \zeta) \, \bar{y}^2 \bar{z} + \tfrac{1}{2} p_{yzz}(\eta, \zeta) \, \bar{y}\bar{z}^2 + \tfrac{1}{6} p_{zzz}(\eta, \zeta) \, \bar{z}^3 \\
& + \tfrac{1}{6} p_{yyyz}(\eta, \zeta) \, \bar{y}^3 \bar{z} + \tfrac{1}{4} p_{yyzz}(\eta, \zeta) \, \bar{y}^2 \bar{z}^2 + \tfrac{1}{24} p_{zzzz}(\eta, \zeta) \, \bar{z}^4 \\
= {}& -47 - 68\bar{y} + 52\bar{z} - 32\bar{y}^2 + 76\bar{y}\bar{z} - 2\bar{z}^2 \\
& - 5\bar{y}^3 + 34\bar{y}^2\bar{z} - 8\bar{y}\bar{z}^2 - 4\bar{z}^3 + 5\bar{y}^3\bar{z} - 2\bar{y}^2\bar{z}^2 + \bar{z}^4 \, . \qquad \square
\end{aligned}$$

For convenience, we will sometimes use the *Fréchet derivative* concept of Functional Analysis to represent results in a more compact notation. Fréchet differentiation is a straightforward generalization of common differentiation to maps between Banach spaces; in our algebraic context, all Banach spaces of interest are finite-dimensional vector spaces.

For a sufficient understanding of our essentially notational use of the concept, we observe that we can interpret the differentiability of a function $f : \mathbb{R} \to \mathbb{R}$ in a domain $D \subset \mathbb{R}$ thus: f is differentiable at $x \in D$ if there exists a *linear map* $u(x) : \mathbb{R} \to \mathbb{R}$ such that

$$\lim_{\Delta x \to 0} \frac{1}{|\Delta x|} \, |f(x + \Delta x) - f(x) - u(x)\,\Delta x| \; = \; 0\,;$$

note that a linear map $\mathbb{R} \to \mathbb{R}$ is given by a real number to be employed as a *factor*. A slightly stronger formulation which is, however, equivalent in our setting is

$$|f(x + \Delta x) - f(x) - u(x)\,\Delta x| \; = \; \mathrm{O}(|\Delta x|^2)\,. \tag{1.10}$$

Thus, differentiation of $f : \mathbb{R} \to \mathbb{R}$ is an operation which maps f into $u : \mathbb{R} \to \mathcal{L}(\mathbb{R} \to \mathbb{R})$, the space of linear maps from $\mathbb{R}$ to $\mathbb{R}$. The derivative u is generally denoted by f' or $\frac{d}{dx}\,f$, etc.

Now, we consider functions or maps from one *vector space* A into another one B and apply the same line of thought: Fréchet differentiation is an operation which maps $f : A \to B$ into a function $u : A \to \mathcal{L}(A \to B)$ such that, for x from the domain $D \subset A$ of differentiability,

$$\|f(x + \Delta x) - f(x) - u(x) \cdot \Delta x \|_B \; = \; \mathrm{O}(\|\Delta x\|_A^2)\,, \tag{1.11}$$

where $\|..\|_A$, $\|..\|_B$ are the norms in A and B, resp., and the $\cdot$ denotes the action of the linear operation $u(x)$. Again, notations like f' etc. are commonly employed for u. Obviously, $f'(x)$ *linearizes* the local variation of f in the neighborhood of x.

A few examples will show the notational power of the Fréchet differentiation concept: Let $A = \mathbb{R}^m$, $B = \mathbb{R}$ (or $\mathbb{C}^m$ and $\mathbb{C}$); i.e. f is a scalar function of m variables. Then the Fréchet derivative $u(x)$ of f at x must satisfy

$$\begin{aligned}
f(x + \Delta x) \; &= \; f(x_1 + \Delta x_1, \ldots, x_m + \Delta x_m) \; = \; f(x) + u(x) \cdot \Delta x + \mathrm{O}(\|\Delta x\|^2) \\
&= \; f(x_1, \ldots, x_m) + \left(\tfrac{\partial f}{\partial x_1}(x), \ldots, \tfrac{\partial f}{\partial x_m}(x) \right) \begin{pmatrix} \Delta x_1 \\ \vdots \\ \Delta x_m \end{pmatrix} + \mathrm{O}(\|\Delta x\|^2)\,;
\end{aligned}$$

thus the Fréchet derivative $u = f'$ of f is given by $x \to \operatorname{grad} f(x) := \left(\frac{\partial f}{\partial x_1}(x), \ldots, \frac{\partial f}{\partial x_m}(x) \right)$, a *row* vector of dimension m.

For a vector of functions, we simply obtain the vector of the Fréchet derivatives. Thus, for $A = \mathbb{R}^m$, $B = \mathbb{R}^n$, and $f : A \to B$,

$$\begin{aligned}
f(x + \Delta x) \; &= \; \begin{pmatrix} f_1(x_1 + \Delta x_1, \ldots, x_m + \Delta x_m) \\ \vdots \\ f_n(x_1 + \Delta x_1, \ldots, x_m + \Delta x_m) \end{pmatrix} = f(x) + u(x) \cdot \Delta x + \mathrm{O}(\|\Delta x\|^2) \\
&= \; \begin{pmatrix} f_1(x) \\ \vdots \\ f_n(x) \end{pmatrix} + \begin{pmatrix} \frac{\partial f_1}{\partial x_1}(x) & \cdots & \frac{\partial f_1}{\partial x_m}(x) \\ \vdots & & \vdots \\ \frac{\partial f_n}{\partial x_1}(x) & \cdots & \frac{\partial f_n}{\partial x_m}(x) \end{pmatrix} \begin{pmatrix} \Delta x_1 \\ \vdots \\ \Delta x_m \end{pmatrix} + \mathrm{O}(\|\Delta x\|^2)\,.
\end{aligned}$$

Now, the Fréchet derivative $u = f'$ of f is $x \to \left(\frac{\partial f_\nu}{\partial x_\mu}(x) \right)$, the $n \times m$ *Jacobian matrix* of f at x.

Higher Fréchet derivatives tend to become less intuitive: When $f' = u$ is a map from A to $\mathcal{L}(A \to B)$, $f'' = u'$ has to be a map from A to $\mathcal{L}(A \to \mathcal{L}(A \to B)) = \mathcal{L}(A \times A \to B)$ or a *bilinear map* from A to B. Thus the Fréchet generalization of the classical second derivative to a scalar function of m variables assigns to each $x \in D$ the *bilinear mapping* $(\Delta x, \overline{\Delta x}) \to \overline{\Delta x}^T f''(x) \Delta x$, with the symmetric $m \times m$ *Hessian matrix* $H(x) := f''(x) = \left(\frac{\partial^2 f}{\partial x_\mu \, \partial x_{\bar\mu}}(x) \right)$. With the use of this higher Fréchet derivative concept, the Taylor expansion of a *system* P of n polynomials p_ν in m variables takes the compact form

$$P(x + \Delta x) = \sum_{\kappa=0}^{\deg(P)} \frac{1}{\kappa!} P^{(\kappa)}(x) (\Delta x)^\kappa .$$

Proposition 1.4. A polynomial $p \in \mathcal{P}^s$ is a holomorphic function on each compact part $D \subset \mathbb{C}^s$. The image $p(D)$ of D is a compact part of $\mathbb{C}$.

Proof: According to Propositions 1.2 and 1.3, each p has a *finite* expansion $\overline{p}(\bar{x}; \xi) = p(\xi + \bar{x})$ in powers of $\bar{x} = x - \xi$ at each $\xi \in \mathbb{C}^s$. $\square$

The fact that bounded domains in $\mathbb{C}^s$ are mapped on *bounded* domains in $\mathbb{C}$ makes it feasible to exclude ∞ as a proper element from $\mathbb{C}$ (or $\mathbb{R}$).

Proposition 1.5. A polynomial $p \in \mathcal{P}^s$ represents the zero mapping if and only if p is the zero polynomial.

Proof: The "if" direction is trivial. Now assume that $p(x) = 0$ for all $x \in \mathbb{C}^s$; we proceed by induction on s:

A univariate polynomial ($s = 1$) cannot vanish for all $x \in \mathbb{C}$ except if it is the zero polynomial: A nonzero polynomial of degree d has at most d zeros. Let the assertion be true for $s \le s_0 - 1$ and consider a polynomial p in s_0 variables: Write p as a univariate polynomial $\overline{p}$ in the variable x_{s_0}, with coefficients which are polynomials in the x_σ, $\sigma = 1(1)s_0 - 1$. By assumption, $\overline{p}$ vanishes for all values of x_{s_0}; hence the coefficients of $\overline{p}$ must vanish for all values of their arguments. By the induction assumption this implies that they are zero polynomials so that p is also the zero polynomial. $\square$

On the other hand, each nonconstant polynomial has zeros in $\mathbb{C}^s$.

Proposition 1.6. A polynomial $p \in \mathcal{P}^s$ has zeros in $\mathbb{C}^s$ except if $p(x) = c$, $c \neq 0$.

Proof: For univariate polynomials, the assertion is well known. For $s > 1$, let p be nonconstant as a function of (say) x_s and consider the univariate polynomial $p(\xi_1, \ldots, \xi_{s-1}, x_s)$. $\square$

According to the inverse function theorem, a function $P : \mathbb{C}^s \to \mathbb{C}^s$ (or $\mathbb{R}^s \to \mathbb{R}^s$) is *invertible* in a neighborhood of some $\xi \in \mathbb{C}^s$ if and only if it is differentiable in the neighborhood and its Fréchet derivative $P'(\xi)$ is a *regular* linear function. (A linear function $L : \mathbb{C}^s \to \mathbb{C}^s$ is regular iff $L x = 0$ implies $x = 0$.) While differentiability is no problem for functions P defined by polynomials, $P'(\xi)$ cannot be uniformly regular for such functions except when P is "essentially linear," i.e. if $\det P'(x) = \text{const}$. An essentially linear system is at most a trivial modification of a linear system, e.g., $p_1(x_1, x_2) = x_1 + x_2^2$, $p_2(x_1, x_2) = x_2$.

Theorem 1.7. For a polynomial (system) $P \in (\mathcal{P}^s)^s$, $P'(\xi)$ must be singular at some $\xi \in \mathbb{C}^s$, except if P is essentially linear.

Proof: Consider $s = 1$ at first, with $P = p_1 \in \mathcal{P}^1$ of degree $d > 1$. According to Proposition 1.2, $P' = \partial_1 p_1$ is a polynomial of positive degree; hence it has at least one zero ξ in $\mathbb{C}$ which implies the singularity of the linear mapping $P'(\xi)$. For $s > 1$, except if the polynomial $\hat{p}(x) := \det P'(x)$ is constant, it is a polynomial of positive degree and must have a zero $\xi \in \mathbb{C}^s$ at which $P'(\xi)$ is singular, according to Proposition 1.6. $\square$

Corollary 1.8. A polynomial (system) $P \in (\mathcal{P}^s)^s$ which is not essentially linear is not invertible over arbitrary domains of the $\mathbb{C}^s$.

Polynomial functions $P : \mathbb{C}^s \to \mathbb{C}^s$ can be invertible only in domains which do not contain a point at which P' is singular. At a singularity of the linear function P', two or more *branches* of the inverse function meet. For $s > 1$, this is a rather complicated situation which is one of the main objects of Singularity Theory. We will be mainly interested in the case where P' is singular at a *zero* ζ of P so that ζ is a *multiple zero* of P; cf. section 8.5.

For $s = 1$, the situation is well known from classical function theory: Without loss of generality, we may assume the singularity to occur at $\xi = 0$ so that $p(x) = a_0 + \sum_{j=m}^{d} a_j x^j$, $m > 1$, and $p(x) = y$ has the m branches $x = (\frac{y - a_0}{a_m})^{1/m} (1 + O(y - a_0)^{1/m})$ in the neighborhood of $y = a_0$ or $x = 0$, respectively.

Exercises

1. (a) Consider a real polynomial $p \in \mathcal{P}^s$ as a function $\mathbb{R}^s \to \mathbb{R}$. For $s = 1$, there are a number of well-known properties of a (sufficiently differentiable) function f; on some connected convex domain $D \subset \mathbb{R}$

f is (strictly) *increasing* if $\eta > \xi \implies f(\eta) \geq (>) f(\xi)$;

f is (strictly) *convex* if $\eta > \xi \implies \frac{(\eta - x) f(\xi) + (x - \xi) f(\eta)}{\eta - \xi} \geq (>) f(x)$ for $x \in (\xi, \eta)$,

or, equivalently (why?), $f''(x) \geq (>) 0$, $x \in D$;

f has a *minimum* at ξ if $f'(\xi) = 0$ and $f''(\xi) > 0$; f has a *turning point* at ξ if $f''(\xi) = 0$.

In which ways can these properties be extended to the multivariate case $s > 1$?

(b) For $s \geq 1$, construct polynomials which have such properties in specified domains or at specified points, respectively.

2. (a) Consider a complex univariate polynomial as a function $\mathbb{R}^2 \to \mathbb{R}^2$, $(u, v) \to (q, r)$, by setting $p(u + iv) = q(u, v) + i r(u, v)$. How do the following properties of p reflect in q and r : $\deg(p) = d$, p even, p odd, other sparsity of p ?

(b) It is well known that the polynomials q and r cannot be chosen arbitrarily but have to satisfy the Cauchy–Riemann equations $\partial_u q - \partial_v r = 0$, $\partial_v q + \partial_u r = 0$, which further imply that q and r individually have to satisfy the "potential equation" $(\partial_{uu} + \partial_{vv}) q = (\partial_{uu} + \partial_{vv}) r = 0$. Verify these relations for a polynomial of some degree $d > 1$ and coefficients $\beta_j + i\gamma_j$.

3. (a) Show that the images under the mapping $p : \mathbb{C} \to \mathbb{C}$, $p(x) = x^2 + a$, cover the entire open complex plane $\mathbb{C}$. At which point(s) is p not invertible?

(b) Take a regular point ξ of p; what is the power series of p about ξ ? What is the beginning of the power series of the inverse function about $\eta = \xi^2 + a$?

4. Take $p_1, p_2 \in \mathcal{P}^2$ from Example 1.3 in section 1.3.2 and consider the mapping $\mathbb{C}^2 \to \mathbb{C}^2$ defined by $(x, y) \to (u = p_1(x, y), v = p_2(x, y))$.

(a) Find the images of the real grid lines $x = m$ and $y = n$, $m, n \in \mathbb{Z}$, in the real u, v-plane. Try to visualize the images of the imaginary grid lines $x = im$, $y = in$ in the complex u, v-plane.

(b) Which points (ξ, η) are mapped into $(0,0)$? Find the preimages of the real coordinate axes ($u = 0$, $v \in \mathbb{R}$), ($u \in \mathbb{R}$, $v = 0$) and of the complex coordinate axes ($u = 0$, $v \in \mathbb{C}$), ($u \in \mathbb{C}$, $v = 0$) in the x, y-plane.

(c) At which points in the x, y-plane is the mapping singular? Take one such point $(\xi, \eta) \neq (0, 0)$ and consider the images of $(\xi + \varepsilon_1, \eta + \varepsilon_2)$ for small ε_i, i.e. neglecting terms of order $O(\varepsilon^2)$. How does the singular situation manifest itself? Compare with the linearized image of the neighborhood of a regular point. What is the special situation at the singular point $(0,0)$?

(d) At a regular point (x_0, y_0) with image (u_0, v_0), find the beginning of the inverse power series $x = x_0 + \alpha_{11}(u - u_0) + \alpha_{12}(v - v_0) +$ quadratic terms, $y = y_0 + \alpha_{21}(u - u_0) + \alpha_{22}(v - v_0) +$ quadratic terms. What happens when (x_0, y_0) approaches a singular point?

1.3 Rings and Ideals of Polynomials

1.3.1 Polynomial Rings

In the set T^s of monomials, we have the natural commutative multiplication

$$x^{j_1} \cdot x^{j_2} = x^{j_1 + j_2}, \tag{1.12}$$

where $j_1 + j_2$ is the vector addition in $\mathbb{N}_0^s$. The distributive law defines a *commutative* product for polynomials $p_i(x) = \sum_{j \in J_i} \alpha_{ij} x^j \in \mathcal{P}^s$, $i = 1, 2$:

$$p_1 \cdot p_2 = \sum_{j_1 \in J_1} \sum_{j_2 \in J_2} \alpha_{1 j_1} \alpha_{2 j_2} x^{j_1 + j_2} \in \mathcal{P}^s. \tag{1.13}$$

The 1-element of this multiplication is the constant polynomial $p(x) = 1$ which is contained in each $\mathcal{P}^s$. Due to the associativity of multiplication in $\mathbb{C}$ and of addition in $\mathbb{N}_0^s$, the product (1.13) is also *associative*.

Therefore, with the multiplication (1.13), the linear spaces $\mathcal{P}^s$ are *commutative rings* of polynomials. The usual notation for a polynomial ring in s variables, with coefficients from $\mathbb{C}$ or $\mathbb{R}$, resp., is $\mathbb{C}[x_1, \ldots, x_s]$ or $\mathbb{R}[x_1, \ldots, x_s]$. We will use this more explicit notation only in cases where the explicit denotation of the variables is important, like in the following paragraph.

The product of two polynomials whose variables do not coincide may be defined in the set of polynomials in the union of the variables: The product of $p_1 \in \mathbb{C}[x_1, \ldots, x_{s_1}]$, $p_2 \in \mathbb{C}[y_1, \ldots, y_{s_2}]$ is defined in $\mathbb{C}[x_1, \ldots, x_{s_1}, y_1, \ldots, y_{s_2}]$ which contains both p_1 and p_2. Naturally, some of the variables in p_1 and p_2 may be common: For example, $p_1(x, y) \cdot p_2(y, z)$ is an element of $\mathbb{C}[x, y, z]$.

In the linear space $\mathcal{P}^s$, the mapping $M_p : \mathcal{P}^s \to \mathcal{P}^s$, with $p \in \mathcal{P}^s$, $M_p q := p \cdot q$, is a *linear mapping*, because

$$p \cdot (\gamma_1 q_1 + \gamma_2 q_2) = \gamma_1 \, p \, q_1 + \gamma_2 \, p \, q_2. \tag{1.14}$$

Hence, the commutative *product mapping* $M : \mathcal{P}^s \times \mathcal{P}^s \to \mathcal{P}^s$ is *bilinear*.

Proposition 1.9. The multiplication (1.13) by a fixed polynomial $p \in \mathcal{P}^s$, $p \neq 0$, is a *regular* linear mapping in $\mathcal{P}^s$, i.e. $q \in \mathcal{P}^s$, $p \cdot q = 0 \Rightarrow q = 0$.
Proof: Let $\alpha_{1j_1^*} x^{j_1^*}$ and $\alpha_{1j_2^*} x^{j_2^*}$ be the *leading* terms of p and q, resp. (cf. section 8.4.1); their product is a term in $p\,q$ which cannot be cancelled because there cannot be another term with the monomial $x^{j_1^*+j_2^*}$. $\square$

Corollary 1.10. The polynomial rings $\mathcal{P}^s$ are *integral domains*, i.e. $p_1 \cdot p_2 = 0$ implies $p_1 = 0$ or $p_2 = 0$ for $p_1,\ p_2 \in \mathcal{P}^s$.

For $p \in \mathcal{P}^s_{d_p}$, $q \in \mathcal{P}^s_{d_q}$, we have $p \cdot q \in \mathcal{P}^s_{d_p+d_q}$. With a monomial basis in each of these linear spaces, the linear mapping $M_p : \mathcal{P}^s_{d_q} \to \mathcal{P}^s_{d_p+d_q}$ must be representable by a matrix acting on the coefficient vector of q. Using rows for coefficient vectors as explained in (1.3), we have

$$(\dots a_{pq}^T \dots) = (\dots a_q^T \dots)\left(M_p \right). \tag{1.15}$$

For $s > 1$, the $d_q \times (d_p + d_q)$ matrices M_p are generally so large and sparse that they are rarely helpful computationally.

For $s = 1$, $p(x) = \sum_{j=0}^{d_p} \alpha_j x^j$, $q(x) = \sum_{j=0}^{d_q} \beta_j x^j$, $p(x)q(x) = \sum_{j=0}^{d_p+d_q} \gamma_j x^j$, we have

$$\left(\gamma_0\, \gamma_1 \dots \dots \gamma_{d_p+d_q} \right) = \left(\beta_0 \dots \beta_{d_q} \right) \begin{pmatrix} \alpha_0 & \alpha_1 & \dots & \alpha_{d_p} & & \\ & \alpha_0 & \dots & \dots & \alpha_{d_p} & \\ & & \ddots & & & \ddots \\ & & & \alpha_0 & \dots & \dots & \alpha_{d_p} \end{pmatrix}, \tag{1.16}$$

which will be used later. Naturally, the roles of p and q in (1.16) may be interchanged.

The representations (1.15)/(1.16) of multiplication in $\mathcal{P}^s$ may also be written in terms of the monomial basis vectors $\mathbf{x}_{d_q}$, $\mathbf{x}_{d_p+d_q}$:

$$p \cdot \mathbf{x}_{d_q} = M_p\, \mathbf{x}_{d_p+d_q} \iff p\,q = p\,a_q^T \mathbf{x}_{d_q} = a_q^T M_p\, \mathbf{x}_{d_p+d_q} = a_{pq}^T \mathbf{x}_{d_p+d_q} = p\,q. \tag{1.17}$$

Note that the same *multiplication matrix* M_p multiplies the basis vector $\mathbf{x}_{d_p+d_q}$ from the left or the coefficient vector a_q^T from the right.

A different kind of polynomial rings in $\mathcal{P}^s$ will also play a central role in our considerations, viz. *quotient rings* or *residue class rings* modulo a polynomial ideal $\mathcal{I} \subset \mathcal{P}^s$. In a quotient ring $\mathcal{R} \subset \mathcal{P}^s_d$, with a fixed basis vector $\mathbf{b}$ of dimension m, the product mapping $M_p : q \to p \cdot q$ maps $\mathcal{R}$ *into itself* and is thus represented by an $m \times m$ matrix for each $p \in \mathcal{R}$. We delay an introduction to quotient rings to section 2.2.

1.3.2 Polynomial Ideals

In linear algebra, *linear combinations* $\sum_{v=1}^{n} \gamma_v \ell_v$ of *linear polynomials* or *functionals* $\ell_v : \mathbb{C}^s \to \mathbb{C}$ play a fundamental role; here, the coefficients γ_v are scalars, i.e. values from $\mathbb{C}$ (or $\mathbb{R}$).

A set of linear functionals forms a *linear space L* if it is *closed under linear combination*. The linear space L has dimension d if more than d elements from L are always linearly dependent; n linear functions are linearly independent iff $\sum_{\nu=1}^{n} \gamma_\nu \ell_\nu = 0$ (the zero functional) implies $\gamma_\nu = 0$, $\nu = 1(1)n$. Any set of d linearly independent elements from L forms a basis of the linear space L. A linear space of linear functionals on $\mathbb{C}^s$ has at most dimension s.

An $n \times s$ matrix A, with rows a_ν^T, $\nu = 1(1)n$, may, e.g., be interpreted as a set of n linear functionals $a_\nu : \mathbb{C}^s \to \mathbb{C}$, with $a_\nu(x) := a_\nu^T \mathbf{x}$. The dimension of the linear space generated by the rows of A is the *rank* of A. The formation of suitable linear combinations is a basic tool in the design of computational algorithms in linear algebra.

In polynomial algebra, likewise, *linear combinations* of polynomials $p_\nu : \mathbb{C}^s \to \mathbb{C}$ are a central object of consideration. Now, however, the coefficients c_ν in $\sum_{\nu=1}^{n} c_\nu p_\nu$ need no longer be scalars—as it was necessary in linear algebra to keep the combination a linear functional.

Definition 1.6. Consider a set of n polynomials $p_\nu \in \mathcal{P}^s$, $\nu = 1(1)n$; any polynomial

$$p = \sum_{\nu=1}^{n} c_\nu \, p_\nu \in \mathcal{P}^s, \quad \text{with arbitrary } \textit{polynomials } c_\nu \in \mathcal{P}^s, \tag{1.18}$$

is a *linear combination* of the p_ν. $\square$

For a numerical analyst, it is very unusual to consider (1.18), with polynomial coefficients c_ν, as a "linear combination." Therefore, whenever there is a danger of confusion between the scalar linear combinations of linear algebra and the polynomial linear combinations of polynomial algebra, we will use the term "polynomial combination" for (1.18).

Since the coefficient polynomials $c_\nu \in \mathcal{P}^s$ may have *arbitrary total degrees*, the potential total degree of a linear combination is *unbounded*, independently of the total degrees of the p_ν. *Example 1.3:* In $\mathcal{P}^2$, let $p_1(x, y) := x^2 + y^2 - 4$, $p_2 := x\,y - 1$. Linear combinations of p_1 and p_2 are, e.g.,

$$p_3 = x \cdot p_1 - y \cdot p_2 = x^3 - 4x + y, \quad p_4 = y \cdot p_1 - x \cdot p_2 = y^3 + x - 4\,y.$$

With polynomial combinations of two or more polynomials, we can always form nontrivial representations of the zero-polynomial: Take, e.g., $p_2 \cdot p_1 + (-p_1) \cdot p_2$, or $p_1 + (y^2 - 4)\, p_2 - x\, p_4$. Compare also Proposition 2.1. $\square$

As is to be expected, sets of polynomials which are *closed under polynomial combination* play a central role in polynomial algebra:

Definition 1.7. A set of polynomials in $\mathcal{P}^s$ which is closed under linear (= polynomial) combination is a *polynomial ideal* in $\mathcal{P}^s$. The ideal which consists of the linear combinations of the polynomials $p_\nu \in \mathcal{P}^s$, $\nu = 1(1)n$, is denoted by

$$\langle p_1, \ldots, p_n \rangle \, ; \tag{1.19}$$

the p_ν form a *basis* of this ideal which they *generate*. $\square$

One of the reasons why ideals are so important in computational polynomial algebra is shown by:

Proposition 1.11. Consider a set Z of points in $\mathbb{C}^s$. The set

$$\mathcal{I}_Z := \{p \in \mathcal{P}^s : p(z) = 0 \ \forall z \in Z\} \tag{1.20}$$

of all polynomials which vanish at each $z \in Z$ is an ideal in $\mathcal{P}^s$.

Proof: Consider $p_\nu \in \mathcal{I}_Z$, $\nu = 1(1)n$, $n \geq 1$. Then, for any $c_\nu \in \mathcal{P}^s$, $p = \sum_{\nu=1}^n c_\nu p_\nu$ obviously vanishes at each $z \in Z$; therefore, $p \in \mathcal{I}_Z$. $\square$

Conversely, assume that a set of polynomials p_ν has a joint zero $z \in \mathbb{C}^s$. Then, by (1.18), all polynomials $p \in \langle p_1, \ldots, p_n \rangle$ vanish at z, i.e. z is a *joint zero* of *all* p in the ideal generated by the p_ν. Thus, joint zeros or zero sets are closely associated with polynomial ideals.

Definition 1.8. $z \in \mathbb{C}^s$ is a *zero of the polynomial ideal* $\mathcal{I} \subset \mathcal{P}^s$ iff $p(z) = 0$ for all $p \in \mathcal{I}$. The *zero set of* $\mathcal{I}$ is denoted by

$$Z[\mathcal{I}] := \{z \in \mathbb{C}^s : z \text{ is a zero of } \mathcal{I}\}. \ \square \tag{1.21}$$

Proposition 1.12. Consider the polynomial system $P = \{p_\nu, \ \nu = 1(1)n\} \subset \mathcal{P}^s$ and the ideal $\langle P \rangle := \langle p_1, \ldots, p_n \rangle$. Then the zero set $Z[\langle P \rangle]$ of the polynomial ideal $\langle P \rangle$ satisfies (cf. (1.4))

$$Z[\langle P \rangle] = \{z \in \mathbb{C}^s : P(z) = 0\} =: Z[P],$$

i.e. it is identical with the zero set $Z[P]$ of the polynomial system P.

Proof: See the argument before Definition 1.8 . $\square$

Example 1.4: Consider s linear polynomials $p_\nu(x) = \alpha_{\nu 0} + a_\nu^T x \in \mathcal{P}_1^s$, $\nu = 1(1)s$, with $a_\nu^T = (\alpha_{\nu 1}, \ldots, \alpha_{\nu s})$, $x = (x_1, \ldots, x_s)^T$; cf. the beginning of this section. As is well known from linear algebra, if the a_ν^T are linearly independent, the p_ν have exactly one joint zero $z \in \mathbb{C}^s$, the solution of the system of linear equations

$$P(x) = \begin{pmatrix} p_1(x) \\ \vdots \\ p_s(x) \end{pmatrix} = 0.$$

The ideal $\mathcal{I}_z := \langle P \rangle$ with zero set $Z[\mathcal{I}_z] = \{z\}$ consists of all polynomial combinations of the p_ν; it has another, simpler, basis consisting of the s univariate linear polynomials

$$b_\sigma(x) := x_\sigma - \zeta_\sigma, \quad \sigma = 1(1)s.$$

To establish $\{b_\sigma, \sigma = 1(1)s\}$ as a basis of $\mathcal{I}_z$ we observe that $p_\nu(x) = \sum_{\sigma=1}^s \alpha_{\nu\sigma}(x_\sigma - \zeta_\sigma) + p_\nu(z) = \sum_{\sigma=1}^s \alpha_{\nu\sigma}(x_\sigma - \zeta_\sigma)$. On the other hand, the $s \times s$ matrix $(\alpha_{\nu\sigma})$ is regular due to the assumed (scalar) linear independence of the p_ν so that, with the elements $\bar{\alpha}_{\sigma\nu}$ of the inverse matrix, $b_\sigma(x) = \sum_\nu \bar{a}_{\sigma\nu} p_\nu(x)$, $\sigma = 1(1)n$.

The ideal $\mathcal{I}_z$ consists of all polynomials $p \in \mathcal{P}^s$ which vanish at z: The Taylor expansion of such a p about z (cf. (1.9)) has no constant term so that each remaining term contains at least one of the $b_\sigma = x_\sigma - \zeta_\sigma$ as a factor. $\mathcal{I}_z$ is also a *prime ideal*, i.e. $p \cdot q \in \mathcal{I}_z \Rightarrow p \in \mathcal{I}_z$ or $q \in \mathcal{I}_z$. $\square$

Proposition 1.12 permits the following approach to the computation of the zeros of a polynomial system $P = \{p_\nu, \nu = 1(1)n\} \subset \mathcal{P}^s$:

Algorithm 1.1. Compute a set of polynomials $b_\kappa \in \mathcal{P}^s$, $\kappa = 1(1)k$, such that

$$\langle P \rangle = \langle b_1, \ldots, b_k \rangle \tag{1.22}$$

and such that the zeros of the system $B = \{b_\kappa, \kappa = 1(1)k\}$ can be readily computed. $\square$
The identity (1.22) establishes the set $\{b_\kappa, \kappa = 1(1)k\}$ as a basis of the ideal $\langle p_\nu, \nu = 1(1)n \rangle$. A
polynomial ideal in $\mathcal{P}^s$ has arbitrarily many different bases, with (generally) varying numbers
of elements; thus k is not necessarily equal to n in (1.22). All this will be considered in much
more detail in Chapter 2 and later chapters.

The algorithmic pattern of Algorithm 1.1 is a generalization of some direct methods of
linear algebra for the solution of systems of linear equations (cf. Example 1.4): The well-known
Gauss–Jordan algorithm *diagonalizes* the matrix of the row vectors a_ν^T which is equivalent to
the formation of the basis $\{b_\nu\}$. It is generally easy to transform a set of linear polynomials such
that the new set of linear polynomials forms a basis of the original ideal and at the same time
is simpler in a specified sense; a good number of algorithms in numerical linear algebra serve
that purpose (though they are generally not formulated in this way). For general polynomial
systems, the same task is much harder; we will concern ourselves with it at great length.

Exercises

1. With p_1, $p_2 \in \mathcal{P}^2$ from Example 1.3, consider the polynomial ideal $\mathcal{I} = \langle p_1, p_2 \rangle$.

(a) Is it possible that $\mathcal{I}$ contains a polynomial from $\mathcal{P}_1^2$? (Hint: Consider Proposition
1.12.)

(b) Find the zero set $Z[\mathcal{I}] = Z[\{p_1, p_2\}]$. Construct polynomials in $\mathcal{P}^2$ which vanish at
each $\zeta \in Z[\mathcal{I}]$ and represent them as polynomial combinations of p_1 and p_2.

(c) Form complex polynomials $p \in \mathcal{I}$ by using complex polynomial coefficients in (1.18);
check that the p vanish on $Z[\mathcal{I}]$.

2. In Example 1.4, take only $s - 1$ linear polynomials p_ν and assume that the a_ν^T, $\nu = 1(1)s - 1$,
are linearly independent.

(a) Show that the zero set $Z[P]$ is one-dimensional, i.e. that there exists an s-vector y
such that $\xi \in Z[P] \Rightarrow \xi + t\,y \in Z[P]$ for each $t \in \mathbb{C}$.

(b) Characterize the ideal $\langle P \rangle$ algebraically and geometrically.

3. Consider a set Z of 4 disjoint points $z_\mu := (\xi_\mu, \eta_\mu) \in \mathbb{R}^2$, $\mu = 1(1)4$.

(a) Let $p_{\mu\nu} \in \mathcal{P}_1^2$ be linear polynomials which vanish at z_μ and z_ν, $\mu.\nu \in \{1, 2, 3, 4\}$.
Construct a prospective basis for the ideal $\mathcal{I}_Z$ consisting of the two quadratic polynomials
$\bar{p}_1 := p_{12}\,p_{34}$, $\bar{p}_2 := p_{13}\,p_{24}$. Prove that, generically, $\langle \bar{p}_1, \bar{p}_2 \rangle = \mathcal{I}_Z$.

(b) Characterize special situations when $\bar{p}_1$, $\bar{p}_2$ are not a basis of $\mathcal{I}_Z$. Find a basis for $\mathcal{I}_Z$
when the z_μ are collinear.

1.4   Polynomials and Affine Varieties

One of the secrets of the success of linear algebra is the fact that many of its objects and
relations may easily be visualized geometrically, at least in two or three real dimensions: The
elements of a linear space in s variables are identified with the points of the affine $\mathbb{R}^s$. Linear

transformations of the linear space become affine transformations of the R^s, and the invariant subspaces under a transformation correspond to the subspaces invariant under the related affine transformation. Most important, the zero sets of a linear mapping of the linear space correspond to linear manifolds in the affine space.

It had been customary for a long time to label introductory courses on linear algebra as "Linear Algebra and Analytic Geometry" because these geometric aspects of linear algebra are not only an important conceptual tool but they also carry a strong modelling potential in their own right. Although the dimensions under consideration in applied linear algebra may be much larger than 2 or 3, we can still use the 3D visualization as a valuable guide for our reasoning. We "see" the zero set of a system of $n \le s$ linear equations in s variables as the intersection of n (hyper)planes. And we can use a geometric language even when we speak about purely algebraic relations.

With the same identification of s variables with the coordinates in s-dimensional affine space, and s-tuples of numbers with the points in that space, we may interpret the zero sets of individual polynomials in $\mathcal{P}^s$ as (hyper)surfaces in affine s-space and the zero sets of systems of such polynomials as the intersections of hypersurfaces.

Definition 1.9. Consider a system P of polynomials $p_\nu \in \mathcal{P}^s$, $\nu = 1(1)n$. The points of the zero set $Z[P] := \{\xi \in \mathbb{C}^s : p_\nu(\xi) = 0, \ \nu = 1(1)n\} \subset \mathbb{C}^s$ form the *affine variety defined by* P. $\square$

According to Proposition 1.12, the zero set $Z[P]$ of the polynomial system $P = \{p_\nu\}$ is also the zero set $Z[\mathcal{I}]$ of the polynomial ideal $\mathcal{I} = \langle P \rangle = \langle p_1, \ldots, p_n \rangle$ generated by the polynomials $p_\nu \in P$ or, equivalently, the set of the *joint* zeros of *all* polynomials in $\mathcal{I}$.

Definition 1.10. The affine variety $Z[P] \subset \mathbb{C}^s$ defined by the polynomial system $P = \{p_\nu, \nu = 1(1)n\} \subset \mathcal{P}^s$ will be denoted by $V[\langle P \rangle]$ or $V[\langle p_1, \ldots, p_n \rangle]$, respectively. $\square$

Affine varieties are the fundamental objects of *algebraic geometry*. Since the main goal of this book is *computational* polynomial algebra, we will not enter into a technical discussion of affine varieties; we will rather employ them in an informal manner as a tool for the geometric visualization of our considerations, in analogy to what is customary in numerical linear algebra.

From our point of view, an important limit for the intuitional potential of affine varieties lies in the distinction between *complex and real domains*. In section 1.2, we have argued that it is advantageous to work in $\mathcal{P}_\mathbb{C}^s$ wherever possible—even when all specified polynomials have real coefficients—because the real subsets of the complex zero sets of systems of real polynomials may be very restricted and even empty. On the other hand, human beings are generally not able to perceive objects in $\mathbb{C}^2$. But a vivid visualization of the case $s = 2$ is definitely a prerequisite for an abstract visualization by analogy for $s > 2$!

Therefore, it has become customary to use the $\mathcal{P}_\mathbb{R}^s$ and the $\mathbb{R}^s$ in examples dealing with affine varieties; cf., e.g., the wonderful text [2.10]. But this may strongly mislead the associative potentials of our brain.

Consider, e.g., the affine variety $U = V[\langle x^2 + y^2 - 1 \rangle]$ defined by the polynomial $p(x, y) = x^2 + y^2 - 1 \in \mathcal{P}^2$. In $\mathbb{R}^2$, U is the well-known unit circle which is a *bounded* variety; it has no points in $\{|\xi| > 1, |\eta| > 1\}$. In $\mathbb{C}^2$, on the other hand, the affine variety U is *unbounded*; there are points (ξ, η) in U for any specified value of $\xi \in \mathbb{C}$, with arbitrarily large

$|\xi|$. If we change the sign of the constant term in p the difference is even more conspicuous: In $\mathbb{R}^2$, $V[\langle x^2 + y^2 + 1\rangle]$ is empty while the complex variety does not change its character with the sign change. At the same time, I have to confess that I cannot truly visualize the complex unit circle in my mind, let alone produce illustrative sketches of it on a 2D paper or screen.

This shortcoming of our geometric intuition of the complex domain in two or more dimensions must be kept in mind throughout this book, particularly in connection with illustrations for $s = 2$ or 3 which are necessarily unable to depict the true situation in $\mathbb{C}^2$ or $\mathbb{C}^3$.

An immediate consequence of Proposition 1.12 is:

Corollary 1.13. For any polynomial ideal $\mathcal{I} \subset \mathcal{P}^s$ and $p_1, \ldots, p_k \in \mathcal{I}$, $k \geq 1$,

$$V[\mathcal{I}] \subset V[\langle p_1, \ldots, p_k\rangle]. \tag{1.23}$$

Intuitively, Corollary 1.13 says that the variety of an ideal is the "smaller" the more polynomials the ideal contains; this is natural since we may expect the fewer *joint* zeros for the ideal.

We are accustomed to this fact in linear algebra: Loosely speaking, the zero set of *one* linear polynomial in s variables is a hyperplane of dimension $s - 1$, that of two linear polynomials is a linear variety of dimension $s - 2$ (e.g., a line in $\mathbb{R}^3$) etc. Finally, a system of s linear polynomials has a joint zero set of dimension 0 which consists of only one point. Here, the term "dimension" has been used in an intuitive sense which is possible for linear varieties: A linear variety in $\mathbb{C}^s$ or $\mathbb{R}^s$ has dimension $k < s$ if it is isomorphic to the $\mathbb{C}^k$ or $\mathbb{R}^k$, respectively. However, even in linear algebra the above is true only if a further polynomial p_{k+1} appended to the current set $p_1, \ldots, p_k$ of generators is *linearly independent* from that set. Otherwise, $\langle p_1, \ldots, p_{k+1}\rangle = \langle p_1, \ldots, p_k\rangle$ and the variety does not change.

Unfortunately, the concept of linear independence does not generalize to polynomial algebra because, in a set of 2 or more polynomials, there are always nontrivial polynomial combinations representing the 0-polynomial; cf. Example 1.3. Therefore, the naive statement that the dimension of the associated variety decreases by one for each polynomial which is appended to a basis of a polynomial ideal has to be used with care. (Also the concept of dimension becomes more subtle for general affine varieties; e.g., they may consist of several components which have different dimensions; cf. also below.) Thus, the naive assumption that the zero set of a system of s polynomials in $\mathcal{P}^s$ consists of isolated points (i.e. is of dimension 0) must be verified carefully in nontrivial situations.

Polynomial systems which behave like systems of linearly independent linear polynomials with respect to the dimensions of their zero sets (varieties), deserve a special name:

Definition 1.11. A polynomial system $P = \{p_1, \ldots, p_n\} \subset \mathcal{P}^s$, $n \leq s$, is called a *complete intersection (system)* iff the variety $V[\langle p_{\nu_1}, \ldots, p_{\nu_k}\rangle]$ of each subsystem of k polynomials from P, $k \leq n$, has dimension $s - k$. $\square$

Example 1.5: The system $\{p_1, p_2\}$ of Example 1.3 is clearly a complete intersection. The system

$$p_1 = z^2 - 2x^2 - 2y^2, \quad p_2 = x - y, \quad p_3 = z - x - y,$$

is not a complete intersection: While any two p_ν have a 1-dimensional variety, the variety $V[\langle p_1, p_2, p_3\rangle]$ is also 1-dimensional; it consists of the "straight line" $y = x$, $z = 2x$. But any generic perturbation of P makes it a complete intersection, e.g., adding a small constant to each (or just one of the) p_ν. $\square$

We will consider complete intersection systems in more detail in section 8.3.

The strong relation between ideals and varieties suggests the following commonly used terminology.

Definition 1.12. An ideal $\mathcal{I} \subset \mathcal{P}^s$ whose zero set (affine variety) $V[\mathcal{I}] \subset \mathbb{C}^s$ has a component of dimension d but none of a higher dimension is called *d-dimensional*. A 0-dimensional ideal has a zero set $Z[\mathcal{I}]$ which consists of isolated points only. The same terminology is used for polynomial systems. $\square$

We are also interested in the local analytic structure of the affine variety of a polynomial ideal: Except at certain *singular points* (cf. section 7.3), the neighborhood of a point $z \in \mathbb{C}^s$ on an affine variety is a homeomorphic image of a Euclidean sphere $\|x - z\| \le r$ of some dimension d, the local dimension of the variety. For example, the points of the unit circle $U = V[\langle x^2 + y^2 - 1 \rangle] \subset \mathbb{R}^2$ near some $z \in U$ are an image of a real 1-dimensional line segment. The same is true when we relate a complex segment of $V[\langle x^2 + y^2 - 1 \rangle] \subset \mathbb{C}^2$ to a segment of a complex 1-dimensional "line."

Such point sets are generally called *manifolds* in analysis and also in linear algebra. Because of the natural embedding of our subject into these areas, we will often use the term "manifold" for the zero set of a polynomial or a polynomial system. Like with the term "affine variety," we will do this in an informal manner; fortunately, in connection with our computational tasks, most of the pathological possibilities which require more refined definitions do not occur.

The visualization of polynomial ideals through the affine varieties of their zero sets raises the following natural question: Is there a unique correspondence between affine varieties and polynomial ideals, or

$$V[\mathcal{I}_1] = V[\mathcal{I}_2] \quad \Rightarrow \quad \mathcal{I}_1 = \mathcal{I}_2 \ ? \tag{1.24}$$

While the answer is "yes" in linear algebra, the fact that it is "no" in polynomial algebra can be established by trivial examples in $\mathcal{P}^1$: The ideals $\mathcal{I}_1 = \langle (x - \alpha_1)(x - \alpha_2) \rangle$ and $\mathcal{I}_2 = \langle (x - \alpha_1)^2 (x - \alpha_2)^3 \rangle$ have the same zero sets $\{\alpha_1, \alpha_2\}$; but they are certainly not identical because the generator of $\mathcal{I}_1$ is not in $\mathcal{I}_2$. At the same time, this example reveals the issue: While $\mathcal{I}_1$ and $\mathcal{I}_2$ have the same zeros, the *multiplicities* of these zeros are not the same.

On the other hand, if we let the V in (1.24) stand for *Visualization* instead of *Variety*, we would definitely wish to have "yes" as an answer. Obviously, this requires the inclusion of multiplicity into our visualization concept. Furthermore, we will soon realize that—in *computational* polynomial algebra—it is absolutely indispensable to *consider the multiplicity of a zero at all times* because an m-fold zero splits into m isolated zeros under almost any perturbation, and analogous statements hold also for positive-dimensional zero sets of a multiplicity > 1.

Therefore, in an abuse of language and notation, we consider multiplicity also in the components of affine varieties in order to be able to use them as suitable visualization tools for polynomial ideals. Thus, in the example above, the "visualization variety" $V[\mathcal{I}_1]$ consists of the two *simple* points α_1 and α_2 while $V[\mathcal{I}_2]$ consists of a *double* point α_1 and a *triple* point α_2; consequently—from this point of view—$V[\mathcal{I}_1] \ne V[\mathcal{I}_2]$. Otherwise, affine varieties would not share the continuity properties of zero sets of polynomials and hence remain unsuitable for the visualization of polynomial ideals.

Exercises

1. (a) The real affine varieties defined by real polynomials from $\mathcal{P}_2^2$ are the so-called "conic sections" of analytic geometry. Recall their well-known categories (ellipses, hyperbolas, parabolas, pairs of lines) and the algebraic criteria for membership in a particular category.

(b) Analogously, the real affine varieties defined by real polynomials in $\mathcal{P}_2^3$ are the well-known "surfaces of second order." There is now a wider selection of categories and associated criteria; also there are interesting partial degeneracies like cones, etc. Try to recover the related information from a source on analytic geometry.

(c) Consider the varieties in $\mathbb{R}^2$ and $\mathbb{R}^3$ which arise as zero sets of ideals of two polynomials in $\mathcal{P}_2^2$ and of two or three polynomials in $\mathcal{P}_2^3$, resp.; cf. Example 1.5. Select polynomial sets which form or do not form complete intersection systems, respectively.

2. Consider two affine varieties $V_1 = V[\langle p_1, \ldots, p_m \rangle]$ and $V_2 = V[\langle q_1, \ldots, q_n \rangle]$ in $\mathbb{C}^s$. Prove the following two relations:

$$V_1 \cap V_2 \; = \; V[\langle p_1, \ldots, p_m, q_1, \ldots, q_n \rangle] \, ; \tag{1.25}$$

$$V_1 \cup V_2 \; = \; V[\langle p_\mu q_\nu, \; \mu = 1(1)m, \; \nu = 1(1)n \rangle] \, . \tag{1.26}$$

Note that these relations imply that the intersection and the union of two affine varieties (or of a finite number of them) is again an affine variety.

3. Compare the exercises for Chapter 1, section 2 in [2.10] which provide a wealth of insight.

1.5   Polynomials in Scientific Computing

The strongest stimulus for my occupation with numerical polynomial algebra came from the fact that polynomials play an increasingly important role in those areas of scientific computing which deal with phenomena in engineering and in the natural sciences, including the biological sciences and medicine. In these areas, polynomials furnish a natural tool for the modelling of relations in the real world which cannot be adequately described by linear models. But the analytic and predictive value of mathematical models carries only as far as our *solving capacity* for the related mathematical problems.

In a good deal of the work in Computer Algebra, it has not been sufficiently considered that "solving"— in a real-world situation—generally means the computational extraction of *numerical* values for quantities which satisfy mathematical relations containing *numerical* data as coefficients. Some or all of these numerical data represent aspects of modelled systems which are only known with a *limited accuracy*. Even in engineering problems, some input quantities may only be known with a low relative accuracy; in other areas, there may only be 1 or 2 meaningful digits in some input data.

Such a low input accuracy may not only stem from the limited accuracy of measurements but also from the inaccuracies introduced by preceding computations. Moreover, in most mathematical models of real-world phenomena, it is unavoidable that some "secondary" effects are not taken into account; this implies that the relations themselves are of limited accuracy. Often, the omission of such terms is vaguely equivalent to a lower accuracy in the coefficients of some terms which are present.

In any case, these circumstances imply that it is not meaningful to determine the requested output values with an arbitrarily high accuracy. It is intuitively clear that result digits which do not remain invariant under changes of the input data well within their ranges of indetermination are meaningless and therefore *need not be computed*. In fact, the *reporting* of such digits may pretend a solution accuracy which is not justified. The key test for the validity of a result is provided by a *backward analysis*: If the computed approximate result values may be interpreted as *exact result values* of a neighboring problem whose data are—within the limited accuracy—*indistinguishable* from the specified ones, then the approximate results are not only valid solutions of the specified problem but they *cannot be improved* under the given information limitations. In Chapter 3, we will introduce a formal basis for these considerations.

This awareness of the *nonexistence of "exact solutions"* in practically all problems of scientific computing also opens the way for the use of floating-point arithmetic in the computational solution of algebraic problems and for the use of approximate methods like truncated iterative methods. Thus, the computational treatment of polynomial problems becomes an intrinsic part of *numerical analysis*, and the concepts and approaches of numerical analysis have to be superimposed on those of polynomial algebra. This mutual penetration of two areas of mathematics which have remained disjoint for a long time, has created a large number of new tasks and challenges. The later parts of this book will be devoted to their elaboration and to their—often preliminary—solution. Many of the insights and results which we will gain should form an indispensible basis for the solution of nonlinear algebraic problems in scientific computing.

1.5.1 Polynomials in Scientific and Industrial Applications

Polynomial models appear in nearly all areas of scientific computing. The recent *Computer Algebra Handbook*[2] devotes nearly 100 pages to an overview of applications of computer algebra. Most of these employ polynomials or systems of polynomials, often in many variables, and require numerical results. Besides applications in other areas of mathematics, the *Handbook* describes applications in Physics, Chemistry, Engineering, and Computer Science. Actually, the use of polynomial models nowadays extends from the Life and Earth Sciences over the whole spectrum of classical scientific activity to the Social Sciences and Finance.

Industrial applications and the relationship between the current state of the art in polynomial systems solving and the industrial needs in this area were studied in a subtask of the European Community Project FRISCO[3] (a Framework for Integrated Symbolic/Numeric Computation, 1996–1999). A report about this subtask may be found at www.nag.co.uk/projects/ FRISCO/frisco/frisco. In the following, we comment on some of the findings in this document.

The report stresses the difficulties which were met in the attempts to obtain sufficiently detailed and meaningful information about the occurrence and particular features of polynomial problems in industrial research and development. This is a well-known phenomenon which inhibits the potential cooperation between academic and applied research in many areas, but, in the case of polynomial algebra, it is aggravated by the abstract appearance of a good deal of the related scientific publications and even of the documentation of the related software. For

---

[2] J. Grabmeier, E. Kaltofen, V. Weispfenning (Eds.): *Computer Algebra Handbook – Foundations, Applications, Systems*; Springer, Berlin, 2003.

[3] Partners in the project were NAg, several universities, and several industrial enterprises.

example, the first sentence of the online description of the `Groebner` package of Maple™ 7 which contains a number of very useful routines for the treatment of polynomial problems, reads: "The `Groebner` package is a collection of routines for doing Groebner basis calculations in skew algebras like Weyl and Ore algebras and in corresponding modules like D-modules." This will not induce an engineer or application scientist in an industrial environment to read any further, even if he/she has been told that this is the place to look for tools for his/her problems.

Hopefully, this book, with its abstinence from unnecessary abstractions and its use of the widely known language and notation of numerical linear algebra, will help a number of application scientists to discover the tools which are available for dealing with the nonlinear polynomial problems which they may have to solve.

Within the scope of the above-mentioned subtask of FRISCO, polynomial algebra problems were uncovered and discussed with practitioners in the following fields:

- Computer Aided Design and Modelling,

- Mechanical Systems Design,

- Signal Processing and Filter Design,

- Civil Engineering,

- Fluids, Materials, and Durability,

- Robotics,

- Simulation.

Typically, the nonlinearities arose because either a linear(ized) model was not feasible or not sufficiently accurate. The problems were either systems of polynomial equations or optimization problems with polynomial objective functions and/or restraints. While the number of polynomial equations and unknowns was anything from moderate (say 16) to very large ($O(10^3)$), their degree was generally quite low (2 to 4) and extreme sparsity was common.

Coefficients were generally real and "in most cases they come from experimental data and thus are known only to a limited accuracy." This supports another main objective of this book and stresses the need for further research in numerical polynomial algebra.

For more details, we refer the reader to the above-mentioned report which is freely available on the Internet.

Exercises

1. If a data value in a problem signifies one of the following quantities, which relative accuracy (how many significant decimal digits) would you attribute to it:

        width of an artery,
        body length,
        depth of a river at a given gauge,
        height of a mountain,
        a city's electric energy consumption in 24 hours,
        national gross income.

Find further quantities with a low relative accuracy which may enter a real-life model computation.

2. Consider a univariate real polynomial p whose constant term is an empirical data value with an absolute tolerance of 10^{-3}. What is the induced indetermination in the location of a real zero of p and on what does it depend? What is the induced indetermination in the location of a double zero of p ?

3. (a) Use a reliable solver to compute the zeros of the polynomials

$$p(x) = x^4 - 2.83088\,x^3 + .00347\,x^2 + 5.66176\,x - 4.00694,$$
$$\tilde{p}(x) = x^4 - 2.83087\,x^3 + .00348\,x^2 + 5.66177\,x - 4.00693.$$

Comment on the result. If the coefficients of p have an absolute tolerance of 10^{-5}, what is a meaningful assertion about the zeros of p in the positive halfplane and about the zero in the left halfplane ?

(b) If we report the zeros of p in the right halfplane with 5 decimal digits after the point, which is the implied accuracy of the coefficients of p ?

Chapter 2

Representations of Polynomial Ideals

Individual polynomials in one or several variables and systems of such polynomials constitute the basic material of numerical polynomial algebra. Many aspects of polynomials are related to the polynomial ideals which they generate. In particular, their zeros which are a central object of computational algebra may be viewed as the joint zeros of all polynomials in the generated ideal; cf. Proposition 1.12. Therefore, it is useful to study the ways in which polynomial ideals may be represented and analyzed. We will find that the representation of a polynomial ideal through its quotient ring and/or its dual space is often more suitable for computational purposes than customary basis representations, even with Groebner bases.

This chapter is meant as an introduction to polynomial algebra for readers without expertise in that subject; this will probably include a majority of those with a numerical analysis background. Our restriction to fundamental relations and our emphasis on linear algebra aspects, with the associated terminology and notation, should help them to gain the understanding of polynomial algebra needed for the main parts of the book. Readers with expertise in polynomial algebra may be surprised how the appearance of the subject changes when it is regarded as an extension of linear algebra and analysis. The central result of this chapter in section 2.4 may be new to them in this form, as will some of the content of section 2.5.

2.1 Ideal Bases

The classical view of polynomial ideals is that of all linear combinations (cf. (1.18)) of a specified set P of polynomials, the *generators* of the ideal; cf. Definition 1.7. However, many different sets of generators may define the same ideal.

Definition 2.1. For a specified polynomial ideal $\mathcal{I} \subset \mathcal{P}^s$, any set of polynomials $G = \{g_\kappa \in \mathcal{P}^s, \ \kappa = 1(1)k\}$ such that

$$\langle g_1, \ldots, g_k \rangle = \mathcal{I} \tag{2.1}$$

is a *basis* of $\mathcal{I}$. $\square$

Example 2.1: In Example 1.4, we have considered a system P of s linearly independent linear polynomials $p_\nu = \alpha_{\nu 0} + a_\nu^T \mathbf{x} \in \mathcal{P}_1^s$, with the joint zero $z \in \mathbb{C}^s$. It was shown that $\langle P \rangle$ has also the basis $\{x_\sigma - \zeta_\sigma, \ \sigma = 1(1)s\}$.

Other bases of $\langle P \rangle$ which are important in the computational determination of z are so-called triangular bases T consisting of the s polynomials

$$t_\sigma := \sum_{\lambda=\sigma}^{s} \gamma_{\sigma\lambda} \, x_\lambda + \gamma_{\sigma 0}, \quad \sigma = 1(1)s,$$

with $\gamma_{\sigma\sigma} \neq 0$, $\sigma = 1(1)s$. The joint zero of a triangular linear system is easily found by a recursive computation of its components $\zeta_s, \zeta_{s-1}, \ldots, \zeta_1$ beginning with the last polynomial t_s (which contains only the variable x_s) and proceeding successively to the first one. This procedure is called *back(ward) substitution* in numerical linear algebra.   □

It is trivial that a given basis G of a polynomial ideal $\mathcal{I}$ may be arbitrarily expanded by the inclusion of other elements from $\mathcal{I}$. On the other hand, it is often not possible to omit one of the generator polynomials in a basis G without an alteration of $\langle G \rangle$.

Definition 2.2. A basis $G = \{g_\kappa, \ \kappa = 1(1)k\}$ of a polynomial ideal $\mathcal{I} = \langle G \rangle$ is called *minimal* if all ideals $\mathcal{I}_l = \langle g_\kappa, \kappa \neq l \rangle$, $l = 1(1)k$, are *proper* subsets of $\mathcal{I}$.   □

For an ideal in $\mathcal{P}^1$, a minimal basis always consists of 1 element and is essentially unique (cf. Proposition 5.2); ideals in $\mathcal{P}^s$, $s > 1$, have many different minimal bases which may have different numbers of elements.

Example 2.2: For $s = 2$, consider $\mathcal{I} = \langle p_1, p_2 \rangle$ with the quadratic polynomials

$$p_1 = x^2 + 4\,x\,y + 4\,y^2 - 4, \qquad p_2 = 4\,x^2 - 4\,x\,y + y^2 - 4;$$

their joint zeros are the 4 intersections of the 2 pairs of parallel lines $p_1 = 0$, $p_2 = 0$,

$$(1.2, .4), \ (-.4, 1.2), \ (-1.2, -.4), \ (.4, -1.2).$$

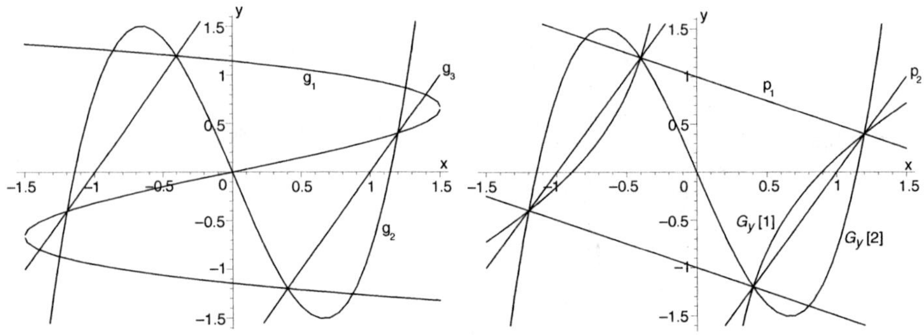

Figure 2.1.

Naturally, $\{p_1, p_2\}$ is a minimal basis; other minimal bases of $\mathcal{I}$ are (cf. Figure 2.1)

$$G_x = \{20\,x\,y + 15\,y^2 - 12, \ 125\,y^3 + 48\,x - 164\,y\},$$

$$G_y = \{20\,x\,y - 15\,x^2 + 12,\; 125\,x^3 - 164\,x - 48\,y\},$$
$$G_{lx} = \{125\,y^3 + 48\,x - 164\,y,\; 625\,y^4 - 1000\,y^2 + 144\},$$
$$G_{ly} = \{125\,x^3 - 164\,x - 48\,y,\; 625\,x^4 - 1000\,x^2 + 144\},$$
$$G_s = \{15\,x^2 - 20\,x\,y - 12,\; 15\,y^2 + 20\,x\,y - 12\}.$$

But $G = \{g_1, g_2, g_3\} = \{125\,y^3 + 48\,x - 164\,y,\; 125\,x^3 - 164\,x - 48\,y,\; 4\,x^2 - 4\,x\,y + y^2 - 4\}$ with its three generators is also a minimal basis of $\mathcal{I}$: $\langle g_1, g_2 \rangle$, $\langle g_2, g_3 \rangle$ and $\langle g_3, g_1 \rangle$ are smaller ideals than $\mathcal{I}$, each of them has more than 4 zeros. $\square$

Since an ideal in $\mathcal{P}^s$ is always an infinite set (except for the trivial ideal $\langle 0 \rangle$), one may ask whether there are polynomial ideals which require an infinite basis for their definition. This is not the case according to

Hilbert's Basis Theorem: Every ideal in $\mathcal{P}^s$ has a finite generating set. (Cf. e.g. [2.10].)

At first sight, one might hope that a minimal basis G of a polynomial ideal $\langle G \rangle$ could be employed in computational polynomial algebra in a similar fashion as bases of linear polynomials are used in linear algebra. This is, however, not the case:

Polynomial ideal bases are *not* bases in the sense of linear algebra.

The distinction arises because the fundamental concept of *linear independence* does not generalize to polynomial combinations which are the building blocks of polynomial ideals.

Proposition 2.1. For a minimal basis $G = \{g_\kappa,\; \kappa = 1(1)k\}$ of an ideal $\langle G \rangle \subset \mathcal{P}^s$, $s > 1$,

$$c_1(x)\,g_1(x) + c_2(x)\,g_2(x) + \ldots + c_k(x)\,g_k(x) = 0 \quad \text{(the zero-polynomial)} \tag{2.2}$$

does *not* imply $c_\kappa(x) = 0$, $\kappa = 1(1)k$, except when $k = 1$.

Proof: For a counterexample, simply choose $g_1 = x^2$, $g_2 = xy$, $g_3 = y^2$; then, e.g., $y \cdot g_1 - x \cdot g_2 = 0$. For $k = 1$, $c\,g = 0$ implies $c = 0$ because $\mathcal{P}^s$ is an integral domain for each value of s; cf. Corollary 1.10. $\square$

Definition 2.3. A nontrivial linear combination (2.2) of the polynomials $g_\kappa \in G$ which equals the zero polynomial is called a *syzygy*[4] in $\langle G \rangle$. Syzygies of the type $g_{\kappa_2} g_{\kappa_1} - g_{\kappa_1} g_{\kappa_2}$ are *trivial*; all other syzygies are *nontrivial*. $\square$

The existence of syzygies in multivariate polynomial ideals presents a major difficulty for computational algorithms. In section 8.3.3, we will see that a basis of s elements for a 0-dimensional ideal in $\mathcal{P}^s$ (a complete intersection system) has only trivial syzygies; cf. Proposition 8.24.

Example 2.3: In Example 2.2, the 2-element bases can only have trivial syzygies. However, the 3-element basis G has the following nontrivial syzygy:

$$(13\,x - 4\,y)\,g_1(x, y) + (16x - 28\,y)\,g_2(x, y) + (-500\,x^2 + 375\,x\,y + 500\,y^2)\,g_3(x, y) = 0.\;\; \square$$

The existence of unique minimal bases makes the *univariate* case exceptional in many respects (details in Chapter 5): Computational polynomial algebra in $\mathcal{P}^1$ is essentially simpler

---

[4]This strange word has originated from the Greek word for "yoke" and has been introduced in astronomy to denote an *alignment* of several heavenly bodies.

than in $\mathcal{P}^s$, $s > 1$. In particular, with respect to any specified polynomial ideal $\mathcal{I} = \langle g \rangle$, deg $g = d \geq 1$, each polynomial $p \in \mathcal{P}^1$ has a *unique decomposition* $p(x) = r(x) + q(x) \cdot g(x)$, with deg $r < d$; cf. (5.16). Obviously, $p \in \mathcal{I}$ iff $r = 0$.

An analogous decomposition for the case $s \geq 2$ and $k \geq 2$ has the form

$$p(x) = r(x) + \sum_\kappa q_\kappa(x) \cdot g_\kappa(x), \quad \text{with } r, q_\kappa \in \mathcal{P}^s. \tag{2.3}$$

Since syzygies may be freely added to a linear combination of the basis elements without altering the represented quantity, it is clear that a decomposition (2.3) of an element $p \in \mathcal{P}^s$ cannot be unique. But we may still hope for a unique remainder r modulo $\langle G \rangle$ of p if the selection of r is suitably restricted; cf. deg $r <$ deg g in the univariate case.

Here, a different ambiguity comes into play: In the univariate case, the restriction deg $r <$ deg g prevents the addition of some multiple of g to r because any multiple of g has a degree $\geq$ deg g. An analogous restriction is not so readily available in the multivariate case as we see from the following example.

Example 2.4:   The ideal $\mathcal{I}$ in Example 2.2 consists of all polynomials in $\mathcal{P}^2$ which vanish on the joint zero set $Z[\mathcal{I}] = \{(1.2, .4), (-.4, 1.2), (-1.2, -.4), (.4, -1.2)\}$. Therefore, any remainder $r \bmod \mathcal{I}$ of some $p \in \mathcal{P}^2$ must copy the values of p at the four points in $Z[\mathcal{I}]$. But a *linear* polynomial r in 2 variables cannot generally assume specified values at 4 disjoint points in $\mathbb{C}^2$; therefore, r must have total degree 2. Since there is a 2-parameter family of $r \in \mathcal{P}_2^2$ which interpolate p on $Z[\mathcal{I}]$, there may be many potential remainders r of total degree 2 for a decomposition (2.3). $\square$

The example indicates that the appropriate restriction for r is not by degree but by requested membership in a particular interpolation space on the zeros of $\mathcal{I}$; this will be confirmed later. The univariate degree restriction may also be interpreted in this way.

Exercises

1. With p_1, p_2 from Example 2.2, the Maple command `gbasis({p1,p2},tdeg(x,y))` yields the (Groebner) basis $G = \{125\,y^3 - 164\,y + 48\,x, \ 20\,xy + 15\,y^2 - 12, \ 5\,x^2 + 5\,y^2 - 8\}$ for $\langle p_1, p_2 \rangle$.

(a) Find a nontrivial syzygy of the 3-element basis G. (Hint: Consider $x\,g_1 - y^2\,g_2$ and add/subtract appropriate multiples of the g_κ to reach 0.)

(b) From that syzygy, conclude that $g_3 \in \langle g_1, g_2 \rangle$ so that G is not minimal.

2.   Consider the ideal of all polynomials in $\mathcal{P}^s$ vanishing at a specified point $z \in \mathbb{C}^s$, with the basis $G = \{x_\sigma - \zeta_\sigma, \ \sigma = 1(1)s\}$; cf. Example 2.1. Show that each polynomial $p \in \mathcal{P}^s$ possesses a unique representation

$$p(x_1, \ldots, x_s) = \rho + q_1(x_1, \ldots, x_s)\,(x_1 - \zeta_1) + q_2(x_2, \ldots, x_s)\,(x_2 - \zeta_2) + \ldots + q_s(x_s)\,(x_s - \zeta_s),$$

with $\rho = p(z) \in \mathbb{C}$, $q_\sigma \in \mathcal{P}^{s-\sigma+1}$, $\sigma = (1)s$. (Hint: Consider $p - \rho$ as a polynomial in x_1 only (with the remaining variables as parameters) and apply the univariate decomposition; continue recursively.)

3. In Example 2.4, $\mathcal{R} = \mathrm{span}\{1, x, y, xy\}$ is a suitable space for interpolation at the zeros of $\mathcal{I}$.

(a) Show that each quadratic polynomial p in x, y has a unique representation $p(x, y) = r(x, y) + \gamma_1\, p_1(x, y) + \gamma_2\, p_2(x, y)$, with $r \in \mathcal{R}$, $\gamma_1, \gamma_2 \in \mathbb{C}$; cf. the note after Example 2.4.

(b) Are there analogous results for polynomials in $\mathcal{P}_3^2$ and $\mathcal{P}_4^2$?

2.2 Quotient Rings of Polynomial Ideals

2.2.1 Linear Spaces of Residue Classes and Their Multiplicative Structure

A more successful approach to the determination of a remainder of a multivariate polynomial p modulo an ideal $\mathcal{I}$ begins by considering *all* potential remainders r in (2.3) *simultaneously.*

Definition 2.4. The set

$$[p]_\mathcal{I} := \{r \in \mathcal{P}^s : p - r \in \mathcal{I}\} \tag{2.4}$$

of all remainders of $p \in \mathcal{P}^s$ modulo the polynomial ideal $\mathcal{I} \subset \mathcal{P}^s$ is the *residue class of p mod $\mathcal{I}$.* $\square$

Obviously, the set $[p]_\mathcal{I}$ is obtained by adding to p each element of $\mathcal{I}$:

$$[p]_\mathcal{I} := \{p\} + \mathcal{I}\,.$$

Thus, $[p]_\mathcal{I}$ has the same cardinality as $\mathcal{I}$; in particular, $p \in \mathcal{I}$ implies $[p]_\mathcal{I} = [0]_\mathcal{I} = \mathcal{I}$.

When p varies within $\mathcal{P}^s$, we obtain an infinite set $\mathcal{R}[\mathcal{I}]$ of residue classes $[p]_\mathcal{I}$. Within this set, we can define the standard operations of a vector space in a straightforward way: For $\alpha \in \mathbb{C}$, $p, q, \in \mathcal{P}^s$,

$$\alpha \cdot [p]_\mathcal{I} := [\alpha\, p]_\mathcal{I}\,, \tag{2.5}$$

$$[p]_\mathcal{I} + [q]_\mathcal{I} := [p + q]_\mathcal{I}\,. \tag{2.6}$$

These definitions are valid because the right-hand sides do not depend on the particular polynomials chosen to represent the residue classes on the left-hand side: Take two different polynomials $p, \bar{p}$ and $q, \bar{q}$ in each of the left-hand side residue classes; then (cf. (2.4))

$$[\alpha\, \bar{p}]_\mathcal{I} = [\alpha\, p]_\mathcal{I} \text{ because } \alpha\bar{p} - \alpha p = \alpha\,(\bar{p} - p) \in \mathcal{I}\,,$$

$$[\bar{p} + \bar{q}]_\mathcal{I} = [p + q]_\mathcal{I} \text{ because } (\bar{p} + \bar{q}) - (p + q) = (\bar{p} - p) + (\bar{q} - q) \in \mathcal{I}\,.$$

Thus, the set $\mathcal{R}[\mathcal{I}]$ of all residue classes mod $\mathcal{I}$ is a *vector space* over $\mathbb{C}$ or $\mathbb{R}$, depending on the coefficient field in $\mathcal{P}^s$. This implies that we must be able to choose a *basis* $\{[b_1], \ldots\}$ in $\mathcal{R}[\mathcal{I}]$ and to represent all elements (= residue classes) in $\mathcal{R}[\mathcal{I}]$ by their components (γ_λ) in that basis:

$$[p]_\mathcal{I} = \sum_\lambda \gamma_\lambda\, [b_\lambda]_\mathcal{I}\,. \tag{2.7}$$

Note that we are in a vector space so that the γ_λ are *scalars*, i.e. complex or real numbers, and that a basis $B = \{[b_\lambda]_\mathcal{I}\} \subset \mathcal{R}[\mathcal{I}]$ must be *linearly independent* in the standard sense:

$$\sum_\lambda \gamma_\lambda\, [b_\lambda]_\mathcal{I} = [0]_\mathcal{I} \quad \Longrightarrow \quad \gamma_\lambda = 0 \ \ \forall \lambda\,. \tag{2.8}$$

This is a first indication that $\mathcal{R}[\mathcal{I}]$ may provide a friendly environment for numerical computations.

In (2.7) and (2.8), we have left open the size m of a basis of $\mathcal{R}[\mathcal{I}]$, i.e. the dimension of $\mathcal{R}[\mathcal{I}]$ as a vector space. Actually the dimension m of $\mathcal{R}[\mathcal{I}]$ depends intimately on the structure of the underlying polynomial ideal $\mathcal{I}$, as shown below.

Consider a set $Z[\mathcal{I}] = \{z_1, \ldots, z_m\} \subset \mathbb{C}^s$ with disjoint z_μ, and the related ideal of (1.20):

$$\mathcal{I} = \{ p \in \mathcal{P}^s : p(z_\mu) = 0, \ \mu = 1(1)m \}.$$

Then (2.4) implies

$$[r]_\mathcal{I} = \{q \in \mathcal{P}^s : q(z_\mu) = r(z_\mu), \ \mu = 1(1)m\}. \tag{2.9}$$

Thus, each residue class $[r]_\mathcal{I} \in \mathcal{R}[\mathcal{I}]$ is fully characterized by the m common values $r(z_\mu) \in \mathbb{C}$ which its elements take at the zeros z_μ of $\mathcal{I}$.

Proposition 2.2. For the above ideal $\mathcal{I} \subset \mathcal{P}^s$ with the m zeros of Z, $\mathcal{R}[\mathcal{I}]$ has dimension m.
Proof: In $\mathcal{R}[\mathcal{I}]$, we introduce a so-called Lagrange basis $\widehat{B} = \{[\hat{b}_\lambda]_\mathcal{I}, \ \lambda = 1(1)m\}$ with $\hat{b}_\lambda \in \mathcal{P}^s$ and

$$\hat{b}_\lambda(z_\mu) = \begin{cases} 0 & \lambda \neq \mu, \\ 1 & \lambda = \mu, \end{cases} \quad \lambda, \mu = 1(1)m. \tag{2.10}$$

Then the values $r(z_\mu)$ are immediately the components of $[r]_\mathcal{I}$ w.r.t. the basis $\widehat{B}$:

$$[r]_\mathcal{I} = \sum_{\mu=1}^{m} r(z_\mu) [\hat{b}_\mu]_\mathcal{I}. \quad \Box \tag{2.11}$$

We will discuss the computational determination of a Lagrange basis for $\mathcal{R}[\mathcal{I}]$ later (cf. section 2.4.1); its existence is trivial.

Proposition 2.3. For each set $Z = \{z_1, \ldots, z_m\} \subset \mathbb{C}^s$ of m disjoint complex s-tuples, there exist polynomials $\hat{b}_\lambda \in \mathcal{P}^s, \lambda = 1(1)m$, which satisfy (2.10).
Proof: Assume that the x_1-components $(\zeta_\mu)_1$ of the $z_\mu \in Z$ are disjoint; for disjoint z_μ, this can always be achieved by a linear transformation of the variables x_σ: for almost all directions in s-space, the projections of the z_μ onto these directions are disjoint. Then, the polynomials

$$\hat{b}_\lambda(x_1, \ldots, x_s) := \prod_{\mu \neq \lambda} (x_1 - (\zeta_\mu)_1) / \prod_{\mu \neq \lambda} ((\zeta_\lambda)_1 - (\zeta_\mu)_1)$$

(the well-known one-dimensional Lagrange basis polynomials) are well-defined and satisfy (2.10). $\Box$

In the above ideal $\mathcal{I}$, all m zeros z_μ are simple. If some of the zeros z_μ of $\mathcal{I}$ are not simple (say z_1), then there exist scalar linear combinations of partial derivative evaluations at z_1 which vanish for all $p \in \mathcal{I}$. The values of a linearly independent set of these derivative evaluations furnish further components for the specification of the residue class $[q]_\mathcal{I}$ of some $q \in \mathcal{P}^s$. The number of these linearly independent derivative evaluations is $m_\mu - 1$ for a zero z_μ of multiplicity m_μ. This will be discussed in detail in section 2.3. Thus we have derived

Theorem 2.4. For a 0-dimensional ideal $\mathcal{I} \subset \mathcal{P}^s$, the vector space $\mathcal{R}[\mathcal{I}]$ of all residue classes mod $\mathcal{I}$ is finite-dimensional; its dimension $m \in \mathbb{N}$ is equal to the number of zeros of $\mathcal{I}$ *counting multiplicities.*

Example 2.5: Consider Example 2.2, with its four isolated zeros. According to the above, the four residue classes,

$$[\frac{(x+.4)(x+1.2)(x-.4)}{1.6 \cdot 2.4 \cdot .8}], \quad [\frac{(x+1.2)(x-.4)(x-1.2)}{.8 \cdot .8 \cdot 1.6}],$$

$$[\frac{(x+.4)(x-.4)(x-1.2)}{-.8 \cdot 1.6 \cdot 2.4}], \quad [\frac{(x+.4)(x+1.2)(x-1.2)}{-.8 \cdot 1.6 \cdot .8}],$$

form a Lagrange basis for $\mathcal{R}[\mathcal{I}]$. Each of these classes contains "nicer" polynomials, i.e. more natural representatives for the classes (cf. also Exercise 2.2-1).

A more natural basis for $\mathcal{R}[\mathcal{I}]$ is $B = \{[1], [x], [y], [xy]\}$; the presence of $[xy]$ is in agreement with the observation at the end of Example 2.4 that there must be remainders mod $\mathcal{I}$ with total degree 2. B is a basis of $\mathcal{R}[\mathcal{I}]$ because it permits interpolation at the $z_\nu = (\xi_\nu, \eta_\nu)$; $\nu = 1(1)4$: The matrix of the linear system $\sum_\mu \alpha_\mu b_\mu(z_\nu) = w_\nu$, $\nu = 1(1)4$, is

$$(b_\mu(z_\nu)) = \begin{pmatrix} 1 & 1 & 1 & 1 \\ 1.2 & -.4 & -1.2 & .4 \\ .4 & 1.2 & -.4 & -1.2 \\ -48 & -.48 & .48 & -.48 \end{pmatrix}, \quad \text{with det} (\ldots) = -6.144 \neq 0. \quad \square$$

For a positive-dimensional ideal, the dimension of the vector space $\mathcal{R}[\mathcal{I}]$ is *infinite*: Consider the 1-dimensional ideal $\mathcal{I} = \langle x_1 \rangle \subset \mathcal{P}^2$. Obviously, $[p(x_1, x_2)]_\mathcal{I} = [p(0, x_2)]_\mathcal{I}$, and the associated vector space $\mathcal{R}[\mathcal{I}]$ is isomorphic to the vector space $\mathcal{P}^1$ of polynomials in one variable which is infinite-dimensional. But each specific computation in $\mathcal{P}^1$ utilizes only a finite-dimensional subset of $\mathcal{P}^1$ so that we may hope that $\mathcal{R}[\mathcal{I}]$ is computationally useful also for positive-dimensional ideals. We will further analyze this situation in section 11.1.

So far, we have considered $\mathcal{R}[\mathcal{I}]$ only as a vector space. But we can also define a *multiplication* for its elements:

$$[p]_\mathcal{I} \cdot [q]_\mathcal{I} := [p\,q]_\mathcal{I} \tag{2.12}$$

which is a valid definition because (cf. the explanations after (2.5) and (2.6))

$$\bar{p}\,\bar{q} - p\,q = \bar{p}\,(\bar{q} - q) + q\,(\bar{p} - p) \in \mathcal{I}.$$

This multiplication is clearly commutative; it is also distributive, and it has the 1-element $[1]_\mathcal{I}$. Hence, with (2.12), $\mathcal{R}[\mathcal{I}]$ is a *commutative ring*.

Definition 2.5. For a polynomial ideal $\mathcal{I} \subset \mathcal{P}^s$, the ring $\mathcal{R}[\mathcal{I}]$, with the operations (2.5), (2.6), and (2.12), is called *quotient ring* or *residue class ring modulo* $\mathcal{I}$; it is often denoted by $\mathcal{P}^s/\mathcal{I}$. $\square$

The term *quotient ring* and the notation $\mathcal{P}^s/\mathcal{I}$ are used because the construction of $\mathcal{R}[\mathcal{I}]$ follows the general recipe for the formation of "quotient spaces" of equivalence classes. Note, however, that $\mathcal{R}[\mathcal{I}]$ is *not an integral domain*; e.g., the product of two elements of a Lagrange basis is $[0]_\mathcal{I}$ because it vanishes at all zeros of $\mathcal{I}$ and is thus in $\mathcal{I}$.

If we want to utilize the multiplication (2.12) computationally, we must express it in terms of the components of the factors and their product w.r.t. a specified basis of $\mathcal{R}[\mathcal{I}]$. Again we restrict ourselves, at first, to 0-dimensional ideals so that the vector space dimension m of the quotient ring $\mathcal{R}[\mathcal{I}]$ is finite. Let $\mathcal{R}[\mathcal{I}] = \mathrm{span}\ \{[b_\mu]_\mathcal{I},\ \mu = 1(1)m\}$.

From $[p_i]_\mathcal{I} = \sum \gamma_{i\mu}[b_\mu]_\mathcal{I},\ i = 1, 2$, we have, due to the distributivity of the product,

$$[p_1]_\mathcal{I}[p_2]_\mathcal{I} = \sum_{\mu=1}^{m}\sum_{\nu=1}^{m} \gamma_{1\mu}\gamma_{2\nu}[b_\mu]_\mathcal{I}[b_\nu]_\mathcal{I} = \sum_{\mu,\nu=1}^{m} \gamma_{1\mu}\gamma_{2\nu}[b_\mu\, b_\nu]_\mathcal{I}. \tag{2.13}$$

The products of the basis elements $[b_\nu]_\mathcal{I}$ must themselves be representable in terms of the basis $\{[b_\mu]_\mathcal{I}\}$:

$$[b_\mu]_\mathcal{I} \cdot [b_\nu]_\mathcal{I} = [b_\mu\, b_\nu]_\mathcal{I} = \sum_{\lambda=1}^{m} \alpha_{\mu\nu\lambda}\, [b_\lambda]_\mathcal{I}, \quad \mu, \nu = 1(1)m\,.$$

To avoid the 3-dimensional array ($\alpha_{\mu\nu\lambda}$), it is customary to subdivide it into m 2-dimensional arrays (= matrices) for fixed μ, $\mu = 1(1)m$. Then one may use standard matrix-vector notation and write

$$[b_\mu]_\mathcal{I} \cdot \begin{pmatrix} [b_1]_\mathcal{I} \\ \vdots \\ [b_m]_\mathcal{I} \end{pmatrix} = \begin{pmatrix} A_{[b_\mu]} \end{pmatrix} \begin{pmatrix} [b_1]_\mathcal{I} \\ \vdots \\ [b_m]_\mathcal{I} \end{pmatrix}, \tag{2.14}$$

where the $m \times m$-matrices $A_{[b_\mu]}$ have the complex numbers $\alpha_{\mu\nu\lambda}$ as elements in their ν-th row and λth column. The set of the m numerical $m \times m$ matrices $A_{[b_\mu]}$ specifies the multiplication in $\mathcal{R}[\mathcal{I}]$ w.r.t. the basis $\{[b_\mu]_\mathcal{I},\ \mu = 1(1)m\}$.

Example 2.6: In Example 2.5, we had at first introduced the Lagrange basis $\widehat{B}$ of $\mathcal{R}[\mathcal{I}]$. By its definition, it is clear that

$$[\hat{b}_\mu]_\mathcal{I}[\hat{b}_\nu]_\mathcal{I} = \begin{cases} [\hat{b}_\mu]_\mathcal{I} & \nu = \mu \\ 0 & \nu \neq \mu \end{cases} \qquad \text{so that} \qquad A_{[\hat{b}_\mu]} = \mathrm{diag}\,(0\ldots \overset{\mu}{1}\ldots 0)\,,$$

and (cf. (2.13)):

$$[p_1]_\mathcal{I}[p_2]_\mathcal{I} = \sum_{\mu,\nu} \gamma_{1\mu}\gamma_{2\nu}[\hat{b}_\mu\hat{b}_\nu]_\mathcal{I} = \sum_{\mu} \gamma_{1\mu}\gamma_{2\mu}[\hat{b}_\mu]_\mathcal{I}\,,$$

which is also intuitively clear.

In Example 2.5, we had then introduced the basis $B_\mathcal{I} = \{[1], [x], [y], [xy]\}$ for the quotient ring $\mathcal{R}[\mathcal{I}]$. The multiplication matrices of $\mathcal{R}[\mathcal{I}]$ w.r.t. this basis are

$$A_{[1]} = I\,, \qquad A_{[x]} = \begin{pmatrix} 0 & 1 & 0 & 0 \\ \frac{4}{5} & 0 & 0 & \frac{4}{3} \\ 0 & 0 & 0 & 1 \\ 0 & \frac{48}{125} & \frac{36}{125} & 0 \end{pmatrix},$$

$$A_{[y]} = \begin{pmatrix} 0 & 0 & 1 & 0 \\ 0 & 0 & 0 & 1 \\ \frac{4}{5} & 0 & 0 & -\frac{4}{3} \\ 0 & \frac{36}{125} & -\frac{48}{125} & 0 \end{pmatrix}, \qquad A_{[xy]} = \begin{pmatrix} 0 & 0 & 0 & 1 \\ 0 & \frac{48}{125} & \frac{36}{125} & 0 \\ 0 & \frac{36}{125} & -\frac{48}{125} & 0 \\ \frac{144}{625} & 0 & 0 & 0 \end{pmatrix}.$$

To find $A_{[x]}$, e.g., we observe that the basis G_s of $\mathcal{I}$ in Example 2.2 implies that, mod $\mathcal{I}$,

$$x^2 = \frac{4}{3}xy + \frac{4}{5}, \quad \text{and} \quad \begin{aligned} xy &= \frac{3}{4}x^2 - \frac{3}{5}, \\ y^2 &= -\frac{4}{3}xy + \frac{4}{5}, \end{aligned}$$

which implies (multiply by y and x, resp., and subtract) $\frac{25}{12}x^2y = \frac{4}{5}x + \frac{3}{5}y$. Now we have

$$[x] \cdot \begin{pmatrix} [1] \\ [x] \\ [y] \\ [xy] \end{pmatrix} = \begin{pmatrix} [x] \\ [x^2] \\ [xy] \\ [x^2y] \end{pmatrix} = \begin{pmatrix} [x] \\ \frac{4}{5}[1] + \frac{4}{3}[xy] \\ [xy] \\ \frac{48}{125}[x] + \frac{36}{125}[y] \end{pmatrix} = \begin{pmatrix} 0 & 1 & 0 & 0 \\ \frac{4}{5} & 0 & 0 & \frac{4}{3} \\ 0 & 0 & 0 & 1 \\ 0 & \frac{48}{125} & \frac{36}{125} & 0 \end{pmatrix} \begin{pmatrix} [1] \\ [x] \\ [y] \\ [xy] \end{pmatrix}.$$

It is clear that these four matrices are not independent. First of all, because of the associativity of multiplication, they must satisfy

$$A_{[x]} \cdot A_{[y]} = A_{[xy]}$$

so that the specification of $A_{[xy]}$ is redundant. Furthermore, because of the commutativity of multiplication, we must have

$$A_{[x]} \cdot A_{[y]} = A_{[y]} \cdot A_{[x]}. \tag{2.15}$$

Actually, in this particular case, after the specification of either $A_{[x]}$ or $A_{[y]}$, the nontrivial rows of the other matrix are uniquely determined by (2.15).

This indicates that the complete information about the polynomial ideal $\mathcal{I} = \langle p_1, p_2 \rangle$ in Example 2.2 is contained in the basis $B_{\mathcal{I}}$ for $\mathcal{R}[\mathcal{I}]$ and the multiplication matrix $A_{[x]}$ and implies that one should be able to determine the zero set $Z[\mathcal{I}]$ of $\mathcal{I}$ from the matrix $A_{[x]}$ alone. We will see in section 2.4 that this is true and how it is done. $\square$

In section 1.3.1, we have observed that—in a ring—multiplication by a fixed element is a *linear mapping*. In the polynomial ring $\mathcal{P}^s$, this is not very helpful because—as a vector space—it has an infinite basis. The quotient ring $\mathcal{R}[\mathcal{I}] = \mathcal{P}^s/\mathcal{I}$, on the other hand, is a vector space of a fixed finite dimension m, for 0-dimensional $\mathcal{I}$. Upon introduction of a basis, all operations in $\mathcal{R}[\mathcal{I}]$ may be represented in terms of standard linear algebra, which is extremely helpful for computational purposes.

It is a standard procedure in mathematics to proceed from a structured space of equivalence classes (modulo some equivalence relation) to the space of a *particular set of representatives*. The operations in that space are derived from those for the equivalence classes.

For the residue class ring $\mathcal{R}[\mathcal{I}] = \mathcal{P}^s/\mathcal{I}$, with a basis $\{[b_\mu]_{\mathcal{I}}, \ \mu = 1(1)m\}$ and elements $[p]_{\mathcal{I}} = \sum_\mu \gamma_\mu [b_\mu]_{\mathcal{I}}$, and with the equivalence relation $\bar{p} \equiv p$ or $[\bar{p}]_{\mathcal{I}} = [p]_{\mathcal{I}}$ iff $\bar{p} - p \in \mathcal{I}$, we may proceed to the vector space span $\{b_\mu, \ \mu = 1(1)m\}$, with the multiplication

$$b_\mu \begin{pmatrix} b_1 \\ \vdots \\ b_m \end{pmatrix} := \begin{pmatrix} A_{[b_\mu]} \end{pmatrix} \begin{pmatrix} b_1 \\ \vdots \\ b_m \end{pmatrix}, \tag{2.16}$$

where the b_μ are arbitrary but *fixed* elements in the residue classes $[b_\mu]_{\mathcal{I}}$. We will also write A_{b_μ} in place of $A_{[b_\mu]}$ under these circumstances. Naturally, the choice of the representatives

$b_\mu \in \mathcal{P}^s$ should be well considered so that the elements of the vector space span $\{b_\mu\}$ appear as a suitable representation of the polynomials in $\mathcal{P}^s$.

Example 2.7: Consider once more the Lagrange basis $\widehat{B}$ in Example 2.5 specified in terms of univariate polynomials of degree 3. If we choose these polynomials as representatives b_μ, then span $\{b_\mu\} = \mathcal{P}_3^1$ as a vector space; this will not lead to a very intuitive representation of the elements of $\mathcal{P}^2$.

On the other hand, if we choose the representatives (cf. Example 2.17)

$$b_1 = \frac{1}{4} + \frac{3}{8}x + \frac{1}{8}y + \frac{25}{48}xy \in [\hat{b}_1]_{\mathcal{I}}, \quad b_2 = \frac{1}{4} - \frac{1}{8}x + \frac{3}{8}y - \frac{25}{48}xy \in [\hat{b}_2]_{\mathcal{I}},$$

$$b_3 = \frac{1}{4} - \frac{3}{8}x - \frac{1}{8}y + \frac{25}{48}xy \in [\hat{b}_3]_{\mathcal{I}}, \quad b_4 = \frac{1}{4} + \frac{1}{8}x - \frac{3}{8}y - \frac{25}{48}xy \in [\hat{b}_4]_{\mathcal{I}},$$

we have span $\{b_\mu\} = $ span $\{1, x, y, xy\}$ which leads to a more natural representation of the $\mathcal{P}^2$ mod $\mathcal{I}$. $\square$

As is customary, we will generally *identify* the vector space span $\{b_\mu\}$, with the multiplication (2.16), with the quotient ring $\mathcal{R}[\mathcal{I}] = \mathcal{P}^s/\mathcal{I}$ and also denote it by $\mathcal{R}[\mathcal{I}]$. This permits a simpler notation, and there is no danger of confusion as long as we remember that equality relations in this setting have to be considered modulo $\mathcal{I}$. Where confusion may arise, we will explicitly append "mod $\mathcal{I}$" to the relation.

Without essential loss of generality, we may assume that none of the monomials x_σ, $\sigma = 1(1)s$, is an element of the polynomial ideal $\mathcal{I} \subset \mathcal{P}^s$. Then, in $\mathcal{R}[\mathcal{I}]$,

$$x_\sigma = \sum_\mu \xi_{\sigma\mu} b_\mu, \quad \sigma = 1(1)s,$$

which implies, for $\sigma = 1(1)s$,

$$x_\sigma \begin{pmatrix} b_1 \\ \vdots \\ b_m \end{pmatrix} = \left(\sum_\mu \xi_{\sigma\mu} b_\mu \right) \begin{pmatrix} b_1 \\ \vdots \\ b_m \end{pmatrix} = \left(\sum_\mu \xi_{\sigma\mu} A_{b_\mu} \right) \begin{pmatrix} b_1 \\ \vdots \\ b_m \end{pmatrix} =: A_\sigma \begin{pmatrix} b_1 \\ \vdots \\ b_m \end{pmatrix}.$$

Definition 2.6. Consider the quotient ring $\mathcal{R} = $ span $\{b_\mu, \ \mu = 1(1)m\}$; cf. the above convention. The $m \times m$ matrices A_σ, $\sigma = 1(1)s$, for which, in $\mathcal{R}$, i.e. mod $\mathcal{I}$,

$$x_\sigma \begin{pmatrix} b_1 \\ \vdots \\ b_m \end{pmatrix} = A_\sigma \begin{pmatrix} b_1 \\ \vdots \\ b_m \end{pmatrix}, \quad \sigma = 1(1)s, \tag{2.17}$$

are called *multiplication matrices* of $\mathcal{R}$ w.r.t. the basis $\{b_\mu, \ \mu = 1(1)m\}$. If $x_\sigma b_\mu = b_{\mu'}$, the μ-th row of A_σ is the unit vector $e_{\mu'}^T$ and a *trivial* row of A_σ; all rows not of this form are called *nontrivial*. $\square$

Definition 2.7. As with monomials, we define

$$A^j := A_1^{j_1} A_2^{j_2} \ldots A_s^{j_s} \quad \text{for} \ j = (j_1, \ldots, j_s) \in \mathbb{N}_0^s. \ \square \tag{2.18}$$

(2.18) is a valid notation because of the associativity of matrix products and

Proposition 2.5. The matrices A^j, $j \in \mathbb{N}_0^s$, derived from the multiplication matrices A_σ for a quotient ring $\mathcal{R} \subset \mathcal{P}^s$ w.r.t. a basis $\{b_\mu, \mu = 1(1)m\}$ and their linear combinations form a *commuting family* $\overline{A}$ of matrices:

$$\Big(\sum_{j\in J_1} \gamma_{1j} A^j\Big)\Big(\sum_{k\in J_2} \gamma_{2k} A^k\Big) = \Big(\sum_{k\in J_2} \gamma_{2k} A^k\Big)\Big(\sum_{j\in J_1} \gamma_{1j} A^j\Big). \tag{2.19}$$

Proof: Let $\mathbf{b} := (b_1 \ldots b_m)^T$. Then, in $\mathcal{R}$,

$$x_{\sigma_1} x_{\sigma_2} \mathbf{b} = \begin{cases} x_{\sigma_1} A_{\sigma_2} \mathbf{b} = A_{\sigma_2} x_{\sigma_1} \mathbf{b} = A_{\sigma_2} A_{\sigma_1} \mathbf{b} \\ x_{\sigma_2} x_{\sigma_1} \mathbf{b} = \cdots = A_{\sigma_1} A_{\sigma_2} \mathbf{b} \end{cases},$$

which implies the commutativity of two individual A_σ; here, $x_{\sigma_1} A_{\sigma_2} = A_{\sigma_2} x_{\sigma_1}$ because x_{σ_1} is a scalar in relation to the numerical matrix A_{σ_2}. The associativity of the matrix products permits the extension to arbitrary A^j; the extension to linear combinations is trivial. $\square$

Proposition 2.5 further implies

Corollary 2.6. Let $p(x) = \sum_{j\in J} \alpha_j x^j \in \mathcal{P}^s$. In a quotient ring $\mathcal{R} \subset \mathcal{P}^s$, with basis $\mathbf{b}$ and multiplication matrices A_σ,

$$p(x)\, \mathbf{b} = \Big(\sum_{j\in J} \alpha_j A^j\Big)\, \mathbf{b} =: p(A)\, \mathbf{b}. \tag{2.20}$$

Theorem 2.7. For a polynomial ideal $\mathcal{I} \subset \mathcal{P}^s$, consider the quotient ring $\mathcal{R}[\mathcal{I}]$, with basis $\mathbf{b}$ and multiplication matrices A_σ. Then, for any $p \in \mathcal{P}^s$,

$$p \in \mathcal{I} \quad \text{iff} \quad p(A) = 0 \quad (\text{zero matrix}). \tag{2.21}$$

Proof: $p \in \mathcal{I}$ implies $p\, b_\mu \in \mathcal{I}, \forall \mu$; therefore, each row of $p(A)$ must vanish according to (2.20). Conversely, $p(A) = 0$ implies $p\, b_\mu \in \mathcal{I}, \forall \mu$, which implies $p\, q \in \mathcal{I}$ for each $q \in \mathcal{R}[\mathcal{I}]$ and $p \in \mathcal{I}$ for $q = 1$. $\square$

Theorem 2.7 establishes that a 0-dimensional polynomial ideal $\mathcal{I} \subset \mathcal{P}^s$ is *fully determined* by its quotient ring $\mathcal{R}[\mathcal{I}]$. Thus we have found that 0-dimensional ideals in $\mathcal{P}^s$ may be represented by some vector space basis $\mathbf{b}$ of $\mathcal{R}[\mathcal{I}]$ and the associated multiplication matrices A_σ, $\sigma = 1(1)s$: p is a member of $\mathcal{I}$ iff $p(A) = 0$. The strong linearity which dominates $\mathcal{R}[\mathcal{I}]$ makes this representation very suitable for computational purposes.

Example 2.8: $\mathcal{R}[\mathcal{I}] = \operatorname{span} \mathbf{b}$, with $\mathbf{b} = (1, x, y, xy)^T$ and

$$A_1 = A_x = \begin{pmatrix} 0 & 1 & 0 & 0 \\ \frac{4}{5} & 0 & 0 & \frac{4}{3} \\ 0 & 0 & 0 & 1 \\ 0 & \frac{48}{125} & \frac{36}{125} & 0 \end{pmatrix}, \quad A_2 = A_y = \begin{pmatrix} 0 & 0 & 1 & 0 \\ 0 & 0 & 0 & 1 \\ \frac{4}{5} & 0 & 0 & -\frac{4}{3} \\ 0 & \frac{36}{125} & -\frac{48}{125} & 0 \end{pmatrix},$$

is a complete representation of the ideal $\mathcal{I} = \langle p_1, p_2 \rangle$; cf. Example 2.6. It is easily checked, e.g., that p_1 and p_2 satisfy

$$p_1(A) = A_1^2 + 4A_1 A_2 + 4A_2^2 - 4I = 0, \quad p_2(A) = 4A_1^2 - 4A_1 A_2 + A_2^2 - 4I = 0.$$

The first and third row of A_1 and the first and second row of A_2 are trivial rows; the remaining rows are nontrivial.　□

　　　Theorem 2.7 raises the following question: If we *choose* a finite-dimensional vector space $\mathcal{R} = \text{span } \mathbf{b} \subset \mathcal{P}^s$, and *specify* commuting multiplication matrices A_σ, $\sigma = 1(1)s$, under what conditions is the set $\{p \in \mathcal{P}^s : p(A) = 0\}$ a 0-dimensional ideal in $\mathcal{P}^s$? We will answer this question in section 8.1.2.

　　　And what about *positive-dimensional* ideals? Here, as we have realized in the beginning of this section, we have to deal with an infinite-dimensional quotient ring $\mathcal{R}[\mathcal{I}]$ and its infinite bases; moreover, the linear mappings which represent multiplication by the x_σ in $\mathcal{R}[\mathcal{I}]$ are between infinite-dimensional spaces and hence represented by "infinite" matrices w.r.t. some particular basis $\mathbf{b}$. If we find a way to handle such mappings computationally, we might be able to use them in an analogous fashion as for finite-dimensional quotient rings. We will return to this question in section 11.1.

2.2.2　Commuting Families of Matrices

Obviously, $\overline{A} := \text{span}_{\mathbb{C}}\{A^j, \ j \in \mathbb{N}_0^s\}$ is the set of all multiplication matrices $p(A)$ which can appear in (2.21). Due to Theorem 2.7, $\overline{A} = \{p(A), \ p \in \mathcal{R}[\mathcal{I}]\}$ because all $\bar{p}$ in the same residue class $[p]_{\mathcal{I}}$ have the same multiplication matrix $\bar{p}(A) = p(A)$. Furthermore, with $\mathcal{R}[\mathcal{I}] = \text{span } \mathbf{b}$, we have $\overline{A} = \text{span } \{A_{b_\mu}, \ \mu = 1(1)m\}$ (cf. (2.14)) so that $\overline{A}$ is an m-dimensional vector space of $m \times m$-matrices.

Definition 2.8. The commuting family of $m \times m$-matrices

$$\overline{A} := \text{span } \{A^j, \ j \in \mathbb{N}_0^s\} = \text{span } \{A_{b_\mu}, \ \mu = 1(1)m\}$$

is the *family of multiplication matrices* of $\mathcal{R}[\mathcal{I}]$ w.r.t. the basis $\mathbf{b}$.　□

　　　Families of commuting matrices have a special distinction in linear algebra; cf., e.g., [2.15]. In particular, such families have *joint eigenvectors* and *joint invariant subspaces*. We list some facts which will be important in what follows. Here, $\overline{B} := \{B_0, B_1, \ldots\} \subset \mathbb{C}^{m \times m}$ denotes a commuting family of $m \times m$ matrices.

Proposition 2.8. Consider $B_0 \in \overline{B}$ and an eigenvalue λ_0 of B_0 with *geometric multiplicity* 1, i.e. rk $(B_0 - \lambda_0 I_0) = m - 1$ and there exists only one eigenvector direction $x \in \mathbb{C}^m$ of B_0 for the eigenvalue λ_0. Then x is a *joint eigenvector* of *all* $B \in \overline{B}$, with associated eigenvalues $\lambda_0(B)$. Furthermore, if B_0 has an invariant subspace $\mathbf{x} = \text{span } (x_1, \ldots, x_{m_0})$, $m_0 > 1$, associated with the eigenvalue λ_0, then $\mathbf{x}$ is a *joint invariant subspace* of *all* $B \in \overline{B}$ associated with the eigenvalues $\lambda_0(B)$.

Proof:　By assumption, the Jordan normal form of B_0 contains exactly the one Jordan block

$$J_0 = \begin{pmatrix} \lambda_0 & 1 & 0 \\ & \ddots & 1 \\ 0 & & \lambda_0 \end{pmatrix} \in \mathbb{C}^{m_0 \times m_0},$$

such that

$$B_0\, X = X\, J_0, \quad \text{with } X = \begin{pmatrix} | & & | \\ x_1 & \cdots & x_{m_0} \\ | & & | \end{pmatrix}.$$

Clearly, x_1 is the only eigenvector of B_0 for the eigenvalue λ_0. For an arbitrary matrix $B \in \overline{B}$, the commutativity implies $B_0\,(B\,X) = B\,B_0 X = B\,X J_0 = (B\,X)\,J_0$ or

$$B_0\, Y = Y\, J_0 \qquad \text{for } Y := BX = (\,y_1 \ldots y_{m_0}\,).$$

From $B_0\, y_1 = \lambda_0\, y_1$, we deduce that y_1 must be a multiple $\lambda\, x_1$ of the only eigenvector x_1; thus $B\, x_1 = y_1 = \lambda\, x_1$ which proves the first assertion. Furthermore, the relation $(B_0 - \lambda_0 I)\, y_2 = y_1 = \lambda\, x_1$ equating the second columns implies $y_2 \in \text{span}\,(x_1, x_2)$: a component of $y_2 \notin \text{span}\,(x_1, x_2)$ would have to be in the kernel of $B_0 - \lambda_0 I$ which consists only of multiples of x_1. This argument can be continued for the remaining y_μ and proves the second assertion:

$$B\, X = X\, T, \quad \text{with } T \text{ upper-triangular with diagonal elements } \lambda. \quad \Box \qquad (2.22)$$

A *nonderogatory* matrix has only eigenvalues of *geometric multiplicity* 1; i.e. it has no eigenspaces of a dimension > 1. Thus, if a commuting family $\overline{B}$ contains a nonderogatory matrix B_0, then the eigenvectors and invariant subspaces of B_0 are *joint* eigenvectors and invariant subspaces of the whole family $\overline{B}$. Unfortunately, the commuting families $\overline{A}$ of multiplication matrices of some quotient ring $\mathcal{R}[\mathcal{I}]$ do not necessarily contain a nonderogatory matrix. However, by Corollary 2.26, they are nonderogatory families.

Definition 2.9. A commuting family of matrices is a *nonderogatory family* iff it has *no joint eigenspace of a dimension > 1*. $\quad \Box$

Proposition 2.9. In a nonderogatory commuting family, for each joint eigenvector $x_{\mu 1}$ there is a unique associated joint invariant subspace span $(\,x_{\mu 1}, x_{\mu 2}, \ldots, x_{\mu m_\mu}\,)$, $m_\mu \geq 1$, such that for each $B \in \overline{B}$ with $B\, x_{\mu 1} = \lambda_\mu(B)\, x_{\mu 1}$, the relation (2.22) holds with $X = (\,x_{\mu 1} \ldots x_{\mu m_\mu}\,)$ and $\lambda = \lambda_\mu(B)$.

Proof: Without loss of generality, consider $B_0 \in \overline{B}$ with an eigenvalue λ_0 of geometric multiplicity 2 and an algebraic multiplicity > 2 so that the Jordan block associated with λ_0 consists of two *separate* Jordan blocks J_1 and J_2 of dimensions $m_1, m_2 \geq 1$ and

$$B_0\, X_1 = X_1\, J_1, \quad B_0\, X_2 = X_2\, J_2,$$

with one eigenvector x_1 in X_1 and the other one x_2 in X_2. By our assumption, there exists a matrix $B_1 \in \overline{B}$ with $B_1\, x_1 = \lambda_0(B_1)\, x_1$ but $B_1\, x_2 \neq \lambda_0(B_1)\, x_2$. There are two possibilities:

(i) x_2 is a joint eigenvector of $\overline{B}$ but the eigenvalue for x_2 is different from that for x_1 for almost all $B \in \overline{B}$. Then the invariant subspaces span X_1 and span X_2 belong to different Jordan blocks and thus to different joint eigenvectors for all $B \in \overline{B}$.

(ii) x_2 is not an eigenvector of B_1. In this case, we combine the columns of X_1 and X_2 into a single matrix X putting the columns x_1 and x_2 into the first two positions. Then the argument in the proof of Proposition 2.8, with obvious variations, establishes the validity of (2.22), and span X is the invariant subspace associated with the only joint eigenvector x_1 in span X. $\quad \Box$

Example 2.9: Consider the polynomial ring $\mathcal{R}$ with basis $\mathbf{b} = (1,\ x,\ y,\ x^2)^T$ and commuting multiplication matrices

$$A_x = \begin{pmatrix} 0 & 1 & 0 & 0 \\ 0 & 0 & 0 & 1 \\ -1 & 1 & 1 & 0 \\ 9 & -12 & -1 & 6 \end{pmatrix}, \quad A_y = \begin{pmatrix} 0 & 0 & 1 & 0 \\ -1 & 1 & 1 & 0 \\ -2 & 0 & 3 & 0 \\ -1 & 0 & 1 & 1 \end{pmatrix}.$$

The eigenanalysis of A_x establishes it as *nonderogatory*:

$$A_x \left(\begin{array}{ccc|c} 1 & 0 & 0 & 1 \\ 2 & 1 & 0 & 1 \\ 1 & 0 & 0 & 2 \\ 4 & 4 & 1 & 1 \end{array} \right) = \left(\begin{array}{ccc|c} 1 & 0 & 0 & 1 \\ 2 & 1 & 0 & 1 \\ 1 & 0 & 0 & 2 \\ 4 & 4 & 1 & 1 \end{array} \right) \left(\begin{array}{ccc|c} 2 & 1 & & \\ & 2 & 1 & 0 \\ & & 2 & \\ & 0 & & 1 \end{array} \right).$$

Thus, by Proposition 2.9, the family of multiplication matrices generated by A_x and A_y has the two joint eigenvectors $(1\ 2\ 1\ 4)^T$, $(1\ 1\ 2\ 1)^T$, and—associated with the first eigenvector—the joint invariant subspace span $\left\{ \begin{pmatrix} 1 \\ 2 \\ 1 \\ 4 \end{pmatrix}, \begin{pmatrix} 0 \\ 1 \\ 0 \\ 4 \end{pmatrix}, \begin{pmatrix} 0 \\ 0 \\ 0 \\ 1 \end{pmatrix} \right\}$. The eigenanalysis of A_y is compatible with this, but it would not have permitted that conclusion since the invariant subspace is a 3-dimensional *eigenspace* for A_y:

$$A_y \left(\begin{array}{ccc|c} 1 & 0 & 0 & 1 \\ 2 & 1 & 0 & 1 \\ 1 & 0 & 0 & 2 \\ 4 & 4 & 1 & 1 \end{array} \right) = \left(\begin{array}{ccc|c} 1 & 0 & 0 & 1 \\ 2 & 1 & 0 & 1 \\ 1 & 0 & 0 & 2 \\ 4 & 4 & 1 & 1 \end{array} \right) \left(\begin{array}{ccc|c} 1 & & & \\ & 1 & & 0 \\ & & 1 & \\ & 0 & & 2 \end{array} \right).$$

With some changes of the multiplication matrices (whose meaning will come to light in the next section), we have, for the same vector space $\mathcal{R}$, a different multiplicative structure:

$$A_x = \begin{pmatrix} 0 & 1 & 0 & 0 \\ 0 & 0 & 0 & 1 \\ -6 & 5 & 2 & -1 \\ 4 & 8 & 0 & 5 \end{pmatrix}, \quad A_y = \begin{pmatrix} 0 & 0 & 1 & 0 \\ -6 & 5 & 2 & -1 \\ 3 & -4 & 2 & 1 \\ -16 & 12 & 4 & -2 \end{pmatrix}.$$

Now neither matrix nor any linear combination or power of them is nonderogatory: Each matrix has *two* eigenvectors for the same eigenvalue, but only *one* is a *joint* eigenvector:

$$A_x \left(\begin{array}{ccc|c} 1 & 0 & 0 & 1 \\ 2 & 1 & 0 & 1 \\ 1 & 0 & 1 & 2 \\ 4 & 4 & 0 & 1 \end{array} \right) = \left(\begin{array}{ccc|c} 1 & 0 & 0 & 1 \\ 2 & 1 & 0 & 1 \\ 1 & 0 & 1 & 2 \\ 4 & 4 & 0 & 1 \end{array} \right) \left(\begin{array}{ccc|c} 2 & 1 & & \\ & 2 & & 0 \\ & & 2 & \\ & 0 & & 1 \end{array} \right),$$

$$A_y \left(\begin{array}{ccc|c} 1 & 0 & 0 & 1 \\ 2 & 1 & 0 & 1 \\ 1 & 0 & 1 & 2 \\ 4 & 4 & 0 & 1 \end{array} \right) = \left(\begin{array}{ccc|c} 1 & 0 & 0 & 1 \\ 2 & 1 & 0 & 1 \\ 1 & 0 & 1 & 2 \\ 4 & 4 & 0 & 1 \end{array} \right) \left(\begin{array}{ccc|c} 1 & & 1 & \\ & 1 & & 0 \\ & & 1 & \\ & 0 & & 2 \end{array} \right).$$

Thus the multiplication matrices form a *nonderogatory commuting family*, with the joint invariant subspace span $\{ \begin{pmatrix} 1 \\ 2 \\ 1 \\ 4 \end{pmatrix}, \begin{pmatrix} 0 \\ 1 \\ 0 \\ 4 \end{pmatrix}, \begin{pmatrix} 0 \\ 0 \\ 1 \\ 0 \end{pmatrix} \}$ associated to the joint eigenvector $(1\ 2\ 1\ 4)^T$.

$\square$

Exercises

1. Consider $\mathcal{I} = \langle p_1, p_2 \rangle$ of Example 2.2 and the Lagrange basis $\hat{\mathbf{b}}$ for $\mathcal{R}[\mathcal{I}]$ of Example 2.5 whose elements $\hat{b}_\mu$ are from span $\{1, x, x^2, x^3\}$.

   (a) Find the Lagrange basis of $\mathcal{R}[\mathcal{I}]$ which consists of elements from span $\{1, x, y, x^2\}$. (The Lagrange basis from span $\{1, x, y, xy\}$ has been displayed in Example 2.7.)

   (b) For $\mathcal{R}[\mathcal{I}]$ with basis $\mathbf{b} = (1, x, y, x^2)^T$, find the multiplication matrices A_x and A_y; cf. the approach in Example 2.6.

2. Consider the specification of an ideal $\mathcal{I} \subset \mathcal{P}^s$ by a quotient ring basis $\mathbf{b} = (b_1, \ldots, b_m)^T$ and the multiplication matrices A_σ. Convince yourself that

$$p_{\sigma\mu}(x) := x_\sigma\, b_\mu(x) - a_{\sigma\mu}^T\, \mathbf{b}(x) \in \mathcal{I}, \quad \sigma = 1(1)s, \ \mu = 1(1)m,$$

where $a_{\sigma\mu}^T$ is the μth row of A_σ.

   (a) For the representation of $\mathcal{I}$ from Example 2.2 by the data in Example 2.8, list the polynomials $p_{\sigma\mu}$. What happens when $a_{\sigma\mu}^T$ is a trivial row?

   (b) The four nontrivial $p_{\sigma\mu}$ constitute a basis of $\mathcal{I}$, but not a minimal one. Refer to Example 2.2 to select a minimal basis.

   (c) For the data of Exercise 1 b) above, find the $p_{\sigma\mu}$ and select a minimal basis.

3. (a) With the aid of suitable software, find the four joint eigenvectors of A_1, A_2 in Example 2.8. Normalize the eigenvectors such that their first components are 1.

   (b) Compare the components of the normalized eigenvectors with the components of the zeros of p_1, p_2 of Example 2.2.

2.3 Dual Spaces of Polynomial Ideals

2.3.1 Dual Vector Spaces

A particularly useful feature of linear spaces or vector spaces is the fact that they have natural *dual counterparts*: the spaces of all *linear functionals* on a given vector space form another vector space which mirrors the properties of the original one. This fundamental insight from linear algebra is also important and useful in polynomial algebra.

Consider the generic $\mathbb{C}^s$, with elements $y = (\eta_1, \ldots, \eta_s)^T$, $\eta_\sigma \in \mathbb{C}$. It is well known from linear algebra that each linear mapping $l : \mathbb{C}^s \to \mathbb{C}$, i.e. each linear functional on $\mathbb{C}^s$, has the form

$$l : y \ \to \ c^T y \ = \ \sum_{\sigma=1}^{s} \gamma_\sigma \eta_\sigma, \quad \gamma_\sigma \in \mathbb{C}.$$

The fact that these functionals form a vector space of dimension s is obvious; the vector space operations are defined by the respective operations on the row vectors $c^T = (\gamma_1, \ldots, \gamma_s)$.

Definition 2.10. The *dual space* $\mathcal{V}^*$ of a vector space $\mathcal{V}$ over $\mathbb{C}$ is the vector space of all linear functionals $l : \mathcal{V} \to \mathbb{C}$, with

$$(\alpha\, l)(y) := \alpha\, l(y)\,; \quad (l_1 + l_2)(y) := l_1(y) + l_2(y)\,. \quad \Box$$

Proposition 2.10. For a finite-dimensional vector space $\mathcal{V}$, $\dim \mathcal{V}^* = \dim \mathcal{V}$.

In this section, we will restrict our considerations to vector spaces $\mathcal{V}$ and $\mathcal{V}^*$ over $\mathbb{C}$ or $\mathbb{R}$ and of a *finite* dimension m.

Just as we have used the *column vector* $\mathbf{b} = (b_1, \ldots, b_m)^T$ as a handy notation for a basis of a vector space $\mathcal{V}$ (of polynomials), we will also use the *row vector* $\mathbf{c}^T = (c_1, \ldots, c_m)$ as an abbreviation for a basis of its dual space $\mathcal{V}^*$. This permits the further abbreviations

$$\mathbf{c}^T(b_\mu) := (c_1(b_\mu), \ldots, c_m(b_\mu))\,, \qquad c_\nu(\mathbf{b}) := \begin{pmatrix} c_\nu(b_1) \\ \vdots \\ c_\nu(b_m) \end{pmatrix}, \qquad (2.23)$$

$$\mathbf{c}^T(\mathbf{b}) := \begin{pmatrix} c_1(b_1) & \ldots & c_m(b_1) \\ \vdots & & \vdots \\ c_1(b_m) & \ldots & c_m(b_m) \end{pmatrix}. \qquad (2.24)$$

Note that, in this notation, each (linear functional) element of the row vector acts on each (polynomial) element of the column vector! Note further that—naturally—linear functionals commute with *scalar* linear combinations: For $a^T = (\alpha_1, \ldots, \alpha_m) \in C^m$,

$$c_\nu(a^T \mathbf{b}) = a^T c_\nu(\mathbf{b})\,, \qquad \mathbf{c}^T(a^T \mathbf{b}) = a^T \mathbf{c}^T(\mathbf{b})\,.$$

As in Proposition 2.10, the following are basic results from linear algebra.

Proposition 2.11. For a basis $\mathbf{b}$ of $\mathcal{V}$ and $c \in \mathcal{V}^*$, $c\,\mathbf{b} = 0$ (0-vector) implies $c = 0$ (0-functional).

Proposition 2.12. If, for some $\mathbf{b} \in (\mathcal{V})^m$ and $\mathbf{c}^T \in (\mathcal{V}^*)^m$, the (numerical) matrix $\mathbf{c}^T(\mathbf{b})$ is *regular*, then $\mathbf{b}$ is a basis of $\mathcal{V}$ and $\mathbf{c}^T$ a basis of $\mathcal{V}^*$.

Definition 2.11. A basis $\mathbf{b} = \{b_\nu\}$ of $\mathcal{V}$ and a basis $\mathbf{c}^T = \{c_\mu\}$ of $\mathcal{V}^*$ which satisfy

$$\mathbf{c}^T(\mathbf{b}) = I \quad \text{(identity matrix)} \qquad (2.25)$$

are called *conjugate bases* of $\mathcal{V}$ and its dual space $\mathcal{V}^*$. $\quad \Box$

Example 2.10: Consider the 3-dimensional vector space $\mathcal{P}_2^1$ of quadratic univariate polynomials. The dual space $(\mathcal{P}_2^1)^*$ consists of all linear functionals $l : \mathcal{P}_2^1 \to \mathbb{C}$. Examples of such functionals are

$$
\begin{aligned}
c_1(p) &:= & p(\xi) && \text{evaluation at } \xi \in \mathbb{C}\,, \\
c_2(p) &:= & p'(\xi) && \text{evaluation of derivative at } \xi\,, \\
c_3(p) &:= & \textstyle\int_{-1}^{+1} p(t)dt && \text{definite integration}\,.
\end{aligned}
$$

With the basis $\mathbf{b} = (1, x, x^2)^T$ in $\mathcal{P}_2^1$, we have

$$c_1(\mathbf{b}) = \begin{pmatrix} 1 \\ \xi \\ \xi^2 \end{pmatrix}, \quad c_2(\mathbf{b}) = \begin{pmatrix} 0 \\ 1 \\ 2\xi \end{pmatrix}, \quad c_3(\mathbf{b}) = \begin{pmatrix} 2 \\ 0 \\ 2/3 \end{pmatrix}.$$

By Proposition 2.12, $\mathbf{c}^T = (c_1, c_2, c_3)$ is a basis of $(\mathcal{P}_2^1)^*$ for all $\xi \in \mathbb{C}$, with the exception of $\xi = \pm\frac{1}{\sqrt{3}}$ for which the matrix $\mathbf{c}^T(\mathbf{b})$ is singular.

The conjugate basis of $\mathbf{b}$ in $(\mathcal{P}_2^1)^*$ consists of the functionals $\hat{c}_\nu$, $\nu = 1(1)3$, which assign to $p \in \mathcal{P}_2^1$ the coefficient of the term with $x^{\nu-1}$. It is obvious that $(\hat{c}_1, \hat{c}_2, \hat{c}_3)$ satisfies (2.25).

Each element in $(\mathcal{P}_2^1)^*$ has a *kernel* of dimension 2: For example, $c_2(p)$ vanishes for $p = \alpha_0 - 2\xi\alpha_2 x + \alpha_2 x^2$, with arbitrary α_0, α_2. $\square$

The fact that the functionals in the conjugate basis retrieve the coefficients in the basis representation of an element of $\mathcal{V}$ is universal

Proposition 2.13. Consider the vector space $\mathcal{V}$ with basis $\mathbf{b}$ and its dual space $\mathcal{V}^*$ with *conjugate* basis $\mathbf{c}^T$. For any element $y = \sum_{\mu=1}^m \alpha_\mu b_\mu = a^T \mathbf{b}$,

$$\mathbf{c}^T(y) = (c_1(y), \ldots, c_m(y)) = a^T. \tag{2.26}$$

Proof: $\quad c_\mu(a^T\mathbf{b}) = a^T c_\mu(\mathbf{b}) = a^T e_\mu = \alpha_\mu. \quad \square$

The relation between a finite-dimensional vector space $\mathcal{V}$ and its dual space $\mathcal{V}^*$ is fully *symmetric*: Consider the space $(\mathcal{V}^*)^*$ of all linear functionals on $\mathcal{V}^*$. Such functionals $\hat{y}$ which map $\mathcal{V}^*$ into $\mathbb{C}$ are obviously provided by the elements of $\mathcal{V}$ with the definition $\hat{y}(l) := l(\hat{y})$. The fact that $(\mathcal{V}^*)^*$ consists *precisely* of the elements $y \in \mathcal{V}$ and may thus be identified with $\mathcal{V}$ is another central theorem of linear algebra.

Theorem 2.14. A finite-dimensional vector space $\mathcal{V}$ over $\mathbb{C}$ is *reflexive*, i.e. the mapping from $\mathcal{V}$ to $(\mathcal{V}^*)^*$ is *bijective*.

Assume now that the vector space $\mathcal{V}$ is also a *commutative ring*, with a multiplication $\mathcal{V} \times \mathcal{V} \to \mathcal{V}$. Then it is useful to introduce also a multiplication $\mathcal{V}^* \times \mathcal{V} \to \mathcal{V}^*$ by

$$(l \cdot u)(y) := l(u\,y). \tag{2.27}$$

The fact that $l \cdot u \in \mathcal{V}^*$ is easily established: $(l \cdot u)(\alpha_1 y_1 + \alpha_2 y_2) = l(\alpha_1 u\, y_1 + \alpha_2 u\, y_2) = \alpha_1 (l \cdot u)(y_1) + \alpha_2 (l \cdot u)(y_2)$. Furthermore, it is an immediate consequence of (2.27) that the mapping $L_u : l \to l \cdot u$ is linear, and that (cf. (1.14))

$$(L_u\, l)(y) = l(M_u\, y). \tag{2.28}$$

Proposition 2.15. Consider a commutative ring $\mathcal{V}$, with basis $\mathbf{b} = (b_1, \ldots, b_m)^T$ and multiplication matrices A_{b_μ} from (2.14), and its dual space $\mathcal{V}^*$, with the conjugate basis $\mathbf{c}^T = (c_1, \ldots, c_m)$. Then

$$(\mathbf{c}^T \cdot b_\mu) = (c_1 \cdot b_\mu, \ldots, c_m \cdot b_\mu) = (c_1, \ldots, c_m)\, A_{b_\mu} = \mathbf{c}^T A_{b_\mu}, \tag{2.29}$$

and, for any element $y = \sum_{\nu=1}^{m} \alpha_\nu \, b_\nu = a^T \mathbf{b}$,

$$(\mathbf{c}^T \cdot b_\mu)(y) \;=\; a^T A_{b_\mu} \,. \tag{2.30}$$

Proof: $(\mathbf{c}^T b_\mu)(\mathbf{b}) = \mathbf{c}^T (b_\mu \mathbf{b}) = \mathbf{c}^T A_{b_\mu} (\mathbf{b})$ implies (2.29) by Proposition 2.11. Further, $(\mathbf{c}^T b_\mu)(a^T \mathbf{b}) = \mathbf{c}^T (a^T A_{b_\mu} \mathbf{b}) = a^T A_{b_\mu} \mathbf{c}^T (\mathbf{b}) = a^T A_{b_\mu}$, by (2.25). $\square$

Example 2.11: We consider the ring $\mathcal{R} \subset \mathcal{P}^1$, with $\mathbf{b} = (1, x, x^2)^T$ and

$$A_x = \begin{pmatrix} 0 & 1 & 0 \\ 0 & 0 & 1 \\ -1 & 1 & 1 \end{pmatrix}, \quad A_{x^2} = \begin{pmatrix} 0 & 0 & 1 \\ -1 & 1 & 1 \\ -1 & 0 & 2 \end{pmatrix}.$$

The conjugate basis $\mathbf{c}^T = (c_1, c_2, c_3)$ of $\mathcal{R}^*$ retrieves the coefficients of $r \in \mathcal{R}$. The coefficients of $x \, r$ and $x^2 r$ are retrieved by the functionals

$$\mathbf{c}^T \cdot x = \mathbf{c}^T A_x = (-c_3, \; c_1 + c_3, \; c_2 + c_3), \quad \mathbf{c}^T \cdot x^2 = \mathbf{c}^T A_{x^2} = (-c_2 - c_3, \; c_2, \; c_1 + c_2 + 2c_3).   \square$$

2.3.2   Dual Spaces of Quotient Rings

After these general observations about dual spaces of finite-dimensional commutative rings, we now consider the dual spaces of quotient rings $\mathcal{R}[\mathcal{I}]$ of 0-dimensional polynomial ideals $\mathcal{I} \subset \mathcal{P}^s$, with $\dim \mathcal{R}[\mathcal{I}] = m$. For $\mathcal{R}[\mathcal{I}]$, with basis $\mathbf{b}$ and related multiplication matrices A_σ (cf. Definition 2.6), there is a unique dual space $(\mathcal{R}[\mathcal{I}])^*$. Multiplication in $(\mathcal{R}[\mathcal{I}])^*$ is defined by (2.27).

Proposition 2.16. In $(\mathcal{R}[\mathcal{I}])^*$, let $\mathbf{c}^T = (c_1, \ldots, c_m)$ be the conjugate basis of the basis $\mathbf{b} = (b_1, \ldots, b_m)^T$ in $\mathcal{R}[\mathcal{I}]$. For $r = \sum_{j \in J} \rho_j x^j \in \mathcal{R}[\mathcal{I}]$,

$$\mathbf{c}^T \cdot r \;=\; \mathbf{c}^T \, r(A) \,. \tag{2.31}$$

Proof: Let $x_\sigma = \sum_\mu \xi_{\sigma\mu} b_\mu$; by Proposition 2.15, $\mathbf{c}^T x_\sigma = \mathbf{c}^T \sum_\mu (\xi_{\sigma\mu} A_{b_\mu}) = \mathbf{c}^T A_\sigma$. Equation (2.31) follows by linear superposition. $\square$

Now we remember that each element $r \in \mathcal{R}[\mathcal{I}]$ is the representative of the residue class $[r]_\mathcal{I} \subset \mathcal{P}^s$. Thus we can extend the domain of the functionals in $(\mathcal{R}[\mathcal{I}])^*$ to all of $\mathcal{P}^s$ by letting them take the same value on all polynomials in the same residue class $[r]_\mathcal{I}$.

Definition 2.12. For a given polynomial ideal $\mathcal{I} \subset \mathcal{P}^s$ with quotient ring $\mathcal{R}[\mathcal{I}] = \mathcal{P}^s/\mathcal{I}$, the *dual space* $\mathcal{D}[\mathcal{I}]$ *of the ideal* $\mathcal{I}$ is the set of linear functionals in $(\mathcal{R}[\mathcal{I}])^*$, with their domain extended to $\mathcal{P}^s$ by

$$l(p) := l(r), \quad r \in \mathcal{R}[\mathcal{I}], \quad p - r \in \mathcal{I}. \quad \square \tag{2.32}$$

Immediate consequences of this definition are shown in:

Proposition 2.17. $\mathcal{D}[\mathcal{I}]$ is a linear space of dimension $\dim(\mathcal{R}[\mathcal{I}])^* = \dim \mathcal{R}[\mathcal{I}]$.

Proof: With the extension (2.32), a basis of $(\mathcal{R}[\mathcal{I}])^*$ is also a basis of $\mathcal{D}[\mathcal{I}]$. $\square$

Proposition 2.18.

$$l(p_1) \;=\; l(p_2) \quad \forall l \in \mathcal{D}[\mathcal{I}] \quad \text{for } p_1 \equiv p_2 \bmod \mathcal{I}, \tag{2.33}$$

and

$$\ker \mathcal{D}[\mathcal{I}] := \{p \in \mathcal{P}^s : l(p) = 0 \ \forall l \in \mathcal{D}[\mathcal{I}]\} = \mathcal{I}. \tag{2.34}$$

Theorem 2.19.

$$\mathcal{D}[\mathcal{I}] = \{l \in (\mathcal{P}^s)^* : l(p) = 0 \ \forall p \in \mathcal{I}\}. \tag{2.35}$$

Proof: $\{ \ldots \} \supset \mathcal{D}[\mathcal{I}]$ by (2.34). On the other hand, $l \in \{ \ldots \}$ implies $l \in (\mathcal{R}[\mathcal{I}])^*$ and hence in $\mathcal{D}[\mathcal{I}]$ by (2.32). $\square$

Theorem 2.19 characterizes the dual space $\mathcal{D}[\mathcal{I}]$ of the polynomial ideal $\mathcal{I}$ as the set of all linear functionals on $\mathcal{P}^s$ whose kernel contains $\mathcal{I}$.

In section 1.3.2, we have seen that we can define a 0-dimensional polynomial ideal $\mathcal{I} \subset \mathcal{P}^s$ with simple zeros by specifying its zero set $Z[\mathcal{I}] = \{z_\mu, \mu = 1(1)m\} \subset \mathbb{C}^s$. Therefore, by Theorem 2.19, the set $Z[\mathcal{I}]$ must also specify $\mathcal{D}[\mathcal{I}]$.

Theorem 2.20. If $\mathcal{I} := \{p \in \mathcal{P}^s : p(z_\mu) = 0, \ z_\mu \in Z \subset C^s, \ \mu = 1(1)m\}$, then

$$\mathcal{D}[\mathcal{I}] = \operatorname{span} \{l_\mu \in (\mathcal{P}^s)^* : l_\mu(p) := p(z_\mu), \ \mu = 1(1)m\} ; \tag{2.36}$$

i.e. the evaluations at the zeros z_μ of $\mathcal{I}$ form a basis of $\mathcal{D}[\mathcal{I}]$.

Proof: We have $l \in \operatorname{span} \{l_\mu\} \Rightarrow l(p) = 0 \ \forall p \in \mathcal{I} \Rightarrow l \in \mathcal{D}[\mathcal{I}]$. Now assume that $\exists \bar{l} \in \mathcal{D}[\mathcal{I}]$ with $\bar{l} \notin \operatorname{span} \{l_\mu\}$; then $\dim \mathcal{D}[\mathcal{I}] = m + 1 > \dim(\mathcal{R}[\mathcal{I}])^* = m$, which contradicts Propositions 2.17 and 2.2. $\square$

Theorem 2.20 suggests that a multiplicity of a zero in $\mathcal{I}$ leads to an extension of $\mathcal{D}[\mathcal{I}]$. In $\mathcal{P}^1$, e.g., if $z_1 \in Z[\mathcal{I}]$ is a 3-fold zero of all $p \in \mathcal{I}$, then $\mathcal{D}[\mathcal{I}]$ must contain the linear functionals $l_{1,1}(p) := p'(z_1)$ and $l_{1,2}(p) := p''(z_1)$ besides those in (2.36); i.e. $l_{1,1}$ and $l_{1,2}$ must appear in an "evaluation basis" of $\mathcal{D}[\mathcal{I}]$. For a multiple zero in $\mathcal{P}^s$, $s > 1$, we expect that $\mathcal{D}[\mathcal{I}]$ will contain evaluations of partial derivatives at a zero z_1.

We introduce the following notation for such functionals on $\mathcal{P}^s$; cf. (1.6) in section 1.2.

Definition 2.13. For $j = (j_1, \ldots, j_s) \in \mathbb{N}_0^s$, with $|j| := \sum_\sigma j_\sigma$, and for $z \in \mathbb{C}^s$, the *differential functional* $\partial_j[z] \in (\mathcal{P}^s)^*$ is defined by

$$\partial_j[z](p) := \frac{1}{j_1! \cdots j_s!} \left(\frac{\partial^{|j|}}{\partial x_1^{j_1} \cdots \partial x_s^{j_s}} \, p \right)(z). \tag{2.37}$$

"$[z]$" may be omitted if it is obvious from the context. $\square$

Example 2.12: In $(\mathcal{P}^3)^*$, at some $z \in \mathbb{C}^3$, we have, e.g.,

$$\partial_{000}[z](p) = p(z), \quad \partial_{020}[z](p) = \tfrac{1}{2} \left(\tfrac{\partial^2}{\partial x_2^2} \, p \right)(z), \quad \partial_{111}[z](p) = \left(\tfrac{\partial^3}{\partial x_1 \partial x_2 \partial x_3} \, p \right)(z),$$

$$\partial_{321}[z](p) = \tfrac{1}{12} \left(\tfrac{\partial^6}{\partial x_1^3 \partial x_2^2 \partial x_3} \, p \right)(z) \quad \text{etc.}$$

If we use the notation x, y, z for the variables in $\mathbb{C}^3$, we may occasionally write ∂_x, ∂_y, ∂_z in place of ∂_{100}, ∂_{010}, ∂_{001} resp., and $\partial_{x^3 y^2 z}$ for ∂_{321}, etc. This and analogous notations should not lead to confusion. $\square$

Definition 2.14. For a 0-dimensional ideal $\mathcal{I} \subset \mathcal{P}^s$, consider a zero $z \in Z[\mathcal{I}]$. If there exists a basis of $\mathcal{D}[\mathcal{I}]$ containing m functionals $\partial_j[z]$ (cf. (2.37)) then z is an *m-fold zero* of $\mathcal{I}$. $\square$

In $\mathcal{P}^1$, the notion of an m-fold zero is classical: z is an m-fold zero of the ideal $\langle p \rangle$ iff p has the factor $(x - z)^m$, or—equivalently—iff $\mathcal{D}[\mathcal{I}]$ contains the functionals $\partial_0[z], \partial_1[z], \ldots, \partial_{m-1}[z]$.

Example 2.13: Let us consider the dual spaces $\mathcal{D} = \mathcal{R}^* \subset (\mathcal{P}^2)^*$ for the rings of Example 2.9. By a relation to be explained in section 2.4.1, the basis $\mathbf{c}^T = (c_1, c_2, c_3, c_4)$ of $\mathcal{D}$ conjugate to the basis $\mathbf{b} = (1, x, y, x^2)^T$ of $\mathcal{R}$ for the first multiplicative structure of $\mathcal{R}$ is found as

$$
\begin{array}{rlllll}
c_1 &=& 2\,\partial_{00}[z_1] & -3\,\partial_{10}[z_1] & +5\,\partial_{20}[z_1] & -\partial_{00}[z_2] \quad, \\
c_2 &=& & \partial_{10}[z_1] & -4\,\partial_{20}[z_1] & \quad, \\
c_3 &=& -\partial_{00}[z_1] & +\partial_{10}[z_1] & -\partial_{20}[z_1] & +\partial_{00}[z_2] \quad, \\
c_4 &=& & & \partial_{20}[z_1] & \quad,
\end{array}
$$

where $z_1 = (2, 1)$, $z_2 = (1, 2)$. The relation (2.25) is easily verified.

Thus, $\mathcal{D}$ is span $\{\partial_{00}[z_1], \partial_{10}[z_1], \partial_{20}[z_1], \partial_{00}[z_2]\}$ since the coefficients of the c_ν form a regular matrix. This shows that the underlying polynomial ideal $\mathcal{I} \subset \mathcal{P}^2$ has a triple zero at z_1 and consists of all polynomials which vanish at z_1 and z_2 and whose first and second x-derivatives vanish at z_1.

For $\mathcal{R}$ with the second multiplicative structure, the conjugate basis of $\mathcal{D}$ is given by

$$
\begin{array}{rlllll}
c_1 &=& -3\,\partial_{00}[z_1] & +2\,\partial_{10}[z_1] & -5\,\partial_{01}[z_1] & +4\,\partial_{00}[z_2] \quad, \\
c_2 &=& 4\,\partial_{00}[z_1] & -3\,\partial_{10}[z_1] & +4\,\partial_{01}[z_1] & -4\,\partial_{00}[z_2] \quad, \\
c_3 &=& & & \partial_{01}[z_1] & \quad, \\
c_4 &=& -\partial_{00}[z_1] & +\partial_{10}[z_1] & -\partial_{01}[z_1] & +\partial_{00}[z_2] \quad.
\end{array}
$$

Now, $\mathcal{D}$ contains the derivative evaluation $\partial_{01}[z_1]$ in place of $\partial_{20}[z_1]$; again $\mathcal{I}$ has a triple zero at z_1, but now it consists of the polynomials with zeros at z_1 and z_2 whose two first derivatives vanish at z_1. $\square$

Note that, in both cases of Example 2.13, the ideal $\mathcal{I}$ has a *triple* zero at z_1 and a simple zero at z_2, but this information is *not sufficient* to specify $\mathcal{I}$.

This example and Theorem 2.20 raise the following question: If we specify an arbitrary vector space $\mathcal{D}$ of linear functionals on $\mathcal{P}^s$, will the intersection of the kernels of all $l \in \mathcal{D}$,

$$\mathcal{I}[\mathcal{D}] \;:=\; \cap_{l \in \mathcal{D}} \{p \in \mathcal{P}^s : l(p) = 0\} \;=\; \{p \in \mathcal{P}^s : l(p) = 0 \ \forall l \in \mathcal{D}\}, \tag{2.38}$$

be a polynomial ideal in $\mathcal{P}^s$?

The only constituent property of ideals (cf. Definition 1.7) which is not evident for $\mathcal{I}[\mathcal{D}]$ is the *closedness w.r.t. multiplication* of an element by an *arbitrary* $q \in \mathcal{P}^s$:

$$l(p) = 0 \ \forall l \in \mathcal{D} \quad \overset{?}{\Rightarrow} \quad (l \cdot q)\,(p) = l(q\,p) = 0 \ \forall l \in \mathcal{D}.$$

This property obviously holds when $\mathcal{D}$ is the span of function evaluations at disjoint $z_\mu \in \mathbb{C}^s$ as in Theorem 2.20. But this implication is *not* true for an arbitrary vector space $\mathcal{D}$ in $(\mathcal{P}^s)^*$ as is easily seen from counterexamples:

In $(\mathcal{P}^1)^*$, let $\mathcal{D} := \mathrm{span}\,\{l_0(p) = p(0), l_2(p) = p''(0)\}$; then $\mathcal{I}[\mathcal{D}] = \{\sum_{\nu=0}^{d} \alpha_\nu x^\nu \in \mathcal{P}^1 : \alpha_0 = \alpha_2 = 0\}$. But $x\,p$ is not in $\mathcal{I}[\mathcal{D}]$ for each $p \in \mathcal{I}[\mathcal{D}]$ with $\alpha_1 \neq 0$.

Definition 2.15. A vector space $\mathcal{D}$ of linear functionals on $\mathcal{P}^s$ is *closed* iff

$$l \in \mathcal{D} \quad \Rightarrow \quad (l \cdot q) \in \mathcal{D} \quad \forall q \in \mathcal{P}^s. \quad \square \tag{2.39}$$

Theorem 2.21. For a *closed* vector space $\mathcal{D}$ of linear functionals on $\mathcal{P}^s$, the set $\mathcal{I}[\mathcal{D}]$ of (2.38) is a *polynomial ideal* in $\mathcal{P}^s$.

Example 2.14: Consider univariate polynomials as above, but let $\mathcal{D} := \mathrm{span}\,\{l_0(p) = p(0), l_1(p) = p'(0)\}$; then $\mathcal{D}$ is closed: For arbitrary $q \in \mathcal{P}^1$, $(l_0 \cdot q)(p) = q(0)\,p(0) = 0$ and $(l_1 \cdot q)(p) = q'(0)\,p(0) + q(0)\,p'(0) = 0$ for all $p \in \mathcal{I}[\mathcal{D}]$. The set $\mathcal{I}[\mathcal{D}]$ is the ideal of all univariate polynomials with a double zero at 0. $\quad\square$

Obviously, the closedness of some vector space of linear functionals on $\mathcal{P}^1$ is easily checked. In the multivariate case, we will consider the checking of closedness in detail in section 8.5.1.

The term "dual space" is also justified by the nice duality between 0-dimensional ideals $\mathcal{I}$ in $\mathcal{P}^s$ and finite-dimensional closed vector spaces $\mathcal{D}$ in $(\mathcal{P}^s)^*$:

$$\mathcal{D}[\mathcal{I}] := \{\, l \in (\mathcal{P}^s)^* : \ l(p) = 0 \quad \forall\, p \in \mathcal{I} \,\},$$

$$\mathcal{I}[\mathcal{D}] := \{\, p \in \mathcal{P}^s : \ l(p) = 0 \quad \forall\, l \in \mathcal{D} \,\}.$$

Furthermore, since $\mathcal{D}[\mathcal{I}]$ is automatically closed by (2.35), $\mathcal{I}[\mathcal{D}[\mathcal{I}]]$ must be an ideal by Theorem 2.21 while, on the other hand, $\mathcal{D}[\mathcal{I}[\mathcal{D}]]$ is a closed vector space in $(\mathcal{P}^s)^*$ for closed $\mathcal{D}$. Actually (without proof)

$$\mathcal{D}[\mathcal{I}[\mathcal{D}]] \ = \ \mathcal{D} \quad \text{and} \quad \mathcal{I}[\mathcal{D}[\mathcal{I}]] \ = \ \mathcal{I}. \tag{2.40}$$

By Proposition 2.18, a 0-dimensional polynomial ideal $\mathcal{I} \subset \mathcal{P}^s$ is *fully determined* by its dual space $\mathcal{D}[\mathcal{I}]$: A polynomial p is a member of $\mathcal{I}$ iff it is in the *kernel* of $\mathcal{D}[\mathcal{I}]$; cf. (2.35) and (2.40). Thus, we have found a further representation for 0-dimensional ideals in $\mathcal{P}^s$, viz. by their dual spaces. The strong linearity of $\mathcal{D}[\mathcal{I}]$ and the duality with $\mathcal{R}[\mathcal{I}]$ make this representation attractive for computational purposes.

In view of Theorem 2.20 and the subsequent remarks, it appears at first that the dual spaces of 0-dimensional ideals provide nothing but a formal description of their zero sets, including the structure of potential multiple zeros. In the case of simple zeros, the knowledge of a basis (2.36) of the dual space $\mathcal{D}[\langle P \rangle]$ of the ideal $\langle P \rangle$ generated by a polynomial system $P \subset \mathcal{P}^s$ is equivalent to the knowledge of the zero set $Z[P]$ of P. Thus it appears that the knowledge of $\mathcal{D}[\langle P \rangle]$ comes as a consequence of solving the system of equations $P = 0$ but cannot be used as a computational tool for that purpose.

- However, as we will see in the following sections, this observation does not account for two important facts:

- Information about $\mathcal{D}[\langle P \rangle]$ may be obtained with reference to a basis which is different from (2.36).

The complete duality between the vector spaces $\mathcal{D}[\langle P \rangle]$ and $\mathcal{R}[\langle P \rangle]$ provides structural relations which may be used computationally.

Example 2.15: Consider the quadratic polynomials $p_1, p_2 \in \mathcal{P}^2$ of Example 2.2. We have found that $\mathcal{R}[\langle p_1, p_2 \rangle]$ has a basis $\{1, x, y, xy\}$ with multiplication matrices A_1, A_2 of Example 2.8. Each polynomial $p \in \mathcal{P}^2$ has a unique representation

$$p(x, y) \equiv c_1(p) + c_2(p)\,x + c_3(p)\,y + c_4(p)\,xy \quad \mathrm{mod}\ \langle p_1, p_2 \rangle;$$

by (2.26), the linear functionals $c_1, \ldots, c_4$ form a *basis* $\mathbf{c}$ of $\mathcal{D}[\langle p_1, p_2 \rangle]$. For given p, the values of these functionals are determined by the first row of $p(A)$ since (cf. Corollary 2.6)

$$p(x) \begin{pmatrix} 1 \\ x \\ y \\ xy \end{pmatrix} \equiv p(A) \begin{pmatrix} 1 \\ x \\ y \\ xy \end{pmatrix} \mod \langle p_1, p_2 \rangle .$$

From the basis $\mathbf{c}$, the *zero revealing basis* $\mathbf{c}_0 := \{l_\mu(p) = p(z_\mu),\ \mu = 1(1)4\}$ should be obtainable by a *basis transformation*.   □

Exercises

1.   For p_1, p_2 of Example 2.2, consider different bases of the 4-dimensional dual space $\mathcal{D}[\langle p_1, p_2 \rangle] \subset \mathcal{P}^2$:

$$\begin{aligned} \mathbf{c}_0^T(p) &:= \text{ evaluations of } p \text{ at the zeros } z_\mu,\ \mu = 1(1)4 , \\ \mathbf{c}^T(p) &:= \text{ coefficients of } r \in [p]_{\langle p_1, p_2 \rangle} \text{ in span } \mathbf{b} = \text{span } (1, x, y, xy)^T ; \end{aligned}$$

cf. Example 2.15. The two bases must be related by a linear transformation

$$\mathbf{c}_0^T = \mathbf{c}^T M_0 , \text{ with } M_0 \in \mathbb{C}^{4 \times 4} .$$

(a) Determine the matrix M_0 from the z_μ (hint: Apply the above relation to $\mathbf{b}$ and use (2.25)). Compare this with the method indicated in Example 2.15.

(b) How can you use M_0 to interpolate on $Z[\langle p_1, p_2 \rangle]$?

(c) How can you use M_0 to find r for $p \notin \text{span } \mathbf{b}$?

2. Consider the following set of differential functionals $\partial_j[z] \in (\mathcal{P}^3)^*$ (the evaluation point z is not denoted):

$$\partial_{000},\ \partial_{100},\ \partial_{110},\ \partial_{101},\ \partial_{011},\ \partial_{020} .$$

(a) Convince yourself that the set is not closed. What is the largest closed subset?

(b) Which functionals have to be appended to make the set closed?

2.4   The Central Theorem of Polynomial Systems Solving

2.4.1   Basis Transformations in $\mathcal{R}$ and $\mathcal{D}$

In the previous two sections, it has become obvious that the choice of the bases in $\mathcal{R}[\mathcal{I}]$ and $\mathcal{D}[\mathcal{I}]$ may play an important role in the computational treatment of ideals $\mathcal{I} = \langle P \rangle$ generated by 0-dimensional systems P of multivariate polynomials. Remember that $\mathcal{R}[\mathcal{I}]$ is a space of representatives so that all equality relations in $\mathcal{R}[\mathcal{I}]$ are actually equivalences mod $\mathcal{I}$.

Let $\dim \mathcal{R}[\mathcal{I}] = m$ and consider two bases of $\mathcal{R}[\mathcal{I}] \subset \mathcal{P}^s$

$$\mathbf{b}_0 = \begin{pmatrix} b_{01} \\ \vdots \\ b_{0m} \end{pmatrix} , \qquad \mathbf{b} = \begin{pmatrix} b_1 \\ \vdots \\ b_m \end{pmatrix} ,$$

and the conjugate bases (cf. Definition 2.11) of $\mathcal{D}[\mathcal{I}] \subset (\mathcal{P}^s)^*$

$$\mathbf{c}_0^T = (c_{01}, \ldots, c_{0m}), \qquad \mathbf{c}^T = (c_1, \ldots, c_m).$$

By Proposition 2.13, a conjugate basis in $\mathcal{D}[\mathcal{I}]$ retrieves the components of the representation of a polynomial p in the basis of $\mathcal{R}[\mathcal{I}]$. Thus, for the above bases and any $p \in \mathcal{P}^s$,

$$p = \mathbf{c}_0^T(p)\,\mathbf{b}_0 = \mathbf{c}^T(p)\,\mathbf{b}. \tag{2.41}$$

In particular, we can express the polynomials of the one basis in terms of the other:

$$\mathbf{b} = \mathbf{c}_0^T(\mathbf{b})\,\mathbf{b}_0 =: M_0\,\mathbf{b}_0, \tag{2.42}$$

where the numerical matrix $\mathbf{c}_0^T(\mathbf{b}) = M_0 \in \mathbb{C}^{m \times m}$ is nonsingular, and (cf. (2.41))

$$\mathbf{c}_0^T = \mathbf{c}^T\,M_0. \tag{2.43}$$

Since the components of $\mathbf{b}_0$ and $\mathbf{b}$ are polynomials, (2.42) is not an ordinary basis transformation as between two bases in $\mathbb{C}^m$. Rather, the quotient ring structure of $\mathcal{R}[\mathcal{I}]$ permits the relation $\mathbf{b}(x) \equiv M_0\,\mathbf{b}_0(x) \bmod \mathcal{I}$ to be written in the simple *numerical* form (2.42). Similarly, the basis change (2.43) in $\mathcal{D}[\mathcal{I}]$ is a *numerical* linear transformation only because both bases vanish on $\mathcal{I}$.

This observation is a key to the computational solution of polynomial systems because it reveals the underlying *linear structure* of 0-dimensional polynomial ideals in $\mathcal{P}^s$:

For $\mathcal{I} \subset \mathcal{P}^s$, basis transformations in $\mathcal{R}[\mathcal{I}]$ and $\mathcal{D}[\mathcal{I}]$ are numerical linear transformations.

Let us now consider the representation of some linear mapping $\mathcal{A} : \mathcal{R}[\mathcal{I}] \to \mathcal{R}[\mathcal{I}]$ w.r.t. the two bases $\mathbf{b}_0$, $\mathbf{b}$:

$$\mathbf{b}_0 \overset{\mathcal{A}}{\to} A^{(0)}\mathbf{b}_0, \qquad \mathbf{b} \overset{\mathcal{A}}{\to} A\,\mathbf{b}.$$

Proposition 2.22.

$$A\,M_0 = M_0\,A^{(0)}. \tag{2.44}$$

Proof: Consider $p \overset{\mathcal{A}}{\to} q$ for p from (2.41). Then, by (2.43) and (2.42),

$$q = \begin{cases} \mathbf{c}_0^T(p)\,A^{(0)}\mathbf{b}_0 &= \mathbf{c}^T(p)\,M_0 A^{(0)}\,\mathbf{b}_0 \\ \mathbf{c}^T(p)\,A\,\mathbf{b} &= \mathbf{c}^T(p)\,A M_0\,\mathbf{b}_0. \end{cases} \qquad \square$$

Example 2.16: Consider the ring $\mathcal{R}[\mathcal{I}]$ of Example 2.9, with the first multiplicative structure. Besides $\mathbf{b} = (1,\ x,\ y,\ x^2)^T$, we consider the basis

$$\mathbf{b}_0 = \begin{pmatrix} -7 + 12\,x - 6\,x^2 + x^3 \\ 6 - 11\,x + 6\,x^2 - x^3 \\ -4 + 8\,x - 5\,x^2 + x^3 \\ 8 - 12\,x + 6\,x^2 - x^3 \end{pmatrix},$$

which spans the quotient ring $\mathcal{R}[\mathcal{I}]$ by a different set of representatives.

Since we know from A_x in Example 2.9 that $x^3 \equiv 9 - 12\,x - y + 6\,x^2 \bmod \mathcal{I}$, we find that

$$\mathbf{b}_0 \;=\; \begin{pmatrix} 2 & 0 & -1 & 0 \\ -3 & 1 & 1 & 0 \\ 5 & -4 & -1 & 1 \\ -1 & 0 & 1 & 0 \end{pmatrix} \mathbf{b} \;=:\; M_0^{-1}\,\mathbf{b},$$

which implies $\mathbf{c}_0^T = \mathbf{c}^T M_0$ for the conjugate dual bases. With $\mathbf{c}^T$ from Example 2.13, we find $\mathbf{c}_0^T = (\,\partial_{00}[z_1],\ \partial_{10}[z_1],\ \partial_{20}[z_1],\ \partial_{00}[z_2]\,)$, which establishes that $\mathbf{b}_0$ is a Lagrange basis of $\mathcal{R}[\mathcal{I}]$, which is conjugate to the dual basis $\mathbf{c}_0^T$ of function and derivative evaluations.

When we transform A_x of Example 2.9 which represents multiplication by x in the basis $\mathbf{b}$ to the basis $\mathbf{b}_0$, we obtain (cf. (2.44))

$$A_x^{(0)} \;=\; M_0^{-1}\,A_x\,M_0 \;=\; \begin{pmatrix} 2 & 1 & 0 & 0 \\ 0 & 2 & 1 & 0 \\ 0 & 0 & 2 & 0 \\ 0 & 0 & 0 & 1 \end{pmatrix}$$

as the representation of multiplication by x in the Lagrange basis $\mathbf{b}_0$. $\square$

In applying these general observations to ideals $\mathcal{I} = \langle P \rangle$, we assume at first that the polynomial system $P \subset \mathcal{P}^s$ has m *disjoint simple* zeros $z_1, \ldots, z_m$. When we choose an associated *Lagrange basis* of $\mathcal{R}[\mathcal{I}]$ as $\mathbf{b}_0$, we have, by (2.36), for an arbitrary $p \in \mathcal{P}^s$:

$$\mathbf{c}_0^T(p) \;=\; (p(z_1), \ldots, p(z_m)) \;=:\; p(\mathbf{z})$$

and, by (2.42), for an arbitrary basis $\mathbf{b}$ of $\mathcal{R}[\mathcal{I}]$:

$$M_0 \;=\; \mathbf{c}_0^T(\mathbf{b}) \;=\; \begin{pmatrix} | & & | \\ \mathbf{b}(z_1) & \cdots & \mathbf{b}(z_m) \\ | & & | \end{pmatrix} \;=:\; \mathbf{b}(\mathbf{z}). \tag{2.45}$$

Equations (2.45) and (2.42) permit the immediate determination of a Lagrange basis for $\mathcal{R}[\mathcal{I}]$ with basis $\mathbf{b}$ and specified dual space $\mathcal{D}_0[\mathcal{I}]$:

$$\mathbf{b}_0 \;=\; M_0^{-1}\,\mathbf{b} \;=\; (\mathbf{c}_0^T(\mathbf{b}))^{-1}\,\mathbf{b}. \tag{2.46}$$

Example 2.17: In Example 2.7, a Lagrange basis in span $\{1, x, y, xy\}$ has been displayed without derivation. The underlying ideal has zeros $(\frac{6}{5}, \frac{2}{5})$, $(-\frac{2}{5}, \frac{6}{5})$, $(-\frac{6}{5}, -\frac{2}{5})$, $(\frac{2}{5}, -\frac{6}{5})$, cf. Example 2.2. With the corresponding $M_0 = \mathbf{c}_0^T(\mathbf{b})$ as displayed in Example 2.5, we have

$$\mathbf{b}_0 \;=\; M_0^{-1}\,\mathbf{b} \;=\; \begin{pmatrix} \frac{1}{4} & \frac{3}{8} & \frac{1}{8} & \frac{25}{48} \\[4pt] \frac{1}{4} & -\frac{1}{8} & \frac{3}{8} & -\frac{25}{48} \\[4pt] \frac{1}{4} & -\frac{3}{8} & -\frac{1}{8} & \frac{25}{48} \\[4pt] \frac{1}{4} & \frac{1}{8} & -\frac{3}{8} & -\frac{25}{48} \end{pmatrix},$$

which confirms the basis polynomials in Example 2.7. $\square$

2.4.2   A Preliminary Version

We are now ready to formulate a preliminary version (viz. for disjoint simple zeros) of our "central theorem."

Theorem 2.23.   Let the 0-dimensional ideal $\mathcal{I} \subset \mathcal{P}^s$ possess m disjoint simple zeros $z_\mu \in \mathbb{C}^s$, $\mu = 1(1)m$. Consider the commuting family $\overline{A} \subset \mathbb{C}^{m \times m}$ of multiplication matrices w.r.t. an arbitrary fixed basis $\mathbf{b}$ of $\mathcal{R}[\mathcal{I}]$. The *joint eigenvectors* of $\overline{A}$ are the m columns $\mathbf{b}(z_\mu)$ of the matrix M_0 of (2.45).

Proof: Take $q \in \mathcal{P}^s$ such that the $q(z_\mu) \in \mathbb{C}$, $\mu = 1(1)m$, are distinct values. Remember (cf. (2.10)) that a Lagrange basis $\mathbf{b}_0$ of $\mathcal{R}[\mathcal{I}]$ satisfies

$$
b_{0\mu}(z_\nu) \;=\; \begin{cases} 0 & \nu \neq \mu, \\ 1 & \nu = \mu. \end{cases}
$$

Therefore, in the basis $\mathbf{b}_0$, multiplication by q must be represented by the *diagonal* matrix

$$
A_q^{(0)} \;=\; \begin{pmatrix} q(z_1) & & 0 \\ & \ddots & \\ 0 & & q(z_m) \end{pmatrix}.
$$

By Proposition 2.22, the matrix A_q which represents multiplication by q in the basis $\mathbf{b}$ satisfies

$$
A_q\, M_0 \;=\; M_0\, A_q^{(0)} \qquad \text{or} \qquad A_q\, \mathbf{b}(z_\mu) \;=\; q(z_\mu)\, \mathbf{b}(z_\mu), \quad \mu = 1(1)m\,;
$$

cf. (2.45). Thus the m columns $\mathbf{b}(z_\mu)$ of M_0 are eigenvectors of A_q; the associated m eigenvalues $q(z_\mu)$ are distinct and simple so that A_q is nonderogatory. Thus, by Proposition 2.8, the column vectors $\mathbf{b}(z_\mu)$ of M_0 are joint eigenvectors of the family $\overline{A}$. Since there are m such vectors, the family $\overline{A}$ has no other joint eigenvectors. $\square$

The importance of this theorem for polynomial systems solving is revealed by

Corollary 2.24.   In the situation of Theorem 2.23, assume that the basis $\mathbf{b}$ of $\mathcal{R}[\mathcal{I}] \subset \mathcal{P}^s$ contains the monomials $1, x_1, x_2, \ldots, x_s$ as elements. Then the joint eigenvectors $\mathbf{b}(z_\mu)$ of the family $\overline{A}$ of multiplication matrices, normalized by 1-component = 1, display *all components* $\zeta_{\mu\sigma}, \sigma = 1(1)s$, of *all zeros* z_μ, $\mu = 1(1)m$, of $\mathcal{I}$:

$$
\text{For } \mathbf{b} = \begin{pmatrix} 1 \\ x_1 \\ x_2 \\ \vdots \\ x_s \\ \vdots \end{pmatrix}, \text{ the normalized joint eigenvectors of } \overline{A} \text{ are } \begin{pmatrix} 1 \\ \zeta_{\mu 1} \\ \zeta_{\mu 2} \\ \vdots \\ \zeta_{\mu s} \\ \vdots \end{pmatrix}, \ \mu = 1(1)m.
$$

$$\tag{2.47}$$

Corollary 2.24 reduces the task of computing all zeros of a 0-dimensional multivariate system P of polynomial equations to the determination of a suitable *monomial basis* of the quotient ring $\mathcal{R}[\langle P \rangle]$ and of its *multiplication matrices*, and to an ordinary *matrix eigenproblem*.

If the hypotheses on the presence of all x_σ in the basis $\mathbf{b}$ of $\mathcal{R}[\mathcal{I}]$ cannot be satisfied, there is an easy way out: Assume that $x_{\sigma'}$ is not an element of $\mathbf{b}$. Then, $A_{\sigma'}\mathbf{b}(z_\mu) = \zeta_{\mu\sigma'}\mathbf{b}(z_\mu)$ so that, with $b_1 = 1$,

$$\zeta_{\mu\sigma'} = (1\ 0\ \ldots\ 0)\, A_{\sigma'}\, \mathbf{b}(z_\mu) =:\ a_{\sigma'1}^T\, \mathbf{b}(z_\mu)\,, \tag{2.48}$$

where $a_{\sigma'1}^T$ is the first row of $A_{\sigma'}$. Thus, the eigenvectors $\mathbf{b}(z_\mu)$ yield all components of the z_μ whenever 1 is an element of the basis $\mathbf{b}$.

The fact that a result like Corollary 2.24 was overlooked until recently (cf. Historical and Bibliographical Notes 2) is the more surprising, as relations like (2.16) and (2.17) "cry" for an interpretation as a matrix eigenproblem: Consider (2.17) which—as an equivalence mod $\mathcal{I}$—we may write as

$$x_\sigma\, \mathbf{b}(x) = A_\sigma\, \mathbf{b}(x) + \mathbf{p}(x) \quad \text{with } \mathbf{p} = (p_\mu(x)),\ p_\mu \in \mathcal{I}.$$

At $x = z_\nu$, $z_\nu \in Z[\mathcal{I}]$, the p_μ vanish and we have

$$\zeta_{\nu\sigma}\, \mathbf{b}(z_\nu) = A_\sigma\, \mathbf{b}(z_\nu), \quad \nu = 1(1)m\,.$$

Example 2.18: In Example 2.8, we represented $\langle P \rangle$ for the quadratic system of Example 2.2 by $\mathbf{b} = (1, x, y, xy)^T$ and

$$A_x = \begin{pmatrix} 0 & 1 & 0 & 0 \\ \frac{4}{5} & 0 & 0 & \frac{4}{3} \\ 0 & 0 & 0 & 1 \\ 0 & \frac{48}{125} & \frac{36}{125} & 0 \end{pmatrix}, \quad A_y = \begin{pmatrix} 0 & 0 & 1 & 0 \\ 0 & 0 & 0 & 1 \\ \frac{4}{5} & 0 & 0 & -\frac{4}{3} \\ 0 & \frac{36}{125} & -\frac{48}{125} & 0 \end{pmatrix}.$$

From either A_x or A_y, we obtain the four normalized eigenvectors

$$\begin{pmatrix} 1 \\ 1.2 \\ 0.4 \\ 0.48 \end{pmatrix}, \quad \begin{pmatrix} 1 \\ -0.4 \\ 1.2 \\ -0.48 \end{pmatrix}, \quad \begin{pmatrix} 1 \\ -1.2 \\ -0.4 \\ 0.48 \end{pmatrix}, \quad \begin{pmatrix} 1 \\ 0.4 \\ -1.2 \\ -0.48 \end{pmatrix},$$

whose second and third components display the four zeros $(1.2, 0.4)$, $(-0.4, 1.2)$, $(-1.2, -0.4)$, $(0.4, -1.2)$ of P.

The above eigenvectors are also eigenvectors of any matrix $p(A)$, $p \in \mathcal{P}^2$. For some p, however, $p(A)$ may have multiple eigenvalues and higher-dimensional eigenspaces which are spanned by two or more of the above eigenvectors; e.g.:

For $p = xy$, $p(A) = A_1 A_2$ has the two eigenvalues ± 0.48, with eigenspaces of dimension 2 spanned by the first/third and the second/fourth eigenvectors, respectively.

For $p = x^2 + y^2$, $p(A) = A_1^2 + A_2^2 = \text{diag}\,(1.6, \ldots, 1.6)$ has one 4-fold eigenvalue 1.6, and each vector in $\mathbb{C}^4$ is an eigenvector.

However, since $\overline{A}$ is a nonderogatory family, the above four vectors are the only *joint* eigenvectors of $\overline{A}$. $\square$

2.4.3 The General Case

We must still analyze the case where the ideal $\mathcal{I} = \langle P \rangle$ has one or more *multiple zeros*. In this case, the vectors $\mathbf{b}(z_\mu)$, $\mu = 1(1)m_0 < m$, cannot form a complete eigenbasis of the multiplication matrices A_q w.r.t. the basis $\mathbf{b}$ of $\mathcal{R}[\mathcal{I}]$. Actually, there cannot be any further *joint* eigenvectors of the family $\overline{A}$ beyond those for the m_0 zeros of $\mathcal{I}$.

Theorem 2.25. For a 0-dimensional ideal $\mathcal{I} \subset \mathcal{P}^s$, consider the commuting family $\overline{A}$ of multiplication matrices $A_q = q(A) \in \mathbb{C}^{m \times m}$ w.r.t. an arbitrary basis $\mathbf{b}$ of $\mathcal{R}[\mathcal{I}]$. *Each joint eigenvector x_μ of $\overline{A}$ has the form $\mathbf{b}(z_\mu)$ for some $z_\mu \in Z[\mathcal{I}]$.*

Proof: Consider $x_\mu \in \mathbb{C}^m$ such that

$$A_q\, x_\mu = \lambda_\mu(q)\, x_\mu\,, \quad \lambda_\mu(q) \in \mathbb{C} \qquad \forall q \in \mathcal{P}^s\,.$$

Assume at first that the basis $\mathbf{b}$ contains the element $b_1 = 1$.

(i) The first component $e_1^T x_\mu$ of x_μ does not vanish: Assume $e_1^T x_\mu = 0$ which implies $e_1^T A_q x_\mu = 0 \ \forall q$. But (cf. (2.20)) $e_1^T A_q \mathbf{b} = e_1^T (q\, \mathbf{b}) = q\, b_1 = q$ so that $e_1^T A_q = \mathbf{c}^T(q)$ contains the coefficients of $[q]_\mathcal{I}$ in the basis $\mathbf{b}$ and $e_1^T A_q x_\mu$ cannot vanish for all q. Therefore, we may assume the joint eigenvector x_μ as normalized by $e_1^T x_\mu = 1$.

(ii) We now establish the existence of a basis $\mathbf{b}_0$ of $\mathcal{R}[\mathcal{I}]$, with conjugate basis $\mathbf{c}_0^T = (c_{01}, \dots)$, such that

$$c_{01}(q\, p) = c_{01}(q)\, c_{01}(p) \qquad \forall q \text{ and } p\,. \tag{2.49}$$

Take x_μ as the first column $c_{01}(\mathbf{b}) = M_0\, e_1$ of a basis transformation $\mathbf{b} = M_0\, \mathbf{b}_0 = \mathbf{c}_0^T(\mathbf{b})\, \mathbf{b}_0$; cf. (2.42). Then $A_q\, x_\mu = c_{01}(A_q\, \mathbf{b}) = c_{01}(q\, \mathbf{b})$. By (2.44), the representation of multiplication by q in the basis $\mathbf{b}_0$ must satisfy $M_0\, A_q^{(0)} = A_q\, M_0$ so that $M_0\, A_q^{(0)}\, e_1 = A_q\, M_0\, e_1 = A_q\, x_\mu = \lambda_\mu(q)\, x_\mu = M_0\, \lambda_\mu(q)\, e_1$, i.e. $A_q^{(0)}\, e_1 = \lambda_\mu(q)\, e_1$.

With $c_{01}(b_1) = 1$ due to the normalization of x_μ, we have $c_{01}(q) = \mathbf{c}_0^T(q\, b_1)\, e_1 = \mathbf{c}_0^T(b_1)\, A_q^{(0)}\, e_1 = \mathbf{c}_0^T(b_1)\, \lambda_\mu(q)\, e_1 = \lambda_\mu(q)$. Analogously, with p in place of b_1, we have $c_{01}(q\, p) = \lambda_\mu(q)\, c_{01}(p) = c_{01}(q)\, c_{01}(p)$.

(iii) Equation (2.49) implies $c_{01}(x^j) = \prod_{\sigma=1}^s (c_{01}(x_\sigma))^{j_\sigma}$, and, for $p = a^T x$,

$$c_{01}(p) = a^T c_{01}(x) = p(c_{01}(x)) = p(z_\mu)\,, \quad \text{with } z_\mu := (c_{01}(x_\sigma),\ \sigma = 1(1)s) \in \mathbb{C}^s\,.$$

Since $c_{01} \in \mathcal{D}[\mathcal{I}]$ implies $c_{01}(p) = p(z_\mu) = 0 \ \forall\, p \in \mathcal{I}$, we must have $z_\mu \in Z[\mathcal{I}]$. So, finally, $x_\mu = c_{01}(\mathbf{b}) = \mathbf{b}(z_\mu)$.

(iv) For an *arbitrary* basis $\widehat{\mathbf{b}} = \widehat{M}\, \mathbf{b}$, we have $\widehat{A}_q = \widehat{M}\, A_q \widehat{M}^{-1}$ and there is a joint eigenvector $\widehat{x}_\mu = \widehat{M}\, x_\mu$ of the family $\widehat{\overline{A}}$ for each joint eigenvector x_μ of $\overline{A}$:

$$\widehat{A}_q\, \widehat{x}_\mu = \widehat{A}_q\, \widehat{M}\, x_\mu = \widehat{M}\, A_q\, x_\mu = \widehat{M}\, \lambda_\mu(q)\, x_\mu = \lambda_\mu(q)\, \widehat{x}_\mu\,,$$

and $\widehat{x}_\mu = \widehat{M}\, \mathbf{b}(z_\mu) = \widehat{\mathbf{b}}(z_\mu)$. $\quad \square$

By this theorem, the set of joint eigenvectors x_μ, $\mu = 1(1)m_0$, of the family $\overline{A}$ of multiplication matrices for $\mathcal{R}[\mathcal{I}]$ w.r.t. an arbitrary basis $\mathbf{b}$ is *identical* to the set $\{\mathbf{b}(z_\mu),\ z_\mu \in Z[\mathcal{I}]\}$ of evaluations of the basis vector $\mathbf{b}$ at the zeros of $\mathcal{I}$.

Corollary 2.26. For a 0-dimensional ideal $\mathcal{I} \subset \mathcal{P}^s$, the family $\overline{A}$ of multiplication matrices for $\mathcal{R}[\mathcal{I}]$ w.r.t. an arbitrary basis is a *nonderogatory commuting family*.

Proof: With $\lambda_\mu(q) = c_{01}(q) = q(z_\mu)$, and Theorem 2.25, there cannot be *two joint* eigenvectors of $\overline{A}$ for the same eigenvalue $q(z_\mu)$. $\quad \square$

By Proposition 2.9, this implies that each joint eigenvector x_μ of $\overline{A}$ has an associated *joint invariant subspace* of a dimension $m_\mu \geq 1$. Collating all results and considerations of this section so far, and considering (2.22), we have finally arrived at our central theorem.

Theorem 2.27 (Central Theorem). Let the 0-dimensional ideal $\mathcal{I} \subset \mathcal{P}^s$ possess m_0 disjoint zeros z_μ, $\mu = 1(1)m_0$. Consider the nonderogatory commuting family $\overline{A} \subset \mathbb{C}^{m \times m}$ of multiplication matrices for $\mathcal{R}[\mathcal{I}]$ w.r.t. an arbitrary basis $\mathbf{b}$. Then the set of the m_0 *joint eigenvectors* of $\overline{A}$, with proper normalization, is identical to the set of the vectors $\mathbf{b}(z_\mu)$, $\mu = 1(1)m_0$. Furthermore, for each zero $z_\mu \in Z[\mathcal{I}]$, there is an associated *joint invariant subspace* span X_μ of $\overline{A}$ of a dimension $m_\mu \geq 1$, $\sum_\mu m_\mu = m$, such that, for each $A_q \in \overline{A}$,

$$A_q \, (X_1 \,|\, X_2 \,|\, \ldots \,|\, X_{m_0}) = (X_1 \,|\, \ldots \,|\, X_{m_0}) \begin{pmatrix} T_{q1} & & 0 \\ & \ddots & \\ 0 & & T_{qm_0} \end{pmatrix}, \qquad (2.50)$$

with *upper-triangular* $m_\mu \times m_\mu$ matrices $T_{q\mu}$ with diagonal elements $q(z_\mu)$.

Note that (2.50) is, generally, *not* a Jordan normal form of A_q because the $T_{q\mu}$ may contain nonzero elements different from a side-diagonal of 1's: cf. Example 2.19 below.

The term "Central Theorem" has not been common in the literature so far; this name appears suitable because Theorem 2.27 clearly points the way to the computational solution of 0-dimensional systems of polynomial equations:

(i) Find a suitable basis $\mathbf{b}$ of $\mathcal{R}[\mathcal{I}]$ and the multiplication matrices A_σ w.r.t. this basis. (A basis is suitable if it contains the 1 and (if feasible) all x_σ as elements; cf. Corollary 2.24.)

(ii) Compute the joint eigenvectors of the family $\overline{A}$ spanned by the A_σ and extract the zeros.

As we will see in section 10.1, task (i) is performed by linear algebra manipulations under the control of polynomial algebra relations; this is displayed by the fact that, for polynomials with rational coefficients, the A_σ have rational elements. Task (ii) is a quadratic problem but is generally considered as part of linear algebra. Thus, the Central Theorem implies (with a grain of salt):

The numerical solution of 0-dimensional systems of polynomial equations
is a task of numerical linear algebra.

If we are only interested in the location of the zeros and in their multiplicity, the extra columns in the X_μ and the above-diagonal elements in the $T_{q\mu}$ of (2.50) are without interest. However, as we have seen in Example 2.13, the location and multiplicity of all zeros of a multivariate polynomial ideal $\mathcal{I}$ do not fully specify the ideal.

The meaning of this additional information may be derived from a comparison of (2.50) with (2.44): The $m \times m$ matrix $T_q := \mathrm{diag}\,(T_{q\mu})$ in (2.50) represents multiplication by q in $\mathcal{R}[\mathcal{I}]$ w.r.t. an expanded Lagrange basis $\mathbf{b}_0$ for whose conjugate basis $\mathbf{c}_0^T$ in $\mathcal{D}[\mathcal{I}]$ we have (cf. (2.42))

$$\mathbf{c}_0^T \, (\mathbf{b}) = (\, X_1 \,|\, X_2 \,|\, \ldots \,|\, X_{m_0} \,) \quad \text{and} \quad A_q^{(0)} = T_q \, ;$$

by (2.17), this implies, for $q = x_\sigma$, $x_\sigma \mathbf{b}_0 = T_{x_\sigma} \mathbf{b}_0$, $\sigma = 1(1)s$, and

$$\mathbf{c}_0^T (x_\sigma \, \mathbf{b}_0) = A_{x_\sigma}^{(0)} \, \mathbf{c}_0^T (\mathbf{b}_0) = T_{x_\sigma}, \qquad \sigma = 1(1)s\,. \qquad (2.51)$$

The relations (2.51) permit the identification of the missing basis functionals $c_{0\nu} \in \mathbf{c}_0$ which complement the function evaluations $\partial_0[z_\mu]$, $\mu = 1(1)m_0$.

We will explain details of this identification in section 8.5.3; at this point, we only apply our newly gained insight to the situation of Example 2.13 (cf. also Example 2.16).

Example 2.19: Obviously, in Example 2.13, the relations (2.50) have been found for $q = x,\ y$ and for the two different multiplicative structures of $\mathcal{R}[\mathcal{I}]$ considered there. In both cases, the joint eigenvectors $\begin{pmatrix} 1 \\ 2 \\ 2 \\ 1 \\ 4 \end{pmatrix}$ and $\begin{pmatrix} 1 \\ 1 \\ 1 \\ 2 \\ 1 \end{pmatrix}$ display the two zeros $(2, 1)$ and $(1, 2)$ of $\mathcal{I}$—as we have already observed in Example 2.16.

Now we consider the first invariant subspace span $\{X_1\}$ of dimension 3 and its associated upper-triangular matrices T_{x1}, T_{y1}. For the first multiplicative structure of $\mathcal{R}[\mathcal{I}]$, we have from (2.51)

$$\mathbf{c}_0^T(x\,\mathbf{b}_0) = T_{x1} = \begin{pmatrix} 2 & 1 & 0 \\ & 2 & 1 \\ 0 & & 2 \end{pmatrix}, \quad \mathbf{c}_0^T(y\,\mathbf{b}_0) = T_{y1} = \begin{pmatrix} 1 & 0 & 0 \\ & 1 & 0 \\ 0 & & 1 \end{pmatrix};$$

this implies $c_{02}(x\,b_{01}) = 1$, $c_{03}(x\,b_{01}) = 0$, $c_{03}(x\,b_{02}) = 1$, and $c_{02}(y\,b_{01}) = c_{03}(y\,b_{01}) = c_{03}(y\,b_{02}) = 0$. We claim that (cf. Example 2.16) the functionals $c_{02} = \partial_{10}[z_1]$ and $c_{03} = \partial_{20}[z_1]$ satisfy these equations. Omitting $[z_1]$ and remembering that $c_{0\nu}(b_{0\mu}) = \delta_{\mu\nu}$ because of (2.25), we have

$$\begin{aligned}
\partial_{10}(x\,b_{01}) &= \partial_{10}(x)\partial_{00}(b_{01}) + \partial_{00}(x)\partial_{10}(b_{01}) = 1 \cdot 1 + 2 \cdot 0 = 1, \\
\partial_{20}(x\,b_{01}) &= \partial_{20}(x)\partial_{00}(b_{01}) + \partial_{10}(x)\partial_{10}(b_{01}) + \partial_{00}(x)\partial_{20}(b_{01}) = 0 \cdot 1 + 1 \cdot 0 + 2 \cdot 0 = 0, \\
\partial_{20}(x\,b_{02}) &= \partial_{20}(x)\partial_{00}(b_{02}) + \partial_{10}(x)\partial_{10}(b_{02}) + \partial_{00}(x)\partial_{20}(b_{02}) = 0 \cdot 0 + 1 \cdot 1 + 2 \cdot 0 = 1;
\end{aligned}$$

and analogous relations for the $y\,b_{0\mu}$.

For the second multiplicative structure, we have

$$\mathbf{c}_0^T(x\,\mathbf{b}_0) = T_{x1} = \begin{pmatrix} 2 & 1 & 0 \\ & 2 & 0 \\ 0 & & 2 \end{pmatrix}, \quad \mathbf{c}_0^T(y\,\mathbf{b}_0) = T_{y1} = \begin{pmatrix} 1 & 0 & 1 \\ & 1 & 0 \\ 0 & & 1 \end{pmatrix}.$$

Obviously, $c_{02} = \partial_{10}[z_1]$ as previously; but now, $c_{03} = \partial_{01}[z_1]$ as is easily checked. $\square$

In retrospect and without all formalism, the Central Theorem states a near-trivial observation:

In the quotient ring $\mathcal{R}[\mathcal{I}]$ of a 0-dimensional ideal $\mathcal{I}$, with m_0 zeros z_μ, the nonderogatory family $\overline{\mathcal{A}}$ of the linear mappings defined by multiplication with some $q \in \mathcal{R}[\mathcal{I}]$ has *one joint eigenelement for each zero* z_μ, viz., the polynomial in $\mathcal{R}[\mathcal{I}]$ which takes the value 1 at z_μ and vanishes at all other zeros. This is clearly the only way for a polynomial to remain *invariant* (except for scaling) mod $\mathcal{I}$ under multiplication with an arbitrary other polynomial.

Naturally, in this form, the Central Theorem would be nonconstructive. It has been turned into a computational tool by the considerations of conjugate basis transformations in the quotient ring $\mathcal{R}[\mathcal{I}]$ and the associated dual space $\mathcal{D}[\mathcal{I}]$.

Exercises

1. Use the procedures `gbasis`, `SetBasis`, `MulMatrix`, `Eigenvectors` of Maple (or analogous procedures of other systems) to compute the zeros of systems of s polynomials in s variables, for $s = 2$ and 3. Compare the results with those of `solve`.

2. Consider the ideal $\mathcal{I} \subset \mathcal{P}^2$ with a triple zero $z_1 = (1, 3)$, a double zero $z_2 = (-1, 1)$, and a simple zero $z_3 = (2, -1)$; both ∂_x and ∂_y vanish for $p \in \mathcal{I}$ at z_1 but only ∂_y vanishes at z_2.

    (a) Form the zero-revealing basis c_0^T of $\mathcal{D}[\mathcal{I}]$. What is the dimension of $\mathcal{R}[\mathcal{I}]$ and $\mathcal{D}[\mathcal{I}]$?

    (b) For $\mathbf{b} = (1, x, y, xy, y^2, xy^2)^T$, form the matrix $c_0^T(\mathbf{b})$. Convince yourself that $\mathbf{b}$ is a basis of $\mathcal{R}[\mathcal{I}]$.

    (c) Determine the Lagrange basis $\mathbf{b}_0$ of $\mathcal{R}[\mathcal{I}]$ in span $\mathbf{b}$ and check its correctness. Use $\mathbf{b}_0$ to interpolate prescribed function and derivative values at the z_μ.

    (d) Determine the basis $\mathbf{c}^T$ of $\mathcal{D}[\mathcal{I}]$ conjugate to $\mathbf{b}$ and check $\mathbf{c}^T(\mathbf{b}) = I$. Use $\mathbf{c}^T$ to find the residuals mod $\mathcal{I}$ in $\mathcal{R}[\mathcal{I}]$ of various polynomials $p \in \mathcal{P}^2$.

3. Consider $\mathcal{R}[\mathcal{I}]$ of Exercise 2 further:

    (a) To find the multiplication matrices A_x, A_y w.r.t. the basis $\mathbf{b}$, you need the residuals mod $\mathcal{I}$ of all monomials in $x\,\mathbf{b}$ and $y\,\mathbf{b}$, respectively. Use $\mathbf{c}^T$ to determine the nontrivial rows of A_x and A_y. Check that they commute.

    (b) Find the multiplication matrices $A_x^{(0)}$, $A_y^{(0)}$ of $\mathcal{R}[\mathcal{I}]$ w.r.t. the Lagrange basis $\mathbf{b}_0$ in two different ways:

        (i) Evaluate $c_0^T(x\,\mathbf{b}_0)$ and $c_0^T(y\,\mathbf{b}_0)$ and use (2.51).

        (ii) Use (2.44): $A_x^{(0)} = M_0^{-1} A_x M_0$, $A_y^{(0)} = M_0^{-1} A_y M_0$.

    (c) Find the multiplication matrices $A_x^{(1)}$, $A_y^{(1)}$ of $\mathcal{R}[\mathcal{I}]$ w.r.t. the basis $\mathbf{b}_1 = (1, x, y, x^2, xy, y^2)^T$: Determine the matrix M_1^{-1} in $\mathbf{b}_1 = M_1^{-1}\mathbf{b}$ (cf. (2.42)); note that all components of $\mathbf{b}_1$ are in $\mathbf{b}$ except x^2, which is in $x\mathbf{b} = A_x\mathbf{b}$. Then use (2.44).

4. (a) From the results of Exercise 3 (b), you can immediately write down the eigenanalysis (2.50) of A_x and A_y: $(X_1\,X_2\,X_3) = M_0$, $T_x = A_x^{(0)}$ and $T_y = A_y^{(0)}$. Why is that so?

    (b) Read the zeros z_μ, $\mu = 1(1)3$, of $\mathcal{I}$ and their multiplicities from the X_μ and T_μ. Try to read also the "differentiation structure" of z_1 and z_2.

    (c) Use linear algebra software to determine the joint eigenvectors of A_x and A_y. Why is it difficult to distinguish them? How do you proceed?

2.5   Normal Sets and Border Bases

2.5.1   Monomial Bases of a Quotient Ring

According to the previous sections, the transformation of a system P of polynomial equations into a matrix eigenproblem requires the construction of a suitable monomial basis $\mathbf{b}$ for the quotient ring $\mathcal{R}[\langle P \rangle]$ and the computation of the associated multiplication matrices.

Example 2.20: Consider the quadratic system of Examples 2.2, 2.8, etc.:

$$P = \begin{cases} p_1(x, y) &= x^2 + 4xy + 4y^2 - 4, \\ p_2(x, y) &= 4x^2 - 4xy + y^2 - 4. \end{cases}$$

By *scalar* linear combination of p_1 and p_2, we can eliminate either x^2 or y^2 to obtain the two polynomials $bb_1, bb_2 \in \langle P \rangle$:

$$bb_1 \;=\; 15\,x^2 - 20\,x\,y - 12\,, \quad bb_2 \;=\; 15\,y^2 + 20\,x\,y - 12\,.$$

This suggests a basis $\mathbf{b} = (1,\,x,\,y,\,xy)^T$ for $\mathcal{R}[\langle P \rangle]$—since we know $\dim \mathbf{b} \le 4$ from the total degrees 2 of p_1 and p_2. To complete the multiplication matrices A_x and A_y w.r.t. this basis $\mathbf{b}$, we must express the monomials in $x\,\mathbf{b} = (x,\,x^2,\,xy,\,x^2y)^T$ and $y\,\mathbf{b} = (y,\,yx,\,y^2,\,y^2x)^T$ in terms of the monomials in $\mathbf{b} \bmod \langle P \rangle$.

This is trivial for x, y, xy which are immediately in $\mathbf{b}$; for x^2 and y^2, the representations are provided by bb_1 and bb_2. In order to obtain the analogous representations for x^2y and y^2x, we form

$$\begin{aligned} y \cdot bb_1 &= 15\,x^2y - 20\,xy^2 - 12\,y\,, \\ x \cdot bb_2 &= 20\,x^2y + 15\,xy^2 - 12\,x \end{aligned}$$

and obtain the two polynomials $bb_3, bb_4 \in \langle P \rangle$:

$$bb_3 \;=\; 125\,x^2\,y - 48\,x - 36\,y\,, \quad bb_4 \;=\; 125\,xy^2 - 36\,x + 48\,y\,,$$

which satisfy our needs. From $\{bb_1,\,bb_2,\,bb_3,\,bb_4\}$, we have directly the multiplication matrices A_x and A_y of Example 2.8.

Note that we could just as well have started by eliminating xy or y^2 :

$$bb_1' = 5\,y^2 + 5\,x^2 - 8\,, \quad bb_2' = 20\,xy - 15\,x^2 + 12\,,$$

which now suggests $\mathbf{b}' = (1,\,x,\,y,\,x^2)^T$ as a basis for $\mathcal{R}[\langle P \rangle]$. Now, $x\,\mathbf{b}' = (x,\,x^2,\,xy,\,x^3)^T$; $y\,\mathbf{b}' = (y,\,yx,\,y^2,\,yx^2)^T$, and we need representations for x^3 and yx^2. The system

$$\begin{aligned} x \cdot bb_1' &= 5\,xy^2 &+ 5\,x^3 & & & -8\,x\,, \\ y \cdot bb_2' &= 20\,xy^2 & & -15\,x^2y & +12\,y\,, \\ x \cdot bb_2' &= & -15\,x^3 & +20\,x^2y & +12\,x \end{aligned}$$

yields $bb_3' = 125\,x^2\,y - 48\,x - 36\,y$, $bb_4' = 125\,x^3 - 164\,x - 48\,y$, and

$$A_x' = \begin{pmatrix} 0 & 1 & 0 & 0 \\ 0 & 0 & 0 & 1 \\ -3/5 & 0 & 0 & 3/4 \\ 0 & 164/125 & 48/125 & 0 \end{pmatrix}, \quad A_y' = \begin{pmatrix} 0 & 0 & 1 & 0 \\ -3/5 & 0 & 0 & 3/4 \\ 8/5 & 0 & 0 & -1 \\ 0 & 48/125 & -36/125 & 0 \end{pmatrix}.$$

Similarly, we could have obtained the multiplication matrices for the basis $(1,\,x,\,y,\,y^2)^T$ and—with slightly more effort—for the bases $(1,\,x,\,x^2,\,x^3)^T$ and $(1,\,y,\,y^2,\,y^3)^T$.

Apparently, *each set of four monomials* from T^2 (cf. Definition 1.1) which has no "holes," i.e. which contains each divisor of an element, can serve as a suitable basis for $\mathcal{R}[\langle P \rangle]$ in this case. $\quad\square$

Definition 2.16. A (nonempty) set $\mathcal{N} = \{x^j,\ j \in J\}$ from T^s is called *convex* or *closed* iff it satisfies

$$x^j \in \mathcal{N} \quad \Rightarrow \quad x^{j'} \in \mathcal{N} \ \ \forall j' : \ x^{j'} | x^j\,, \tag{2.52}$$

where $x^{j'} \mid x^j$ is a common shorthand for "$x^{j'}$ divides x^j". $\square$

The following intuitive notion will prove handy in the formulation of many considerations.

Definition 2.17. $x^{j'} \in T^s$ is a $\begin{cases} negative \\ positive \end{cases}$ *neighbor of* $x^j \in T^s$ *iff*

$$ j' = \begin{cases} j - e_\sigma, \\ j + e_\sigma, \end{cases} \quad \text{for some } \sigma \in \{1, \ldots, s\} \text{ with } e_\sigma := (0, .., \overset{\sigma}{1}, .., 0). \quad \square$$

With this notion, Definition 2.16 reads: $\mathcal{N}$ is closed iff it contains all negative neighbors of its elements.

Example 2.21: Clearly, each closed set of monomials contains 1. When we illustrate sets of monomials in T^2 by their exponents in $\mathbb{N}_0^2$, the following sets in the upper row are closed while those in the lower row are not:

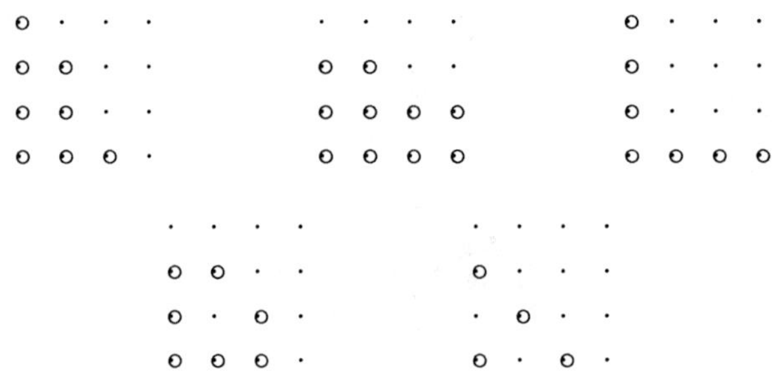

In principle, since the elements of $\mathcal{R}[\mathcal{I}]$ are only *representatives* of residue classes, we could regard monomial bases for $\mathcal{R}[\mathcal{I}]$ which are not closed. For example, $(1, x, x\,y, x\,y^2)^T$ or rather $([1], [x], [xy], [xy^2])^T$ would be a valid basis for $\mathcal{R}[\mathcal{I}]$ in Example 2.19. But the computational use of such bases is awkward; therefore we will *not* consider nonconvex sets of monomials as bases of a quotient ring.

Definition 2.18. The set of all closed subsets of T^s with m elements will be denoted by $T^s(m)$. $\square$

The set $T^s(m)$ is the "reservoir" for the potential monomial bases of an m-dimensional quotient ring. The magnitude of the sets $T^s(m)$ increases very rapidly with the number s of variables and the number m of elements. In three variables, e.g., there are already 48 different potential bases for a quotient ring of dimension 6.

For a *generic m-dimensional quotient ring* $\mathcal{R} \subset \mathcal{P}^s$, *each* set from $T^s(m)$ can serve as a basis. For a *specified m-dimensional quotient ring* $\mathcal{R}[\mathcal{I}]$, on the other hand, it is generally *not* true that *each* set from $T^s(m)$ is valid as a basis **b**: Let $\mathbf{c}_0^T$ be a basis for the dual space $\mathcal{D}[\mathcal{I}]$; then, according to Proposition 2.12, $\mathbf{c}_0^T(\mathbf{b})$ must be a *regular $m \times m$* matrix. This may or may not be a restriction for the selection of **b**, the column vector of the basis monomials.

Definition 2.19. Consider a 0-dimensional polynomial ideal $\mathcal{I} \subset \mathcal{P}^s$. A closed set of monomials from T^s is a *normal set* $\mathcal{N}[\mathcal{I}]$ of $\mathcal{I}$ iff the monomials in $\mathcal{N}[\mathcal{I}]$ form a *basis* of $\mathcal{R}[\mathcal{I}]$. If we wish to emphasize the validity of a set from $T^s(m)$ as a basis of some $\mathcal{R}$, we will call it a *feasible normal set for* $\mathcal{R}$. $\square$

The parallel use of the two notations and terms $\mathcal{N}[\mathcal{I}]$ and $\mathbf{b}$ for a basis of a quotient ring $\mathcal{R}[\mathcal{I}]$ is a little awkward. But, first of all, it is quite widespread in the literature; moreover, we will strictly consider the normal set $\mathcal{N}$ as a *set*, i.e. an unordered collection of its elements, while $\mathbf{b}$ is a *column vector* whose components are arranged in a specified order and to which linear algebra operations can be applied.

Example 2.22: $T^2(4) = \{\,\{1, y, y^2, y^3\},\ \{1, x, y, y^2\},\ \{1, x, y, xy\},\ \{1, x, y, x^2\},\ \{1, x, x^2, x^3\}\,\}$ with $x_1 = x$, $x_2 = y$. For P from Example 2.20, each element from $T^2(4)$ may be used as a normal set $\mathcal{N}[\langle P \rangle]$.

On the other hand, for $\bar{P} = \{x^2 + y^2 - 2,\ x^2 - y^2 - 1\}$, the only element from $T^2(4)$ which can serve as a normal set of $\langle \bar{P} \rangle$ is $\{1, x, y, xy\}$, it is the only *feasible* normal set in this case. $\square$

To understand the different situation for the two systems in Example 2.22, we consider, at first, the case where the ideal $\mathcal{I} \subset \mathcal{P}^s$ has m disjoint *simple* zeros z_μ, $\mu = 1(1)m$.

Proposition 2.28. For a polynomial ideal $\mathcal{I} \subset \mathcal{P}^s$ with m simple zeros, the monomials in a set $\mathcal{N} \subset T^s(m)$ can form a basis $\mathbf{b}$ of $\mathcal{R}[\mathcal{I}]$ iff

$$s_\mathcal{N}(z_1, \ldots, z_m) := \det\big(\mathbf{b}(z_\mu),\ \mu = 1(1)m\big) =: \det(\mathbf{b}(\mathbf{z})) \neq 0. \qquad (2.53)$$

Proof: By Theorem 2.20, $\mathbf{c}_0^T = (\partial_{0..0}[z_\mu],\ \mu = 1(1)m)$ is a basis of $\mathcal{D}[\mathcal{I}]$; thus, (2.53) is an immediate consequence of Proposition 2.12. $\square$

Since (2.53) is only a special version of

$$s_\mathcal{N}(c_1, \ldots, c_m) := \det\big(\mathbf{c}_0^T\,(\mathbf{b}(x))\big) \neq 0, \qquad (2.54)$$

Proposition 2.28 can immediately be extended to polynomial ideals with multiple zeros when we know the basis elements c_μ of the associated dual space $\mathcal{D}[\mathcal{I}]$, cf. section 2.3.2. Details will be considered in section 8.5.2.

Proposition 2.29. For a polynomial ideal $\mathcal{I} \subset \mathcal{P}^s$ with m zeros counting multiplicities, a set $\mathcal{N} \subset T^s(m)$ is a *feasible normal set* iff (2.53) (or (2.54)) holds where $\mathbf{b}$ is the associated normal set vector.

If we consider the components $\zeta_{\mu\sigma}$ of the z_μ in (2.53) as indeterminates, $s_\mathcal{N}$ is a polynomial in $\mathbb{C}[\zeta_{\mu\sigma},\ \mu = 1(1)m,\ \sigma = 1(1)s]$ and $s_\mathcal{N} = 0$ describes a manifold $S_\mathcal{N}$ of codimension 1 in the $\mathbb{C}^{ms}$ of the $\zeta_{\mu\sigma}$. A specified element $\mathcal{N}$ from $T^s(m)$ with the associated basis $\mathbf{b}$ is a feasible normal set of $\mathcal{I}$ iff the zeros of $\mathcal{I}$ *do not lie on this manifold* $S_\mathcal{N}$. This shows also that each $\mathcal{N} \subset T^s(m)$ is a feasible normal set for *almost all* ideals $\mathcal{I} \subset \mathcal{P}^s$ with m zeros—if we consider these ideals parametrized by their zeros in $\mathbb{C}^s$.

Generally, $S_\mathcal{N}$ will contain constellations of zeros which display certain *symmetries* or *degeneracies*. However, in a higher-dimensional complex space, it is virtually impossible to characterize all such exceptional constellations geometrically.

Example 2.22, continued: For the system $\bar{P}$, $Z[\bar{P}] = \{(\sqrt{\tfrac{3}{2}}, \sqrt{\tfrac{1}{2}}),\ (-\sqrt{\tfrac{3}{2}}, \sqrt{\tfrac{1}{2}}),\ (-\sqrt{\tfrac{3}{2}}, -\sqrt{\tfrac{1}{2}}),\ (\sqrt{\tfrac{3}{2}}, -\sqrt{\tfrac{1}{2}})\}$. Thus, the set $Z[\bar{P}]$ is invariant under many mappings of the $\mathbb{C}^2$: reflection at the x-axis, reflection at the y-axis, reflection at the origin, etc. For $\mathcal{N} = \{1, x, y, y^2\}$, we have

$$s_{\mathcal{N}}(z_1, z_2, z_3, z_4) \;=\; \det \begin{pmatrix} 1 & 1 & 1 & 1 \\ \sqrt{\frac{3}{2}} & -\sqrt{\frac{3}{2}} & -\sqrt{\frac{3}{2}} & \sqrt{\frac{3}{2}} \\ \sqrt{\frac{1}{2}} & \sqrt{\frac{1}{2}} & -\sqrt{\frac{1}{2}} & -\sqrt{\frac{1}{2}} \\ \frac{1}{2} & \frac{1}{2} & \frac{1}{2} & \frac{1}{2} \end{pmatrix} \;=\; 0,$$

while $\mathcal{N} = \{1, x, y, xy\}$ yields

$$s_{\mathcal{N}}(z_1, z_2, z_3, z_4) \;=\; \det \begin{pmatrix} 1 & 1 & 1 & 1 \\ \sqrt{\frac{3}{2}} & -\sqrt{\frac{3}{2}} & -\sqrt{\frac{3}{2}} & \sqrt{\frac{3}{2}} \\ \sqrt{\frac{1}{2}} & \sqrt{\frac{1}{2}} & -\sqrt{\frac{1}{2}} & -\sqrt{\frac{1}{2}} \\ \frac{\sqrt{3}}{2} & -\frac{\sqrt{3}}{2} & \frac{\sqrt{3}}{2} & -\frac{\sqrt{3}}{2} \end{pmatrix} \;=\; -12.$$

The zeros of the system P above form a square and also have many symmetries; but, in terms of x, y, they are concealed by the rotation of the principal axes of the two conics. $\square$

For ideals with multiple zeros, the missing elements $\partial_{0..0}[z_\mu]$ in the basis $\mathbf{c}_0^T$ of $\mathcal{D}[\mathcal{I}]$ may be replaced with appropriate derivative evaluations at the multiple zeros, cf. sections 2.3.2 and 2.4. We will see in sections 6.3 and 9.3 how m_μ-fold zeros with a specified derivative structure may be interpreted as limiting constellations of m_μ individual zeros. In this sense, the multiple zeros fill the gaps which would otherwise remain in the $\mathbb{C}^{ms}$ of the components of sets of m disjoint zeros.

Each $p \in \mathcal{P}^s$ belongs to a unique residue class $[p]_{\mathcal{I}}$ mod $\mathcal{I}$ and each residue class has a unique representative in the quotient ring $\mathcal{R}[\mathcal{I}]$; cf. section 2.2.1.

Definition 2.20. For $p \in \mathcal{P}^s$, the *normal form* $NF_{\mathcal{I}}[p]$ is the unique polynomial in $\mathcal{R}[\mathcal{I}]$ which satisfies

$$p - \mathrm{NF}_{\mathcal{I}}[p] \in \mathcal{I}. \tag{2.55}$$

With a specified basis vector $\mathbf{b}$ of $\mathcal{R}[\mathcal{I}]$ and the conjugate basis $\mathbf{c}^T$ of $\mathcal{D}[\mathcal{I}]$, Proposition 2.13 implies

$$\mathrm{NF}_{\mathcal{I}}[p] \;=\; \mathbf{c}^T(p)\,\mathbf{b}(x) \;=\; \sum_{\mu=1}^{m} c_\mu(p)\,b_\mu(x). \tag{2.56}$$

2.5.2 Border Bases of Polynomial Ideals

In order to complete step (i) of the procedure described below Theorem 2.27, we must also find the multiplication matrices A_σ of $\mathcal{R}[\mathcal{I}]$ with respect to the basis $\mathbf{b}$.

Definition 2.21. For a specified closed set $\mathcal{N} \subset \mathcal{T}^s$, we define the following sets in $\mathcal{T}^s$:

(i) The *corner set*

$$C[\mathcal{N}] \;:=\; \{\, x^j \in \mathcal{T}^s \;:\; x^j \notin \mathcal{N},\ \text{but } each \text{ negative neighbor of } x^j \in \mathcal{N} \,\}; \tag{2.57}$$

(ii) the *border set*

$$B[\mathcal{N}] \;:=\; \begin{cases} \{\, x^j \in \mathcal{T}^s \;:\; x^j \notin \mathcal{N},\ \text{but } some \text{ negative neighbor of } x^j \in \mathcal{N} \,\}, \\ \{\, x^j \in \mathcal{T}^s \;:\; x^j \notin \mathcal{N},\ x^j \text{ a positive neighbor of some } x^{j'} \in \mathcal{N} \,\}; \end{cases} \tag{2.58}$$

(iii) the family of *hull sets* $H_i[\mathcal{N}]$, $i = 0, 1, 2, \ldots$, defined by

$$H_0[\mathcal{N}] := \mathcal{N}, \qquad H_{i+1}[\mathcal{N}] := H_i[\mathcal{N}] \cup B[H_i[\mathcal{N}]]. \qquad \square \tag{2.59}$$

For any closed set $\mathcal{N} \subset T^s$, we have

$$C[\mathcal{N}] \subset B[\mathcal{N}], \qquad \mathcal{N} \subset H_1[\mathcal{N}] \subset H_2[\mathcal{N}] \subset \ldots .$$

Example 2.23: $s = 2$:

$$
\begin{array}{lll}
\mathcal{N} &= \circ \\
C[\mathcal{N}] &= \triangleleft \\
B[\mathcal{N}] &= \diamond \\
H_1[\mathcal{N}] &= \circ \cup \diamond \\
H_2[\mathcal{N}] &= \circ \cup \diamond \cup \triangleright
\end{array}
$$

Let $\mathcal{N} \subset T^s$ be a normal set of $\mathcal{I} \subset \mathcal{P}^s$, let $\mathbf{b}_{\mathcal{N}} = \{b_\mu, \ \mu = 1(1)m\}$ be the associated basis of $\mathcal{R}[\mathcal{I}]$, and let $\mathbf{c}_{\mathcal{N}}^T$ be the *conjugate* basis of $\mathcal{D}[\mathcal{I}]$. The rows of the multiplication matrices A_σ contain the representations of the monomials $x_\sigma \cdot x^{j_\mu}$, $x^{j_\mu} \in \mathcal{N}$, in terms of $\mathbf{b}_{\mathcal{N}}$ for $\sigma = 1(1)s$. From (2.13), we have

$$A_\sigma = \begin{pmatrix} \ldots & \mathbf{c}_{\mathcal{N}}^T(x_\sigma b_1) & \ldots \\ & \vdots & \\ \ldots & \mathbf{c}_{\mathcal{N}}^T(x_\sigma b_m) & \ldots \end{pmatrix}, \qquad \sigma = 1(1)s . \tag{2.60}$$

By (2.58), the products $x_\sigma b_\mu$, $b_\mu \in \mathcal{N}$, are either in $\mathcal{N}$ or in $B[\mathcal{N}]$. For $x_\sigma b_\mu = b_{\mu'} \in \mathcal{N}$, the row vectors $\mathbf{c}_{\mathcal{N}}^T(x_\sigma b_\mu) = e_{\mu'}^T$ are trivial. Therefore, the specification of the A_σ requires the determination of the row vectors $\mathbf{c}_{\mathcal{N}}^T(x^j) \in \mathbb{C}^m$, $\forall x^j \in B[\mathcal{N}]$.

Proposition 2.30. For a 0-dimensional ideal $\mathcal{I} \subset \mathcal{P}^s$ and any feasible normal set $\mathcal{N}$ of $\mathcal{I}$, the polynomials

$$bb_\kappa := x^{j_\kappa} - \mathbf{c}_{\mathcal{N}}^T(x^{j_\kappa})\,\mathbf{b}_{\mathcal{N}}, \qquad \forall\, x^{j_\kappa} \in B[\mathcal{N}], \tag{2.61}$$

form an ideal basis of $\mathcal{I}$.

Proof: From $x^{j_\kappa} = \mathbf{c}_{\mathcal{N}}^T(x^{j_\kappa})\,\mathbf{b}_{\mathcal{N}}$ mod $\mathcal{I}$, we have $bb_\kappa \in \mathcal{I}$, $\kappa = 1(1)k := |B[\mathcal{N}]|$. We must show that each $p \in \mathcal{I}$ may be written as $\sum_\kappa d_\kappa(p)\,bb_\kappa$, $d_\kappa \in \mathcal{P}^s$. We show that each $p \in \mathcal{P}^s$ may be written as $p = \mathbf{c}_{\mathcal{N}}^T(p)\mathbf{b}_{\mathcal{N}} + \sum_\kappa d_\kappa(p)\,bb_\kappa$; then $p \in \mathcal{I}$ implies $\mathbf{c}_{\mathcal{N}}^T(p) = 0$. It suffices to show that each monomial $x^j \notin \mathcal{N}$ may be written in the form

$$x^j = \mathbf{c}_{\mathcal{N}}^T(x^j)\,\mathbf{b}_{\mathcal{N}} + \sum_\kappa d_\kappa(x^j)\,bb_\kappa . \tag{2.62}$$

Note that we have to verify (2.62) as an equality in $\mathcal{P}$.

Let $x^j \in H_r[\mathcal{N}]$, $r \in \mathbb{N}$; then, by the recursive definition (2.59) of H_r, x^j permits a representation (generally not unique)

$$x^j = x_{\sigma_r} \ldots x_{\sigma_1} b_\mu, \qquad b_\mu \in \mathcal{N}, \ \sigma_\rho \in \{1, \ldots, s\} .$$

We have

$$x_{\sigma_1} b_\mu = (a_{\sigma_1}^T)_\mu \, \mathbf{b} + \text{some } bb_\kappa \quad \text{and} \quad x_{\sigma_\rho} \mathbf{b} = A_{\sigma_\rho} \mathbf{b} + \mathbf{bb}_{\sigma_\rho} \,, \quad \rho = 2(1)r \,,$$

where $\mathbf{bb}_{\sigma_\rho}$ is a column vector of components which are either 0, for the trivial rows of A_{σ_ρ}, or one of the bb_κ, for the nontrivial rows. Because the x_{σ_ρ} commute with the A_{σ_ρ}, we obtain

$$x_{\sigma_r} \dots x_{\sigma_1} b_\mu = (a_{\sigma_1}^T)_\mu A_{\sigma_2} \dots A_{\sigma_r} \mathbf{b}$$
$$+ (a_{\sigma_1}^T)_\mu A_{\sigma_2} \dots A_{\sigma_{r-1}} x_{\sigma_r} \mathbf{bb}_{\sigma_r} + (a_{\sigma_1}^T)_\mu A_{\sigma_2} \dots A_{\sigma_{r-2}} x_{\sigma_{r-1}} x_{\sigma_r} \mathbf{bb}_{\sigma_{r-1}} + \dots \,,$$

which is of the requested form. $\square$

Definition 2.22. For a 0-dimensional ideal $\mathcal{I} \subset \mathcal{P}^s$ with a feasible normal set $\mathcal{N}$, the polynomials (2.61) form the $\mathcal{N}$-*border basis* $\mathcal{B}_{\mathcal{N}}[\mathcal{I}]$ of $\mathcal{I}$. $\square$

Example 2.23: The four polynomials $bb_1, \dots, bb_4$ in Example 2.19 form the border basis for $\langle P \rangle$ for the normal set $\{1, x, y, xy\}$. Analogously, the four polynomials $bb_1', \dots, bb_4'$ form the border basis $\mathcal{B}_{\mathcal{N}'}[\langle P \rangle]$ for $\mathcal{N}' = \{1, x, y, x^2\}$. $\square$

We summarize our considerations about representations of polynomial ideals.

Definition 2.23. A representation of a 0-dimensional ideal $\mathcal{I} \subset \mathcal{P}^s$ is called (in our context) a *normal set representation* of $\mathcal{I}$ if it consists of
   - a normal set basis $\mathbf{b}_{\mathcal{N}}$ of the quotient ring $\mathcal{R}[\mathcal{I}]$,
   - the multiplication matrices A_σ of $\mathcal{R}[\mathcal{I}]$ w.r.t. $\mathbf{b}_{\mathcal{N}}$ and the border basis $\mathcal{B}_{\mathcal{N}}$. $\square$

For computational purposes, this is our *standard representation* of a 0-dimensional polynomial ideal $\mathcal{I}$. Trivially, the "and" in the second item may be replaced by "or"; cf. (2.60) and (2.61). From such a representation, a standard matrix eigenproblem generates the complete zero set $Z[\mathcal{I}]$ of $\mathcal{I}$.

The design of a constructive algorithm which *determines* a feasible normal set $\mathcal{N}[\langle P \rangle]$ and the border basis $\mathcal{B}_{\mathcal{N}}[\langle P \rangle]$ for a specified polynomial system P has not yet been discussed. It will be an important topic in later parts of the book, particularly in Chapter 10.

2.5.3   Groebner Bases

We cannot finish this chapter without a word on *Groebner bases*, which have played such a dominant role in computational polynomial algebra for decades; cf. Historical and Bibliographical Notes 2. How do they figure in our approach to polynomial algebra?

From our point of view, a *Groebner basis* for an ideal $\langle P \rangle \subset \mathcal{P}^s$ generates just *one of the large variety* of possible representations of the kind described above. It is the border basis $\mathcal{B}_{\mathcal{N}}[\langle P \rangle]$ for a very special normal set $\mathcal{N}[\langle P \rangle]$ which is uniquely determined by the polynomial system P after a so-called *term order* has been specified. A term order arranges the monomials $x^j \in T^s$ in a linear order which begins with 1 as the lowest element and which is consistent with multiplication; we will discuss it in section 8.4.1. The introduction of a term order permitted Buchberger to formulate a strategy for finding this particular $\mathcal{N}$ and the associated border basis $\mathcal{B}_{\mathcal{N}}[\langle P \rangle]$ for a given system P and to prove that the algorithm based on that strategy *comes to an end after finitely many steps.*

Actually, for a Groebner basis one can show that a subset of the border basis which contains only the representations (2.61) of the *corner* monomials of $\mathcal{N}$ already constitutes a complete basis for the ideal; this subset is usually denoted as the *reduced* Groebner basis or, quite often, as *the* Groebner basis. But this distinction is not so important because, for almost all computational purposes, the full border basis or, equivalently, the multiplication matrices are needed anyway.

We will deal with Groebner bases in section 8.4 and other places. In this introductory chapter, we have avoided the introduction of Groebner bases for three reasons:

(1) The introduction of a strict linear order between the variables $x_1, \ldots, x_s$ and their power products is an artificial technical tool which has no intuitive foundation for most polynomial systems. While it brings certain formal advantages, it may also carry severe penalties.

(2) Everything which can be done with a Groebner basis representation of an ideal $\mathcal{I}$ can also be done with an arbitrary other border basis representation of $\mathcal{I}$. Thus, an emphasis on Groebner bases obscures the view of the great flexibility which prevails in the representation of polynomial ideals.

(3) Representations based on different normal sets $\mathcal{N}$ may have very different qualities when used in connection with data of limited accuracy (cf. Chapter 3) and with approximate numerical computation (cf. Chapter 4). The consideration of this important aspect is also obscured by a fixation on one very particular normal set.

Thus, we do not believe that the understanding of the fundamental relations in polynomial algebra profits from the introduction of term order and Groebner bases at an early stage.

2.5.4 Polynomial Interpolation

Polynomial interpolation is—in a sense—the inverse task to finding the zeros of a polynomial ideal; it is also best understood by considering it in the context of a dual space and its quotient ring. Therefore, we include a principal account of it here, as a basis for later consideration in section 5.4 for the univariate and section 9.6 for the multivariate case.

Definition 2.24. The following task is a *polynomial interpolation problem*:
Given m linear functionals $l_\mu : \mathcal{P}^s \to \mathbb{C}$ and associated values $w_\mu \in \mathbb{C}$, $\mu = 1(1)m$, find $p^* \in \mathcal{P}^s$ such that $l_\mu(p^*) = w_\mu$, $\mu = 1(1)m$. $\quad\square$

An interpolation problem is well-defined iff $\mathcal{D} := \mathrm{span}\,\{l_\mu\} \subset (\mathcal{P}^s)^*$ is a *closed* vector space of linear functionals (cf. Definition 2.15); then it defines an ideal $\mathcal{I}[\mathcal{D}] := \{p \in \mathcal{P}^s : l(p) = 0 \,\forall l \in \mathcal{D}\} \subset \mathcal{P}^s$ (cf. Theorem 2.21). Obviously, a polynomial interpolation problem can only be solved mod $\mathcal{I}[\mathcal{D}]$, i.e. in the quotient ring $\mathcal{R}[\mathcal{D}]$. For a chosen normal set basis $\mathbf{b}$ of $\mathcal{R}[\mathcal{D}]$, we must transform the basis $\mathbf{c}_0^T = \{l_\mu, \mu = 1(1)m\}$ of $\mathcal{D}$ into the basis $\mathbf{c}^T = \{c_\mu, \mu = 1(1)m\}$ which is conjugate to $\mathbf{b}$, i.e. which furnishes the coefficients of the interpolation polynomial $p^* \in \mathcal{R}[\mathcal{D}]$; cf. section 2.4.1. The complete solution of a polynomial interpolation problem is the set of all polynomials in the residue class $[p^*]_{\mathcal{I}[\mathcal{D}]}$.

By (2.43), we have

$$\mathbf{c}^T = \mathbf{c}_0^T M_0^{-1} \qquad \text{with } M_0 := \mathbf{c}_0^T(\mathbf{b}) . \tag{2.63}$$

From a computational point of view, $\mathbf{b}$ should be chosen such that the matrix M_0 is sufficiently well-conditioned. In the univariate case, with $l_\mu(p) = p(z_\mu)$ and disjoint z_μ, M_0 is the

Vandermonde matrix. Note that the l_μ may be more general than just evaluation functionals; cf. Example 2.10.

Example 2.24: Consider the zero set $Z := \{(1.2, 0.4), (-0.4, 1.2), (-1.2, -0.4), (0.4, -1.2)\}$ of the polynomial system $P(x, y)$ of Examples 2.2 and the following ones, and let $l_\mu(p) := p(z_\mu)$, $\mu = 1(1)4$. With the basis $\mathbf{b} := (1, x, y, xy)^T$ of the associated quotient ring $\mathcal{R}$, we have

$$M_0 \; = \; \begin{pmatrix} | & & | \\ \mathbf{b}(z_1) & \cdots & \mathbf{b}(z_4) \\ | & & | \end{pmatrix} \; = \; \begin{pmatrix} 1 & 1 & 1 & 1 \\ 1.2 & -.4 & -1.2 & .4 \\ -4 & 1.2 & -.4 & -1.2 \\ .48 & -.48 & .48 & -.48 \end{pmatrix},$$

with $\mathrm{cond}_{\max}(M_0) \approx 5$; cf. section 3.2.2.

Consider the interpolation problem: Find $p^*(x, y) \in \mathrm{span}\,\mathbf{b}$ with the values $.1, -.15, .5, .9$ at the z_μ, $\mu = 1(1)4$. We obtain $\mathbf{c}^T = (.1, -.15, .5, .9)\, M_0^{-1} = (.3375, -.01875, -.44375, -.078125)$ or

$$p^*(x, y) \; = \; .3375 - .01875\, x - .44375\, y - .078125\, xy\,.$$

Note that we may add *any* polynomial from $\langle P \rangle$ to p^* and obtain a correct interpolation polynomial.   □

Exercises

1. Try to develop an expression for the size of $T^2(m)$. What is the size of $T^3(4)$, $T^3(5)$?

2. (a) In Example 2.19, from linear combinations of suitable multiples of p_1 and p_2, find the border bases and multiplication matrices for the remaining three normal sets in $T^2(4)$. How many different multiples of p_1, p_2 (including the polynomials themselves) are necessary to determine the full border bases?

(b) How much savings is possible if only one multiplication matrix is determined? For the "univariate" normal sets, how are the other components of the zeros obtained?

(c) Compute the condition numbers of $\mathbf{c}_0^T(\mathbf{b})$ for the five different normal sets in $T^2(4)$. What do you conclude?

3) (a) In $\mathbb{R}^2$, assume various symmetric or otherwise degenerate positions of m zeros and try to predict which normal sets from $T^2(m)$ they admit or exclude, respectively. Confirm your predictions by computing $\det \mathbf{b}(\mathbf{z})$; cf. (2.53).

(b) For $s = 2$ and small m, find the function $s_\mathcal{N}(z_1, \ldots, z_m)$ in terms of the components of the z_μ. Try to describe the manifolds $S_\mathcal{N}$ in terms of the zero positions. Note that a "singularity" manifold exists even for $m = 2$ for any choice of the two zeros. This implies that, for $s \geq 2$, $m \geq 2$, there is no normal set in $T^s(m)$ which is *uniformly feasible* for all ideals in $\mathcal{P}^s(m)$.

4. Use the results of Exercise 2 and 3 from section 2.4 to solve the following interpolation task: Find $p^* \in \mathcal{P}^2$ such $p^*(1, 3) = .8$, $\partial_x p^*(1, 3) = -.5$, $\partial_y p^*(1, 3) = 2.2$, $p^*(-1, 1) = 1.7$, $\partial_y p^*(-1, 1) = -1.4$, $p^*(2, -1) = .2$. Use various natural choices for $\mathbf{b}$ and compare the results.

Historical and Bibliographical Notes 2

Algebraic models have been a source of computational problems since the beginning of scientific computing. The Babylonians were able to solve special polynomial systems 2500 years ago. In China, at the beginning of the 14th century, the solving of polynomial systems was explained by Zhu Shiejie in his "Jade Mirror of the 4 Unknowns" [2.1]. The development of algebra in Europe from the late Middle Ages to 1900 may be found in any text on the history of mathematics. At that time, "algebra" was strongly algorithmic and essentially synonymous with the study of solving systems of polynomial equations. This changed radically during the 20th century: "The rabbit is put into a hat, and attention focused on the hat" (O. Perron).

Buchberger's seminal work on "Groebner Bases" appeared just when it was realized that the new electronic computers are an ideal tool for symbol manipulation. His Ph.D. thesis *"An Algorithm for Finding a Basis for the Residue Class Ring of a Zero-Dimensional Polynomial System"* ([2.2], 1965) and subsequent publications ([2.3], [2.4], and others) have fuelled and dominated the development of polynomial computer algebra; his central concept of a term order has been generally adopted. For the computation of the zeros, Buchberger [2.3] suggested the derivation of a "triangular" ideal basis, with a univariate polynomial at the bottom of the recursion. This led to a dominance of lexicographic term ordering for that purpose; the more efficiently computable total degree Groebner bases were transformed into lexicographical bases by the FGLM-algorithm [2.5]. (The fact that this often turned a well-conditioned representation into an ill-conditioned one was rarely minded.)

Initially, the representation of the multiplicative structure of the *quotient ring* $\mathcal{R}[\langle P \rangle]$ of a polynomial system P had been the center of attention (cf. [2.2], [2.3]), but the emphasis soon shifted to the basis of the ideal $\langle P \rangle$, the Groebner basis, and its algorithmic use. It was not before Maple7 (2001) that the black-box commands for the generation of the normal set and the associated multiplication matrices appeared in Maple! Even less attention was given to the *dual space* of the quotient rings which had been introduced in [2.6] in 1991.

The fundamental relation between the eigenelements of multiplication in the quotient ring and the zeros of the ideal must have been known to algebraists of the late 19th and early 20th centuries, in the language of their time. An elaboration may have been held back by the infeasibility of a numerical solution of nontrivial matrix eigenproblems. There are quotations of a theorem by Stickelberger from the 1920s, which is equivalent to Theorem 2.27, but its relevance has remained concealed. In the 1980s, there are new allusions in the direction of the Central Theorem, notably in [2.7], but only w.r.t. eigen*values*. It appears that the rather informal report [2.8] of 1988 contained the first explicit demonstration of the full content of Corollary 2.24; curiously, the authors were led to their insight through the presentation of resultants in the algebra textbook [2.9] of 1931! Due to its unfortunate medium, [2.8] remained unnoticed for several years, and independent discoveries of the Central Theorem occurred. Even so, its use for the numerical computation of the zeros of a polynomial system became standard only around 2000; cf. the late arrival in Maple quoted above.

The proposed use of *arbitrary* closed monomial sets as bases for $\mathcal{R}[\langle P \rangle]$ requires the abandonment of a term order as uppermost principle. With the multiplicative structure of $\mathcal{R}[\langle P \rangle]$ at the heart of the zero structure of $\langle P \rangle$, it appears natural to use a normal set representation $(\mathcal{N}, \mathcal{B}_\mathcal{N})$ as the *standard* representation of a 0-dimensional polynomial ideal; this will be further confirmed in Part III of this book. Such an approach is still rather unusual.

For readers not familiar with polynomial algebra, there are several textbooks of a more introductory kind which emphasize computational aspects. Common to all of them is the fact that only (small) parts of their contents are needed as a background for our own text. Reference [2.10] is very intuitive and stresses geometric interpretations; the same is true for [2.11] by the same authors. Another very intuitive text is [2.12], which reads like a conversation with the authors and stimulates the reader to further explore his/her understanding by numerous "tutorials." Although its title indicates a more specialized content, [2.13] also provides a broad and very concise introduction to polynomial algebra. The relevance of [2.14] shown by its title is borne out by its content. Naturally, this is not a complete list.

References

[2.1] Zhu Shiejie: Jade Mirror of the Four Unknowns (Chinese), 1303; cf., e.g., Jock Hoe: The Jade Mirror of the 4 Unknowns - some reflections, Math. Chronicle **7** (1978), part 3, 125–156.

[2.2] B. Buchberger: An Algorithm for Finding a Basis for the Residue Class Ring of a Zero-Dimensional Polynomial Ideal (German), Ph.D. Thesis, Univ. Innsbruck, 1965.

[2.3] B. Buchberger: An Algorithmic Criterion for the Solvability of Algebraic Systems of Equations (German), Aequationes Mathematicae **4** (1970), 374–383.

[2.4] B. Buchberger: Gröbner Bases: An Algorithmic Method in Polynomial Ideal Theory, in: N.K. Bose (ed.): Multidimensional Systems Theory, D. Reidel, Dordrecht, 1985, 184–232.

[2.5] J.Ch. Faugère, P. Gianni, D. Lazard, T. Mora: Efficient Computation of Zero-Dimensional Groebner Bases by Change of Ordering, J. Symbol. Comp. **16** (1993), 329–344.

[2.6] M.G. Marinari, H.M. Moeller, T. Mora: Groebner Bases of Ideals Given by Dual Bases, in Proceed. ISSAC 91, ACM, New York, 55–63.

[2.7] D. Lazard: Resolution des Systèmes Algébriques, Theor. Comput. Sci. **15** (1981), 77–110.

[2.8] W. Auzinger, H.J. Stetter: An Elimination Algorithm for the Computation of all Zeros of a System of Multivariate Polynomial Equations, in: Conf. in Numerical Analysis, ISNM **80**, Birkhäuser, Basel, 1988, 11–30.

[2.9] O. Perron: Algebra, vol.1, 2nd Ed., (German), Walter de Gruyter, Berlin, 1931.

[2.10] D. Cox, J. Little, D. O'Shea: Ideal, Varieties, and Algorithms, 2nd Ed., Springer, New York, 1996.

[2.11] D. Cox, J. Little, D. O'Shea: Using Algebraic Geometry, Springer, New York, 1998.

[2.12] M. Kreuzer, L. Robbiano: Computational Commutative Algebra 1, Springer, New York, 2000.

[2.13] Th. Becker, V. Weispfenning: Gröbner Bases—A Computational Approach to Commutative Algebra, Springer, New York, 1993.

[2.14] B. Mishra: Algorithmic Algebra, Springer, New York, 1993.

[2.15] R.A. Horn, Ch.R. Johnson: Matrix Analysis, Cambridge Univ. Press, Cambridge (UK), 1985.

Chapter 3

Polynomials with Coefficients of Limited Accuracy

In section 1.5, we have considered the use of polynomials as modelling functions in scientific computing. We have realized that, generally, some coefficients of such polynomials have a limited accuracy only, and we have briefly looked at potential sources for this indetermination. In this chapter, we introduce a formal framework for dealing with this indetermination in the context of polynomial algebra.

In this formalization, the vague character of this indetermination must be preserved. When polynomials contain coefficients of limited accuracy, questions like: "Is $\xi = 3.52$ a zero of p?" or "Do p_1 and p_2 possess a nontrivial common divisor?" cannot have a yes or no answer in many situations. Instead, we answer such questions with a "validity value," a positive real number δ which provides a *continuous transition* between clearly positive ($\delta \ll 1$) and clearly negative ($\delta \gg 1$) answers; the interpretation of values ≈ 1 must be left to the expert for the model. This is achieved by the association of a one-parametric *family of neighborhoods* with a data value of limited accuracy.

3.1 Data of Limited Accuracy

Situations with geometric aspects are a typical source of polynomial systems; e.g., many problems in *robotics* permit such a formulation. Here, some coefficients represent lengths and angles of robotic agents and relative positions of objects to be manipulated; it is obvious that those data have only a limited meaningful accuracy even in high precision tools. In other areas of scientific computing, like biology or economics, the indetermination of some coefficients in polynomial models may amount to several percent! Nevertheless, one wishes to have a reliable mathematical approach for the qualitative and quantitative prediction of the model behavior. The formalism explained in this section provides an appropriate tool for the handling of algebraic problems with an inherent indetermination.

3.1.1 Empirical Data

In most models of real-life situations, the coefficient 4.865 in a polynomial like

$$p(x, y) := x^3 + 4.865\, xy^2 - y^3 \tag{3.1}$$

does not signify that precise rational number but rather *any real number* from some small neighborhood of 4.865. When we associate with that coefficient a *tolerance* of 10^{-3}, we do not mean to restrict the potential values to a precise interval like [4.864, 4.866]. The appropriate interpretation is rather that there is an *indetermination of order* 10^{-3}, i.e. that the last digit is uncertain, with a deviation by more than few units highly improbable though not fully impossible.

On the other hand, there will also be coefficients in such polynomials which are to be considered as *exact*; they may, e.g., be implied by normalization or by some identity. Generally, such coefficients will be *integers*, like 1, -2, etc., or simple fractions, like $\frac{1}{2}$, etc. In particular, a *vanishing* coefficient is often of that nature: The fact that a certain monomial does *not* occur in a polynomial modelling function is intrinsic rather than coincidental. More generally, the *sparsity pattern* of a polynomial is, generally, an intrinsic part of the underlying model. Thus, the coefficients -1 and 0, resp., of the monomials y^3 and x^2y, resp., in the polynomial (3.1) will normally not contain an indetermination, but they will signify the exact integers -1 and 0.

Definition 3.1. Numerical data of an algebraic problem fall in two categories:

- *intrinsic data* α represent an *exact value* $\in \mathbb{R}$ or $\mathbb{C}$ in the sense of classical mathematics;

- *empirical data* $(\bar{\alpha}, \varepsilon)$ have a *specified value* $\bar{\alpha} \in \mathbb{R}$ or $\mathbb{C}$ and a *tolerance* $\varepsilon \in \mathbb{R}_+$ which indicate the *range of potential values* for that quantity in the following way:

We assume that the tolerance ε of an empirical data item indicates the *order of magnitude* of the indetermination of that quantity—which is all that is generally known about such an indetermination. Thus, typical values for tolerances are 10^{-4}, $5 \cdot 10^{-6}$, $2 \cdot 10^{-3}$, etc.; if the indetermination is specified *relative* to the size of the empirical value, the tolerance may have a value like $|\bar{\alpha}| \cdot 10^{-4}$ etc., where $\bar{\alpha}$ is the associated specified value of the empirical quantity. In other words: If an empirical quantity

$$\text{has} \quad \begin{array}{l} m \text{ meaningful decimal digits after its point} \\ m \text{ meaningful decimal digits} \end{array} \quad \text{then} \quad \begin{array}{l} \varepsilon \approx 10^{-m}, \\ \varepsilon \approx |\bar{\alpha}| \cdot 10^{-m}. \end{array} \qquad \square$$

Intuitively, this means that any value $\tilde{\alpha}$ with $|\tilde{\alpha} - \bar{\alpha}| \le \varepsilon$ should be considered as a *valid instance* for the empirical quantity $(\bar{\alpha}, \varepsilon)$. But it also means that valid instances $\tilde{\alpha}$ with a larger distance from $\bar{\alpha}$ may well occur, though this is assumed to be less and less likely as $\frac{|\tilde{\alpha} - \bar{\alpha}|}{\varepsilon}$ increases. It is important to understand that 1 is *not a strict boundary* for $\frac{|\tilde{\alpha} - \bar{\alpha}|}{\varepsilon}$ but rather a mark for the interpretation of potential numerical values of $(\bar{\alpha}, \varepsilon)$.

Tolerances of that kind have been widely used in engineering and technology for a long time, with the above understanding.

Example 3.1:

(a) The specification of the electric voltage supplied to our homes has a tolerance which is commonly interpreted in this way: "Almost always," the deviation of the momentary voltage from the specified one will be less than the specified tolerance, but, occasionally, the deviation may exceed this tolerance somewhat. Also, it is assumed that such an excessive deviation will be less and less likely the larger it is, not in the sense of probability theory but in the intuitive sense of everyday life.

(b) When the level of some large body of water is reported as "5.16 m" it is clear that the reporting of more decimal digits would be meaningless since the surface of the water is never motionless. One will automatically assume that several measurements of the water level, at the same location and within a short time interval, will produce deviating values; most of these may lie between 5.15 and 5.17, but one will not be surprised also to obtain a value like 5.19. In this case, the water level would be adequately described by the empirical quantity (5.16 m, 1 cm). □

When we consider the indetermination introduced into an algebraic problem by the indetermination of its empirical data, the *intrinsic* data are considered as *fixed*. Therefore, it will generally be convenient to consider their specification as a *part of the specification of the algebraic problem* and reserve the term "data" to the empirical data of the problem.

Formally, with an empirical data quantity we associate a *family of neighborhoods*, parametrized by a positive real parameter δ in the following way:

Definition 3.2. The empirical quantity $(\bar{\alpha}, \varepsilon)$, with the *specified value* $\bar{\alpha} \in \mathbb{C}$ or $\mathbb{R}$ and the *tolerance* $\varepsilon > 0$, defines the *family of neighborhoods*

$$N_\delta(\bar{\alpha}, \varepsilon) := \{\tilde{\alpha} : |\tilde{\alpha} - \bar{\alpha}| \le \delta\varepsilon\}, \quad \delta \in \mathbb{R}_0; \tag{3.2}$$

for *real* $\bar{\alpha}$, it must be specified (or clear from the context) whether the $\tilde{\alpha}$ in $N_\delta(\bar{\alpha}, \varepsilon)$ may be in $\mathbb{C}$ or whether they are restricted to the real line.   □

Naturally, one could replace the word "neighborhood" in the above by the word "interval." However, we believe that the term neighborhood conveys more of the *vagueness* which is inherent in empirical data: Beyond a certain accuracy, their value is simply not defined, and the boundary is not sharp (as it is in an interval) but *blurred*; this is why one must consider a *family* of such neighborhoods rather than one interval. Perhaps the notion of a "data cloud" is best suited to describe that situation.

Obviously, our concept of empirical data is closely related to the concept of "fuzzy data" which plays a useful role in various engineering applications; cf. the related literature. In a simplified view, the introduction of a probability distribution for δ would turn our empirical data items into fuzzy data items. We have abstained from using this approach for two reasons:

- probabilities are foreign to the algebraic world and they would have introduced an awkward formalism;

- in most of the real-world applications, the choice of a probability distribution for an empirical quantity cannot be based on substantial information.

Rather than dealing with formal probabilities, we associate with the parameter δ in (3.2) a *validity scale*: The values $\tilde{\alpha} \in N_\delta(\bar{\alpha}, \varepsilon)$, with $\delta = O(1)$, are considered as *valid instances* of the empirical quantity $(\bar{\alpha}, \varepsilon)$. (Here and throughout the book, we abuse the symbol $O(1)$ to denote real numbers "of the order 1," without any relation to a limit process!) Generally, one may visualize the relation between values of the parameter δ and an intuitive concept of validity by the scale

$$\delta : \quad 0 \quad \dots \quad 1 \quad \dots \quad 3 \quad \dots \quad 10 \quad \dots \quad 30 \quad \dots$$

valid	probably valid	possibly valid	probably invalid	invalid

$$\tag{3.3}$$

In a particular context, a more precise adjustment of the validity scale to values of δ may be possible for the experts who have modelled the situation. In any case, while the validity of $\tilde{\alpha}$

decreases with an increase of the value of δ necessary to achieve $\tilde{a} \in N_\delta(\bar{a}, e)$, the value $\delta = 1$ is not a bound for the validity of a numerical value $\tilde{\alpha}$ but only a normalization unit on a continuous validity scale; cf. Example 3.1.

It is straightforward to extend this concept of empirical data and their validity to *structured quantities* with complex or real components, like vectors or matrices; cf. Figure 3.1:

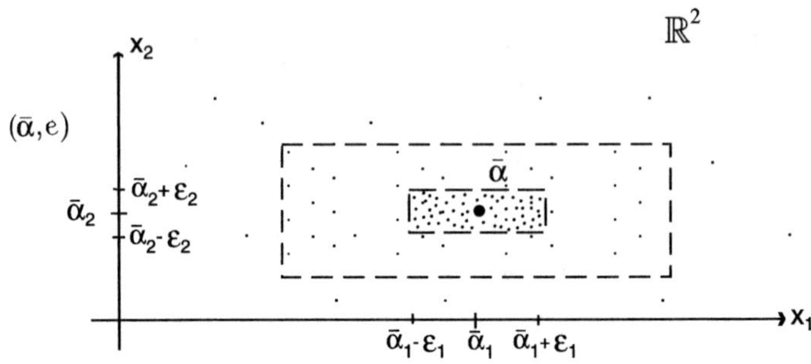

Figure 3.1.

Definition 3.3. The empirical vector $(\bar{a}, e)$, with *specified value*

$$\bar{a} = (\bar{\alpha}_1, \ldots, \bar{\alpha}_M) \in \mathbb{C}^M \text{ or } \mathbb{R}^M \text{ and } \textit{tolerance } e := (\varepsilon_1, \ldots, \varepsilon_M), \quad \varepsilon_j > 0 \qquad (3.4)$$

defines a *family* N_δ, $\delta > 0$, *of neighborhoods*

$$N_\delta(\bar{a}, e) := \{\tilde{a} : \|\tilde{a} - \bar{a}\|_e^* \le \delta\}, \quad \text{where } \|\tilde{a} - \bar{a}\|_e^* := \|(\ldots, \tfrac{|\tilde{\alpha}_j - \bar{\alpha}_j|}{\varepsilon_j}, \ldots)\|^*. \qquad (3.5)$$

For $\bar{a}$ with *real* components $\bar{\alpha}_j$, it must be specified (or clear from the context) whether the $\tilde{a} \in N_\delta(\bar{p}, e)$ are restricted to *real* values or not. $\square$

The norm $\|..\|^*$ in (3.5) is a vector norm in the *dual vector space* $(\mathbb{R}^M)^*$ of $\mathbb{R}^M$, i.e. the vector space of the linear functionals on $\mathbb{R}^M$. Consider an *absolute norm* on $\mathbb{R}^M := \{u = (u_1, \ldots, u_M)^T\}$, i.e. a norm with $\| |u| \| = \|u\|$. Then $\|..\|^*$ denotes the associated *dual norm* or *operator norm*, i.e.

$$\|v^T\|^* := \sup_{u \ne 0} \frac{|v^T u|}{\|u\|} = \sup_{\|u\|=1} |v^T u|. \qquad (3.6)$$

In our finite-dimensional vector spaces, the sup in (3.6) is always attained. The $\|..\|_e$ norm directly associated with the e-weighted dual norm $\|..\|_e^*$ of (3.5) is

$$\| u \|_e := \| (\varepsilon_1 |u_1|, \ldots, \varepsilon_M |u_M|)^T \|. \qquad (3.7)$$

The following norms (often called p-norms) are most commonly used in linear algebra:

$$
\begin{aligned}
\|u\|_{(1)} &:= \textstyle\sum_j |u_j|, & \|v^T\|_{(1)}^* &:= \max_j |v_j|; \\
\|u\|_{(\infty)} &:= \max_j |u_j|, & \|v^T\|_{(\infty)}^* &:= \textstyle\sum_j |v_j|; \\
\|u\|_{(2)} &:= (\textstyle\sum_j |u_j|^2)^{\frac{1}{2}}, & \|v^T\|_{(2)}^* &:= (\textstyle\sum_j |v_j|^2)^{\frac{1}{2}}.
\end{aligned}
\qquad (3.8)
$$

With (3.5), the $\|..\|_{(1)}^*$ norm appears most appropriate as it requires

$$|\tilde{\alpha}_j - \bar{\alpha}_j| \leq \varepsilon_j \delta, \quad j = 1(1)M, \tag{3.9}$$

for $a \in N_\delta(\bar{a}, e)$; it is the only norm where the requirements on the individual components of $\tilde{a}$ remain *separated*. Throughout this book, we will use this norm if not stated otherwise.

3.1.2 Empirical Polynomials

The primary data objects in this book are polynomials. To specify a particular polynomial, we must specify its structure which is represented by its *support* (cf. Definition 1.2) and the *numerical values of the coefficients*. Throughout, we assume that the structural information about a polynomial is intrinsic and fixed within a computational algebraic task. Some or all of the coefficients, on the other hand, may be empirical. If this is the case, the algebraic problems with these polynomial data require a treatment which differs in important ways from what is usual in computer algebra, where all data are automatically assumed to be intrinsic. Our concept of empirical polynomials provides a mathematical framework for the treatment of such problems which constitute the vast majority of all algebraic problems arising in scientific computing.

Definition 3.4. For an *empirical polynomial* $(\bar{p}, e)$, with specified polynomial $\bar{p} \in \mathcal{P}^s$, $s \geq 1$,

$$\bar{p}(x) = \sum_{j \in J} \bar{\alpha}_j x^j, \quad J \subset \mathbb{N}_0^s, \quad \bar{\alpha}_j \in \mathbb{C} \text{ or } \mathbb{R}, \tag{3.10}$$

we let

$$\emptyset \neq \tilde{J} := \{j \in J : \alpha_j \text{ is an empirical coefficient of } (\bar{p}, e)\} \subset J$$

be the *empirical support* of $(\bar{p}, e)$, with $|\tilde{J}| = M$. The components ε_j of the *tolerance e* of the empirical polynomial $(\bar{p}, e)$ refer only to subscripts in the empirical support $\tilde{J}$:

$$e := (\varepsilon_j > 0, \ j \in \tilde{J}) \in \mathbb{R}_+^M. \tag{3.11}$$

The *neighborhoods* $N_\delta(\bar{p}, e)$, $\delta > 0$, contain the polynomials $\tilde{p} \in \mathcal{P}^s$, $\tilde{p}(x) = \sum_{j \in J} \tilde{\alpha}_j x^j$, with

$$\|\tilde{p} - \bar{p}\|_e^* := \|(\ldots, \frac{|\tilde{\alpha}_j - \bar{\alpha}_j|}{\varepsilon_j}, \ldots, j \in \tilde{J})\|^* \leq \delta, \text{ and } \tilde{\alpha}_j = \bar{\alpha}_j, \ j \in J \setminus \tilde{J}. \tag{3.12}$$

That is, $\tilde{p} \in \mathcal{P}^s$ is a δ-*neighbor* of $(\bar{p}, e)$ iff it has the *same support* as $\bar{p}$ and if its coefficients satisfy (3.12). For $\bar{p}$ with *real* coefficients $\bar{\alpha}_j$, it must be specified (or clear from the context) whether $N_\delta(\bar{p}, e)$ is restricted to *real* polynomials or not. □

It is trivial that $N_{\delta_1}(\bar{p}, e) \subset N_{\delta_2}(\bar{p}, e)$ for $\delta_1 < \delta_2$. In agreement with our assumptions about the validity of individual data objects (cf. (3.3)), we say that

$$\text{all } \tilde{p} \in N_\delta(\bar{p}, e) \text{ with } \delta = O(1) \text{ are } \textit{valid instances} \text{ of } (\bar{p}, e). \tag{3.13}$$

Remember that, because of (3.9), we use the max-norm (i.e. the dual of the 1-norm) for the coefficient vector almost exclusively.

Example 3.2: Consider the empirical polynomial $(\bar{p}, e)$ with

$$\bar{p}(x, y) \;=\; x^3 + 4.865\,xy^2 - y^3 + 2.902\,x^2 + 0.0\,xy - 8.389\,x + 2\,y - 17.54\,,$$

with

$$J \;=\; \{(3, 0), (1, 2), (0, 3), (2, 0), (1, 1), (1, 0), (0, 1), (0, 0)\}\,,$$

$$\tilde{J} \;=\; \{(1, 2), (2, 0), (1, 1), (1, 0), (0, 0)\}\,,$$

$$e \;=\; (10^{-3},\, 5 \cdot 10^{-4},\, 10^{-4},\, 10^{-3},\, 5 \cdot 10^{-3})\,.$$

With respect to the max-norm, the following polynomials are valid instances of the empirical polynomial $(\bar{p}, e)$:

$$\tilde{p}_1(x, y) \;=\; x^3 + 4.8642\,xy^2 - y^3 + 2.9025\,x^2 - 8.3888\,x + 2\,y - 17.541\,,$$

$$\tilde{p}_2(x, y) \;=\; x^3 + 4.865\,xy^2 - y^3 + 2.9018\,x^2 - 0.00006\,xy - 8.3896\,x + 2\,y - 17.536\,.$$

Note that we have assumed that the xy-term has an *empirical* coefficient $(0, \varepsilon_{11})$ in $(\bar{p}, e)$ as $j = (1, 1)$ appears in the supports J and $\tilde{J}$. Therefore, an xy-term may appear with a sufficiently small coefficient in a valid instance of $(\bar{p}, e)$; cf. $\tilde{p}_2$.

The polynomial

$$\tilde{p}_3(x, y) \;=\; x^3 + 4.866\,xy^2 - y^3 + 2.901\,x^2 - 8.385\,x + 2\,y - 17.53$$

may be considered as a (probably) valid instance of $(\bar{p}, e)$ while

$$\tilde{p}_4(x, y) \;=\; x^3 + 4.85\,xy^2 - y^3 + 2.90\,x^2 + 0.001\,xy - 8.38\,x + 2\,y - 17.5$$

should be considered as a (probably) invalid instance; cf. (3.3).   □

More generally, an empirical polynomial is a polynomial function in $s \geq 1$ variables, with one or several empirical parameters $(\bar{\alpha}_\nu, \varepsilon_\nu)$ and possibly some intrinsic parameters. The neighborhood definition (3.12) is not affected by this extension, except that there may no longer be a well-defined support. Since this extension is straightforward but formally tedious, we will generally consider empirical polynomials to have the form (3.10) throughout this book.

Definition 3.4 of an empirical polynomial is naturally extended to systems $(\bar{P}, E)$ of n empirical polynomials.

Definition 3.5. The empirical polynomials $(\bar{p}_\nu, e_\nu)$, with empirical supports $\tilde{J}_\nu$ and tolerances $e_\nu \in \mathbb{R}_+^{M_\nu}$, $\nu = 1(1)n$, define an *empirical system* $(\bar{P}, E)$, with $\bar{P} = \{\bar{p}_\nu,\, \nu = 1(1)n\}$, $E = \{e_\nu,\, \nu = 1(1)n\}$. A system

$$\widetilde{P} \;:=\; \{\tilde{p}_1, \ldots, \tilde{p}_n\} \in N_\delta(\bar{P}, E) \qquad \text{iff } \tilde{p}_\nu \in N_\delta(\bar{p}_\nu, e_\nu),\, \nu = 1(1)n\,. \;\;\square \qquad (3.14)$$

The principal algebraic structure associated with one or several polynomials in $\mathcal{P}^s$, $s \geq 1$, is that of an *ideal*; cf. section 1.3.2 and Chapter 2. What happens if one or all of the *generating polynomials* of an ideal are empirical ?

Consider the ideal $\mathcal{I} = \langle \bar{P} \rangle = \langle \bar{p}_1, \ldots, \bar{p}_s \rangle \in \mathcal{P}^s$ and assume that $\mathcal{I}$ is a complete intersection; cf. Definition 1.11. When we replace the $\bar{p}_\nu$ by empirical polynomials $(\bar{p}_\nu, e_\nu)$, we may interpret the result as a family of sets $\mathcal{I}_\delta$ of ideals:

$$\mathcal{I}_\delta := \{ \langle \tilde{p}_1, \ldots, \tilde{p}_s \rangle \ : \ \tilde{p}_\nu \in N_\delta(\bar{p}_\nu, e_\nu) \}, \quad \text{for a fixed } \delta > 0.$$

Is $\mathcal{I}_\delta$ a meaningful mathematical quantity?

At first, it is important to note that each *member* $\widetilde{\mathcal{I}} \in \mathcal{I}_\delta$ is a proper polynomial ideal with a well-defined generating basis. Thus, each $\widetilde{\mathcal{I}}$ can be subject to the usual algebraic and computational manipulations, under the usual rules. Without much loss of generality, we may further assume that all $\widetilde{\mathcal{I}} \in \mathcal{I}_\delta$ are 0-dimensional. Then, each $\widetilde{\mathcal{I}}$ defines an m-dimensional quotient ring $\widetilde{\mathcal{R}} := \mathcal{R}[\widetilde{\mathcal{I}}]$ and dual space $\widetilde{\mathcal{D}} := \mathcal{D}[\widetilde{\mathcal{I}}]$, and a zero set $Z[\widetilde{\mathcal{I}}]$ with m zeros counting muliplicities. Also, membership of some polynomial $u \in \mathcal{P}^s$ in a particular $\widetilde{\mathcal{I}}$ is well-defined.

One could argue that the ideals $\widetilde{\mathcal{I}} \in \mathcal{I}_\delta$ are *close to each other*, e.g., w.r.t their zero sets. Since we will not need the concept of ideal neighborhoods in the following, we refrain from a further discussion.

3.1.3 Valid Approximate Results

It is our goal to solve algebraic problems with polynomial data of limited accuracy in a meaningful way. Since we admit dense infinite sets $N_\delta(\bar{a}, e) \subset \mathcal{A}$, $\delta = O(1)$, of *valid* input data for such a problem, we must generally expect dense infinite sets of result values which have to be regarded as *valid results* of the given problem. We will now formalize this situation.

For an empirical algebraic problem, we consider the mapping from the space $\mathcal{A}$ of the empirical data to the result space $\mathcal{Z}$ which assigns to a particular input value $\tilde{a} \in N_{\bar{\delta}}(\bar{a}, e)$ the value $\tilde{z} \in \mathcal{Z}$ of the *exact* result of the algebraic problem for $\tilde{a}$. Here, $\tilde{z}$ may be a real or a complex number or a set of such numbers. When there are several independent result quantities, we may restrict our mapping to one particular result quantity.

Definition 3.6. For an empirical algebraic problem, with empirical data $(\bar{a}, c)$, the mapping

$$F : \quad \tilde{A} \subset \mathcal{A} \ \longrightarrow \ \mathcal{Z} \tag{3.15}$$

which assigns to each data value $\tilde{a} = (\tilde{\alpha}_j)$ in some domain $\tilde{A} \supset N_{\bar{\delta}}(\bar{a}, e)$, $\bar{\delta} > 1$, the *exact* result $z \in \mathcal{Z}$ of the algebraic problem with these data, is called the *data→result mapping* for that situation.   □

Here, $\bar{\delta} > 1$ denotes a value beyond which we do not wish to consider the indetermination of our input data with limited accuracy; cf. (3.5) and (3.3). As previously agreed (cf. the remark below Example 3.1), we consider the intrinsic data as a fixed part of the specification of our algebraic problem and thus also of the associated data→result mapping.

Definition 3.7. For a data→result mapping (3.15), the sets

$$Z_\delta(\bar{a}, e) := \{ z = F(a), \ a \in N_\delta(\bar{a}, e) \}, \quad 0 < \delta \leq \bar{\delta}, \tag{3.16}$$

are called (δ-)*pseudoresult sets* of (3.15). In the case of *real* specified data $\bar{a}$, it must be clear whether the data neighborhoods N_δ are restricted to the real domain or not.   □

Note that each result value $\tilde{z} \in Z_\delta(\bar{a}, e)$ is the *exact result* of the algebraic problem for some input data $\tilde{a} \in N_\delta(\bar{a}, e)$. In particular, for $\delta = O(1)$, each result value in Z_δ is the exact result of a valid instance of the algebraic problem for the model under discussion.

Definition 3.8. In the situation described by a data$\rightarrow$result mapping (3.15), the values in a pseudoresult set $Z_\delta(\bar{a}, e)$ with $\delta = O(1)$ are *valid (approximate) results* of the empirical algebraic problem. $\square$

The word "approximate" in Definition 3.8 is really not necessary; we use it occasionally to emphasize that, by the notion of an empirical problem, any result value $\tilde{z} \in Z_\delta$ (including $\bar{z} = F(\bar{a})$)—while being the exact result $F(\tilde{a})$ of the algebraic problem for some particular $\tilde{a} \in N_\delta(\bar{a}, e)$—can only represent an *approximate solution* for the problem at hand. In fact, in a decimal representation of the values $\tilde{z} \in Z_\delta$ only those digits which coincide after rounding to that digit can contain safe information about the problem. The vague notion of $O(1)$ is necessary because of the equally vague character of tolerance specifications; cf. (3.3).

On the other hand, it must be emphasized once more that the *mathematical structure of the algebraic problem* and the definition of a *solution* have not been altered at all by our approach: The data$\rightarrow$result mapping F fully embodies these aspects in their classical meanings; cf. Definition 3.6.

Example 3.3: Consider the univariate monic polynomial

$$\bar{p}(x) := x^4 - 2.83088\,x^3 + 0.00347\,x^2 + 5.66176\,x - 4.00694 \tag{3.17}$$

and assume that it is the specified polynomial of an empirical polynomial with tolerance vector $e = (\varepsilon_j = 10^{-5},\ j = 0(1)3)$. Thus, in all coefficients except the leading one, the trailing digit is not well determined; we assume that this indetermination is restricted to the real domain. We wish to determine the zeros of the empirical polynomial $(\bar{p}, e)$. As $\bar{p}$ is a perturbed version of

$$p(x) = (x - \sqrt{2})^3(x + \sqrt{2}) \approx x^4 - 2.82843\,x^3 + 5.65685\,x - 4.00000\,,$$

we expect one real zero in the left halfplane (near $-\sqrt{2}$) and three zeros in the right halfplane (near $\sqrt{2}$). For simplicity, we consider only $\delta = 1$ in the following.

Let us first look at the zero in the left halfplane: With a considerate shifting of the coefficients to appropriate corners of the domain N_1 and with the help of a zerofinding code, we find that the negative zero can take values in the interval $[-1.4142168, -1.4142104]$ (rounded to 7 digits) for $a \in N_1$. Thus, each of the 6-digit values $-1.414217, \ldots, -1.414210$ is a valid approximate zero $\tilde{z}$ of $(\bar{p}, e)$, and it is meaningful to specify $\tilde{z}$ to 6 digits after the decimal point, with a potential indetermination of a few units in the last place, although the coefficients of $(\bar{p}, e)$ have an indetermination in the 5th digit.

This changes dramatically when we consider the three zeros in the right halfplane which form a cluster. For the specified polynomial $\bar{p}$, a zerofinder yields the three close real zeros (rounded to 5 digits)

$$1.41421,\ 1.41481,\ 1.41607.$$

However, the equally valid polynomial

$$\tilde{p}(x) := x^4 - 2.83087\,x^3 + 0.00348\,x^2 + 5.66177\,x - 4.00693 \in N_1(\bar{p}, e)$$

has the zeros (rounded to 5 digits)

$$1.38583 \quad \text{and} \quad 1.42963 \pm 0.02578\, i\, !$$

This shows at first that it is not possible to attribute individual pseudozero sets Z_1 to each of the three zeros in the right halfplane since the conjugate pair can only arise after two real zeros have merged into a double zero for some polynomial "between" $\bar{p}$ and $\tilde{p}$. Thus, we must treat the three zeros as *one* result which may as well consist of three real zeros as of one real zero and a complex conjugate pair.

Quantitatively, it appears that a pseudozero set Z_1 *for the zero triple* in the right halfplane has a diameter of nearly $6 \cdot 10^{-2}$! Nevertheless, each $\tilde{z} \in Z_1$ is an exact zero of some $\tilde{p} \in N_1(\bar{p}, e)$ and there are exactly three zeros (counting potential multiplicities) in Z_1 for each $\tilde{p} \in N_1(\bar{p}, e)$. We will show later that it is possible to give a better quantitative description of the *cluster* although the *location* of the individual zeros is so indeterminate (cf. the end of section 3.4 and section 6.3). □

Example 3.4: In [3.13], the authors consider a perturbed model for the conformations of the cyclohexane molecule described by the polynomial system

$$P(x) := \begin{pmatrix} 1.313\, x_2^2 x_3^2 + .959\, x_2^2 + 1.389\, x_2 x_3 + .774\, x_3^2 - .310 \\ 1.269\, x_3^2 x_1^2 + .755\, x_3^2 + 1.451\, x_3 x_1 + .917\, x_1^2 - .365 \\ 1.352\, x_1^2 x_2^2 + .837,\, x_1^2 + 1.655\, x_1 x_2 + .838\, x_2^2 - .413 \end{pmatrix} = 0\,;$$

the tolerance of all entries is apparently $.5 \cdot 10^{-3}$. They display numerically computed approximations for the 4 real zeros to 10 digits. In section 3.3, we will establish that the approximate zeros obtained by *rounding* their numerical results *to 3 digits* are valid approximate results under the above tolerance. Thus, it is not meaningful—and even misleading—to display more than 3 or 4 digits of the zeros of P, no matter how they have been computed. □

Since the definition of pseudoresults uses the unknown data→result mapping F (cf. Definition 3.6), it is generally not possible to compute an approximate *representation* of a pseudoresult set Z_δ, except for demonstrative purposes in simple situations and with a high computational effort. Also, how should one visualize a domain in some high-dimensional $\mathbb{C}^m$? The only aspect of importance of the pseudoresult sets Z_δ with $\delta = O(1)$ which define the valid results of an empirical problem (cf. Definition 3.8) is their approximate *diameter* in terms of a suitable metric in the result space $\mathcal{Z}$. This quantity determines the *meaningful number of digits* for the specification of a valid result. The following sections will be devoted to estimations of this quantity.

Exercises

1. (a) Consider the real-life quantities of Exercise 1.5-1; associate tolerances in the sense of Definition 3.2 with these empirical data.

(b) Consider an empirical data item representing the length of an object, with specified value 1.20 m. Describe situations where the following tolerances may be meaningful: .1 m, .01 m, .001 m, .0001 m.

2.   For a univariate quadratic polynomial $p = x^2 + \alpha_1 x + \alpha_0$, the data→result mapping $F : \mathbb{C}^2 \to \mathbb{C}^2$ which maps the coefficients α_1, α_0 into the two zeros z_1, z_2 is explicitly known. Consider the empirical polynomial with real coefficients $(\bar{\alpha}_1, \varepsilon)$, $(\bar{\alpha}_0, \varepsilon)$.

(a) Choose some $\bar{a} = (\bar{\alpha}_1, \bar{\alpha}_0)$ such that the zeros in $Z_1(\bar{p}, e)$ are real. Try to plot $Z_1(\bar{p}, e)$ in the z_1, z_2-plane: Evaluate F along the boundaries of $N_1(\alpha_1, \varepsilon) \times N_1(\alpha_0, \varepsilon)$ for a sufficiently large value of ε. Vary $\bar{a}$ such that $\bar{z}_1$ gets close to $\bar{z}_2$ and observe the effect on the pseudozero set Z_1.

(b) Choose $\bar{a}$ such that there is a conjugate complex pair of zeros and plot an image of $Z_1(\bar{p}, e)$ in the (Re z_ν, Im z_ν)-plane. Vary $\bar{a}$ such that $|\text{Im } z_\nu|$ is close to zero.

3. In Example 9.3, obtain the approximate zeros of u and verify that u is a multiple of $\tilde{p}$. Also find the remainder of the division of u by $\tilde{p}$; why is it not possible to derive the same conclusion in this fashion directly?

3.2   Estimation of the Result Indetermination

3.2.1   Well-Posed and Ill-Posed Problems

So far, we have tacitly assumed some basic *regularity* of the algebraic problem: We expect that the domain $\widehat{A}$ of the data→result mapping F is *open* in the data space $\mathcal{A}$ so that results exist for *all* $\tilde{a}$ in a neighborhood $N_\delta(\bar{a}, e)$, $\delta < \bar{\delta}$. We also expect that F is *surjective* so that the family $\{Z_\delta\}$ of pseudoresult sets consists of *compact connected* sets in the natural topology of the result space $\mathcal{Z}$ and a *full neighborhood* of $\bar{z} = F(\bar{a})$ consists of approximate results $F(\tilde{a})$. These are natural assumptions for a problem whose results are to be determined by approximate computations.

Proposition 3.1. For an algebraic problem with empirical data, assume that

$$M := \dim \mathcal{A} \geq \dim \mathcal{Z} =: m \geq 1 ; \qquad (3.18)$$

furthermore assume that the data→result mapping F is *continuous* on $\widehat{A} \supset N_{\bar{\delta}}(\bar{a}, e)$ for some $\bar{\delta} > 1$, and that it is *surjective*, with an image $F(\widehat{A})$ of dimension m in $\mathcal{Z}$. Then, the associated δ-pseudoresult sets Z_δ are compact connected sets in $\mathcal{Z}$.
Proof: By (3.2) and (3.5), the sets N_δ are compact connected sets in the data space $\mathcal{A}$. A continuous surjective mapping maps a compact connected set onto a compact connected set in the image space. $\square$

Definition 3.9. An algebraic problem with empirical data whose data→result mapping satisfies the hypotheses of Proposition 3.1 is called *well-posed*; otherwise it is called *ill-posed*. $\square$

There are 4 principal types of ill-posed algebraic problems:

1. The dimension of the result space $\mathcal{Z}$ is 0 :

This happens when the problem has only *discrete* result values, e.g. integers or truth values. If there are at least two *different* result values for data in $N_{\bar{\delta}}(\bar{a}, e)$, F cannot be continuous, and $Z_{\bar{\delta}}$ is not a connected set. Only if F is constant in $N_{\bar{\delta}}$ so that all Z_δ consist only of this one value, the problem is well posed.

In certain cases, it is possible to deal computationally with such problems in a meaningful way; cf., e.g., section 6.1.

Example 3.5: If the number of real roots of a real polynomial takes different values within $N_\delta(\bar{p}, e)$, the question for this number is obviously ill posed; cf. Example 3.3. □

2. The dimension of the domain of F is $< M$:

This happens when a result of the algebraic problem is only defined for data on an algebraic manifold S of dimension $d < M$ in A; cf. Figure 3.2. Such problems are called *overdetermined* or *singular*. Assume that $\bar{a} \notin S$. If the intersection of S with $N_\delta(\bar{a}, e)$ is empty, the pseudoresult sets Z_δ, $\delta < \bar{\delta}$, are empty and no valid instance of the algebraic problem has an exact solution. On the other hand, if the domain of F on the manifold S has a nonempty intersection with the neighborhoods $N_\delta(\bar{a}, e)$ for $\delta \geq \delta_0 > 0$, $\delta_0 = O(1)$, we may restrict F to $\widehat{A_0} := N_{\bar{\delta}} \cap S$. If the hypotheses of Proposition 3.1 apply to the data→result mapping $F_0 : \widehat{A_0} \to Z$, we have a well-posed empirical problem with valid results. Note that the *exact* problem with the *specified* data $\bar{a}$ has no solution in this case!

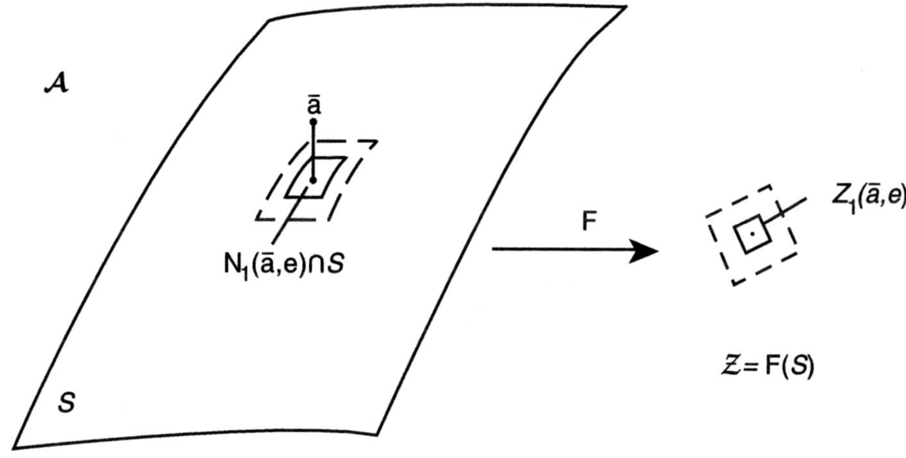

Figure 3.2.

In the later parts of this book, we will meet a great variety of overdetermined or singular empirical problems; we will show how valid solutions may be computed.

Example 3.6: In the family of monic polynomials in P_4^1, the map F which assigns to the 4 coefficients $\alpha_j \in \mathbb{C}$, $j = 0(1)3$, of some p the inflection point z of the graph of p, or more formally, $z \in \mathbb{C}$ such that $p(z) = p'(z) = p''(z) = 0$, has a 2-dimensional manifold $S \subset A = \mathbb{C}^4$ as its domain. Clearly, the polynomial system $P = \{p, p', p''\}$ is an overdetermined system which has a solution only for data $a \in S$. Also, if some $p(x; a)$ with an inflection point, i.e. with $a \in S$, is subject to an arbitrarily small generic perturbation, it loses the inflection point; in this sense, the problem is singular.

For an empirical polynomial $(p(x; \bar{a}), e)$ in P_4^1, with $\bar{a} \notin S$, on the other hand, the neighborhoods $N_\delta(\bar{a}, e)$ will always intersect with S for sufficiently large values of δ. If the smallest $\delta = \delta_0$ for which this happens is O(1), we have valid instances $\tilde{p}$ of $(p(x; \bar{a}), e)$ which possess an inflection point $\tilde{z}$, and the pseudoresult set Z_δ of F is not empty for $\delta \geq \delta_0$. The empirical polynomial of Example 3.3 is an example of this situation, as we will see in section 6.3.4.   □

3. The dimension of the image of F is $< m$:

This happens when the result components of an algebraic problem naturally satisfy some (nonlinear) constraints: they lie in an algebraic manifold $\overline{S} \subset Z$ of dimension $\bar{d} < m$ so that the data→result mapping F is not surjective. This implies that each arbitrarily small neighborhood in Z of a valid approximate result $\tilde{z} \in \overline{S}$ contains infinitely many values $z \in Z$ which cannot be interpreted as exact results of the algebraic problem for whatever values of the empirical data. The unavoidable perturbations induced by numerical computation will generally prevent a computed result to be an element of $\overline{S}$ although it may be very close to a valid result.

We do not attempt to "repair" this situation by a more general definition of valid results. Instead, we will consider such situations as they arise in the context of particular algebraic problems and deal with them in specific appropriate ways where it is meaningful to do so.

Example 3.7: Consider a system P of two generic quadratic polynomials in two variables; as $Z[\langle P \rangle]$ consists of four zeros, the quotient ring $\mathcal{R}[\langle p \rangle]$ has dimension 4. With respect to the basis $\mathbf{b} = (1, x, y, x^2)^T$, the multiplication matrix A_x of $\mathcal{R}$ has 2 nontrivial rows, with 8 elements, while A_y has 3 nontrivial rows, with 12 elements. Generically, both matrices are nonderogatory; i.e. the eigenvectors of A_x as well as those of A_y are the joint eigenvectors of the commuting family $\overline{A}$ of multiplication matrices and hence equal. Therefore, the 12 nontrivial elements of A_y cannot represent 12 independent result components but only 8; they must lie on an 8-dimensional manifold $\overline{S} \subset \mathbb{C}^{12}$.

   When P has floating-point coefficients and the elements of A_y are computed by floating-point computation, the resulting elements will generally not lie on $\overline{S}$; this implies that the computed $\tilde{A}_y$ is not a multiplication matrix for some system $\tilde{P}$ close to P but not a multiplication matrix at all. Such a situation is frequently met in numerical polynomial algebra; in section 9.2, we will explain how it may be handled.   □

4. The data→result mapping F is *discontinuous*:

This may happen when the neighborhoods $N_\delta(\bar{a}, e)$ are not *fully contained* inside the domain $\widehat{A}$ on which the data→result mapping F is a *continuous* map from A to Z, for $\delta \geq \hat{\delta}$, $\hat{\delta} = $ O(1). For example, the structure of the results may change as the data a approach some manifold $\tilde{S}$ in the data space, say some result component may *diverge to* ∞. In this case, all pseudoresult set $Z_\delta(\bar{a}, e)$ for $\delta \geq \hat{\delta}$ extend to infinity so that there exist valid results of arbitrarily large modulus. Often, this will imply that, formally, there exists a disconnected part of these pseudoresult sets.

For such problems, it is important to *recognize* the discontinuity. In certain situations, it may constitute a natural phenomenon. Again, the computational treatment of empirical problems of this kind must be considered in case studies; cf., e.g., section 9.4.

Example 3.8: Consider an empirical polynomial system $(\bar{P}, E)$, with $\bar{P} = \{x^2 y - y^2, \, x y + \bar{\alpha} x - 1\}$, with $|\bar{\alpha}| \neq 0$, small, with an associated tolerance $\varepsilon > |\bar{\alpha}|$. The specified system $\bar{P}$ has three zeros which are close to $(\frac{1}{w}, w)$ where w is a third root of unity, and a fourth zero $(\frac{1}{\bar{\alpha}}, 0)$. Obviously, the data→result mapping for this zero is discontinuous at $\alpha = 0$, which is inside the data set N_δ for $\delta \geq |\bar{\alpha}|/\varepsilon$ so that the pseudozero set Z_1 for this zero extends to ∞ and "beyond." $\quad\square$

Ill-posed problems abound in polynomial algebra; thus, their computational treatment is of particular interest and presents a challenge for the further development of Numerical Polynomial Algebra. In many cases, the approximate computational solution of such problems becomes feasible only through the consideration of empirical data; cf. Example 3.6. In various parts of this book, we will meet ill-posed problems of all kinds and attempt an appropriate treatment.

3.2.2 Condition of an Empirical Algebraic Problem

In Numerical Analysis, the "condition" of a computational mathematical problem plays an important role: It denotes the *sensitivity* of the results of the problem against *perturbations of its data*; cf. any text on Numerical Analysis. A problem with a *low* sensitivity is called "well-conditioned," one with a *high* sensitivity "ill-conditioned." Various types of *condition numbers* are introduced to permit a quantitative assessment of the condition of a problem. It is important to note that "condition" refers to the *mathematical problem* under consideration—*not to a computational procedure for its numerical solution.*

Generally, in analyzing the condition of a mathematical problem, one considers perturbations in *all* data of the problem. With our algebraic problems with data of limited accuracy, we have agreed to regard the intrinsic data of the problem as a fixed part of the *specification of the problem*. Therefore, it is consistent to consider only perturbations in the *empirical* data in a discussion of the condition of such a problem; we will do so in this section and in the remainder of the book.

Definition 3.10. For an algebraic problem with empirical data $(\bar{a}, e) = ((\bar{\alpha}_j, e_j), \ j = 1(1)M)$, let $\bar{z}$ be the result of the specified problem. A number $C > 0$ is a *condition number* for the problem if the *condition estimate*

$$\|\tilde{z} - \bar{z}\| \ \leq \ C \, \|\tilde{a} - \bar{a}\| \tag{3.19}$$

holds for the exact results $\tilde{z}$ of the problems with data $\tilde{a} \in N_{\bar{\delta}}(\bar{a}, e) \subset \mathcal{A}$, $\bar{\delta} > 1$ fixed. (The norms in (3.19) are arbitrary fixed vector norms in $\mathcal{Z}$ and $\mathcal{A}$. If necessary, the notation of C should refer to these norms, but it is generally assumed that the choice of norms is specified or clear from the context.)

In the same situation, C_j is a *condition number with respect to the data item* α_j if

$$\|\tilde{z} - \bar{z}\| \ \leq \ C_j \, |\tilde{\alpha}_j - \bar{\alpha}_j| \tag{3.20}$$

holds for the *exact* results $\tilde{z}$ of the specified problem with only $\bar{\alpha}_j$ replaced by some $\tilde{\alpha}_j$ for which $\tilde{a}$ remains in $N_{\bar{\delta}}$.

Naturally, in condition estimates like (3.19) and (3.20), (near-)minimal values should be used for the condition numbers to make the estimates realistic. $\quad\square$

In Numerical Analysis, perturbation estimates are often formulated in terms of *relative* perturbations because this conforms better with the fixed relative accuracy of floating-point computations; cf. section 4.3. It is easy to convert (3.19) and (3.20) into *relative condition estimates*:

$$\frac{\|\tilde{z} - \bar{z}\|}{\|\bar{z}\|} \le (C \, \frac{\|\bar{a}\|}{\|\bar{z}\|}) \, \frac{\|\tilde{a} - \bar{a}\|}{\|\bar{a}\|}, \tag{3.21}$$

$$\frac{\|\tilde{z} - \bar{z}\|}{\|\bar{z}\|} \le (C_j \, \frac{|\bar{a}_j|}{\|\bar{z}\|}) \, \frac{|\tilde{a}_j - \bar{a}_j|}{|\bar{a}_j|}. \tag{3.22}$$

Such relative condition estimates are particularly widespread in numerical linear algebra. The well-known condition number $\mathrm{cond}(A) := \|A\|\|A^{-1}\|$ of a *matrix* $A \in \mathbb{R}^{m \times m}$ or $\mathbb{C}^{m \times m}$ permits relative condition estimates like

$$\frac{\|\tilde{A}x - Ax\|}{\|Ax\|} \le \mathrm{cond}(A) \, \frac{\|\tilde{A} - A\|}{\|A\|}, \qquad \frac{\|\tilde{A}^{-1} - A^{-1}\|}{\|\tilde{A}^{-1}\|} \le \mathrm{cond}(A) \, \frac{\|\tilde{A} - A\|}{\|A\|},$$

or, for the sensitivity of the solution z of the linear system $A\,x = b$ with respect to variations in the right-hand side b,

$$\frac{\|\tilde{z} - z\|}{\|z\|} \le \mathrm{cond}(A) \, \frac{\|\tilde{b} - b\|}{\|b\|}.$$

In numerical polynomial algebra, there are markedly fewer situations where the use of relative condition estimates is more advantageous than the use of absolute estimates. For this reason, we will generally employ absolute condition estimates of type (3.19) and (3.20), except in a few special situations. Furthermore, (3.21) and (3.22) show that absolute estimates can easily be turned into relative ones, with the added advantage that the normalizing denominators may be suitably chosen.

Since condition refers to the behavior of the *exact* solutions of the problem, it is completely determined by the data$\rightarrow$result mapping F; cf. Definition 3.6. Actually, the condition estimate (3.19) is nothing but a *Lipschitz condition* for the data$\rightarrow$result mapping F. Moreover, if F is *differentiable* with respect to the *empirical data* $\tilde{a}$ on $N_{\tilde{\delta}}$, its (Fréchet) derivative F' furnishes condition estimates for the problem. They follow from the following general result in multivariate analysis.

Theorem 3.2. Let $F : \mathbb{C}^M \to \mathbb{C}^m$ have a continuous Fréchet derivative in some convex domain $A \subset \mathbb{C}^M$; then, for $a, a + \Delta a \in A$,

$$\|F(a + \Delta a) - F(a)\| \le \sup_{\lambda \in (0,1)} \|F'(a + \lambda \, \Delta a)\| \cdot \|\Delta a\|. \tag{3.23}$$

If the Fréchet derivative F' is Lipschitz continuous, with Lipschitz constant L', then

$$F(a + \Delta a) - F(a) = F'(a) \, \Delta a + r(a, \Delta a) \quad \text{with } \|r(a, \Delta a)\| \le \frac{L'}{2} \|\Delta a\|^2. \tag{3.24}$$

Proof: By the multivariate mean-value theorem, we have

$$F(a + \Delta a) - F(a) = \int_0^1 F'(a + \lambda \, \Delta a) \, d\lambda \cdot \Delta a, \tag{3.25}$$

which implies (3.23). For Lipschitz continuous F',

$$F(a + \Delta a) - F(a) - F'(a)\,\Delta a \;=\; \int_0^1 [F'(a + \lambda\,\Delta a) - F'(a)]\,d\lambda \cdot \Delta a\,,$$

$$\Big\| \int_0^1 [F'(a + \lambda\,\Delta a) - F'(a)]\,d\lambda \cdot \Delta a \Big\| \;\leq\; \int_0^1 L'\,\lambda \|\Delta a\|\,d\lambda \cdot \|\Delta a\| \;\leq\; \tfrac{L'}{2}\,\|\Delta a\|^2\,. \qquad \Box$$

Corollary 3.3. Consider a well-posed algebraic problem with data of limited accuracy and let the associated data$\rightarrow$result mapping F be continuously (Fréchet) differentiable in its domain $\tilde{A}$. Then the condition estimate (3.19) holds with

$$C \;=\; \sup_{a \in N_{\tilde{\delta}}(\bar{a},e)}\; \| F'(a) \|\,, \tag{3.26}$$

where the norm in (3.26) is the norm for a linear map from $\mathcal{A}$ to $\mathcal{Z}$ induced by the norms in these spaces. Moreover, for Lipschitz continuous F' we have

$$C \;=\; \|F'(\bar{a})\| + \mathrm{O}(\bar{\delta}^2)\,. \tag{3.27}$$

The condition estimate (3.20) with respect to a particular data item holds with

$$C_j \;=\; \sup_{a \in N_{\tilde{\delta}}(\bar{a},e)}\; \Big\| \frac{\partial F}{\partial \alpha_j}(a) \Big\| \;=\; \Big\| \frac{\partial F}{\partial \alpha_j}(\bar{a}) \Big\| + \mathrm{O}(\bar{\delta}^2)\,. \tag{3.28}$$

There is no restriction on the (finite) dimensions of $\mathcal{A}$ and $\mathcal{Z}$ in the fundamental result (3.23) from multivariate analysis. If there is an explicit representation

$$F(a) \;=\; \begin{cases} F_1(\alpha_1, \ldots, \alpha_M) \\ \quad\vdots \\ F_m(\alpha_1, \ldots, \alpha_M) \end{cases} \tag{3.29}$$

then (cf. section 1.2)

$$F'(a) \;=\; \begin{pmatrix} \dfrac{\partial F_\mu}{\partial \alpha_j}(a) \\[4pt] {\scriptstyle \mu=1(1)m} \\ {\scriptstyle j=1(1)M} \end{pmatrix} \;\in\; \mathbb{C}^{m \times M} \text{ or } \mathbb{R}^{m \times M}, \text{ resp.,} \tag{3.30}$$

and

$$\frac{\partial F}{\partial \alpha_j}(a) \;=\; \begin{pmatrix} \dfrac{\partial F_\mu}{\partial \alpha_j}(a) \\[4pt] {\scriptstyle \mu=1(1)m} \end{pmatrix} \;\in\; \mathbb{C}^m \text{ or } \mathbb{R}^m, \text{ resp.} \tag{3.31}$$

For $m = 1$, the $1 \times M$ matrix $F'(a)$ is the *gradient* of F.

There are well-known expressions for the norms of linear maps induced by particular norms in the domain and range spaces (cf. any text on Numerical Linear Algebra): For example, for the max norm (without weights) in both $\mathcal{A}$ and $\mathcal{Z}$,

$$\| F'(a) \| \;=\; \max_{\mu=1(1)m} \sum_{j=1}^{M} \Big| \frac{\partial F_\mu}{\partial \alpha_j}(a) \Big|\,; \tag{3.32}$$

for the Euclidean norm (without weights) in $\mathcal{A}$ and $\mathcal{Z}$,

$$\| F'(a) \| \; = \; \max \; (\text{singular values of } F'(a)) \, . \tag{3.33}$$

Example 3.9: Consider the interpolation problem for values a_ν at the corners $c_\nu := (\cos \varphi_\nu,\ \sin \varphi_\nu)$, $\varphi_\nu = \frac{\nu\pi}{3}$, $\nu = 1(1)6$, of a unit hexagon. As we will see in section 9.6, a possible interpolation polynomial is

$$F(a; x, y) \; = \; \frac{1}{6}\,(a_1, \ldots, a_6) \begin{pmatrix} 2 + 4x + \sqrt{3}\,y - 2x^2 + 2\sqrt{3}\,xy - 4x^3 \\ 2 - 4x + \sqrt{3}\,y - 2x^2 - 2\sqrt{3}\,xy + 4x^3 \\ -1 + x + 4x^2 - 4x^3 \\ 2 - 4x - \sqrt{3}\,y - 2x^2 + 2\sqrt{3}\,xy + 4x^3 \\ 2 + 4x - \sqrt{3}\,y - 2x^2 - 2\sqrt{3}\,xy - 4x^3 \\ -1 - x + 4x^2 + 4x^3 \end{pmatrix} =: a^T q(x, y) \, . $$

(3.34)

In (3.34), there are two kinds of data for a value $v = F(a; \xi, \eta)$ of the interpolation polynomial: the specified values a_ν at the interpolation knots and the coordinates (ξ, η) at which we evaluate the interpolation polynomial. Thus, we have $M = 6 + 2$, $m = 1$.

The condition of v with respect to a perturbation or indetermination in the a_ν, $\nu = 1(1)6$, is displayed by $\frac{\partial}{\partial a^T} F(a; \xi, \eta) = q(\xi, \eta)$; the Fréchet derivative $\frac{\partial}{\partial a^T} F(a; x, y)$ is a *column* vector because we differentiate with respect to a row vector argument. The condition with respect to individual a_ν is shown by the moduli of the respective components of $q(\xi, \eta)$; a comprehensive condition number is the norm of $q(\xi, \eta)$ dual to the norm in which we measure a perturbation in a^T. At the origin, e.g., we obtain individual condition numbers $\frac{1}{3}, \frac{1}{3}, \frac{1}{6}, \frac{1}{3}, \frac{1}{3}, \frac{1}{6}$, and $C = \frac{5}{3}$ for the maximum norm in $\mathcal{A}$.

The condition of v with respect to a perturbation or indetermination in the evaluation point (ξ, η) is displayed by $\frac{\partial}{\partial(\xi,\eta)} F(a; \xi, \eta) =$

$$a^T \begin{pmatrix} \vdots & \vdots \\ \frac{\partial}{\partial \xi} q(\xi, \eta) & \frac{\partial}{\partial \eta} q(\xi, \eta) \\ \vdots & \vdots \end{pmatrix} = \frac{1}{6} a^T \begin{pmatrix} 4 - 4\xi + 2\sqrt{3}\,\eta - 12\,\xi^2 & \sqrt{3} + 2\sqrt{3}\,\xi \\ -4 - 4\xi - 2\sqrt{3}\,\eta + 12\,\xi^2 & \sqrt{3} - 2\sqrt{3}\,\xi \\ 1 + 8\xi - 12\,\xi^2 & 0 \\ -4 - 4\xi + 2\sqrt{3}\,\eta + 12\,\xi^2 & -\sqrt{3} + 2\sqrt{3}\,\xi \\ 4 - 4\xi - 2\sqrt{3}\,\eta - 12\,\xi^2 & -\sqrt{3} - 2\sqrt{3}\,\xi \\ -1 + 8\xi + 12\,\xi^2 & 0 \end{pmatrix} ;$$

the effect of some $\Delta\xi$, $\Delta\eta$ is given by $\Delta v = \frac{\partial}{\partial(\xi,\eta)} F(a; \xi, \eta) \begin{pmatrix} \Delta\xi \\ \Delta\eta \end{pmatrix}$. Again, one can consider effects of particular perturbations or form a comprehensive condition number by taking the appropriate norm. $\square$

For most nontrivial algebraic problems, explicit expressions for the data$\rightarrow$result mapping F are not available. Rather, a well-posed algebraic problem with data $a \in \mathbb{C}^M$ and results $z \in \mathbb{C}^m$ may often be formulated as a system of m equations

$$G(x; a) = 0 \, , \qquad G : \mathbb{C}^m \times \mathbb{C}^M \longrightarrow \mathbb{C}^m \, , \tag{3.35}$$

and analogously in the real case. E.g., (3.35) may be a system of m polynomial equations in the m unknown *components* ζ_μ, $\mu = 1(1)m$, of *one particular result* $z \in \mathbb{C}^m$ which is to be computed. We will meet many interesting situations representable in the form (3.35) which defines the associated data→result mapping in an implicit way:

$$G\left(F(a);\, a\right) \equiv 0 \qquad \text{for } a \in N_{\bar{\delta}}\,. \tag{3.36}$$

Proposition 3.4. Consider a well-posed algebraic problem, with data $a = (\alpha_1, \ldots, \alpha_M) \in \mathbb{C}^M$ and results $z = (\zeta_1, \ldots, \zeta_m) \in \mathbb{C}^m$, which can be represented in the form (3.35) for $a \in N_{\bar{\delta}}(\bar{a}, e)$ and $x \in Z_{\bar{\delta}}(\bar{a}, e)$. Assume that G is Fréchet differentiable with respect to x and a in $Z_{\bar{\delta}} \times N_{\bar{\delta}}$. Then the Fréchet derivative $F'(a)$ of the data→result mapping $F : a \to z$ implicitly defined by (3.36) satisfies, for $a \in N_{\bar{\delta}}$,

$$\frac{\partial G}{\partial x}\left(F(a);\, a\right) \cdot F'(a) + \frac{\partial G}{\partial a}\left(F(a);\, a\right) = 0\,. \tag{3.37}$$

Proof: The proposition is proved by total differentiation of G with respect to a in (3.36). $\qquad\square$

Corollary 3.5. Under the hypotheses of Proposition 3.4, if the $m \times m$ matrix function $\frac{\partial G}{\partial x}(F(a);\, a)$ is regular in $N_{\bar{\delta}}$,

$$\| F'(a) \| \leq \| [\tfrac{\partial G}{\partial x}(F(a);\, a)]^{-1} \| \cdot \| \tfrac{\partial G}{\partial a}(F(a);\, a) \|\,. \tag{3.38}$$

Thus, we may obtain a condition number C by differentiating an equation (a system of equations) for the result quantity z with respect to the data.

The implicit algebraic problem (3.35) remains unaffected by multiplication from the left with an arbitrary constant *regular $m \times m$* matrix M. While this also does not affect (3.36) and (3.37), it may well affect the right-hand side of (3.38) since, for matrices A, B, $\|(M\,A)^{-1}\|\,\|(M\,B)\| \neq \|A^{-1}\|\,\|B\|$ in general.

It is well known from numerical linear algebra that a considerate normalization and pre-processing ("preconditioning") of a system of linear equations may have a significant influence on the subsequent solution procedure, including the sensitivity to round-off. This must be equally true for systems of nonlinear equations. In this text, we do not consider these effects; we assume that (3.35) is the appropriate formulation of the algebraic problem to be solved. But the analysis of this aspect poses an interesting research topic in computational polynomial algebra. For *empirical* problems, there is a further important aspect; cf. section 3.3.2.

Example 3.10: Consider an empirical univariate polynomial $(\bar{p}, e)$, with the specified polynomial $\bar{p} = p(x; \bar{a}) = \sum_{j=0}^{n} \bar{\alpha}_j x^j$ and tolerances ε_j, $j \in \tilde{J} \subset \{0, \ldots, n\}$. Consider one particular simple zero ζ of $p(x; a)$ and denote its dependence on the empirical coefficients by $\zeta(a)$, with $a := (\alpha_j, j \in \tilde{J})$ so that

$$p(\zeta(a);\, a) \equiv 0\,.$$

By differentiation with respect to a, we obtain

$$p'(\zeta(a);\, a) \cdot \frac{d\zeta(a)}{da} + \frac{\partial p}{\partial a}(\zeta(a);\, a) = 0\,.$$

At $\bar{a}$, with $\zeta(\bar{a}) =: \bar{\zeta}$, this implies, for each $j \in \tilde{J}$,

$$\frac{\partial \zeta}{\partial \alpha_j}(\bar{a}) = -\frac{\bar{\zeta}^{\,j}}{\bar{p}'(\bar{\zeta})}\,. \tag{3.39}$$

Thus, at least for small variations in the α_j, we may use $\left|\frac{\bar{\zeta}^j}{\bar{p}'(\zeta)}\right|$, $j \in \tilde{J}$ for C_j in

$$|\tilde{\zeta} - \bar{\zeta}| \leq \sum_{j \in \tilde{J}} C_j \, |\tilde{\alpha}_j - \bar{\alpha}_j| \leq \left(\sum_{j \in \tilde{J}} C_j\right) \max_{j \in \tilde{J}} |\tilde{\alpha}_j - \bar{\alpha}_j| . \quad \Box \tag{3.40}$$

As is to be expected, the condition of a simple zero of a univariate polynomial is inversely proportional to the modulus of the derivative at the zero. This holds also for zeros of arbitrary univariate functions which are differentiable in a neighborhood of the zero. If the derivative vanishes at or very near the zero—as for a multiple zero or for a zero of a dense cluster, resp.—there does not exist a condition estimate of the type (3.19).

3.2.3 Linearized Estimation of the Result Indetermination

In connection with empirical problems, condition estimates are mainly used for the estimation of the indetermination in the pseudoresults due to the indetermination in the empirical data. That indetermination is precisely what is captured in the pseudoresult sets of Definition 3.7, but while these sets are an important conceptual tool, they are not suitable for computational handling. What we actually want is an estimate of the extension of a pseudoresult set Z_δ with $\delta = O(1)$ in the directions of the various result components. This relates immediately to the natural question: How many digits are meaningful in the numerical specification of a valid result $\tilde{z}$? Clearly, if a result component varies by a few units of 10^{-r} inside Z_1, then it would be meaningless and even misleading to specify more than r decimal digits (after the decimal point) of a valid approximation for this result component.

Fortunately, the linearization result (3.24) for Lipschitz continuously differentiable F (a condition often satisfied in the algebraic context) permits an alternate description of the pseudoresult sets Z_δ: It is true that it is only approximate; but it is very much simpler and more intuitive than the original description by Definition 3.7, and—in the present context—we are only interested in an order-of-magnitude answer anyway.

Proposition 3.6. For a data$\rightarrow$result mapping F which satisfies the hypotheses of Corollary 3.3 let

$$\dot{Z}_\delta(\bar{a}, e) := \{\tilde{z} = F(\bar{a}) + F'(\bar{a}) \, (\tilde{a} - \bar{a}) , \; \tilde{a} \in N_\delta(\bar{a}, e)\} . \tag{3.41}$$

Then, with Z_δ from (3.16),

$$\mathrm{dist}\,(\dot{Z}_\delta(\bar{a}, e), Z_\delta(\bar{a}, e)) \leq \frac{L'}{2} \, (\delta \|e\|^*)^2 , \tag{3.42}$$

where, as usual for sets in metric spaces,

$$\mathrm{dist}\,(S_1, S_2) := \max \left(\max_{s_1 \in S_1} \min_{s_2 \in S_2} d(s_1, s_2), \; \max_{s_2 \in S_2} \min_{s_1 \in S_1} d(s_1, s_2)\right) .$$

Proof: Let $\dot{z}(a) := F(\bar{a}) + F'(\bar{a})(a - \bar{a})$ and $z(a) := F(a)$. Then, by (3.24),

$$\mathrm{dist}\,(\dot{Z}_\delta(\bar{a}, e), Z_\delta(\bar{a}, e)) \leq \max_{\tilde{a} \in N_\delta(\bar{a},e)} \|\dot{z}(\tilde{a}) - z(\tilde{a})\| \leq \max_{\tilde{a} \in N_\delta(\bar{a},e)} \frac{L'}{2} \|\tilde{a} - \bar{a}\|^2 = \frac{L'}{2} (\delta \|e\|^*)^2 . \quad \Box$$

Proposition 3.6 shows that we may safely use $\dot{Z}_\delta$ from (3.41) in place of Z_δ from (3.16) for $\delta = O(1)$, sufficiently small tolerances e, and a moderate L'. $\dot{Z}_\delta(\bar{a}, e)$ is the image in $\mathcal{Z}$ of $N_\delta(\bar{a}, e)$ under the *linear map* $F'(\bar{a})$ and can be more easily described, handled, and visualized than Z_δ. Therefore, we will often use $\dot{Z}_\delta$ in place of Z_δ in this book.

Proposition 3.7. For a well-posed empirical algebraic problem with a data→result mapping $F : \mathbb{C}^M \to \mathbb{C}^m$, with components F_μ, $\mu = 1(1)m$, which satisfies the hypotheses of Corollary 3.3, we have for the components ζ_μ, $\mu = 1(1)m$, of the result quantity $z \in \mathbb{C}^m$,

$$\max_{z_1, z_2 \in \dot{Z}_\delta(\bar{a}, e)} |\zeta_{\mu 1} - \zeta_{\mu 2}| \;=\; 2 \max_{\tilde{z} \in \dot{Z}_\delta} |\tilde{\zeta}_\mu - \bar{\zeta}_\mu| \;=\; 2 \max_{\tilde{a} \in N_\delta(\bar{a}, e)} |F'_\mu(\bar{a}) \, (\tilde{a} - \bar{a})|$$

$$\leq 2\delta \sum_{j=1}^M |\tfrac{\partial F_\mu}{\partial \alpha_j}(\bar{a})| \, \varepsilon_j \;\leq\; 2\delta \, (\sum_{j=1}^M |\tfrac{\partial F_\mu}{\partial \alpha_j}(\bar{a})|) \, \max_j \varepsilon_j \;=\; 2\delta \, \|F'_\mu(\bar{a})\| \|e\|^* \,,$$
$$\tag{3.43}$$

where $F'(\bar{a})$ is the gradient of F at $\bar{a}$ and $\|..\|^*$ is the norm in the data space; cf. Definition 3.3. For a result space of a dimension $m > 1$, we may use the condition estimate (3.19) with $C = \|F'(\bar{a})\|$ (cf. (3.27) and (3.32)) to obtain a bound for the diameter of $\dot{Z}_\delta$ in terms of the norm in $\mathcal{Z}$:

$$\operatorname{diam} \dot{Z}_\delta \;:=\; 2 \max_{\tilde{z} \in \dot{Z}_\delta} \|\tilde{z} - \bar{z}\| \;\leq\; 2 \max_{\tilde{a} \in N_\delta(\bar{a}, e)} \|F'(\bar{a})\| \|\tilde{a} - \bar{a}\| \;\leq\; 2\delta \, \|F'(\bar{a})\| \|e\|^* . \tag{3.44}$$

But since the extension of Z_δ in the directions of the individual result components may vary considerably, the componentwise estimate (3.43) is often preferable.

For the purpose of estimating the indetermination in the approximate results of our empirical problem, (3.43) and (3.44) need only be evaluated within the *correct order of magnitude* (and with $\delta = 1$) because the original indetermination in the data is only known by its order of magnitude in virtually all cases. Furthermore, it is not really important whether we specify one decimal digit more or less of a result, but it is important *not to specify* 10 digits in a situation where diam $Z_1 \approx 10^{-3}$!

Example 3.10, continued: Consider the zero $\bar{\zeta} \approx -1.414213$ of $\bar{p}$ of (3.17) in Example 3.3. From (3.39) and (3.43), we have for $\tilde{\zeta} \in Z_1(\bar{a}, e)$

$$|\tilde{\zeta} - \bar{\zeta}| \;\leq\; (\sum_{j=0}^{3} |\bar{\zeta}|^j)/|\bar{p}'(\bar{\zeta})| \cdot 10^{-5} \;\approx\; 7.24/22.64 \cdot 10^{-5} \;\leq\; 3.2 \cdot 10^{-6} .$$

This yields an indetermination of $\approx 6 \cdot 10^{-6}$ in the zero, in excellent agreement with our previous observations. There are no comparable estimates for the clustered zeros in the right halfplane since p' is not bounded away from zero there.   □

So far, we have silently assumed that the exact result values are *isolated points* in some m-dimensional space. In dealing with multivariate polynomial problems, we will also face the situation where there are positive-dimensional *result manifolds*. We will address the condition problem for such data→result mappings in section 7.2.3.

Exercises

1. The following algebraic problems are ill-posed; associate them with one of the categories in section 3.2.1.

(a) Find valid (pseudo)factors of an empirical multivariate polynomial.

(b) Determine whether an empirical univariate polynomial is "stable" (all zeros z_μ satisfy $\mathrm{Re}\, z_\mu < 0$).

(c) Find a valid greatest common divisor of two empirical univariate polynomials.

(d)[5] Find the coefficients of a valid Groebner basis with more than s elements for a regular empirical polynomial system in $\mathcal{P}^s$.

(e) Find the coefficients of a valid Groebner basis for Example 3.8.

2. Consider the *real* empirical polynomial $(\bar{p}, e)$ with

$$x^6 - 4.751\, x^5 + 13.014\, x^4 + 14.144\, x^3 - 2.282\, x^2 - 41.137\, x + 23.099$$

and $\varepsilon_j = .5 \cdot 10^{-3}$ for $j = 0(1)5$. $\bar{p}$ has 3 pairs of complex zeros $\zeta_\mu = \xi_\mu \pm \mathrm{i}\,\eta_\mu$, $\mu = 1(1)3$. Analyze the indetermination of the real and imaginary parts ξ_μ and η_μ due to the indetermination in the coefficients.

(a) For each of the ζ_μ, split (3.39) into real and imaginary parts and form separate estimates (3.40). How many digits are meaningful in a specification of the ξ_μ, η_μ?

(b) Can you expect that all polynomials in $N_1(\bar{p}, e)$ have no real zeros? Verify your answer experimentally.

3.3 Backward Error of Approximate Results

The following is a fundamental task in the context of empirical problems: Consider a well-posed empirical algebraic problem with a data$\rightarrow$result function $F : \mathcal{A} \rightarrow \mathcal{Z}$; given an approximate result value $\tilde{z} \in \mathcal{Z}$, *from whatever source*, determine whether $\tilde{z}$ is a valid result of the problem.

According to Definitions 3.7 and 3.8, $\tilde{z}$ is a valid approximate result if there exist data $\tilde{a} \in N_\delta(\bar{a}, e)$, $\delta = O(1)$, such that $\tilde{z}$ is the *exact* result of the algebraic problem with data $\tilde{a}$. In the verification of this condition, the set of *all* data $\tilde{a} \in \mathcal{A}$ for which this condition holds plays an important role.

Definition 3.11. For an empirical algebraic problem with data$\rightarrow$result function $F : \mathcal{A} \rightarrow \mathcal{Z}$, and for a *given* approximate result $\tilde{z} \in \mathcal{Z}$, the *equivalent-data set (for z)* is defined by

$$\mathcal{M}(\tilde{z}) := \{\tilde{a} \in \mathcal{A} : F(\tilde{a}) = \tilde{z}\} . \tag{3.45}$$

For algebraic problems, the equivalent-data set is generally an algebraic manifold in the empirical data space $\mathcal{A}$; therefore, we will often call $\mathcal{M}(\tilde{z})$ the equivalent-data manifold. $\square$

Example 3.11: Consider an empirical polynomial $(\bar{p}, e)$, with $\bar{p}(x) = \sum_{j \in J} \bar{\alpha}_j x^j$ and empirical support $\tilde{J} \subset J \subset \mathbb{N}_0^s$; cf. Definition 3.3. In order that a specified $\tilde{z}$ is a zero of $\tilde{p}(x) = \sum_{j \in J} \tilde{\alpha}_j x^j$, the coefficients $\tilde{a} = (\tilde{\alpha}_j)$ must satisfy

$$\tilde{p}(\tilde{z}) = \sum_{j \in J} \tilde{\alpha}_j \tilde{z}^j = \sum_{j \in J} (\tilde{\alpha}_j - \bar{\alpha}_j)\, \tilde{z}^j + \bar{p}(\tilde{z}) = 0 .$$

---

[5]Assumes familiarity with Groebner bases.

Since $\tilde{\alpha}_j = \bar{\alpha}_j$ for $j \in J \setminus \tilde{J}$,

$$\mathcal{M}(\tilde{z}) := \{a \in \mathcal{A} : \sum_{j \in \tilde{J}} (\tilde{\alpha}_j - \bar{\alpha}_j)\, \tilde{z}^j + \bar{p}(\tilde{z}) = 0\}. \tag{3.46}$$

Thus, the equivalent-data set is a *linear manifold* in the space $\mathcal{A}$ of the empirical coefficients; its representation requires merely the computation of the *residual* $\bar{p}(\tilde{z})$. □

As in this example, explicit or implicit representations for the equivalent-data manifold can generally be obtained for computational algebraic problems without great difficulty; moreover, $\mathcal{M}(z)$ is often a *linear* manifold in the space $\mathcal{A}$. In particular, since polynomials are *linear in their coefficients*, this happens when the empirical data are coefficients of *polynomials* which occur in the problem in a linear fashion.

The verification task

$$\text{Given } \tilde{z} \in \mathcal{Z} : \quad \exists\, \tilde{a} \in N_\delta(\bar{a}, e) : F(\tilde{a}) = \tilde{z} ? \tag{3.47}$$

is now reduced to the following two steps:

(a) Determine the equivalent-data manifold $\mathcal{M}(\tilde{z})$;

(b) Check whether $\mathcal{M}(\tilde{z})$ has a nonempty intersection with $N_\delta(\bar{a}, e)$.

Since we know that—for a well-posed problem—(b) will always have a positive answer for a sufficiently large value of δ, our interest is rather in answering the question:

$$\text{What is the smallest value of } \delta \text{ for which (3.47) holds ?} \tag{3.48}$$

or—equivalently—in solving the task:

(b') Find the shortest distance $\delta(\tilde{z})$ of $\mathcal{M}(\tilde{z})$ from $\bar{a}$ in the metric (3.5) .

Naturally, this approach breaks down if the equivalent-data set $\mathcal{M}(\tilde{z})$ turns out to be *empty*, i.e. if $\tilde{z}$ does not lie in the image $F(\widehat{A}) \subset \mathcal{Z}$ of the domain $\widehat{A}$ of F. For *well-posed* problems, as we consider them here, this restriction generally presents no problem; cf. section 3.2.1. For the *ill-posed* problems of type 3, on the other hand, this is very nontrivial; we will return to such cases in later parts of the book. For now, we assume that $\mathcal{M}(\tilde{z}) \neq \emptyset$.

Proposition 3.8. For a well-posed empirical problem, the shortest distance of an equivalent-data set $\mathcal{M}(\tilde{z})$ to $\bar{a}$ in the metric (3.5) is uniquely defined.

Proof: For F satisfying the hypotheses of Proposition 3.1, the set $\mathcal{M}(\tilde{z})$ defined by (3.45) is closed and its intersection with the ball $\{a \in \mathcal{A} : \|a - \bar{a}\| \leq \bar{\delta}\}$ is closed and bounded. On that intersection, the continuous function $\|a - \bar{a}\|_e^*$ attains a unique minimum.   □

Note that, for a nonstrict norm like the max-norm, the assertion of Proposition 3.8 does not imply that there is a unique *data point a* at which the shortest distance is assumed.

Definition 3.12. In the situation previously described, with $\|.\|_e^*$ from (3.5),

$$\delta(\tilde{z}) := \min_{a \in \mathcal{M}(\tilde{z})} \|a - \bar{a}\|_e^* \tag{3.49}$$

is the *backward error* of the approximate result $\tilde{z}$ for the empirical algebraic problem with data $(\bar{a}, e)$.   □

The term "backward error" was introduced by J. Wilkinson in [3.5]. His idea to interpret the deviation (or "error" in numerical analysis terminology) of an approximate result $\tilde{z}$ as the effect of a deviation in the data of the original problem has become a central tool in numerical analysis and, more generally, in applied mathematics. Definition 3.12 follows directly the original idea of Wilkinson: Our backward error $\delta(\tilde{z})$ is the norm of the minimal correction which must be applied to the specified data $\bar{a}$ in order that $\tilde{z}$ becomes an exact solution of the problem. Our individual weights for the empirical data components make the concept more flexible, our normalization by the individual tolerances permits a comparison of the backward errors for different problem types.

3.3.1 Determination of the Backward Error

We consider the task of finding the shortest distance (3.49) between a linear or algebraic manifold and a fixed point in a finite-dimensional vector space, with reference to a metric induced by a *weighted dual norm* $\|\ldots\|_e^*$; cf. (3.5). To simplify the notation and without loss of generality, we *shift the origin* of the data space $\mathcal{A}$ to the specified data point $\bar{a}$; intuitively, this means that we use the *deviations* $\Delta\alpha_j := \alpha_j - \bar{\alpha}_j$ as variables. We denote the data space with the shifted origin by $\Delta\mathcal{A}$.

For a *linear manifold*, finding its shortest norm distance from the origin is a classical task; we assemble a few relevant results.

Let the linear manifold $\mathcal{M}$ in the M-dimensional vector space $\Delta\mathcal{A}$ be given by

$$\Delta a^T C = c^T , \tag{3.50}$$

where $\Delta a^T = (\Delta\alpha_1 \ldots \Delta\alpha_M)$, $C = (\gamma_{\mu\kappa}) \in \mathbb{C}^{M\times k}$, $c^T = (\gamma_{0\kappa}) \in \mathbb{C}^k$, $1 \le k \le M$. Without loss of generality, we assume $\operatorname{rank} C = k$; otherwise, $\mathcal{M}$ could be specified by fewer than k linear equations. Thus, the codimension of $\mathcal{M}$ is k, its dimension $M - k$. In $\Delta\mathcal{A}$, the dual norm $\|\ldots\|_e^*$, with $0 < e \in \mathbb{R}^M$, takes the form

$$\|\Delta a^T\|_e^* := \|(\ldots \frac{|\Delta\alpha_j|}{\varepsilon_j} \ldots)\|^* ; \tag{3.51}$$

the columns of the matrix C are assessed by the associated vector norm $\|..\|$; cf. the discussion of norms in section 3.1.1. We want to find $\min_{\Delta a^T C = c^T} \|\Delta a^T\|_e^*$; cf. (3.49). This is easy for $k = 1$ and $k = M$.

Proposition 3.9. For $k = 1$, with $C = (\gamma_1, \ldots, \gamma_M)^T \in \mathbb{C}^M$, $c^T = \gamma_0 \in \mathbb{C}$, and for any norm,

$$\min_{\Delta a^T C = \gamma_0} \|\Delta a^T\|_e^* = |\gamma_0| / \left\| \begin{pmatrix} \gamma_1 \\ \vdots \\ \gamma_M \end{pmatrix} \right\|_e . \tag{3.52}$$

Proof: By (3.6), we have $|\gamma_0| = |\Delta a^T C| \le \|\Delta a^T\|_e^* \|C\|_e$. In our finite dimensions, equality is attained. $\square$

For the 1-norm and the 2-norm in $\mathbb{R}^M$ (cf. section 3.1.1), the minimizing Δa may be explicitly specified:

Proposition 3.10. Consider $u \in \mathbb{C}^M$ with
$$\left.\begin{array}{rcl} \|u\|_{(1)} & = & \sum_j |u_j| \\[4pt] \|u\|_{(2)} & = & (\sum_j |u_j|^2)^{1/2} \end{array}\right\} = 1\,;$$

then $v^T = (v_1, \ldots, v_M) \in \mathbb{C}^M$ satisfies $\|v^T\|_{(i)}^* = 1$ and $|v^T u| = 1$ for $v_j = \begin{cases} \rho \dfrac{u_j^*}{|u_j|} \\[8pt] \rho u_j^* \end{cases}$,

with an arbitrary $\rho \in \mathbb{C}$, $|\rho| = 1$, where u^* denotes complex conjugation.

Proof: By straightforward verification. $\square$

Example 3.11, continued: For an approximate zero $\tilde{z} \in \mathbb{C}^s$, $s \geq 1$, of an empirical polynomial $(\bar{p}, e)$ with empirical support $\tilde{J}$, $|\tilde{J}| = M$, we had obtained the equivalent-data manifold as

$$\mathcal{M}(\tilde{z}) = \Big\{ \Delta a \in \Delta\mathcal{A} \ : \ \sum_{j \in \tilde{J}} \Delta \alpha_j \, \tilde{z}^j + \bar{p}(\tilde{z}) = 0 \Big\}, \tag{3.53}$$

which implies $\gamma_j = \tilde{z}^j$, $j \in \tilde{J}$, $\gamma_0 = -p(\tilde{z})$ in Proposition 3.9; cf. (3.46). Thus, by (3.52), the backward error of the approximate zero $\tilde{z}$ is, for the i-norm in $\mathbb{C}^M$,

$$\delta(\tilde{z}) = \frac{|\bar{p}(\tilde{z})|}{\|(\tilde{z}^j)\|_e} = \begin{cases} |\bar{p}(\tilde{z})| \,/\, \sum_{j \in \tilde{J}} \varepsilon_j |\tilde{z}|^j & (i = 1) \\[8pt] |\bar{p}(\tilde{z})| \,/\, (\sum_{j \in \tilde{J}} (\varepsilon_j |\tilde{z}|^j)^2)^{1/2} & (i = 2) \end{cases}, \tag{3.54}$$

and it is attained for

$$\rho \cdot \Delta \alpha_j^* = \delta(\tilde{z}) \cdot \begin{cases} \varepsilon_j \dfrac{(z^j)^*}{|z^j|} & (i = 1) \\[8pt] \varepsilon_j^2 \, (z^j)^* \,/\, (\sum_{j \in \tilde{J}} (\varepsilon_j |\tilde{z}|^j)^2)^{1/2} & (i = 2) \end{cases} \quad j = 1(1)M\,, \tag{3.55}$$

where $\rho \in \mathbb{C}$, $|\rho| = 1$, must be chosen so that $\Delta a^* \in \mathcal{M}(\tilde{z})$. $\square$

In Proposition 3.10, for a vector u with *real* components the minimizing dual vector v is also real. On the other hand, for a u with some nonreal components, the minimizing v must also have nonreal components. Therefore, if we restrict the variations Δa to the real domain, we can use (3.52) in Proposition 3.9 *only if* the components γ_j of C are also real. Otherwise we have to form separate equations (3.50) for the real and imaginary parts which may double the dimension k of C. Thus, the expressions (3.54) for the backward error of an approximate zero of an empirical polynomial and (3.55) for the associated modification of the coefficients do not apply to a nonreal zero of a real empirical polynomial whose indetermination can only be real. (Compare section 5.2.1 for the backward error in this case.)

Proposition 3.11. For $k = M$, the only point $\Delta a_0^T = c^T C^{-1}$ of the equivalent-data manifold $\mathcal{M}$ has distance $\|\Delta a_0^T\|_e^*$ from the origin.

Example 3.12: Consider a monic univariate empirical polynomial $(\tilde{p}, e)$ of degree M, with all coefficients but the leading one empirical, and assume that a set of M disjoint approximate zeros $\tilde{z}_\mu \in \mathbb{C}$, $\mu = 1(1)M$, has been computed. We want to find the backward error of the *combined set of approximate zeros*, i.e. the e-distance (3.12) between $\bar{p}$ and the polynomial $\tilde{p} = \prod_{\mu=1}^M (x - \tilde{z}_\mu)$ which has *all* the $\tilde{z}_\mu$ as exact zeros.

The deviations $\Delta \alpha_j$ for the coefficients of $\tilde{p}$ have to lie on the M linear manifolds $\mathcal{M}(\tilde{z}_j)$ from (3.53) simultaneously; i.e. they coincide with the intersection Δa_0 of these manifolds.

With the columns of C and the elements of c_0 as in Example 3.11 above, we have

$$\delta(\tilde{z}_1, \ldots, \tilde{z}_M) = \|\Delta a_0^T\|_e^* = \|\tilde{p} - \bar{p}\|_e^* \qquad \text{with}$$

$$\Delta a_0^T = (\Delta\alpha_0, \ldots, \Delta\alpha_{M-1}) = -(\bar{p}(\tilde{z}_1), \ldots, \bar{p}(\tilde{z}_M)) \begin{pmatrix} 1 & \cdots & 1 \\ \tilde{z}_1 & & \tilde{z}_M \\ \vdots & & \vdots \\ \tilde{z}_1^{M-1} & \cdots & \tilde{z}_M^{M-1} \end{pmatrix}^{-1} . \quad \square$$

For values of k between 1 and M, the determination of the backward error $\delta(\tilde{z})$ depends more explicitly on the choice of the norm. We consider only our standard norm in detail where

$$\delta(\tilde{z}) := \min_{\Delta a^T \in \mathcal{M}(\tilde{z})} \|\Delta a^T\|_e^T = \min_{\Delta a^T \in \mathcal{M}(\tilde{z})} \max_j \frac{|\Delta\alpha_j|}{\varepsilon_j} . \tag{3.56}$$

When the $(M{-}k)$-dimensional linear manifold $\mathcal{M}$ is restricted to the *real domain*, (3.56) becomes a *linear program* in the real variables $\Delta\alpha_j$ and δ:

$$\min \delta \quad \text{with the constraints (3.50) and} \quad \begin{array}{c} \Delta\alpha_j \le \varepsilon_j\delta \\ -\Delta\alpha_j \le \varepsilon_j\delta \end{array}, \quad j = 1(1)M . \tag{3.57}$$

Such standard linear minimization tasks can be solved by widely available packaged software, e.g., by Maple's `simplex` package.

Often it may be advantageous to solve the k linear equality conditions (3.50) for k of the $\Delta\alpha_j$ (say $\Delta\alpha_{j_1}, \ldots, \Delta\alpha_{j_k}$) so that (3.57) becomes a minimization problem in the $M - k + 1$ variables δ and $\Delta\alpha_{j_{k+1}}, \ldots, \Delta\alpha_{j_M}$ only. We may write it as

$$\delta = \min f(\Delta\alpha_{j_{k+1}}, \ldots, \Delta\alpha_{j_M}) , \tag{3.58}$$

where the piecewise linear convex function f is defined in terms of the coefficients $c_{\kappa\nu}$ originating in solving (3.50) for the $\Delta\alpha_{j_\kappa}$, $\kappa = 1(1)k$:

$$f(\Delta\alpha_{j_{k+1}}, \ldots, \Delta\alpha_{j_M}) :=$$
$$\max \left(|c_{\kappa 0} + \sum_{\nu=k+1}^M c_{\kappa\nu}\Delta\alpha_{j_\nu}|/\varepsilon_{j_\kappa} , \ \kappa = 1(1)k ; \ |\Delta\alpha_{j_\kappa}|/\varepsilon_{j_\kappa} , \ \kappa = k+1(1)M \right) . \tag{3.59}$$

For (3.58) and (3.59), one can use a method of descent along edges of the graph of f.

When $\mathcal{M}$ is a *complex* linear manifold in the complex data space $\mathcal{A}$, this function f remains a piecewise smooth, *convex* function so that the minimization problem (3.56) for (3.50) defines a unique minimal value for δ as in the real case. Software for convex optimization in a complex space is also available.

Another possibility for the determination of δ is a parametrization of $\mathcal{M}$ and the determination of $\min_{\Delta a \in \mathcal{M}} \|\Delta a\|_e^*$ in terms of the parameter(s).

Example 3.13: We consider the empirical polynomial (3.17) of Example 3.3 and choose a cubic polynomial $\tilde{s}$ close to $(x - \sqrt{2})^3$ as a tentative *approximate divisor* of our empirical polynomial $(\bar{p}, e)$ to represent the zero cluster in the positive halfplane; we want to determine the backward error of $\tilde{s}$, i.e. the norm of the smallest variation of $\bar{p}$ which leads to a polynomial $\tilde{p}$ which has $\tilde{s}$ as an exact divisor. Let

$$\tilde{s}(x) = x^3 - 4.2451\,x^2 + 6.0069\,x - 2.8333 ;$$

thus, our "result" $z \in \mathbb{R}^3$ consists of the 3 coefficients σ_μ, $\mu = 0, 1, 2$, of $\tilde{s}$, and we have $M = 4$, $m = 3$. The equivalent-data manifold $\mathcal{M}(\tilde{s})$ in the data space $\mathcal{A} = \mathbb{C}^4$ is 1-dimensional and contains all monic 4-th degree polynomials $\tilde{p}$ (or rather their coefficients $\tilde{\alpha}_j$, $j = 0(1)3$) which are exact multiples of $\tilde{s}$. Here, we have a natural parametrization of $\mathcal{M}(\tilde{s})$ by the fourth zero ζ of $\tilde{p}$ and we know that we are only interested in values of ζ near $-\sqrt{2}$. From

$$\tilde{p}(x) = (x - \zeta)\, \tilde{s}(x) = x^4 + (\sigma_2 - \zeta)\, x^3 + (\sigma_1 - \zeta\, \sigma_2)\, x^2 + (\sigma_0 - \zeta\, \sigma_1)\, x - \zeta\, \sigma_0 ,$$

we have $\quad \Delta p(x) := \tilde{p}(x) - \bar{p}(x)$

$$= (\sigma_2 - \zeta - \bar{\alpha}_3)\, x^3 + (\sigma_1 - \zeta\, \sigma_2 - \bar{\alpha}_2)\, x^2 + (\sigma_0 - \zeta\, \sigma_1 - \bar{\alpha}_1)\, x + (-\zeta\, \sigma_0 - \bar{\alpha}_0) =: \sum_{j=0}^{3} \Delta\alpha_j\, x^j .$$

For the max-norm in $\Delta\mathcal{A}$, we have

$$\delta(\tilde{s}) = \min_\zeta \| \Delta p \|_e^* = \min_\zeta \max_{j=0(1)3} \left(\frac{|\Delta\alpha_j(\zeta)|}{\varepsilon_j} \right) .$$

Near $\zeta = \sqrt{2}$, the minimum is attained for $|\Delta\alpha_2| = |\Delta\alpha_0|$, or $\zeta \approx -1.4142$, with $\delta \approx 4.84$. Thus $\tilde{s}$ is not quite a safely valid divisor of $(\bar{p}, e)$; cf. (3.3). $\quad\square$

The previous discussion has shown that it is straightforward to obtain a numerical value for the backward error $\delta(\tilde{z})$ if the set $\mathcal{M}(\tilde{z})$ of data $a \in \Delta\mathcal{A}$ for which the algebraic problem has the exact result $\tilde{z}$ is a *linear* manifold in $\Delta\mathcal{A}$. Fortunately, in polynomial algebra, this covers a great deal of the interesting situations as we will see.

If the set $\mathcal{M}(\tilde{z})$ is a nonempty *nonlinear algebraic manifold*, the situation is not so simple. However, we do not really have to find the shortest distance to the origin on a general algebraic manifold; we can relax this requirement in various ways:

First of all, we may generally assume that a reasonably computed approximate result $\tilde{z}$ will generate an equivalent-data manifold which has a minimal distance (perhaps a local one) from the origin in $\Delta\mathcal{A}$ *for rather small values of the* $\Delta\alpha_j$. This restricts our search to a compact domain about the origin, say $\|\Delta a\|_e^* \leq 100$, and also excludes irrelevant components of the manifold.

Furthermore, if we are able to locate *some* point on $\mathcal{M}(\tilde{z})$ with a norm distance O(1) from the origin, we are finished: then $\tilde{z}$ is a valid result. Thus, the considerate selection of a few points on $\mathcal{M}(\tilde{z})$ and the determination of their norms may complete the task. On the other hand, if we can establish that the minimal distance is sufficiently larger than 1, we know that $\tilde{z}$ is *not* a valid approximate result. If $\mathcal{M}$ is given as the intersection of several higher-dimensional manifolds, it is sufficient to establish $\delta > $ O(1) for *one* of them.

Almost always, $\mathcal{M}(\tilde{z})$ is smooth in the domain of interest. When we replace $\mathcal{M}$ by its *tangential manifold* at some Δa_0 near the minimum, the minimal distance from the origin of this linear manifold will not differ much from that of $\mathcal{M}$.

If $\|a\|_e^*$ is *convex* on $\mathcal{M}$ in a sufficiently large domain about the minimal a, established software for nonlinear convex optimization is normally able to find the value of the minimum when given the origin $\Delta a = 0$ as a starting location. We may often stop such an algorithm prematurely, because our task has been solved; cf. the considerations above.

Example 3.14: Consider the exponential "polynomial"

$$q(x; a) = 1.56 \exp(-.74\, x) - .26 \exp(.45\, x),$$

with all coefficients empirical and $\varepsilon = .005$. A plot shows that $\bar{q}$ has its only zero near 1.5. Is $\tilde{z} = 1.5$ a valid zero?

The equivalent-data manifold $\mathcal{M}(z) \subset \mathbb{R}^4$ is given by $q(z; \bar{a} + \Delta a) = 0$, in which two of the four $\Delta\alpha_j$ occur nonlinearly. But we can restrict the modifications to the two linearly occurring coefficients and try to satisfy

$$(1.56 + \Delta\alpha_1) \exp(-.74 \cdot 1.5) + (-.26 + \Delta\alpha_2) \exp(.45 \cdot 1.5) = 0$$

with minimal $|\Delta\alpha_j|$. This leads to $\Delta\alpha_1 = \Delta\alpha_2 \approx -.0015$ and a norm distance to 0 of $\approx .3$. Thus we have verified the validity of 1.5 as a zero of the empirical function q.

We can also linearize $q(1.5; \bar{a} + \Delta a)$ and solve the arising linear minimization problem. This leads to modifications of all 4 coefficients by $\approx .0009$ and thus to an approximate backward error of .18. The smallness of the modifications suggests that the exact solution of the nonlinear minimization would not deviate significantly from the solution of the linear approximation. $\quad\square$

3.3.2 Transformations of an Empirical Polynomial

In dealing with polynomials as algebraic objects, we are used to "transforming" them freely, i.e. changing the basis for their representation according to need. In particular, a *shift of the origin* of the coordinate system is considered a trivial operation: For $p \in \mathcal{P}^s$, $s \geq 1$, a shift of the origin to $c \in \mathbb{C}^s$ transforms $p(x) = \sum_{j \in J} \alpha_j x^j$ into

$$\vec{p}(x) \;=\; p(c + x) \;=\; \sum_{j \in J} \alpha_j (c + x)^j \;=\; \sum_{j \in J} \vec{\alpha}_j(c) \, x^j, \tag{3.60}$$

where the $\vec{\alpha}_j(c)$ are *scalar products* of the α_j and vectors of powerproducts of the components of c. For a univariate polynomial $p = \sum_{\nu=0}^n \alpha_\nu x^\nu$, we have simply

$$\vec{\alpha}_\nu(c) \;=\; \frac{1}{\nu!} \, p^{(\nu)}(c) \;=\; \sum_{\nu'=0}^{\nu} \alpha_{\nu'} \binom{\nu}{\nu'} c^{\nu - \nu'}, \quad \nu = 0(1)n;$$

the computation is generally performed with the *extended Horner algorithm*.

More generally, from the Taylor expansion of $p(c + x)$ in (3.60), we have immediately:

Proposition 3.12. For $p \in \mathcal{P}^s$ of total degree d, the shifted polynomial $\vec{p}$ of (3.60) has the coefficients (cf. (1.6) and (2.37) for the notation)

$$\vec{\alpha}_j(c) \;=\; \partial_j[c]\,p \;:=\; \frac{1}{j_1! \cdots j_s!} \, \frac{\partial^{|j|}}{\partial_{x_1}^{j_1} \cdots \partial_{x_s}^{j_s}} \, p(c), \quad |j| \leq d. \tag{3.61}$$

The *sparsity pattern* of $\vec{p}$ may differ strongly from that of p even for univariate polynomials; e.g., an even polynomial will be turned into a dense one. For multivariate polynomials, which are generally very sparse, the loss of sparsity can be dramatic. Of course, with relevant information about the structure of p, a transformation (3.60) may also be used to *gain* sparsity; we will not pursue this aspect further.

All this assumes that the coefficients α_j of p are known exactly and that the arithmetic operations in (3.60) are performed exactly. This is important because it is well known that the result of a scalar product operation may be very sensitive to small changes in the components of the factors. A scalar product is strongly ill-conditioned if its factors are nearly orthogonal (or unitary, resp.), i.e. if the modulus of the result is much smaller than the product of the norms of the factors. In the situation of (3.60), this happens when p and/or a number of derivatives of p nearly vanish at c; cf. (3.61).

In an *empirical polynomial* from some real-life situation, the information about the inherent indetermination of p generally refers to the coefficients in a *particular representation* of the polynomial. Our silent assumption that this representation employs a monomial basis will often not be satisfied; but almost all of our considerations so far hold for an arbitrary basis as long as this basis is *used throughout*. When we now consider a *change* of basis for an empirical polynomial, it is not so obvious how the indetermination may be characterized in the new representation. Clearly, the continuous map (3.60) from the coefficient set $(\alpha_j, \ j \in J)$ to the shifted set $(\bar{\alpha}_j, \ j \in \bar{J})$ transforms the family of neighborhoods $N_\delta(\bar{p}, e)$, $\delta > 0$, into a family of neighborhoods $\vec{N}_\delta$ of the polynomial $\vec{p}$. Also one should be able to find *bounds* $\bar{\varepsilon}_j$ for the potential variation of the coefficients $\vec{\alpha}_j$ within a particular neighborhood $\vec{N}_\delta$.

Let us consider the determination of these tolerances $\vec{\varepsilon}_j$: With an empirical polynomial $(\bar{p}, e)$ in $s \geq 1$ variables, we consider the transformation (3.60) for each $\tilde{p} \in N_1(\bar{p}, e)$; we use $j = (j_1, \ldots, j_s)$ as a multisubscript and multiexponent as usual. From (3.61), we have

$$|\vec{\tilde{\alpha}}_j(c) - \vec{\bar{\alpha}}_j(c)| \ \leq \ \max_{\tilde{p} \in N_1(\bar{p}, e)} | \partial_j[c]\, \tilde{p} - \partial_j[c]\, \bar{p} | \,. \tag{3.62}$$

For $p = \sum_{k \in J} \alpha_k x^k$ (note the subscript change),

$$\partial_j[c]\, p \ = \ \sum_{k \in J} \alpha_k\, \partial_j[c] x^k \ = \ \sum_{k \in J,\, k \geq j} \alpha_k \binom{k}{j} c^{k-j} \,, \tag{3.63}$$

where $k \geq j$ denotes the componentwise relation $k_\sigma \geq j_\sigma$, $\forall \sigma$ and $\binom{k}{j} := \prod_\sigma \binom{k_\sigma}{j_\sigma}$. With (3.63) and with the max-norm in the coefficient space, (3.62) implies

$$|\vec{\tilde{\alpha}}_j(c) - \vec{\bar{\alpha}}_j(c)| \ \leq \ \max_{|\tilde{\alpha}_k - \bar{\alpha}_k| \leq \varepsilon_k} \ \Big|\, \sum_{k \in J,\, k \geq j} (\tilde{\alpha}_k - \bar{\alpha}_k) \binom{k}{j} c^{k-j} \Big| \ \leq \ \sum_k \varepsilon_k \binom{k}{j} |c|^{k-j} \ =: \ \vec{\varepsilon}_j \,.$$

The fact that the second inequality may be attained follows from the interpretation of the sum in the max as a scalar product $u^T v$, with $u^T = (\ldots (\tilde{\alpha}_k - \bar{\alpha}_k) \ldots)$ and $v = (\ldots \binom{k}{j} c^{k-j} \ldots)^T$. With the weighted dual norms $\|..\|_e^*$ and $\|..\|_e$ from (3.5), we have $\max_{\|u^T\|_e^* \leq 1} |u^T v| \leq \|v\|_e$, and the well-known attainment of the inequality for some u^T. Thus we have proved

Proposition 3.13.　Within the polynomial neighborhood family $\vec{N}_\delta(c) \subset \mathcal{P}^s$, $\delta > 0$, which results by the transformation (3.60) from the family $N_\delta(\bar{p}, e)$, the variations of the coefficients $\vec{\alpha}_j(c)$ are bounded by

$$|\vec{\tilde{\alpha}}_j(c) - \vec{\bar{\alpha}}_j(c)| \ \leq \ \vec{\varepsilon}_j \cdot \delta \ = \ \Big[\sum_{k \in J,\, k \geq j} \varepsilon_k \binom{k}{j} |c|^{k-j} \Big] \cdot \delta, \quad j \in \bar{J}. \tag{3.64}$$

Since different u are needed to attain the bound for different $\vec{\alpha}_j$, it is generally not possible to attain the tolerances $\vec{\varepsilon}_j$ simultaneously for all $j \in \tilde{J}$: In the data space $\mathcal{A}$ of the coefficient vectors $\vec{a}$, the domain of the potential variations of the $\vec{\alpha}_j$ is only a subset of the Cartesian product of the componentwise domains (3.64); cf. Figure 3.1. Note that another silent assumption in Definition 3.3, viz. the *mutual independence* of the indeterminations in the individual empirical coefficients, need not really be true in practical applications. For the $\vec{N}_\delta(\vec{p}, \vec{e})$, we *know* that this assumption is not true and that we may thus overestimate the effect of the indetermination.

Nevertheless, it may be meaningful to consider the family $N_\delta(\vec{p}, \vec{e})$ with $\vec{e}$ from (3.64) as the result of the shift operation (3.60), particularly for small shifts c. For *large shifts* c, however, the sheer size of the $\vec{\varepsilon}_j$ will generally be prohibitive. Such shifts destroy the meaning of tolerances for the coefficients in the representation of the shifted polynomial.

Example 3.15: Consider a univariate empirical polynomial $(\vec{p}, e)$ of degree n, with all coefficients empirical with $\varepsilon_j = \varepsilon$. For *small shifts*, with $|c| \ll 1$, we have $\vec{\varepsilon}_j = (1 + j \cdot O(|c|))\, \varepsilon$, i.e. a small increase in the tolerances. For *moderate shifts* with $|c| = O(1)$, we have $\vec{\varepsilon}_j = O(\binom{n+1}{j+1})\, \varepsilon$; this will change the order of magnitude for larger n. For *large shifts* with $|c| > O(1)$, we have $\vec{\varepsilon}_j = O(|c|^n)\, \varepsilon$ for smaller values of j so that the $\vec{\varepsilon}_j$ become meaningless for large $|c|$ and n. $\square$

In any case, one should consider the transformation of an empirical polynomial only if it offers definite advantages of some kind. This is true for the simple shift of the origin in (3.60); for more intricate transformations—which we have not considered—it holds as well.

By (3.62), we may give a different interpretation to (3.64): When we regard $c \in \mathbb{C}^s$ as an approximate zero of the derivative $\partial_j p$ of the empirical polynomial $(\vec{p}, e)$, then there exists a $\tilde{p} \in N_1(\vec{p}, e)$ with $\partial_j[c]\, \tilde{p} = 0$ iff (3.62) permits the vanishing of $\vec{\alpha}_j(c)$ with $\vec{\vec{\alpha}}_j = \partial_j[c]\, \vec{p}$. Thus we have:

Proposition 3.14. For an empirical polynomial $(\vec{p}, e)$ of total degree d in s variables, $s \geq 1$, with empirical support $\tilde{J}$, the backward error of an approximate zero $\tilde{z} \in \mathbb{C}^s$ of the derivative $\partial_j p$, $|j| \leq d$, is given by

$$\delta(\tilde{z}) = |\partial_j[\tilde{z}]\, \vec{p}| \,/\, [\, \sum_{k \in \tilde{J},\, k \geq j} \varepsilon_k \binom{k}{j} |\tilde{z}|^{k-j} \,], \quad j \in \tilde{J}, \tag{3.65}$$

with the previous notational conventions. Note that the max-norm part of the expression (3.54) for the backward error of an approximate zero of an empirical polynomial is the special case $j = 0$ of (3.65).

Exercises

1. Consider the situation of Example 3.3, but with the task of finding a valid inflection point of $(\vec{p}, e)$.

(a) Given an approximate inflection point $\tilde{z}_{infl}$, find a representation of the equivalent-data manifold $\mathcal{M}(\tilde{z}_{infl}) \subset \Delta\mathcal{A} = \mathbb{C}^4$
    - by a set of equations in the $\Delta\alpha_j$,
    - by a parameter representation.

(b) Determine the (inflection point) backward error δ_{infl} of $\tilde{z}_{infl} = 1.41421$.

(c) By numerical experimentation, modify $\tilde{z}_{infl}$ such that δ_{infl} decreases. Can you find a $\tilde{z}$ with $\delta_{infl}(\tilde{z}) \leq 1$?

2. In the x, y-plane, consider the empirical ellipse which is the variety of $(\bar{p}, e)$, with $\bar{p}(x, y) = 3.02\, x^2 - 2.87\, x + 1.93\, y^2 + .66\, y - 5.31$ and $e = (.005, \ldots, .005)$, and the straight line $t(x, y) = 1.1\, x + 2.1\, y - 4.2 = 0$. We want to verify that t is a valid tangent of the empirical ellipse. (What is meant by "valid tangent"? Plot the situation.)

(a) Represent the ellipses in the neighborhood of $\bar{p}$ by

$$\tilde{p}(x, y) = \bar{p}(x, y) + \Delta\alpha_{20}\, x^2 + \Delta\alpha_{10}\, x + \Delta\alpha_{02}\, y^2 + \Delta\alpha_{01}\, y + \Delta\alpha_{00}.$$

In $\Delta\mathcal{A} = \mathbb{C}^5$, determine the quadratic equation in the $\Delta\alpha_j$ which represents the equivalent-data manifold $\mathcal{M}(t)$. (Hint: Solve the system $\tilde{p}(x, y) = t(x, y) = 0$ and request that the two zeros coincide.)

(b) Determine upper bounds for the minimal weighted norm distance of $\mathcal{M}(t)$ from the origin by finding various small $\Delta a \in \mathcal{M}(t)$. Can you establish the validity of t in this way ?

(c) Linearize the equation of $\mathcal{M}(t)$ and find the minimal norm distance from the origin of the tangential hyperplane. Convince yourself that this distance represents the backward error $\delta(t)$ sufficiently well.

3. Consider once more the empirical polynomial $(\bar{p}, e)$ of Example 3.3. Shift the origin of the x-space to $c = 1.41$.

(a) Find the tolerances $\vec{\varepsilon}_j$ for the coefficients of the shifted polynomial $\vec{p}$ which guarantee that all shifted polynomials from $N_\delta(\bar{p}, e)$ are in $N_\delta(\vec{p}, \vec{e})$.

(b) For which tasks related to $(\bar{p}, e)$ may this shift be helpful, for which is it not ?

3.4   Refinement of Approximate Results

If we have found an approximate result $\tilde{z}$ for an empirical algebraic problem whose backward error $\delta(\tilde{z}) > O(1)$, we would like to compute a *correction* Δz such that $\tilde{z} + \Delta z$ is, hopefully, a valid approximate result or has, at least, a significantly reduced backward error. This task has hardly been considered in classical algebra, because the mere concept of an approximate result and its improvement belongs to analysis rather than to algebra. In numerical analysis, on the other hand, the refinement of an approximate result is one of the most fundamental tasks; it has been at the center of attention in all areas of computational mathematics, including numerical linear algebra, and it is also a central task in numerical polynomial algebra.

The only situation where iterative improvement has been used in polynomial algebra for centuries is the numerical computation of zeros of a univariate polynomial. Since Abel's famous result it has been well known that zeros of polynomials of a degree higher than 4 cannot—generally—be represented in a closed form which permits numerical evaluation for numerically specified coefficients. And although such representations exist for 3rd- and 4th-degree polynomials, their numerical evaluation is not convenient and requires approximate computation. Also, in the real and complex domain, a natural metric is provided by the modulus.

Therefore, a great number of approaches for the iterative improvement of an approximate value for a particular zero of a univariate polynomial of an arbitrary degree were developed.

Many of them, in one way or other, subdivide a domain in $\mathbb{C}$ or $\mathbb{R}$ which is known to contain all zeros into subdomains containing only a certain set of zeros. These are then further refined until a sufficiently accurate approximation for one particular zero has been obtained. These approaches have been presented in classical textbooks, and we will not attempt to characterize them here. Most of these approaches are tied to the determination of a zero of a univariate polynomial; they cannot readily be extended to the iterative refinement of approximate solutions of more general tasks in polynomial algebra. Furthermore, in agreement with the assumption of exact coefficients, the emphasis is on fast convergence to "arbitrarily" accurate values for the zeros.

With empirical data, on the other hand, high accuracy is generally meaningless; cf. section 3.2. What is needed is the transition from an approximate but moderately invalid result, with a backward error in the 10's or 100's, to a valid result, or from a borderline valid result to a safely valid one. In this context, we rarely wish to take more than one or two steps of such a refinement procedure. And we need a scheme which can be adapted to a great variety of situations and tasks.

Such a scheme is provided by *local linearization* or *Newton's method* which is a very general method in function space. It can be applied whenever the exact result of a mathematical task (not necessarily an algebraic one) may be characterized by the vanishing of some functional image of the result, with weak assumptions on the mapping which defines that image. The simplest and best-known example is, of course, the iterative improvement of an approximate zero of some function $f : \mathbb{C} \to \mathbb{C}$; here, z^* is an exact zero iff $f(z^*) = 0$. The multivariate analog is immediate: For $f : \mathbb{C}^s \to \mathbb{C}^s$, an exact zero $z^* \in \mathbb{C}^s$ satisfies $f(z^*) = 0$.

As an example of a less immediate task of this kind, consider the determination of a *divisor* of a univariate polynomial $p \in \mathcal{P}_n^1$: Here, the function $f : \mathbb{C}^m \to \mathbb{C}^m$ assigns to a monic $s \in \mathcal{P}_m^1$, $m < n$, the remainder $r \in \mathcal{P}_{m-1}^1$ in

$$p(x) \; = \; s(x)\, q(x) + r(x)\,. \tag{3.66}$$

A polynomial s^* of degree m is an *exact* divisor of p iff f maps s^* into the zero polynomial. This situation and its multivariate generalizations will be treated at their appropriate place in this book; cf., e.g., sections 6.2.3, 9.2.2, and 9.3.2.

As a general setting for the application of the *Newton refinement scheme*, we assume—as in section 3.2.2—that our well-posed problem with data $a \in \mathbb{C}^M$ and results $z \in \mathbb{C}^m$ may be formulated as a system of m equations (cf. (3.35))

$$G(x; a) \; = \; 0\,, \qquad G : \mathbb{C}^m \times \mathbb{C}^M \to \mathbb{C}^m\,,$$

where G is Fréchet differentiable w.r.t. both x and a in $Z_{\bar{\delta}}(\bar{a}, e) \times N_{\bar{\delta}}(\bar{a}, e)$; cf. (3.35)–(3.38) and Proposition 3.4. For empirical data $(\bar{a}, e)$ and an approximate result $\tilde{z}$, we have

$$G(\tilde{z}; \bar{a}) \; = \; r \; \in \; \mathbb{C}^m\,,$$

with r not sufficiently small to ascertain the validity of $\tilde{z}$. We want to determine a correction Δz such that

$$0 \; \approx \; G(\tilde{z} + \Delta z;\, \bar{a}) \; = \; G(\tilde{z};\, \bar{a}) + \frac{\partial G}{\partial x}(\tilde{z};\, \bar{a}) \cdot \Delta z + \mathrm{O}(\|\Delta z\|^2)\,. \tag{3.67}$$

The core idea of a Newton refinement step is to linearize (3.67) by neglecting the quadratic term and to solve the *linear system*

$$\frac{\partial G}{\partial x}(\tilde{z}; \bar{a}) \cdot \Delta z \;=\; - G(\tilde{z}; \bar{a}) \;=\; -r \tag{3.68}$$

for Δz, assuming the regularity of the $m \times m$ matrix $\frac{\partial G}{\partial x}(\tilde{z}; \bar{a})$, as in Corollary 3.5.

With a slightly more stringent definition of well-posedness than in Definition 3.9, the nonsingularity of the linear mapping $\frac{\partial G}{\partial x}(\tilde{z}; \bar{a}) \,:\, \mathbb{C}^m \to \mathbb{C}^m$ would be implied for all $\tilde{z} \in Z_\delta(\bar{a}, e)$ with sufficiently small δ. But in our context, we would not employ Newton's approach if the backward error of $\tilde{z}$ were small. Therefore, we rather make the explicit assumption about the regularity of the Jacobian which comes to light in the solution of (3.68) anyway.

Proposition 3.15. Assume that $\frac{\partial G}{\partial x}(\tilde{z}; \bar{a})$ is regular, with $\|[\frac{\partial G}{\partial x}(\tilde{z}; \bar{a})]^{-1}\| =: K$, and that $\frac{\partial G}{\partial x}(x; \bar{a})$ satisfies a Lipschitz condition with respect to x with Lipschitz constant L', in a sufficiently large neighborhood of $\tilde{z}$. Then, with Δz from (3.68),

$$\|G(\tilde{z} + \Delta z; \bar{a})\| \;\le\; \frac{L'}{2}\, K^2\, \|r\|^2 \,. \tag{3.69}$$

Proof: The proof follows immediately from (3.24) in Theorem 3.2 applied to G and from $\|\Delta z\| \le K\, \|r\|$. The norms are from the image space of G; accordingly, the operator norm for the inverse of the Jacobian is for linear maps from $\mathcal{Z}$ to the image of G. $\quad\square$

The bound (3.69) for the reduced residual after one Newton refinement step displays the potential *obstacles* for success:

- The *residual r* of the initial approximate result $\tilde{z}$ may be too large. (A scaling which reduces $\|r\|$ will generally increase K by the same factor.)

- The *Jacobian $\frac{\partial G}{\partial x}$* may be near-singular so that K is too large. (Furthermore, the linear system (3.68) is ill-conditioned in this case.)

- The situation may be *strongly nonlinear* so that the Jacobian $\frac{\partial G}{\partial x}$ changes rapidly with x and L' is too large.

These are the standard restrictions for Newton's method; they must be excluded wherever Newton's method can be applied.

Example 3.16: A successful refinement of an approximate zero $\tilde{z}$ of a *univariate* polynomial p depends on a small deviation between the zeros of p and of its tangent at $\tilde{z}$. This deviation may become large if

- the residual $r = p(\tilde{z})$ is large,

- the slope $p'(\tilde{z})$ of the tangent is small,

- the variation of $p'(x)$ near $\tilde{z}$ is large. $\quad\square$

In a well-posed empirical algebraic problem, if the approximate result $\tilde{z}$ has a moderate backward error $\delta(\tilde{z})$, then one Newton correction Δz will generally reduce $\delta(\tilde{z} + \Delta z)$ to O(1). On the other hand, if no significant reduction is achieved, the situation is probably not suitable for the use of linearization which is the basis of Newton refinement.

Exercises

1. Assume that the coefficients of $\bar{p}$ in (3.17) are exact and attempt to determine highly accurate approximations of the 3 positive zeros of $\bar{p}$ by Newton iteration from some chosen initial approximation $\tilde{z}_0$ near $\sqrt{2}$.

(a) Vary $\tilde{z}_0$ and observe the generated sequence of approximates.

(b) Vary the number of digits used in a) and observe potential effects.

(c) For some $\tilde{z}_0$, determine r, K, and L' of Proposition 3.15 and compare (3.69) with the computed value.

2. Consider the situation of Example 3.13.

(a) For $n = 4$, $m = 3$, determine the function $f : \mathbb{C}^3 \to \mathbb{C}^3$ which maps s into r in (3.66). What are the argument and result components in this case? What are $\tilde{z}$ and $\bar{a}$ in this situation?

(b) Perform one Newton refinement step to correct $\tilde{s}$ of Example 3.13 into a valid divisor of $(\bar{p}, e)$. Compute the backward error of $\tilde{s} + \Delta s$.

Historical and Bibliographical Notes 3

From its very beginnings, the computational solution of application problems has had to deal with data of limited accuracy. A formalization became necessary when reliable answers were sought as in Astronomy and Surveying. C. F. Gauss was active in both areas; he made statistical assumptions on the indetermination of measured data: the famous normal distribution. Fuzzy sets are a more recent generalization of this approach; cf., e.g., [3.1]. Intervals with sharply defined bounds as in [3.2] are rarely adequate models.

Our deformalized statistical model of families of neighborhoods and validity values (section 3.1) follows publications like [3.3], [3.4]; it keeps the focus on the algebraic aspects but embeds them into analysis by means of the data→result mapping. Considerations of the well-posedness of algebraic tasks and of their condition w.r.t. various input quantities are a natural consequence; they have been standard in numerical analysis since the pioneering work [3.5] of J. Wilkinson and are found in any text on numerical analysis. Their use in numerical algebraic computation is an absolute prerequisite, but as yet the exception rather than the rule.

Pseudoresult sets have been introduced and used in interval mathematics since its beginnings; cf., e.g., [3.2]. In the family-of-neighborhoods model, they become more realistic and less demanding; for algebraic tasks, they have been introduced, e.g., by [3.3], [3.4]. The linearized estimation of their extension is another standard approach in numerical analysis.

The formal introduction of the backward error is due to [3.5]. Specific versions of the expressions in Propositions 3.9 and 3.10 may be found with many authors. In 1964, Oettli and Prager established (3.52) for an approximate zero of a linear system, with the 1-norm, in their seminal publications [3.6]. For a zero of a univariate polynomial and the 1-norm, it is found in [3.3]; the 2-norm version is used in [3.4]. In the volume [3.7], these results have been put into the more general framework of a posteriori backward error analysis. Our approach shows that explicit expressions exist for all norms if the equivalent-data manifold is linear and of codimension 1. In algebraic problems, there is often a set of results (like zeros, coefficients) for the same data; analyses of the backward error of such sets (as in Example 3.13) have not come to my attention so far.

The refinement of approximate solutions of nonlinear problems with the aid of local linearization is one of the oldest techniques in applied mathematics as indicated by its "patron" Newton. Convergence proofs for its application to systems of equations appeared in the 1930s. A breakthrough was the extension to functional equations in a Banach space by L.V. Kantorovich in [3.8], 1948.

For readers from the computer algebra community who wish to get an introductory overview of numerical analysis, I would suggest the combination of [3.9] and [3.10]: [3.9] is restricted to numerical linear algebra; it introduces the subject in 40 "lectures," in a very explicit and intuitive language. Reference [3.10] is restricted to topics from analysis; it also features an explicit and intuitive style. Naturally, there are scores of more advanced and technical texts, like [3.11] for numerical linear algebra and [3.12] for ordinary differential equations.

References

[3.1] G.J. Klir, U. St. Clair, B. Yuan: Fuzzy Set Theory - Foundation and Applications, Pearson Education, POD, 1997.

[3.2] R.E. Moore: Interval Analysis, Prentice-Hall, Englewood Cliffs NJ, 1966. and

R.E. Moore: Methods and Applications of Interval Analysis, SIAM, Philadelphia, 1979.

[3.3] R.G. Mosier: Root Neighborhoods of a Polynomial, Math. Comp. **47** (1986), 265–273.

[3.4] K.-C. Toh, L.N. Trefethen: Pseudozeros of Polynomials and Pseudospectra of Companion Matrices, Numer. Math. **68** (1994), 403–425.

[3.5] J. Wilkinson: Rounding Errors in Algebraic Processes, Prentice-Hall, Englewood Cliffs, NJ, 1963.

[3.6] W. Oettli, W. Prager: Compatibility of Approximate Solutions of Linear Equations with Given Error Bounds for Coefficients and Right Hand Sides, Numer. Math. **6** (1964), 405–409 and

W. Oettli: On the Solution Set of a Linear System with Inaccurate Coefficients, J. Soc. Indust. Appl. Math. Ser. B Numer. Anal. **2** (1965), 115–118.

[3.7] F. Chaitin-Chatelin, V. Fraysse: Lectures on Finite Precision Computations, SIAM, Philadelphia, 1996.

[3.8] L.V. Kantorovich: Functional Amalysis and Applied Mathematics (Russian), Uspekhi Mat. Nauk **3** (1948).

[3.9] L.N. Trefethen, D. Bau: Numerical Linear Algebra, SIAM, Philadelphia, 1997.

[3.10] W. Gautschi: Numerical Analysis - An Introduction, Birkhäuser, Berlin, 1997.

[3.11] G.H. Golub, Ch.F. Van Loan: Matrix Computations, 3rd Ed., John Hopkins Univ. Press, Baltimore, 1996.

[3.12] E. Hairer, S.P. Nørsett, G.Wanner: Solving Ordinary Differential Equations I, 2nd Ed.,
Springer, New York, 1993 and

E. Hairer, G. Wanner: Solving Ordinary Differential Equations II, 2nd Ed., Springer, New
York, 1996.

[3.13] I.Z. Emiris, B. Mourrain: Computer Algebra Methods for Studying and Computing
Molecular Conformations, Algorithmica **25** (1999), 372–402.

Chapter 4

Approximate Numerical Computation

4.1 Solution Algorithms for Numerical Algebraic Problems

A *numerical algebraic problem* assigns to a set of numerical data from a data domain A in the *data space* $\mathcal{A}$ a set of numerical results in the *result space* $\mathcal{Z}$. More formally (cf. Definition 3.6), it defines a data$\rightarrow$result mapping F from a domain in $\mathcal{A}$ into $\mathcal{Z}$, where both the data space $\mathcal{A}$ and the result space $\mathcal{Z}$ are product spaces of real or complex numbers. Generally, this data$\rightarrow$result mapping can only be specified *implicitly*, e.g., by a polynomial whose coefficients are the data and whose zeros are the results of the algebraic problem; cf. (3.35) in section 3.2.2. Nevertheless, many mathematical properties (algebraic, analytic, numerical) of the numerical algebraic problem, or of its data$\rightarrow$result mapping, resp., may be derived from this implicit formulation. But generally, it does not permit the immediate numerical evaluation of the image $z \in \mathcal{Z}$ of specified numerical data $a \in \mathcal{A}$.

When we consider this numerical evaluation of the data$\rightarrow$result mapping F for specified numerical data, we must clarify what we mean by numerical data.

Definition 4.1. For the purpose of this book, *numerical data* are ordered sets (vectors) of real or complex numbers α_μ which are *finite decimal fractions*. (This includes integers and finite binary fractions, like floating-point numbers in a standard representation.) $\square$

Note that this *excludes* general rational numbers, like 3/7, and general algebraic numbers, like the `RootOf ( .. )` numbers of Maple. For specified values of empirical data this is no restriction. If a problem formulation contains intrinsic data of this kind which enter the numerical computation, they must either be approximated by floating-point numbers or entered through the backdoor by a reformulation of the problem, e.g., by multiplication of an equation by the least common denominator of its rational coefficients or by appending some polynomials to the problem formulation. An approximation modifies the particular data value and the effect of this may have to be taken into account.

The results of empirical algebraic problems always carry some indetermination (cf. Definition 3.8); therefore it is sufficient to compute approximate values for them and to establish their validity (cf. section 3.3.1). Generally, approximate results with a backward error moderately too large may be refined into valid results; cf. section 3.4. This suggests that many algebraic problems with numerical data may be solved by an algorithmic procedure which follows the following pattern.

Algorithmic Scheme 4.1.

Step 1: Compute a (possibly crude) *approximate result* $\tilde{z}$ by some approximate solution procedure for the problem.

Step 2: Compute the *backward error* $\delta(\tilde{z})$ of $\tilde{z}$; **if** $\delta(\tilde{z}) = O(1)$ **then** report $\tilde{z}$; **stop** .

Step 3: Compute a *correction* Δz by a refinement step; set $\tilde{z} := \tilde{z} + \Delta z$; **go to** step 2 .

Naturally, this scheme must be supplemented by appropriate control procedures which terminate the algorithm in case of nonconvergence.

In many situations, Step 1 serves to analyze the *global structure* of the specified problem and to reach a situation where the *local structure* becomes dominant for the further computation. The precise values at which we arrive in this "local domain" are often irrelevant. Thus, it appears particularly unreasonable to employ exact computation in the *beginning phase* of the computational solution of an algebraic problem when we are still far away from the situation in which the final approach to a sufficiently accurate result takes place.

(When we drive to a distant geographic location, the precise route which we initially take is not important as long as it leads us into a proper vicinity of our goal. *There*, we must be increasingly careful to take the right turns if we wish to arrive at the correct site.)

In Step 2, the checking of the quality of a provisional approximate result must naturally be done against the *original specification* of the problem. Perturbations which have been introduced into the computation by approximations of various kinds will come to light in Step 2; in all but singularly sensitive situations, they will be compensated in the subsequent Step 3. But for the refinement to be effective, the residuals or remainders computed in Step 2 and used in Step 3 must be reliable; it may be necessary to use a higher precision for their computation. In Step 3 itself, the necessary precision depends on the circumstances. We will return to this point in section 4.3.4.

This computational model is in contrast to the mainstream of computer algebra, where the attention is focused on algorithms which generate exact results for exact data. This focus introduces a classification of problems into

(i) Problems with *exact solution algorithms*: These problems admit algorithms which generate the *exact results* of the algebraic problem, for data from an appropriate data domain.

(ii) Problems with *asymptotic solution algorithms*: These problems admit algorithms such that—for data from an appropriate data domain and for a specified $\Delta > 0$—the result $\tilde{z}$ generated by the algorithm satisfies $\|\tilde{z} - F(a)\| \leq \Delta$ where $\|..\|$ is a norm in the result space $\mathcal{Z}$. The number of operations in the algorithm generally tends to ∞ for $\Delta \to 0$.

The following are well-known examples of algebraic problems of category (i) and (ii), resp.:

(i) systems of linear equations, greatest common divisors of univariate polynomials, border bases of multivariate polynomial systems (e.g., Groebner bases);

(ii) polynomial zeros, matrix eigenvalue problems, singular-value decomposition of matrices.

But for most purposes, the distinction of the problems in category (i) is *irrelevant*:

- Although an exact solution algorithm will generate the exact rational result value z, the *representation* of z may be impractical.

- The *computational cost* of the exact solution algorithm may be unduly large while a reasonable approximate result $\tilde{z}$ can be obtained cheaper with an asymptotic solution algorithm.

Example 4.1: Consider the linear system

$$
\begin{pmatrix}
-85 & -55 & -37 & -35 & 97 & 50 & 79 & 56 & 49 & 63 \\
57 & -59 & 45 & -8 & -93 & 92 & 43 & -62 & 77 & 66 \\
54 & -5 & 99 & -61 & -50 & -12 & -18 & 31 & -26 & -62 \\
1 & -47 & -91 & -47 & -61 & 41 & -58 & -90 & 53 & -1 \\
94 & 83 & -86 & 23 & -84 & 19 & -50 & 88 & -53 & 85 \\
49 & 78 & 17 & 72 & -99 & -85 & -86 & 30 & 80 & 72 \\
66 & -29 & -91 & -53 & -19 & -47 & 68 & -72 & -87 & 79 \\
43 & -66 & -53 & -61 & -23 & -37 & 31 & -34 & -42 & 88 \\
-76 & -65 & 25 & 28 & -61 & -60 & 9 & 29 & -66 & -32 \\
78 & 39 & 94 & 68 & -17 & -98 & -36 & 40 & 22 & 5
\end{pmatrix}
\; x =
\begin{pmatrix}
-88 \\ -43 \\ -73 \\ 25 \\ 4 \\ -59 \\ 62 \\ -55 \\ 25 \\ 9
\end{pmatrix}
$$

$$(4.1)$$

whose integer matrix elements and right-hand sides have been generated by a random procedure. The *exact* solution is

$$
z = \left(\frac{7828344912434084 76131}{871457446318467875527}, \; \frac{-2485679712713251977 81}{871457446318467875527}, \; \frac{-1141741239586916224104}{871457446318467875527}, \; \dots \right)^T ;
$$

but for almost all purposes, an approximate result like

$$
\tilde{z} = (.8983049, -.2852325, -1.3101515, \dots, -1.6168161)^T
$$

is fully satisfactory, and one would not compute the exact result to round it to the approximate one. $\square$

Example 4.2: The determination of a border basis is, computationally, a linear process; therefore the exact coefficients of the basis polynomials are rational numbers, and they may be determined by exact computation. The following system of two quadratic equations in two variables describes the intersection of two ellipses (cf. Figure 4.1 in section 4.2.3); the coefficients were obtained from trigonometric function values by highly accurate rational approximation:

$$
p_1(x, y) \; := \; -4 + 3 \cdot \left(\tfrac{172966043}{174178537} x - \tfrac{42176556}{358072327} y \right)^2 + \left(\tfrac{1}{3} + \tfrac{42176556}{358072327} x + \tfrac{172966043}{174178537} y \right)^2 ;
$$

$$
p_2(x, y) \; := \; -4 + \left(\tfrac{1}{3} - \tfrac{42176556}{358072327} y + \tfrac{172966043}{174178537} x \right)^2 + 4 \cdot \left(\tfrac{172966043}{174178537} y + \tfrac{42176556}{358072327} x \right)^2 .
$$

$$(4.2)$$

For this system, the exact unnormalized `plex(y,x)` Groebner basis, with integer coefficients, is

$$
\begin{aligned}
g_1(x) \; = \; & 44876973556839568016 \dots \quad \dots 31345109387246215125 \, x^4 - \\
& 60424891988994024351 \dots \quad \dots 84140992854205425900 \, x^3 - \\
& 94567705076758430878 \dots \quad \dots 34282401394672776250 \, x^2 + \\
& 10875251578559783152 \dots \quad \dots 26595945775784882084 \, x + \\
& 45288589154165595222 \dots \quad \dots 89025539054727808429 \, , \\[1em]
g_2(x, y) \; = \; & 86258840277318527495 \dots \quad \dots 86362495534440858984 \, y - \\
& 11689400385422846400 \dots \quad \dots 52400754665553496625 \, x^3 - \\
& 10696741293834117608 \dots \quad \dots 76330609374373593000 \, x^2 + \\
& 13404023887808939206 \dots \quad \dots 55701166384938121263 \, x + \\
& 11237804869699497732 \dots \quad \dots 67607599906195374410 \, .
\end{aligned}
$$

In the position of the dots, there are between 63 and 80 (!) further digits. Only by counting the number of digits in the individual coefficients, one may discover that the coefficients of the x-powers in g_2 are by $O(10^{16})$ larger than the coefficient of y, and one can, at best, divine the first digit of a *normalized* decimal representation of this Groebner basis, which is (rounded)

$$\tilde{g}_1(x) \approx x^4 - .134645648\,x^3 - 2.10726565\,x^2 + .242334781\,x + 1.00917209\,;$$

$$\tilde{g}_2(x, y) \approx y - (1.35515390\,x^3 + 1.24007479\,x^2 - 1.55393045\,x - 1.30280037) \cdot 10^{16}\,.$$

$$(4.3)$$

Considering the fact that the system (4.2) is really quasi-empirical because of the preceding substitution of approximating rational numbers for exact irrational data and the fact that the zeros of g_1 can only be computed approximately in any case, the generation of the altogether 1180 digits in g_1, g_2 appears as an unreasonable waste of computation. □

Thus, even in the rare event of an intrinsic algebraic problem with an exact solution algorithm, it is generally not meaningful to compute the exact rational results; for many purposes, these must be approximated by decimal fractions anyway. For intrinsic problems without exact solution algorithms, like zeros of polynomials, exact numerical results cannot be computed. For empirical problems which constitute the overwhelming majority of numerical algebraic problems, exact results are *not defined*. Therefore we summarize:

It is the goal of a solution algorithm for a numerical algebraic problem to generate *sufficiently good approximations* for the results of the specified algebraic problem.

This relaxation of the strict mathematical task of finding exact solutions for algebraic problems permits a more relaxed attitude towards the numerical execution of the solution algorithm: We may use *approximate* operations in place of exact ones! In particular, this means that we may use *floating-point* computation in place of rational (= integer) computation very widely in solution algorithms for numerical algebraic problems.

It is not from a lack of insight but due to common sense that 99.999...% of all systems of linear equations, from very small to extraordinarily large ones, are solved in floating-point arithmetic although they *could* be solved in exact rational arithmetic. On the other hand, the predominance of rational computation in the solution of computational algebraic problems, even when the data are purely numerical, is mainly due to historical reasons. Presently, most of the more advanced procedures in current computer algebra software systems do not *admit* floating-point data and they make *no use* of floating-point computation.

In Example 4.2, e.g., no current computer algebra system permits us to compute an approximate Groebner basis directly for the approximate system $(\tilde{p}_1, \tilde{p}_2)$ from (4.2)

$$1.027748\,y^2 - 0.467871\,xy + 2.972252\,x^2 + 0.662026\,y + 0.0785252\,x - 3.888889\,,$$
$$3.958378\,y^2 + 0.701807\,xy + 1.041622\,x^2 - 0.0785252\,y + 0.662026\,x - 3.888889\,,$$

$$(4.4)$$

by calling some appropriate procedure.

In the further chapters of this book, we will try to explain how a great number of algebraic problems with numerical results may be solved in floating-point arithmetic, with full reliability and to whatever accuracy is needed or meaningful. The fact that most numerical computations require only a moderate accuracy in their results makes the use of *approximate computation* so natural, even in situations where exact computation is available. The assessment of approximate results through their backward errors does not at all depend on how they have been obtained. Generally, the effects of approximate computation are fully absorbed into the natural indetermination inherent in the problem and do not affect the validity of the results; they are thus *just as acceptable* as results obtained by exact computation.

Exercises

1. The quo/rem combination of procedures in Maple computes the quotient and the remainder of univariate polynomial division: For specified polynomials p and s, it generates q and r such that $p = q \cdot s + r$. It works for decimal fraction coefficients as well as for rational coefficients.

(a) Choose some p and s with rational coefficients (with nontrivial denominators) and compute the exact q, r. For various choices of Digits, round p, s into decimal approximations $\tilde{p}$, $\tilde{s}$ and apply quo/rem once more. Compare the results to the rounded exact results.

(b) When we regard only $\tilde{p}$ as the specified part of an empirical polynomial but $\tilde{s}$ as intrinsic, it is easy to find the backward error of the computed decimal results; why? Convince yourself that the backward error of the decimal results is O(1), or find an example where this does not hold. Why must the backward error be smaller when s is also considered as empirical?

2. Consider a 3×3 system of linear equations $(A_0 + w\, A_1)\, \mathbf{x} = b_0 + w\, b_1$ with an indeterminate w and chosen decimal fractions in the elements of the matrices and right-hand sides. What kind of expressions do you expect for the components of the solution $\mathbf{x}(w)$?

(a) Find the solution by Maple's solve. How can you verify that the solution is correct within round-off? Find several ways.

(b) Assume that the elements in the matrices and right-hand sides are the specified values of empirical quantities, with appropriate tolerances. How would you define the backward error of a computed result of the kind found in a).

(c) Find the formal power series in w for $(A_0 + w\, A_1)^{-1}$ and form the corresponding power series for $\mathbf{x}(w)$. Evaluate the first few terms numerically and compare the result with the Taylor expansion of $\mathbf{x}(w)$ from a). From the numerical evidence, for which values of w do you expect the Taylor series to converge?

4.2 Numerical Stability of Computational Algorithms

4.2.1 Generation and Propagation of Computational Errors

Throughout computational mathematics, the word "error" is used as a *technical term*: It does *not* refer to a faulty action by a human being or a computer; rather it denotes a deviation of a computed value from a mathematically defined "true" value which occurs because of the deviation of an action of a (human or electronic) computer from a mathematically defined action. Since we assume that both actions are strictly deterministic, errors in the sense of computational mathematics are fully *reproducible*; they will arise in precisely the same fashion whenever a particular implementation of a particular algorithm is activated with a specified set of data.

For example, when we replace $\sqrt{2}$ by 1.41421 in some computation about a geometrical object, we commit a computational error. The action is deliberate: To obtain a numerical answer, we must perform this (or an analogous) replacement. As a consequence, the final numerical result will not coincide with the mathematically defined result. It is clear that this is not an "error" in the sense of conversational language. We will always use the term *error* with that meaning; cf. Definition 4.2.

In dealing with computational errors, one has to distinguish clearly between the generation of an error at some point of an algorithm and the errors (= deviations) which appear as a consequence of this error during the further execution of the algorithm. When we replace $\sqrt{2}$ by 1.41421 we *generate* an error in the quantity which takes that value; by continuing the computation with this value, we *propagate* that error. The effect which it has on further intermediate and final results depends strongly on the use of that quantity in the remainder of the algorithm: It may change many digits of some result quantity, or just a few digits, or it may leave all meaningful digits of that result unaltered.

For a formal treatment of computational errors, we must subdivide the flow of the computation during the execution of a particular algorithm into *computational steps*. Each such step has numerical *input* and *output* quantities; the computation within the step transforms (maps) specified input data into well-defined output data.

Definition 4.2. A *computational step* in a numerical algorithm is a map φ from a set of *input quantities* x_ν (from specified domains) to a set of *output quantities* y_μ; this map constitutes the *exact operation* of that computational step. An *implementation* of this computational step specifies an *approximate operation*, i.e. another map $\tilde{\varphi}$ of the input to the output quantities, often on subdomains of the original domains only. The *generated computational error e* in that computational step is the *difference* between the images of these two maps for arguments in the joint domains:

$$e(x_1, \ldots, x_n) := \tilde{\varphi}(x_1, \ldots, x_n) - \varphi(x_1, \ldots, x_n) \, . \tag{4.5}$$

The sign of the generated computational error is a matter of choice, and there is no universally agreed convention; we use the above sign in this book.   □

Example 4.3: In some algorithm, we may consider the computation of the Euclidean norm $\|x\|$ for $x \in \mathbb{R}^{100}$ as one computational step. Denote by $\mathbb{R}_{sing}$ the single-precision floating-point numbers; cf. section 4.3.1. In an implementation, the exact operation $\|x\|$ may be replaced by the mapping which assigns to $x \in \mathbb{R}^{100}_{sing}$ the result of the floating-point computation

$$\tilde{s} := 0; \qquad \textbf{for } \nu = 1(1)100 \textbf{ do } \tilde{s} := \tilde{x}_\nu \,\tilde{\times}\, \tilde{x}_\nu \,\tilde{+}\, \tilde{s}; \qquad \tilde{y} := \widetilde{\sqrt{\tilde{s}}} \, ;$$

(cf. section 4.3). The generated computational error in this computational step is, for $\tilde{x} \in \mathbb{R}^{100}_{sing}$, $e(\tilde{x}) := \tilde{y}(\tilde{x}) - \|\tilde{x}\|$.

It is well known (cf. section 4.3.2) that this approximate operation is *not* symmetric in the components $\tilde{x}_\nu$ of $\tilde{x}$ but that its result depends on the sequence in which the approximate squares of the components are added. Also there are vectors $\tilde{x} \in \mathbb{R}^{100}_{sing}$ for which our approximate operation *fails* due to exponent overflow; this is not of interest in the present context.   □

The ambiguity of the subdivision of an algorithm into computational steps is well displayed by Example 4.3 : The considered step could be further subdivided into 201 elementary computational steps, viz. 100 muliplications, 100 additions, and one squareroot. On the other hand, the norm computation may be contained in a section of the algorithm which is a natural choice for *one* computational step. The specification of the subdivisions of a computational procedure depends on the purpose of the analysis.

If we choose the maximally coarse subdivision by considering the whole procedure as one step we may not be able to reach any conclusions because of its complexity. If we choose each elementary operation as a separate step, we will be drenched in uninteresting detail. The considerate choice of the subdivision is essential for the derivation of meaningful assertions, particularly with respect to the propagated error.

When we consider the complete sequence of computational steps which compose some given algorithm, a number of conceptual difficulties arise. In a strict sense, the above definition of a generated computational error makes sense only for the first computational step; in the following steps, the arguments of the exact and of the approximate operation will generally no longer coincide; also, the results of the exact operations may not be in the domain of the approximate operation. It is customary to disregard these difficulties except in some very particular investigations; instead, the following assumption is made.

The *properties* of the computational procedure along the path taken by the approximate computation and the path taken by the exact computation coincide sufficiently so that they need not be distinguished.

This assumption which appears unrealistic and naive at first sight is justified for the widely practiced form of error analysis: The properties of the computational steps are considered not for a specific set of input data but for rather wide domains, and *bounds* for the errors are considered rather than specific error values. This recourse to bounds in place of values is also necessary because the actual behavior of computational errors is often very erratic which makes a more detailed analysis hopeless. Error bounds, on the other hand, are generally insensitive against small variations in the computational path along which they are considered. Naturally, this assumes that the data→result mapping represented by the algorithm is Lipschitz continuous in a sufficiently large neighborhood of the given data.

Such an approach is also consistent with the fact that the analysis of the *error propagation* has to be restricted to a *first order analysis* in general. This means that the propagation of the generated error from each individual computational step is regarded independently and that cross-effects between the errors from different computational steps are disregarded. For the analysis of small perturbations, this is standard practice throughout mathematics. Again, this relies on the assumption made above that the behavior of the algorithmic procedure is not extremely sensitive to the precise computational path taken.

For the complete algorithm and its implementation, we have the following situation:

$$\begin{array}{lccl}
\text{computational step 1 generates an error } e_1, & & & \\
\cdots \quad 2 & \cdots & e_2, & \text{and propagates } e_1 \\
\cdots \quad 3 & \cdots & e_3, & \text{and propagates } e_1, e_2 \\
& \cdots & & \\
\cdots \quad N & \cdots & e_N, & \text{and propagates } e_1, \ldots, e_{N-1}
\end{array}$$

We denote the propagated effect of the generated error e_ν at the end of step n by $e_{\nu n}$ and identify $e_{\nu\nu}$ with e_ν. Then, at the end of step n, the first order difference between the values $\tilde{y}_n$ generated in the implementation and the exact values y_n is

$$\tilde{y}_n - y_n = \sum_{\nu=1}^{n} e_{\nu n} . \tag{4.6}$$

Note that (4.6) is rather a symbolic statement than an equation because the sets of input and output quantities may vary from step to step and so will the meaning of the $e_{\nu n}$. A strict formal treatment taking all this into account would obscure the situation rather than clarify it. In each particular case, it is straightforward to establish the correct meaning of the above description. This holds as well for the further statements about the propagated error.

4.2.2   Numerical Stability

At the end of the computational procedure, we have

$$\tilde{y}_N - y_N = \sum_{\nu=1}^{N} e_{\nu N} \quad \text{and} \quad \|\tilde{y}_N - y_N\| \leq \sum_{\nu=1}^{N} \|e_{\nu N}\| . \tag{4.7}$$

In order to make use of (4.7), we need a quantitative estimate of the contributions $e_{\nu N}$ of the individual generated computational errors e_ν to the total error of the final computed result $\tilde{y}_N$, i.e. its deviation from the exact result y_N. For this purpose, we consider the data$\rightarrow$result mappings F_ν (cf. Definition 3.6) which map the exact results y_ν of step ν into the exact final result y_N of the algorithm. If the steps $\nu + 1$ through N use the values of quantities which were generated in steps prior to step ν but did not appear in step ν, we include these quantities formally into the input and output of step ν. Thus, we consider the following data transformations:

$$
\begin{array}{ll}
\text{data } a & y_N = F(a) \\
\varphi_1 : \ \downarrow & \\
\quad \text{results } y_1 & y_N = F_1(y_1) \\
\varphi_2 : \ \downarrow & \\
\quad \text{results } y_2 & y_N = F_2(y_2) \\
\qquad \cdots & \qquad \cdots \\
\varphi_\nu : \ \downarrow & \\
\quad \text{results } y_\nu & y_N = F_\nu(y_\nu) \\
\varphi_{\nu+1} : \ \downarrow & \\
\qquad \cdots & \qquad \cdots \\
\quad \text{results } y_{N-1} & y_N = F_{N-1}(y_{N-1}) \\
\varphi_N : \ \downarrow & \\
\quad \text{results } y_N &
\end{array}
$$

Now we consider an approximate realization $\tilde{\varphi}_\nu$ of some particular step ν, under the assumption that all previous steps have been executed exactly; this will generate values $\tilde{y}_\nu = y_\nu + e_\nu$ in place of y_ν. We interpret the generated error e_ν as a *data perturbation* of the "data" y_ν of the data$\rightarrow$result mapping F_ν. This shows that the effect $e_{\nu N}$ of the generated error e_ν on the final result y_N depends on the *condition numbers* C_ν of the mappings F_ν:

$$\|e_{\nu N}\| \leq C_\nu \cdot \|e_\nu\| ,$$

with norms and condition numbers properly adjusted (cf. Definition 3.10). With our restriction to first order effects in (4.7), this yields

$$\|\tilde{y}_N - y_N\| \le \sum_{\nu=1}^{N} \|e_{\nu N}\| \le \sum_{\nu=1}^{N-1} C_\nu \cdot \|e_\nu\| + \|e_N\| . \tag{4.8}$$

Again, the bound (4.8) is not intended for an actual quantitative estimation of the total effect of the computational errors within an implementation of an algorithm; since all quantities in (4.8) represent worst case bounds, their superposition would generally give a bound which is unrealistically large, often by orders of magnitude. Rather, (4.8) displays the qualitative structure of the error generation and propagation in an approximate realization of an algorithm: The contribution of the error generated in the ν-th computational step to the error of the final computed result $\tilde{y}_N$ depends on the *condition* of the particular data$\rightarrow$result mapping F_ν defined above.

While the sizes of the generated computational errors e_ν may be controlled by a considerate *implementation* of the computational steps, their propagation to the final result depends only on the structure of the *algorithm*: For a particular algorithm, with specified data a, the condition numbers C_ν of the various data$\rightarrow$result mappings F_ν are well determined. Thus, it must be a primary goal of good algorithmic design to ensure that the mappings F_ν are well-conditioned.

At this point, it may seem that the result of our error propagation analysis may strongly depend on the primary subdivision of the algorithm into computational steps. However, this is not the case: Perturbation sensitivity is essentially governed by the derivatives of data$\rightarrow$result mappings (cf. section 3.2.2), and derivatives obey the *chain rule* which says that the derivative of a composite map is the product of the derivatives of the component maps.

This chain rule structure of the perturbation sensitivity in composite mappings is also the reason why it cannot be expected that the conditions of *all* intermediate data$\rightarrow$result mappings F_ν are markedly better than the overall condition of the complete algorithm. Rather, the condition C of the overall data$\rightarrow$result mapping F must serve as a reference level for the assessment of the sensitivities of the mappings F_ν with respect to *their* data y_ν.

Definition 4.3. An algorithm for the solution of a numerical problem is called *numerically stable* if *none* of its partial data$\rightarrow$result mappings F_ν has a significantly worse condition than the data$\rightarrow$result mapping F of the problem, within the domain of data for which the algorithm is supposed to be used. Otherwise, i.e. if there exists at least one F_ν whose condition is markedly worse than that of F, the algorithm is called *numerically unstable.* $\square$
(Like in our validity scale (3.3), the inequalities which appear verbally in this definition are to be interpreted in an order-of-magnitude sense: For numerical stability, none of the condition numbers C_ν should exceed C by a factor $> O(1)$, and there is a *continuous scale* of decreasing numerical stability toward full instability as this factor increases.)

More important is another aspect of Definition 4.3 and the discussion preceeding it: Should we use the sensitivity to errors in the *absolute* or the *relative* sense; cf. section 3.2.2. The choice must depend on whether we expect to bound the absolute or the relative sizes of the generated computational errors e_ν in the implementation of the algorithm. For computational errors which are mainly caused by the use of floating-point arithmetic, it is the *relative error* which is more appropriate; cf. section 4.3.2. Therefore, in numerical linear algebra, the definition of numerical stability is generally based on *relative condition*.

In the context of algebraic algorithms, the choice is not always so clear. Also, since numerical stability is a qualitative rather than a quantitative concept and since it is a really *bad* numerical instability which we have to avoid, it is often not crucial whether absolute or relative perturbation sensitivity is considered. On the other hand, an important aspect to take into account is the type of accuracy which we wish to achieve in the result: For result values near the origin, e.g., a high relative accuracy will generally be much harder to achieve than a high absolute accuracy.

4.2.3   Causes for Numerical Instability

In place of formal derivations, we discuss an example which displays the two main causes of numerical instability in algebraic algorithms.

Example 4.4: We consider the two ellipses of Example 4.2 defined by (4.2) with exact rational coefficients. From Figure 4.1, we see that all 4 intersection points are *well-conditioned*: Small changes in the positions of the ellipses can only have small effects on the position of the intersections. Formally (cf. Corollary 3.5), the condition of the intersection coordinates is determined by the inverse of the Jacobian $\begin{pmatrix} \frac{\partial p_1}{\partial x} & \frac{\partial p_1}{\partial y} \\ \frac{\partial p_2}{\partial x} & \frac{\partial p_1}{\partial y} \end{pmatrix}$. For p_1, p_2 from (4.2), none of the elements of the inverse Jacobian at an intersection point has a modulus above .2.

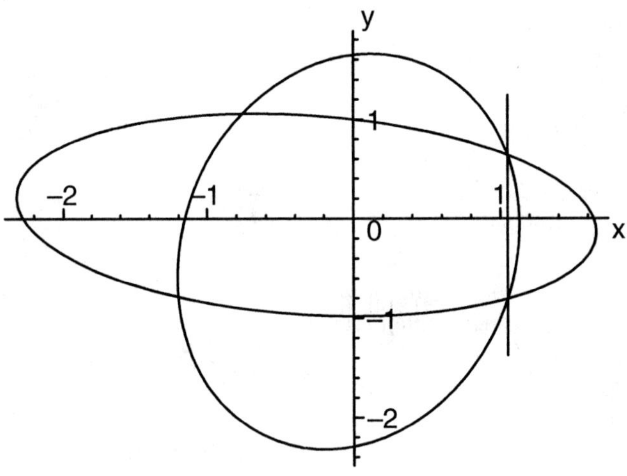

Figure 4.1.

Our proposed algorithm for the computation of approximations of the 4 zeros of (4.2) consists in solving (4.3) which has been obtained from the *exact* plex(y,x) Groebner basis $\{g_1, g_2\}$ by normalization and rounding to 10 digits: We use $\tilde{g}_1$ to obtain the four x-coordinates ξ_μ and then $\tilde{g}_2$ for the associated y-coordinates η_μ, $\mu = 1(1)4$. An execution of this algorithm with the 10-digit decimal floating-point arithmetic of Maple yields the coordinate pairs (rounded)

$$(-1.2044, 2 \cdot 10^7), \ (-.7604, 0), \ (1.04972, -2.1942 \cdot 10^{12}), \ (1.04973, 2.1939 \cdot 10^{12}) \ !$$

While the ξ_μ are reasonable approximations of the exact values, the η_μ are meaningless.

Actually, the two negative ξ_μ-values have 10 correct digits while the nearly coinciding positive ξ_μ have 5 correct digits. But the computation of the η_μ from $\tilde{g}_2$ (or from the exact g_2 as well) amplifies any error in a ξ_μ by $\approx 10^{17}$. Only with more than 17 correct digits in ξ_μ can we expect to get any correct digits in η_μ!

g_2 is an example of extreme *cancellation of leading digits* : In the y-normalized version (4.3), the value of y appears as the sum of 4 terms of $O(10^{16})$, for an $O(1)$ value of x. After substitution of an exact ξ_μ (which is inaccessible), the first 16 or 17 digits of that sum would cancel and the remaining ones represent the exact value of η_μ. The condition number of the map from the "intermediate result" ξ_μ to the "final result" η_μ (cf. (4.8)) is $O(10^{17})$ and the algorithm is extremely numerically unstable; cf. Definition 4.3. Note that the cause for this instability lies in the *exact* polynomial g_2: To *solve for* y in the situation just described must lead to an explosive amplification of any perturbations, from whatever source.

Since we have previously computed the 1180 digits of the exact polynomials g_1, g_2, we can easily repeat the algorithm with `Digits:=20`. As expected, we obtain 3 correct η_μ-digits for the negative ξ_μ; but the close pair of positive ξ_μ has turned into a conjugate-complex pair with tiny (10^{-10}) imaginary parts and the associated η_μ have remained meaningless.

This time the fault is with g_1: The exact positive zeros of g_1 *coincide* in their leading 16 digits. This implies that their positions are extremely ill-conditioned; they react to a perturbation ε in g_1 like $\sqrt{\varepsilon}$! This means that we need some 40 correct digits in the coefficients of $\tilde{g}_1$ to obtain the values of the two positive ξ_μ with ≈ 20 correct digits. But, as we have seen, this is necessary to get a few correct digits for the η_μ from g_2! Actually, when we repeat the algorithm with `Digits:=40`, we obtain 6 correct digits of these η_μ.

The propagation of a computational error (here the rounding of the exact g_1) through *densely clustered zeros* is another main reason for numerical instability in an algorithm for a well-conditioned problem. Obviously, it does not occur for linear problems; but algebraic problems are prone to this fallacy.

In this example[6], the problem cannot be held responsible for either of the two catastrophic instabilities in our solution algorithm for (4.2). When we form the 10-digit rounded normalized version of the exact `tdeg(y,x)` Groebner basis of (4.2), we obtain

$$\begin{aligned}
\tilde{b}_1(x,y) &= x^3 + .9150804097\, x^2 - .738 \cdot 10^{-16}\, y - 1.146681904\, x - .9613670932\,, \\
\tilde{b}_2(x,y) &= y^2 + x^2 + .1662751961\, y + .1417843036\, x - 1.767676768\,, \\
\tilde{b}_3(x,y) &= y\, x - 4.156064804\, x^2 - 1.049726058\, y + .1436149083\, x + 4.428914553\,,
\end{aligned}$$

$$(4.9)$$

which we can easily complete to a full $\mathcal{N}$-border basis for the normal set $\mathcal{N} = \{1, x, y, x^2\}$ by

$$\tilde{b}_4(x,y) = x^2 y - .4159811307\, x^2 - 1.101924797\, y - .186013438\, x + .653643075\,;$$

cf. section 2.5.2. The normalized eigenvectors of the 10-digit multiplication matrix A_y w.r.t. the basis vector $\mathbf{b} = (1, x, y, x^2)^T$ (cf. section 2.4.2) yield the 4 zeros of (4.2) with an accuracy

---

[6]I owe this example to my student W. Windsteiger.

of 8 to 10 decimal digits in *both* components. But even for this basis, when we use the matrix A_x whose theoretically identical eigenvectors are determined via the eigenvalues ξ_μ, the clustered positive ξ_μ corrupt the computation of the associated eigenvectors and hence of the associated η_μ. $\square$

Almost all cases of numerical instability of algebraic algorithms can be traced to the two situations which we have just met:

(i) an expression is *unnecessarily* solved for a quantity which occurs with a *tiny coefficient*;

(ii) a zero (or eigenvalue) *cluster* is introduced *unnecessarily*.

We have stressed the missing necessity because both situations may be unavoidable in the algorithmic solution of a badly *ill-conditioned* algebraic problem. According to Definition 4.3, this does not constitute numerical instability but simply displays the inherent excessive sensitivity of the given task. In a *well-conditioned* problem, however, it must be possible to avoid both situations which corrupt the computational solution unnecessarily.

Both instability generating situations can be *algorithmically diagnosed* during the execution of the algorithm: We can *test* for small "leading" coefficients and for close zeros or eigenvalues, and we can *provide an algorithmic escape route*.

In Example 4.3, it is the use of the term order `plex(y,x)` which unnecessarily steers the computation through the 2-cluster of the projections of the zeros on the x-axis although the intersections of the two algebraic varieties are well-separated in x, y-space. The tiny coefficient property of g_2 is an automatic consequence: For two x-values with an $O(\varepsilon)$ difference, the associated y-values can only have an $O(1)$ difference iff the expression for y has $O(\varepsilon^{-1})$ coefficients ! A change of the term order eliminates the instability.

In the numerical solution of systems of linear equations by elimination algorithms, the occurrence of the situation (i) is largely suppressed by the well-known *pivoting* techniques: For the elimination of a variable, an equation is chosen in which the coefficient of that variable has maximal modulus. For well-conditioned linear systems, this guarantees numerical stability; cf. texts on numerical linear algebra.

It is important to note that, generally, the imminent numerical instability of an algorithm can *only* be detected *during the actual execution* of the algorithm because its occurrence depends also on the *data* with which the algorithm is executed. Typically, a numerical instability in an algebraic algorithm appears only for data from a particular domain which may be a small subdomain of the data domain for which the algorithm is designed. In Example 4.3, no numerical instability will arise for the same algorithm if the configuration in Figure 4.1 is rotated a bit.

We conclude this section with a principal observation about numerical algorithms:

> In order to preserve numerical stability for a wide domain of data,
> an algorithm must provide alternative routes whose activation depends on the data
> and on intermediate results generated in the execution of the algorithm.

Exercises

1.   Choose a polynomial of degree ≥ 4 with 5-decimal-digit coefficients and analyze its evaluation at a chosen 5-decimal-digit argument for two different algorithms:

Algorithm 1: Formation of the powers, formation of the terms, summation;

Algorithm 2: Horner's algorithm.

(a) Define "steps" of the algorithms in various ways and compare the stepwise results arising from a 10-digit decimal arithmetic and from an exact arithmetic. Try to distinguish between generated and propagated errors. Compare their orders of magnitude.

(b) Try to determine bounds for the error propagation and compare them with computational results.

(c) What are differences in the behavior of the two algorithms? Check their behavior for very large arguments, very small arguments, and for values close to a zero of the polynomial.

(d) Evaluate an approximate zero of the polynomial. Which parts of the computed residual are due to the approximations in the zero and in the evaluation, respectively?

2. Consider the "elliptically distorted" unit circle $p_1(x, y) = x^2 + \alpha\, xy + y^2 - 1 = 0$, $\alpha \in \mathbb{R}$, $|\alpha|$ small, and the three straight lines through the origin $p_2(x, y) = y^3 - 3\,x^2 y = 0$. (Plot.)

(a) Determine explicit expressions for the 6 zeros z_μ of (p_1, p_2) in terms of α. Convince yourself that the problem is well-conditioned w.r.t to variations in α.

(b) By suitable manipulations of p_1, p_2, find the border basis of $\langle p_1, p_2 \rangle$ for $\mathcal{N} = \{1, y, x, y^2, xy, xy^2\}$ and form the multiplication matrices A_x and A_y. Why must we expect an instability in the computation of the normalized eigenvectors of either A_x or A_y for very small $|\alpha|$ and for which zeros? Choose various α and compare the zeros from the eigenvectors with those from (a).

(c) Find a path from the border basis to the z_μ which is numerically stable for very small $|\alpha|$. (Hint: Consider A_{x+y}.) Check.

4.3 Floating-Point Arithmetic

It is not the purpose of this section to give a comprehensive account of the features of floating-point arithmetic; a majority of the readers of this book will be familiar with it anyway. Rather, this section is addressed to those readers from the computer algebra community who do not have a coherent knowledge of and computational experience with floating-point arithmetic. It explains the main features of floating-point arithmetic which are relevant in its use for algebraic computations. Floating-point arithmetic has been standardized in [4.4].

General-purpose interactive software products for scientific computing implement a *decimal* floating-point arithmetic on the underlying binary processors. In many of these systems (notably in the Maple releases from 6 up), this arithmetic follows essentially the same principal rules which the standard requires for binary arithmetic. Therefore, we will explain these principles in terms of decimal arithmetic; for convenience, we refer to Maple's floating-point implementation for illustration.

The IEEE floating-point arithmetic standard [4.4] regulates two important aspects of numerical computation:

(i) the *numbers* which are available,

(ii) the *arithmetic operations* within this set of numbers.

4.3.1   Floating-Point Numbers

The numbers of a *floating-point number set* $\mathbb{F}$ constitute a *finite discrete* subset of the set of real numbers: For each $x \in \mathbb{F}$, there is a well-defined positive distance to the preceeding and to the succeeding number in $\mathbb{F}$. As a finite automaton, a microprocessor cannot deal with a dense infinite set of numbers; it must *know* a number in order to be able to handle it properly. For reasons explained above, we consider only *decimal* floating-point number sets $\mathbb{F}(D)$ in the following.

Definition 4.4. For a specified *mantissa length* $D \in \mathbb{N}$ and integers $E_{\min} < 0 < E_{\max}$, the floating-point number set $\mathbb{F}(D) \subset \mathbb{R}$ consists of the numbers

$$s \cdot M \cdot 10^E \quad \text{and} \quad 0, \tag{4.10}$$

where the *mantissa*[7] M, $10^{-1} \le |M| < 1$ is a decimal fraction of D digits and the *exponent* E is an integer $\in [E_{\min}, E_{\max}]$; the *sign* s is $+1$ or -1.
(This does not exclude customary *notations* like -12.3456 or $.0005$ in place of $-.123456 \cdot 10^2$ or $+.500000 \cdot 10^{-3}$, e.g., for numbers in $\mathbb{F}(6)$.) $\square$

By (4.10), each set $\mathbb{F}(D)$ consists of *finitely many* numbers; but $E_{\max}$ and $|E_{\min}|$ are generally so large that, for practical purposes, $\mathbb{F}(D)$ extends to $\pm\infty$ and the "gap" around 0 is negligible. Therefore, in this text, we disregard the potential appearance of floating-point results with an exponent *outside* $[E_{\min}, E_{\max}]$; this phenomenon is commonly called "overflow" or "underflow," respectively. Readers interested in this phenomenon and the related safeguarding measures may consult, e.g., [4.3].

The true restriction is in the mantissa length D which governs the "grid density" of the numbers in $\mathbb{F}(D)$ on $\mathbb{R}$. Denote the mantissa M by $.m_1 \ldots m_D$, where $m_\nu \in \{0, 1, 2, \ldots, 9\}$ but $m_1 \neq 0$; for $m_1 = 0$, $.m_2 \ldots m_D 0 \cdot 10^{E-1}$ is the standard representation of $M \cdot 10^E$. For each fixed E, there are exactly $9 \cdot 10^{D-1}$ positive and as many negative numbers in $\mathbb{F}(D)$ which lie in $\pm[.1 \cdot 10^E, .999..9 \cdot 10^E]$, at a *constant spacing* $\Delta_E := 10^{E-D}$. After a further step of 10^{E-D}, at $10^E = .1 \cdot 10^{E+1}$, the spacing jumps to $\Delta_{E+1} = 10^{E+1-D}$! Thus, the positive part of the set $\mathbb{F}(D)$ consists of successive *intervals* $[.1 \cdot 10^E, 10^E]$ *of equidistant numbers*; but from one interval to the next, the spacing jumps by a factor of 10! This implies that, globally or "from a distance," there is a constant *relative* spacing of the floating-point numbers while, locally or "at close view," there is a constant absolute spacing.

This—at first sight strange—"semi-logarithmic" spacing has advantages and disadvantages for scientific computing.

Large and small numbers are represented with the same *relative* accuracy. In real-life applications, this makes the choice of the units irrelevant: Whether some length is expressed in meters or kilometers or millimeters, it can always be entered into a computation with the same accuracy. Thus, floating-point numbers model the experience that tolerances of realistic data are, almost universally, relative to the size of the data.

On the other hand, after some computation, this may have changed: The small intermediate or final result may originate from large data and vice versa; cf., e.g., Exercise 4.2-1. Shifts of the origin may change the numerical environment of a point drastically; cf. also section 3.3.2.

---

[7]The IEEE standard had called M the *significand*, but this term is not widely used.

These effects are more likely to play a role in nonlinear than in linear algebraic computations and must be kept in mind in the design of algebraic algorithms. If necessary, their occurrence must be monitored and alternate algorithmic paths provided; cf. section 4.2.3.

In the processor-based binary floating-point sets $\mathbb{F}_2(D)$, the mantissa length D is determined by the hardware and restricted to a few fixed values ("single-precision," "double-precision," "quad-precision"). In the decimal floating-point sets $\mathbb{F}(D)$ of many computer algebra systems, implemented on top of this hardware, the mantissa length D can be chosen freely, up to values beyond any practical need. A customary default value is $D = 10$ which provides a precision between single and double precision of the binary standard; it is well-suited for almost all practical problems with data of limited accuracy. For the demonstrative examples in this book, we will often use a shorter mantissa length.

4.3.2 Arithmetic with Floating-Point Numbers

As floating-point numbers are real numbers, the arithmetic operations $\pm$, $*$, and $/$ are automatically defined for them. However, the *discreteness* of the floating-point number sets $\mathbb{F}(D)$ prevents their closedness with respect to these operations. In order to remain within $\mathbb{F}(D)$ while performing arithmetic operations, we must define *arithmetic pseudo-operations* which model the genuine operations as closely as possible but map two floating-point operands into a floating-point result. It is the outstanding feature of the IEEE binary floating-point standard that it has prescribed a strict design principle for these pseudo-operations.

Definition 4.5. A *rounding* is a map $\square$ from $\mathbb{R}$ to some $\mathbb{F}(D)$ which satisfies

$$\square\,(-x) \;=\; -\,\square\,(x) \qquad \text{and} \qquad x_1 < x_2 \;\Rightarrow\; \square\,(x_1) \leq \square\,(x_2)\,. \qquad \square \qquad (4.11)$$

The transitivity in (4.11) implies:

Proposition 4.1. For $x \in \mathbb{F}(D)$, $\square\,x = x$. For $x \in \mathbb{R}, \notin \mathbb{F}(D)$, let $x_\downarrow$ and $x_\uparrow$ be the two neighbors of x in $\mathbb{F}(D)$; then $\exists\, x_\square \in [x_\downarrow, x_\uparrow]$ such that

$$\square\,x \;:=\; \begin{cases} x_\downarrow & \text{for } x < x_\square\,, \\ x_\uparrow & \text{for } x > x_\square\,, \end{cases} \qquad (4.12)$$

while $\square\,x_\square$ must be specified. The choice of $x_\square$ defines the map $\square$.
There are four standard choices for $x_\square$:

$$x_\square \;:=\; \begin{cases} (x_\downarrow + x_\uparrow)/2 & \text{round-to-nearest} & \\ \mathrm{sign}(x)\,|x|_\downarrow & \text{round-to-0} & (\text{"truncate"}) \\ x_\uparrow & \text{round-to-}+\infty & (\text{"round-up"}) \\ x_\downarrow & \text{round-to-}-\infty & (\text{"round-down"}) \end{cases} \qquad (4.13)$$

Round-to-nearest is the default, with $\square\,x_\square :=$ the neighbor with an even last digit. The other rounding modes have $x_\square \in \mathbb{F}$.

Example 4.5: In $\mathbb{F}(5)$, in the above sequence of the modes,

$$\Box\, .987635 \cdot 10^E \;=\; \left\{ \begin{array}{l} .98764 \\ .98763 \\ .98764 \\ .98763 \end{array} \right\} \cdot 10^E . \qquad \Box$$

Now we can define pseudo-operations $\tilde{\nabla}$ in $\mathbb{F}(D)$ for arithmetic operations ∇ in $\mathbb{R}$ by a *uniform* rule.

Definition 4.6. For a binary operation $\nabla : \mathbb{R} \times \mathbb{R} \to \mathbb{R}$, the associated (pseudo-)operation $\tilde{\nabla} : \mathbb{F}(D) \times F(D) \to \mathbb{F}(D)$ is defined by

$$x_1 \,\tilde{\nabla}\, x_2 \;:=\; \Box\,(x_1 \nabla x_2) . \tag{4.14}$$

For a unary operation $\nabla : \mathbb{R} \to \mathbb{R}$, the associated (pseudo-)operation $\tilde{\nabla} : \mathbb{F}(D) \to \mathbb{F}(D)$ is

$$\tilde{\nabla}\, x \;:=\; \Box\,(\nabla x_2) . \tag{4.15}$$

Division by 0 is prohibited. $\Box$

The only unary operation whose implementation according to (4.15) was required in the Standard is $\sqrt{x}$. But in most systems, (4.15) holds also for operations like trigonometric and hyperbolic functions, exponentials, logarithms, etc., except perhaps for extreme arguments; cf. Exercise 4.3-3.

Definition 4.7. A set $\mathbb{F}(D)$ of floating-point numbers (4.10) together with a rounding $\Box$ from (4.13) defines a *floating-point arithmetic* $(\mathbb{F}(D), \Box)$ with its arithmetic operations defined by (4.14)/(4.15). $\Box$

Thus, in Maple, e.g., we do not have "a" decimal floating-point arithmetic, but 4 copies for each natural number $D < D_{\max}$. The default arithmetic is $(\mathbb{F}(10), \text{round-to-nearest})$.

Naturally, (4.14)/(4.15) are not an instruction for the implementation of floating-point arithmetic. The need for a pseudo-operation has arisen precisely from the fact that, generally, $x_1 \nabla x_2$ cannot be formed. Thus, (4.14) is to be read as the *requirement* that the result must be identical with the rounded result of the real operation.

Floating-point arithmetic differs from real arithmetic in important details:

Proposition 4.2. Floating-point arithmetic preserves

- the *commutativity* of addition and multiplication:

$$x_1 \,\tilde{+}\, x_2 \;=\; x_2 \,\tilde{+}\, x_1 , \qquad x_1 \,\tilde{*}\, x_2 \;=\; x_2 \,\tilde{*}\, x_1 ; \tag{4.16}$$

- the relations between addition/subtraction and the minus sign:

$$x_1 \,\tilde{+}\, (-x_2) \;=\; x_1 \,\tilde{-}\, x_2 , \qquad x_1 \,\tilde{-}\, (-x_2) \;=\; x_1 \,\tilde{+}\, x_2 . \tag{4.17}$$

But identities with *two or more operations* are generally not preserved: For example,[8]

---

[8]The $\neq$ signs signify that inequality *may* happen.

- addition and multiplication are *not associative*:

$$(x_1 \,\tilde{+}\, x_2) \,\tilde{+}\, x_3 \;\neq\; x_1 \,\tilde{+}\, (x_2 \,\tilde{+}\, x_3)\,, \qquad (x_1 \,\tilde{*}\, x_2) \,\tilde{*}\, x_3 \;\neq\; x_1 \,\tilde{*}\, (x_2 \,\tilde{*}\, x_3)\,; \tag{4.18}$$

- addition/subtraction and multiplication/division are *not inverse operations*:

$$(x_2 \,\tilde{+}\, x_1) \,\tilde{-}\, x_1 \;\neq\; x_2 \;\neq\; (x_2 \,\tilde{-}\, x_1) \,\tilde{+}\, x_1\,, \qquad (x_2 \,\tilde{*}\, x_1) \,\tilde{/}\, x_1 \;\neq\; x_2 \;\neq\; (x_2 \,\tilde{/}\, x_1) \,\tilde{*}\, x_1\,;$$

- addition/subtraction are *not distributive* with multiplication:

$$(x_1 \,\tilde{\pm}\, x_2) \,\tilde{*}\, x_3 \;\neq\; (x_1 \,\tilde{*}\, x_3) \,\tilde{\pm}\, (x_2 \,\tilde{*}\, x_3)\,;$$

- squaring and squareroot are *not inverse operations*:

$$\widetilde{\sqrt{x \,\tilde{*}\, x}} \;\neq\; x \;\neq\; \widetilde{\sqrt{x}} \,\tilde{*}\, \widetilde{\sqrt{x}}\,.$$

Proof: a) The preservation of properties which refer to *one* operation only follows directly from (4.14).

b) Equation (4.14) requests that the result of each individual operation is rounded *immediately*, before the next operation. For examples, cf. Exercises 4.3-1.   □

Corollary 4.3.   Arithmetic expressions whose interpretation assumes associativity and distributivity must be arranged (by parentheses) into a sequence of binary operations to make their floating-point evaluation well-defined. Different arrangements may often lead to differing evaluations.

But interactive systems do accept expressions like $x_1 + x_2 + x_3$. Yes—but they have a fixed rule how they introduce the parentheses! Maple, e.g., interprets this as $(x_1 + x_2) + x_3$ ("sequential precedence"). It is important to know the rules for the system one is using.

There is a more subtle effect which easily escapes attention: We all use tools like Maple's expand freely, with the clear intention of changing the structure of an expression by the use of associativity and distributivity. Generally, this changes the arrangement as a sequence of binary operations severely. Therefore, we must not be surprised if the substitution of floating-point numbers before and after expand yields different results.

Example 4.6: As a trivial example, consider the floating-point evaluations of $(x + y) * (x - y)$ and $x^2 - y^2$ in $\mathbb{F}(6)$, with $x = 3.14159$ and $y = 3.14127$:

$$(x \,\tilde{+}\, y) \,\tilde{*}\, (x \,\tilde{-}\, y) \;=\; 6.28286 \,\tilde{*}\, .000320000 \;=\; .00201052\,;$$

$$x \,\tilde{*}\, x \,\tilde{-}\, y \,\tilde{*}\, y \;=\; 9.86959 \,\tilde{-}\, 9.86758 \;=\; .00201000\,.$$

The exact result is $.0020105152 \notin \mathbb{F}(6)$; the evaluations of the two expressions are not only different, but while the first one equals $\square$(exact result), the second one is not even the exact result rounded to 3 digits.

For $y = .00314127$, on the other hand, both expressions yield the same result 9.86959.   □

In an expression like $x^T y = \sum_{\nu=1}^{100} \xi_\nu \eta_\nu$, there is an immense number of potential ways to arrange the 100 terms in some order and to insert parentheses. For almost all floating-point data sets, the evaluation of the different arrangements will give many different results. Which of these is closest to the exact result depends on the data in an intricate fashion which prevents an a priori selection of a particular arrangement. But there are some principal observations which aid in the design of algorithms which will not behave particularly badly in floating-point.

4.3.3   Floating-Point Errors

By (4.14), any deviation between the floating-point and the exact result of an operation with floating-point operands may be interpreted as the effect of the rounding operator $\square$.

Definition 4.8. For a specified floating-point arithmetic $(\mathbb{F}(D), \square)$, the function $\varepsilon : \mathbb{R} \to \mathbb{R}$

$$\varepsilon(x) := \square(x) - x \tag{4.19}$$

is the *elementary (absolute) round-off error*. The *elementary relative round-off error* is

$$\rho(x) := \frac{\varepsilon(x)}{x}, \quad x \neq 0. \tag{4.20}$$

(There is no universal agreement about the sign of the elementary round-off error; in this book, we will adhere to (4.19).) $\square$

Trivially, $\varepsilon(x)$ vanishes for $x \in \mathbb{F}$. For $x \in \mathbb{R}$, $x \notin \mathbb{F}$, it is important to know a *bound* for the elementary round-off error.

Proposition 4.4. Let Δ_E be the local spacing of the numbers in $\mathbb{F}(D)$ (cf. section 4.3.1). For $x \in \mathbb{R}$, $|x| \in (.1 \cdot 10^E, \, 10^E)$,

$$|\varepsilon(x)| \begin{cases} \leq \frac{1}{2}\Delta_E = .5 \cdot 10^{E-D} & \text{for round-to-nearest,} \\ < \quad \Delta_E = 10^{E-D} & \text{for the other standard roundings,} \end{cases} \tag{4.21}$$

$$|\rho(x)| < \begin{cases} .5 \cdot 10^{-D+1} & \text{for round-to-nearest,} \\ 10^{-D+1} & \text{for the other standard roundings.} \end{cases} \tag{4.22}$$

Due to (4.14), the same bounds, with $x = x_1 \nabla x_2$, apply to the round-off errors $|x_1 \tilde{\nabla} x_2 - x_1 \nabla x_2|$ and $|(x_1 \tilde{\nabla} x_2 - x_1 \nabla x_2)/x_1 \nabla x_2|$ of floating-point operations.

Proof: Equation (4.21) follows directly from the lengths of the spacings Δ_E; (4.22) follows from $|x| > 10^{E-1}$; cf. (4.10). $\square$

Proposition 4.4 contains the most important *qualitative characterization* of floating-point arithmetic:

> The elementary *relative* round-off error and the relative error generated
> in one floating-point operation are *uniformly bounded* on $\mathbb{R}$.

The value of this important bound for a specified floating-point arithmetic is often denoted by eps or macheps when it refers to hardware arithmetic. According to the IEEE binary standard, we have for round-to-nearest:

$$\text{macheps} = \begin{cases} 2^{-24} \approx .6 \cdot 10^{-7} & \text{for single-precision,} \\ 2^{-53} \approx 1.1 \cdot 10^{-16} & \text{for double-precision.} \end{cases}$$

Thus, IEEE single-precision floating-point arithmetic is comparable in accuracy with a 7-digit decimal floating-point arithmetic and double-precision with 16 decimal digits.

By (4.19) and (4.20), we can formally express the result of a floating-point operation $\tilde{\nabla}$ by the exact result and its elementary round-off error. Due to (4.14),

$$x_1 \,\tilde{\nabla}\, x_2 \;=\; x_1 \,\nabla\, x_2 + \varepsilon(x_1 \nabla x_2) \quad \text{and} \quad x_1 \,\tilde{\nabla}\, x_2 \;=\; (x_1 \,\nabla\, x_2)\,(1 + \rho(x_1 \nabla x_2))\,. \tag{4.23}$$

Thus we can recursively represent the floating-point result of a composite expression in terms of exact operations and elementary round-off errors. With the uniform bound (4.22), this permits some rough quantitative assessment of potential round-off error effects.

Example 4.6, continued: By (4.23), $(x \,\tilde{+}\, y) \,\tilde{*}\, (x \,\tilde{-}\, y) =$

$$((x+y)(1+\rho_+) * (x-y)(1+\rho_-))\,(1+\rho_*) \;=\; (x^2 - y^2)\,(1 + \rho_+ + \rho_- + \rho_* + O(\rho^2))$$

while $x \,\tilde{*}\, x \,\tilde{-}\, y \,\tilde{*}\, y =$

$$(x^2(1+\rho_1) - y^2(1+\rho_2))\,(1+\rho_3) \;=\; (x^2 - y^2)\,(1 + \tfrac{x^2}{x^2-y^2}\rho_1 - \tfrac{y^2}{x^2-y^2}\rho_2 + \rho_3 + O(\rho^2))\,.$$

For the product form, we can be certain that the total generated error of a floating-point evaluation will not exceed 3 eps for arbitrary values of x, y. For the difference of squares, on the other hand, the relative error may grow explosively for small results; in our example, the bound yields ≈ 1500 eps which explains the loss of 3 significant digits. For $|x| \gg |y|$, the bound is only about 2 eps; this displays the missing "robustness" of this form of the expression w.r.t. to floating-point evaluation. $\quad \square$

Generally, significant differences in accuracy between different forms of one and the same mathematical expression arise only when the data$\to$result mapping of the expression is *ill-conditioned*. In this case, we must expect that the result is very sensitive not only to perturbations of the data but also to perturbations of the results of intermediate operations as they are introduced by the floating-point evaluation; cf. section 4.2.2 and the concept of numerical stability.

Typically, ill-conditioned expressions are characterized by results which are much smaller in modulus than the data; cf. Example 4.6. Other common cases of ill-conditioned evaluations are the computation of the scalar product of two near-orthogonal vectors and the evaluation of a polynomial near a zero.

Example 4.7: Consider

$$x^T = (2.39255, -3.96742, 4.21645, -3.67804),$$
$$y^T = (7.46803, -4.00862, -2.19400, 6.66637)$$

with components from $\mathbb{F}(6)$; their exact scalar product is .0014475221. When we evaluate the expression $((\xi_1 * \eta_1 + \xi_2 * \eta_2) + \xi_3 * \eta_3) + \xi_4 * \eta_4$ in $\mathbb{F}(6)$, we obtain

$$((2.39255 \,\tilde{*}\, 7.468803 \,\tilde{+}\, 3.96742 \,\tilde{*}\, 4.00862) \,\tilde{-}\, 4.21645 \,\tilde{*}\, 2.19400) \,\tilde{-}\, 3.67804 \,\tilde{*}\, 6.66637$$
$$= ((17.8676 \,\tilde{+}\, 15.9039) \,\tilde{-}\, 9.25089) \,\tilde{-}\, 24.5192 = .0014\,.$$

This is a reasonable result because it shows the small absolute size of the scalar product correctly. But because of its dramatically reduced *relative* accuracy, its use in further computations may be dangerous. In this case, a rearrangement of the terms would not have helped because the later loss in relative accuracy originates in the rounding after the multiplications: The digits "rounded away" at this point cannot be recovered later. $\square$

An analogous situation is the evaluation of a univariate or multivariate *polynomial near a zero*: The final result is much smaller in modulus than the data from which it is computed. For reasons of efficiency, the accumulation of a polynomial value $p(z)$ is generally done in a Horner type algorithm

$$ p(z) = \sum_{\nu=0}^{n} \alpha_\nu z^\nu = (\ldots ((\alpha_n * z + \alpha_{n-1}) * z + \alpha_{n-2}) * z + \ldots + \alpha_1) * z + \alpha_0 . \quad (4.24) $$

Like in the accumulation of a scalar product, the loss of significant digits occurs when intermediate results are so large that trailing digits are discarded which the small final result $p(z)$ cannot recover; cf. Example 4.8.

4.3.4   Local Use of Higher Precision

In section 3.3 we have seen how the *backward error* is the main tool for checking the validity of a computed result; its value is independent of how that result has been obtained. It is sufficient to determine the order of magnitude of the backward error; this is fortunate because its computation is always based on some sort of *residual*, e.g., the value of a polynomial at an approximate zero, and expressions for residuals tend to be ill-conditioned.

On the other hand, we have also seen (cf. section 3.4) that such residuals may be employed to compute *corrections* of an approximate result by a Newton-type approach. Here, the relative accuracy of the correction is generally limited by the relative accuracy of the residuals from which it has been computed. Thus, for that purpose, we want a reasonable relative accuracy in the residuals in spite of their ill-conditioning. In this section, we will introduce a few possible approaches to achieve that goal.

A natural protection against the effects of an excessive sensitivity of an expression w.r.t. intermediate errors is a *reduction of the generated intermediate errors* by the use of a higher precision *during* its evaluation. The application of this tool in the computation of an ill-conditioned scalar product is particularly instructive.

Assume that the components of the vectors x, y are given in some floating-point format $\mathbb{F}(D)$ and that the result $x^T y$ is also required in that format. When we convert the components into an *extended format* $\mathbb{F}(D_+)$ (filling their mantissas with trailing zeros) and form their products $\xi_\nu \eta_\nu$ in the arithmetic of $F(D_+)$, we lose fewer trailing digits (none for $D_+ \geq 2 D$) of these products. After the accumulation of the products in their specified order in $\mathbb{F}(D_+)$, the result is converted back to $\mathbb{F}(D)$. In the accumulation, a severe cancellation of leading digits may occur; but now there are *meaningful back-up digits* to take the trailing positions of the final result which would have been filled by zeros otherwise.

Example 4.7, ctd: Take $D_+ = 10$ for the extended precision. Then our computation takes the following form:

$$ ((2.392550000 \mathbin{\tilde{*}} 7.4688030000 \mathbin{\tilde{+}} \ldots) \mathbin{\tilde{-}} \ldots) \mathbin{\tilde{-}} 3.678040000 \mathbin{\tilde{*}} 6.666370000 $$
$$ = ((17.86763518 \mathbin{\tilde{+}} 15.90387916) \mathbin{\tilde{-}} 9.250891300) \mathbin{\tilde{-}} 24.51917551 = .00144753 . $$

This is not the correct 6-digit result .00144752, but considerably enhanced in comparison with the $\mathbb{F}(6)$ result .0014. □

Since scalar products are omnipresent in scientific computations, interactive software packages contain procedures for the evaluation of scalar products which employ this technique. Their use is a simple precaution against a potential loss of accuracy.

Example 4.8: The specified polynomial (3.17) of Example 3.3 shows extreme cancellation of leading digits in the computation of a residual at a location within the 3-cluster of zeros.

$$p(x) = x^4 - 2.83088\,x^3 + .00347\,x^2 + 5.66176\,x - 4.00694$$

has coefficients in $\mathbb{F}(6)$. We wish to test the approximate zero $\tilde{z} = 1.41418 \in \mathbb{F}(6)$ which we have somehow obtained.

When we evaluate (cf. (4.24)) $p(\tilde{z}) = ((1.0 * \tilde{z} - 2.83088) * \tilde{z} + \ldots) * \tilde{z} - 4.00694$ in $\mathbb{F}(6)$ arithmetic, we obtain the residual 0 which tells us nothing about the exact residual, except perhaps that it is small.

When we evaluate the same expression in our extended arithmetic of $\mathbb{F}(10)$, we obtain $.1 \cdot 10^{-8}$; the exact value of $p(\tilde{z})$ is $\approx -.113344 \cdot 10^{-9}$. While the $\mathbb{F}(10)$ value should suffice for a validity test, it will not provide any useful information for a refinement of the zero. To obtain a value of the residual with 6 significant digits, an intermediate $\mathbb{F}(16)$ arithmetic is needed. □

If such a failure of the residual computation is discovered a posteriori (this could even be done algorithmically), it is still possible to utilize the unsatisfactory computation as a basis for a refinement: We determine the arithmetic errors generated in its computational steps with a *local* higher precision and then propagate these residuals through the algorithm. Because these residuals have lower exponents, this second run through the algorithm may use the *original* precision; it will generate an approximation to the trailing digits which were lost in the original result. If necessary the procedure can be repeated; this has led to the name "staggered corrections" for it. We show its use for the Horner algorithm in Example 4.8.

We regard the evaluation algorithm (4.24) for $p(z)$ as a *recursion*

$$\begin{aligned}
y_n &:= \alpha_n, \\
y_\nu &:= y_{\nu+1} * z + \alpha_\nu, \quad \nu = n-1(-1)0, \\
p(z) &:= y_0,
\end{aligned}$$

or (equivalently) as a *linear system* for the vector y of the exact intermediate results y_ν:

$$\begin{pmatrix} 1 & & & \\ -z & 1 & & 0 \\ & \ddots & \ddots & \\ 0 & & \ddots & 1 \\ & & & -z & 1 \end{pmatrix} \begin{pmatrix} y_n \\ y_{n-1} \\ \vdots \\ y_1 \\ y_0 \end{pmatrix} - \begin{pmatrix} \alpha_n \\ \alpha_{n-1} \\ \vdots \\ \alpha_1 \\ \alpha_0 \end{pmatrix} = 0, \qquad \text{or}$$

$$Z \cdot y - a = 0. \tag{4.25}$$

The vector $\tilde{y} = y + \Delta y$ of the *computed* intermediate results $\tilde{y}_\nu$ from the evaluation of p leaves a residual

$$Z \cdot \tilde{y} - a =: d \tag{4.26}$$

in (4.25) which implies

$$Z \cdot \Delta y = d . \tag{4.27}$$

If the residual vector d is computed from (4.26) in a higher precision, it will cushion the cancellation of leading digits and retain meaningful digits. After normalization, d will have exponents smaller than those in a; therefore, the same may be expected for the correction vector Δy relative to y. Thus, even if Δy is computed from (4.27) in the original lower precision, it will contribute meaningful information for a correction of y. Note that the evaluation of (4.27) is nothing than the original Horner recursion; no linear system has to be solved.

Example 4.8, continued: Since we were not satisfied with the evaluation of the residual in $\mathbb{F}(10)$, we attempt a correction. With the intermediate results $\tilde{y}_\nu$ from the evaluation of $p(1.41418)$ in $\mathbb{F}(10)$, we compute the residual vector d of (4.26) in $\mathbb{F}(16)$ and obtain

$$d \approx \tilde{Z} \begin{pmatrix} \tilde{y}_4 \\ \tilde{y}_3 \\ \vdots \\ \tilde{y}_1 \\ \tilde{y}_0 \end{pmatrix} - \begin{pmatrix} \alpha_4 \\ \alpha_3 \\ \vdots \\ \alpha_1 \\ \alpha_0 \end{pmatrix} = \begin{pmatrix} 0 \\ 0 \\ 0 \\ .46908 \\ .44998 \end{pmatrix} \cdot 10^{-9} ;$$

for the correction Δy_0 of the residual $\tilde{y}_0$, we obtain from (4.27) evaluated in $\mathbb{F}(10)$,

$$\Delta y_0 = [.46908 * \tilde{z} + .44998] \cdot 10^{-9} = .1113343554 \cdot 10^{-8} .$$

With $\tilde{y}_0 = .1 \cdot 10^{-8}$, this yields the improved residual $\tilde{y}_0 - \Delta y_0 = -.113343554 \cdot 10^{-9}$ which has 3 more correct digits than the residual directly obtained from (4.24) in $\mathbb{F}(16)$. $\quad\square$

This technique of regarding an evaluation procedure as a (recursive) system of equations and applying a Newton step to it, with a residual computed in higher precision, may be successfully applied in various situations. It is usually computationally cheaper than a repetition of the complete procedure in the higher precision arithmetic.

Exercises

1. (a) In the 3-argument floating-point sums and products of (4.18), how many distinct values are to be expected for the 6 different arrangements? Find values for x_1, x_2, x_3, such that the maximal number of distinct values is obtained.

(b) Find values for the arguments in the other expressions in Proposition 4.2 such that the mathematically equivalent expressions have distinct floating-point evaluations.

2. In which fashion should the truncated Taylor-expansion $s(x) := \sum_{\nu=0}^{13} x^\nu / \nu!$ be evaluated in a floating-point arithmetic?

(a) Find intervals on $\mathbb{R}$ of values of x which require different evaluation sequences for reasonable floating-point results.

(b) Try to formulate an algorithm which chooses the optimal floating-point evaluation scheme for specified $x \in \mathbb{R}$. Is such an algorithm dependent on the value of the mantissa length D?

(c) The relative error of the exact value of $s(x)$ as an approximation of $\exp(x)$ is less than $.115 \cdot 10^{-10}$; why? Are there $\mathbb{F}(10)$-floating-point values of x such that the optimal $\mathbb{F}(10)$-evaluation of $s(x)$ yields $\square \exp(x)$?

3. (a) Consider $\bar{x} = 3125 \cdot 10^6$ as exact. To find a 10-digit correct value of $\sin(\bar{x})$,

- call $\sin(\bar{x})$ in Maple (or some similar system),

- compute $x_r := \bar{x} - \lfloor \frac{\bar{x}}{2\pi} \rfloor * 2\pi \in [0, 2\pi)$ and call $\sin(x_r)$.

Comment on the result. Which precision do you need in the computation of x_r to obtain 10 correct digits of $\sin(\bar{x})$ from $\sin(x_r)$? Verify.

(b) Consider the empirical quantity $(\bar{x}, \varepsilon)$. What is the admissible size of ε if .9 is to be a valid value of $\sin((\bar{x}, \varepsilon))$?

(c) Consider $(\bar{x}, \varepsilon)$ with $\varepsilon = .5 \cdot 10^6$. What is the set of valid values of $\sin((\bar{x}, \varepsilon))$?

(d) With (a)–(c) in mind, can you think of a situation where the evaluation of $\sin(\bar{x})$ makes sense?

4.4 Use of Intervals

4.4.1 Interval Arithmetic

In our considerations of empirical algebraic problems, we have introduced sets of valid results or pseudoresult sets; cf. Definition 3.4 : The set $Z_\delta(\bar{a}, e)$ contains all results $z \in Z$ which can arise as exact results $F(\tilde{a})$ of the problem for data $\tilde{a}$ from the neighborhood $N_\delta(\bar{a}, e) \subset A$, where F is the data$\rightarrow$result mapping, and A and Z are the data and result spaces of our problem; cf. (3.16).

We have assumed that F is *continuous* on some sufficiently large dense data domain $\bar{A} \supset N_{\bar{\delta}}(\bar{a}, e)$; thus, we may extend the data$\rightarrow$result mapping F into a mapping $\widehat{F}$ of the subsets of $\bar{A}$ into Z so that $Z_\delta = \widehat{F}(N_\delta)$. A *direct evaluation* of this set extension $\widehat{F}$ of the data$\rightarrow$result mapping F would be attractive from various points of view, particularly in a computer algebra setting where the historical quest for "exact results" prevails.

Arithmetic operations with sets of numbers are defined by a simple rule:

Definition 4.9. Let $X_1, X_2, \ldots$ be compact sets in $\mathbb{R}$ or $\mathbb{C}$ and ∇ an arithmetic operation. Then the *set extension* $\widehat{\nabla}$ of ∇ is defined by

$$X_1 \widehat{\nabla} X_2 := \cup_{x_1 \in X_1, x_2 \in X_2} (x_1 \nabla x_2) ; \tag{4.28}$$

in a division, X_2 must not contain 0. For arithmetic functions on $\mathbb{R}$ or $\mathbb{C}$, the analogous rule is

$$\widehat{F}(X) := \cup_{x \in X} f(x) \qquad \text{for } X \subset \text{domain of } F. \quad \Box \tag{4.29}$$

For individual operations with *real* arguments and results, everything appears deceptively simple: On the real line $\mathbb{R}$, compact sets are contained in bounded *intervals*; the restriction to *closed* intervals is no loss of generality in our context. When we exclude division by an interval containing 0, the result of an arithmetic operation with interval operands is an interval so that it suffices to compute the bounds of the result interval from the bounds of the operands. It is also straightforward to implement the above operations in a *floating-point interval arithmetic* by the introduction of an *outward rounding* of intervals: round-down for the lower and round-up for the upper bound.

Since the arithmetic operations with the rounding modes round-up and round-down are available in all general-purpose processors, an implementation of floating-point interval arithmetic operations meets no principal problems. A slight difficulty appears for the implementation of (4.29) when f is not monotone inside X. Interval operations are available in systems like Fortran-SC ([4.9]).

A first *principal obstacle* for an efficient use of such an interval arithmetic appears in the evaluation of arithmetic expressions $F(x_1, \ldots, x_n)$ which contain some arguments x_ν *more than once*. The evaluation of (4.28) requires the simultaneous substitution of identical values from the X_ν at all occurrences of x_ν. But if we evaluate $F(X_1, \ldots, X_n)$ in a sequence of interval arithmetic operations, this identification is lost. This leads to the inclusion of values in the result interval which are not in the right-hand side of (4.28) and hence to an *artificial widening* of the result interval.

This effect may happen already in the evaluation of powers: When we evaluate X^2 as $X * X$ for an X which contains 0, the product $X * X := \bigcup_{x_1, x_2 \in X} x_1 * x_2$ contains negative values which cannot appear in $X^2 := \bigcup_{x \in X} x^2$. Thus the evaluation of univariate polynomials p at interval arguments X cannot be based on an interval arithmetic evaluation of (4.24); for short intervals in which p is monotonic, we can simply use the values at the bounds of X. But the accurate evaluation of nontrivial multivarate polynomials for interval arguments will generally present serious difficulties.

This artificial widening of result intervals also occurs when intermediate result intervals are used more than once in the further computation. Thus, the naive use of interval arithmetic operations in the evaluation of a nontrivial expression will generally lead to result intervals which are considerably larger then the true images of the argument intervals under the data→result mapping. Note that this effect is independent of the use of floating-point arithmetic in the evaluation of the result bounds; it occurs also when exact rational arithmetic is used throughout. The round-off errors of section 4.3.3 play no significant role in this context.

A second *principal limitation* for the use of intervals in numerical computation appears when we have *multi-dimensional results* $z = (z_1, \ldots, z_m)$. For a data→result mapping $F : \mathbb{R}^n \to \mathbb{R}^m$, $m > 1$, $\widehat{F}(X_1, \ldots, X_n) = \bigcup_{x_\nu \in X_\nu} F(x_1, \ldots, x_n)$ is generally *not a multi-dimensional interval* $Z = Z_1 \times \ldots \times Z_m \subset \mathcal{Z} = \mathbb{R}^m$, i.e. the Cartesian product of univariate intervals Z_μ. The best interval inclusion of $F(X_1, \ldots, X_n) =: (F_\mu(X_1, \ldots, X_n), \mu = 1(1)m)$ which we can hope to determine is the Cartesian product of the

$$\overline{F}_\mu(X_1, \ldots, X_n) := [\, \min_{x_\nu \in X_\nu} F_\mu(x_1, \ldots, x_n), \, \max_{x_\nu \in X_\nu} F_\mu(x_1, \ldots, x_n) \,]; \qquad (4.30)$$

and, generally, the *computed* univariate intervals Z_μ will only be crude, sometimes grossly widened inclusions of these $\overline{F}_\mu$.

But, more important, the true multi-dimensional set $\widehat{F}(X_1, \ldots, X_n)$ is generally only a small subset of the Cartesian product of the $\overline{F}_\mu$ intervals. Although its extremal points lie on the boundary hyperplanes of the product of the $\overline{F}_\mu$, the product may, intuitively and quantitatively, be a gross misrepresentation of the set $\widehat{F}(X_1, \ldots, X_n)$.

Example 4.9: Consider the real empirical polynomial $(\bar{p}, e)(x) = x^2 - (2, .01)\, x + (.9775, .01)$. For the two zeros z_μ, $\mu = 1, 2$, we obtain pseudoresult sets (for $\delta = 1$, rounded outwards) $Z_1 = [.7987, .9448]$, $Z_2 = [1.0452, 1.2112]$. Often, this situation is illustrated by the Cartesian product of the two intervals (the rectangle in Figure 4.2) which suggests that this is the set of potential results, but the true $\delta = 1$ result set is as shown in Figure 4.2.

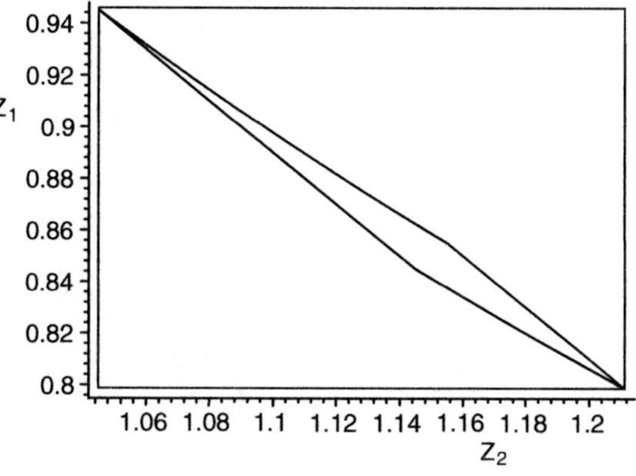

Figure 4.2.

Obviously, the zero pairs (z_1, z_2) which can occur for arbitrary coefficient data from $N_1(\bar{a}, e)$ occupy only a very small fraction of the rectangle formed by the two intervals. $\square$

Naturally, this misrepresentation is particularly harmful when such result multi-intervals are used in further computation. A well-known and geometrically intuitive example is the following:

Example 4.10: Consider the mapping $F : \mathbb{R}^2 \to \mathbb{R}^2$ defined by

$$F(x, y) = \frac{\sqrt{2}}{2} \begin{pmatrix} 1 & -1 \\ 1 & 1 \end{pmatrix} \begin{pmatrix} x \\ y \end{pmatrix}$$

which represents a 45° rotation of the x,y-plane about the origin. When a two-dimensional interval, i.e. an axi-parallel rectangle, is mapped by $\widehat{F}$, its area remains unchanged but its sides are turned by 45° so that it is no longer a two-dimensional interval; hence, by (4.30), it must be included in an axi-parallel rectangle (a square, in fact). This result interval has at least *twice* the original area as is easily checked. After 8 iterations of the mapping $\widehat{F}$ we have returned each point and also each two-dimensional interval to its original position. The 8-fold iteration of the stepwise procedure, with an interval enclosure after each step, will transport the center of the interval to its original location, but the area of the interval will have multiplied by 256 or more! $\square$

Intervals in $\mathbb{C}$ are essentially equivalent with intervals in $\mathbb{R}^2$; therefore, this problem of an artificial expansion of the result set by interval enclosure occurs in each computational step which is not simply an addition/subtraction. For example, the product of two complex intervals is *not* a complex interval but a set in $\mathbb{C}$ whose boundaries are conics. Thus even the efficient computation of the smallest enclosing interval presents a challenge. Therefore, complex interval arithmetic is even more prone to misrepresentations than real interval arithmetic.

4.4.2   Validation Within Intervals

If we accept the unavoidable inclusion of nontrivial multi-dimensional pseudoresult sets $Z_\delta \in \mathcal{Z}$ by axi-parallel intervals, there is an appropriate use of interval arithmetic for the *validation* of such inclusions, from whatever source they may be. It is based on a clever modification of the classical Brouwer's Fixed Point Theorem which states that a continuous function $g : \mathbb{R}^s \to \mathbb{R}^s$ has a *fixed point* (not necessarily unique) in a convex, closed, and bounded set $\overline{Z} \subset \mathbb{R}^s$ if $g(\overline{Z}) \subset \overline{Z}$, i.e. if g maps $\overline{Z}$ into itself. How can we choose g such that its fixed points are solutions of our algebraic problem?

As in section 3.2.2 we assume that our problem is well-posed and may be formulated as a system of m equations $G(x; a) = 0$ so that its result $z(a)$ for data a satisfies $G(z(a); a) = 0$ for $a \in N_{\tilde{a}}$; cf. (3.35) and (3.36). Let R be a fixed, regular $m \times m$ matrix. Then, a fixed point z of $g(x; a) := x - R \cdot G(x; a)$ satisfies $G(z; a) = 0$. However, the condition

$$g(\overline{Z}; a) := \{z - R \cdot G(z; a),\ z \in \overline{Z}\} \subset \overline{Z} \tag{4.31}$$

cannot be checked computationally in this form: The set $\overline{Z} - R \cdot \{G(z; a),\ z \in \overline{Z}\}$ is always a *superset* of $\overline{Z}$ (except in trivial situations). Therefore, we must rewrite the representation of $g(\overline{Z}; a)$ in a clever way.

At first, the inclusion set $\overline{Z}$ of the result $z(a)$ is replaced by an approximation $\tilde{z}$ plus an inclusion set $\Delta\overline{Z} := \overline{Z} - \tilde{z}$ for the correction $z(a) - \tilde{z}$. This permits the use of the *Mean-Value Theorem* in $\mathbb{R}^s$ (cf. (3.25)):

$$G(z; a) = G(\tilde{z}; a) + [\int_0^1 \frac{\partial G}{\partial x}(\tilde{z} + \lambda(z - \tilde{z});\ a)\, d\lambda]\, (z - \tilde{z})$$

with its associated *set version*

$$G(\overline{Z}; a) \subset G(\tilde{z}; a) + \frac{\partial G}{\partial x}(\overline{\tilde{z} \cup \overline{Z}};\ a) \cdot \Delta\overline{Z}, \tag{4.32}$$

where $\frac{\partial G}{\partial x}(Y; a)$, $Y \subset \mathbb{R}^s$, is the Jacobian of G w.r.t. x with its elements evaluated for the set Y.

Furthermore, the substitution $\overline{Z} = \tilde{z} + \Delta\overline{Z}$ permits a rearrangement of the terms in (4.31). Since the intended application is for $\tilde{z} \in \overline{Z}$ (or $\tilde{z}$ very close to $\overline{Z}$), and for $R \approx (\frac{\partial G}{\partial x}(\zeta; a))^{-1}$ with $\zeta \in \overline{\tilde{z} \cup \overline{Z}}$, the set matrix $I - R \cdot \frac{\partial G}{\partial x}(\overline{\tilde{z} \cup \overline{Z}};\ a)$ has *small* elements around 0, and the same is true for the components of $\Delta\overline{Z}$. Thus, when we rewrite (4.31) with the aid of (4.32) as

$$g(\overline{Z}; a) := \tilde{z} - R\,G(\tilde{z}; a) - [I - R\,\frac{\partial G}{\partial x}(\overline{\tilde{z} \cup \overline{Z}};\ a)]\,\Delta\overline{Z} \subset \overline{Z},$$

the left-hand side becomes a point value plus a small set close to the origin. This makes a *computational confirmation* of the requested inclusion feasible by standard interval arithmetic.

When we now subtract $\tilde{z}$ from both sides, we can write the inclusion condition fully in terms of the *correction interval* $\Delta\overline{Z} := \overline{Z} - \tilde{z}$.

Theorem 4.5. In the situation previously described, if $\tilde{z} \in \mathcal{Z}$, $\Delta\overline{Z} \subset \mathcal{Z}$, and the regular matrix R have been chosen such that

$$-R\,G(\tilde{z}; a) - [\,I - R\,\frac{\partial G}{\partial x}(\overline{\tilde{z} \cup \tilde{z} + \Delta\overline{Z}};\ a)\,]\,\Delta\overline{Z} \subset \Delta\overline{Z}, \tag{4.33}$$

then there exists a zero $z(a)$ of $G(x; a) = 0$ in $\tilde{z} + \Delta \overline{Z}$.

This theorem has emerged during a period of intensive consideration of the use of interval analysis in numerical computation in the 1970s. An early version is due to R. Krawczyk; its final formulation contains important contributions by S. Rump (cf. [4.8] and the references given there).

Since we have carried the data vector a through all the above considerations, we can now immediately extend Theorem 4.5 to empirical algebraic problems: Suppose that we can choose $\tilde{z}$, $\Delta \overline{Z}$, and R such that (4.33) holds for all $a \in N_\delta(\bar{a}, e)$; then the results $z(a)$ for all these data must be contained in $\tilde{z} + \Delta \overline{Z}$.

Corollary 4.6. In the situation of Theorem 4.5, the inclusion

$$-R\, G(\tilde{z};\, N_\delta(\bar{a}, e)) - [\, I - R\, \frac{\partial G}{\partial x} \overline{(\tilde{z} \cup \tilde{z} + \Delta \overline{Z};\, N_\delta(\bar{a}, e))}\,]\, \Delta \overline{Z} \subset \Delta \overline{Z}, \qquad (4.34)$$

implies $Z_\delta(\bar{a}, e) \subset \tilde{z} + \Delta \overline{Z}$ for the pseudoresult sets of $G(x;\, (\bar{a}, e)) = 0$.

For nontrivial empirical algebraic problems, the appropriate choice of $\tilde{z}$ and of the correction multi-interval $\Delta \overline{Z}$ requires that an approximate solution has been found with great accuracy—if it is possible at all. At this point, a confirmation of the validity of that solution by the computation of its backward error (cf. section 3.3.1) is generally considerably simpler. It is true that the successful confirmation of (4.34) provides more information than the backward error. Only if this information (with all its shortcomings) is really meaningful (cf. section 4.4.3), the approach of this section should be considered.

Example 4.11: We consider the empirical quadratic polynomial $p(x;\, (\bar{a}, e))$ of Example 4.9. We want to use (4.34) to find a validated inclusion for $Z_\delta(\bar{a}, e)$: cf. Figure 4.2. There are two natural choices for $G(x; a)$:

(i) We consider the two zeros individually ($m = 1$) and take $G(x; a + \Delta a) = p(x; a + \Delta a)$.

(ii) We consider the two zeros jointly ($m = 2$) and take

$$G((x_1, x_2);\, a + \Delta a) = \begin{pmatrix} x_1 + x_2 - 2 + \Delta \alpha_1 \\ x_1 x_2 - .8775 - \Delta \alpha_0 \end{pmatrix}.$$

For both choices, it turns out that (4.34) cannot be satisfied for $\delta = 1$ (the situation in Figure 4.2). Therefore, we keep δ as an indeterminate and try to find the largest δ for which $\Delta \overline{Z} = [-\Delta, +\Delta]$ or $\begin{pmatrix} [-\Delta_1, +\Delta_1] \\ [-\Delta_2, +\Delta_2] \end{pmatrix}$, resp., exists which satisfies (4.34). All intervals in the following will be symmetric to 0; therefore we denote them shortly by $\pm \beta := [-\beta, +\beta]$.

(i) We take $\tilde{z} = z_1 = .85$ and the 1×1 matrix $R = r = 1/p'(\tilde{z}; \bar{a}) = -1/.3$. Thus we have the requirement

$$p(.85;\, N_\delta)/.3 + [\,1 + p'(.85 \pm \Delta;\, N_\delta)/.3\,] \cdot (\pm \Delta) \subset \pm \Delta.$$

Multiplication by .3 and evaluation yields

$$\pm .0185\, \delta + \pm (2\, \Delta^2 + .01\, \delta\, \Delta) \subset \pm\, .3\, \Delta;$$

the largest feasible Δ is the smaller zero of $2\Delta^2 - (.3 - .01\,\delta)\,\Delta + .0185\,\delta$. The zeros turn complex for $\delta \stackrel{>}{\approx} .5846$, the associated $\Delta \approx .0735$. Thus, (4.34) provides a proof that (rounded to 4 digits) $Z_{.5846} \subset [.7765, .9235]$. It is easy to find that, actually, $Z_{.5846} \approx [.8179, .8931]$. An analogous situation arises for $\tilde{z} = z_2 = 1.15$, with a slightly smaller maximal δ.

(ii) With $\tilde{z} = (.85, 1.15)$, we choose

$$R = \left(\frac{\partial G}{\partial x}(\tilde{z}; \bar{a})\right)^{-1} = \begin{pmatrix} 1 & 1 \\ 1.15 & .85 \end{pmatrix}^{-1} = \begin{pmatrix} -17/6 & 10/3 \\ 23/6 & -10/3 \end{pmatrix}$$

and obtain

$$R \cdot \begin{pmatrix} \pm.0185 \\ \pm.0215 \end{pmatrix} \delta + \left[I - R \cdot \begin{pmatrix} 1 & 1 \\ 1.15 \pm \Delta_2 & .85 \pm \Delta_1 \end{pmatrix} \right] \cdot \begin{pmatrix} \pm\Delta_1 \\ \pm\Delta_2 \end{pmatrix} \subset \begin{pmatrix} \pm\Delta_1 \\ \pm\Delta_2 \end{pmatrix}$$

$$\text{or} \quad \begin{pmatrix} \pm.185 \\ \pm.215 \end{pmatrix} \delta + \begin{pmatrix} \pm 10\,\Delta_2 & \pm 10\,\Delta_1 \\ \pm 10\,\Delta_2 & \pm 10\,\Delta_1 \end{pmatrix} \begin{pmatrix} \pm\Delta_1 \\ \pm\Delta_2 \end{pmatrix} \subset 3 \begin{pmatrix} \pm\Delta_1 \\ \pm\Delta_2 \end{pmatrix}.$$

For specified δ, the maximal Δ_1, Δ_2 must satisfy $.185\,\delta + 20\,\Delta_1\Delta_2 - 3\Delta_1 = .215\,\delta + 20\,\Delta_1\Delta_2 - 3\Delta_2 = 0$. With $\Delta_2 = .01\,\delta - \Delta_1$ and $20\,\Delta_1^2 - (3 - .2\,\delta)\,\Delta_1 + .185\,\delta = 0$, we find a maximal $\delta \approx .5633$ beyond which Δ_1 becomes complex and associated maximal $\Delta_1 \approx .0722$, $\Delta_2 \approx .0778$. This yields the validated inclusion

$$Z_{.5633} \subset \begin{pmatrix} [.7778, .9222] \\ [1.0722, 1.2278] \end{pmatrix},$$

which is once more a considerable overestimation, cf. Figure 4.2.

The misrepresentation of the true situation is clearly displayed when we compute the backward error of extreme approximate zero pairs which are inside this inclusion. The backward error of $\tilde{z}_1 = .9222$, $\tilde{z}_2 = 1.2278$, as simultaneous zeros of $p(x; [\bar{a}, e))$, is ≈ 15.5.　$\square$

This example which has dealt with a near-trivial situation, points to various shortcomings of the interval approach to the validation of inclusions of pseudoresult sets $Z_\delta(\bar{e})$ for nonlinear empirical algebraic problems:

The applicability may be restricted to small indeterminations in the empirical data.

The validated inclusion may be a considerable overestimation of the true pseudoresult set.

In multidimensional problems, the inclusion consists of multidimensional intervals which may represent the true pseudoresult set poorly. Together with the overestimation, this may convey severely misleading information.

In nontrivial situations, the elaboration of (4.34) may be rather involved and lead to widenings of the intervals which prevent the verification of the inclusion.

It is often emphasized that the power of the interval approach lies in the fact that the verification of (4.34)—through the embedded fixed point theorem—implies a mathematical proof of the inclusion. But the information conveyed by the backward error is just as mathematically rigid: There exists a data set $\tilde{a}$ in $N_\delta(\bar{a}, e)$ for which the approximate result $\tilde{x}$ is the exact result.

4.4.3　Interval Mathematics and Scientific Computing

In the 1970s, intervals were introduced into numerical computation for two reasons:

(i) To permit the determination of *exact results*, in the form of enclosing intervals, for computational problems with exact data by means of floating-point computation;

(ii) To permit the numerical solution of computational problems with data of *limited accuracy* by substitution of intervals for such data and computing enclosing intervals for the solution sets.

Objective (i) is of a purely mathematical nature. If the emphasis is put on a provably correct inclusion of a result which consists of one point or of several separate points in $\mathbb{R}^s$ or $\mathbb{C}^s$, irrespective of the width of the inclusion, this goal is achievable for a large number of problems in polynomial algebra. However, it may require a high precision of the floating-point arithmetic employed and a huge computational effort to obtain an inclusion at all.

This objective may also be interpreted thus: Since the determination of exact numerical results is principally impossible for practically all truly nonlinear algebraic tasks, one is satisfied with provably correct inclusions whose width can (theoretically) be made arbitrarily narrow if the computational effort is raised high enough.

From the mathematical point of view, it was natural to consider objective (ii) as a generalization of objective (i). Exact data are replaced by exact interval data while the general goal remains the same: To compute provably correct inclusions of the solution sets defined by the interval data through the interval extension $\widehat{F}$ of the data$\rightarrow$result mapping F; cf. the beginning of section 4.4.1. Clearly, the computational effort had to be higher than for point data and increasing with increasing widths of the data intervals. Furthermore, a successful computation of an inclusion could not be guaranteed a priori.

From the point of view of Scientific Computing, this approach contains a *fundamental mistake*: The empirical data of almost all real-world application problems are points with an indetermination which is, at best, known by its order of magnitude; such data *cannot* be suitably represented by intervals with *specified bounds*. As we have discussed in section 3.1, empirical problems cannot have exact results but only results of some degree of validity as displayed by the scale (3.3). Furthermore, to obtain a valid result of some problem with empirical data must require *less* computational effort than with strict intrinsic data. Also, in principle, this effort must *decrease* with an increasing indetermination of the data, because fewer meaningful digits can be determined.

This situation has not been widely recognized for a long time because the tasks considered were almost exclusively of a linear nature. In a linear context, it is often not so difficult to treat objective (ii) successfully as a generalization of (i), at least for sufficiently narrow data intervals. In numerical polynomial algebra, the inherent nonlinearity and increased complexity of the tasks makes this approach infeasible except for trivial situations. This has led the way to the natural emphasis on backward errors and linearized estimates of result indeterminations shown in sections 3.2. and 3.3.

As some of the readers may know, the author of this text also supported the "mathematical" point of view for a while, in the 1980s, before he began to concern himself with nonlinear algebraic problems in the context of Scientific Computing. But it was not really the mounting computational difficulty which made him change his mind but the principal insight:

> It cannot be meaningful to solve vaguely defined empirical problems
> by formulating them as strictly defined intrinsic problems.

It is for this reason that interval mathematical methods and computations will not be used in the remaining parts of this text.

Exercises

1. For $w_\downarrow, w_\uparrow \in \mathbb{C}$, $\operatorname{Re} w_\downarrow \le \operatorname{Re} w_\uparrow$, $\operatorname{Im} w_\downarrow \le \operatorname{Im} w_\uparrow$, the complex interval $[w_\downarrow, w_\uparrow] \subset \mathbb{C}$ is defined by

$$[w_\downarrow, w_\uparrow] := \{w \in \mathbb{C} : \operatorname{Re} w \in [\operatorname{Re} w_\downarrow, \operatorname{Re} w_\uparrow], \ \operatorname{Im} w \in [\operatorname{Im} w_\downarrow, \operatorname{Im} w_\uparrow]\}.$$

Binary and unary operations with complex intervals are defined by (4.28) and (4.29).

(a) For the arithmetic operations $\pm$, $*$, $/$, find the definitions of $w_\downarrow, w_\uparrow$ in the tightest inclusion

$$[w_\downarrow, w_\uparrow] \supset [u_\downarrow, u_\uparrow] \nabla [v_\downarrow, v_\uparrow]$$

(exclude $0 \in [v_\downarrow, v_\uparrow]$ for division). Under what restrictions is equality attainable for multiplication and division?

(b) For the functions $f(u) = u^2$, $\exp(u)$, $\log(u)$, $cos(u)$, find the definitions of $w_\downarrow, w_\uparrow$ in the tightest inclusion $[w_\downarrow, w_\uparrow] \supset f([u_\downarrow, u_\uparrow])$.

(c) Draw conclusions for the use of complex interval arithmetic.

2. For $\bar{p}$ from (3.17), try to find equivalent expressions whose interval arithmetic evaluation for $X = [1.35, 1.47]$ is only a moderate overestimation of the true value $\bar{p}(X) \approx [-.000462, .000261]$.

3. Consider the application of (4.34) to the empirical linear system

$$G((x_1, x_2); (\bar{a}, e)) = \left\{ \begin{array}{l} (\bar{\alpha}_{11}, \varepsilon_{11})\, x_1 + (\bar{\alpha}_{12}, \varepsilon_{12})\, x_2 + (\bar{\alpha}_{10}, \varepsilon_{10}) \\ (\bar{\alpha}_{21}, \varepsilon_{21})\, x_1 + (\bar{\alpha}_{22}, \varepsilon_{22})\, x_2 + (\bar{\alpha}_{20}, \varepsilon_{20}) \end{array} \right\} = 0.$$

(a) What are the natural choices for $\tilde{z}$ and R? Formulate (4.34) for the linear system with this choice.

(b) Besides the regularity of the matrix $\begin{pmatrix} \bar{\alpha}_{11} & \bar{\alpha}_{12} \\ \bar{\alpha}_{21} & \bar{\alpha}_{22} \end{pmatrix}$, what is a necessary and sufficient condition on δ for the feasibility of an inclusion for Z_δ?

(c) Choose an empirical system and find the inclusion. To what extent is it an overestimation?

Historical and Bibliographical Notes 4

The *dynamical* algorithmic view of mathematics is at least as old as the *static* axiomatic view, which was moreover restricted to Greek geometry for a long time. "Recipes" for the calculation of results for application problems have been common since the beginnings of arithmetic computation in all cultures; they continue to appear in today's elementary arithmetic texts. The necessity of a formalized representation of algorithms appeared when the recipe was to be executed by an electronic rather than a human computer. Early Fortran ([4.1]) and Algol (see [4.2]) were the first widely used "programming languages" permitting the concise description of a numerical algorithm such that it could be automatically "compiled" into the machine code of an electronic computer. Interactive software systems, like Maple or Mathematica®, have retained the stepwise dynamical character of an algorithm more clearly than the highly developed programming languages necessary for the specification of large, sophisticated computational tasks.

The classical text on the stability of numerical algorithms is [4.3]; its first chapter is an excellent, well-readable introduction to the subject. At its start, it cites a quotation from Gauss's Theoria Motus which shows that he was well aware of the problem. Its full, potentially disastrous impact was only realized when computations with thousands of elementary arithmetic operations became commonplace.

In the 1960s and '70s, each manufacturer's mainframes featured a distinct floating-point arithmetic, to the dismay of numerical analysts. After the revolutionary advent of the *microprocessor* around 1980, the standard [4.4] was finally approved in 1985 and recognized as an American National Standard; it became a standard of the International Standards Organization (ISO) not much later. In 1986, Intel produced the first microprocessor chip which implemented the complete standard. In the beginning of the 1990s, "full IEEE arithmetic" was still mentioned in advertisements for PCs. A few years later, each general-purpose microprocessor had a standard-conforming arithmetic. Reference [4.3] contains an excellent description of floating-point arithmetic. The refinement technique described in section 4.3.4 has been described independently by various authors; a very general presentation may be found in [4.5].

Interval analysis has been mentioned in Hist.Bibl.Notes 3; the seminal text was [3.2]. The approach was pushed by scholars at the University of Karlsruhe who also developed hardware and software suitable for the purpose. The volumes [4.6] and the journal [4.7] convey an overview of the features and achievements of interval computation; [4.8] gives a thorough treatment of its application to systems of linear and nonlinear equations.

References

[4.1] Fortran (Formula Translator), developed during the later 1950s; cf. Ann. Hist. Comput. **6** (1984), (1).

[4.2] A.J. Perlis, K. Samelson (Eds.): Report on the Algorithmic Language Algol, Numer. Math. **1** (1959), 41–60.

[4.3] N.J. Higham: Accuracy and Stability of Numerical Algorithms, 2nd ed., SIAM, Philadelphia, 2002.

[4.4] IEEE Standard for Binary Floating-Point Arithmetic, ANSI/IEEE Standard 754-1985; cf. SIGPLAN Notices **22** (1987), (2) 9–25.

[4.5] H.J. Stetter: Sequential Defect Correction in High-Accuracy Floating-Point Algorithms, in: Numerical Analysis (Proceedings, Dundee 1983), Lecture Notes in Math. vol. 1066, Springer, Berlin, 1984, 186–202.

[4.6] U. Kulisch, H.J. Stetter (eds.): Scientific Computation with Automatic Result Verification, Computing Suppl. **6**, Springer, Vienna, 1988.

R. Albrecht, G. Alefeld, H.J. Stetter (eds.): Validation Numerics - Theory and Application, Computing Suppl. **9**, Springer, Vienna, 1993.

[4.7] Reliable Computing, an International Journal devoted to Reliable Mathematical Computations based on Finite Representations and Guaranteed Accuracy, Kluwer Publ., Dordrecht (NL).

[4.8] A. Neumaier: Interval Methods for Systems of Equations, Cambridge University Press, Cambridge (UK), 1990.

[4.9] J.H. Bleher, et al.: Fortran-SC - A Study of a FORTRAN Extension for Engineering/Scientific Computation with Access to ACRITH, Computing **39** (1987), 93–110.

Part II

Univariate Polynomial Problems

Chapter 5

Univariate Polynomials

We consider univariate polynomials

$$p(x; a) = \sum_{\nu=0}^{n} \alpha_\nu x^\nu = a^T \mathbf{x} \in \mathcal{P}_n^1 =: \mathcal{P}_n, \quad a^T \in \mathbb{C}^{n+1}, \quad \mathbf{x} := (1, x, \ldots, x^n)^T; \quad (5.1)$$

cf. (1.3) and Definition 1.4. Only when we explicitly say so, we assume $p(x; a)$ to be *monic*, i.e. to have $\alpha_n = 1$. In the data space $\mathcal{A} = \mathbb{C}^{n+1}$, our standard norm $\|..\|^*$ is a weighted maximum norm; cf. section 3.1.1.

Intrinsic univariate polynomials (whose coefficients are assumed as exact; cf. Definition 3.1) have been an object of mathematical investigation for centuries; their analytic and algebraic properties are well known, and even their numerical analysis is rather complete. We only recall some basic facts about such polynomials in the following section; then we devote our attention to aspects which come to light when some or all of the coefficients are *empirical*.

5.1 Intrinsic Polynomials

5.1.1 Some Analytic Properties

From the analytic point of view, univariate polynomials with real or complex coefficients are uniformly analytic or *holomorphic functions* from $\mathbb{C}$ to $\mathbb{C}$, where—as agreed in section 1.2—$\mathbb{C}$ denotes the *affine* complex plane, i.e. $\mathbb{C} := \{\xi + i\eta, \ \xi, \eta \in \mathbb{R}\}$: They are holomorphic in each point of $\mathbb{C}$; cf. Proposition 1.4. For $p \in \mathcal{P}_n$, $p^{(n+1)}$ is the zero polynomial, and the Taylor expansion of p at any $x \in \mathbb{C}$ is finite and at most of length n. Even if $p(x; a)$ has real coefficients, it is often helpful to regard it as a function from $\mathbb{C}$ to $\mathbb{C}$.

Since $p' \in \mathcal{P}_{n-1}$, there are at most $n - 1$ points ζ'_ν where the conformity of the map $p : \mathbb{C} \to \mathbb{C}$ can be disturbed: For $p \in \mathcal{P}_n$, the map $x \to p(x)$ is conformal except at the at most $n - 1$ branch points ζ'_ν where $p'(\zeta'_\nu) = 0$. Therefore, p is bijective in each open disk around some $\xi \in \mathbb{C}$ with $p'(\xi) \neq 0$, with radius $\rho = $ distance to the nearest branch point. Naturally, p can be analytically continued on a suitable Riemann surface, but, in this book, we are interested in computational algebra rather than computational complex analysis.

For each $p \in \mathcal{P}_n$ there exist n not necessarily distinct numbers $\zeta_\nu \in \mathbb{C}$, $\nu = 1(1)n$, such that

$$p(x) \;=\; \alpha_n \prod_{\nu=1}^{n} (x - \zeta_\nu)\,. \tag{5.2}$$

Thus, counting multiplicities, each $p \in \mathcal{P}_n$ has exactly n zeros $\zeta_\nu \in \mathbb{C}$. ζ is of (exact) multiplicity m iff $p(\zeta) = p'(\zeta) = \ldots = p^{(m-1)}(\zeta) = 0$, $p^{(m)} \neq 0$.

Important from our point of view is the consideration of a polynomial $p(x;a) \in \mathcal{P}_n$ as a function of its *coefficients* $a = (\alpha_0, \alpha_1, \ldots, \alpha_n) \in \mathbb{C}^{n+1}$: $p(x;a) \in \mathcal{P}_n$ is a *linear* function of a.

Now we consider some *particular* zero ζ of $p(x;a) \in \mathcal{P}_n$ as a function $F : \mathbb{C}^{n+1} \to \mathbb{C}$ of its coefficients a. (This F is a data$\to$result mapping in the sense of Definition 3.6.) For a *simple* zero ζ, the mapping $F : a \to \zeta(a)$ is differentiable and hence *Lipschitz continuous*; cf. Example 3.10 (3.39)–(3.40). Equation (3.40) implies that the condition of a zero with respect to a variation in some coefficient α_ν is bad if $|p'(\zeta)|$ is small. At a multiple zero, $p'(\zeta(a)) = 0$ and $\zeta(a)$ is no longer Lipschitz-continuous for that value of a.

Proposition 5.1. For an m-fold zero $(m > 1)$ ζ of $p(x;a) \in \mathcal{P}_n$, the mapping $F : a \to \zeta(a)$ is *Hölder continuous*, with an exponent $\frac{1}{m}$, at the value of the coefficient vector a which generates the multiple zero.

Proof: At the m-fold zero $\zeta(a)$, $p^{(\mu)}(\zeta(a);a) = 0$ for $\mu = 0(1)m-1$, while $p^{(m)}(\zeta(a);a) \neq 0$. Let $\zeta(a + \Delta a) = \zeta + \Delta\zeta$; then

$$
\begin{aligned}
0 \;&=\; p(\zeta(a + \Delta a); a + \Delta a) - p(\zeta(a); a) \\
&=\; p(\zeta + \Delta\zeta; a + \Delta a) - p(\zeta + \Delta\zeta; a) + p(\zeta + \Delta\zeta; a) - p(\zeta; a) \\
&=\; \textstyle\sum_\nu \Delta\alpha_\nu (\zeta + \Delta\zeta)^\nu + \frac{1}{m!}\, p^{(m)}(\tilde\zeta; a)\, \Delta\zeta^m\,,
\end{aligned}
$$

with $\tilde\zeta := \zeta + \lambda\Delta\zeta$, $\lambda \in (0,1)$. Let $|p^{(m)}(\zeta + \Delta\zeta; a)| \geq l_m$ for $|\Delta\zeta| \leq \rho$; then we have, for Δa such that $|\Delta\zeta| \leq \rho$,

$$
|\Delta\zeta|^m \;=\; \frac{m!}{|p^{(m)}(\tilde\zeta; a)|} \cdot \Big| \sum_\nu \Delta\alpha_\nu (\zeta + \Delta\zeta)^\nu \Big| \;\leq\; \Big(\frac{m!}{l_m} \max_\nu \big((|\zeta| + \rho)^\nu\big) \Big) \cdot \sum_\nu |\Delta\alpha_\nu|\,. \qquad \square
$$

This displays the well-known fact that, under a sufficiently small generic perturbation of p, an m-fold zero ζ splits into m simple zeros which differ from ζ by $O(\|\Delta a\|^{\frac{1}{m}})$. These cluster zeros are also extremely sensitive to changes of the coefficients; cf. Example 3.3. We will return to this situation in section 6.3.

On the other hand, we have this positive observation: For a univariate polynomial, the data$\to$zero mapping $F : a \to \zeta(a)$ is always *continuous*, except when $\zeta(a)$ diverges to ∞ which can only happen when $\alpha_n \to 0$.

When the coefficients of a univariate polynomial follow a continuous path $a(s) \in \mathbb{C}^{n+1}$, $0 \leq s \leq 1$, with $\alpha_n(s) \neq 0$, then each zero of p follows a continuous path in $\mathbb{C}$. In particular, if $\zeta(a(0))$ and $\zeta(a(1))$ are on *different sides of a closed curve* $\gamma \subset \overline{\mathbb{C}}$, there must exist (at least) one $s^* \in (0,1)$ with $\zeta(a(s^*)) \in \gamma$; i.e. the path $\zeta(a(s))$ cannot "jump" across a boundary.

Example 5.1: Polynomials in $\mathbb{R}[x]$ which have all their zeros ζ_ν in the *left halfplane* $\mathrm{Re}\, x < 0$ are generally called *stable* because this implies that the functions $\sum_\nu c_\nu \exp(\zeta_\nu t)$ tend to zero as the real parameter $t \to +\infty$. For a stable polynomial $\bar{p} = p(x; \bar{a}) \in \mathcal{P}_n$, it may be important to know the largest open neighborhood $N_{\delta^*}^\circ(\bar{p}) := \{p(x; a) : \|a - \bar{a}\|^* < \delta^*\}$ which contains only stable polynomials. If $p(x; a^*)$, with $\|a^* - \bar{a}\|^* = \delta^*$, is *not* stable it must have a zero on the imaginary axis. Compare section 6.1.3.   □

Example 5.2: Consider two polynomials $p_0 = p(x; a_0)$ and $p_1 = p(x; a_1)$ with *real* coefficients a_0, a_1 resp. Assume that p_1 has two more *real zeros* than p_0 and consider a real coefficient path $a(t)$ from a_0 to a_1. For the two conjugate complex zeros ζ, ζ^* which turn into two disjoint real zeros as $a_0 \to a_1$, the zero paths $\zeta(a(t))$, $\zeta^*(a(t))$ must remain conjugate-complex since $p(x; a(t))$ has real coefficients. Thus, $\zeta(a(t))$ real for $t \geq \bar{t}$ implies $\zeta^*(a(\bar{t})) = \zeta(a(\bar{t}))$ so that $p(x; a(t))$ must have a real *double* zero at some $\bar{t} \in (0, 1)$. Cf. section 6.1.2.   □

5.1.2 Spaces of Polynomials

The set $\mathcal{P} = \mathbb{C}[x]$ of *all* univariate polynomials with complex coefficients is an infinite-dimensional vector space, and each subset $\mathcal{P}_n$ of polynomials of maximal degree n is a vector space of dimension $n+1$. The study of *polynomial sequences (systems)* $\mathbf{p} = \{p_0, p_1, \ldots, p_n, \ldots\}$ $\subset \mathcal{P}$ such that

$$\mathcal{P}_n = \mathrm{span}\,(p_0, \ldots, p_n) \quad \forall n \tag{5.3}$$

has been an important subject of classical analysis.

The most natural such system is the system $\mathbf{p_0} := \{p_\nu(x) := x^\nu, \ \nu \in \mathbb{N}_0\}$ of *powers*; this system is also predominantly used in computational algebra to represent and handle polynomials in $\mathcal{P}$; cf. (1.3) and (5.1). We have all been conditioned to the use of $\mathbf{p_0}$ by its notational simplicity and by its overwhelming presence in theoretical work and also in many practical applications. Our own presentation in this book will be no exception from this rule.

However, one must be aware of the fact that, for the solution of computational problems, there may be severe disadvantages connected with the power basis for polynomials. On $[0, 1]$, e.g., the graphs of the powers x^n become less and less distinct with growing n; this is an indication of the modest suitability of $\mathbf{p_0}$ for the representation of polynomials of higher degrees in computational tasks. It is well known, e.g., that the interpolation of data on the $n+1$ equidistant knots $t_{n\nu} = \frac{\nu}{n}$, $\nu = 0(1)n$, by a polynomial $p = a^T \mathbf{x} \in \mathcal{P}_n$ leads to a linear system for the coefficients α_ν whose condition grows exponentially with n; cf. section 5.4.2.

A guiding principle for the design of other polynomial systems $\mathbf{p}$ which satisfy (5.3) is the introduction of a *scalar product* $[p_1, p_2]$ in $\mathcal{P}$; this supplies each vector space $\mathcal{P}_n$ with a Euclidean norm $\|p\| := [p, p]^{1/2}$ which makes the choice of a system of *orthogonal or unitary bases* with the property (5.3) an attractive choice.

Definition 5.1. For a specified scalar product $[p, q]$ in $\mathcal{P}$, a sequence of polynomials $p_\nu \in \mathcal{P}_\nu$ which satisfies (5.3) and the orthogonality condition

$$[p_\nu, p_\mu] = 0 \qquad \text{for } \nu \neq \mu \tag{5.4}$$

is a *system of orthogonal polynomials*. If $[p_\nu, p_\nu] = 1$ for all ν, the system is called *orthonormal*. The term *unitary* is used for systems of polynomials with complex coefficients which satisfy (5.4).   □

An immediate consequence of the orthogonality (5.4) is

Proposition 5.2. The *representation* of some $p \in \mathcal{P}_n$ in the orthogonal basis $\mathbf{p}$ is

$$p(x) = \sum_{\nu=0}^{n} \frac{[p, p_\nu]}{[p_\nu, p_\nu]} \, p_\nu(x) \, ; \tag{5.5}$$

the *best approximation* of some polynomial $q \in \mathcal{P}^N$, $N > n$, with respect to the scalar product norm of $\mathbf{p}$ by a polynomial $p_n^* \in \mathcal{P}_n$ is

$$p_n^*(x) = \sum_{\nu=0}^{n} \frac{[q, p_\nu]}{[p_\nu, p_\nu]} \, p_\nu(x) \, , \tag{5.6}$$

which satisfies $[p_n^*, q - p_n^*] = 0$.

Systems of orthogonal polynomials play a considerable role in computational analysis, in analogy to the role of orthogonal and unitary bases in the $\mathbb{R}^n$ and $\mathbb{C}^n$, resp., in computational linear algebra. The systematic use of such systems in computational *polynomial algebra* has not yet been investigated, it seems. We will not fill this white spot in this book; but some of our results are formulated in a form that their extension to more general bases is immediate.

All systems $\mathbf{p}$ of orthogonal polynomials, with respect to a scalar product of the form

$$[p, q] := \int_a^b p(\xi)\, q(\xi)\, w(\xi)\, d\xi \, , \quad \text{with a weight function } w(x) > 0 \, , \ x \in (a, b), \tag{5.7}$$

have a few fundamental properties in common which we list for easy reference:

1) The polynomials $p_\nu \in \mathbf{p}$ are connected by a *three-term recurrence relation*. With a normalization $p_n(x) = x^n + \dots$, it has the form

$$p_{\nu+1}(x) = (x - \beta_\nu)\, p_\nu(x) - \gamma_\nu\, p_{\nu-1}(x), \quad \nu \geq 1 \, . \tag{5.8}$$

2) p_n has n *real disjoint* zeros in (a, b); there are no zeros in $\mathbb{C}$ outside that interval.

3) Any $p \in \mathcal{P}$ with $[p, p_\nu] = 0$, $\nu = 0(1)n$, has at least $n + 1$ zeros with a sign change in (a, b).

We sketch the proof for 3) because it is a prototype for similar arguments and remarkably simple: For a contradiction, assume $p(x) = \prod_{\nu=1}^{m}(x - \zeta_\nu)^{\alpha_\nu}\, \bar{p}(x)$, with $m \leq n$, $\zeta_\nu \in (a, b)$, α_ν odd, and $\bar{p}(x) \neq 0$ in (a, b). Let $q(x) := \prod_{\nu=1}^{m}(x - \zeta_\nu) \in \mathcal{P}_m$; then $p(x)\, q(x) \not\equiv 0$ is without a sign change on (a, b) so that $[p, q] \neq 0$. But this contradicts the assumption on p and (5.5).

For computational purposes, the most important system of orthogonal polynomials are the *Chebyshev polynomials* T_ν which are intimately connected with the *trigonometric functions*:

$$T_\nu(\cos\varphi) = \cos(\nu\varphi) \quad \text{or} \quad T_\nu(x) = \cos(\nu \arccos x) \, , \quad \text{for all } \nu \, . \tag{5.9}$$

They may also be defined as the orthogonal system for the scalar product

$$[p, q] := \int_{-1}^{1} \frac{p(\xi)\, q(\xi)\, d\xi}{\sqrt{1 - \xi^2}} \, , \quad \text{with the normalization } T_\nu(1) = 1 \text{ for all } \nu \, . \tag{5.10}$$

The first few Chebyshev polynomials are

$$T_0(x) = 1 \, , \ T_1(x) = x \, , \ T_2(x) = 2x^2 - 1 \, , \ T_3(x) = 4x^3 - 3x \, , \ T_4(x) = 8x^4 - 8x^2 + 1 \, , \quad \text{etc.}$$

From (5.9), we obtain many of their properties directly: Their 3-term recurrence (5.8) is

$$T_{v+1}(x) = 2x\,T_v(x) - T_{v-1}(x)\,, \qquad v \geq 1\,; \tag{5.11}$$

the n *zeros* of T_n in $(-1, +1)$ are

$$\xi_v = \cos\left(\tfrac{2v-1}{2n}\,\pi\right), \quad v = 1(1)n\,; \tag{5.12}$$

the $n + 1$ *extrema* of T_n in $[-1, +1]$—including those at the end-points—lie at

$$\eta_v = \cos\left(\tfrac{v}{n}\,\pi\right), \quad v = 0(1)n\,. \tag{5.13}$$

Obviously, the values at the extrema are, alternatingly, $+1$ and -1 so that

$$\|T_n\|_{\max} := \max_{x \in [-1,+1]} |T_n(x)| = 1\,, \qquad \text{for all } n\,. \tag{5.14}$$

For a polynomial $p \in \mathcal{P}_n$ which is specified by its coefficients b^T with respect to the Chebyshev basis, i.e. $p(x) = \sum_{v=0}^n \beta_v T_v(x)$, it appears that its evaluation at some $\xi \in \mathbb{R}$ would require the evaluations $T_v(\xi)$ and linear combinations. However, due to the recurrence (5.11), there is a more efficient way of performing the evaluation of $p(\xi)$:

Algorithm 5.1 (Clenshaw-Curtis). The recursion

$$\begin{aligned}
\beta_{n-1} &:= \beta_{n-1} + 2\xi\,\beta_n\,, \\
\beta_v &:= \beta_v + 2\xi\,\beta_{v+1} - \beta_{v+2}\,, \qquad \text{for } v = n-2(-1)1\,, \\
\beta_0 &:= \beta_0 + \xi\,\beta_1 - \beta_2
\end{aligned}$$

yields $p(\xi) := \beta_0$.

This is checked by using (5.11) recursively on the terms of $p(x) = \sum_{v=0}^n \beta_v T_v(x)$, beginning at the top end. Similar algorithms exist for the evaluation of p from its coefficients with respect to an arbitrary orthogonal system of polynomials, due to the 3-term recurrence (5.8).

From (5.14), we have the bound for

$$\|p\|_{\max} := \max_{x \in [-1,+1]} |p(x)| = \max_{x \in [-1,+1]} \left| \sum_{v=0}^n \beta_v T_v \right| \leq \sum_{v=0}^n |\beta_v|. \tag{5.15}$$

Naturally, there is the analogous bound $\|\sum_{v=0}^n \alpha_v x^v\|_{\max} \leq \sum_{v=0}^n |\alpha_v|$ with the coefficients a^T with respect to the power basis, but the moduli of the β_v are generally smaller than those of the α_v for increasing n.

Example 5.3: Consider a truncated Taylor expansion for $\exp(-x^2)$, say

$$p(x) := 1 - x^2 + \tfrac{1}{2}x^4 - \tfrac{1}{6}x^6 + \tfrac{1}{24}x^8 - \tfrac{1}{120}x^{10},$$

whose representation in terms of Chebyshev polynomials is

$$p(x) := \tfrac{19807}{30720} - \tfrac{1925}{6144}T_2(x) + \tfrac{59}{1536}T_4(x) - \tfrac{41}{12288}T_6(x) + \tfrac{1}{6144}T_8(x) - \tfrac{1}{61440}T_{10}(x)\,.$$

The power expansion yields the bound $\tfrac{163}{60} \approx 2.7167$ for the maximum value of $|p(x)|$ in $[-1, +1]$ while the Chebyshev expansion gives the correct bound 1. $\quad\square$

By an affine transformation of the independent variable x, Chebyshev basis polynomials can be defined for arbitrary finite basis intervals $[a, b] \subset \mathbb{R}$ in place of $[-1, 1]$.

5.1.3   Some Algebraic Properties

Most computational problems with univariate polynomials can be solved without appeal to their algebraic structure. Nevertheless, in the context of this book, we will emphasize the algebraic background of such computational procedures. In particular, we will introduce a number of considerations and techniques which will prove crucial in dealing with *multivariate* polynomials.

Let us first list a number of well-known properties which distinguish ideals in $\mathcal{P}$ from those in $\mathcal{P}^s$, $s > 1$; cf. section 2.1:

- Each ideal $\mathcal{I} \subset \mathcal{P}$ can be *generated by one single polynomial s* which is unique except for scalar factors. (Each $\mathcal{I}$ is a "principal ideal.") Thus it is no restriction to write $\langle s \rangle$ for an arbitrary ideal in $\mathcal{P}$.

- Each ideal $\langle s \rangle \subset \mathcal{P}$ (except, trivially, $\langle 0 \rangle$) is *zero-dimensional*; for deg $s = n$, it has exactly n zeros counting multiplicities.

- Each ideal $\langle s \rangle \subset \mathcal{P}$ defines a *unique decomposition* for each polynomial $p \in \mathcal{P}$, of the form

$$p(x) = q(x)\,s(x) + r(x), \qquad \text{with } r \in \mathcal{P}_{n-1}, \tag{5.16}$$

which can be determined by a *division algorithm*; cf. section 5.3.

Some immediate consequences of these properties are:

Since an ideal $\mathcal{I} \subset \mathcal{P}$ consists of all polynomial multiples of its generator s, it is clear that the generator is a polynomial of lowest degree in $\mathcal{I}$. Therefore, no polynomial of degree $< n$ can be an element of $\langle s \rangle$ with deg $s = n$.

Due to (5.16), each *residue class* $[p]_{\langle s \rangle}$ contains precisely one polynomial $r \in \mathcal{P}_{n-1}$. These polynomials may be used as representatives of the residue classes in the *quotient ring* $\mathcal{R}[\langle s \rangle]$ (cf. section 2.2); thus, as a vector space, $\mathcal{R}[\langle s \rangle]$ is isomorphic with $\mathcal{P}_{n-1}$ and therefore of dimension n. With this identification, $\{1, x, \ldots, x^{n-1}\}$ is a *normal set* $\mathcal{N}[\langle s \rangle]$ of the ideal $\langle s \rangle$, with the associated normal set vector

$$\mathbf{b}(x) := (1, x, \ldots, x^{n-1})^T; \tag{5.17}$$

cf. Definition 2.19. It is also the *only* closed set of n monomials from $\mathcal{T}^1$ so that the normal set is *unique*. The unique remainder r in the polynomial division (5.16) of a polynomial $p \in \mathcal{P}$ by s represents the *normal form* $\mathrm{NF}_{\langle s \rangle}[p]$ of p with respect to the basis $\mathbf{b}$ of $\mathcal{R}[\langle s \rangle]$; cf. Definition 2.20. Thus, polynomial division is a fundamental operation with univariate polynomials; we will consider its properties, in particular for empirical polynomials, in section 5.3.

One can also "expand" univariate polynomials "in powers of s":

Proposition 5.3. For $s \in \mathcal{P}_n$, $p \in \mathcal{P}_N$, $k - 1 \leq N/n < k$, there exist unique coefficient vectors $d_\kappa^T = (\delta_{\kappa 0}, \ldots, \delta_{\kappa,n-1}) \in \mathbb{C}^n$, $\kappa = 0(1)k - 1$, such that, with the normal set vector $\mathbf{b}(x)$ of (5.17),

$$p(x) = (d_0^T \mathbf{b}(x)) + (d_1^T \mathbf{b}(x))\,s(x) + (d_2^T \mathbf{b}(x))\,(s(x))^2 + \ldots + (d_{k-1}^T \mathbf{b}(x))\,(s(x))^{k-1}. \tag{5.18}$$

Proof: From (5.16), we have $d_0^T \mathbf{b} := r$; recursive division of q by s yields $q = q_1 s + r_1$, $q_1 = q_2 s + r_2$, etc. so that $d_\kappa^T \mathbf{b} := r_\kappa$, $\kappa = 1(1)k - 1$. $\square$

Note that (5.18) reduces to an ordinary Taylor-expansion for $s(x) = x - \zeta \in \mathcal{P}_1$, where it takes the form

$$p(x) \;=\; \delta_0 + \delta_1\,(x - \zeta) + \delta_2\,(x - \zeta)^2 + \ldots + \delta_N\,(x - \zeta)^N\,.$$

Like this Taylor-expansion, we may split (5.18) into a main part and a remainder part. If we terminate the main part after the $(s(x))^m$ term, the remainder part behaves like $O(|x - \zeta_\nu|^{m+1})$ in the vicinity of a zero ζ_ν, $\nu = 1(1)n$, of s, in analogy to the Taylor-expansion. Thus, the main part is a good approximation of p in the vicinity of all zeros of s *simultaneously*.

Example 5.4: Take $s(x) = x^2 - 1$, $p(x) = 3\,x^6 - 7\,x^5 + 2\,x^4 + x^3 - 5\,x^2 + 4\,x$; an easy computation yields

$$p(x) \;=\; -\,2\,x + (8 - 13\,x)\,(x^2 - 1) + (11 - 7\,x)\,(x^2 - 1)^2 + 3\,(x^2 - 1)^3\,.$$

When we approximate p by the first two terms, the remaining two terms combine into $(3\,x^2 - 7\,x + 8)\,(x - 1)^2(x + 1)^2$; thus the remainder has *double* zeros at ± 1 and is quadratically small at both of these locations. Compare Figure 5.1. $\square$

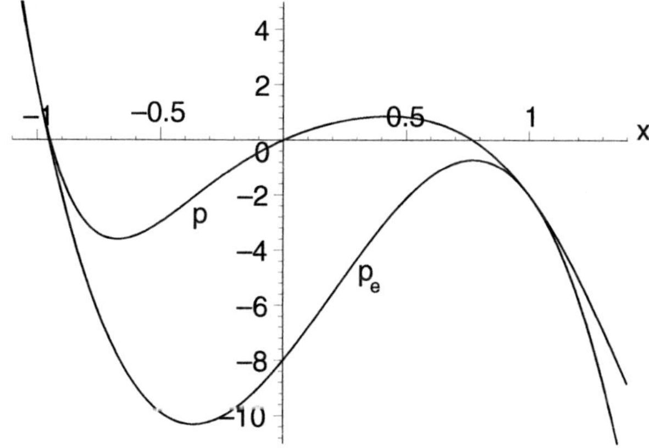

Figure 5.1.

If $\mathcal{I} \subset \mathcal{P}$ has the disjoint zeros $\zeta_\nu \in \mathbb{C}$, $\nu = 1(1)m$, with multiplicities $m_\nu \geq 1$, $\sum_\nu m_\nu = n$, the generator of $\mathcal{I}$ is

$$p(x) := \prod_{\nu=1}^{m}(x - \zeta_\nu)^{m_\nu}\,, \tag{5.19}$$

and $\langle p \rangle$ has the *primary decomposition*

$$\langle p \rangle \;=\; \cap_{\nu=1}^{m}\,\langle (x - \zeta_\nu)^{m_\nu} \rangle\,. \tag{5.20}$$

By Theorem 2.20, the *zeros* of $p \in \mathcal{P}_n$ define the conjugate *Lagrange basis* $\mathbf{c}_0^T$ of *the dual space* $\mathcal{D}[\langle p \rangle]$: For each zero ζ_ν there is a basis element $c_{\nu 0}$ with $c_{\nu 0}(f) := f(\zeta_\nu)$; cf. (2.36).

For an m_ν-fold zero, $m_\nu > 1$, there are $m_\nu - 1$ further basis elements $c_{\nu\mu}$, $\mu = 1(1)m_\nu - 1$, with $c_{\nu\mu}(f) = \partial^\mu[\zeta_\nu](f) := \frac{1}{\mu!}\frac{\partial^\mu}{\partial x^\mu} f(\zeta_\nu)$ so that

$$\mathbf{c}_0^T := \{c_{\nu\mu}^{(0)} \in \mathcal{P}_n^* : c_{\nu\mu}^{(0)} = \partial^\mu[\zeta_\nu], \ \mu = 0(1)m_\nu - 1, \ \nu = 1(1)m\}. \tag{5.21}$$

This follows from the primary decomposition (5.20) of $\langle p \rangle$.

The functionals c_ν, $\nu = 0(1)n - 1$, which assign to a polynomial $f \in \mathcal{P}$ its *normal form coefficients*, i.e. the coefficients of its remainder $r(x) = \sum_{\nu=0}^{n-1} c_\nu(f) x^\nu$ upon division by p, also form a basis $\mathbf{c}^T = (c_0, \ldots, c_{n-1})$ of $\mathcal{D}[\langle p \rangle]$. $\mathbf{c}$ is the *conjugate basis* of the normal set basis $\mathbf{b}$ of $\mathcal{R}[\langle p \rangle]$; cf. Proposition 2.13 and (2.26).

On the other hand, there is the *Lagrange basis* $\mathbf{b}_0$ of $\mathcal{R}[\langle p \rangle]$ which is conjugate to the basis $\mathbf{c}_0^T$ of (5.21) of the dual space; cf. Definition 2.11. For p from (5.19), its elements $b_{\nu\mu}^{(0)} \in \mathcal{P}_{n-1}$, $\mu = 0(1)m_\nu - 1$, $\nu = 1(1)m$, have to satisfy (cf. (2.25))

$$c_{\nu\mu}^{(0)}(b_{\nu\mu}^{(0)}) = 1, \quad c_{\nu'\mu'}^{(0)}(b_{\nu\mu}^{(0)}) = 0 \quad \text{for } \nu'\mu' \neq \nu\mu. \tag{5.22}$$

According to our considerations in section 2.4, the matrix $M_0 = \mathbf{c}_0^T(\mathbf{b})$ is the key to all relations between these fundamental bases of $\mathcal{R}[\langle p \rangle]$ and $\mathcal{D}[\langle p \rangle]$, resp.: We have (cf. (2.42) and (2.43))

$$\mathbf{b} = M_0\,\mathbf{b}_0, \quad \mathbf{b}_0 = M_0^{-1}\,\mathbf{b}, \tag{5.23}$$

$$\mathbf{c}_0^T = \mathbf{c}^T\,M_0, \quad \mathbf{c}^T = \mathbf{c}_0^T\,M_0^{-1}. \tag{5.24}$$

From (5.21), we have, e.g., with $m_1 = 3$ and $m_\mu = 1$ for $\mu > 1$,

$$M_0 = \begin{pmatrix} 1 & 0 & 0 & 1 & \cdots & 1 \\ \zeta_1 & 1 & 0 & \zeta_2 & \cdots & \zeta_m \\ \zeta_1^2 & 2\zeta_1 & 1 & \zeta_2^2 & \cdots & \zeta_m^2 \\ & \cdots & & & \cdots & \\ \zeta_1^{n-1} & (n-1)\zeta_1^{n-2} & \binom{n-1}{2}\zeta_1^{n-3} & \zeta_2^{n-1} & \cdots & \zeta_m^{n-1} \end{pmatrix}. \tag{5.25}$$

Since M_0 transforms one basis of $\mathcal{R}$ into another one, it must be *regular*.

Definition 5.2. For a polynomial ideal $\langle p \rangle$, $p \in \mathcal{P}_n$, with the monomial basis $\mathbf{b}$ for its quotient ring $\mathcal{R}[\langle p \rangle]$ (cf. (5.17)) and the Lagrange basis $\mathbf{c}_0^T$ for its dual space $\mathcal{D}[\langle p \rangle]$ (cf. (5.21)), the matrix $M_0 := \mathbf{c}_0^T(\mathbf{b})$ of (5.23)–(5.25) is called *Vandermonde matrix* and denoted by $V[\zeta_1, \ldots, \zeta_n]$ (or the like). With some of the multiplicities $m_\nu > 1$, the associated matrix M_0 (cf., e.g., (5.25)) is called generalized or *confluent* Vandermonde matrix and denoted by (e.g.) $V[\zeta_1, \zeta_1, \zeta_1, \zeta_2, \ldots]$ because an m_ν-fold zero ζ_ν may be regarded as a confluence of m_ν disjoint zeros. $\quad\square$

By (5.23)–(5.24), the rows and columns of the generalized Vandermonde matrix M_0 and its inverse M_0^{-1} have an intuitive meaning.

Corollary 5.4. In the situation under consideration,

(i) the columns $c_{\nu\mu}^{(0)}(\mathbf{b})$ of M_0 are the evaluations of the basis vector $\mathbf{b}(x)$ (and its derivatives) at the ζ_ν;

(ii) the rows $\mathbf{c}_0^T(b_\nu)$ of M_0 are the values of a particular basis element $b_\nu(x)$ (and its derivatives) at the ζ_μ;

(iii) the rows $\mathbf{c}^T(b_\nu^{(0)})$ of M_0^{-1} are the coefficients of the ν-th Lagrange basis element;

(iv) the columns $c_\nu(\mathbf{b}_0)$ of M_0^{-1}, due to (5.22), are the coefficients of x^ν in the various Lagrange basis elements.

Example 5.5: Consider $p(x) = (x-1)^3 x (x+1) = x^5 - 2x^4 + 2x^2 - x \in \mathcal{P}_5$. Obviously, there is a triple zero $\zeta_1 = 1$ and simple zeros $\zeta_2 = 0$, $\zeta_3 = -1$. Thus, the "evaluation basis" $\mathbf{c}_0^T$ of $\mathcal{D}[\langle p \rangle]$ has the basis $(\partial^0[1], \partial^1[1], \partial^2[1], \partial^0[0], \partial^0[-1])$; cf. (5.21). With the normal set basis $\mathbf{b} = (1, x, \ldots, x^4)^T$, we have (cf. (5.25))

$$M_0 = \mathbf{c}_0^T(\mathbf{b}) = \begin{pmatrix} 1 & 0 & 0 & 1 & 1 \\ 1 & 1 & 0 & 0 & -1 \\ 1 & 2 & 1 & 0 & 1 \\ 1 & 3 & 3 & 0 & -1 \\ 1 & 4 & 6 & 0 & 1 \end{pmatrix}, \quad M_0^{-1} = \begin{pmatrix} 0 & 17/8 & -3/8 & -13/8 & 7/8 \\ 0 & -5/4 & 3/4 & 5/4 & -3/4 \\ 0 & 1/2 & -1/2 & -1/2 & 1/2 \\ 1 & -2 & 0 & 2 & -1 \\ 0 & -1/8 & 3/8 & -3/8 & 1/8 \end{pmatrix}.$$

The 2nd column, e.g., is the evaluation of $\partial\,\mathbf{b}(x)$ at 1, while the 3rd row, e.g., is the evaluation of the third basis monomial x^2 and its derivatives at the various zeros.

By (iii), the rows of M_0^{-1} display the Lagrange basis $\mathbf{b}_0$ for $\mathcal{R}[\langle p \rangle]$:

$$b_{10}^{(0)}(x) = \tfrac{17}{8}x - \tfrac{3}{8}x^2 - \tfrac{13}{8}x^3 + \tfrac{7}{8}x^4, \quad b_{11}^{(0)}(x) = -\tfrac{5}{4}x + \tfrac{3}{4}x^2 + \tfrac{5}{4}x^3 - \tfrac{3}{4}x^4,$$

$$b_{12}^{(0)}(x) = \tfrac{1}{2}x - \tfrac{1}{2}x^2 - \tfrac{1}{2}x^3 + \tfrac{1}{2}x^4,$$

$$b_{20}^{(0)}(x) = 1 - 2x + 2x^3 - x^4, \quad b_{30}^{(0)}(x) = -\tfrac{1}{8}x + \tfrac{3}{8}x^2 - \tfrac{3}{8}x^3 + \tfrac{1}{8}x^4.$$

By (iv), the columns of M_0^{-1} yield the coefficients of the remainder polynomial mod p of degree 4 of a polynomial f from its values at the ζ_μ :

$$r(x) = (f(1), f'(1), f''(1)/2, f(0), f(-1))\, M_0^{-1}\, \mathbf{b}(x).$$

Naturally, r is the interpolation polynomial of degree 4 for the values of f at the ζ_μ. $\quad\square$

5.1.4 The Multiplicative Structure

Let us now consider the *multiplicative structure* of the rings $\mathcal{R}[\langle p \rangle]$ and $\mathcal{D}[\langle p \rangle]$ for $p \in \mathcal{P}_n$; cf. section 2.2. With the monomial basis $\mathbf{b} = (1, x, \ldots, x^{n-1})^T$ of $\mathcal{R}[\langle p \rangle]$, multiplication of the b_ν by x yields a result outside span $\mathbf{b}$ (before reduction mod $\langle p \rangle$) only for $b_n = x^{n-1}$:

$$x \cdot x^{n-1} = x^n \equiv -\frac{1}{\alpha_n} \sum_{\nu=0}^{n-1} \alpha_\nu x^\nu \mod \langle p \rangle.$$

Therefore, the *multiplication matrix* $A = A_x$ for $\mathcal{R}[\langle p \rangle]$ which satisfies $x \cdot \mathbf{b}(x) \equiv A\,\mathbf{b}(x) \mod \langle p \rangle$ (cf. Definition 2.6) is

$$A = \begin{pmatrix} 0 & 1 & & & 0 \\ 0 & 0 & 1 & & \\ & & & \ddots & \\ & & & & 1 \\ -\frac{\alpha_0}{\alpha_n} & -\frac{\alpha_1}{\alpha_n} & \cdots & \cdots & -\frac{\alpha_{n-1}}{\alpha_n} \end{pmatrix}. \tag{5.26}$$

This matrix (or its transpose or some other permutation) is often called the *Frobenius matrix* or *companion matrix* of p. In agreement with our notational conventions, we will always assume the form (5.26) for the Frobenius matrix of a polynomial p.

The multiplication matrix w.r.t. $\mathbf{b}$ for an arbitrary polynomial $q \in \mathcal{P}$, with $\mathrm{NF}_{\langle p \rangle}[q] = \mathbf{c}^T[q]\,\mathbf{b} =: a^T\mathbf{b}$, is, by (2.20),

$$A_q = \sum_{\nu=0}^{n-1} a_\nu A^\nu .$$

Thus the family $\overline{A}$ of all multiplication matrices w.r.t. the basis $\mathbf{b}$ of $\mathcal{R}[\langle p \rangle]$ simply consists of the polynomials in the Frobenius matrix A, and it is immediate that it is commuting.

For other bases of $\mathcal{R}[\langle p \rangle]$, the representation of multiplication mod $\langle p \rangle$ may be either obtained directly (cf. Exercise 5.1-1) or via (2.44).

From the "Central Theorem" (Theorem 2.27) we have:

Theorem 5.5. For $p \in \mathcal{P}_n$ and an arbitrary basis $\mathbf{b}$ of $\mathcal{R}[\langle p \rangle]$, let A represent multiplication by x in $\mathcal{R}[\langle p \rangle]$. If p has m disjoint zeros ζ_ν, $\nu = 1(1)m$, with multiplicities m_ν, then (cf. (2.50))

$$A\,(\,X_1\mid \dots \mid X_m\,) = (\,X_1 \mid \dots \mid X_m\,) \begin{pmatrix} T_1 & & 0 \\ & \ddots & \\ 0 & & T_m \end{pmatrix} \tag{5.27}$$

with

$$X_\nu = \begin{pmatrix} \vdots & \vdots & & \vdots \\ \mathbf{b}(\zeta_\nu) & \partial\mathbf{b}(\zeta_\nu) & \dots & \partial^{m_\nu-1}\mathbf{b}(\zeta_\nu) \\ \vdots & \vdots & & \vdots \end{pmatrix} \in \mathbb{C}^{n \times m_\nu},$$

$$T_\nu = \begin{pmatrix} \zeta_\nu & 1 & & 0 \\ & \ddots & \ddots & \\ & & \ddots & 1 \\ 0 & & & \zeta_\nu \end{pmatrix} \in \mathbb{C}^{m_\nu \times m_\nu}. \tag{5.28}$$

Proof: Application of the elements $c_{\nu\mu}^{(0)}$ of the basis $\mathbf{c}_0$ of $\mathcal{D}[\langle p \rangle]$ of (5.21) to $A \cdot \mathbf{b}(x) \equiv x \cdot \mathbf{b}(x)$ mod $\langle p \rangle$ yields, for $\nu = 1(1)m$,

$$\begin{aligned}
\partial^0[\zeta_\nu] : \quad & A\,\mathbf{b}(\zeta_\nu) = \zeta_\nu\,\mathbf{b}(\zeta_\nu), \\
\partial^1[\zeta_\nu] : \quad & A\,\partial\mathbf{b}(\zeta_\nu) = \zeta_\nu\,\partial\mathbf{b}(\zeta_\nu) + \mathbf{b}(\zeta_\nu), \\
& \quad \vdots \qquad\qquad\qquad \vdots \\
\partial^{m_\nu-1}[\zeta_\nu] : \quad & A\,\partial^{m_\nu-1}\mathbf{b}(\zeta_\nu) = \zeta_\nu\,\partial^{m_\nu-1}\mathbf{b}(\zeta_\nu) + \partial^{m_\nu-2}\mathbf{b}(\zeta_\nu). \quad \square
\end{aligned}$$

For any $p \in \mathcal{P}_n$, with arbitrary $n \in \mathbb{N}$, the $n \times n$ matrix A is immediately available; cf. (5.26). Therefore, (5.28) shows that the determination of the location and multiplicity of *all zeros* of p requires only the determination of the *eigenvalues* of A. Contrary to the general multivariate case (cf. section 2.4.3), the eigenvectors and the Jordan normal form of A carry no supplementary information. We return to the determination of zeros in section 5.1.5.

By (2.44) and (5.25), the $n \times n$ matrix $(X_1 \mid \ldots \mid X_m)$ in (5.27) is identical with M_0 and the $n \times n$ matrix diag (T_ν) with the T_ν from (5.28) is the multiplication matrix of $\mathcal{R}[\langle p \rangle]$ w.r.t the Lagrange basis $\mathbf{b}_0$. Thus, all the elements which play a role in the Central Theorem have an intuitive interpretation for univariate polynomial ideals.

Example 5.5, continued: For p of Example 5.5, we have

$$
A = \begin{pmatrix} 0 & 1 & & & \\ & 0 & 1 & & \\ & & 0 & 1 & \\ & & & 0 & 1 \\ 0 & 1 & -2 & 0 & 2 \end{pmatrix} \qquad \text{with eigenvalues } 1 \text{ (triple)}, 0, -1.
$$

The eigendecomposition (5.28) of A is

$$
A \begin{pmatrix} 1 & 0 & 0 & 1 & 1 \\ 1 & 1 & 0 & 0 & -1 \\ 1 & 2 & 1 & 0 & 1 \\ 1 & 3 & 3 & 0 & -1 \\ 1 & 4 & 6 & 0 & 1 \end{pmatrix} = \begin{pmatrix} 1 & 0 & 0 & 1 & 1 \\ 1 & 1 & 0 & 0 & -1 \\ 1 & 2 & 1 & 0 & 1 \\ 1 & 3 & 3 & 0 & -1 \\ 1 & 4 & 6 & 0 & 1 \end{pmatrix} \begin{pmatrix} 1 & 1 & 0 & 0 & 0 \\ 0 & 1 & 1 & 0 & 0 \\ 0 & 0 & 1 & 0 & 0 \\ 0 & 0 & 0 & 0 & 0 \\ 0 & 0 & 0 & 0 & -1 \end{pmatrix}.
$$

The right-hand matrix is the multiplication matrix w.r.t. the Lagrange basis. $\square$

Corollary 5.6. In the situation of Theorem 5.5, let A_q represent multiplication by $q \in \mathcal{P}$ in $\mathcal{R}[\langle p \rangle]$. Then we have

$$
A_q \, (X_1 \mid \ldots \mid X_m) = (X_1 \mid \ldots \mid X_m) \begin{pmatrix} T_1^{(q)} & & 0 \\ & \ddots & \\ 0 & & T_m^{(q)} \end{pmatrix}
$$

with X_ν as in (5.28) and

$$
T_\nu^{(q)} = \begin{pmatrix} q(\zeta_\nu) & \partial q(\zeta_\nu) & \cdots & \partial^{(m_\nu - 1)} q(\zeta_\nu) \\ & \ddots & \ddots & \vdots \\ & & \ddots & \partial q(\zeta_\nu) \\ 0 & & & q(\zeta_\nu) \end{pmatrix} \in \mathbb{C}^{m_\nu \times m_\nu}. \tag{5.29}
$$

The eigenvalues of A_q are the values of q at the zeros ζ_ν of p, with their respective multiplicities. *Proof*: By Proposition 2.9, *all* A_q have the same invariant subspaces X_ν. The $T_\nu^{(q)}$ follow by application of $\mathbf{c}_0$ to $A_q \cdot \mathbf{b}(x) \equiv q(x) \cdot \mathbf{b}(x) \bmod \langle p \rangle$ and (1.8). $\square$

Example 5.5, continued: For $q = (x+1)^2$, we have $A_q = A^2 + 2A + I = \begin{pmatrix} 1 & 2 & 1 & 0 & 0 \\ 0 & 1 & 2 & 1 & 0 \\ 0 & 0 & 1 & 2 & 1 \\ 0 & 1 & -2 & 1 & 4 \\ 0 & 4 & -7 & -2 & 9 \end{pmatrix}$

and

$$A_q \begin{pmatrix} 1 & 0 & 0 & 1 & 1 \\ 1 & 1 & 0 & 0 & -1 \\ 1 & 2 & 1 & 0 & 1 \\ 1 & 3 & 3 & 0 & -1 \\ 1 & 4 & 6 & 0 & 1 \end{pmatrix} = \begin{pmatrix} 1 & 0 & 0 & 1 & 1 \\ 1 & 1 & 0 & 0 & -1 \\ 1 & 2 & 1 & 0 & 1 \\ 1 & 3 & 3 & 0 & -1 \\ 1 & 4 & 6 & 0 & 1 \end{pmatrix} \begin{pmatrix} 4 & 4 & 1 & 0 & 0 \\ 0 & 4 & 4 & 0 & 0 \\ 0 & 0 & 4 & 0 & 0 \\ 0 & 0 & 0 & 1 & 0 \\ 0 & 0 & 0 & 0 & 0 \end{pmatrix},$$

in accordance with (5.27), (5.28), and (5.29). The eigenvalues of A_q are $4 = q(\zeta_1)$ (triple), $1 = q(\zeta_2)$, and $0 = q(\zeta_3)$.   □

5.1.5 Numerical Determination of Zeros of Intrinsic Polynomials

In principle, this problem is settled by Theorem 5.5: The zeros are the eigenvalues of the Frobenius matrix (5.26), with the same multiplicity. Since highly efficient and accurate software for the numerical determination of eigenvalues is widely available, this solves the problem in almost all practically relevant situations. Moreover, interactive packages like Maple and others, contain procedures (e.g., `solve`) which provide the zeros on the "push of a button."

It is true that the eigenvalue problem for (5.26) is nonnormal; but we know the condition of the individual eigenvalues from (3.39). (The condition of a result does not depend on how it is obtained.) The only two situations where an ill-conditioning must be expected are
- dense clustering or a multiplicity greater than 1,
- very large modulus.

For polynomials with real or complex coefficients, a multipicity ≥ 1 is a *singular* phenomenon: an m-fold zero disappears under infinitesimal perturbations and turns into an m-cluster. Thus, one can reasonably deal with zero clusters/multiple zeros numerically only in the context of empirical polynomials, which we will do in section 6.3. Of course, one can claim that—for an *intrinsic polynomial*—the distinction between a multiple zero and a cluster is well defined and that it must therefore be possible to determine the distinction and the location of zeros in a dense cluster to any desired accuracy. There is specialized software which achieves that goal (see below); but the goal is essentially academic and very rarely plays a role in scientific computing.

Zeros with a very large modulus may have a poor absolute condition by (3.39). They can arise only when the modulus of the leading coefficient α_n is tiny relative to other coefficients in p. With an intrinsic polynomial, we can determine the reciprocal polynomial (5.40) to any desired accuracy, which takes the zero close to the origin. Again, the highly accurate determination of such a zero is generally academic; in a practical context, the polynomial is empirical and the neighborhoods N_δ for $\delta = O(1)$ may contain polynomials of a lower degree so that ∞ is a valid zero. Compare section 5.2.3.

A third kind of potentially hard polynomial zero problems arises from polynomials of a *very high degree*, say $O(100)$ and more. With a grain of salt, one can say that such polynomials do not occur in scientific computing where the prevailing degrees are below 10 and two-digit degrees are quite rare. By their very nature, polynomials of a very high degree do not constitute reasonable models for real-life phenomena, from the approximation and from the handling point-of-view.

In any case, software has been designed which successfully computes approximations of a specified accuracy for all zeros of an intrinsic polynomial with floating-point coefficients, of an arbitrary degree. Naturally, such software must employ a multi-precision package which permits the use of higher and higher floating-point precision as it becomes necessary. For efficiency reasons, it is necessary to restrict this increase selectively to very ill-conditioned zeros or to very stringent requirements. For readers interested in this aspect of zero computation, we recommend the package MPsolve of Bini and Fiorentini which is well documented in [5.5] and incorporates a number of mathematical and numerical niceties. In this book, we do not deal with this highly specialized subject.

The only aspect of such procedures which we shortly describe now is the *simultaneous refinement* of all zeros of a univariate polynomial which utilizes the information on the current approximations of the other zeros in the refinement of a particular zero. We observe that, for p with exact zeros ζ_ν, $p'(x) = \sum_{\nu=1}^{n} \prod_{\nu' \neq \nu} (x - \zeta_{\nu'})$ and replace the derivative evaluation in Newton's method at an approximation $\tilde{\zeta}_\nu$ by $\prod_{\nu' \neq \nu} (\tilde{\zeta}_\nu - \tilde{\zeta}_{\nu'})$ where the $\tilde{\zeta}_{\nu'}$ are the currently available approximations for the other zeros. This leads to the *simultaneous refinement step*

$$\tilde{\zeta}_\nu \;\rightarrow\; \tilde{\zeta}_\nu - \frac{p(\tilde{\zeta}_\nu)}{\prod_{\nu' \neq \nu} (\tilde{\zeta}_\nu - \tilde{\zeta}_{\nu'})}\,, \qquad \nu = 1(1)n\,. \tag{5.30}$$

This idea has already appeared in Weierstrass' work ([5.2]) and been rediscovered several times; the most commonly used name for the procedure (5.30) appears to be *Durand-Kerner method* (cf. [5.3]).

It is interesting that (5.30) may also be interpreted as a vectorial Newton step for the Vieta system

$$\zeta_1 \ldots \zeta_n \;=\; (-1)^n \alpha_0\,,$$
$$\zeta_2 \ldots \zeta_n + \zeta_1 \zeta_3 \ldots \zeta_n + \ldots + \zeta_1 \ldots \zeta_{n-1} \;=\; (-1)^{n-1} \alpha_1\,,$$
$$\ldots$$
$$\zeta_1 + \zeta_2 + \ldots + \zeta_n \;=\; -\alpha_{n-1}\,,$$

for the n zeros ζ_ν of a monic polynomial with coefficients α_ν, $\nu = 0(1)n - 1$. Therefore, it converges quadratically from sufficiently close initial approximations. Theoretically, it works also for multiple zeros and clusters, as long as all approximations $\tilde{\zeta}_\nu$ remain disjoint, but numerical difficulties arise from close $\tilde{\zeta}_\nu$ and the quadratic convergence is lost. This shows once more that the numerical determination of clustered zeros requires special attention.

Another idea is *implicit deflation* with current approximations: In the simultaneous refinement step, we may derive the correction of $\tilde{\zeta}_\nu$ not from p but from the *rational function*

$$\tilde{p}_\nu(x) \;:=\; \frac{p(x)}{\prod_{\nu' \neq \nu} (x - \tilde{\zeta}_{\nu'})}\,, \qquad \nu = 1(1)n\,.$$

The classical Newton refinement for $\tilde{p}_\nu$ becomes

$$\tilde{\zeta}_\nu \;\rightarrow\; \tilde{\zeta}_\nu - \frac{p(\tilde{\zeta}_\nu)/p'(\tilde{\zeta}_\nu)}{1 - \frac{p(\tilde{\zeta}_\nu)}{p'(\tilde{\zeta}_\nu)} \cdot \sum_{\nu' \neq \nu} \frac{1}{\tilde{\zeta}_\nu - \tilde{\zeta}_{\nu'}}}\,. \tag{5.31}$$

Again, this procedure has been suggested—with varying arguments—independently by several authors; it is now commonly called *Aberth's method* ([5.4]). The simultaneous refinement (5.31) for $\nu = 1(1)n$ has a local convergence rate which is *cubic* for the case of disjoint zeros. Numerical experience indicates that Aberth's method is *globally convergent* for almost all initial approximations $\zeta_{0\nu}$, but no proof of this remarkable property has as yet been obtained.

Exercises

1. In section 2.4.2, we have considered the relations between two different bases of a quotient ring $\mathcal{R}[\mathcal{I}]$ and the conjugate bases of $\mathcal{D}[\mathcal{I}]$; cf. (2.42) and (2.43). Apply this for $\mathcal{R}[\langle p \rangle]$, $p \in \mathcal{P}_n$, with the monomial basis $\mathbf{b}$ and (in place of $\mathbf{b}_0$) the Chebyshev basis $\mathbf{b}_T$ (cf. section 5.1.2).

(a) For various values of n, determine the transformation matrices M_T and M_T^{-1} of $\mathbf{b} = M_T\,\mathbf{b}_T$; cf. (5.23). What are the associated transformations (5.24) for the conjugate dual space bases $\mathbf{c}$ and $\mathbf{c}_T$. Interpret the rows and columns of M_T and M_T^{-1}; cf. Corollary 5.4.

(b) Use a) to represent p of Example 5.5 in terms of the Chebyshev basis of $\mathcal{P}_5$.

(c) For this p, determine the multiplication matrix A_T of $\mathcal{R}[\langle p \rangle]$ w.r.t. $\mathbf{b}_T$: Like in (5.26), all rows except the last one are independent of p and determined by (5.11); for the last row, use the result of b). Another approach is through the use of (2.44).

(d) Compute the eigendecomposition (5.27) of A_T and interpret the result.

2. Consider the polynomial $p \in \mathcal{P}_{10}$ of Example 5.3.

(a) Compute the zeros ζ_μ of p from the multiplication matrices A and A_T of $\mathcal{R}[\langle p \rangle]$ w.r.t. $\mathbf{b}$ and $\mathbf{b}_T$, respectively. Plot the zeros in $\mathbb{C}$.

(b) Are the ζ_μ approximations of zeros of $\exp(-x^2)$? What do you conclude?

5.2 Zeros of Empirical Univariate Polynomials

We recall the framework which we have introduced in section 3.1 for the consideration of univariate polynomials with some coefficients of limited accuracy:

- An *empirical quantity* $(\bar{a}, e)$, with the *specified value* $\bar{a}$ and the *tolerance* e, defines a *family of neighborhoods* $N_\delta(\bar{a}, e)$ in the data space $\mathcal{A}$; cf. Definition 3.3.

- An *empirical polynomial* $(\bar{p}, e)$ has one or more empirical coefficients $(\bar{\alpha}_j, \varepsilon_j)$, $j \in \tilde{J}$; it defines a *family of polynomial neighborhoods* $N_\delta(\bar{p}, e)$, cf. Definition 3.4. $\tilde{J}$ is the *empirical support* of $(\bar{p}, e)$.

Remember that the concept of an "empirical polynomial" does not denote one "blurred" polynomial but a *family* of neighboring polynomials, each of which is a perfectly normal (exact) polynomial, with all standard analytic and algebraic properties. The parameter $\delta > 0$ indicates the *degree of validity*, cf. (3.3): data within N_δ, $\delta = O(1)$, are considered as *valid instances* for the situation under consideration.

In general considerations, we leave the choice of the *norm* in the space $\mathcal{A}$ or $\Delta\mathcal{A}$, resp., of the empirical coefficients open and formulate results in terms of a *tolerance-weighted dual norm* $\|..\|_e^*$; cf. (3.5). The associated vector norm $\|..\|_e$ in $\mathbb{C}^n$ is (3.7). In examples, we shall generally use a weighted max-norm for $\|..\|_e^*$.

In Chapter 3, it has become obvious that the aim of a computational task with empirical polynomials can only be the determination of a *valid result*; cf. Definition 3.8. All results within a *pseudoresult* set Z_δ with a sufficiently small δ of $O(1)$ must be considered as *equally acceptable* in the context of the task; cf. Definition 3.7. Therefore, in our Algorithmic Scheme 4.1 for the solution of a computational empirical algebraic problem, the computation of the *backward error* $\delta(\tilde{z})$ of an approximate result $\tilde{z}$ plays a central role; cf. Definition 3.12. If it is sufficiently small, we have obtained a *valid approximate result* and are finished. In well-behaved problems, this may happen without a refinement step in the Algorithmic Scheme 4.1.

Regarding the actual result indetermination, the influence of the *condition* of the algebraic problem has to be kept in mind; this has been explained and discussed in section 3.2 In the case of an *ill-conditioned problem*, a small backward error may well be associated with a poor determination of the solution. Clustered zeros represent such a case; they will be treated in section 6.3. Compare also Example 5.7.

In contrast to a backward error analysis, a *detailed forward error analysis* of the effects of the indetermination in the empirical data is either expensive or, most often, infeasible. In the case of zeros of univariate polynomials, it would require the explicit computation of inclusion sets for pseudozero sets in $\mathbb{C}^n$; in sections 4.4.2 and 4.4.3, we have discussed some principal limitations for this task. Generally, it is only the approximate *size* of the pseudoresult sets which is an important piece of information because it determines the precision with which the approximate results may reasonably be reported; cf. section 3.2.3.

In the following two subsections, we consider the two main tools in the solution of empirical algebraic problems, backward error and pseudoresult sets, for the task of determining zeros of empirical univariate polynomials.

5.2.1 Backward Error of Polynomial Zeros

We consider an empirical polynomial $(\bar{p}, e)$ of degree n, with empirical support $\tilde{J} \subset \{0, 1, .., n\}$, $|\tilde{J}| = M \le n + 1$, and empirical coefficients $(\bar{\alpha}_j, \varepsilon_j)$ for $j \in \tilde{J}$; cf. Definition 3.4. Here, $\bar{p}(x) = \sum_{\nu=0}^{n} \bar{\alpha}_\nu x^\nu$, $\bar{\alpha}_\nu \in \mathbb{C}$, $e = (\varepsilon_j > 0, j \in \tilde{J})$. The *intrinsic coefficients* $\bar{\alpha}_\nu$, $\nu \notin \tilde{J}$, of $(\bar{p}, e)$ which are *invariant* over all polynomials in $N_\delta(\bar{p}, e)$ will always be assumed to be known and fixed in a given context.

With a tolerance-weighted norm $\|..\|_e^*$ (cf. (3.5)), the polynomial neighborhoods $N_\delta(\bar{p}, e)$ are

$$N_\delta(\bar{p}, e) := \{\tilde{p}(x) = \sum_{\nu=0}^{n} \tilde{\alpha}_\nu x^\nu, \ \tilde{\alpha}_\nu \in \mathbb{C} : \ \|(\ldots, \Delta\alpha_j, \ldots; \ j \in \tilde{J})\|_e^* \le \delta; \ \tilde{\alpha}_\nu = \bar{\alpha}_\nu, \ \nu \notin \tilde{J}\};$$

$$(5.32)$$

cf. (3.12). For a $\bar{p}$ with *real* coefficients $\bar{\alpha}_\nu$, we must specify whether the $\tilde{\alpha}_\nu$ are also to be restricted to $\mathbb{R}$. Except when explicitly noted otherwise, we assume that $\varepsilon_n \ll |\alpha_n|$ if $n \in \tilde{J}$; this implies that *all polynomials in $N_\delta(\bar{p}, e)$ have the same degree n*. As in (5.32), we will often choose $(\bar{\alpha}_j, \ j \in \tilde{J})$ as the origin of the shifted data space $\Delta\mathcal{A}$, with the *deviations* $\Delta\alpha_j := \tilde{\alpha}_j - \bar{\alpha}_j$, $j \in \tilde{J}$, as components.

In section 3.3.1, we have used the case of a univariate empirical polynomial $(\bar{p}, e)$ to visualize the concepts introduced there. Therefore, we can immediately refer to Example 3.11

for the situation of *one* approximate zero $\tilde{z} \in \mathbb{C}$ of $(\bar{p}, e)$. Its backward error $\delta(\tilde{z})$ is the minimal δ such that there exists a polynomial p in $N_\delta(\bar{p}, e)$ which has $\tilde{z}$ as an *exact* zero; cf. Definition 3.12.

In the empirical data space $\Delta \mathcal{A} = \mathbb{C}^M$, the equivalent-data manifold $\mathcal{M}(\tilde{z})$ consists of those M-tuples $\Delta a = \tilde{a} - \bar{a}$ for which $p(\tilde{z}, \tilde{a}) = 0$; cf. Definition 3.11. Thus, $\mathcal{M}(\tilde{z})$ is the *linear* manifold (3.53)

$$\mathcal{M}(\tilde{z}) = \{\Delta a \in \Delta \mathcal{A} : \sum_{j \in \tilde{J}} \Delta \alpha_j \, \tilde{z}^j + \bar{p}(\tilde{z}) = 0\},$$

The linearity of $\mathcal{M}(\tilde{z})$ and its codimension 1 permit the explicit solution of the minimization problem (3.49) for $\delta(\tilde{z})$; cf. Proposition 3.9. For the weighted max-norm in $\Delta \mathcal{A}$, we have obtained

$$\delta(\tilde{z}) = \frac{|\bar{p}(\tilde{z})|}{\|(\tilde{z}^j)\|_e} = |\bar{p}(\tilde{z})| / \sum_{j \in \tilde{J}} \varepsilon_j |\tilde{z}|^j \tag{5.33}$$

in Example 3.11; cf. (3.54). This value is attained for

$$\rho \cdot \Delta \alpha_j^* = \delta(\tilde{z}) \, \varepsilon_j \, \frac{(\tilde{z}^j)^*}{|\tilde{z}^j|}, \quad j = \tilde{J}, \tag{5.34}$$

with $\rho \in \mathbb{C}$, $|\rho| = 1$, such that $\Delta a^* \in \mathcal{M}(\tilde{z})$; cf. Proposition 3.10 and (3.55).

Example 5.6: For our well-known polynomial (3.17), the value of $\bar{p}$ at $\tilde{z} = 1.43244$ is $\approx$ .000015. With the assumed tolerance 10^{-5} for (3.17), this yields, by (3.54),

$$\delta(\tilde{z}) \approx \frac{1.5 \cdot 10^{-5}}{(1 + \tilde{z} + \tilde{z}^2 + \tilde{z}^3) \cdot 10^{-5}} \approx .2$$

so that $\tilde{z}$ is a valid approximate zero of (3.17); cf. also Example 6.4. However, $\tilde{z}$ is sufficiently removed from the center of the cluster that it is no longer a valid zero of $\bar{p}'$: When we use (3.54) to compute the backward error of $\tilde{z}$ as an approximate zero of $\bar{p}'$, we obtain

$$\delta_1(\tilde{z}) = \frac{|\bar{p}'(\tilde{z})|}{(1 + 2\tilde{z} + 3\tilde{z}^2) \cdot 10^{-5}} \approx 26.$$

For $\tilde{z} = -1.41430$, on the other hand, which is quite close to the exact zero at ≈ -1.41421, we obtain $p(\tilde{z}) \approx .00196$ and $\delta(\tilde{z}) \approx 27$ which should exclude $\tilde{z}$ from being considered as a pseudozero of $(\bar{p}, e)$; cf. (3.3). This shows how the well-conditioned negative zero of $(\bar{p}, e)$ is far less affected by the indetermination in the coefficients.   □

As to be expected, the backward error of one zero of a univariate polynomial is simply a *weighted residual* $|p(\tilde{z})|$. The denominator $\|(\tilde{z}^j)\|_e$ in (3.54) shows that the size of the backward error depends critically on the choice of the *origin* for the x-axis: For large $|\tilde{z}|$, even a large residual $p(\tilde{z})$ can be annihilated by a small change in the coefficients! Vice versa, after a shift of the x-origin, the empirical coefficients of the new polynomial will have quite different tolerances; cf. section 3.3.2. Therefore, if tolerances are to be realistic, they must also account for the choice of the origin of a monomial basis for the empirical polynomial.

Example 5.7: It is widely known that a relative perturbation of only 1 bit in the single-precision floating-point representation of the coefficient $\alpha_{19} = 210$ of the Wilkinson polynomial $p_W(x) := \prod_{\mu=1}^{20}(x - \mu)$ induces huge changes $\Delta\zeta_\mu$ in the zeros $\zeta_\mu = \mu$ for $\mu \geq 10$; e.g., ζ_{20} becomes $\tilde{\zeta}_{20} \approx 23.549$. But when we consider this 1-bit perturbation as an indetermination in α_{19}, the huge residual $p_W(\tilde{\zeta}_{20}) \approx 1.78 \cdot 10^{21}$ is fully compatible with the pseudozero property of $\tilde{\zeta}_{20}$ because, with $\varepsilon_{19} = 2^{-16}$ as the only tolerance, $\varepsilon^{19}\,\tilde{\zeta}_{20}^{19}$ is just as large so that $\delta(\tilde{\zeta}_{20}) = 1$. $\quad\square$

The preceding discussion is in no way dependent on the basis used for the representation of $\bar{p}$; it generalizes immediately to an *arbitrary basis*. For example, when $\bar{p}$ is represented in terms of Chebyshev polynomials T_ν (cf. section 5.1.2), then the denominator in (3.54) simply changes to $\left\| \left(T_j(\tilde{\zeta})\right) \right\|_e$, as may be derived from the discussion in section 3.3.1.

Assume now that we have computed *several* approximate zeros $\tilde{\zeta}_1, \ldots, \tilde{\zeta}_m$ of an empirical polynomial $(\bar{p}, e)$ which are to be used concurrently. Then it would be deceptive to employ only individual validations via (3.54): Even for $m = 2$, it may well happen that $\tilde{\zeta}_1$ and $\tilde{\zeta}_2$ are valid approximate zeros while there is no $p \in N_\delta(\bar{p}, e)$, $\delta = O(1)$, which can have *both zeros simultaneously*. In this case, we must rather consider the set $\{\tilde{\zeta}_\mu\}$ of m zeros as *one result* $\tilde{z} \in \mathbb{C}^m$ whose backward error is to be determined.

The associated procedure is straightforward: The equivalent-data manifold $\mathcal{M}(\tilde{\zeta}_1, \ldots, \tilde{\zeta}_m)$ is defined by

$$\sum_{j\in\tilde{J}} \Delta\alpha_j\, \tilde{\zeta}_\mu^j + \bar{p}(\tilde{\zeta}_\mu) = 0, \quad \mu = 1(1)m, \tag{5.35}$$

with codimension m (except in some degenerate situations); the determination of the backward error $\delta(\tilde{\zeta}_1, \ldots, \tilde{\zeta}_m)$ via the minimization of $\|\Delta a\|_e^*$ over $\mathcal{M}$ is standard; cf. (3.57) in section 3.3.1. In section 6.2.1, we will find a natural *parameter representation* of $\mathcal{M}$ which generally leads to a simpler formulation of the minimization problem; cf. (6.20).

From (5.35), it is clear that $\delta(\tilde{\zeta}_1, \ldots, \tilde{\zeta}_m) \geq \max_\mu \delta(\tilde{\zeta}_\mu)$; almost always, the inequality is strict since the minimal distance to the origin of the intersection of linear manifolds is generally larger than each individual minimal distance.

Example 5.6, continued: Although a valid real zero of (3.17) may lie anywhere between (approx.) 1.385 and 1.445 and although there are always 3 zeros near $\sqrt{2}$ which may all be real, the values 1.40, 1.41, 1.42 cannot *simultaneously* be zeros of a polynomial in a small neighborhood of $\bar{p}$: The backward error $\delta(\tilde{z})$ of the result quantity $\tilde{z} = \{1.40, 1.41, 1.42\}$ with respect to $(\bar{p}, e)$ is > 3400; cf. also Example 6.4. $\quad\square$

The expression (3.54) for the backward error of an approximate zero cannot be used for a *complex* approximate zero $\tilde{\zeta}$ of a *real* empirical polynomial if only real variations of the coefficients are permitted. The reason is that the underlying Proposition 3.9 is based on the relation (3.6) which assumes that u and v are from matching dual spaces, i.e. *both* vectors must either be in $\mathbb{R}^M$ or in $\mathbb{C}^M$! Thus, the naive validation of $\tilde{\zeta} \in \mathbb{C}$ by (3.54) would be misleading because the implied nearest polynomial with exact zero $\tilde{\zeta}$ is complex. Obviously, we must verify that $\tilde{\zeta}$ *and its complex conjugate* $\tilde{\zeta}^*$ are valid simultaneous zeros! By taking real and imaginary parts of the complex manifold (3.53), with $\Delta\alpha_j$ *real*, we obtain

$$\sum_{j\in\tilde{J}} \Delta\alpha_j\, \mathrm{Re}\,(\tilde{\zeta}^j) + \mathrm{Re}\,\bar{p}(\tilde{\zeta}) = 0\,, \qquad \sum_{j\in\tilde{J}} \Delta\alpha_j\, \mathrm{Im}\,(\tilde{\zeta}^j) + \mathrm{Im}\,\bar{p}(\tilde{\zeta}) = 0\,,$$

as the specification of the equivalent-data manifold $\mathcal{M}(\tilde{\zeta}, \tilde{\zeta}^*)$ of codimension 2 in $\mathbb{R}^M$, and $\delta(\tilde{\zeta}, \tilde{\zeta}^*)$ is easily found from there.

Example 5.6, continued: Consider once more our polynomial (3.17) and $\tilde{\zeta} = 1.414 + .029\,i$ for which (3.54) yields a backward error $\delta(\tilde{\zeta}) \approx .96$ so that $\tilde{\zeta}$ is a valid approximate zero for *complex deviations* $\Delta\alpha_\nu$. But from the *real* representation

$$\Delta\alpha_0 + \Delta\alpha_1 \operatorname{Re} \tilde{\zeta} + \Delta\alpha_2 \operatorname{Re} \tilde{\zeta}^2 + \Delta\alpha_3 \operatorname{Re} \tilde{\zeta}^3 = -\operatorname{Re} \bar{p}(\tilde{\zeta}),$$

$$\Delta\alpha_1 \operatorname{Im} \tilde{\zeta} + \Delta\alpha_2 \operatorname{Im} \tilde{\zeta}^2 + \Delta\alpha_3 \operatorname{Im} \tilde{\zeta}^3 = -\operatorname{Im} \bar{p}(\tilde{\zeta}),$$

of the manifold $\mathcal{M}(\tilde{\zeta}, \tilde{\zeta}^*)$ with real codimension 2, we find $\delta(\tilde{\zeta}, \tilde{\zeta}^*) \approx 54$ which shows that $\tilde{\zeta}$ cannot be a zero of a *real* polynomial within the tolerance neighborhood of $(\bar{p}, e)$; cf. also Example 6.4. $\quad\square$

5.2.2   Pseudozero Domains for Univariate Polynomials

In section 3.1.3, we have introduced the concept of *data→result mappings* as a basis for the definition of sets of valid results or *pseudoresult sets* Z_δ of empirical algebraic problems; cf. Definitions 3.6 to 3.8. For the problem of finding *some* zero $\tilde{z}$ of the empirical univariate polynomial $(\bar{p}, e)$, we obtain an explicit expression for the *δ-pseudozero set* directly from the expression (3.54):

$$Z_\delta(\bar{p}, e) \;=\; \{\zeta \in \mathbb{C} \,:\, |\bar{p}(\zeta)| \leq \|(\zeta^\nu)\|_e \cdot \delta\} \subset \mathbb{C}. \tag{5.36}$$

In this section, we restrict our considerations to the case where the empirical polynomial $(\bar{p}, e)$, $\bar{p} \in \mathcal{P}_n$, has n well-separated zeros; here, "well-separated" means that they remain separated for each $\tilde{p} \in N_\delta(\bar{p}, e)$, $\delta < \bar{\delta} = O(1)$. (The important case of *clustered zeros* will be treated in section 6.3.) Under this assumption, we have n separate data→result mappings $F_\nu : a \rightarrow \zeta_\nu$, $\nu = 1(1)n$. According to our convention, $a \in \mathcal{A} = \mathbb{C}^M$ is the vector of the $M \leq n + 1$ *empirical* coefficients of the empirical polynomial $(\bar{p}, e)$; the (fixed) intrinsic coefficients are incorporated into the definition of $(\bar{p}, e)$ and of the mappings F_ν.

By Definition 3.7, each data→result mapping F_ν defines a pseudozero set $Z_{\delta,\nu}(\bar{p}, e)$.

Definition 5.3. For an empirical polynomial $(\bar{p}, e)$, $\bar{p} \in \mathcal{P}_n$, with well-separated zeros, the pseudoresult sets for the individual zeros ($\nu = 1(1)n$)

$$Z_{\delta,\nu}(\bar{p}, e) := \{\zeta \in \mathbb{C} \,:\, \zeta = F_\nu(a), \, a \in N_\delta(\bar{a}, e)\} \subset \mathbb{C}, \quad 0 < \delta \leq \bar{\delta}, \tag{5.37}$$

are the *(δ-)pseudozero domains* of $(\bar{p}, e)$. $\quad\square$

Proposition 5.7. For an empirical univariate polynomial $(\bar{p}, e)$ with well-separated zeros and for sufficiently small $\delta > 0$, each pseudozero domain $Z_{\delta,\nu}(\bar{p}, e)$ contains exactly one zero of each polynomial $\tilde{p} \in N_\delta(\bar{p}, e)$.

Proof: Since the zeros of $\bar{p}$ are disjoint and since the zeros are continuous functions of the coefficients (cf. section 5.1.1), there must exist $\bar{\delta} > 0$ such that the sets $Z_{\delta,\nu}(\bar{p}, e)$ remain separated for $\delta < \bar{\delta}$. For $\tilde{p} \in N_\delta(\bar{p}, e)$, $\delta < \bar{\delta}$, consider $p(x; t) := (1 - t)\,\bar{p}(x) + t\,\tilde{p}(x), t \in [0, 1]$. The one zero $\zeta_\nu(0)$ of $\bar{p}$ in $Z_{\delta,\nu}$ is the beginning of a path $\zeta_\nu(t)$ of zeros of $p(x; t)$ which leads to a zero $\zeta_\nu(1)$ of $\tilde{p}$ and remains in $Z_{\delta,\nu}$ because the $p(x; t)$ remain in $N_\delta(\bar{p}, e)$. If there were a further zero of $\tilde{p}$ in $Z_{\delta,\nu}$, the reverse argument would imply a further disjoint zero of $\bar{p}$ in $Z_{\delta,\nu}(\bar{p}, e)$. $\square$

For special norms, pseudozero domains of univariate polynomials have been suggested and analyzed by Mosier. It is remarkable that he has already introduced the idea of considering *families* of pseudozero domains in his seminal paper [3.3]. Pseudozero domains represent the potential variation of the individual zeros due to the indetermination in $(\bar{p}, e)$. For an empirical polynomial with well-separated zeros, the domains $Z_{\delta,\nu}(\bar{p}, e)$ are the connected components of the pseuodzero set (5.36).

If we restrict attention to the real domain for real polynomials, it suffices to find the end points of the real intervals which compose the set

$$Z_\delta(\bar{p}, e) := \{\xi \in \mathbb{R} : |\bar{p}(\xi)| \le \|(\ldots, \varepsilon_j|\xi|^j, \ldots)\| \cdot \delta\} \subset \mathbb{R};$$

cf. (5.36). Since absolute values $|\xi|$ of real quantities ξ may be segmentwise replaced by $+\xi$ or $-\xi$, this is a straightforward computation. In the complex domain, one has to employ contour-following techniques for the tracing of the boundary of a $Z_{\delta,\nu} \subset \mathbb{C}$. This is feasible; but it is generally expensive, particularly compared with the effort for the computation of an approximate zero $\tilde{\zeta}_\nu$ and of its backward error. As explained in section 3.2.3, the (approximate) computation of the condition numbers of the individual zeros ζ_ν is straightforward and permits the estimation of the approximate sizes of the $Z_{\delta,\nu}$.

In Example 3.10 in section 3.2.2, we have applied this approach to the indetermination of the zeros of univariate empirical polynomials and obtained the estimates (3.39) for the condition number w.r.t. the perturbation of an individual coefficient and (3.40) for a combined condition number; this yields the estimates

$$Z_{\delta,\nu} \overset{\subset}{\approx} \{w \in \mathbb{C} : |w - \bar{\zeta}_\nu| \le \frac{\|\left(\bar{\xi}_\nu^j\right)\|_e}{|\bar{p}'(\bar{\zeta}_\nu)|} \cdot \delta\}, \qquad \text{diam } Z_{\delta,\nu} \approx 2\frac{\|\left(\bar{\xi}_\nu^j\right)\|_e}{|\bar{p}'(\bar{\zeta}_\nu)|} \cdot \delta; \qquad (5.38)$$

cf. (3.44). For reasonably small tolerances (cf. Proposition 3.6), the information in (5.38) is just as valuable as a plot of the pseudozero domains. It also indicates whether the zeros of $(\bar{p}, e)$ are well-separated in the above sense.

Moreover, a plot of the complete collection of pseudozero domains conveys the same miscomprehension as the collection of the backward errors $\{\delta(\tilde{\zeta}_1), \ldots, \delta(\tilde{\zeta}_n)\}$ for a set of approximate zeros $\tilde{\zeta}_\nu$, $\nu = 1(1)n$, of $(\bar{p}, e)$: While we may have $\tilde{\zeta}_\nu \in Z_{\delta,\nu}$ or—equivalently— $\delta(\tilde{\zeta}_\nu) \le \delta$ for each $\nu = 1(1)n$, there will, generally, *not exist* a polynomial $\tilde{p} \in N_\delta(\bar{p}, e)$ with $\tilde{p}(\tilde{\zeta}_\nu) = 0$ for each $\nu = 1(1)n$! In section 5.2.1, this has led us to consider $m > 1$ particular zeros $\zeta_{\nu_1}, \ldots, \zeta_{\nu_m}$ as *one* result $\tilde{z}$ of a corresponding data→result mapping $F_{\nu_1\ldots\nu_m} : \mathcal{A} \to \mathbb{C}^m$.

Definition 5.4. For an empirical polynomial $(\bar{p}, e)$, $\bar{p} \in \mathcal{P}_n$, with well-separated zeros, the *simultaneous (δ-)pseudozero domain* of m zeros ζ_{ν_μ}, $\mu = 1(1)m$, is

$$Z_{\delta,\nu_1\ldots\nu_m} := \{(\zeta_{\nu_1}, \ldots, \zeta_{\nu_m}) = F_{\nu_1\ldots\nu_m}(a), \ a \in N_\delta(\bar{a}, e)\} \subset \mathbb{C}^m. \quad \square \qquad (5.39)$$

Naturally, $Z_{\delta,v_1\ldots v_m} \subset Z_{\delta,v_1} \times \ldots \times Z_{\delta,v_m}$; in fact, the Z_{δ,v_μ} are the *projections* of $Z_{\delta,v_1\ldots v_m}$ onto the m component spaces $\mathbb{C}$. But, for the same reasons as explained in section 5.2.1, $Z_{\delta,v_1\ldots v_m}$ is generally a *proper* subset of the Cartesian product of the Z_{δ,v_μ}: The choice of a particular value for a $\tilde{\zeta}_v$ restricts the choice of values for the remaining zeros if they are to be zeros of the *same* neighboring polynomial.

If we assign values for $m = M$ approximate zeros, then there is generally[9] a unique relation between the m-tuple $(\zeta_{v_1}, \ldots, \zeta_{v_M})$ and a point $\Delta a \in \Delta \mathcal{A} = \mathbb{C}^M$; i.e. the equivalent-data manifold reduces to that one point. Thus, $Z_{\delta,v_1\ldots v_m}$ consists precisely of those points in $\mathbb{C}^M$ for which the associated Δa are in $N_\delta(\bar{a}, e)$. For $m > M$, the equivalent-data manifold and the simultaneous pseudozero domain are generally empty. If we consider a complete set $(\tilde{\zeta}_1, \ldots, \tilde{\zeta}_n)$ of approximate zeros, the associated domain $Z_{\delta,1\ldots n} \subset \mathbb{C}^n$ is generally nonempty iff there are at least n empirical coefficients (out of $n+1$) in $(\bar{p}, e)$. Each n-tuple $(\zeta_1, \ldots, \zeta_n) \in Z_{\delta,1\ldots n} \subset \mathbb{C}^n$ is the complete exact zero set $Z_0[\tilde{p}]$ of some polynomial $\tilde{p} \in N_\delta(\bar{p}, e)$.

In Example 4.9 with Figure 4.2, we have determined the simultaneous pseudozero domain of the two real zeros of a quadratic empirical polynomial; we have seen that $Z_{\delta,1} \times Z_{\delta,2}$ is *not* a realistic description of the indetermination in the zero set. For higher degrees and more than two zeros, this effect may become much more extreme. For two (and more) complex zeros or more than three real zeros, a graphical representation of the simultaneous pseudozero domain is not feasible. But the concept is important as a tool for understanding how the actual indetermination in the complete zero set of a polynomial may be strongly exaggerated by the Cartesian product of the domains $Z_{\delta,v}$ in $\mathbb{C}$. At the same time, the determination of the *simultaneous backward error* $\delta(\tilde{\zeta}_1, \ldots, \tilde{\zeta}_m)$ from (5.35) is a safe check for the simultaneous validity of several approximate zeros of an empirical polynomial.

5.2.3 Zeros with Large Modulus

The accurate computation of zeros with a very large modulus may often present numerical difficulties: The computation of the residual of p at such a zero ξ and other evaluations involving ξ will generally involve a very strong cancellation of leading digits. Rescaling of the variable may help in cases where the reason for the large moduli has been an original ill-chosen scaling. If there are only one or a few large zeros, it is generally advisable to compute these zeros from the *reciprocal polynomial*

$$q(y) := y^n\, p\left(\frac{1}{y}\right) = \sum_{v=0}^{n} \alpha_{n-v}\, y^v \qquad \text{for } \alpha_0 \neq 0 ; \tag{5.40}$$

since q contains the original coefficients α_v of p, their tolerances are *unaffected* by this transformation which is important in the case of empirical polynomials. Trivially, to each zero η_μ of q there corresponds a zero $\xi_\mu = 1/\eta_\mu$ of p. Due to the assumption $\alpha_0 \neq 0$, $0 \notin Z[p]$.

Proposition 5.8. For an empirical polynomial $(\bar{p}, e)$, consider the reciprocal empirical polynomial $(\bar{q}, e)$ obtained from (5.40), with the tolerances of the coefficients unchanged. If η is a valid approximate zero of $(\bar{q}, e)$, then $\xi = 1/\eta$ is a valid approximate zero of $(\bar{p}, e)$.

---

[9]The word "generally" in many places refers to the fact that special symmetric positions of the $\tilde{\zeta}_v$ and/or the coefficients $\bar{\alpha}_v$ may render the standard dimension counts invalid.

Proof: Consider the backward errors

$$\delta_p(\xi) \;=\; \frac{|\sum_\nu \alpha_\nu \xi^\nu|}{\sum_\nu \varepsilon_\nu |\xi|^\nu} \;=\; \frac{|\sum_\nu \alpha_\nu (\frac{1}{\eta})^\nu|}{\sum_\nu \varepsilon_\nu |\frac{1}{\eta}|^\nu} \;=\; \frac{|\sum_\nu \alpha_{n-\nu} \eta^\nu|}{\sum_\nu \varepsilon_{n-\nu} |\eta|^\nu} \;=\; \delta_q(\eta). \qquad \square$$

Zeros with large moduli occur when p has one or several tiny leading coefficients. In this case, the reciprocation of p is not only numerically beneficial but also improves the efficiency of the Newton refinement of a large zero. Consider

$$p(x) \;:=\; \varepsilon\, x^{n+1} + \alpha_n\, x^n + \alpha_{n-1} x^{n-1} + \ldots + \alpha_0, \qquad \text{with } |\varepsilon| \ll \alpha_n.$$

Then there is one large zero $\xi_0 \approx \hat{\xi} = -\alpha_n/\varepsilon$ of p which corresponds to a tiny zero $\eta_0 \approx \hat{\eta} = -\varepsilon/\alpha_n$ of the reciprocal polynomial

$$q(y) \;=\; \varepsilon + \alpha_n\, y + \alpha_{n-1} y^2 + \ldots + \alpha_0\, y^{n+1}.$$

It is easily found that one Newton step for q from $\hat{\eta}$ leads to the approximation

$$\tilde{\eta}_0 \;=\; \hat{\eta} - \frac{\alpha_{n-1}}{\alpha_n}\, \hat{\eta}^2 + \left(\frac{\alpha_{n-2}}{\alpha_n} - 2\left(\frac{\alpha_{n-1}}{\alpha_n}\right)^2\right) \hat{\eta}^3 + O(\varepsilon^4) \;=\; \eta_0\,(1 + O(\varepsilon^3))$$

of the exact tiny zero η_0 of q, and correspondingly to the excellent relative approximation $\tilde{\xi}_0 := 1/\tilde{\eta}_0 = \xi_0\,(1 + O(\varepsilon^3))$ of the huge zero ξ_0 of p. One Newton step from $\hat{\xi}$ for p, on the other hand, leads to a value $\hat{\xi}_\infty = \xi_0\,(1 + O(\varepsilon^2))$.

Naturally, computation and the refinement of zeros of p with a modulus of $O(1)$ must be performed with the polynomial p, irrespective of the tiny leading coefficient; the reciprocal polynomial is only to be used for the exceptional very large zeros.

When an empirical polynomial $(\bar{p}, e)$ has a tiny leading coefficient $(\bar{\alpha}_{n+1}, \varepsilon_{n+1})$ whose tolerance ε_{n+1} is larger than $|\alpha_{n+1}|$, then 0 is a valid value for α_{n+1}. This implies that 0 is a valid zero of the reciprocal polynomial $(\bar{q}, e)$ and ∞ is a valid zero of $(\bar{p}, e)$; thus the pseudozero domain $Z_\delta(\bar{p}, e)$ which contains the large zero of $\bar{p}$ is *not bounded*. In $\mathbb{C}$, this domain will be a connected set about the complex point ∞; its restriction to $\mathbb{R}$ will separate into unbounded intervals on both ends of the real line. In some applications, it may be more reasonable to observe that the degree n polynomial $\tilde{p}(x) := \sum_{\nu=0}^{n} \bar{\alpha}_\nu x^\nu \in N_\delta(\bar{p}, e)$ is a *valid instance* of the empirical polynomial $(\bar{p}, e)$ and thus of the modelled situation. This indicates that the huge zero is "spurious" and has no meaning for the analysis.

More interesting, from the mathematical as well as from the numerical point of view, is the case when the reciprocal polynomial has a *cluster of zeros about 0*, or—correspondingly—the polynomial p has a zero cluster about ∞, i.e. several related zeros with a very large modulus. This case is discussed separately in section 6.3.5.

Exercises

1. Consider the empirical polynomial $(\bar{p}, e)$ with

$$\bar{p}(x) \;:=\; x^5 - .552\,x^4 - 5.616\,x^3 + 4.630\,x^2 + 3.693\,x - 1.611$$

and $\varepsilon_j = .0005$, $j = 0(1)4$, and only real deviations $\Delta\alpha_j$.

(a) Plot the backward error $\delta(z)$ of an individual approximate zero at z for $-3 \leq z \leq 2.5$. What can you conclude about the location and the condition of the zeros? Are the zeros well-separated?

(b) Since the zeros of $\bar{p}$ are disjoint, there is a smallest $\bar{\delta} > 0$ so that the pseudozero domains $Z_{\delta,\nu}$ are disjoint for $\delta < \bar{\delta}$. Determine $\bar{\delta}$ by plotting $\delta(z)$ between the two close zeros of $\bar{p}$, or by solving $\delta'(z) = 0$. Give a qualitative distinction of $N_{\bar{\delta}}(\bar{p}, e)$ and of the N_δ, $\delta > \bar{\delta}$, in terms of the zeros.

(c) For the zeros $\bar{\zeta}_\nu$ of $\bar{p}$, compare the estimates for the domains $Z_{1,\nu}$ from (5.38) with the values obtained from the solutions of $\delta(z) = 1$. What happens for the two close zeros? Is there a meaningful interpretation of the inclusions from (5.38)?

(d) Assume that we fix an approximate zero of $(\bar{p}, e)$ at some value $\hat{\zeta}_\nu$ in some $Z_{\delta,\nu}$. Convince yourself that the simultaneous 1-pseudozero domains for another zero $\zeta_{\nu'}$ together with $\hat{\zeta}_\nu$ are smaller than the original $Z_{1,\nu'}$. (Compute $\delta(\zeta_{\nu'}, \hat{\zeta}_\nu)$ at the boundaries of $Z_{1,\nu'}$.)

(e) Fix $\hat{\zeta}_5 = 1.66$ which lies inside the domain for the two close zeros. Find (experimentally) the range of values for the other zero ζ_4 such that ζ_4 and $\hat{\zeta}_5$ are simultaneously valid for $(\bar{p}, e)$. Comment.

(f) Fix all zeros $\bar{\zeta}_\nu$ of $\bar{p}$ except $\bar{\zeta}_5 \approx 1.6556$. Find the 1-pseudozero domain for $\tilde{\zeta}_5$ in this situation, i.e. the interval $\{\tilde{\zeta}_5 \in \mathbb{C} : \delta(\bar{\zeta}_1, \bar{\zeta}_2, \bar{\zeta}_3, \bar{\zeta}_4, \tilde{\zeta}_5) \leq 1\}$.

2. With the polynomial p_W of Example 5.7, consider the empirical polynomial (p_W, e) with the only empirical coefficient $(\bar{\alpha}_{19}, \varepsilon_{19}) = (210, 2^{-16})$, with only real deviations. Which precision of a decimal floating-point arithmetic is required to represent p_W and $(p_W \pm 2^{-16})$ correctly?

(a) Compute the condition of the zeros ζ_μ of p_W w.r.t. to absolute and relative perturbations of α_{19}. What do you conclude?

(b) Compute the esimate (5.38) for the real pseudozero domains $Z_{1,\mu}$. Which zeros of (p_W, e) are separated for $\delta = 1$? Find $\bar{\varepsilon}_{19}$ such that *all* $Z_{1,\mu}$ are disjoint for $|\Delta\alpha_{19}| < \bar{\varepsilon}_{19}$.

(c) Let $p_W^+(t) := p_W + \varepsilon_{19}\, t\, x^{19}$, $p_W^-(t) := p_W - \varepsilon_{19}\, t\, x^{19}$, and $\zeta_\mu^+(t)$, $\zeta_\mu^-(t)$ their zeros, with $\zeta_\mu^\pm(0) = \mu$. Describe the expected paths $\zeta_\mu^\pm(t)$ for $0 \leq t \leq 1$, using the information of a) and b). Check by computation.

(d) Represent p_W by fewer decimal digits and try to determine bounds on the potential effects, from a) and the coefficients of p_W. Watch the actual influence on the zeros; from how many digits onward are all zeros real and reasonable approximations?

3. (a) Confirm the assertions in section 5.2.3 about the results of a Newton step from $\hat{\eta}$ and $\hat{\xi}$.

(b) Consider the polynomial

$$\bar{p}(x) := .00001\, x^6 - 2.345\, x^5 + 5.318\, x^4 - 3.852\, x^3 + 4.295\, x^2 - 1.972\, x + 5.321 \,.$$

From $\hat{\xi} = 234500$, perform one Newton step to obtain $\tilde{\xi}_\infty$. From $\hat{\eta} = .00001/2.345$, perform one Newton step in the reciprocal polynomial $\bar{q}(y)$ to obtain $\tilde{\xi}_0 = 1/\tilde{\eta}_0$. Compare the results with an accurate value for the large zero ξ_0 of $\bar{p}$. Form the residuals of $\hat{\xi}$ and the other approximations of ξ_0; why is the large size of the residual compatible with the accuracy of the approximations?

(c) Assume that all coefficients in $\bar{p}$ except the leading tiny one have a tolerance of .0005 and compute the backward errors of the approximations for ξ_0 computed in b). Are the

Newton steps meaningful for the empirical polynomial $(\bar{p}, e)$ with these tolerances? Find the approximate extension along the real axis of the pseudozero domain Z_1 containing ξ_0. How many digits of an approximation for ξ_0 are meaningful?

(d) Compute an approximation for the real zero ξ_1 of $\bar{p}$ near 2. Determine the condition of ξ_1 w.r.t changes of the coefficient of x^6. How much will ξ_1 change when the leading coefficient is set to zero? Confirm by computation.

5.3   Polynomial Division

In section 5.1.3, we have observed that the operation of polynomial division plays a fundamental role in the ring $\mathbb{C}[x] = \mathcal{P}^1$ of univariate polynomials over $\mathbb{C}$: For two polynomials p and $s \neq 0$ of degrees $n \geq m$, there exist uniquely two polynomials, the *quotient* $q \in \mathcal{P}_{n-m}$ and the *remainder* $r \in \mathcal{P}_{m-1}$ such that

$$p(x) \;=\; q(x)\, s(x) + r(x) \,. \tag{5.41}$$

Proposition 5.9. In (5.41), r is the *normal form* $\mathrm{NF}_{\langle s \rangle}[p]$ of p mod $\langle s \rangle$ and the *interpolation polynomial* of p at the zeros of s.
Proof: The first statement follows from the remarks after (5.16); the second one is obvious from (5.41). $\square$

For an intrinsic *dividend* p and *divisor* s and with exact computation, q and r are determined by the well-known

Algorithm 5.2 (Division Algorithm).
$\qquad q(x) := 0; \quad r(x) := p(x);$

$\qquad$ **while** $\quad \deg(r) \geq \deg(s) \quad$ **do**

$\Delta q(x) := \; \mathrm{l.t.}(r)/\mathrm{l.t.}(s); \quad q(x) := q(x) + \Delta q(x); \quad r(x) := r(x) - \Delta q(x)\, s(x); \quad$ **od**

where l.t. denotes the leading term.

5.3.1   Sensitivity Analysis of Polynomial Division

The relation (5.41) defines a map from the data space $\mathcal{A}$ of the $n + m + 2$ coefficients of $p \in \mathcal{P}_n$ and $s \in \mathcal{P}_m$ to the result space $\mathcal{Z}$ of the $n + 1$ coefficients of q and r. We want to understand the sensitivity of the result with respect to small changes in the data; this is of importance for a floating-point execution of the division algorithm and for the interpretation of the result of polynomial division with empirical polynomials.

In linear algebra notation, with a row vector $\mathbf{a}^T = (\alpha_0, \ldots, \alpha_n)$ for the coefficients of $p(x) = \sum_{\nu=0}^{n} \alpha_\nu x^\nu = a^T \mathbf{x}$ and row vectors $c^T \in \mathbb{C}^{m+1}$, $b^T \in \mathbb{C}^{n-m+1}$, $r^T \in \mathbb{C}^m$ for the coefficients of s, q, r, the relation (5.41) takes the form

$$(\alpha_0 \ldots \alpha_{m-1} \mid \alpha_m \ldots \alpha_n) \cdot \mathbf{x} =$$

$$[\,(\beta_0 \ldots \beta_{n-m})
\begin{pmatrix}
\gamma_0 & \cdot\cdot & \gamma_{m-1} & \mid & \gamma_m & 0 & & \cdot\cdot & \\
0 & \gamma_0 & \cdot\cdot & \mid & \cdot\cdot & \gamma_m & 0 & \cdot\cdot & \\
\cdot\cdot & 0 & \gamma_0 & \mid & \cdot\cdot & \cdot\cdot & \gamma_m & 0 & \cdot\cdot \\
 & & & \mid & \gamma_0 & \cdot\cdot & \cdot\cdot & \gamma_m & 0 \\
 & 0 & & \mid & 0 & \gamma_0 & \cdot\cdot & & \cdot\cdot \\
 & & & \mid & \cdot\cdot & 0 & \gamma_0 & \cdot\cdot & \gamma_m
\end{pmatrix}
+ (\rho_0 \ldots \rho_{m-1} \mid 0 \ldots) \,] \cdot \mathbf{x}$$

or

$$(a_2^T \mid a_1^T) = b^T (S_2 \mid S_1) + (r^T \mid 0) , \tag{5.42}$$

with an obvious definition of the partitions. If $n - m + 1 \leq m + 1$, i.e. $m \geq \frac{n}{2}$, S_1 is a full lower triangular matrix of $n - m + 1$ rows and columns and there are no empty rows in S_2.

γ_m is the leading coefficient of s and hence $\neq 0$ so that S_1 is nonsingular. Therefore, we may represent the result vectors b, r as

$$b^T = a_1^T S_1^{-1} \qquad \in \mathbb{C}^{n-m+1} , \tag{5.43}$$

$$r^T = a_2^T - b^T S_2 \qquad \in \mathbb{C}^m . \tag{5.44}$$

From this representation it is obvious that the absolute condition of the coefficient vectors b^T and r^T of q and r depends on $\| S_1^{-1} \|$.

Proposition 5.10. For small perturbations of p and s in the polynomial division (5.41), the generated perturbations of q and r satisfy

$$\| \Delta b^T \|^* \leq (\| \Delta a_1^T \|^* + \| a_1^T S_1^{-1} \Delta S_1 \|^*) \, \| S_1^{-1} \|^* + \mathrm{O}(\| \Delta a_1 \| \, \| \Delta S_1 \| + \| \Delta S_1 \|^2) ,$$

$$\| \Delta r^T \|^* \leq \| \Delta a_2^T \|^* + \| b^T \| \, \| \Delta S_2 \|^* + \| \Delta b^T \| \, \| S_2 \|^* + \mathrm{O}(\| \Delta b \| \, \| \Delta S_2 \|) .$$

Proof: For a regular square matrix X and ΔX such that $\| X^{-1} \Delta X \| < 1$,

$$(X + \Delta X)^{-1} = [X (I + X^{-1} \Delta X)]^{-1} = (I + X^{-1} \Delta X)^{-1} X^{-1} = (I - X^{-1} \Delta X + \mathrm{O}(\| \Delta X \|^2)) \, X^{-1} .$$

Application to the perturbed relations (5.43) and (5.44) yields the bounds. $\square$

Proposition 5.9 shows that, for small perturbations of the dividend and the divisor, the effects on the quotient and the remainder are essentially proportional to the data perturbations. An ill-conditioning can only arise from an unduly large value of $\| S_1^{-1} \|^*$.

Proposition 5.11. For the triangular Toeplitz matrix

$$S_1 =
\begin{pmatrix}
\gamma_m & & & & \\
\vdots & \ddots & & 0 & \\
\gamma_0 & & \gamma_m & & \\
 & \ddots & & \ddots & \\
0 & & \gamma_0 & \cdots & \gamma_m
\end{pmatrix}
\in \mathbb{C}^{M \times M} , \qquad \gamma_m \neq 0 ,$$

$| \gamma_m | \ll \| S_1 \|$ implies that $\| S_1^{-1} \|$ may be as large as $\mathrm{O}(\| S_1 \|^{M-1} / \gamma_m^M)$. (For $M < m + 1$, only the γ_μ with $\mu \geq m + 1 - M$ appear in S_0.)

Proof: For a lower triangular matrix X with a *zero diagonal*, it is well known that $(I - X)^{-1} = I + X + X^2 + \ldots + X^{M-1}$ (since $X^M = 0$). With $I - X := \frac{1}{\gamma_m} S_1$, we have $S_1^{-1} = \frac{1}{\gamma_m}(I - X)^{-1}$ and $\|X\| \le 1 + \frac{1}{|\gamma_m|}\|S_1\| = O(\frac{1}{|\gamma_m|}\|S_1\|)$. $\quad\square$

Proposition 5.10 shows that perturbations in the data of a polynomial division can be strongly amplified if the *leading coefficient of the divisor is small* relative to its other coefficients since $\|S_1\|^* = \sum_{\mu=0}^{m} |\gamma_\mu|$ for $\|b^T\|^* = \max_\nu |\beta_\nu|$. Furthermore, $M = n - m + 1 = \deg(q) + 1$; this displays that the growth of the perturbation occurs as more and more coefficients of q are computed, with a small $\mathrm{LT}(s)$ in the denominator; cf. the Division Algorithm.

Clearly, the expressions in Propositions 5.9 and 5.10 are only bounds for the effects of a perturbation in p and/or s. It is well known from situations in linear algebra analogous to the one in (5.43) and (5.44) that a moderate error propagation can occur for special data and perturbations even in cases which are principally ill conditioned. On the other hand, if the leading coefficient of the divisor $\gamma_m = O(\|S_1\|)$, the resulting q and r are *well-conditioned*; we will call such a divisor well-behaved. In floating-point arithmetic, one may try to obtain such a situation by a suitable scaling, if this is feasible and meaningful.

By Proposition 5.8, the remainder r in the division of p by s is the normal form of p in the ideal $\langle s \rangle$. Thus, an ill-conditioning of division by s means that the *normal forms* mod $\langle s \rangle$ are poorly determined.

Proposition 5.12. If s is not well-behaved (see above), the determination of the membership of a polynomial p in the ideal $\langle s \rangle$ becomes exponentially ill-conditioned with increasing degree of p.

Proof: Membership of p in $\langle s \rangle$ is equivalent to $\mathrm{NF}_{\langle s \rangle} = 0$. In Proposition 5.10, $M = n - m + 1$, where $n = \deg p$. $\quad\square$

Example 5.8: With $s = x - \zeta \in \mathcal{P}_1$, we have $r = p(\zeta) \in \mathcal{P}_0$ and $q(x) = \frac{p(x) - p(\zeta)}{x - \zeta} \in \mathcal{P}_{n-1}$; the Division Algorithm becomes the well-known Horner Algorithm for the evaluation of p (cf. (4.24)). In our linear algebra notation (5.42), we have

$$S_1 = \begin{pmatrix} 1 & & & \\ -\zeta & 1 & & 0 \\ & \ddots & \ddots & \\ 0 & & -\zeta & 1 \end{pmatrix} \in \mathbb{C}^{n \times n}, \quad \text{with} \quad S_1^{-1} = \begin{pmatrix} 1 & & & \\ \zeta & 1 & & 0 \\ \vdots & \ddots & \ddots & \\ \zeta^{n-1} & \cdots & \zeta & 1 \end{pmatrix}.$$

For large $|\zeta|$, 1 is small relative to $|\zeta|$ and $\|S_1^{-1}\| = O(|\zeta|^{n-1})$. For small $|\zeta|$, $\|S_1^{-1}\| = O(1)$. Proposition 5.11 shows that the evaluation of $p(\zeta)$ for $|\zeta| \gg 1$ becomes exponentially ill-conditioned with increasing $\deg p$. $\quad\square$

While Propositions 5.9 and 5.10 relate to the absolute sensitivity of the division result to perturbations of the dividend and divisor, we have observed another detrimental effect for the Horner Algorithm in Example 4.8 in section 4.3.4: In a floating-point evaluation of $p(\zeta)$, the relative accuracy of the result of the Horner Algorithm is jeopardized if $p(\zeta)$ is much smaller in modulus than the data and intermediate results; the algorithm is *numerically unstable* when ζ is nearly a zero of p. For the general division algorithm, the analogous situation arises when the zeros of s are nearly zeros of p so that s is nearly a factor of p. In this case, a naive floating-point execution of polynomial division will lead to a cancellation of leading digits in

(5.44). As explained in section 4.3.4, the intermediate use of a higher floating-point precision will generally be an effective remedy.

5.3.2 Division of Empirical Polynomials

By now, we know how to define valid approximate results for problems with empirical data: Assume that we have a dividend $(\bar{p}, e_p)$ of degree n and a divisor $(\bar{s}, e_s)$ of degree $m \leq n$. We consider the polynomial division of $(\bar{p}, e_p)$ by $(\bar{s}, e_s)$.

Definition 5.5. A polynomial $\tilde{r}$ is a *valid approximate remainder* if $\deg \tilde{r} \leq m - 1$ and if there exist, for $\delta = O(1)$, polynomials $\tilde{p} \in N_\delta(\bar{p}, e_p)$, $\tilde{s} \in N_\delta(\bar{s}, e_s)$ and $q \in \mathcal{P}_{n-m}$ (arbitrary) such that

$$\tilde{p}(x) = q(x) \cdot \tilde{s}(x) + \tilde{r}(x) . \tag{5.45}$$

A polynomial $\tilde{q}$ is a *valid approximate quotient* if $\deg \tilde{q} \leq n - m$ and if there exist, for $\delta = O(1)$, polynomials $\tilde{p} \in N_\delta(\bar{p}, e_p)$, $\tilde{s} \in N_\delta(\bar{s}, e_s)$ and $r \in \mathcal{P}_{m-1}$ (arbitrary) such that

$$\tilde{p}(x) = \tilde{q}(x) \cdot \tilde{s}(x) + r(x) . \tag{5.46}$$

The polynomials $\tilde{q} \in \mathcal{P}_{n-m}$ and $\tilde{r} \in \mathcal{P}_{m-1}$ are a *valid approximate quotient/remainder pair* if there exist, for $\delta = O(1)$, polynomials $\tilde{p} \in N_\delta(\bar{p}, e_p)$, $\tilde{s} \in N_\delta(\bar{s}, e_s)$ such that

$$\tilde{p}(x) = \tilde{q}(x) \cdot \tilde{s}(x) + \tilde{r}(x) . \quad \square \tag{5.47}$$

It is obvious that a valid remainder is supplemented to a valid quotient/remainder pair by q in (5.45), and similarly a valid quotient is supplemented by r in (5.46). But $\tilde{r}$ and $\tilde{q}$ may be valid as a remainder and quotient separately without forming a valid pair.

Candidates for valid remainders and/or quotients are the results of performing the division algorithm on $\bar{p}$ and $\bar{s}$ in floating-point arithmetic. For a well-behaved divisor (cf. the previous section), the generated polynomials $\tilde{r}$ and $\tilde{q}$ should be very close to the exact results of the division and thus well within the validity bounds—with a floating-point precision which conforms with the tolerances. This validity may also occur for ill-conditioned cases: Now, the $\tilde{q}$ and $\tilde{r}$ may differ substantially from the q and r in $\bar{p} = q \cdot \bar{s} + r$; but these deviations may be interpretable as the effect of small changes in p and s which are within the tolerance limits. This may be different when there are only few empirical coefficients in p and s, perhaps due to sparsity. Then it is important to establish that numerically computed remainders/quotients can be exact results for nearby data in the restricted sense.

We now consider the computation of the backward error of an approximate remainder; the other two cases in Definition 5.5 may be treated similarly and will only be sketched.

Assume that we have—somehow—obtained approximate expressions $\tilde{q}$ and $\tilde{r}$ for the quotient and remainder of $(\bar{p}, e_p)$ and $(\bar{s}, e_s)$. We want to establish (5.45) for some $\tilde{p} = \bar{p} + \Delta p$ and $\tilde{s} = \bar{s} + \Delta s$, with $\|\Delta p\|_{e_p}^* \leq \delta$, $\|\Delta s\|_{e_s}^* \leq \delta$. The empirical data space $\Delta \mathcal{A}$ of the Δp, Δs, with its norm composed of the norms $\|..\|_{e_p}^*$, $\|..\|_{e_s}^*$, has a dimension M equal to the sum of the sizes of the empirical supports of p and s. In $\Delta \mathcal{A}$, we need the equivalent-data manifold $\mathcal{M}(\tilde{r})$ of those corrections Δp, Δs, for which, with an *arbitrary* $\Delta q \in \mathcal{P}_{n-m}$,

$$\bar{p} + \Delta p = (\tilde{q} + \Delta q) \cdot (\bar{s} + \Delta s) + \tilde{r} , \qquad \text{or} \tag{5.48}$$

$$\Delta p - \tilde{q} \cdot \Delta s - \Delta q \cdot \bar{s} = -(\bar{p} - \tilde{q}\,\bar{s} - \tilde{r})\,. \tag{5.49}$$

In (5.49), we have neglected the quadratic correction term $\Delta q\,\Delta s$; the effect of this linearization may be checked a posteriori and—if necessary—accommodated in a refinement step. In the following, we use $\mathcal{M}$ for the linearized manifold.

We claim that the codimension of $\mathcal{M}(\tilde{r})$ is m. Assume at first that $(\bar{p}, e_p)$ and $(\bar{s}.e_s)$ are intrinsically monic, and so is $\tilde{q}$. Then, (5.49) represents n linear equations for the M corrections Δp, Δs. However, $\Delta q \in \mathcal{P}_{n-m-1}$ constitutes a set of $n - m$ *free parameters* which raises the codimension of the linear manifold $\mathcal{M}(\tilde{r})$ to m. In the nonmonic case, we have $n + 1$ linear equations but $\Delta q \in \mathcal{P}_{n-m}$ so that the codimension remains the same. Thus we must have $M \geq m$ empirical coefficients in $(\bar{p}, e_p)$, $(\bar{s}, e_s)$ to have a well-posed problem; cf. section 3.2.1.

In this case, we can compute the closest distance of $\mathcal{M}(\tilde{r})$ from the origin of $\Delta\mathcal{A}$:

$$\delta(\tilde{r}) = \min_{\Delta p, \Delta s \in \mathcal{M}(\tilde{r}),\, \Delta q} \left(\max(\|\Delta p\|_{e_p}^*, \|\Delta s\|_{e_s}^*) \right)\,. \tag{5.50}$$

Note that in this minimization task, the coefficients $\Delta\beta_\nu$ of Δq appear only in the equality constraints (5.49) but not in the modulus constraints $|\Delta\alpha_\nu| \leq \delta\,\varepsilon_{p,\nu}$, $|\Delta\gamma_\mu| \leq \delta\,\varepsilon_{s,\mu}$; cf. (3.57).

Due to the omission of the quadratic term in (5.49), the minimizing Δp, Δs, Δq will not satisfy (5.48) exactly. But the residual $\Delta q\,\Delta s$ will generally be so small that it can be absorbed into the tolerance of $(\bar{p}, e_p)$.

For the computation of the backward error $\delta(\tilde{q})$ of an approximate quotient, the roles of $\tilde{r}$ and $\tilde{q}$ are exchanged: Now, we want $\bar{p} + \Delta p = \tilde{q} \cdot (\bar{s} + \Delta s) + (\tilde{r} + \Delta r)$ or

$$\Delta p - \tilde{q} \cdot \Delta s - \Delta r = -(\bar{p} - \tilde{q}\,\bar{s} - \tilde{r})\,. \tag{5.51}$$

Here, $\mathcal{M}(\tilde{q})$ is automatically linear; the arbitrary $\Delta r \in \mathcal{P}_{m-1}$ contributes m free parameters and the codimension of $\mathcal{M}(\tilde{q})$ is $n - m$ or $n - m + 1$ for the monic and nonmonic case, respectively.

Finally, in the computation of the backward error $\delta(\tilde{q}, \tilde{r})$ of an approximate quotient/ remainder pair, we have no linearization and no free parameters in

$$\Delta p - \tilde{q} \cdot \Delta s = -(\bar{p} - \tilde{q}\,\bar{s} - \tilde{r})\,, \tag{5.52}$$

and there are n or $n+1$ equations, resp.; this requires at least that number of empirical coefficients in the dividend and divisor.

From the codimensions of the equivalent-data manifold it appears that all three cases are well-posed for an empirical dividend $(\bar{p}, e_p)$ with all coefficients empirical, even with an intrinsic divisor. On the other hand, this may not be the case if only the divisor is empirical.

A particularly interesting situation arises when we expect to have a valid remainder 0 for the division of $(\bar{p}, e_p)$ by $(\bar{s}, e_s)$. This is equivalent to the assertion that there exist pairs $(\tilde{p}, \tilde{s})$ in the respective tolerance neighborhoods such that $\tilde{s}$ exactly divides $\tilde{p}$ or (vice versa) $\tilde{p}$ is an exact multiple of $\tilde{s}$. Note that we have recognized the division algorithm as unstable in this case at the end of section 5.3.1. We will analyze this important case in more detail in section 6.2.

Example 5.9: Consider $\bar{p}(x) = x^3 - 6.25\,x^2 + 11.14\,x - 5.83$ and $\bar{s}(x) = x^2 - 3.87\,x + 3.12$, with tolerances .005 in all coefficients (except the leading 1's). Assume that we have somehow obtained $\tilde{q} = x - 2.39$ and $\tilde{r} = -1.20\,x + 1.60$ and wish to check the validity of $\tilde{r}$ as a remainder of the polynomial division. Equation (5.49) yields 3 equations (from the coefficients of $1, x, x^2$), with one free parameter $\Delta\beta_0$ from the arbitrary quotient correction Δq. Elimination of $\Delta\beta_0$ leaves two equality conditions for the five coefficients in Δp, Δs, and the minimization of $(\|\Delta p\|_{e_p}^*, \|\Delta s\|_{e_s}^*)$ yields $\delta(\tilde{r}) \approx .36$ and thus validity of $\tilde{r}$. The remaining quadratic residual in (5.48) is $\approx 10^{-5}\,(1 - x)$ and may be neglected.

When we now consider $\tilde{q}$, $\tilde{r}$ as an approximate quotient/remainder pair, we lose the free parameter and have 3 equality constraints for the $\Delta\alpha_\nu$ and $\Delta\beta_\nu$. The minimal e-norm distance of their intersection from the origin of $\Delta\mathcal{A}$ is raised to ≈ 1.58. This means that we must modify some of the coefficients (three in fact) in $\bar{p}$ and $\bar{s}$ by $\approx .008$ to accommodate $\tilde{q}$ and $\tilde{r}$ as *exact* quotient and remainder. The admissibility of this must be decided. $\square$

Since the exact quotient/remainder pair q, r can be found for the dividend $\bar{p}$ and divisor $\bar{s}$ from the linear system (5.42), it must be possible to *correct* some approximate but invalid quotient/remainder pair $\tilde{q}$, $\tilde{r}$ by solving a linear system. Consider

$$\bar{p}(x) = (\tilde{q}(x) + \Delta q(x))\,\bar{s}(x) - (\tilde{r}(x) + \Delta r(x))$$

$$\text{or}\qquad \bar{p}(x) - \tilde{q}(x)\,\bar{s}(x) - \tilde{r}(x) = \Delta b^T\,(\bar{S}_2\,|\,\bar{S}_1) + (\Delta r^T\,|\,0),\qquad (5.53)$$

where $\Delta q(x) = \Delta b^T \mathbf{x}$, $\Delta r(x) = \Delta r^T \mathbf{x}$, and $\bar{S}_1$, $\bar{S}_2$ are the matrices in (5.42) for $\bar{s}$. The linear system (5.53) has as many equations as unknowns, its exact solution Δb^T, Δr^T refines $\tilde{q}$, $\tilde{r}$ into the exact quotient and remainder of $\bar{p}, \bar{s}$.

For empirical dividends and/or divisors, such an accurate refinement is not meaningful. It will generally suffice to compute and use the first few significant digits of Δq, Δr. The backward error of $\tilde{q}, \tilde{r}$ is an indication of the refinement which is meaningful. Compare Exercise 5.3-2 below.

Exercises

1. By Propositions 5.9 and 5.10, the condition of polynomial division by s is ill conditioned if $|\gamma_m| \ll \sum |\gamma_\mu|$. Form a polynomial $s(x, \zeta)$ with a few fixed zeros of $O(1)$ and one indeterminate zero ζ.

(a) Visualize the behavior of $\|S_1^{-1}\|^*$ by values and plots
- for increasing $|\zeta|$ and fixed large M,
- for increasing M and fixed large $|\zeta|$.

(b) Take some polynomial p of high degree and compute the remainder r of $p(x)/s(x, \zeta)$ for some large $|\zeta|$. Observe the effects on r of small changes in p.

(c) Form a high degree multiple p_0 of $s(x, \zeta)$, $|\zeta|$ large. Compute the normal form (= remainder) of $p_0 \bmod s$. Observe the dependence on the floating-point precision used in the computation.

2. For $(\bar{p}, e_p)$ with $\bar{p}(x) = x^5 - .552\,x^4 - 5.616\,x^3 + 4.630\,x^2 + 3.693\,x - 1.611$ and tolerances $\varepsilon_{p,\nu} = .0005$, $\nu = 1(1)4$, and $(\bar{s}, e_s)$ with $\bar{s}(x) = x^2 - 3.28\,x + 2.69$ and $\varepsilon_{s,\mu} = .005$, $\mu = 0, 1$, consider the approximate quotient $\tilde{q}(x) = x^3 + 2.73\,x^2 + .64\,x - .60$ and remainder $\tilde{r}(x) = -.012\,x + .011$.

(a) Determine the backward errors $\delta(\tilde{r})$, $\delta(\tilde{q})$, and $\delta(\tilde{q}, \tilde{r})$. Comment.

(b) Compute corrections Δq and Δr by a refinement step (5.53). Refine $\tilde{q}$, $\tilde{r}$ by appending only *one* more correct (rounded-to-nearest) digit to each coefficient. Determine the backward errors of a) for the refined quantities.

(c) What do you conclude from a) and b) about the meaningful accuracy of a quotient and remainder for $(\bar{p}, e_p)/(\bar{s}, e_s)$?

5.4 Polynomial Interpolation

In section 2.5.4, we have introduced the general polynomial interpolation problem in Definition 2.23: Given m linear functionals $l_\mu : \mathcal{P}^s \to \mathbb{C}$ and associated values $w_\mu \in \mathbb{C}$, $\mu = 1(1)m$, find $r^* \in \mathcal{P}^s$ such that $l_\mu(r^*) = w_\mu$, $\mu = 1(1)m$. For the interpolation problem to be well-defined, $\mathcal{D} = \operatorname{span} \{l_\mu\} \subset (\mathcal{P}^s)^*$ must be a *closed m-dimensional vector space* of linear functionals so that it defines an ideal $\mathcal{I}[\mathcal{D}]$ and a quotient ring $\mathcal{R}[\mathcal{D}]$.

Take the basis $\mathbf{c}_0^T = (l_\mu, \ \mu = 1(1)m)$ of $\mathcal{D}$ and some basis $\mathbf{b}$ of $\mathcal{R}[\mathcal{D}]$, and let $w^T = (w_\mu, \ \mu = 1(1)m)$. Then any polynomial $r \in \mathcal{P}^s$ in the residue class mod $\mathcal{I}[\mathcal{D}]$ of

$$r^*(x) \ = \ w^T \, (\mathbf{c}_0^T(\mathbf{b}))^{-1} \, \mathbf{b}(x) \ = \ w^T \, \mathbf{b}_0(x) \, , \tag{5.54}$$

with $\mathbf{b}_0$ the conjugate basis of $\mathbf{c}_0^T$, is a solution of the interpolation problem; cf. (2.63) in section 2.5.4.

In the univariate case $s = 1$, the situation is more intuitive because any reasonable basis of $\mathcal{R}[\mathcal{D}]$ contains only polynomials of degree $\leq m - 1$ and each residue class has one unique element in $\mathcal{P}_{m-1}$; cf. section 5.1.3. Thus, (5.54) defines *the* unique interpolation polynomial $r^* \in \mathcal{P}_{m-1}$ of lowest degree.

In the classical situation, $\mathbf{c}_0^T$ is spanned by a set of *evaluation* functionals; i.e. $\mathbf{c}_0^T$ is a *Lagrange basis* of $\mathcal{D}$, and the matrix $M_0 = \mathbf{c}_0^T(\mathbf{b})$ in (5.54) is a *Vandermonde matrix*; compare Definition 5.2.

Definition 5.6. Given a Lagrange basis $\mathbf{c}_0^T = (c_1^{(0)}, \ldots, c_m^{(0)})$ of an m-dimensional dual space $\mathcal{D}$, i.e. a set of m evaluation functionals $c_\mu^{(0)}$, $\mu = 1(1)m$, which are linearly independent on $\mathcal{P}_{m-1}$, and a set w of *data values* $w_\mu \in \mathbb{C}$, $\mu = 1(1)m$, we call the unique polynomial $r \in \mathcal{P}_{n-1}$ which satisfies

$$c_\mu^{(0)}(r) \ = \ w_\mu \, , \quad \mu = 1(1)m \, , \tag{5.55}$$

the *interpolation polynomial* for w on $\mathbf{c}_0^T$. When w satisfies $w_\mu = c_\mu^{(0)}(f)$ for some polynomial $f \in \mathcal{P}$—or more generally for some arbitrary *function* $f : \mathbb{C} \to \mathbb{C}$—we call r the *interpolation polynomial of f on $\mathbf{c}_0^T$*. $\square$

Proposition 5.13. For $f \in \mathcal{P}$, the unique normal form $r = \mathrm{NF}_{\mathcal{I}[\mathcal{D}]}[f]$ is the interpolation polynomial of f on $\mathbf{c}_0^T$, where $\mathcal{D} = \operatorname{span} \mathbf{c}_0^T$.
Proof: $f - r \in \mathcal{I}[\mathcal{D}]$ implies $\mathbf{c}_0^T(f - r) = 0$. $\square$

5.4.1 Classical Representations of Interpolation Polynomials

The study of univariate polynomial interpolation has a very old tradition; in this section, we display some of the classical results in the context of our conceptual framework, without proofs which can be found in older textbooks of numerical or applied analysis. Almost exclusively, the case of evaluation functionals at a set of points ζ_ν, $\nu = 1(1)n$ for the function and perhaps some derivatives has been considered. Special representations for the interpolation polynomial have been developed which permit a simple computation—with pencil and paper(!)—of its coefficients and an inexpensive Horner-like evaluation at an argument $\xi \neq \zeta_\nu$. Note that, for specified functionals $\mathbf{c}_0^T$ and data w, the unique interpolation polynomial $r \in \mathcal{P}_{n-1}$ may be *represented* in a number of different ways.

When $\mathbf{c}_0^T$ consists only of function evaluations at a set of *interpolation nodes* $\zeta_\nu \in \mathbb{C}$, an explicit representation of the associated Lagrange basis $\mathbf{b}_0 = (b_\nu^{(0)})$ is

$$b_\nu^{(0)}(x) := \prod_{\nu' \neq \nu} \frac{(x - \zeta_{\nu'})}{(\zeta_\nu - \zeta_{\nu'})} = \frac{s(x)}{(x - \zeta_\nu)\, s'(\zeta_\nu)}, \tag{5.56}$$

with $s(x) := \prod_\nu (x - \zeta_\nu)$; it is easily verified that these $b_\nu^{(0)}$ satisfy $b_\nu^{(0)}(\zeta_{\nu'}) = \delta_{\nu\nu'}$. Also, the inverse of the Vandermonde matrix can be explicitly expressed in terms of the ζ_ν.

For univariate polynomial interpolation, the *recursive* determination of r is of particular interest: One may not wish to fix the complete set of ζ_ν a priori. The so-called *Newton representation* of the interpolation polynomial r uses the basis $\mathbf{b}^N = (b_\nu^N, \nu = 1(1)n)$ of the $\mathcal{P}_{n-1}$, with the $b_\nu^N \in \mathcal{P}_{\nu-1}$ defined by

$$b_1^N(x) := 1, \quad b_\mu^N(x) := (x - \zeta_{\mu-1})\, b_{\mu-1}^N(x) = \prod_{\nu < \mu}(x - \zeta_\nu), \quad \mu = 2, 3, \dots . \tag{5.57}$$

For disjoint ζ_ν, the b_μ^N, $\mu \leq \nu$, form a basis of $\mathcal{P}_{\nu-1}$ because the matrix $M_N := \mathbf{c}_0(\mathbf{b}^N)$ is upper triangular, with nonvanishing diagonal elements.

For the same reason, the coefficients $w^T (\mathbf{c}_0(\mathbf{b}^N))^{-1}$ of r in this basis (cf. (5.54)) can be formed recursively. They are the *divided differences* of the data w_ν which are recursively defined by

$$\begin{aligned}
w[\zeta_\nu] &:= w_\nu, & \nu &= 1(1)\dots, \\
w[\zeta_\nu, \zeta_{\nu+1}] &:= \frac{w[\zeta_{\nu+1}] - w[\zeta_\nu]}{\zeta_{\nu+1} - \zeta_\nu}, & \nu &= 1(1)\dots, \\
&\quad \dots \\
w[\zeta_\nu, \dots, \zeta_{\nu+\ell}] &:= \frac{w[\zeta_{\nu+1}, \dots, \zeta_{\nu+\ell}] - w[\zeta_\nu, \dots, \zeta_{\nu+\ell-1}]}{\zeta_{\nu+\ell} - \zeta_\nu}, & \nu &= 1(1)\dots, \\
&\quad \dots
\end{aligned} \tag{5.58}$$

In (5.58), $w[\zeta_\nu, \dots, \zeta_{\nu+\ell}]$ is called a *divided difference of order ℓ*. Divided differences are *invariant under arbitrary permutations* of the ζ_ν as can be easily established recursively.

Proposition 5.14. The polynomials

$$r_\mu(x) := w[\zeta_1]\, b_1^N(x) + w[\zeta_1, \zeta_2]\, b_2^N(x) + \dots + w[\zeta_1, \dots, \zeta_\mu]\, b_\mu^N(x), \quad \mu = 1, 2, \dots, \tag{5.59}$$

satisfy $r_\mu(\zeta_\nu) = w_\nu$, $\nu = 1(1)\mu$.

The *evaluation* of (5.59) at some ξ can also be performed recursively after rewriting it in the Horner-like form (cf. (5.57))

$$\begin{aligned}
r_\mu(x) = (\dots (w[\zeta_1, .., \zeta_\mu]\,(x - \zeta_{\mu-1}) + w[\zeta_1, .., \zeta_{\mu-1}])\,(x - \zeta_{\mu-2}) \\
+ \dots + w[\zeta_1, \zeta_2]\,)\,(x - \zeta_1) + w[\zeta_1]
\end{aligned} \tag{5.60}$$

which reduces the evaluation effort.

If one does not need a representation for the interpolation polynomial r but only its *value(s)* at one (or very few) point(s), one can use the so-called *Neville Algorithm* which is based on

Proposition 5.15. (Aitken): The interpolation polynomials $r_{\nu\ldots\nu+\ell}$ at $\zeta_\nu, \ldots, \zeta_{\nu+\ell}$ satisfy the recursion

$$r_{\nu\ldots\nu+\ell}(x) \;=\; \frac{(x - \zeta_\nu)\, r_{\nu+1\ldots\nu+\ell}(x) - (x - \zeta_{\nu+\ell})\, r_{\nu\ldots\nu+\ell-1}(x)}{\zeta_{\nu+\ell} - \zeta_\nu}. \tag{5.61}$$

Neville's Algorithm uses the recursion (5.61) at the specified evaluation argument ξ; thus, it simply forms linear combinations of the original data values in a recursive fashion to obtain the value of the interpolation polynomial at ξ.

The data values w_ν specified at the interpolation nodes ζ_ν may be from any source and completely without relation to each other. If they are the values at the ζ_ν of a *real function w* which is *sufficiently smooth*, the divided differences of order ℓ are closely related to the ℓ-th derivative of w.

Proposition 5.16. For a function $w : \mathbb{R} \to \mathbb{R}$ with n continuous derivatives,

$$w[\zeta_1, \ldots, \zeta_n] \;=\; \frac{1}{(n-1)!}\, w^{(n-1)}(\tilde{\zeta}), \quad \text{with } \tilde{\zeta} \in (\min_\nu \zeta_\nu, \max_\nu \zeta_\nu). \tag{5.62}$$

The relation (5.62) establishes the smooth behavior of divided differences under a *confluence* of the nodes. Obviously, for w as in (5.58),

$$\lim_{\zeta_\nu \to \bar{\zeta}} w[\zeta_1, \ldots, \zeta_n] \;=\; \frac{1}{(n-1)!}\, w^{(n-1)}(\bar{\zeta}). \tag{5.63}$$

This permits the extension of the Newton representation (5.59) of the interpolation polynomial to data for *derivatives*. Assume that we have specified values for w' and w'' at ζ_1, in addition to function values at all the ζ_ν. Then we may simply use three copies of ζ_1 in (5.59):

$$r(x) \;=\; w[\zeta_1] + w[\zeta_1, \zeta_1]\,(x - \zeta_1) + w[\zeta_1, \zeta_1, \zeta_1]\,(x - \zeta_1)^2 + w[\zeta_1, \zeta_1, \zeta_1, \zeta_2]\,(x - \zeta_1)^3 + \ldots,$$

where

$$w[\zeta_1, \zeta_1] = w'(\zeta_1), \quad w[\zeta_1, \zeta_1, \zeta_1] = \frac{1}{2} w''(\zeta_1), \quad w[\zeta_1, \zeta_1, \zeta_2] = \frac{w[\zeta_1, \zeta_2] - w[\zeta_1, \zeta_1]}{\zeta_2 - \zeta_1},$$

$$w[\zeta_1, \zeta_1, \zeta_1, \zeta_2] = \frac{w[\zeta_1, \zeta_1, \zeta_2] - w[\zeta_1, \zeta_1, \zeta_1]}{\zeta_2 - \zeta_1}, \quad \text{etc.}$$

Note, however, that the differentiation functionals must form a proper basis of a dual space of a polynomial ideal; i.e. they must form a *closed set*; cf. Definition 2.15. It is not feasible, e.g., to specify $\partial^0[\zeta]$ and $\partial^2[\zeta]$ without $\partial^1[\zeta]$: There is no polynomial of degree 1 which can take these values (except for $w''(\zeta) = 0$).

If (and only if) the data w are from a smooth function, it is a natural question to ask for the *interpolation error*, i.e. for the potential deviation between the function w and its interpolation polynomial r in an interval $[a, b]$ containing the interpolation nodes.

Proposition 5.17. For a function $w : [a, b] \to \mathbb{C}$ with n continuous derivatives and its interpolation polynomial $r \in \mathcal{P}_{n-1}$ at $\zeta_1, \ldots, \zeta_n$ in $[a, b]$, there holds, at any $\xi \in [a, b]$,

$$
r(\xi) - w(\xi) \; = \; -w[\zeta_1, \ldots, \zeta_n, \xi] \cdot \prod_{\nu=1}^{n} (\xi - \zeta_\nu) \; = \; -\frac{w^{(n)}(\tilde{\xi})}{n!} \cdot \prod_{\nu=1}^{n} (\xi - \zeta_\nu), \quad \text{with } \tilde{\xi} \in (a, b).
$$

$$(5.64)$$

Both expressions for the interpolation error cannot be evaluated in the usual sense: The divided difference form would require the knowledge of $w(\xi)$ while the derivative form would require the knowledge of $\tilde{\xi}$. But both forms permit the computation of upper/lower bounds or of estimates for the deviation $r(\xi) - w(\xi)$ between the (unknown) function w and its interpolation polynomial r.

Equation (5.64) displays the great crux of polynomial interpolation for a larger number of data: Assume that $w^{(n)}$ has little variation in $[a, b]$; then the deviation behaves essentially like $s(x) := \prod_{\nu=1}^{n} (x - \zeta_\nu)$, i.e. it *oscillates strongly*. For equidistant ζ_ν, the amplitudes of these oscillations increase substantially towards the outer parts of the interval covered by the ζ_ν. Obviously, such an interpolant is not a feasible approximation for w ! On the other hand, (5.64) indicates how these oscillations of the error may be diminished by adding to r a suitable constant or polynomial multiple of s: This does not affect the interpolation property but may notably influence the oscillations. Compare Exercise 5.4-1.

On [-1,+1], with $\zeta_\nu = \xi_\nu$ of (5.12), the zeros of the *Chebyshev polynomial* T_n, we have $\prod_{\nu=1}^{n} (x - \zeta_\nu) = \frac{1}{2^{n-1}} T_n(x)$ and the oscillations have a *uniform amplitude* of $\frac{1}{2^{n-1}}$. With an affine transformation of x, this distribution of nodes may be moved to an arbitrary finite interval. While this establishes once more the distinction of the Chebyshev polynomials which rests on their property (5.14), it is also clear that such unevenly spaced, irrational nodes are generally not suitable for practical applications.

For this reason, large-scale interpolation in scientific computing, e.g., for the purpose of *plotting*, has developed in a different direction. It has, almost exclusively, been based on *polynomial splines*, i.e. *piecewise polynomial functions*, rather than on high degree polynomials covering large intervals. For a discussion of spline functions, we must refer to the relevant literature.

5.4.2　Sensitivity Analysis of Univariate Polynomial Interpolation

In polynomial interpolation, there are two kinds of numerical data: The values w^T which may be values $\mathbf{c}_0^T(f)$ of a function f and possibly its derivatives at the interpolation nodes, and the values ζ_μ of these nodes. We should know, qualitatively at least, how the interpolation polynomial reacts upon perturbations of these data.

By (5.54), the interpolated data w^T enter the interpolation polynomial in a linear fashion; hence, the effect of perturbations in these values is described by the interpolation polynomial

$\Delta r(x) = \Delta w^T \mathbf{b}_0(x)$ of the perturbations:

$$|\Delta r(x)| \leq \|\Delta w^T\|^* \|\mathbf{b}_0\| \leq \max_\nu |\Delta w_\nu| \cdot \sum_\nu |b_\nu^{(0)}(x)| =: \max_\nu |\Delta w_\nu| \cdot B(x). \quad (5.65)$$

In the case of no derivative data, we have $\sum_\nu b_\nu^{(0)}(x) \equiv 1$ (use $w^T = (1, .., 1)$) and hence $B(x) \geq 1$ for all x. By (5.56), the Lagrange basis polynomials $b_\mu^{(0)}$ with their zeros at the nodes ζ_ν, $\nu \neq \mu$, *oscillate* strongly in the domain where the nodes are located. Thus, $B(x)$ can assume substantial values between the nodes; for equidistant real nodes, the maxima of B increase rapidly toward the boundary nodes. Outside the domain of the interpolation nodes, $B(x)$ grows like x^{n-1} for n data. For a larger number of equidistant nodes, the interpolation polynomial remains well-conditioned w.r.t perturbations of the data w^T only in the central part of the interpolation interval; cf. Figure 5.2 and Example 5.10.

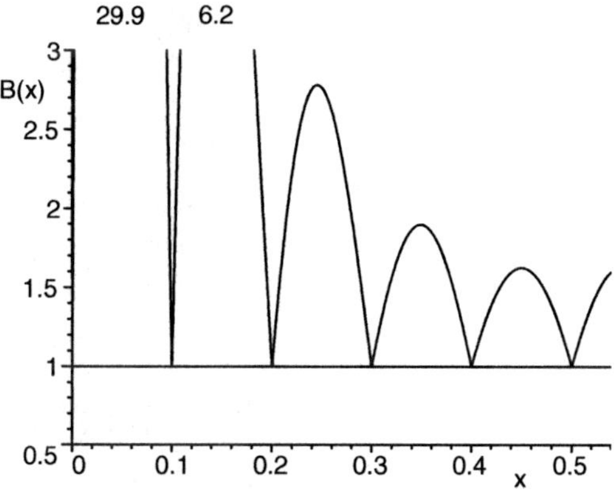

Figure 5.2.

Example 5.10: Consider the set $\{\zeta_\mu = \mu/10, \ \mu = 0(1)10\}$ of interpolation nodes. At the points $\eta_\mu := (\mu - .5)/10$ halfway between the nodes ($\mu = 1(1)10$) and outside the nodes ($\mu \leq 0$ and ≥ 11), we obtain the following values (rounded):

η_μ	$-.05$	.05	.15	.25	.35	.45	..	.95	1.05	1.15
$|b_0^{(0)}(\eta_\mu)|$	3.7	.18	.01	.002	.001	.0003	..	.01	.18	3.7
$|b_3^{(0)}(\eta_\mu)|$	63	4.2	1.1	.98	.46	.11	..	1.6	30	601
$|b_5^{(0)}(\eta_\mu)|$	85	4.9	1.0	.41	.32	.67	..	4.9	85	1650
$B(\eta_\mu)$	385	25	5.9	2.8	1.9	1.6	..	25	385	7190

This shows clearly, that the effect of perturbations ε in the data does not remain $O(\varepsilon)$ in the boundary parts of the interpolation interval and that it grows excessively outside that interval. $\quad\square$

Like the interpolation error (5.64), the condition function $B(x)$ has a more uniform behavior over the interpolation interval if the nodes are denser near the boundaries. Again the

Chebyshev nodes, i.e. the n zeros (5.12) of T_n or their affine images, are suitable nodes from this point of view. But the practical objections against these nodes remain.

Example 5.10, continued: We move the 11 zeros of T_{11} to $[0, 1]$ by the affine map $\xi \to \frac{1}{2} + \frac{1}{2}\xi$ and compute the entries of the previous table for interpolation at these nodes; note that the two extreme nodes are not 0 and 1 as above but .005 and .995. For $\mu = 1(1)10$, the η_ν are the midpoints between the Chebyshev nodes, outside $[0, 1]$, we have taken the same values as above. The values are generally rounded to two digits.

η_μ	$-.05$	$.025$	$.084$	$.18$	$.29$	$.43$	$..$	$.975$	1.05	1.15
$\lvert b_0^{(0)}(\eta_\mu)\rvert$	.41	.006	.007	.008	.009	.011	..	.30	7.7	86
$\lvert b_3^{(0)}(\eta_\mu)\rvert$	3.1	.05	.05	.06	.08	.11	..	.17	9.0	207
$\lvert b_5^{(0)}(\eta_\mu)\rvert$	5.4	.09	.11	.14	.22	.65	..	.09	5.4	144
$B(\eta_\mu)$	65	2.0	2.0	2.0	2.0	2.0	..	2.0	65	1340

The uniformity of the (near-)maxima of B is striking and the slight increase in the central part, due to the increased spacing of the nodes there, negligible. $\square$

The effects of perturbations in the interpolation nodes can also be assessed. When the data are some independent values w_ν, one has to determine how the perturbation affects the $b_\nu^{(0)}$. We consider only the case without derivative specifications so that

$$\prod_{\mu \neq \nu}(\zeta_\nu - \zeta_\mu) \cdot b_\nu^{(0)}(x) = \prod_{\mu \neq \nu}(x - \zeta_\mu);$$

cf. (5.56). Differentiation w.r.t some $\zeta_\lambda \neq \zeta_\nu$ deletes the factors $(\zeta_\nu - \zeta_\lambda)$ and $(x - \zeta_\lambda)$, resp., so that

$$\frac{\partial b_\nu(x)}{\partial \zeta_\lambda} = \frac{1}{\zeta_\nu - \zeta_\lambda}\left(b_\nu(x) + \frac{\prod_{\mu \neq \lambda,\nu}(x - \zeta_\mu)}{\prod_{\mu \neq \lambda,\nu}(\zeta_\nu - \zeta_\mu)}\right);$$

the right fraction is $b_\nu^{(0)}$ for the node set without ζ_λ. Thus, the *relative* perturbation of $b_\nu^{(0)}$ is $O(\lvert\zeta_\nu - \zeta_\lambda\rvert^{-1})$, which will generally not be excessive. When we differentiate w.r.t. ζ_ν, the right-hand side vanishes; when we write the product on the left-hand side as $s'(\zeta_\nu)$ (cf. (5.56)), we obtain

$$\frac{\partial b_\nu(x)}{\partial \zeta_\nu} = -\frac{s'(\zeta_\nu)\, b_\nu(x)}{s''(\zeta_\nu)};$$

thus, again, the relative perturbation is generally moderate.

The situation is different if we assume that the w_ν are function evaluations $f(\zeta_\nu)$ so that the value changes systematically with a perturbation of ζ_ν. If $f \in \mathcal{P}_{n-1}$ so that it is reproduced by the interpolation, there is no effect at all. Otherwise, the effect is proportional to the difference of the derivatives of r and f at ζ_ν.

5.4.3   Interpolation Polynomials for Empirical Data

Now we consider the interpolation of data with limited accuracy; we will also use the term "approximate interpolation" for this task. At first, we assume that we have exact (i.e. intrinsic) interpolation abscissae ζ_ν, $\nu = 1(1)n$, but that *empirical values* $(\bar{w}_\nu, \varepsilon_\nu)$ are specified there.

Definition 5.7. A polynomial $\tilde{r} \in \mathcal{P}_k$, $k \leq n - 1$, is a *valid (approximate) interpolation polynomial* for the empirical data $(\bar{w}^T, e)$ on the ζ_ν, $\nu = 1(1)n$, if

$$\tilde{r}(\zeta_\nu) \in N_\delta(\bar{w}_\nu, \varepsilon_\nu), \quad \nu = 1(1)n, \quad \text{with } \delta = O(1). \tag{5.66}$$

As usual, $\varepsilon_\mu = 0$ for some $\mu \in \{1, \ldots, n\}$ implies the requirement $\tilde{r}(\zeta_\mu) = \bar{w}_\mu$. $\quad\square$

The essential liberty which we gain from the relaxation of (5.55) is the choice of a degree for $\tilde{r}$ which may be *lower* than $n - 1$. Actually, since there exists a unique polynomial $r \in \mathcal{P}_{n-1}$ which satisfies $r(\zeta_\nu) = \bar{w}_\nu$ for all ν, approximate interpolation makes sense *only if* we attempt to use a lower degree.

Definition 5.8. For specified interpolation nodes ζ_ν and data $(\bar{w}_\nu, \varepsilon_\nu)$, $\nu = 1(1)n$, the *backward interpolation error* of a polynomial $\tilde{r}$ is

$$\delta(\tilde{r}) := \|r(z) - w^T\|_e^* = \max_\nu |\tilde{r}(\zeta_\nu) - \bar{w}_\nu|/\varepsilon_\nu . \tag{5.67}$$

The *backward interpolation error for degree k* is

$$\delta(k) := \min_{r \in \mathbf{P}_k} \delta(r), \quad \text{with } \delta(r) \text{ from (5.64)}. \quad\square \tag{5.68}$$

With this concept, there are approximate interpolation polynomials of all degrees $k \geq 0$ for a specified data set; but there can be valid interpolation poynomials of degree k only if the backward interpolation error for degree k is $O(1)$. Since $\mathcal{P}_{k-1} \subset \mathcal{P}_k$, we have

$$\min_{\rho_0} (\max_\nu \frac{|\rho_0 - \bar{w}_\nu|}{\varepsilon_\nu}) = \delta(0) \geq \delta(1) \geq \ldots \geq \delta(n-2) \geq \delta(n-1) = 0;$$

so there must be some $k \leq n-1$ with $\delta(k) = O(1)$. Due to our vague definition of $O(1)$, the minimal degree k^* for which valid approximate interpolation is possible may not be unambiguously defined. $\quad\square$

For a specified data set, the determination of $\delta(k)$ and of the associated minimizing $r \in \mathcal{P}_k$ is a standard *linear minimization problem*: Let $r(x) = \sum_{\kappa=0}^k \rho_\kappa x^\kappa$; then we have to solve

$$\min_{\rho_0, \ldots, \rho_k} \delta : \left| \sum_{\kappa=0}^k \rho_\kappa \zeta_\nu^\kappa - \bar{w}_\nu \right| \leq \varepsilon_\nu \delta, \quad \nu = 1(1)n . \tag{5.69}$$

It is more customary to address the task of approximate interpolation with a (weighted) 2-norm in the space of the empirical data: The max norm in (5.67) is simply replaced by the Euclidean norm. Now, the relation between (5.66) and (5.67) is no longer so immediate; there appear factors $\sqrt{n}$ in both directions which may be absorbed into the $O(1)$ for small k only. The determination of $\delta(k)$ and the minimizing r become straightforward *least squares problems*; therefore, this approach is usually called "interpolation by least squares." As at other places in this book, we refrain from an explicit analysis of this analogous approach because it is well-known and because we feel that the use of the maximum norm is often more appropriate in connection with empirical data.

Independently of the $\mathbb{R}^n$-norm used in (5.67), approximate interpolation comes often under the name of "smoothing" or "smoothing interpolation"; this refers to the fact that the approximate interpolating polynomials of a degree $< n - 1$ generally display less oscillation than the genuine interpolation polynomial. This is particularly evident in the case $k = 1$.

Example 5.11: Consider the set $\{\zeta_\nu = \nu/10,\ \nu = 0(1)10\}$ of interpolation nodes. We take the values $\bar{w}_\nu$ at the ζ_ν from the functions $\exp(x)$ and $\exp(-x)$, resp., but assign tolerances $\varepsilon_\nu = 10^{-6}$ to them. For a minmax smoothing interpolation on the above grid, we obtain the following approximate backward errors for degree k:

$$\exp(x) \qquad\qquad \delta(3) \approx 510\,, \quad \delta(4) \approx 26\,, \quad \delta(5) \approx .96\,, \quad \delta(6) \approx .032\,,$$

$$\exp(-x) \qquad\qquad \delta(3) \approx 190\,, \quad \delta(4) \approx 9.6\,, \quad \delta(5) \approx .35\,, \quad \delta(6) \approx .012\,.$$

Thus, if we are satisfied to interpolate the exponential functions to 6 decimal digits on the above grid, we may use a polynomial of degree 5 in place of the genuine interpolation polynomial of degree 10. $\square$

Exercises

1. Consider the polynomial interpolation of $\cos(x)$ on the 7 nodes $\zeta_\nu = (\nu - 1)\,\pi/6,\ \nu = 1(1)7$.

   (a) Form the Vandermonde matrix $V(\zeta_1, \ldots, \zeta_7)$ and compute the coefficients of the interpolation polynomial $r \in \mathcal{P}_6$. Why is r only of degree 5? Find r by interpolating $-\sin(x)$ on $\zeta_2, \zeta_3, \zeta_4$ by an odd polynomial $\bar{r} \in \mathcal{P}_5$ and setting $r(x) = \bar{r}(x - \pi/2)$. Give the reason. Use the symmetry of $\cos(x)$ and of the node set also in the following tasks.

   (b) Determine the interpolation error $r(\eta_\nu) - \cos(\eta_\nu)$ at the $\eta_\nu = (\nu - 1/2)\,\pi/6,\ \nu = 1(1)6$. Comment.

   (c) Form $s(x) := \prod_{\nu=1}^{7}(x - \zeta_\nu)$ and $\hat{r}(x, \rho) := r(x) + \rho\,s(x) \in \mathcal{P}_7$ which also interpolates $\cos(x)$ on the ζ_ν. Determine $\rho \in \mathbb{R}$ such that $\max_\nu |\hat{r}(\eta_\nu, \rho) - \cos(\eta_\nu)|$ is minimal. Compare the minimizing ρ with the values of $\frac{1}{7!}\frac{\partial^7}{\partial x^7}\cos(x)$ in $(0, \pi)$.

   (d) Compare the approximation qualities of r and $\hat{r}$. By how much have the oscillations of the interpolation error been reduced?

   (e) Replace the exact values at the ζ_ν, $\nu = 2, 3, 5, 6$, by empirical quantities $(\cos(\zeta_\nu), .5 \cdot 10^{-5})$ and try to find an approximate interpolation polynomial with a lower degree than 6. Which is the minimal k which permits a valid approximate interpolation of degree k ?

2. Use least squares smoothing in Example 5.11, i.e. use the Euclidean norm in place of the maximum norm in (5.69). Compare various aspects of the results.

Historical and Bibliographical Notes 5

Univariate polynomials have been used and their zeros have been determined for centuries; an early numerical analysis text of high quality is [5.1]. With the arrival of functional analysis, orthogonal systems of polynomials were widely studied. As a polynomial counterpart to trigonometric polynomials and Fourier expansions, Chebyshev polynomials (cf. (5.9)) play a central role in polynomial approximation and expansion; cf. (5.6). Details may be found in most texts on applied and/or numerical analysis, e.g., in [3.10].

   Because of their simple algebraic structure, ideals of univariate polynomials are used as examples in most texts on polynomial algebra; cf. [2.10]–[2.14]. Yet the fundamental role of the Vandermonde matrix for the relations (5.23) and (5.24) and of the Frobenius matrix for Theorem 5.5 (the univariate version of the Central Theorem) is rarely elaborated although it is the basis for the computational treatment of the multivariate case. The numerically interesting expansion (5.18) of a polynomial in terms of one of lower degree is rarely discussed.

The algorithmic flavor of mathematics in the late 19th century is shown by Weierstrass's use of (5.30) in his proof of the "fundamental theorem of algebra" ([5.2]). With the advent of the computer, algorithms for the approximate computation of polynomial zeros flourished; Weierstrass's method was rediscovered, e.g., in [5.3]. The approach in [5.4] is still considered as superior with respect to its convergence properties; MPSolve of [5.5] uses (5.31) for refinement. Interactive systems (like MATLAB®) use Theorem 5.5 in their polynomial rootfinders. According to comparative runs at Bell Labs ([5.6]), MPSolve beats computation via eigenvalues on well-conditioned polynomials, particularly of a very high degree, and for sparse polynomials, while it is inferior for dense, ill-conditioned polynomials.

The potential extreme sensitivity of apparently well-separated zeros has been spectacularly exhibited by Wilkinson's example in [3.5, §9]; cf. Example 5.7 and Exercise 5.2-2. The example also shows that the power basis representation is often not well-suited for the computation of polynomial zeros and an unnecessary transformation to a power basis may be ill-advised. A representation of the potential indetermination of polynomial zeros has been a goal of interval analysis from its start (cf. [3.2], [4.8]), and of pseudozero domains (cf., e.g., [3.3], [3.4]). The fact that the potential locations of the individual zeros under given perturbations of the coefficients are strongly tied has not received much attention so far; the same is true for the simultaneous backward error of a *set* of approximate zeros.

Polynomial *division* is an example of a crucial algebraic operation whose numerical analysis appears to be rather untouched so far. Polynomial *interpolation* on the other hand, because of its importance in applications, has been a dominant subject of computational analysis; the names attached to various algorithmic approaches (Lagrange, Newton, Hermite, etc.) bear evidence of that. References [3.10] and [5.7] give a solid background for our superficial account in section 5.4.

References

[5.1] A.S. Householder: The Numerical Treatment of a Single Nonlinear Equation, McGraw-Hill, New York, 1970.

[5.2] K. Weierstrass: Neuer Beweis des Satzes, dass jede ganze rationale Function einer Veränderlichen dargestellt werden kann als Product aus linearen Functionen derselben Veränderlichen, Sitzungsber. Königl. Akad. Wiss., Berlin, 1891.

[5.3] I.O. Kerner: Ein Gesamtschritt-Verfahren zur Bestimmung der Nullstellen von Polynomen, Numer. Math. **8** (1966), 290–294.

[5.4] O. Aberth: Iteration Method for Finding All Zeros of a Polynomial Simultaneously, Math. Comp. **27** (1973), 319–344.

[5.5] D.A. Bini, G. Fiorentino: Design, Analysis, and Implementation of a Multiprecision Polynomial Rootfinder, Numer. Algorithms **23** (2000), 127–173.

[5.6] `http://cm.bell-labs.com/who/sjf/eigensolveperformance.html`

[5.7] A.M. Ostrowski: Solution of Equations and Systems of Equations (2nd Ed.), Academic Press, New York, 1966.

Chapter 6

Various Tasks with Empirical Univariate Polynomials

In this chapter, we address a number of different tasks of practical importance which involve univariate polynomials. The common trait of these tasks is that they acquire a different character when they are posed for empirical polynomials: Classically, their results depend *discontinuously* on the polynomial data; thus, they cannot generally be solved or evaluated numerically with floating-point computation, at least not when the data are close to a discontinuity manifold of the problem. For empirical polynomials, these problems become continuous when they are posed in the set topology of the associated polynomial neighborhoods; this makes their approximate treatment meaningful and feasible.

A number of the tasks discussed in this chapter may be formulated in terms of predicates on sets of polynomials. Others are related to the determination of valid divisors, or—equivalently—valid sets of zeros, of one or several empirical polynomials.

6.1 Algebraic Predicates

6.1.1 Algebraic Predicates for Empirical Data

In many mathematical models, assertions about the location of the complete zero set of some univariate polynomial in the model are of prime importance: The modeled system has some characteristic property if all zeros of a certain polynomial p lie in some specified part of $\mathbb{C}$. We may abbreviate this by saying: "p has property Π" or "the assertion (predicate) Π is true for p." Clearly, this is a *discontinuous* situation: Generally, there must be polynomials which lose or gain the property upon an arbitrarily small perturbation.

More generally, we consider *predicates* which assign a *truth value* (**true** or **false**) to a given polynomial $p \in \mathcal{P}$, or rather to the set a of coefficients of the polynomial $p(x; a)$.

Definition 6.1. The mapping $\Pi : \mathcal{P}^1 \to \mathbb{T} := \{\textbf{true}, \textbf{false}\}$ is an *algebraic predicate* on $\mathcal{P}^1$ iff

- the truth value $\Pi(p)$ is a function of the coefficients of p ;

- the *truth domain* $Q(\Pi) := \{a \in \mathcal{A} : \Pi(p(x; a)) = \textbf{true}\}$ of Π in the space $\mathcal{A}$ of the coefficients a of $p(x; a) \in \mathcal{P}_n^1$ is semi-algebraic.

We call the algebraic predicate Π *noncritical* if the dimension of $Q(\Pi)$ in $\mathcal{A}$ equals $\dim \mathcal{A}$, otherwise Π is *critical*. $\square$

Example 6.1: Over the real univariate polynomials, the predicate $\Pi(p) =$ "p is positive over $\mathbb{R}$" is algebraic: Trivially, the truth value of $\Pi(p(x; a))$ depends only on the coefficients a of p. Also, for all polynomials $p(x; a)$ with $\Pi(p(x; a)) = \textbf{true}$, the discriminant of p must have a fixed sign and some further algebraic inequalities must hold; thus the truth domain $Q(\Pi)$ is semi-algebraic. For degree 2, e.g., $\Pi(p(x; a)) = \textbf{true}$ iff $s(\alpha_0, \alpha_1, \alpha_2) = \alpha_1^2 - 4\alpha_0 \alpha_2 < 0$ and $\alpha_0 > 0$. Since these are open domains in the coefficient space $\mathcal{A}$, the predicate is noncritical. $\square$

The reason for the distinction between noncritical and critical predicates comes to light when we attempt to evaluate algebraic predicates for polynomials which are empirical—as they will generally be in a modelling situation: A *critical* predicate Π may be extended to empirical polynomials in a manner which we have used previously; cf. Chapters 3 and 5. For the critical predicate in disguise "z is a zero of p," e.g., we have defined the extension "z is a pseudozero of $(\bar{p}, e)$" as valid if there exist polynomials $\tilde{p} \in N_\delta(\bar{p}, e)$, $\delta = O(1)$, with $\tilde{p}(z) = 0$. In the same way, we can generally define an extension $\widetilde{\Pi}$ of a critical predicate Π as valid if there exist polynomials $\tilde{p} \in N_\delta(\bar{p}, e)$, $\delta = O(1)$, with $\Pi(\tilde{p}) = \textbf{true}$; cf. Figure 6.1.

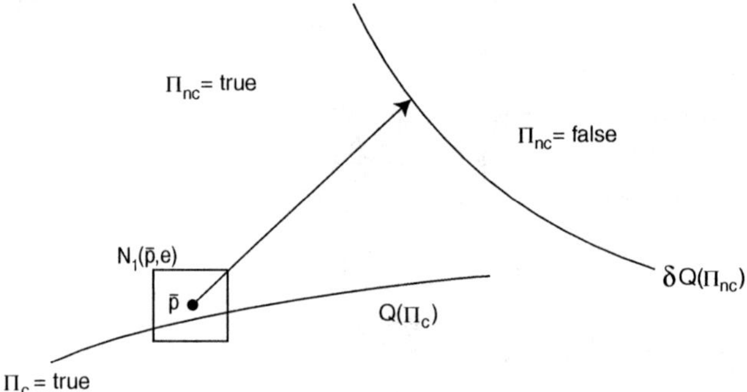

Figure 6.1.

For a noncritical predicate Π like "p is positive over $\mathbb{R}$," on the other hand, we have a different situation: For $\widetilde{\Pi}((\bar{p}, e))$ to be valid, we want to be certain that Π is true not only for $\bar{p}$ but for *all* $\tilde{p}$ in the neighborhoods $N_\delta(\bar{p}, e)$, $\delta = O(1)$. This is equivalent to the requirement that the e-nearest $\hat{p}$ with $\Pi(\hat{p}) = \textbf{false}$ or $\neg\Pi(\hat{p}) = \textbf{true}$ is *not* in an N_δ neighborhood with $\delta = O(1)$. Then, the violation of Π by a neighboring polynomial would require a perturbation well above the specified tolerance level.

We may reduce the extension and evaluation of a noncritical predicate for an empirical polynomial $(\bar{p}, e)$, $\bar{p} = p(x; \bar{a})$, to our established procedure for a critical predicate as follows: For the noncritical predicate Π with its truth domain $Q(\Pi)$ in the empirical data space $\mathcal{A}$ of $(\bar{p}, e)$, we consider the *boundary set* $\partial Q(\Pi) \subset \mathcal{A}$ of that component of $Q(\Pi)$ which contains $\bar{a}$; obviously, $\partial Q(\Pi)$ has a positive codimension. Then we introduce the critical predicate $\partial\Pi$ with truth domain $\partial Q(\Pi)$. Now we may call the extension $\widetilde{\Pi}$ valid at $(\bar{p}, e)$ if $\Pi(\bar{p}) = \textbf{true}$ and if $\widetilde{\partial\Pi}((\bar{p}, e))$ is *not valid*; cf. Figure 6.1.

When a critical algebraic predicate Π is applied to an empirical polynomial $(\bar{p}, e)$, we can associate with Π a *backward error* $\delta(\Pi) \in \mathbb{R}_+$ in the standard way as the minimal e-norm distance of $Q(\Pi)$ from $\bar{a} \in \mathcal{A}$ or of the shifted truth domain $\Delta Q(\Pi)$ from the origin in $\Delta\mathcal{A}$, respectively. When we consider $\delta(\Pi)$ as the *image* of the extended predicate $\widetilde{\Pi}$, we have turned the original discontinuous map $\Pi : \mathcal{P}^1 \to \mathbb{T}$ into a map $\widetilde{\Pi}$ which maps the empirical polynomials in $\mathcal{P}^1$ to the nonnegative reals and which is *continuous* with respect to $\bar{a}$ and e. Via the "boundary predicate" $\partial\Pi$ of a noncritical predicate, we have extended this approach also to noncritical predicates Π.

Definition 6.2. For a *critical* algebraic predicate Π on $\mathcal{P}^1$, the value of its extension $\widetilde{\Pi}$ for an empirical polynomial $(\bar{p}, e)$ is defined as

$$\widetilde{\Pi}((\bar{p}, e)) := \min_{\Delta a \in \Delta Q(\Pi)} \|\Delta a\|_e^* . \tag{6.1}$$

Π is *valid* for $(\bar{p}, e)$ if $\widetilde{\Pi}((\bar{p}, e)) \leq O(1)$.

For a *noncritical* algebraic predicate Π on $\mathcal{P}^1$, we consider the boundary set $\partial\Delta Q(\Pi)$ of $\Delta Q(\Pi)$—also denoted as *critical manifold* of Π—and define

$$\partial\widetilde{\Pi}((\bar{p}, e)) := \min_{\Delta a \in \partial\Delta Q(\Pi)} \|\Delta a\|_e^* . \tag{6.2}$$

Π is *valid* for $(\bar{p}, e)$ if $0 \in \Delta Q(\Pi)$ and $\partial\widetilde{\Pi}((\bar{p}, e)) > O(1)$. $\square$

Example 6.1, continued: Consider the quadratic empirical polynomial $(\bar{p}, e)$, with $\bar{p}(x) = 2.47\,x^2 - 7.81\,x + 6.27$ and $e = (.5, .5, .5) \cdot 10^{-2}$; $s(6.27, -7.81, 2.47) = -.9515$ and $6.27 > 0$ so that $\Pi(\bar{p}) = $ **true**. In the space $\Delta\mathcal{A} = \mathbb{R}^3$ of the coefficient deviations $\Delta\alpha_i$, $i = 0, 1, 2$, the critical manifold $\partial\Delta Q(\Pi)$ is given by

$$s(6.27 + \Delta\alpha_0, -7.81 + \Delta\alpha_1, 2.47 + \Delta\alpha_2) = (-7.81 + \Delta\alpha_1)^2 - 4\,(6.27 + \Delta\alpha_0)\,(2.47 + \Delta\alpha_2) = 0;$$

since $6.27 + \Delta\alpha_0 > 0$ is always satisfied for small perturbations, we may disregard this condition. The minimal distance of $\partial\Delta Q(\Pi)$ from the origin, in terms of the norm $\|\Delta a\|_e^* = \max_{i=0,1,2}(\frac{|\Delta\alpha_i|}{.005})$ in $\Delta\mathcal{A}$, is found as $\delta(\partial\widetilde{\Pi}) \approx 3.77$ for $\Delta\alpha_0 = \Delta\alpha_1 = \Delta\alpha_2 \approx -.01884$.

Thus, $\partial\widetilde{\Pi}((\bar{p}, e)) > 1$ and Π is a valid assertion for $(\bar{p}, e)$ in a restricted sense: All polynomials in $N_1(\bar{p}, e)$ are positive over the reals. With our more relaxed concept of empirical quantities, however, we may hesitate to make that assertion: The closest polynomial with coefficients on the critical manifold is $\tilde{p}(x) \approx 2.4512\,x^2 - 7.8288\,x + 6.2512$, with $\|\tilde{p} - \bar{p}\|_e^* \approx 3.77$, so that we may well regard $\tilde{p}$ as a polynomial in an $O(1)$ tolerance neighborhood of $\bar{p}$. This is a typical case where a scrutiny of the context will be necessary for a wise decision. $\square$

For a computational determination of the validity of an algebraic predicate for an empirical polynomial, we must be able to evaluate the minima in (6.1) or (6.2), resp.; for this purpose, we must have a characterization of the critical manifolds $\Delta Q(\Pi)$ or $\partial\Delta Q(\Pi)$, resp., in algebraic terms, like in Example 6.1 above. For notational simplicity, we will, at this point, denote either critical manifold by $s(\Delta a) = 0$. In many cases, the critical manifolds will be nonlinear and the determination of their minimal distance from the origin in $\Delta\mathcal{A}$ may present serious difficulties, in any norm. General approaches and algorithms for *global nonlinear optimization* may be found in the special literature on that subject; cf. also [6.8]. Their immediate applicability will depend on the particular form of the critical manifold under consideration.

A relief in this situation is the fact that we may often not need to find the actual minimum: If we have located some point Δa on the critical manifold with $\|\Delta a\|_e^* = O(1)$, we have already reached a decision about the validity of $\widetilde{\Pi}$ for $(\bar{p}, e)$: a positive one for a critical and a negative one for a noncritical Π. In the following, we outline two *heuristic* techniques which may be used in the present situation and which may often be successful.

We assume, at first, that the critical manifold of codimension m passes closely by the origin $\Delta a = 0$ of $\Delta \mathcal{A} = \mathbb{C}^M$ and that its functional representation is sufficiently regular there. If the critical manifold is represented by a polynomial system $q(\Delta a) = 0$, $q : \mathbb{C}^M \to \mathbb{C}^m$, sufficient regularity means that the Fréchet derivative $q'(\Delta a)$, a linear map from $\mathbb{C}^M$ to $\mathbb{C}^m$, exists and is regular and slowly varying as a function of Δa near $\Delta a = 0$. Then, the intersection of the orthogonal linear manifold at the origin, with parameter representation $\Delta a = \sum_{\mu=1}^m t_\mu \, (q'(0))_\mu$, and the critical manifold $q(\Delta a) = 0$ may yield a point Δa^0 on the critical manifold not too far from its closest point to the origin. (The dimensions of these two manifolds add up to M so that their intersection will generally consist of isolated points.) If $\|\Delta a^0\|_e^*$ is of moderate size, we may move *along* the critical manifold in directions of decreasing $\|\Delta a\|_e^*$ until we have reached a sufficiently small value or an approximate minimum. Under the above assumptions, this should be the global minimum of $\|\Delta a\|_e^*$ on the critical manifold; cf. Figure 6.2a.

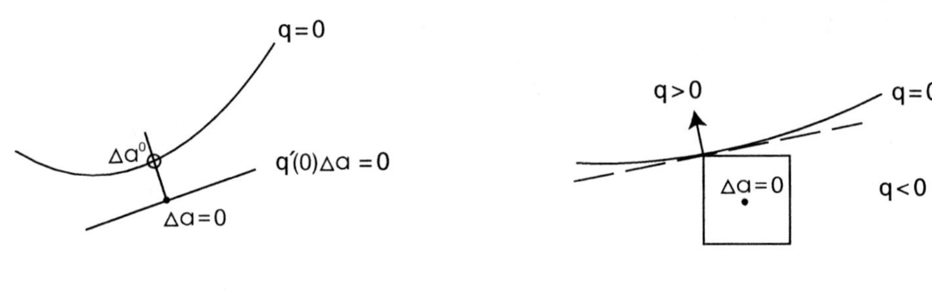

Figure 6.2.

a   b

If the codimension of the critical manifold is 1, we may use the following fact in the search for minima of the norm distance to the origin.

Proposition 6.1. On a generic algebraic manifold $q(y) = 0$ of codimension 1 in $\mathbb{R}^n$, with $q(0) \neq 0$, a point $y^* = (\eta_1^*, \ldots, \eta_n^*)$ has a locally minimal max-norm distance from the origin iff it satisfies

$$|\eta_\nu^*| = \|y^*\|_{\max} \quad \text{and} \quad \text{sign}\left(\frac{\partial q}{\partial \eta_\nu}(y^*) \Big/ \eta_\nu^*\right) = -\text{sign}\, q(0) \quad \text{for all } \nu = 1(1)n. \quad (6.3)$$

Proof: For y^* to be a local minimum of $\|y\|_{\max}$ on the generic manifold $q(y) = 0$, it is necessary and sufficient that y^* is the only common point of the tangential hyperplane $h(y; y^*) = \text{grad}\, q(y^*) \cdot (y - y^*) = 0$ of the manifold at y^*—which exists due to genericity—and the hypercube $\{y : \|y\| \leq \|y^*\|\}$. Otherwise, there exist neighboring points of y^* on the manifold which are in the interior of the hypercube and thus have a smaller norm; the case that the intersection lies in a lower-dimensional linear manifold on the surface of the hypercube may be dismissed because of the genericity assumption.

In a generic situation, with all components of grad $(y^*) \neq 0$, the above condition requires that y^* is a *corner* of the hypercube, i.e. the first condition in (6.3). In this case, the hyperplane has no further points in common with the hypercube iff its normal vector lies in the same orthant as $\pm y^*$, i.e. if the left-hand signs in the second condition in (6.3) are equal. If no further component of $q = 0$ passes between y^* and the origin, this sign must equal -sign $q(0)$; cf. Figure 6.2b. $\square$

Thus, for a generic manifold of codimension 1, it suffices to know the orthant of y^* with the *global* minimal max-norm distance from the origin and to solve the univariate polynomial equation $q(\pm t, \ldots, \pm t) = 0$ for t, with the appropriate combination of signs. To determine the orthant, we may assume that the sign distribution in grad q remains invariant between the manifold and the origin (cf. the considerations above); then the vector grad $q(0)$ points into the orthant in which $\pm y^*$ lies. The correctness of the second part of (6.3) may be checked a posteriori by the evaluation of grad $q(y^*)$. If it is not satisfied we must investigate further orthants for local minima of the norm distance between the manifold $q(y) = 0$ and the origin.

This procedure may be applied immediately to the critical manifold if it has codimension 1 in $\Delta\mathcal{A}$. Naturally, we must consider the weighting (3.5) of the max-norm introduced by the tolerance vector e of the empirical polynomial; this changes the first condition in (6.3) to

$$|\Delta\alpha_j^*| = \varepsilon_j \, \|\Delta a^*\|_e^*, \quad \text{for all } j \in \tilde{J}. \tag{6.4}$$

Thus, we have to solve the equation $q(\pm\varepsilon_1\delta, \ldots, \pm\varepsilon_M\delta) = 0$ with the appropriate sign distribution. If we obtain a real zero $\delta \leq O(1)$, a decision about the validity has been reached; in this case it is not necessary to determine the actual minimal distance.

When we do not readily find a $\delta \leq O(1)$, the situation is not so simple: Since we must make a decision between $\delta \leq O(1)$ and $\delta > O(1)$, we must actually determine the minimal distance of the critical manifold from the origin with sufficient reliability. Proposition 6.1 requires the search of 2^{M-1} orthant cones for potential candidates which satisfy (6.4), but nongeneric situations may also arise. Therefore, one must strive to use special information about the predicate Π to restrict the search. Fortunately, in some important cases, it is possible to restrict the search to *two* of the 2^{M-1} orthant cones which are known a priori to contain the potential minima; cf. Propositions 6.2 and 6.3 in the following subsections.

Another approach has been proposed and applied by Karmarkar/Lakshman and Hitz/Kalt-ofen in several papers; cf. [6.7]. For suitable situations, they introduce a real parameter ζ to *parametrize* the critical manifold as $q(\Delta a(\zeta)) = 0$; thus, the distance $\|\Delta a(\zeta)\|_e^*$ becomes a function $\delta(\zeta)$ and one may determine min $\delta(\zeta)$ over the feasible locations of ζ. If $\delta(\zeta)$ may be expressed explicitly in terms of ζ, this permits a straightforward determination of the minimal value. If $\delta(\zeta)$ can only be determined numerically for specified values of ζ, one must attempt to collect sufficient information to make the decision about the validity of Π.

Example 6.1, continued: The boundary predicate $\partial\Pi$ for our noncritical predicate $\Pi(p)$ (p is positive over $\mathbb{R}$) is "p has a multiple zero on $\mathbb{R}$"; cf. section 6.1.2. Let $\zeta \in \mathbb{R}$ be the point at which $\tilde{p}(x, \bar{a} + \Delta a) = \bar{p}(x) + (x^2 \Delta\alpha_2 + x \Delta\alpha_1 + \Delta\alpha_0)$ and its derivative have a common zero on the real axis, which implies

$$\zeta^2 \Delta\alpha_2 + \zeta \Delta\alpha_1 + \Delta\alpha_0 + \bar{p}(\zeta) = 2\zeta \Delta\alpha_2 + \Delta\alpha_1 + \bar{p}'(\zeta) = 0.$$

For specified $\bar{p}$ and $\zeta \in \mathbb{R}$, this defines a manifold $\mathcal{M}(\zeta)$ of codimension 2 in $\Delta\mathcal{A}$; its minimal e-norm distance $\delta(\zeta)$ from the origin can readily be computed for each specific value of ζ, as we have seen in section 3.3.1. But it is not possible to express $\delta(\zeta)$ as a manageable function of ζ, even in this simple 2nd degree case. $\quad\square$

It is obvious that algebraic predicates for *multivariate* polynomials may be extended to empirical multivariate polynomials in the same way as we have done it for univariate polynomials in this section, except that the technical details become more involved. We will return to this subject in Part III of the book.

6.1.2 Real Polynomials with Real Zeros

We consider the predicate $\Pi(p) = $ "$p \in \mathbb{R}[x]$ has only real zeros," with $\neg\Pi(p) = $ "p has a pair of conjugate-complex zeros." This property of a real univariate polynomial is of interest in various application contexts. The predicate Π is clearly noncritical because there are p for which a full coefficient neighborhood lies in the truth domain $Q(\Pi)$. Hence we must find the critical manifold $\partial Q(\Pi)$ which constitutes the boundary between $Q(\Pi)$ and $Q(\neg\Pi)$.

Proposition 6.2. For $\Pi(p)$ as above,

$$\partial Q(\Pi) = \{p \in \mathbb{R}[x] \;:\; p \in Q(\Pi) \text{ has a multiple real zero}\}.$$

Proof: For $p \in Q\Pi$ and $\hat{p} \in Q(\neg\Pi)$, consider $p(t) := t\, p + (1-t)\,\hat{p}$: When t moves from 0 to 1, the conjugate-complex pair(s) of zeros of $\hat{p}$ must turn real; for real $p(t)$, this is only possible through a confluence on the real line. $\quad\square$

Consider an empirical real polynomial $(\bar{p}, e)$ whose specified polynomial $\bar{p}$ has only real zeros, and its associated empirical data space $\Delta\mathcal{A} = \mathbb{R}^M$ of deviations $\Delta\alpha_j$ of the empirical coefficients. By Proposition 6.2, the boundary manifold $\partial Q \subset \Delta\mathcal{A}$ of the component of $Q(\Pi)$ about the origin consists of values Δa for which $p(x; \bar{a} + \Delta a)$ has a multiple real zero, or— equivalently—for which $p(x; \bar{a} + \Delta a)$ and its derivative $p'(x; \bar{a} + \Delta a)$ have a *common* real zero. This manifold ∂Q is algebraically defined by the *Sylvester matrix* $S(p, p')$ of p and p'; cf. section 6.2. With $\tilde{p}(x; \bar{a} + \Delta a) = \bar{p}(x) + \sum_{j \in \tilde{\jmath}} \Delta\alpha_j x^j =: \sum_{v=0}^{n} \tilde{\alpha}_v x^v$,

$$s(\Delta a) := \det S(\tilde{p}, \tilde{p}') = \det \begin{pmatrix} \tilde{\alpha}_0 & \cdots & \cdots & \tilde{\alpha}_n & & & \\ & \ddots & & & & \ddots & \\ & & & \tilde{\alpha}_0 & \cdots & \cdots & \tilde{\alpha}_n \\ \tilde{\alpha}_1 & 2\tilde{\alpha}_2 & \cdots & n\tilde{\alpha}_n & & & \\ & \ddots & & & & \ddots & \\ & & \tilde{\alpha}_1 & \cdots & \cdots & \cdots & n\tilde{\alpha}_n \end{pmatrix} = 0 \qquad (6.5)$$

is a necessary and sufficient condition for $\tilde{p}$ and $\tilde{p}'$ to have a common zero. Here, $S(\tilde{p}, \tilde{p}') \in \mathbb{R}^{(2n-1)\times(2n-1)}$, with the coefficients of $\tilde{p}$ occupying the first $n-1$ rows and those of $\tilde{p}'$ the remaining n rows. Thus, $s(\Delta a)$ is a polynomial of degree $\leq 2n - 1$ in the M variables of the data space $\Delta\mathcal{A}$ in which the critical manifold

$$\partial Q := \{\Delta a \;:\; \exists \zeta \in \mathbb{R} \;:\; p(\zeta; \bar{a} + \Delta a) = p'(\zeta; \bar{a} + \Delta a) = 0\} = \{\Delta a \;:\; s(\Delta a) = 0\} \qquad (6.6)$$

has codimension 1.

Proposition 6.3. The minima of $\|\Delta a\|_e^*$ on the critical manifold (6.6) lie in the orthants $\pm(+, +, +, \ldots)$ and $\pm(+, -, +, -, \ldots)$.

Proof: Consider an empirical real polynomial $(\bar{p}, e)$, where $\bar{p}$ has only disjoint real zeros. Let $\bar{p}(x) + \Delta p(x)$ be the closest polynomial, in terms of $\|..\|_e^*$, with a multiple real zero, which we assume to lie at $\xi \in \mathbb{R}$. For specified ξ, we know from (5.34) that

$$\Delta\alpha_\nu \;=\; -\frac{\bar{p}(\xi)}{\sum_\nu \varepsilon_\nu |\xi|^\nu} \, \varepsilon_\nu \,\text{sign}\,\xi^\nu\,, \quad \nu = 0(1)n\,, \tag{6.7}$$

determines the Δp with the smallest e-norm such that $(p + \Delta p)(\xi) = 0$; note that ξ enters only through the common factor and sign ξ.

The case $\xi = 0$, with $\Delta\alpha_0 = -\bar{\alpha}_0$, is trivial. For $\xi \neq 0$, we consider the common factor whose modulus is $\|\Delta a\|_e^*$, as a parameter $t \in \mathbb{R}$ and realize that the minimizing Δa lies on the line

$$\Delta\alpha_\nu \;=\; t \cdot \begin{cases} \varepsilon_\nu & \text{for } \xi > 0\,, \\ (-1)^\nu \varepsilon_\nu & \text{for } \xi < 0\,. \end{cases} \qquad \square \tag{6.8}$$

Hence, *without a priori knowledge about* ξ, we need only intersect the two lines (6.8) with the manifold ∂Q of (6.6) to obtain the minimizing $\Delta a \in \partial Q$ and their e-norms; this requires the determination of the real zeros of smallest modulus for the two univariate polynomials $s(t\,e)$ and $s(t\,e_\pm)$ where $e_\pm := (\varepsilon_0, -\varepsilon_1, \varepsilon_2, -\varepsilon_3, \ldots)$.

The parametrization approach also works in this case. A parametrization is provided by the real value ζ at which $\tilde{p}$ and $\tilde{p}\,'$ vanish simultaneously for a $\tilde{p} \in N_\delta(\bar{p}, e)$. The potential candidates for $\tilde{p}$ are characterized by (6.7); with Δp_e and Δp_o for the even and odd parts of Δp, they are $(j = 0, 1)$

$$\tilde{p}_j(\zeta; t) \;=\; \bar{p}(\zeta) + t \cdot [\Delta p_e(\zeta) + (-1)^j \Delta p_o(\zeta)] \;=\; 0\,,$$

with derivatives

$$\tilde{p}_j\,'(\zeta; t) \;=\; \bar{p}\,'(\zeta) + t \cdot [\Delta p_e\,'(\zeta) + (-1)^j \Delta p_o\,'(\zeta)] \;=\; 0\,.$$

These two linear equations in t have a solution iff

$$\bar{p}(\zeta) \cdot [\Delta p_e\,'(\zeta) + (-1)^j \Delta p_o\,'(\zeta)] - \bar{p}\,'(\zeta) \cdot [\Delta p_e(\zeta) + (-1)^j \Delta p_o(\zeta)] \;=\; 0\,. \tag{6.9}$$

For each real zero ζ of this determinant, we obtain a unique $|t| = \delta(\zeta)$; thus, $\neg\widetilde{\Pi}((\bar{p}, e)) = \min_\zeta \delta(\zeta)$.

Example 6.2: Consider the empirical polynomial $(\bar{p}, e)$ with

$$\bar{p}(x) \;:=\; 1.00\,x^4 - 1.29\,x^3 - 2.01\,x^2 + 1.26\,x + .85\,,$$

with zeros at $\approx -1.04411, -.46815, .92762, 1.87465$, with real tolerances $\varepsilon_\nu = .005$ for all ν, including $\nu = 4$. We would like to know whether we may safely assume that $(\bar{p}, e)$ has only real zeros.

The polynomial $s(\Delta a)$ of (6.5) is of degree 7 in the 5 variables $\Delta\alpha_\nu$, $\nu = 0(1)4$; according to (6.8) we replace them by $t\,\varepsilon_\nu$ or $(-1)^\nu t\,\varepsilon_\nu$, resp. The resulting two 7th degree polynomials in t have 5 real zeros each, the minimal modulus ones are ≈ 19.6 and 24.2, respectively. Thus,

$\neg \tilde{\Pi}((\bar{p}, e)) = 19.6$; the associated polynomial $\tilde{p}$ has a double zero at ≈ 1.3007 and two negative zeros. The backward error is sufficiently large to confirm that the empirical polynomial $(\bar{p}, e)$ has only real zeros.

In the parametrized approach, we form the polynomial (6.9) for $j = 0$ and 1, resp.; these are polynomials of degree 6 in ζ (the ζ^7 terms cancel), with 4 real zeros each. For each real zero, we form $t(\zeta) = -\bar{p}(\zeta) / [\Delta p'_e(\zeta) + (-1)^j \Delta p'_o(\zeta)]$ and select the minimal modulus value. From $\tilde{p}_0$, we obtain the minimal $|t(\zeta)| \approx 19.6$ for $\zeta \approx 1.3007$ as above. $\square$

6.1.3   Stable Polynomials

The solutions of the homogeneous linear differential equation with constant real coefficients

$$\partial^n y(t) + \sum_{\nu=0}^{n-1} \alpha_\nu \, \partial^\nu y(t) = 0 \tag{6.10}$$

tend to 0 for $t \to \infty$ iff all zeros of the associated characteristic polynomial $p(x) := x^n + \sum_{\nu=0}^{n-1} \alpha_\nu x^\nu$ have *negative real parts*; therefore, such polynomials are often called *stable*. We consider the predicate $\Pi(p) = $ "p is stable" which is again noncritical.

Because of the practical importance of the stability of differential equations, various characterizations for the truth domain $Q(\Pi)$ have been developed; the best-known is the

Routh-Hurwitz criterion. For $p(x) := \sum_{\nu=0}^{n} \alpha_\nu x^\nu$, with $\alpha_n > 0, \alpha_0 \neq 0$, we form the $n \times n$ matrix $H(a)$ by arranging the coefficients $\alpha_{n-1}, \ldots, \alpha_0$ along the main diagonal and filling the rows by further α_ν: to the right with increasing ν, to the left with decreasing ν. For $n = 5$, e.g., we obtain

$$H(a) = \begin{pmatrix} \alpha_4 & \alpha_5 & 0 & 0 & 0 \\ \alpha_2 & \alpha_3 & \alpha_4 & \alpha_5 & 0 \\ \alpha_0 & \alpha_1 & \alpha_2 & \alpha_3 & \alpha_4 \\ 0 & 0 & \alpha_0 & \alpha_1 & \alpha_2 \\ 0 & 0 & 0 & 0 & \alpha_0 \end{pmatrix}. \tag{6.11}$$

The polynomial p is stable iff all principal minors of $H(a)$ are positive. For $n = 5$, e.g., this requires $H_1 = \alpha_4 > 0$,

$$H_2 = \begin{vmatrix} \alpha_4 & \alpha_5 \\ \alpha_2 & \alpha_3 \end{vmatrix} > 0, \quad H_3 = \begin{vmatrix} \alpha_4 & \alpha_5 & 0 \\ \alpha_2 & \alpha_3 & \alpha_4 \\ \alpha_0 & \alpha_1 & \alpha_2 \end{vmatrix} > 0, \quad H_4 = \begin{vmatrix} \alpha_4 & \alpha_5 & 0 & 0 \\ \alpha_2 & \alpha_3 & \alpha_4 & \alpha_5 \\ \alpha_0 & \alpha_1 & \alpha_2 & \alpha_3 \\ 0 & 0 & \alpha_0 & \alpha_1 \end{vmatrix} > 0,$$

and $H_5 = H_4 \cdot \alpha_0 > 0$. Thus, the evaluation of $\Pi(p)$ for an intrinsic polynomial p is straightforward, but the boundary set $\partial Q(\Pi)$ appears quite unmanageable. Therefore, we proceed like in section 6.1.2.

Proposition 6.4. For $\Pi(p)$ as above,

$$\partial Q(\Pi) = \{p \in \mathbb{R}[x] : p \text{ has no zeros in } \mathbb{C}_+ \text{ but at least one zero on the imaginary axis}\}.$$

Proof: The proof is analogous to the proof of Proposition 6.3. $\square$

For real p, a zero ζ with $\mathrm{Re}\,\zeta = 0$ must either be 0 or one of a conjugate-complex pair. With $p(x;a) = \sum_{\nu=0}^{n} \alpha_\nu x^\nu \in \mathbb{R}[x]$, a purely imaginary zero $\pm i\,\eta$ must satisfy

$$\hat{p}(\eta;a) := p(i\eta;a) = \sum_{\nu=0}^{n} \alpha_\nu (i\eta)^\nu = \sum_{\nu=0}^{\lfloor n/2 \rfloor} (-1)^\nu \alpha_{2\nu}\,\eta^{2\nu} + i\eta \sum_{\nu=0}^{\lfloor (n-1)/2 \rfloor} (-1)^\nu \alpha_{2\nu+1}\,\eta^{2\nu}$$

$$=: \quad p_r(\eta^2; a_0) \quad + \quad i\eta\, p_i(\eta^2; a_1) \quad = \quad 0\,, \tag{6.12}$$

where a_0, a_1 denote the sets of coefficients with even and odd subscripts, respectively. Thus, the critical manifold ∂Q in the empirical data space $\Delta\mathcal{A}$ of $(\bar{p}, e)$ is given by

$$\partial Q := \{\Delta a \in \Delta\mathcal{A} : \exists \hat{\eta} \in \mathbb{R}_+ : p_r(\hat{\eta}; \bar{a}_0 + \Delta a_0) = p_i(\hat{\eta}; \bar{a}_1 + \Delta a_1) = 0\} \cup \{\Delta\alpha_0 = -\alpha_0\}\,. \tag{6.13}$$

As in section 6.1.2, we have to determine the minimal norm perturbation of $\bar{p}$ for which two real polynomials related to $\bar{p}$ have a common real zero and, again, we can characterize the orthants in $\Delta\mathcal{A}$ where this happens.

Proposition 6.5. The minima of $\|\Delta a\|_e^*$ on the critical manifold (6.13) lie in the orthants $\pm(+, +, -, -, +, +, -, -, \ldots)$ and $\pm(+, -, -, +, +, -, -, +, \ldots)$.

Proof: At first, we observe that the coefficient sets of p_r and p_i are *disjoint*. For fixed $\hat{\eta} > 0$, we can tell from (5.34) in which orthants of $\Delta\mathcal{A}$ the minimizing Δa_0 and Δa_1 may lie; this depends only on the known *sign* of $\hat{\eta}$. When we refer this back to the original coefficients α_ν in (6.12) and consider the potential combinations of orthants for p_r and p_i, we obtain the assertion. $\square$

Remark: Proposition 6.5 is equivalent to *Kharitonov's Theorem* ([6.6]) for the stability of linear differential equations with constant coefficients; in our language it says the following:

Consider an empirical real polynomial $(\bar{p}, e)$, where $\bar{p}$ of degree n has all zeros in $\mathbb{C}_-$. Let $\Delta p_0(y; e)$ and $\Delta p_1(y; e)$ be the even and odd parts of

$$\Delta p(y;e) := \sum_{\nu=0}^{n} \varepsilon_\nu y^\nu = \sum_{\nu=0}^{\lfloor n/2 \rfloor} \varepsilon_{2\nu}\,y^{2\nu} + \sum_{\nu=0}^{\lfloor (n-1)/2 \rfloor} \varepsilon_{2\nu+1}\,y^{2\nu+1} =: \Delta p_e(y; e) + \Delta p_o(y; e)\,.$$

Iff the 4 polynomials $(i, j = 0, 1)$, with $\delta > 0$, $\delta\,\varepsilon_n < |\bar{\alpha}_n|$,

$$\tilde{p}_{ij}(y;\delta) := \bar{p}(y) + \delta \cdot [(-1)^i \Delta p_e(y; e) + (-1)^j \Delta p_o(y; e)]$$

are stable, then all $\tilde{p} \in N_\delta(\bar{p}, e)$ are stable.

With the information from Proposition 6.5, we may use equation (6.13) and form the two polynomials

$$q_j(t) := \det S(p_r(\hat{\eta}; \bar{a}_0 + t\,e_0), p_i(\hat{\eta}; \bar{a}_1 + (-1)^j t\,e_1))\,, \quad j = 0, 1\,, \tag{6.14}$$

with $e_0 := (\varepsilon_0, -\varepsilon_2, \varepsilon_4, \ldots)$, $e_1 := (\varepsilon_1, -\varepsilon_3, \varepsilon_5, \ldots)$. Then

$$\neg\tilde{\Pi}((\bar{p}, e)) = \min_{j=0,1}\,\min\{|t| : t \in \mathbb{R},\ q_j(t) = 0\}\,.$$

For a second approach, we consider the real nonnegative parameter $\hat{\eta}$ in (6.13) and determine the backward error of $\hat{\eta}$ as a simultaneous zero of the empirical polynomials $\bar{p}_r(\hat{\eta})$ and $\bar{p}_i(\hat{\eta})$, resp., with the associated tolerances. Since the coefficient sets of $\bar{p}_r$ and $\bar{p}_i$ are *disjoint*, we may take the individual backward errors $\delta_r(\hat{\eta})$ and $\delta_i(\hat{\eta})$ and form

$$\delta(\hat{\eta}) \; = \; \max \; (\delta_r(\hat{\eta}), \, \delta_i(\hat{\eta})) \; = \; \max \; \left(\frac{|\bar{p}_r(\hat{\eta})|}{\sum_\nu \varepsilon_{2\nu} \, \hat{\eta}^\nu} \, , \; \frac{|\bar{p}_i(\hat{\eta})|}{\sum_\nu \varepsilon_{2\nu+1} \, \hat{\eta}^\nu} \right) .$$

The desired value of $\neg\Pi((\bar{p}, e))$ is $\min_{\hat{\eta} \geq 0} \delta(\hat{\eta})$. The evaluation of the minimum requires a piecewise analysis which is, generally, greatly simplified by the available contextual information.

There appears to be a third independent approach based on the Routh-Hurwitz criterion. Since we know the orthants in which the minimal perturbations leading to a violation of the criterion may lie, we can form (cf. (6.11)) the two $n \times n$ matrices

$$H^{(j)}(\delta) \; := \; H(\bar{a} + \delta \cdot (e_0 + (-1)^j e_1)) \, , \qquad j = 0, \, 1 \, , \tag{6.15}$$

and their principal minors $H_\nu^{(j)}(\delta)$, $\nu = 1(1)n$. Since all these $H_\nu^{(j)}(\delta)$ must be positive if all polynomials in $N_\delta(\bar{p}, e)$ are to be stable, we must determine the smallest modulus real zero over the polynomial equations

$$H_\nu^{(j)}(\delta) \; = \; 0 \, , \qquad \nu = 1(1)n \, , \quad j = 0, \, 1 \, .$$

A closer inspection of the matrices in (6.14) and (6.15), resp., reveals that the principal minors $H_{n-1}^{(j)}(\delta)$ are *identical* with the polynomials $-q_j(t)$ of (6.14). For $n = 5$, e.g., we have

$$H_4^{(j)}(\delta) = \begin{vmatrix} \alpha_4 + \delta & \alpha_5 \pm \delta & 0 & 0 \\ \alpha_2 - \delta & \alpha_3 \mp \delta & \alpha_4 + \delta & \alpha_5 \pm \delta \\ \alpha_0 + \delta & \alpha_1 \pm \delta & \alpha_2 - \delta & \alpha_3 \mp \delta \\ 0 & 0 & \alpha_0 + \delta & \alpha_1 \pm \delta \end{vmatrix} , \quad q_j(\delta) = \begin{vmatrix} \alpha_0 + \delta & \alpha_2 - \delta & \alpha_4 + \delta & 0 \\ 0 & \alpha_0 + \delta & \alpha_2 - \delta & \alpha_4 + \delta \\ \alpha_1 \pm \delta & \alpha_3 \mp \delta & \alpha_5 \pm \delta & 0 \\ 0 & \alpha_1 \pm \delta & \alpha_3 \mp \delta & \alpha_5 \pm \delta \end{vmatrix} .$$

$H_n^{(j)}$ adjoins the factor $(\alpha_0 + \delta)$ to $H_{n-1}^{(j)}(\delta)$ which represents the appended component of ∂Q in (6.13). Thus, surprisingly, it appears that the polynomials $H_\nu^{(j)}(\delta)$, $\nu < n - 1$, cannot yield a smaller $|\delta|$ than the $H_{n-1}^{(j)}$.

Example 6.3: We consider the empirical polynomial $(\bar{p}, e)$ with

$$\bar{p}(x) \; := \; x^4 + 3.38 \, x^3 + 6.44 \, x^2 + 8.19 \, x + 7.85$$

and tolerances $\varepsilon_\nu = .01$ for all coefficients except the leading one. The zeros of $\bar{p}$ are $\approx$ $-1.5704 \pm .9322 \, i$, $-.1196 \pm 1.5295 \, i$ so that $\bar{p}$ is stable. We want to know whether it is safe to declare the empirical polynomial $(\bar{p}, e)$ stable. We display the application of the approaches explained above.

In the first approach, we form $p_r(\hat{\eta}; \bar{a}_0) = \hat{\eta}^2 - 6.44 \, \hat{\eta} + 7.85$ and $p_i(\hat{\eta}; \bar{a}_1) = -3.38 \, \hat{\eta} + 8.19$. Their Sylvester matrix is only 3×3 and the polynomials (6.14) become

$$q_{0,1}(t) \; = \; \begin{vmatrix} 7.85 + .01 \, t & -6.44 + .01 \, t & 1 \\ 8.19 \pm .01 \, t & -3.38 \pm .01 \, t & 0 \\ 0 & 8.19 \pm .01 \, t & -3.38 \pm .01 \, t \end{vmatrix}$$

$$= \; \begin{cases} -21.515728 + .333970 \, t + .000372 \, t^2 \, , \\ -21.515728 + .448162 \, t + .002686 \, t^2 \, . \end{cases}$$

Their zeros are approximately -958, 60.4 and -206, 38.9, resp., which clearly establishes $\neg\widetilde{\Pi}((\bar{p}, e)) = \textbf{false}$.

In the second approach, we have

$$\delta(\hat{\eta}) \;=\; \max\; \left(\frac{|\hat{\eta}^2 - 6.44\,\hat{\eta} + 7.85|}{.01\,(1 + \hat{\eta})} \,,\; \frac{|-3.38\,\hat{\eta} + 8.19|}{.01\,(1 + \hat{\eta})} \right).$$

For small $\hat{\eta}$, the second term dominates; equality occurs at $\hat{\eta} \approx 2.07$ with $\delta \approx 39$ and this value remains valid for all larger $\hat{\eta}$. $\square$

Exercises

1. Solve the question in the continuation of Example 6.1 by the approach of section 6.1.2.

2. Consider the empirical real polynomial

$$\bar{p}(x) := x^5 + .327\,x^4 + 3.920\,x^3 + .773\,x^2 + 3.502\,x + .208,$$

with tolerances .001 on each of the coefficients except the leading one.

(a) Prove that the predicate "$(\bar{p}, e)$ is stable" is valid. Find the e-nearest polynomial to $\bar{p}$ which is not stable.

(b) With the help of a), determine an empirical polynomial $(\bar{p}_1, e)$ of degree 5 with the same tolerances for which the backward error of the above predicate is 1. Which predicate for $(\bar{p}_1, e)$ is valid under these circumstances?

3. For a real empirical polynomial, find a procedure to determine the backward error of the predicate $\Pi_{uc}(\bar{p}, e) := $ "$(\bar{p}, e)$ has all zeros inside the complex unit circle":

(a) The critical manifold $\Delta\mathcal{S}_{uc}$ of Π_{uc} in the data space $\Delta\mathcal{A}$ of $(\bar{p}, e)$ consists of the coefficients of the polynomials which have a zero (generally a conjugate-complex pair of zeros) *on* the unit circle. For $\zeta = \exp(i\,\varphi)$, determine its backward error $\delta(\zeta)$ as a zero of $(\bar{p}, e)$. Convince yourself that—for *real* variations of the α_ν—the equivalent-data manifolds of ζ and of its conjugate value ζ^* are identical so that $\delta(\zeta) = \delta(\zeta, \zeta^*)$.

(b) Form $\delta(\zeta)^2$ as a function of φ. (Remember the relations between the exponentials of purely imaginary arguments and the trigonometric functions.) You can then find $\delta(\Pi_{uc}) = \min_\varphi \delta(\zeta)$ from the real zeros of the φ-derivative of $\delta(\zeta)^2$.

(c) Form real polynomials p with their zeros inside but close to the unit circle. Determine the tolerances e on the coefficients for which the predicate $\Pi_{uc}(p, e)$ is valid.

6.2   Divisors of Empirical Polynomials

6.2.1   Divisors and Zeros

The fact that a polynomial s divides a polynomial p without a remainder ("s is a divisor of p") is commonly denoted by $s \,|\, p$ (cf. Definition 2.16):

$$s \,|\, p \qquad \Longleftrightarrow \qquad \exists q \,:\, p(x) = q(x) \cdot s(x). \tag{6.16}$$

For univariate polynomials, one may assume s to be *monic* without loss of generality so that the leading coefficients of p and q are equal. $s(x) = p(x)/\mathrm{lc}(p)$ and $s(x) = 1$ are the *trivial divisors* of p.

From (6.16), one has the immediate consequence

$$s \mid p \quad \Longleftrightarrow \quad \text{all zeros of } s \text{ are zeros of } p \quad \Longleftrightarrow \quad p \in \langle s \rangle . \qquad (6.17)$$

For a polynomial $p \in \mathbb{C}[x]$ of degree n, (6.17) implies that there are at most 2^n different monic divisors since each of the n zeros of p may be a zero of s or not. For disjoint zeros, the bound is assumed; in the case of multiple zeros of p, it cannot be realized. If p is *real* and if we admit only *real* divisors, the total number of potential divisors may be much smaller.

In section 5.3.2, we have considered the division of empirical polynomials and defined valid quotients and remainders; the definition of valid divisors follows the same pattern.

Definition 6.3. A monic polynomial $\tilde{s}$ is a *valid approximate divisor* or *pseudodivisor* of the empirical polynomial $(\bar{p}, e)$ if there exist, for $\delta = O(1)$, polynomials $\tilde{p} \in N_\delta(\bar{p}, e)$ and $q \in \mathcal{P}$ (arbitrary) such that

$$\tilde{p}(x) = \bar{p}(x) + \Delta p(x) = q(x) \cdot \tilde{s}(x) . \quad \Box \qquad (6.18)$$

Proposition 6.6. $\tilde{s}$ is a pseudodivisor of $(\bar{p}, e)$ iff the zeros of $\tilde{s}$ (with consideration of their potential multiplicity) are *simultaneous* pseudozeros of $(\bar{p}, e)$.

Proof: By (6.18), all zeros of $\tilde{s}$ are zeros of one and the same $\tilde{p} \in N_\delta(\bar{p}, e)$, with $\delta = O(1)$, and hence simultaneous pseudozeros of $(\bar{p}, e)$. Reversely, the request that the zeros of $\tilde{s}$ are simultaneous pseudozeros of $(\bar{p}, e)$ demands the existence of a $\tilde{p} \in N_\delta(\bar{p}, e)$ which satisfies (6.18). Compare section 5.2.1. $\quad \Box$

As a consequence of (6.17) and of Proposition 6.6, many statements about (pseudo)zeros, in particular about sets of (pseudo)zeros, can also be formulated in terms of (pseudo)divisors. This flexibility is often an asset for the understanding as well as for computational purposes. Although (6.17) clearly holds also for *multivariate* polynomials, the consequences are less immediate because the zero set of s is always infinite. In this part of the book, we consider univariate polynomials only.

To verify that some monic polynomial $\tilde{s}$ of degree m is a pseudodivisor of the empirical polynomial $(\bar{p}, e)$, we form its backward error

$$\delta(\tilde{s}) = \min_{q \in \mathbf{P}_{n-m}} \|\Delta p\|_e^* = \min_{q \in \mathbf{P}_{n-m}} \| q(x) \cdot \tilde{s}(x) - \bar{p}(x) \|_e^* . \qquad (6.19)$$

The associated equivalent-data manifold $\mathcal{M}(\tilde{s})$ in the empirical data space $\Delta \mathcal{A}$ of $(\bar{p}, e)$ is linear and parametrized by the $n - m + 1$ free coefficients of q. For $\tilde{s}$ of degree m, the manifold $\mathcal{M}(\tilde{s})$ must be identical with $\mathcal{M}(\tilde{\zeta}_1, \ldots, \tilde{\zeta}_m)$, where the $\tilde{\zeta}_\mu$ are the zeros of $\tilde{s}$ (with multiplicities appropriately considered; cf. section 6.3).

In section 5.2.1, we had represented the linear manifold $\mathcal{M}(\tilde{\zeta}_1, \ldots, \tilde{\zeta}_m)$ of codimension m by the m linear equations (5.35). Now, we have a parameter representation

$$\Delta \alpha_\nu = \sum_{\mu=0}^{n-m} c_{\nu\mu} \beta_\mu - \bar{\alpha}_\nu , \quad \nu = 0(1)n , \qquad (6.20)$$

of the same manifold by the $n - m + 1$ coefficients β_μ of $q(x) = \sum_{\mu=0}^{n-m} \beta_\mu x^\mu$; the $c_{\nu\mu}$ are simple linear expressions in the coefficients γ_μ of $\tilde{s}$. If *all* coefficients of $(\bar{p}, e)$ are empirical, the dimension M of $\Delta\mathcal{A}$ is $n + 1$ and the $n - m + 1$ parameters β_μ imply the codimension m (degenerate situations are disregarded). If $M < n+1$, then $n+1 - M$ of the $\Delta\alpha_\nu$ in (6.20) must vanish and the number of *free* parameters drops to $M - m$ so that the codimension m is retained. As we have also seen in section 5.2.1, $M \geq m$ is necessary for a validity of m simultaneous approximate zeros as well as of an approximate divisor of degree m, except if they happen to be exact.

Computationally, the parameter representation (6.20) leads directly to the linear minimization task

$$\text{minimize } \delta \quad \text{with} \quad \pm \left(\sum_\mu c_{\nu\mu} \beta_\mu - \bar{\alpha}_\nu \right) \leq \delta\,\varepsilon_\nu\,, \quad \nu = 0(1)n\,, \tag{6.21}$$

which also accommodates intrinsic coefficients of $(\bar{p}, e)$ through $\varepsilon_\nu = 0$. Thus, for $m > 1$, the divisor formulation is computationally more convenient than the simultaneous-zero formulation.

We also note that the divisor formulation does not require the *evaluation* of $\bar{p}$ at the pseudozeros $\tilde{\zeta}_\mu$. Instead, (6.21) checks the individual steps in the recursive evaluation of $\bar{p}(\tilde{\zeta}_\mu)$ by a Horner algorithm, which avoids the inherent numerical instability of the residual computation.

Example 6.4: We repeat the 3 parts of Example 5.6 in section 5.2.1 in terms of divisors.

For the approximate zero $\tilde{\zeta} = 1.43244$, we check the approximate divisor $\tilde{s}(x) = x - 1.43244$. The parameter representation (6.20) of $\mathcal{M}(\tilde{\zeta})$ becomes

$$\Delta\alpha_0 = -\tilde{\zeta}\,\beta_0 - \bar{\alpha}_0, \;\; \Delta\alpha_1 = \beta_0 - \tilde{\zeta}\,\beta_1 - \bar{\alpha}_1, \;\; \Delta\alpha_2 = \beta_1 - \tilde{\zeta}\,\beta_2 - \bar{\alpha}_2, \;\; \Delta\alpha_3 = \beta_2 - \tilde{\zeta} - \bar{\alpha}_3\,,$$

and (6.21) yields $\delta(\tilde{s}) \approx .20$.

For the 3 simultaneous approximate zeros 1.40, 1.41, 1,42 or the divisor $\tilde{s}(x) = \prod_\mu (x - \zeta_\mu) = x^3 - 4.23\,x^2 + 5.9642\,x - 2.80308$, the quotient q is only $x - \beta_0$ and $\mathcal{M}(\tilde{s})$ is represented by

$$\Delta\alpha_0 = -2.80308\,\beta_0 - \bar{\alpha}_0, \qquad \Delta\alpha_1 = -2.80308 + 5.9642\,\beta_0 - \bar{\alpha}_1,$$
$$\Delta\alpha_2 = 5.9642 - 4.23\,\beta_0 - \bar{\alpha}_2, \quad \Delta\alpha_3 = -4.23 + \beta_0 - \bar{\alpha}_3\,;$$

(6.21) yields $\delta(\tilde{s}) \approx 3400$.

For the complex approximate zero $\tilde{\zeta} = 1.414 + .029\,i$, the divisor $\tilde{s}(x) = (x - \tilde{\zeta})(x - \tilde{\zeta}^*) = x^2 - 2.828\,x + 2.000237$, the quadratic quotient $q(x) = x^2 + \beta_1 x + \beta_0$ introduces 2 parameters and (6.21) yields $\delta(\tilde{s}) \approx 54$.

All these values are, naturally, in complete agreement with those obtained in Example 5.6 in terms of zeros; the set-up of the minimization is rather simpler than there. $\quad\square$

6.2.2 Sylvester Matrices

Consider an exact factorization of $p \in \mathcal{P}^n$: Let

$$\sum_{\nu=0}^{n} \alpha_\nu x^\nu \;=\; p(x) \;=\; q(x) \cdot s(x) = \sum_{j=0}^{n-m} \beta_j x^j \cdot \sum_{\mu=0}^{m} \gamma_\mu x^\mu\,,$$

where, at this point, we do not require s to be monic but assume some fixed normalization of the leading coefficients, with $\alpha_n = \beta_{n-m}\, \gamma_m$. The linearized effects of a small perturbation Δp may be found from $p + \Delta p = (q + \Delta q)\,(s + \Delta s)$, with

$$\Delta s \cdot q + \Delta q \cdot s \; = \; \Delta p \; [\, -\Delta q \cdot \Delta s \,] . \tag{6.22}$$

With $\Delta p = \sum_{\nu=0}^{n-1} \Delta\alpha_\nu x^\nu$, $\Delta q = \sum_{j=0}^{n-m-1} \Delta\beta_j x^j$, $\Delta s = \sum_{\mu=0}^{m-1} \Delta\gamma_\mu x^\mu$, the linearization of (6.22) may be written as

$$(\Delta\gamma_0, .., \Delta\gamma_{m-1}, \Delta\beta_0, .., \Delta\beta_{n-m+1})
\begin{pmatrix}
\beta_0 & \cdots & \cdots & \beta_{n-m} & & & \\
 & \ddots & & & \ddots & & \\
 & & \ddots & & & \ddots & \\
 & & & \beta_0 & \cdots & \cdots & \beta_{n-m} \\
\gamma_0 & & \cdots & & \gamma_m & & \\
 & \ddots & & & & \ddots & \\
 & & \gamma_0 & & \cdots & & \gamma_m
\end{pmatrix}
\begin{pmatrix} 1 \\ x \\ \vdots \\ x^{n-1} \end{pmatrix}$$

$$\tag{6.23}$$

$$= (\Delta\alpha_0, \ldots, \Delta\alpha_{n-1}) \begin{pmatrix} 1 \\ \vdots \\ x^{n-1} \end{pmatrix} , \quad \text{or} \quad (\Delta c \; \Delta b) \cdot S(q,s) \cdot \mathbf{x} = (\Delta a) \cdot \mathbf{x} .$$

Definition 6.4. For two univariate polynomials p_1, p_2 of degrees n_1, n_2, the matrix $S(p_1, p_2) \in \mathbb{C}^{(n_1+n_2)\times(n_1+n_2)}$ which contains the coefficients of p_1, p_2 in a staggered arrangement as in (6.23) is called the *Sylvester matrix* or *resultant matrix* of p_1 and p_2.   □

There exist various other conventions for arranging the coefficients of p_1, p_2 in a matrix with $n_1 + n_2$ rows and columns. All these matrices are commonly called Sylvester matrices and, naturally, serve the same purpose in a slightly different notation. In this book, we use the form (6.23) because it matches our other notational conventions.

The *rank* of $S(p_1, p_2)$ is intimately connected with the relative positions of the zeros of p_1 and p_2. The content of the following two theorems and various patterns for their proof have been known for a long time. We spell out linear algebra-oriented proofs because they shed further light on our uses of Sylvester matrices.

Theorem 6.7. $S(p_1, p_2)$ is regular iff p_1 and p_2 have no zeros in common, or—equivalently— iff they are *relatively prime*, i.e. have no common factor of a positive degree. $S(p_1, p_2)$ has a rank deficiency $d \leq \min(n_1, n_2)$ iff p_1 and p_2 have exactly d zeros in common (counting multiplicities), or—equivalently—have a common factor of degree d.

Proof: Let $\mathbf{x} := (1, x, \ldots, x^{n_1+n_2-1})^T$ and $\mathbf{x}(z) := (1, z, \ldots, z^{n_1+n_2-1})^T \in \mathbb{C}^{n_1+n_2}$. Let $z_\nu^{(i)}$, $\nu = 1(1)n_i$, be the zeros of p_i, $i = 1, 2$. In the case of an m-fold zero z, we supplement $\mathbf{x}(z)$ by the vectors $(\partial^\mu \mathbf{x})(z)$, $\mu = 1(1)m - 1$, so that there are always exactly $n_1 + n_2$ vectors $\mathbf{x}(z_\nu^{(i)})$, with a suitable numbering. Furthermore, let $S_1 \in \mathbb{C}^{n_2 \times (n_1+n_2)}$ and $S_2 \in \mathbb{C}^{n_1 \times (n_1+n_2)}$ be the upper and lower Toeplitz submatrices of $S(p_1, p_2)$. The kernel of S_i is spanned by the vectors $\mathbf{x}(z_\nu^i)$, $\nu = 1(1)n_i$, $i = 1, 2$.

Case 1: p_1, p_2 have no common zeros. Assume that $S(p_1, p_2)$ is singular so that there exists a vector $\mathbf{z} \in \mathbb{C}^{n_1+n_2}$ with $S(p_1, p_2)\,\mathbf{z} = 0$. Clearly, z must be in $\ker S_1$ and $\ker S_2$:

$$\mathbf{z} = \sum_{\nu=1}^{n_1} w_\nu^{(1)}\mathbf{x}(z_\nu^{(1)}) = \sum_{\nu=1}^{n_2} w_\nu^{(2)}\mathbf{x}(z_\nu^{(2)}), \qquad \text{or}$$

$$\left(\cdots \ \ \mathbf{x}(z_\nu^{(1)}) \ \ \cdots \ \bigg| \ \cdots \ \ \mathbf{x}(z_\nu^{(2)}) \ \ \cdots \right) \begin{pmatrix} \vdots \\ w_\nu^{(1)} \\ \vdots \\ -w_\nu^{(2)} \\ \vdots \end{pmatrix} = 0 .$$

But the Vandermonde matrix of the $n_1 + n_2$ disjoint zeros $z_\nu^{(1)}$, $z_\nu^{(2)}$ is regular so that this implies $w_\nu^{(i)} = 0$, $\nu = 1(1)n_i$, $i = 1, 2$, and the nonexistence of $\mathbf{z}$.

Case 2: p_1 and p_2 have exactly d common zeros (counting multiplicities) so that $p_i(x) = \hat{p}_i(x) \cdot \hat{g}(x)$, $i = 1, 2$, with $\hat{g}(x) = \sum_{\mu=0}^{d} \hat{\gamma}_\mu x^\mu$.

a) The rank deficiency of $S(p_1, p_2)$ is at least d:

$$\begin{pmatrix} p_1(x) \\ \vdots \\ x^{n_2-1}p_1(x) \\ p_2(x) \\ \vdots \\ x^{n_1-1}p_2(x) \end{pmatrix} = \begin{pmatrix} \hat{\alpha}_0^{(1)} & \cdots & \hat{\alpha}_{n_1-d}^{(1)} & & \\ & \ddots & & \ddots & \\ & & \hat{\alpha}_0^{(1)} & \cdots & \hat{\alpha}_{n_1-d}^{(1)} \\ \hat{\alpha}_0^{(2)} & \cdots & \hat{\alpha}_{n_2-d}^{(2)} & & \\ & \ddots & & \ddots & \\ & & \hat{\alpha}_0^{(2)} & \cdots & \hat{\alpha}_{n_2-d}^{(2)} \end{pmatrix} \begin{pmatrix} \hat{\gamma}_0 & \cdots & \hat{\gamma}_d & & \\ & \ddots & & \ddots & \\ & & \hat{\gamma}_0 & \cdots & \hat{\gamma}_d \end{pmatrix} \mathbf{x},$$

or $S(p_1, p_2)\mathbf{x} = B\,C\,\mathbf{x}$, with $B \in \mathbb{C}^{(n_1+n_2)\times(n_1+n_2-d)}$, $C \in \mathbb{C}^{(n_1+n_2-d)\times(n_1+n_2)}$. Since B and C have only $n_1 + n_2 - d$ columns or rows, resp., their product must have rank deficiency at least m.

b) The rank deficiency of $S(p_1, p_2)$ is at most d (cf. Case 1) : Assume $S(p_1, p_2)$ has rank deficiency $\bar{d} > d$ and $\ker S(p_1, p_2)$ spanned by $\mathbf{z}_1, \ldots, \mathbf{z}_{\bar{d}} \in \mathbb{C}^{n_1+n_2}$. Each $\mathbf{z}_\mu$ must be in $\ker S_1$ and $\ker S_2$:

$$\mathbf{z}_\mu = \sum_{\nu=1}^{n_1} w_{\nu\mu}^{(1)}\mathbf{x}(z_\nu^{(1)}) = \sum_{\nu=1}^{n_2} w_{\nu\mu}^{(2)}\mathbf{x}(z_\nu^{(2)}), \quad \mu = 1(1)\bar{d}, \qquad \text{or}$$

$$\left(\cdots \ \ \mathbf{x}(z_\nu^{(1)}) \ \ \cdots \ \bigg| \ \cdots \ \ \mathbf{x}(z_\nu^{(2)}) \ \ \cdots \right) \begin{pmatrix} \vdots & & \\ \cdots & w_{\nu\mu}^{(1)} & \cdots \\ \vdots & & \\ \cdots & -w_{\nu\mu}^{(2)} & \cdots \\ \vdots & & \end{pmatrix} = 0 .$$

Since d of the columns in the Vandermonde matrix are duplicated while the remaining ones are disjoint, its rank deficiency is exactly d and there can be at most d linearly independent columns in the second matrix. Thus the dimension $\bar{d}$ of $\ker S(p_1, p_2)$ is at most d. $\qquad \square$

Theorem 6.8. If p_1, p_2 have a common divisor g of degree $d > 0$ so that $S(p_1, p_2)$ has rank deficiency d, then the last $n_1 + n_2 - d$ columns of $S(p_1, p_2)$ are linearly independent. Furthermore, $S(p_1, p_2)$ may be factored into a regular $(n_1 + n_2) \times (n_1 + n_2)$ matrix M and a lower triangular matrix L such that the upper d rows of L vanish; then the $(d + 1)$st row of L contains the coefficients of (a scalar multiple of) g.

Proof: a) Consider the factorization $S(p_1, p_2) = B C$ in the proof of Theorem 6.7. By Theorem 6.7, B has rank $n_1 + n_2 - d$ so that its columns are linearly independent. Since g has degree d, $\gamma_d \neq 0$ in C so that the lower triangular matrix of the last $n_1 + n_2 - d$ columns of C is regular. This implies the linear independence of the last $n_1 + n_2 - d$ columns of $B C$.

b) We form a triangularization $S = M L$ of $S(p_1, p_2)$, with regular M and lower-triangular L, columnwise from *right to left* and *bottom to top*. Due to the linear independence of the last $n_1 + n_2 - d$ columns of $S(p_1, p_2)$, no column interchanges are necessary within the triangularization of these columns (row interchanges do not affect the independence of the columns). Thus we have $\bar{S} = \bar{M} \bar{L}$, where $\bar{S}$ and $\bar{M}$ contain the $n_1 + n_2 - d$ rightmost columns of S and M and $\bar{L}$ is the lower right triangle of L. The columns of $\bar{M}$ span the range of S; thus the remaining d first columns of S lie in the column space of $\bar{M}$. If we complete $\bar{M}$ into a regular square matrix M, its first d columns cannot figure in the representation of S so that the top d rows of L must remain empty.

$$
\begin{array}{ccccc}
S & = & M & \cdot & L \quad ,
\end{array}
$$

$$
\left(\begin{array}{c|c}
\overset{d}{\mid} & \overset{n_1+n_2-d}{\bar{S}} \\
\mid & \\
\mid &
\end{array} \right)
=
\left(\begin{array}{c|c}
\mid & \\
\mid & \bar{M} \\
\mid &
\end{array} \right)
\cdot
\left(\begin{array}{c|cc}
\mid & & 0 \\
\mid & & \\
\mid & \bar{L} &
\end{array} \right) .
$$

Now consider the d zeros $z_1, \ldots, z_d$ of g; the associated vectors $\mathbf{x}(z_\mu)$, $\mu = 1(1)d$, are in ker S_1 and ker S_2 and hence in ker $S(p_1, p_2) = $ ker L. The $(d + 1)$st row of L contains the coefficients of a d-th degree polynomial which therefore vanishes at the z_μ and thus coincides with g or a scalar multiple of it. Details of the algorithm indicated in the proof will be considered in the following sections. $\square$

Example 6.5: With $p_1 = \prod_{i=1}^{3}(x - i) = x^3 - 6x^2 + 11 x - 6$ and $p_2 = (x - 1)(x - 4) = x^2 - 5 x + 4$, we have $d = 1$ and $S = $

$$
\begin{pmatrix}
-6 & 11 & -6 & 1 & 0 \\
0 & -6 & 11 & -6 & 1 \\
4 & -5 & 1 & 0 & 0 \\
0 & 4 & -5 & 1 & 0 \\
0 & 0 & 4 & -5 & 1
\end{pmatrix}
=
\begin{pmatrix}
1 & 1/4 & -1 & 1 & 0 \\
0 & 1 & 2 & -1 & 1 \\
0 & 0 & 1 & 0 & 0 \\
0 & 0 & 0 & 1 & 0 \\
0 & 0 & 0 & 0 & 1
\end{pmatrix}
\begin{pmatrix}
0 & 0 & 0 & 0 & 0 \\
-8 & 8 & 0 & 0 & 0 \\
4 & -5 & 1 & 0 & 0 \\
0 & 4 & -5 & 1 & 0 \\
0 & 0 & 4 & -5 & 1
\end{pmatrix} ,
$$

with $-8 + 8 x = 8 (x - 1) = 8 g(x)$. $\square$

6.2.3 Refinement of an Approximate Factorization

After recalling these well-known facts about Sylvester matrices, let us now analyze an approximate factorization, with monic $\tilde{s} \in \mathcal{P}^m$, $\tilde{q} \in \mathcal{P}^{n-m}$:

$$
\bar{p}(x) \approx \tilde{q}(x) \cdot \tilde{s}(x) ; \tag{6.24}
$$

we try to obtain corrections Δq and Δs such that

$$\bar{p} = (\tilde{q} + \Delta q) \cdot (\tilde{s} + \Delta s) \, ,$$

or, disregarding the quadratic terms in the corrections,

$$\Delta s \cdot \tilde{q} + \Delta q \cdot \tilde{s} = \bar{p} - \tilde{q}\,\tilde{s} =: r \, . \tag{6.25}$$

A comparison of coefficients in (6.25) yields $n+1$ linear equations for the $n-m+1$ coefficients of Δq and the m coefficients of Δs. If (6.25) is regular, $\tilde{s} + \Delta s$ will be an exact divisor of $\bar{p} - \Delta q \Delta s$ which should generally lie in $N_\delta(\bar{p}, e)$.

The matrix of the linear system (6.25) is the Sylvester matrix $S(\tilde{q}, \tilde{s})$, cf. (6.23). From Theorem 6.7, we know that it is regular iff $\tilde{q}$ and $\tilde{s}$ have no zeros in common. From the numerical point of view, a *near-singularity* of $S(\tilde{q}, \tilde{s})$ is just as bad; this means that *closely adjacent* zeros must not be attributed to different divisors. This is particularly important for the zeros in a cluster which correspond to a perturbed multiple zero: All zeros in a zero cluster of p must go into the same factor of a factorization of p; cf. section 6.3.4.

Let us now consider the computational solution of (6.25). Assume at the moment that we are only interested in the correction $\Delta s \in \mathcal{P}_{m-1}$ which turns $\tilde{s}$ of (6.24) into a valid approximate divisor. (In the following, we omit the $\tilde{\ }$ on s and q.) When we take remainders modulo the ideal $\langle s \rangle$ in (6.25), we obtain

$$\mathrm{NF}_{\langle s \rangle}(q \cdot \Delta s) = \mathrm{NF}_{\langle s \rangle}\, r \, .$$

With the monomial basis $(1, \ldots, x^{m-1})^T$ for the quotient ring $\mathcal{R}[\langle s \rangle]$ and $\Delta s = \sum_{\mu=0}^{m-1} \Delta \gamma_\mu x^\mu$, $\mathrm{NF}_{\langle s \rangle} r = \sum_{\nu=0}^{n-1} \rho_\nu x^\nu$, and with the *multiplication matrix* $A_q \in \mathbb{C}^{m \times m}$ representing multiplication by $q \bmod \langle s \rangle$, this yields the linear system

$$\Delta c^T \cdot A_q = (\ldots \Delta \gamma_\mu \ldots) \cdot A_q = (\ldots \rho_\mu \ldots) =: (r^*)^T \, . \tag{6.26}$$

The matrix A_q and the vector $(\ \rho_\mu\ )$ are obtained in a simple fashion: We may subdivide the Sylvester matrix $S(q, s)$ of (6.23) into the blocks

$$S(q, s) = \begin{pmatrix} S_{11} & S_{12} \\ S_{21} & S_{22} \end{pmatrix} ,$$

where $S_{22} \in \mathbb{C}^{(n-m) \times (n-m)}$ is a lower triangular matrix with a unit diagonal (for monic s). We may annihilate the block S_{12} from *right to left* by subtracting suitable multiples of the shifted identical rows of γ_μ in $(S_{21}\ S_{22})$; this begins with the elimination of the complete β_{n-m} diagonal in S_{12} (cf. (6.23)) by a formal multiplication from the left of $S(q, s)$ with

$$M_1 = \begin{pmatrix} I & (-\beta_{n.m}\,\hat{I}) \\ 0 & I \end{pmatrix} ,$$

where $\hat{I}$ is a suitably shifted diagonal of 1's. There is *only one* row combination to be computed because $(S_{11}\ S_{12})$ also consists of shifted identical rows. After $n-m$ such operations, we arrive at

$$M_{n-m} \cdots M_1\, S(q, s) = \begin{pmatrix} I & -B \\ 0 & I \end{pmatrix} S(q, s) = \begin{pmatrix} S_{11}^* & 0 \\ S_{21} & S_{22} \end{pmatrix} . \tag{6.27}$$

Proposition 6.9. S_{11}^* is the multiplication matrix A_q in $\mathcal{R}[\langle s \rangle]$.

Proof: We note that

$$(S_{11}\ S_{12}) \begin{pmatrix} 1 \\ x \\ \vdots \\ x^{m-1} \end{pmatrix} = \begin{pmatrix} q(x) \\ x\,q(x) \\ \vdots \\ x^{m-1}q(x) \end{pmatrix} \quad \text{while} \quad A_q \begin{pmatrix} 1 \\ x \\ \vdots \\ x^{m-1} \end{pmatrix} = \mathrm{NF}_{\langle s \rangle} \begin{pmatrix} q(x) \\ x\,q(x) \\ \vdots \\ x^{m-1}q(x) \end{pmatrix}.$$

Our operations subtract multiples of $s(x)$ from the elements of $(q(x),\ x\,q(x),\ \dots,\ x^{m-1}q(x))^T$ until each element is in $\mathcal{P}_{m-1}$; this implies that each element has been reduced to its normal form mod $\langle s \rangle$. $\square$

In the same fashion, we may eliminate the entries of the row vector of the coefficients of $r = r^T\mathbf{x}$ in (6.25) from right to left by subtracting multiples of the appropriate rows in $(S_{21}\ S_{22})$; thus we obtain (cf. (6.26)) $(r^*)^T =: r^T - \bar{r}^T(S_{21}\ S_{22}) \in \mathbb{C}^m$. Due to the Toeplitz structure of the upper and lower parts of $S(q, s)$, the total number of arithmetic operations is only $O(n^2)$.

If $\deg q \geq \deg s$, $r_1 := \mathrm{NF}_{\langle s \rangle} q \neq q$ and $A_q = A_{r_1}$. r_1 is the remainder of the division of q by s; it is the coefficient of the second term $r_1\,s$ in the expansion (5.18) of p in powers of s:

$$p(x) = r(x) + r_1(x)\,s(x) + r_2(x)\,(s(x))^2 + \dots . \tag{6.28}$$

To form A_{r_1} from r_1, we must reduce the multiples $r_1 \cdot (1,\ x,\ \dots,\ x^{m-1})^T$ mod s. The partial triangularization of $S(q, s)$ in (6.27) *combines* the computation of r_1 with this reduction; it is therefore more economic except when r_1 is already available; cf., e.g., section 6.3.5.

We write the procedure in linear algebra terms: With $\begin{pmatrix} I & -B \\ 0 & I \end{pmatrix}^{-1} = \begin{pmatrix} I & B \\ 0 & I \end{pmatrix}$, we have from (6.25) and (6.27)

$$(\Delta c^T\ \Delta b^T)\,S(q, s) = (\Delta c^T\ \Delta b^T) \begin{pmatrix} I & B \\ 0 & I \end{pmatrix} \begin{pmatrix} S_{11}^* & 0 \\ S_{21} & S_{22} \end{pmatrix} = r^T = (r^*)^T + \bar{r}^T(S_{21}\ S_{22})$$

$$\text{or} \quad (\Delta c^T\ \ \Delta c^T B + \Delta b^T) \begin{pmatrix} S_{11}^* & 0 \\ S_{21} & S_{22} \end{pmatrix} = ((r^*)^T + \bar{r}^T S_{21}\ \ \bar{r}^T S_{22}). \tag{6.29}$$

Because of the regularity of S_{22}, this implies

$$\Delta c^T B + \Delta b^T - \bar{r}^T = 0 \tag{6.30}$$

and (6.25) so that Δq can also be obtained.

Example 6.6: Consider the empirical polynomial $(\bar{p}, e)$ with

$$\bar{p} = .2345\,x^5 - .3204\,x^4 - 1.5086\,x^3 + 2.2478\,x^2 + 1.6565\,x - 2.6163$$

and $\varepsilon_\nu = .5 \cdot 10^{-4}$, $\nu = 0(1)5$. $\bar{p}$ has two negative real zeros near -2.28 and -1.20, two clustered real zeros near 1.42 and a zero near 2.00. Grouping the two negative zeros into $\tilde{s}$ and the three positive zeros into $\tilde{q}$, we obtain an approximate factorization (6.24)

$$\bar{p}(x) = (.2345\,x^3 - 1.14\,x^2 + 1.81\,x - .95)\,(x^2 + 3.49\,x + 2.74)$$
$$+ (.001195\,x^4 + .01747\,x^3 + .0045\,x^2 + .0126\,x - .0133).$$

The linear minimization (6.21) yields a backward error $\delta(\tilde{s}) \approx 180$; we proceed immediately to a refinement of $\tilde{s}$. The algorithmic procedure described above reduces the Sylvester matrix $S(\tilde{q}, \tilde{s})$, appended with a top row of the coefficients of r,

$$
\begin{pmatrix}
-.0133 & .0126 & .0045 & .01747 & .001195 \\
-.95 & 1.81 & -1.14 & .2345 & 0 \\
0 & -.95 & 1.81 & -1.14 & .2345 \\
2.74 & 3.49 & 1 & 0 & 0 \\
0 & 2.74 & 3.49 & 1 & 0 \\
0 & 0 & 2.74 & 3.49 & 1
\end{pmatrix}
$$

$$
\text{into} \quad
\begin{pmatrix}
.1105 & .1339 & 0 & 0 & 0 \\
4.4160 & 8.0023 & 0 & 0 & 0 \\
-21.9263 & -23.5120 & 0 & 0 & 0 \\
2.74 & 3.49 & 1 & 0 & 0 \\
0 & 2.74 & 3.49 & 1 & 0 \\
0 & 0 & 2.74 & 3.49 & 1
\end{pmatrix}
$$

(rounded). From

$$
(\Delta\gamma_0, \Delta\gamma_1)
\begin{pmatrix}
4.4160 & 8.0023 \\
-21.9263 & -23.5120
\end{pmatrix}
= (.1105, .1339) , \tag{6.31}
$$

we obtain $\Delta s(x) \approx .0047 - .0041\, x$ and $\tilde{s}_{\text{new}}(x) = x^2 + 3.4859\, x + 2.7447$, with a backward error of $\approx .26$. Thus, $\tilde{s}_{\text{new}}$ is a valid approximate divisor of $(\bar{p}, e)$. $\quad\square$

Due to the unit diagonal in the right lower block S_{22} of $S(s, q)$, the reduction of $S(s, q)$ to A_q can always be performed without numerical problems. A potential near-singularity of $S(s, q)$ is thus distilled into the matrix A_q. From Corollary 5.6, we know that the eigenvalues of A_q are the values of q at the zeros z_μ of s, $\mu = 1(1)m$. Although A_q is, generally, nonnormal so that its eigenvalues do not fully characterize $\text{cond}(A)$, it is obvious that eigenvalues of small modulus in A_q, i.e. small absolute values of q at the zeros of s, will lead to an ill-conditioned system (6.26). Such values are most likely to occur when zeros of q and s are adjacent. We will analyze this further in section 6.3.4.

This insight is further emphasized by another interpretation of (6.25): If we evaluate this relation at the m zeros z_μ, $\mu = 1(1)m$, of s, we obtain

$$
q(z_\mu) \cdot \Delta s(z_\mu) = r(z_\mu) , \quad \mu = 1(1)m . \tag{6.32}
$$

(In the case of a multiple zero $\bar{z}$ in s, we substitute $\bar{z}$ also in differentiated versions of (6.25)). This shows that Δs is the *interpolation polynomial* of degree $m - 1$ of the values $r(z_\mu)/q(z_\mu)$ at the nodes z_μ, $\mu = 1(1)m$; small moduli of some of the $q(z_\mu)$ are likely to lead to a high sensitivity of Δs to small changes in r.

This interpretation also explains why it is *not* destabilizing to have a complete cluster of zeros in one and the same factor of p. It is true that the Vandermonde matrix for clustered interpolation knots is ill-conditioned, but from (6.32) we may assume that the values $(r/q)(z_\mu)$ have a smooth behavior and hardly vary at all between the zeros of a cluster. For such values, the interpolation problem is not ill-conditioned although its matrix has a large inverse; this can be shown by writing the interpolation in terms of divided differences of the data (cf. section 5.4.1). The determination of a correction of $\tilde{s}$ via interpolation may be more economic than via the reduction procedure. The latter one does not require the explicit computation of zeros, however.

Example 6.6, continued: The two zeros of $\tilde{s}$ are -2.2973, -1.1927 (rounded) and the values of r/q at these points are $.0141$, $.0096$ (rounded). Linear interpolation yields $\Delta s \approx .0047 - .0041\, x$ as previously. $\quad\square$

In the partial triangularization of $S(q, s)$, a numerical instability may arise if the leading coefficient of s is tiny relative to other coefficients in s, i.e. if s has some huge zero, because large intermediate values may appear in the elimination. Again the interpolation view makes it plausible that a mixture of huge and ordinary zeros in s is unfortunate for manipulations with s as a divisor.

In the case of disjoint zeros of $\bar{p}$, there are $\binom{n}{m}$ exact divisors of $\bar{p}$ of degree m. Thus, in an ill-conditioned situation, the pseudodivisor obtained from (6.25) as a Newton correction of the approximate factorization (6.24) of $\bar{p}$ may not necessarily contain the zeros which one had in mind.

6.2.4 Multiples of Empirical Polynomials

Let us return to the beginning of section 6.2 but exchange the attributes of s and p (cf. Definition 6.3): Assume that s is empirical and we want to find whether a given $p \in \mathcal{P}$ is a valid multiple of $(\bar{s}, e)$.

Definition 6.5. A polyomial p is a *valid approximate multiple* or *pseudomultiple* of $(\bar{s}, e)$ if there exist, for $\delta = O(1)$, polynomials $\tilde{s} \in N_\delta(\bar{s}, e)$ and $q \in \mathcal{P}$ (arbitrary) such that

$$p(x) \;=\; q(x) \cdot \tilde{s}(x)\,. \quad\square \tag{6.33}$$

This definition immediately implies the analogous result to Proposition 6.6.

Proposition 6.10. p is a pseudomultiple of $(\bar{s}, e)$, with $\bar{s} \in \mathcal{P}_m$, iff there exists a set of m zeros of p which are simultaneous pseudozeros of $(\bar{s}, e)$ (with consideration of their potential multiplicity).

For agreement with previous considerations, we assume at first that $(\bar{s}, e)$ is monic, i.e. that all $\tilde{s} \in N_\delta(\bar{s}, e)$ are monic. We proceed as in the beginning part of section 6.2.3, with a slightly different notation and a different interpretation; cf. (6.24) and (6.25): We divide p by $\bar{s}$ and obtain a remainder r, then we determine corrections of the quotient q and of $\bar{s}$ such that the remainder disappears. This yields

$$p \;=\; q \cdot \bar{s} + r \;=\; (q + \Delta q) \cdot (\bar{s} + \Delta s)\,;$$

neglecting the quadratic terms in the corrections, we have

$$\Delta s \cdot q + \Delta q \cdot \bar{s} \;=\; r\,. \tag{6.34}$$

With monic $(\bar{s}, e)$, $\Delta s \in \mathcal{P}_{m-1}$ and $\Delta q \in \mathcal{P}_{n-m-1}$, r is in $\mathcal{R}[\langle \bar{s} \rangle]$ and has degree $m - 1$ so that the last $n - m$ of the n linear equations (6.34) are homogeneous. Otherwise, we have precisely the situation of (6.25).

Therefore, we may proceed as in section 6.2.3 and take remainders modulo $\bar{s}$ to obtain $\mathrm{NF}_{\langle\bar{s}\rangle}(q \cdot \Delta s) = r$; with $\Delta s = \sum_{\mu=0}^{m-1} \Delta\gamma_\mu x^\mu$, $r = \sum_{\mu=0}^{m-1} \rho_\mu x^\mu$, and with the *multiplication matrix* $A_q \in \mathbb{C}^{m\times m}$ representing multiplication by $q \bmod \langle\bar{s}\rangle$, we arrive once more at the m by m linear system (6.26). The computational reduction of the $n \times n$ Sylvester matrix $S(q, \bar{s})$ into the $m \times m$ multiplication matrix $A_{\bar{q}} = S_{11}^*$ has been explained in the previous section. If $\|\Delta s\|_e^* = O(1)$ but only moderately so, one may wish to iterate the procedure, with $\tilde{s} := \bar{s} + \Delta s$ in place of $\bar{s}$, to be sure that the neglect of the quadratic terms in (6.34) has been admissible.

As in section 6.2.2, we may also substitute the m zeros z_μ, $\mu = 1(1)m$, of $\bar{s}$ into (6.34) and obtain equation (6.32) for the (linearized) correction Δs which leads to $\tilde{s}$ with an exact multiple $p - \Delta q \Delta s$. From (6.32), Δs is directly obtained by interpolation.

This approach no longer depends on assuming that the polynomials in $(\bar{s}, e)$ are monic; in (6.32), we may readily consider Δs as a polynomial of degree m. Then the interpolation problem has a one-dimensional set of solutions of which we may select the one with the smallest e-weighted max-norm.

Example 6.7: We want to reuse the computations in Example 6.6; therefore we choose $\bar{s} = x^2 + 3.49\,x + 2.74$ and $\bar{q} = .2345\,x^3 - 1.14\,x^2 + 1.81\,x - .95$ as in Example 6.6 and adjust p such that its remainder at division by $\bar{s}$ is $.1339\,x + .1105$. This yields the following task:

Given $p := .2345\,x^5 - .321595\,x^4 - 1.52607\,x^3 + 2.2433\,x^2 + 1.7778\,x - 2.4925$ and the empirical polynomial $(\bar{s}, e)$, with $\bar{s}$ as above and a tolerance of $.005$ on the coefficients of x and 1, is p a pseudomultiple of $(\bar{s}, e)$?

The linear system (6.34) agrees with the system (6.31) in Example 6.6 and yields $\Delta s \approx .0047 - .0041\,x$. Thus (within the linearization) there is an $\tilde{s} \in N_\delta(\bar{s}, e)$ with $\delta \approx .94$ such that p is an exact multiple of $\tilde{s}$. As p leaves a remainder of $O(.0001)$ upon division by $\bar{s} + \Delta s$ and the inverse of the above matrix is $O(1)$, the effect of the quadratic terms must be negligible.

With the use of (6.32), we proceed like in the continuation of Example 6.6 and obtain the same result. We can now also admit that the leading coefficient 1 of $\bar{s}$ may be subject to a perturbation: If we use a *quadratic* Δs in (6.32) and solve $\min \|\Delta s\|_e^*$, we obtain $\Delta s \approx .0044 - .0044\,x - .0001\,x^2$ and a slightly smaller backward error.

Finally, if we know the zeros of p, we may directly use Proposition 6.10 to verify that p is a pseudomultiple of $(\bar{s}, e)$. The 5 zeros of p are $-2.2843, -1.2016, 1.0579, 1.8997 \pm .2267\,i$ (rounded). The first two of these are close to the zeros $-2.2973, -1.1927$ (rounded) of $\bar{s}$. The backward error of the two zeros of p as simultaneous approximate zeros of $\bar{s}$ is found to be $\approx .96$; in this case, we simply have to form $\tilde{s} = (x + 2.2843)(x + 1.2016) \approx x^2 + 3.4859\,x + 2.7448$ and consider its e-weighted distance from $\bar{s}$. $\square$

Exercises

1. Consider a *linear* divisor $s = x - \tilde{z}$, with $\tilde{z}$ a crude approximate zero of $\bar{p}$. Use (6.32) to refine the divisor s. Convince yourself that the result is identical with the result of a Newton step for the refinement of $\tilde{z}$ as a zero of $\bar{p}$.

2. (a) Form the $(m + 1) \times (m + 1)$ Sylvester matrix for $q(x) = x$ and an arbitrary, not necessarily monic, $s \in \mathcal{P}_m$ and reduce it as in (6.27). Show that S_{11}^* is the Frobenius matrix A of s.

(b) Convince yourself that, for an arbitrary $q = \sum_{\nu=0}^{n} \beta_\nu x^\nu$, the reduction (6.27) is equivalent to the evaluation of $S_{11}^* = A_q = q(A)$ by the Horner algorithm

$$\sum_{\nu=1}^{n} = (\dots((\beta_n A + \beta_{n-1} I)\, A + \beta_{n-2} I)\, A + \dots + \beta_1 I)\, A + \beta_0 I \,.$$

3. Consider Example 6.6:

(a) Use Proposition 6.6 to establish that $\tilde{s}$ is not a valid divisor of $(\bar{p}, e)$ but that $\tilde{s}_{\text{new}}$ is.

(b) Form a rough approximate factorization which separates the two clustered zeros of $\bar{p}$ and find the condition number of the corresponding Sylvester matrix $S(q, s)$. Compute the matrix S_{11}^* and determine its condition number. Refine the approximate factorization by the use of (6.32); how does the ill-conditioning come to light now?

(c) Form a rough approximate factorization which puts the two clustered zeros of $\bar{p}$ into the quadratic factor s. Verify that $S(q, s)$ is not ill-conditioned and use it to refine the factorization. In using (6.32) for the refinement, consider and verify the remarks below (6.32).

4. For $(\bar{s}, e)$ of Example 6.7, with real variations of its coefficients γ_0, γ_1 only, characterize the domain in the real z_1, z_2-plane in which two zeros of a polynomial p must lie in order that p is a valid multiple of $(\bar{s}, e)$. (Hint: Consider $\tilde{s}(x) = (x - z_1)(x - z_2)$.)

6.3   Multiple Zeros and Zero Clusters

The multiplicity of zeros has been an algebraic and analytic concept for a long time while the closely related concept of a "zero cluster" has only appeared with numerical computation. We have used both concepts previously in our general considerations and in examples. We will now discuss them in connection with empirical univariate polynomials and consider their algorithmic aspects.

6.3.1   Intuitive Approach

We begin by exhibiting some characteristic phenomena with the help of our polynomial $(\bar{p}, e)$ of (3.17) which we have used in many examples:

$$\bar{p}(x) = x^4 - 2.83088\, x^3 + 0.00347\, x^2 + 5.66176\, x - 4.00694 \,, \tag{6.35}$$

with $e = (1, 1, 1, 1) \cdot 10^{-5}$. The 3 zeros of $\bar{p}$ in the right half-plane are, to 5 decimal digits,

$$1.41421, \quad 1.41481, \quad 1.41607.$$

The following polynomials are all in $N_1(\bar{p}, e)$ and thus valid instances of $(\bar{p}, e)$:

$$\begin{aligned}
p_+(x) &= x^4 - 2.83087\, x^3 + 0.00348\, x^2 + 5.66177\, x - 4.00693 \,, \\
p_-(x) &= x^4 - 2.83089\, x^3 + 0.00346\, x^2 + 5.66175\, x - 4.00695 \,, \\
p_3(x) &= x^4 - 2.83088\, x^3 + 0.003472486\, x^2 + 5.66175549657\, x - 4.00693860488 \,, \\
p_2(x) &= x^4 - 2.830876\, x^3 + .00346308594\, x^2 + 5.6617555046\, x - 4.0069311266 \,,
\end{aligned}$$

but their respective zeros in the right half-plane vary widely (rounded to the digits specified):

$$p_+ \; : \quad 1.38583, \; 1.42963 - .025776\,\mathrm{i}, \; 1.42963 + .025776\,\mathrm{i},$$
$$p_- \; : \quad 1.40014 - .025272\,\mathrm{i}, \; 1.40014 + .025272\,\mathrm{i}, \; 1.44482,$$
$$p_3 \; : \quad 1.415031, \; 1.415031, \; 1.415031,$$
$$p_2 \; : \quad 1.41393, \; 1.41558, \; 1.41558.$$

Note that—in this very ill-conditioned situation—we must specify p_3 and p_2 to many digits in order to obtain the multiplicities in their zeros, at least within rounding accuracy.

Under these circumstances, it is clearly meaningless to specify the *location* of the 3 zeros of the *empirical* polynomial $(\bar{p}, e)$ in the right half-plane more accurately than by stating that they are "clustered about 1.41." On the other hand, the existence in $N_1(\bar{p}, e)$ of a polynomial like p_3 with an *exact triple zero* would also permit the statement that $(\bar{p}, e)$ "possesses a valid 3-fold zero" near 1.41503.

Now we regard the 3rd degree polynomials which have one of the above sets of clustered zeros as their zeros. Rounded to 7 decimal digits we obtain

$$
\begin{aligned}
s(x) &= x^3 - 4.2450936\,x^2 + 6.0069389\,x - 2.8333344, \\
s_+(x) &= x^3 - 4.2450841\,x^2 + 6.0069378\,x - 2.8333263, \\
s_-(x) &= x^3 - 4.2451030\,x^2 + 6.0069399\,x - 2.8333426, \\
s_3(x) &= x^3 - 4.2450900\,x^2 + 6.0069288\,x - 2.8333273, \\
s_2(x) &= x^3 - 4.2450936\,x^2 + 6.0069389\,x - 2.8333344.
\end{aligned}
$$

Apparently, the coefficients of these "cluster polynomials" hardly reflect the violent variations in the locations of their zeros; they are *well-conditioned* functions of the coefficients of our 4th degree polynomials in $N_1(\bar{p}, e)$. Thus, the 3rd degree empirical polynomial $(\bar{s}, e_s)$ with

$$\bar{s}(x) = x^3 - 4.24509\,x^2 + 6.00694\,x - 2.83333 \tag{6.36}$$

and $e_s = (1, 1, 1) \cdot 10^{-5}$ appears as an appropriate description of the 3-cluster of $(\bar{p}, e)$.

In which sense does $\bar{s}$ provide a *characterization* of the cluster when it is clear that the zero locations for $(\bar{s}, e_s)$ are just as fluctuating as for $(\bar{p}, e)$? At first, we note that the coefficient of x^2 is the negative sum of the zeros; its mild variation shows that the *arithmetic mean* of the cluster zeros is a well-conditioned function of the coefficients in $(\bar{p}, e)$. For the 5 polynomials under consideration, the arithmetic means of the clustered zeros in the right half-plane are (to the digits shown)

$$p \; : \; 1.415031, \quad p_+ \; : \; 1.415028, \quad p_- \; : \; 1.415034, \quad p_3 \; : \; 1.415031, \quad p_2 \; : \; 1.415030.$$

Similarly, the arithmetic mean of the *squares* of the cluster zeros (their "2nd moment") is well-conditioned:

$$p \; : \; 2.002314, \quad p_+ \; : \; 2.002288, \quad p_- \; : \; 2.002340, \quad p_3 \; : \; 2.002313, \quad p_2 \; : \; 2.002311.$$

Apparently, some "statistical information" about the zeros in a cluster remains stable against the indetermination within an empirical polynomial and may be obtained from an associated cluster polynomial like $\bar{s}$.

Also, $\bar{s}$ is an *exact divisor* of some polynomial $\tilde{p} \in N_1(\bar{p}, e)$, and we may compute it from this property; cf. section 6.2. Its zeros represent a *potential constellation* of the cluster zeros for some $(\bar{p}, e)$. The lower degree of a cluster polynomial makes the numerical computation of its zeros easier than for the full polynomial which may have a much higher degree.

From these observations, we draw the following conclusion: For a cluster of $m > 1$ zeros—i.e. zeros which lie in one and the same pseudozero domain (cf. the next section)—we should generally not attempt to determine individual zero locations but valid approximate coefficients of a cluster polynomial of degree m; these coefficients yield the meaningful information about the potential location of the zeros. We will now turn to a more formal treatment of zero clusters and multiple zeros.

6.3.2 Zero Clusters of Empirical Polynomials

An exact m-fold zero ζ_0 of a univariate polynomial p satisfies

$$p(\zeta_0) \;=\; p'(\zeta_0) \;=\; \ldots \;=\; p^{(m-1)}(\zeta_0) \;=\; 0 \quad \text{or} \quad (x - \zeta_0)^m \,|\, p \,. \tag{6.37}$$

Under a *generic* small perturbation Δp of p, ζ_0 splits into m disjoint simple zeros ζ_μ, $\mu = 1(1)m$, with $|\zeta_\mu - \zeta_0| = O(\|\Delta p\|^{\frac{1}{m}})$ for $\|\Delta p\| \to 0$. More precisely,

$$\zeta_\mu \;=\; \zeta_0 + \left[\frac{-\Delta p(\zeta_0)}{\partial^m p(\zeta_0)} \right]_\mu^{\frac{1}{m}} (1 + O(\|\Delta p\|^{\frac{1}{m}})), \quad \mu = 1(1)m\,, \tag{6.38}$$

where $[..]_\mu^{\frac{1}{m}}$ denotes the m different values of the m-th root in $\mathbb{C}$; cf. Proposition 5.1. Equation (6.38) shows that

- a multiple zero is not persistent under generic perturbations of p;

- small perturbations of p may lead to large variations in the zero positions.

Example 6.8: p_3 of section 6.3.1 has a genuine 3-fold zero at $\zeta_0 = 1.415031$. The polynomial $p_3(x) + 10^{-5}(x^3 + x^2 + x + 1)$ has its 3 zeros in the positive halfplane at $\approx 1.38584,\ 1.42962 \pm .02578\,\mathrm{i}$, at a distance from ζ_0 of $\approx .02919$ and $.02963$, resp.; the value of the bracket in (6.38) is $.02948$ in this case. $\square$

Proposition 6.11. The zeros originating from the splitting of an m-fold zero of a nearby polynomial are very ill-conditioned although they are simple.

Proof: By (3.39), the condition of a simple zero ζ with respect to a change in the coefficient α_j is given by $|\zeta|^j / |p'(\zeta)|$. At $\zeta_\mu = \zeta_0 + \Delta\zeta_\mu$, we have

$$p'(\zeta_\mu) \;=\; m\, \partial^m p(\zeta_0)\, (\Delta\zeta_\mu)^{m-1} + O(|\Delta\zeta_\mu|^m)$$

so that the condition of ζ_μ is $O(|\Delta\zeta_\mu|^{1-m}) = O(\|\Delta p\|^{\frac{1}{m}-1})$ as $\|\Delta p\|, |\Delta\zeta_\mu| \to 0$. $\square$

Thus, even for exact polynomials, approximations for a multiple zero or for simple zeros in a cluster are difficult to determine numerically. Most iterative refinement procedures (cf. section 5.1.5) rely, directly or indirectly, on $p' \neq 0$ in a vicinity of the zero; in the situation presently discussed where $|p'|$ is extremely small and vanishes near the zero(s), they either converge only linearly or not at all. If the multiplicity of the zero is known, one can repair some of these deficiencies; e.g., the adapted Newton correction $\Delta\zeta := -m\, p(\zeta)/p'(\zeta)$ converges quadratically to a nearby exact m-fold zero of p (if it exists). But the evaluation of the numerator and denominator involves heavier and heavier cancellation of leading digits. (For a different view on the computation of multiple zeros, cf. the end of section 6.3.4.)

On the other hand, for most of the iterative methods which refine all zeros simultaneously (cf., e.g., (5.30) and (5.31) in section 5.1.5), one can show that the convergence of the zeros in a cluster is *uniform* so the refinement of their arithmetic mean $\bar{\zeta} := \frac{1}{m}\sum_{\text{cluster}}\zeta_\mu$ converges *quadratically*. Since this quantity is well-determined even for a zero cluster of an empirical polynomial (cf. sections 6.3.1 and 6.3.4), these iterative methods may be safely used for the determination of $\bar{\zeta}$. An alternative method for the determination of $\bar{\zeta}$ will be explained in section 6.3.3.

To understand zero clusters of *empirical* polynomials, let us at first consider the case of real coefficients, with real variations, and assume that all zeros ζ_ν of the specified polynomial $\bar{p}$ in $(\bar{p}, e)$ are real, simple, and $\neq 0$. Consider a graph of the backward error $\delta(\xi)$ as a function of $\xi \in \mathbb{R}$:

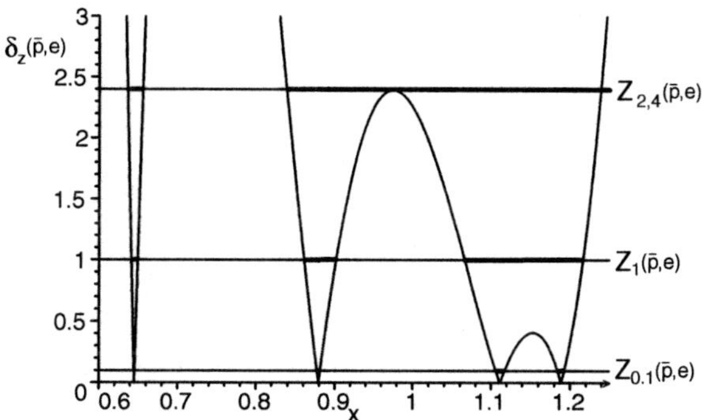

Figure 6.3.

The pseudozero intervals $Z_{\delta,\nu}(\bar{p}, e) \subset \mathbb{R}$ of the zeros ζ_ν are obtained as the intervals with $\delta(\xi) \leq \delta$. For a sufficiently small threshold δ, all extrema of $\delta(\xi)$ will be larger than δ and the generated intervals will each contain *one* zero ζ_ν of $\bar{p}$; cf. Figure 6.3. As we increase the threshold δ we will reach values $\hat{\delta}$ where

$$\hat{\delta} = \max_{\xi \in [\zeta_\nu, \zeta_{\nu+1}]} \delta(\xi) =: \delta(\hat{\xi}_\nu) \tag{6.39}$$

so that the intervals $Z_{\delta,\nu}$ and $Z_{\delta,\nu+1}$ *merge* at their common endpoint $\hat{\xi}_\nu$. The combined interval now contains the two zeros ζ_ν and $\zeta_{\nu+1}$ of $\bar{p}$. It is easy to see that—in the same fashion—new pseudozero intervals with more and more zeros of $\bar{p}$ arise whenever the threshold δ reaches another one of the extrema of the backward error function; cf. (6.39). If we would let δ increase further and further, we would finally arrive at one pseudozero interval containing all zeros of $\bar{p}$—except in the case when $\alpha_0 \neq 0$ is intrinsic: Here, $\delta(0) = \infty$ and the pseudozero intervals in $\mathbb{R}_+$ and $\mathbb{R}_-$ remain separated.

In the case of complex coefficients and zeros, the analogous transition from pseudozero domains containing only one zero each—for sufficiently small δ—to larger and larger pseudozero domains containing more and more zeros of $\bar{p}$ can be made.

Realistically, we are only interested in $\delta = O(1)$. For $\delta = O(1)$, the (real or complex) pseudozero domains $Z_{\delta,\nu}$ may be all separated; this is the case of "well-separated" zeros where each $Z_{\delta,\nu}$ contains exactly one zero of each polynomial $\tilde{p} \in N_\delta(\bar{p}, e)$; cf. Proposition 5.7. Here, we consider the situation where there exist O(1)-pseudozero domains containing more than one zero of $\bar{p}$.

Definition 6.6. A δ-pseudozero domain for $(\bar{p}, e)$, $\delta = O(1)$, which contains $m > 1$ zeros of $\bar{p}$, is an *m-cluster domain* (for tolerance level δ).　　$\square$

Proposition 6.12. An m-cluster domain $Z_\delta(\bar{p}, e)$ contains exactly m zeros (counting multiplicities) of each $\tilde{p} \in N_\delta(\bar{p}, e)$.

Proof: Like in the proof of Proposition 5.7, one need only follow the paths issuing from the m zeros of $\bar{p}$ with $p(x; t)$ defined as previously. If there appears a μ-fold zero along a path for $t = \hat{t}$, there must be μ paths entering that zero for $t \to \hat{t}$ and as many leaving it for $t > \hat{t}$. If $\bar{p}$ itself has a multiple zero, then there are multiple paths issuing from that zero.　　$\square$

Definition 6.7. For each $\tilde{p} \in N_\delta(\bar{p}, e)$, $\delta = O(1)$, the m-tuple of zeros $(\zeta_1, \ldots, \zeta_m)$ of $\tilde{p}$ in an m-cluster domain is a *valid m-cluster of zeros* for $(\bar{p}, e)$ at tolerance level δ.　　$\square$

Superficially, a valid m-cluster appears to be the analogue of a pseudozero for $(\bar{p}, e)$; but there is an important distinction: While each point in a pseudozero domain $Z_{\delta,\nu}$ of $(\bar{p}, e)$ is a pseudozero, an m-tuple of points in an m-cluster domain is a valid m-cluster for $(\bar{p}, e)$ iff these points are *simultaneous* zeros of some specific $\tilde{p} \in N_\delta(\bar{p}, e)$; cf. section 5.2.2. For example, if $\bar{p}$ has an m-fold zero ζ_0 and δ is very small, $\zeta_1, \ldots, \zeta_m$ must lie approximately on a circle in $\mathbb{C}$ about ζ_0, at approximately equal angular distances; cf. (6.38) and Example 6.8.

How can we tell that m zeros $\tilde{\zeta}_\mu$ of some $\tilde{p} \in N_\delta(\bar{p}, e)$ are contained in the same m-cluster domain $Z_\delta(\bar{p}, e)$? In this case, there must exist curves in $\mathbb{C}$ joining any two of these $\tilde{\zeta}_\mu$ which fully remain in Z_δ, i.e. the backward error along these curves must remain below δ. Conversely, if some $\tilde{\zeta}_{\mu_1}$ and $\tilde{\zeta}_{\mu_2}$ do not belong to the same Z_δ, the backward error along *any* curve connecting $\tilde{\zeta}_{\mu_1}$ and $\tilde{\zeta}_{\mu_2}$ must exceed δ somewhere. A natural choice for a test is the collection of the straight line segments between the $\tilde{\zeta}_\mu$ and their arithmetic mean $\bar{\zeta}$; this implies that the size of the backward error at $\bar{\zeta}$ gives an indication of the presence or nonpresence of a cluster domain. In the following subsection, we will consider this further.

6.3.3　Cluster Polynomials

In section 6.3.1, we have found that the coefficients of an appropriate divisor $\bar{s}(x)$ of degree 3 of the empirical polynomial $(\bar{p}, e)$ of (3.17) are *well-determined* within the tolerance of the polynomial—in contrast to the locations of the 3 zeros of $\bar{s}$ which form a 3-cluster of $(\bar{p}, e)$. We have seen that $\bar{s}$ characterizes the cluster in a statistical sense. This approach can be used generally for clusters of univariate polynomials.

Assume that we have found a set of m valid zeros $\tilde{\zeta}_\mu$, $\mu = 1(1)m$, an empirical polynomial $(\bar{p}, e)$ which—intuitively—form a cluster and assume that their arithmetic mean

$\bar{\zeta} := \frac{1}{m} \sum_{\mu=1}^{m} \tilde{\zeta}_\mu$ also has a backward error of $O(1)$. A refinement of the individual zeros fails because of the indetermination in the residuals $\bar{p}(\tilde{\zeta}_\mu)$ and the smallness of the $|\bar{p}'(\tilde{\zeta}_\mu)|$. Thus we expect that there exists an m-cluster domain at the location of the $\tilde{\zeta}_\mu$, i.e. that the zeros lie in one connected pseudozero domain $Z_\delta(\bar{p}, e)$; cf. Definition 6.6. We wish to confirm this expectation and to obtain reliable quantitative information about the cluster.

For this purpose, we determine an m-th degree *valid divisor* $\bar{s}(x)$ of $(\bar{p}, e)$ which is a perturbation of $(x - \bar{\zeta})^m$; thus it must have the form

$$\bar{s}(x) =: \hat{s}(x - \bar{\zeta}) := (x - \bar{\zeta})^m + \sum_{\mu=0}^{m-1} \hat{\sigma}_\mu \, (x - \bar{\zeta})^\mu \,, \tag{6.40}$$

with *small* coefficients $\hat{\sigma}_\mu$, $\mu = 0(1)m - 1$. According to (6.38), the m zeros ζ_μ of s satisfy

$$\zeta_\mu = \bar{\zeta} + (-\hat{\sigma}_0)^{\frac{2\pi i \mu}{m}} \left(1 + O(\|\hat{\sigma}\|^{\frac{1}{m}})\right), \quad \mu = 1(1)m \,, \tag{6.41}$$

where $\hat{\sigma}$ denotes the vector of the $\hat{\sigma}_\mu$.

Definition 6.8. A polynomial of the form (6.40) is an *m-cluster polynomial* if $\|\hat{\sigma}\|^{1/m} = (\max_\mu |\hat{\sigma}_\mu|)^{1/m} \ll 1$; cf. (6.41). It is an *$m$-cluster polynomial for* $(\bar{p}, e)$ if s is a valid divisor of $(\bar{p}, e)$. $\quad\square$

In our situation, there are two natural choices for $\bar{\zeta}$:

- if reasonable approximations ζ_μ for the zeros in the cluster are available, we may use their arithmetic mean;

- we may use the zero of $\bar{p}^{(m-1)}$ which is closest to the cluster location.

The determination of the approximate divisor $\hat{s}$ of $(\bar{p}, e)$ by the procedure in section 6.2.3 is naturally started with $s_0 = (x - \bar{\zeta})^m$. If s_0 itself is not a valid divisor of $(\bar{p}, e)$, its refinement will yield some valid divisor s whenever there are m zeros of $\bar{p}$ reasonably close to $\bar{\zeta}$ while the other zeros of $\bar{p}$ are well-separated from $\bar{\zeta}$.

Essentially, a cluster polynomial s is simply a valid divisor of an empirical polynomial $(\bar{p}, e)$. Its sensitivity to variations in p like that of any other divisor is mainly determined by the condition of the Sylvester matrix $S(q, s)$. This implies that a cluster polynomial of a degree *lower* than the number of zeros in the cluster cannot be well defined. If there is a doubt about the appropriate degree for s, it is better to take it too large than too small! When we include a zero $\hat{\zeta}$ which is not actually a proper part of the cluster, the associated cluster polynomial will be $s(x) (x - \hat{\zeta})$. If $\hat{\zeta}$ is sufficiently well-separated from the remaining zeros of p, it will be a well-conditioned zero and its inclusion will not disturb the well-conditioning of the cluster polynomial.

An algorithmic question in the numerical determination of a cluster polynomial is the following: Should one determine it as a polynomial $\hat{s}$ in $\Delta x := x - \bar{\zeta}$, as in (6.40), or as a polynomial s in x? Moving the origin in $\mathbb{C}$ to $\bar{\zeta}$ requires the transformation of $\bar{p}$ which is a critical numerical operation; cf. section 3.3.2. On the other hand, once this has been done with sufficient care, the subsequent computation of the $\hat{\sigma}_\mu$ deals with small quantities and may be less sensitive to round-off errors than the computation of the coefficients of $s(x)$. The computation of the zeros from $\hat{s}$ should generally be simpler than from s. But for small m, there appears to be no general advantage in the one or the other choice. Also the reaction to a small change in $\bar{p}$ must be the same for s and $\hat{s}$, except for the differing representation.

Example 6.9: Compare section 6.3.1. For demonstration purposes, we choose $\bar{\zeta} = 1.415$ as a rough meanvalue of various approximate zero triples and find a backward error of ≈ 21 for $\bar{s}(x) = (x - 1.415)^3$ as divisor of $\bar{p}$. In this case, $\mathcal{M}(\bar{s})$ is linear of dimension 1 which makes the determination of the backward error straightforward. For a refinement of the cluster polynomial, we may either work in powers of $\Delta x := (x - 1.415)$ or of x.

a) Refinement in terms of Δx: $\hat{s}_0(\Delta x) = \Delta x^3$. $\bar{p}$ transforms to (rounded)

$$\hat{p}(\Delta x) := \bar{p}(\Delta x + 1.415) = \Delta x^4 + 2.82912\,\Delta x^3 - .0002656\,\Delta x^2 - .0000025\,\Delta x\,,$$

which shows that $\bar{\zeta}$ is a valid double zero but not a triple zero. With $\hat{q}_0(\Delta x) = (\Delta x + 2.82912)$,

$$\hat{p}(\Delta x) \;=\; \hat{q}_0(\Delta x) \cdot \Delta x^3 - (26.56\,\Delta x^2 + .25\,\Delta x) \cdot 10^{-5}\,,$$

$$A_{\hat{q}} = \begin{pmatrix} 2.82912 & 1 & 0 \\ 0 & 2.82912 & 1 \\ 0 & 0 & 2.82912 \end{pmatrix} \bmod \hat{s}_0, \quad \mathrm{NF}_{\langle \hat{s}_0 \rangle} \hat{r} = \hat{r} = (26.56\,\Delta x^2 + .25\,\Delta x) \cdot 10^{-5}.$$

From the linear system $(\; \Delta \hat{\sigma}\;)\, A_{\hat{q}} = (\;\hat{\rho}\;)$, we obtain

$$\Delta \hat{s}(\Delta x) \approx (-9.36\,\Delta x^2 - .09\,\Delta x) \cdot 10^{-5} \quad \text{and} \quad \hat{s}(\Delta x) = \Delta x^3 - .0000936\,\Delta x^2 - .0000009\,\Delta x\,,$$

with a backward error $\ll 1$. Backtransformation and rounding to 5 digits gives (6.36).

b) Refinement in terms of x : $s_0(x) = (x - 1.415)^3$. With $q_0(x) = x + 1.414232$ (from the determination of the backward error), we have (rounded)

$$\bar{p}(x) \;=\; q_0(x) \cdot s_0(x) - (.000112\,x^3 - .0002098e\,x^2 - .0000764\,x + .0002109)\,.$$

After reduction of $x^2 q_0$ (third row of A_{q_0}) and $r \bmod s_0$, we have

$$(\;\Delta\sigma\;) \begin{pmatrix} 1.414232 & 1 & 0 \\ 0 & 1.414232 & 1 \\ 2.8331484 & -6.006675 & 5.659232 \end{pmatrix} = (\,-.0005282\; .0007491\; -.0002656\,)$$

which yields $\Delta s(x) \approx -.000094\,x^2 + .000264\,x - .000186$ and (6.36). Naturally, this simple example gives no indication about the differing rounding effects in a) and b). $\square$

With an m-cluster polynomial $\hat{s}$ for $(\bar{p}, e)$, we may now confirm the existence of an m-cluster domain of $(\bar{p}, e)$ about $\bar{\zeta}$: We bound the variation Δs which we must permit so that the straight line segments from each ζ_μ to $\bar{\zeta}$, $\mu = 1(1)m$, consist of exact zeros for polynomials $\hat{s} + \Delta s$; this defines a neighborhood of $\hat{s}$. Then we transfer this neighborhood to a neighborhood of $\bar{p}$: From the divisor property of $\hat{s}$, we have

$$q \cdot \hat{s} \;=\; \tilde{p} \;=:\; \bar{p} + \Delta p\,, \quad \text{with } \|\Delta p\|_e^* =: \delta_0\,.$$

With a maximal $\|\Delta s\|^*$ for the above neighborhood of $\hat{s}$, we obtain

$$q \cdot (\hat{s} + \Delta s) \;=\; \tilde{p} + q \cdot \Delta s \;=:\; \bar{p} + \Delta \bar{p}\,, \quad \text{with } \|\Delta \bar{p}\|_e^* \le \delta_0 + \|q\,\Delta s\|_e^* =: \bar{\delta}\,.$$

Thus, all points on the m straight line segments are zeros of some $\tilde{p} + q \cdot \Delta s = \bar{p} + \Delta\bar{p}$, with $\|\Delta\bar{p}\|_e^* \leq \bar{\delta}$. If $\bar{\delta} = O(1)$, we know that the m cluster zeros of $(\bar{p}, e)$ lie in one and the same pseudozero domain $Z_{\bar{\delta}}(\bar{p}, e)$ which is thus an m-cluster domain for $(\bar{p}, e)$.

Example 6.10: We use $\bar{s}$ of (6.36) which is a valid cluster polynomial for $(\bar{p}, e)$, with a backward error $\delta_0 \approx .64$. The zeros of $\bar{s}$ are $\zeta_1 \approx 1.39157$ and $\zeta_{2,3} \approx 1.422676 \pm .02057\,$i; their mean $\bar{\zeta}_s = 1.41503$ becomes a zero of $\bar{s}$ for a perturbation Δs with $\|\Delta s\|^* = \bar{s}(\bar{\zeta}_s) / (1 + \bar{\zeta}_s + \bar{\zeta}_s^2) \approx .3 \cdot 10^{-5}$ (cf. (3.54)). When we form the corresponding weighted residuals $r(\zeta) := |\bar{s}(\zeta)| / (1 + |\zeta| + |\zeta|^2)$ for ζ along the straight line segments between $\bar{\zeta}_s$ and ζ_1, $\zeta_{2,3}$ resp., we find that $r(\zeta)$ decreases monotonically to 0 as ζ moves from $\bar{\zeta}_s$ to one of the zeros so that the above bound holds throughout. Therefore,

$$\bar{\delta} = \delta_0 + \|q\,\Delta s\|_e^* \leq \delta_0 + \|q\|\,\|\Delta s\|^*/10^{-5} \approx .64 + 2.4 \times .3 = 1.36 = O(1)\,.$$

We have thus established that the 3 zeros of $(\bar{p}, e)$ in the positive halfplane are contained in one 3-cluster domain for $(\bar{p}, e)$. As is to be expected from (6.41), the s-residual at the meanvalue is the maximal value taken inside the cluster. $\square$

Since an m-cluster polynomial s is an exact divisor of some $\tilde{p} \in N_\delta(\bar{p}, e)$ with $\delta = O(1)$, its zeros form a valid m-cluster of zeros for $(\bar{p}, e)$ at tolerance level δ; cf. Definition 6.7. As we have observed in section 6.3.1, the coefficients of s express valid quantitative information about the potential positions of the ill-conditioned zeros in the cluster relative to $\bar{\zeta}$ and each other.

Proposition 6.13. The first m moments—relative to $\bar{\zeta}$ or to the origin, resp.—of the zeros within an m-cluster are (well-known) fixed polynomials in the coefficients of an associated cluster polynomial $\hat{s}$ or s, resp.

Proof: By Vieta, the $\hat{\sigma}_\mu$ and the σ_μ are the elementary symmetric functions of the $\Delta\zeta_\mu$ and the ζ_μ, resp., in the cluster:

$$\sum_{\mu=1}^m \Delta\zeta_\mu = -\hat{\sigma}_{m-1}\,, \qquad\qquad \sum_{\mu=1}^m \zeta_\mu = -\sigma_{m-1}\,,$$
$$\sum_{\mu=2}^m \sum_{\mu'<\mu} \Delta\zeta_\mu \Delta\zeta_{\mu'} = \hat{\sigma}_{m-2}\,, \quad \text{resp.,} \quad \sum_{\mu=2}^m \sum_{\mu'<\mu} \zeta_\mu \zeta_{\mu'} = \sigma_{m-2}\,,$$
$$\text{etc.} \qquad\qquad\qquad\qquad\qquad \text{etc.}$$
$$\prod_{\mu=1}^m \Delta\zeta_\mu = (-1)^m \hat{\sigma}_0\,, \qquad\qquad \prod_{\mu=1}^m \zeta_\mu = (-1)^m \sigma_0\,.$$

It is well known from classical algebra that the *moments* of a set of quantities may be expressed as polynomials in their symmetric fundamental functions; e.g.,

$$\sum_{\mu=1}^m \Delta\zeta_\mu^2 = \hat{\sigma}_{m-1}^2 - 2\,\hat{\sigma}_{m-2}\,, \quad \text{resp.,} \quad \sum_{\mu=1}^m \zeta_\mu^2 = \sigma_{m-1}^2 - 2\,\sigma_{m-2}\,. \quad \square$$

Example 6.11: Compare section 6.3.1. Consider the cluster polynomials $\hat{s}$ and s obtained in Example 6.9. With $\Delta\hat{s}$, we can refine the arithmetic mean $\bar{\zeta}$ of the 3 cluster zeros:

$$\Delta\bar{\zeta} = \frac{1}{3}\sum_{\mu=1}^3 \Delta\zeta_\mu = \frac{1}{3}\sum_\mu \zeta_\mu - \bar{\zeta} = -\frac{1}{3}\hat{\sigma}_2 \approx .000031\,.$$

This is also the zero of $\partial^2 \bar{p}$; cf. the remark below Definition 6.8. The square sum of the deviations $\Delta\zeta_\mu$ of the zeros from $\bar{\zeta}$ is obtained as $\approx -2\,\hat{\sigma}_1 \approx .0000018$ which is the correct value for the zeros of $\bar{p}$. (Note, however, that this implies small $|\Delta\zeta_\mu|$ only when the ζ_μ are *real*; for complex $\Delta\zeta_\mu$, there may be considerable cancellation in a sum of squares!)

From the coefficients of s, we obtain the same improved arithmetic mean and, e.g.,

$$\frac{1}{3}\sum_{\mu=1}^{3}\zeta_\mu^2 \;=\; \frac{1}{3}\,(\sigma_2^2 - 2\,\sigma_1) \;\approx\; 2.002314\,,$$

which is the "expected value" for that average when p varies in $N_1(\bar{p}, e)$. $\square$

6.3.4 Multiple Zeros of Empirical Polynomials

For $m > 1$, the algebraic predicate Π^m : "p has an m-fold zero" is a critical predicate in the sense of Definition 6.1: In an arbitrarily small neighborhood of p^* with $\Pi^m(p^*)$ =**true**, there are $\tilde{p}$ with $\Pi^m(\tilde{p})$ =**false**. For empirical polynomials $(\bar{p}, e)$, we extend Π^m such that its range becomes $\mathbb{R}_+$; cf. section 6.1.1.

$\zeta \in \mathbb{C}$ is an exact m-fold zero of $p \in \mathcal{P}$ if it satisfies (6.37). In analogy with other definitions, we define a valid m-fold zero of an *empirical* polynomial by

Definition 6.9. For a univariate empirical polynomial $(\bar{p}, e)$, a value $\zeta \in \mathbb{C}$ is a *valid m-fold zero* if there exists $\tilde{p} \in N_\delta(\bar{p}, e), \delta = O(1)$, such that $(x - \zeta)^m \,|\, \tilde{p}$. $\square$

Definition 6.10. For a univariate empirical polynomial $(\bar{p}, e)$, a connected set

$$Z_\delta^m(\bar{p}, e) \;:=\; \{\,\zeta \in \mathbb{C} \;:\; \exists\,\tilde{p} \in N_\delta(\bar{p}, e) \;:\; (x - \zeta)^m \,|\, \tilde{p}\,\} \tag{6.42}$$

is an *m-fold pseudozero domain* of $(\bar{p}, e)$. $\square$

Example 6.12: For our empirical polynomial $(\bar{p}, e)$, $\zeta = 1.415031$ is a valid 3-fold zero because

$$\bar{p} \;=\; (x + 1.414213)\,(x - 1.415031)^3 + (-.2486\,x^2 + .4503\,x - .1395)\cdot 10^{-5}$$

so that there exists $\tilde{p} \in N_\delta(\bar{p}, e)$ with a 3-fold zero ζ for $\delta \geq .45$. This shows that $Z_\delta^3(\bar{p}, e)$ is not empty for $\delta \geq \underline{\delta}$, with some $\underline{\delta} < .45$; in Example 6.13 below, we will see that $\underline{\delta} \approx .06$. On the other hand, we know that $\bar{p}$ does not possess an exact 3-fold zero. Therefore, $Z_\delta^3(\bar{p}, e)$ must be empty for sufficiently small positive δ. This also establishes that the existence of a 3-cluster for $(\bar{p}, e)$ does not imply the existence of a 3-fold zero in $N_\delta(\bar{p}, e)$ for all $\delta > 0$.

In section 6.3.1, we have also found $\zeta = 1.41558$ to be a valid 2-fold zero of $(\bar{p}, e)$. For each δ, the associated domain Z_δ^2 must enclose the domain Z_δ^3 and lie inside the 3-cluster domain Z_δ of $(\bar{p}, e)$. However, the sizes of these domains behave very differently for small $\bar{e} := \|e\|$: While $Z_\delta = O(e^{1/3})$, $Z_\delta^2 = O(e^{1/2})$ and $Z_\delta^3 = O(e)$; cf. the end of this section. $\square$

The backward error $\delta^m(\zeta) := \min_{\Delta\alpha\in\mathcal{M}^m(\zeta)} \|\Delta\alpha\|_e^*$ of an approximate m-fold zero ζ of $(\bar{p}, e)$ is determined by the associated equivalent-data manifold $\mathcal{M}^m(\zeta)$ which derives from the requirements (6.37) for ζ as an exact m-fold zero of a polynomial $\bar{p} + \Delta p$. Thus

$$\mathcal{M}^m(\zeta) := \{\Delta a \in \Delta\mathcal{A} : \sum_{\nu=0}^{n}\Delta\alpha_\nu\zeta^\nu + \bar{p}(\zeta) = 0, .., \sum_{\nu=m-1}^{n}\tbinom{\nu}{m-1}\Delta\alpha_\nu\zeta^{\nu-m+1} + \partial^{m-1}\bar{p}(\zeta) = 0\},$$

$$\tag{6.43}$$

with codimension m; thus $\mathcal{M}^m(\zeta)$ needs at least m empirical coefficients in $(\tilde{p}, e)$ to be nonempty.

Equation (6.37) is clearly an *overdetermined system* of m univariate polynomial equations so that it can have a solution ζ only if the coefficients of $\tilde{p}$ satisfy some *consistency conditions* $S^m(\tilde{a}) = 0$ which define an algebraic manifold $\mathcal{S}^m \subset A$. $\mathcal{S}^m$ is the truth domain $Q(\Pi^m)$ of the predicate Π^m; cf. Definition 6.1. The determination of a multiple zero of an intrinsic polynomial is an *ill-posed* problem of type 2) in section 3.2.1. Obviously, $(\tilde{p}, e)$ has a valid m-fold zero iff $\mathcal{S}^m \cap N_\delta(\tilde{p}, e) \neq \emptyset$ for $\delta \geq \underline{\delta} = O(1)$, where

$$\underline{\delta}(\tilde{p}, e) := \min_{a \in \mathcal{S}^m} \|a - \bar{a}\|_e^* . \tag{6.44}$$

Proposition 6.14. The codimension of the manifold $\mathcal{S}^m$ is $m - 1$.

Proof: Without loss of generality, we consider monic polynomials of degree n. For each $\tilde{a} \in \mathcal{S}^m \subset A = \mathbb{C}^n$, there exists a monic polynomial q of degree $n - m$ and $\zeta \in \mathbb{C}$ with

$$p(x; \tilde{a}) =: \tilde{p} = q(x) \cdot (x - \zeta)^m .$$

A variation of the $n - m$ coefficients of q and of ζ keeps $\tilde{p}$ on $\mathcal{S}^m$; these are $n - m + 1$ independent parameters. Thus the codimension of $\mathcal{S}^m$ is $m - 1$. $\square$

If the linear manifold $\mathcal{M}^m(\zeta)$ is nonempty for a candidate value ζ, we can determine its backward error $\delta^m(\zeta)$. If $\delta^m(\zeta) = O(1)$, we have found a valid m-fold zero. But if $\delta^m(\zeta)$ is not sufficiently small, we cannot proceed as usual since, generally, $\tilde{p}$ does not possess an m-fold zero. Instead, we take the following recourse: Assume that we know, from the determination of $\delta^m(\zeta)$,

$$\tilde{p}(x) = q(x) \cdot (x - \zeta)^m + r(x) .$$

What we hope to find is

$$\tilde{p}(x) = (q(x) + \Delta q(x)) \cdot (x - (\zeta + \Delta\zeta))^m \in \mathcal{S}^m , \quad \text{with } \|\tilde{p} - \bar{p}\|_e^* \stackrel{>}{\approx} \underline{\delta}(\tilde{p}, e) .$$

Now we subtract and linearize:

$$\tilde{p}(x) - \bar{p}(x) \doteq \Delta q(x) \cdot (x - \zeta)^m - q(x)\, m\, (x - \zeta)^{m-1}\, \Delta\zeta - r(x) =: \sum_{\nu=0}^{n-1} \Delta\alpha_\nu(\Delta q, \Delta\zeta) ;$$

then we solve the minimization problem

$$\min_{\Delta q, \Delta\zeta} \|\Delta a\|_e^* . \tag{6.45}$$

Due to the linearization, the resulting $q + \Delta q$ and $\zeta + \Delta\zeta$ will not realize the minimal distance (6.44) precisely. But this is not necessary: Either they define a $\tilde{p}$ close enough to $\bar{p}$, then $\zeta + \Delta\zeta$ is a valid m-fold zero; or $\|\tilde{p} - \bar{p}\| > O(1)$, then there are (most probably) no valid m-fold zeros accessible from our candidate value ζ.

Example 6.13: From Example 6.12, we have for our standard $\bar{p}$ and $\zeta = 1.415031$, $q(x) = x + 1.414213$ and $r(x) = (-.2486\, x^2 + .4503\, x - .1395) \cdot 10^{-5}$. Thus,

$$\Delta p(x) = \Delta\beta_0 \cdot (x - 1.415031)^3 - (x + 1.414213) \cdot 2 (x - 1.415031)^2 \, \Delta\zeta - r(x) \approx$$

$$(-2.8333 \, \Delta\beta_0 - 5.6634 \, \Delta\zeta + .0000014) + (6.0069 \, \Delta\beta_0 + 4.0000 \, \Delta\zeta - .0000045) \, x$$
$$+ (-4.2451 \, \Delta\beta_0 + 2.8317 \, \Delta\zeta + .0000025) \, x^2 + (\Delta\beta_0 - 2 \, \Delta\zeta) \, x^3 \,.$$

Minimization of the maximum modulus of the coefficients yields a backward error (6.45) of $\approx .06 \approx \underline{\delta}$ which shows that $\bar{p}$ is actually very close to having an exact 3-fold zero. $\qquad\square$

Let us consider the *condition* of a valid m-fold zero ζ, with $(x - \zeta)^m \mid p$, $p \in N_1(\bar{p}, e) \cap \mathcal{S}^m$. Since we may only admit variations of p in the consistency manifold $\mathcal{S}^m$, the associated cluster polynomials remain of the form $(x - \tilde{\zeta})^m$ and their coefficients $\tilde{\sigma}_{m-1} = m \, \tilde{\zeta}$. Thus the potential variation of $\tilde{\zeta}$ for a variation of $\tilde{p}$ within $N_1(\bar{p}, e)$ is bounded by the potential variation of $\tilde{\sigma}_{m-1}$ which we have found to be $O(\|e\|^*)$ for a sufficiently well-conditioned Sylvester matrix $S(q, s)$; cf. the end of section 6.2.3. Thus, if it is nonempty, the size of Z_δ^m can only be $O(\|e\|^*)$ and the potential m-fold zeros are *well-conditioned* functions of the coefficients of $(\bar{p}, e)$.

This view of the situation remains relevant as $e \to 0$: If the intrinsic polynomial $\bar{p}$ has an m-fold zero $\bar{\zeta}$ and if we consider only perturbations of $\bar{p}$ which *retain an m-fold zero*, $\bar{\zeta}$ is *well-conditioned*; cf. also [6.2]. When we have an approximation ζ of $\bar{\zeta}$, then (6.45) will yield a $\Delta\zeta$ such that $|(\zeta + \Delta\zeta) - \bar{\zeta}| = O(|\zeta - \bar{\zeta}|^2)$. The potential iteration of this procedure provides a stable, quadratically convergent algorithm for the computation of $\bar{\zeta}$.

6.3.5 Zero Clusters about Infinity

In section 5.2.3, we have found it advisable to consider zeros ξ_ν with a very large modulus as reciprocals of zeros η_ν of the reciprocal polynomial (5.40). Such zeros occur when one or several leading coefficients are tiny relative to other coefficients in $p(x) = \sum_{\nu=0}^n \alpha_\nu x^\nu$. For the empirical polynomial $(\bar{p}, e)$, assume that

$$|\bar{\alpha}_{n-\mu}| \ll \|\bar{a}^T\|^* \quad \text{for} \quad \mu = 0(1)m - 1 \quad \text{and} \quad |\bar{\alpha}_{n-m}| = O(\|\bar{a}^T\|^*) \,.$$

Then the reciprocal polynomial $(\bar{q}, e)$ with $\bar{q} = y^n \, \bar{p}(\frac{1}{y})$ has y^m as a near-divisor, which indicates an m-cluster of zeros about 0.

If, for $\mu = 0(1)m - 1$, $0 \in N_\delta(\bar{\alpha}_{n-\mu}, \varepsilon_{n-\mu})$ with $\delta = O(1)$, then 0 is a valid m-fold zero of $(\bar{q}, e)$ and ∞ a valid m-fold zero of $(\bar{p}, e)$; equivalently, the lower degree polynomial $\tilde{p}(x) = \sum_{\nu=0}^{n-m} \bar{\alpha}_\nu x^\nu \in \mathcal{P}_{n-m}$ is in $N_\delta(\bar{p}, e)$ and thus a valid instance of $(\bar{p}, e)$. In this case, the empirical polynomial $(\bar{p}, e)$ is essentially of degree $n - m$ and its m huge zeros should be disregarded.

We assume now that $0 \notin N_\delta(\bar{\alpha}_{n-\mu}, \varepsilon_{n-\mu})$, $\delta = O(1)$, for some $\mu < m$. Then we may determine the cluster polynomial s_q associated with the near-divisor $\tilde{s}_q(y) := y^m$

$$s_q(y) = \tilde{s}_q(y) + \Delta s(y) = y^m + \sum_{\mu=0}^{m-1} \hat{\sigma}_\mu \, y^\mu$$

such that it is a valid divisor of $\bar{q}(y)$. Since an expansion (6.28) of $\bar{q}$ in powers of $\tilde{s}_q$ is trivial, we know the coefficient $r_1(y)$ in (6.28) and we can compute the matrix for the refinement (6.25) from r_1; cf. the remark after Proposition 6.9.

The zeros η_μ, $\mu = 1(1)m$, of a valid cluster polynomial s_q are valid zeros of $(\bar{q}, e)$; by Proposition 5.8, this implies that the $\xi_\mu = \frac{1}{\eta_\mu}$ are valid zeros of $(\bar{p}, e)$. Since the cluster domain of $(\bar{q}, e)$ contains the origin, the cluster domain of $(\bar{p}, e)$ will contain $\infty \in \mathbb{C}$.

Example 6.14: The *Mignotte polynomials* have the structure

$$p_{\text{Mign}}(x) = x^n + (a\,x + 1)^{n-m}, \qquad \text{with } n \gg 1,\ |a| \gg 1. \tag{6.46}$$

Obviously, p_{Mign} has a cluster of $n - m$ small zeros near 0 and a cluster of m large zeros about infinity. The reciprocal polynomial q is

$$q(y) = 1 + y^m\,(y + a)^{n-m} = 1 + y^m\,(a^{n-m} + (n - m)\,a^{n-m-1}y + \ldots).$$

The remainder $r \bmod y^m$ is 1, the coefficient polynomial r_1 in (6.28) consists of the first m terms in the factor of y^m above, which is the whole term $(y + a)^{n-m}$ if $n - m \le m$. The matrix A_{r_1} which contains the remainders mod y^m of $r_1\,y^\mu$, $\mu = 0(1)m - 1$, is simply the upper-triangular Toeplitz matrix with the coefficients of r_1. The coefficients of $\Delta s(y)$ are in the first row of the inverse of A_{r_1}.

For a numerical example, we take $n = 10$, $m = 6$, $a = 10$; the fact that we take an intrinsic polynomial is of no avail here. Thus, $p(x) = x^{10} + (10\,x + 1)^4$ and

$$q(y) = 1 + y^6\,(10000 + 4000\,y + 600\,y^2 + 40\,y^3 + y^4) = r(y) + y^6 r_1(y).$$

The first row of the inverse of the 6×6 matrix

$$A_{r_1} = \begin{pmatrix} 10000 & 4000 & 600 & 40 & 1 & 0 \\ 0 & 10000 & 4000 & 600 & 40 & 1 \\ 0 & 0 & 10000 & 4000 & 600 & 40 \\ 0 & & \ddots & \ddots & & \ddots \end{pmatrix}$$

is $(.0001, -.00004, .00001, -.000002, .00000035, -.000000056)$, which are the coefficients of Δs. The reciprocals ξ_μ of the 6 zeros η_μ of $s(y) = y^6 + \Delta s(y)$ are (rounded)

$$4.085384 \pm 2.321360\,\mathrm{i},\ .0666266 \pm 4.642784\,\mathrm{i},\ -3.952011 + 2.321429\,\mathrm{i}\,;$$

the exact values coincide with the exact zeros of p in 8 decimal digits, after a computation in 10-digit floating-point arithmetic. $\quad\square$

Exercises

1. Consider the empirical polynomial $(\bar{p}, e)$ with

$$\bar{p}(x) = x^8 - 4.150\,x^7 + 6.279\,x^6 - 4.736\,x^5 + 4.542\,x^4 - 6.271\,x^3 + 4.983\,x^2 - 1.859\,x + .262,$$

and a tolerance $.5 \cdot 10^{-3}$ on each coefficient except the leading one.

(a) Analyze the zero cluster of $(\bar{p}, e)$ in all the respects that we have applied throughout section 6.3 to the example in section 6.3.1.

(b) Has $(\bar{p}, e)$ a valid 4-fold zero?

2. Consider the empirical polynomial $(\bar{p}, e)$ with $\bar{p}(x) =$

$$.000143\,x^8 - .00110\,x^7 - .00173\,x^6 + .00186\,x^5 - 2.130\,x^4 + 1.212\,x^3 + 1.180\,x^2 - .417\,x - 1.000,$$

with tolerances specified by assuming that all coefficients have been rounded to the decimal digits shown. Determine the zeros of $\bar{p}$.

(a) Convince yourself that $(\bar{p}, e)$ has a 4-cluster about ∞: Determine a polynomial $\hat{s}(y) = y^4 + \sum_{\nu=0}^{3} \hat{\sigma}_\nu y^\nu$, with tiny coefficients $\hat{\sigma}_\nu$, which is a valid divisor of the reciprocal empirical polynomial $(y^8 \bar{p}(1/y), e)$.

(b) Compare the zeros of $\hat{s}$ and their reciprocals with the zeros of $\bar{q} := y^8 \bar{p}(1/y)$ and those of $\bar{p}$, respectively.

(c) Establish the poor condition of the cluster zeros of $\bar{q}$ by computing their condition numbers (3.40) as well as by introducing perturbations into $\bar{q}$. Observe the effects on the "cluster zeros" of $\bar{p}$.

6.4   Greatest Common Divisors

For a set $\{p_1, \ldots, p_k\}$ of two or more univariate polynomials, the following two predicates are equivalent:

"The $p_1, \ldots, p_k$ have *common zeros*" and "the $p_1, \ldots, p_k$ have a nontrivial *common divisor*."

A common divisor of degree $m > 1$ is equivalent to m common zeros (counting multiplicities). Thus, many questions can be posed from either point of view; cf. the remark after Proposition 6.6.

For a set of $k > 1$ intrinsic polynomials, the determination of common zeros or divisors is an *ill-posed problem* of class 2) in the sense of section 3.2.1: In the joint data space $\mathcal{A}$ of two or more polynomials of given degrees, the coefficients $(a_1, a_2, \ldots)$ which permit a common divisor of a degree d or d common zeros lie on an algebraic manifold $\mathcal{S}_d$ of a dimension less than dim $\mathcal{A}$ and the data$\rightarrow$result mapping is only defined for data on $\mathcal{S}_d$, the truth domain Q of the above predicates. For a set of *empirical* univariate polynomials, the discrete **true–false** values are replaced by a *continuous* result in $\mathbb{R}_+$ as we have explained in section 6.1: For pseudozeros or pseudodivisors, the predicate is always "true" but the minimal achievable backward error may be so large that it is effectively false. It is this *smooth transition* between **true** and **false** which makes the problem accessible to a solution by approximate computation.

Greatest common divisors have attracted the attention of algebraists for a very long time, in their theoretical aspects as well as their algorithmic ones. This is also one of the few areas in computational algebra where numerical aspects have been considered more thoroughly. We will not be able to relate the results of all these approaches; like in our treatment of univariate polynomial zero finding, we will rather attempt to expose those ideas which contribute to the goal of this book.

6.4.1   Intrinsic Polynomial Systems in One Variable

In this section, we recall some facts about common divisors and zeros. Since any polynomial $p \in \mathcal{P}$ of a positive degree n defines exactly n zeros $\zeta_\nu \in \mathbb{C}$ (counting multiplicities), a *system*

of two or more univariate polynomials must be *overdetermined* with respect to zero finding:
The system

$$P(x) \; = \; (\, p_1(x), \; p_2(x), \; \ldots, \; p_k(x)\,)^T \; = \; 0 \tag{6.47}$$

of k polynomial equations of degrees n_κ can have a common zero ζ only if the p_κ are interrelated
so that ζ is a zero of each individual p_κ. We have met such a system in section 6.3.4: An m-
fold zero of a univariate polynomial has to satisfy the system (6.37). In this special case, the
coefficients of the polynomials in P come from one and the same set $\{\alpha_0, \alpha_1, \ldots, \alpha_n\}$; this will
not be assumed in the further discussion.

The zero problem for (6.47) may also be posed in terms of *common divisors*:

$$\text{For } d > 0, \quad \exists\,?\,g \in \mathcal{P}_d \,:\; p_\kappa(x) \; = \; q_\kappa(x)\,g(x)\,, \quad \kappa = 1(1)k\,, \tag{6.48}$$

and—if yes—what is the g with the highest degree, the "greatest" common divisor $\gcd(P)$.
Clearly, the zeros of $\gcd(P)$ constitute the complete solution set of the system (6.47) and vice
versa. Algebraically, both formulations are versions of the problem:

$$\text{Given } P = \{\, p_1, \ldots, p_k \,\} \subset \mathcal{P}, \; k \geq 2, \text{ what is the basis polynomial } g \text{ of the ideal } \langle P \rangle \,? \tag{6.49}$$

This points immediately to an algorithmic way for the determination of g: Without loss of
generality, we assume that the p_κ are monic and ordered by degrees, i.e. $n_1 \leq n_2 \leq \ldots \leq n_k$.
We reduce $p_2, \ldots, p_k$ by p_1 to polynomials $r_{\kappa 1}$ of degree at most $n_1 - 1$ by polynomial division
(cf. section 5.3):

$$p_\kappa(x) \; = \; q_{\kappa 1}(x) \cdot p_1(x) + r_{\kappa 1}(x)\,, \quad \kappa = 2(1)k\,. \tag{6.50}$$

Iff all $r_{\kappa 1}$ vanish, $g = p_1$ and we are finished. Otherwise, we take one of the $r_{\kappa 1}$ of lowest
positive degree (say r_{21}), normalize it to be monic, and reduce the remaining $r_{\kappa 1}$ by it, obtaining
polynomials $r_{\kappa 2}$ of lower degree than that of r_{21}. Iff all new remainders vanish, $g = r_{21}$;
otherwise ..., and the recursive continuation is clear. Since the degree of the reductor decreases
at least by one in each recursive step, the procedure must end, either with a g of positive degree,
or with $g = 1$.

For *two* polynomials p_1, p_2, $\deg p_1 \leq \deg p_2$, this is the well-known Euclidean Algorithm
which successively divides the last-before-last remainder by the last one.

Algorithm 6.1 (Euclidean Algorithm). $\quad r_0 := p_1\,, \quad r_1 := \mathrm{rem}(p_2, p_1)\,, \quad i := 2\,,$

$\qquad$ **while** $r_{i-1} \neq 0$ **do** $r_i := \mathrm{rem}(r_{i-2}, r_{i-1})\,, \quad i := i + 1$ **od** ;

$\qquad g := r_{i-2}\,.$

The attempt to express this procedure (where we have omitted the normalization requested
in the text to conform with the standard formulation) in terms of row operations on a matrix
leads to the Sylvester matrix $S(p_2, p_1)$, cf. section 6.2.2. The reason why n_1 copies of the
coefficients of p_2 and n_2 copies of the coefficients of p_1 are needed is seen thus: To generate the
coefficients of $r_1 = \mathrm{rem}(p_2, p_1)$ by subtracting multiples of rows containing the coefficients of
p_1 from a row with the coefficients of p_2, we need rows representing $x^\lambda\, p_1$ for $\lambda = 0(1)n_2 - n_1$.
Analogously, if $\deg r_{i-1} = \deg r_{i-2} - 1$, we need 2 rows of r_{i-1} coefficients to generate the
coefficients of r_i, and more if the difference in degrees is greater. The recursion process leads
to a total of n_2 shifted rows for p_1 and n_1 rows for p_2, as they appear in $S(p_2, p_1)$, cf. (6.23).

The Sylvester matrix $S(p_2, p_1)$ also appears when we pose the problem (6.49) in a quantitative form: Find the maximal d such that

$$\exists u_1 \in \mathcal{P}_{n_2-1},\ u_2 \in \mathcal{P}_{n_1-1}\ :\ u_2(x)\, p_2(x) - u_1(x)\, p_1(x) = g(x) \in \mathcal{P}_d, \tag{6.51}$$

which may be written as

$$(\ldots u_2^T \ldots \mid \ldots u_1^T \ldots)\, (\, S(p_2, p_1)\,)\, (\mathbf{x}) = (\gamma_0, \ldots, \gamma_{d-1}, 1, 0, \ldots)\, (\mathbf{x}). \tag{6.52}$$

As a square matrix, $S(p_2, p_1)$ has generically a trivial kernel. By Theorem 6.7, the existence of a kernel of dimension $d > 0$ is equivalent to the existence of a common divisor of degree d of p_1 and p_2. By the elimination procedure in section 6.2.3, we may transform $S(p_2, p_1) = \begin{pmatrix} S_{11} & S_{12} \\ S_{21} & S_{22} \end{pmatrix}$ into $M\, S(p_2, p_1) = \begin{pmatrix} S_{11}^* & 0 \\ S_{21} & S_{22} \end{pmatrix}$, with (cf. (6.27))

$$S_{11}^* \begin{pmatrix} 1 \\ x \\ \vdots \\ x^{n_1-1} \end{pmatrix} \overset{\langle p_1 \rangle}{\equiv} \begin{pmatrix} p_2(x) \\ x\, p_2(x) \\ \vdots \\ x^{n_1-1} p_2(x) \end{pmatrix}; \tag{6.53}$$

cf. Proposition 6.9. S_{22} is lower triangular with a unit diagonal and thus nonsingular; hence, the rank deficiency of $S(p_2, p_1)$ must fully transfer to S_{11}^* and the kernel must remain the same.

Proposition 6.15. p_1 and p_2 have a common divisor of degree d iff the rank deficiency of S_{11}^* in (6.53) is d. The kernel of S_{11}^* is spanned by the m vectors $\mathbf{z}_\mu = (1, \zeta_\mu, \ldots, \zeta_\mu^{n_1-1})^T \in \mathbb{C}^{n_1}$, where the ζ_μ are the common zeros of p_1 and p_2. (In the case of a multiple common zero, $\mathbf{z}_\mu$ is supplemented by further vectors in the well-known way.)

Proof: The assertion follows from (6.53) when we represent it with respect to the Lagrange basis $\mathbf{b}_0$ of $\mathcal{R}[\langle p_1 \rangle]$. Then the multiplication matrix becomes $A_{p_2}^{(0)} = \operatorname{diag}(p_2(\zeta_1), \ldots, p_2(\zeta_{n_1}))$ and, by (2.42)/(2.44), we have

$$A_{p_2}\, M_0 = M_0\, A_{p_2}^{(0)}, \qquad \text{with } M_0 = \mathbf{c}_0^T(\mathbf{b}) = \begin{pmatrix} 1 & \cdots & 1 \\ \zeta_1 & \cdots & \zeta_{n_1} \\ \vdots & & \vdots \\ \zeta_1^{n_1-1} & \cdots & \zeta_{n_1}^{n_1-1} \end{pmatrix} \quad \text{or}$$

$$S_{11}^*\, M_0 = M_0\, \operatorname{diag}(p_2(\zeta_1), \ldots, p_2(\zeta_{n_1})). \tag{6.54}$$

At the *common* zeros ζ_μ, $p_2(\zeta_\mu) = 0$. Modifications for multiple zeros are as usual. $\square$

The one direction (degree of gcd $\leq$ rank deficiency) of Proposition 6.15 may be generalized immediately to the case of more than 2 polynomials. Consider the situation of (6.50), with $k \geq n_1 + 1$, and

$$r_\kappa(x) := \operatorname{rem}_{p_1} p_\kappa(x) =: \sum_{\nu=0}^{n_1-1} \rho_{\kappa,\nu} x^\nu, \qquad \kappa = 2(1)k. \tag{6.55}$$

Proposition 6.16. If the polynomial set $P = \{p_1, p_2, \ldots, p_k\}$ has d common zeros (counting multiplicities) or a nontrivial gcd $g \in \mathcal{P}_d$, resp., then the matrix $R := \begin{pmatrix} \rho_{2,0} & \cdots & \rho_{2,n_1-1} \\ \vdots & & \vdots \\ \rho_{k,0} & \cdots & \rho_{k,n_1-1} \end{pmatrix}$

has rank deficiency at least d. Equivalently, if rk $R = n_1 - d$, P has at most d common zeros (counting multiplicities), i.e. a potential gcd of P has at most degree d. In particular, if R is nonsingular, there are no common zeros and gcd $P = 1$; i.e. the p_κ are coprime.

Proof: Since the assumption implies $p_1(x) = s_1(x)\, g(x)$, a reduction by p_1 leaves divisibility by g invariant; thus the assumption implies that $\overline{P} = \{p_1, r_2, \ldots, r_k\}$ also has the nontrivial gcd $g(x)$, i.e. $r_\kappa(x) = s_\kappa(x)\, g(x)$, with deg $s_\kappa \leq n_1 - d - 1$, $\kappa = 2(1)k$. This implies

$$
R = \begin{pmatrix} \sigma_{2,0} & \cdots & \sigma_{2,n_1-d-1} \\ \vdots & & \vdots \\ \sigma_{k,0} & \cdots & \sigma_{k,n_1-d-1} \end{pmatrix} \begin{pmatrix} \gamma_0 & \cdots & 1 \\ & \ddots & & \ddots \\ & & \gamma_0 & \cdots & 1 \end{pmatrix}.
$$

The two matrix factors have $n_1 - d$ columns or rows, resp.; hence rk $R \leq n_1 - d$. $\quad\square$

The assumption $k \geq n_1 + 1$ appears quite restrictive, but it is necessary to accumulate at least n_1 rows in R with its n_1 columns. Note that this is also the number of columns and rows in S_{11}^*; this points to a suitable way for $2 < k \leq n_1$: We supplement the system P by polynomials which cannot introduce spurious common zeros. When we interpret the classical case $k = 2$ as using the system $P = \{p_1, p_2, x\, p_2, \ldots, x^{n_1-1} p_2\}$ with $n_1 + 1$ members, we realize that we may simply replace sufficiently many of the extra polynomials there by the p_κ, $\kappa = 3(1)k$.

The other direction (rank deficiency $\leq$ degree of gcd) holds only if the remainders r_κ are linearly independent, or—equivalently—the values of the p_κ, $\kappa \geq 2$, on the zeros of p_1 are linearly independent. Because of (6.54), this independence exists for the polynomial set $\{p_1, p_2, x\, p_2, \ldots, x^{n_1-1} p_2\}$. If the linear independence has thus been secured for n_1 remainders r_κ, further remainders or polynomials, resp., may be introduced without restriction.

Proposition 6.17. Consider $P = \{p_1, p_2, \ldots, p_k\}$, $k > 2$, ordered by increasing degree. Form the remainders mod p_1

$$
r_{2,0} := \text{rem } p_2, \quad \ldots, \quad r_{2,n_1-1} := \text{rem } x^{n_1-1} p_2, \quad r_\kappa := \text{rem } p_\kappa, \quad \kappa = 3(1)k .
$$

Let $r_\kappa(x) =: \sum_{\nu=0}^{n_1-1} \rho_{\kappa,\nu}\, x^\nu$ and

$$
\overline{R} := \begin{pmatrix} \rho_{2,0,0} & \cdots & \rho_{2,0,n_1-1} \\ \vdots & & \vdots \\ \rho_{2,n_1-1,0} & \cdots & \rho_{2,n_1-1,n_1-1} \\ \rho_{3,0} & \cdots & \rho_{3,n_1-1} \\ \vdots & & \vdots \\ \rho_{k,0} & \cdots & \rho_{k,n_1-1} \end{pmatrix}. \tag{6.56}
$$

The system P has m common zeros (counting multiplicities) and a gcd of degree m iff $\overline{R}$ has rank deficiency m.

Proof: The $n_1 \times n_1$ matrix R_2 of the upper n_1 rows of $\overline{R}$ is the matrix S_{11}^* of (6.53); its rank deficiency equals deg gcd(p_1, p_2). The rank deficiency of $\overline{R}$ and the degree of gcd(P) cannot be larger than that of R_2. If rk $R_2 = n_1 - m < n_1$, the kernel of R_2 is spanned by the vectors $\mathbf{z}_\mu$, $\mu = 1(1)m$, in the proof of Theorem 6.7. Since $r_\kappa(\zeta_\mu) = p_\kappa(\zeta_\mu)$, the rank deficiency of $\overline{R}$ and the degree of gcd(P) remain at m iff the $\mathbf{z}_\mu$, $\mu = 1(1)m$, also annihilate the lower rows r_κ, $\kappa = 3(1)k$, of $\overline{R}$. If $r_\kappa^T \mathbf{z}_\mu \neq 0$ for some μ and κ, the rank deficiency *and* the degree of the gcd(P) are reduced by 1. $\quad\square$

The structure of the matrix (6.56) suggests:

Definition 6.11. For $k > 2$ polynomials in P, a *generalized Sylvester matrix* $S(p_k, \ldots, p_2, p_1)$ is given by

$$S(p_k, \ldots, p_2, p_1) := \begin{pmatrix} \alpha_{k,0} & \alpha_{k,1} & \cdots & & \alpha_{k,n_k} \\ \cdots & & \cdots & & \\ \alpha_{3,0} & \alpha_{3,1} & \cdots & \alpha_{3,n_3} & 0 \\ & & S(p_2, p_1) & & \end{pmatrix}. \qquad (6.57)$$

If the polynomials are ordered by increasing degrees n_κ and if $n_k \leq n_1 + n_2$, $S(p_1, p_2, \ldots, p_k)$ has $n_1 + n_2 + k - 2$ rows and $n_1 + n_2$ columns. Otherwise, for one of the polynomials in the Sylvester matrix proper, further shifted rows of coefficients must be appended to reach n_k columns. $\square$

It is an immediate consequence of Proposition 6.17 that, for $k > 2$ polynomials in P, the rank deficiency of a generalized Sylvester matrix determines the existence and degree of a common divisor or the existence and number of common zeros, respectively.

Example 6.15:   For disjoint $\zeta_\nu \in \mathbb{C}$, $\nu = 1(1)5$, let p_κ, $\kappa = 1(1)4$, be the polynomials with zeros (ζ_1, ζ_2), $(\zeta_1, \zeta_3, \zeta_4)$, $(\zeta_1, \zeta_4, \zeta_5)$, $(\zeta_1, \zeta_3, \zeta_5)$, respectively. The linear polynomials $r_\kappa := \mathrm{rem}_{p_1} p_\kappa(x) =: \rho_{\kappa,0} + \rho_{\kappa,1} x$, $\kappa = 2, 3, 4$, satisfy $r_\kappa(\zeta_1) = 0$ so that the first column of the 4×2 matrix R is $-\zeta_1$ times the second column and R has rank deficiency 1. This implies a gcd of degree 1, viz. $g(x) = x - \zeta_1$.

The generalized Sylvester matrix $S(p_4, \ldots, p_1)$ has one row each with the coefficients of p_4 and p_3 (and a 0 in the last column), two shifted rows for p_2, and three shifted rows for p_1, which yields a 7×5 matrix; its partial triangularization from right to left gemerates the matrix R. $\square$

Example 6.16: An m-fold zero ζ of $p \in \mathcal{P}_n$ must be a common zero of the system (6.37). For $m \geq 2$, we take $p_1 = p^{(m-1)}$, $p_2 = p^{(m-2)}$, and $p_\mu = p^{(m-\mu)}$ for $\mu = 3(1)m$; then, the matrix $\overline{R}$ of (6.56) has n rows and $n - m + 1$ columns. The generalized Sylvester matrix $S(p, p', \ldots, p^{(m-1)})$ has $2n - m + 1$ rows and $2n - 2m + 3$ columns if $n - 2m + 2 \geq 0$; otherwise there are $n + m - 1$ rows and $n + 1$ columns. $\square$

When the rank deficiency of S_{11}^* or an analogous matrix for more than 2 polynomials is 1, the kernel must be of the form $\mathbf{z}_\mu = (1, \zeta_\mu, \ldots, \zeta_\mu^{n_1-1})^T$; hence the common zero and the gcd $g(x)$ are explicitly displayed by the kernel. With a rank deficiency $d > 1$, the algorithmic determination of the kernel will generally produce *some* basis $\mathbf{v}_1, \ldots, \mathbf{v}_d$ of the kernel from which g and the common zeros have to be determined.

Assume $d < n_1$ and let

$$g(x) = x^d + \sum_{\mu=0}^{d-1} \gamma_\mu x^\mu = \prod_{\mu=1}^{d} (x - \zeta_\mu); \qquad (6.58)$$

multiple zeros are possible. With the appropriate definition of the $\mathbf{z}_\mu$ for a multiple zero, span $(\mathbf{z}_1, \ldots, \mathbf{z}_d)$ and span $(\mathbf{v}_1, \ldots, \mathbf{v}_d)$ are both representations of the kernel, so there must be a nonsingular $d \times d$ matrix W so that

$$\begin{pmatrix} \vdots & & \vdots \\ \mathbf{z}_1 & \cdots & \mathbf{z}_d \\ \vdots & & \vdots \end{pmatrix} = \begin{pmatrix} \vdots & & \vdots \\ \mathbf{v}_1 & \cdots & \mathbf{v}_d \\ \vdots & & \vdots \end{pmatrix} W \, .$$

From (6.58), we have (in the case of disjoint ζ_μ)

$$0 = (\zeta_1^d \, \cdots \, \zeta_m^d) + (\gamma_0 \, \cdots \, \gamma_{d-1}) \begin{pmatrix} 1 & \cdots & 1 \\ \zeta_1 & \cdots & \zeta_d \\ \vdots & & \vdots \\ \zeta_1^{d-1} & \cdots & \zeta_d^{d-1} \end{pmatrix}$$

$$= \left[(v_{1m} \, \cdots \, v_{mm}) + (\gamma_0 \, \cdots \, \gamma_{-1}) \begin{pmatrix} v_{10} & \cdots & v_{m0} \\ \vdots & & \vdots \\ v_{1,m-1} & \cdots & v_{m,m-1} \end{pmatrix} \right] W \, .$$

Thus, the coefficient vector c^T of g is obtained from the regular linear system

$$c^T \cdot V + v^T = 0 \, , \tag{6.59}$$

with V and v^T from the relation above, and the zeros of g are the common zeros of P. The case $d = n_1$ is trivial since it implies $r_\kappa \equiv 0$, $\kappa = 2(1)k$, and $g(x) = p_1(x)$.

Let us conclude this discussion of intrinsic systems of univariate polynomials by a natural observation.

Proposition 6.18. If the system P contains only real polynomials, a potential gcd P must be real.

Proof: Since each p_κ can have complex zeros only in conjugate pairs, potential common complex zeros must also consist of conjugate pairs. $\quad\square$

6.4.2 Empirical Polynomial Systems in One Variable

The fact that the characterization and the determination of gcd's is dominated by the rank of certain matrices has once more displayed the *discontinuous* character of the associated data→result mappings and the ill-posedness of the task for intrinsic polynomials. This changes when we pose the same problem (6.47) for a set $(\overline{P}, E)$ of empirical polynomials:

$$(\overline{P}(x), \, E) = ((\bar{p}_1(x), e_1), \ldots, (\bar{p}_k(x), e_k))^T \, , \tag{6.60}$$

with our customary concept of empirical polynomials; cf. Definition 3.4 and (3.12). Remember that, in the case of *real $\bar{p}_\kappa$*, we have to specify whether the indetermination in the coefficients is restricted to the real domain or not.

Now, we ask for a *common pseudozero* of $(\overline{P}, E) = 0$, i.e. for a value $\tilde{\zeta} \in \mathbb{C}$ such that, for $\delta = O(1)$,

$$\exists \, \tilde{p}_\kappa \in N_\delta(\bar{p}_\kappa, e_\kappa) \quad \text{with} \quad \tilde{p}_\kappa(\tilde{\zeta}) = 0 \, , \quad \kappa = 1(1)k \, . \tag{6.61}$$

With the max-norm $\|..\|^*$ in the definition (3.12) of the tolerance neighborhoods, this is equivalent with the requirement (cf. (5.36))

$$|\bar{p}_\kappa(\tilde{\zeta})| \;\le\; \delta \cdot \sum_{\nu=0}^{n_\kappa} \varepsilon_{\kappa\nu}|\tilde{\zeta}|^\nu \quad \text{for } \kappa = 1(1)k \quad \text{with } \delta = O(1)\,; \tag{6.62}$$

the *backward error* $\delta(\tilde{\zeta})$ of $\tilde{\zeta}$ as a common pseudozero of $(\overline{P}, E)$ is thus defined by

$$\delta(\tilde{\zeta}) \;:=\; \max_\kappa \; \frac{|\bar{p}_\kappa(\tilde{\zeta})|}{\sum_{\nu=0}^{n_\kappa} \varepsilon_{\kappa\nu}|\tilde{\zeta}|^\nu} \,. \tag{6.63}$$

Note that this holds only if the $(\bar{p}_\kappa, e_\kappa)$ have no common empirical data so that the backward errors of $\tilde{\zeta}$ as a pseudozero of the individual polynomials are unrelated.

Through (6.63), we may assign a backward error $\delta(\zeta)$ to *any* $\zeta \in \mathbb{C}$ and the question for the existence of a common pseudozero—or a valid approximate common zero—is no longer of a qualitative but of a *quantitative* nature.

Proposition 6.19. The empirical polynomial system (6.60) has valid common zeros iff

$$\delta(\overline{P}, E) \;:=\; \min_{\zeta \in \mathbb{C}} \delta(\zeta) \;=\; \min_{\zeta \in \mathbb{C}} \max_\kappa \; \frac{|\bar{p}_\kappa(\zeta)|}{\sum_{\nu=0}^{n_\kappa} \varepsilon_{\kappa\nu}|\zeta|^\nu} \;=\; O(1)\,. \tag{6.64}$$

The empirical system has no valid common zeros iff $\delta(\overline{P}, E) > O(1)$.

In terms of predicates (cf. section 6.1.1), the first statement is about the validity of the critical predicate $\Pi_{cz}(P) :=$ "P has common zeros" while the second statement is about the validity of the noncritical predicate $\neg\,\Pi_{cz}(P) =$ "P has no common zeros." In the data space $\Delta\mathcal{A}$ of the system $\overline{P}$, the truth domain $\Delta Q(\Pi_{cz})$ of Π_{cz} is at the same time the boundary set $\partial \Delta Q(\neg\Pi_{cz})$ of the truth domain $\Delta Q(\neg\Pi_{cz})$ of $\neg\Pi_{cz}$, which explains why the same quantity $\delta(\overline{P}, E)$ determines the validity of either predicate.

When we assume that the tolerances E in $(\overline{P}, E)$ have been specified such that our validity scale (3.3) applies, and when we consider our usage of the symbol $O(1)$, we find that there is a "gray zone" between the empirical systems for which Π_{cz} is valid and those for which $\neg\Pi_{cz}$ is valid.

For $\delta(\overline{P}, E) \overset{<}{\approx} 3$ (say), we will assert that Π_{cz} is valid for $(\overline{P}, E)$, and for $\delta(\overline{P}, E) \overset{>}{\approx} 10$ (say), we will assert that $\neg\Pi_{cz}$ is valid. But for empirical systems with a δ value in between, we must admit that the case is open. This reflects the unavoidable degree of arbitrariness in the specification of tolerances for most empirical quantities; the above margin (3–10) may even be too small in many applications. This has to be kept in mind in the following.

Candidates for approximate common zeros may be found by (visual or algorithmic) inspection of the zero sets of the individual $\bar{p}_\kappa$: A potential common δ-pseudozero ζ of (6.60) can only lie in the intersection of the δ-pseudozero sets of the individual $(\bar{p}_\kappa, e_\kappa)$.

Proposition 6.20. In the situation under discussion, for $\delta > 0$,

$$\delta(\overline{P}, E) \;\le\; \delta \qquad \Leftrightarrow \qquad \cap_{\kappa=1}^{k} Z_\delta(\bar{p}_\kappa, e_\kappa) \;\neq\; \emptyset\,. \tag{6.65}$$

Proof: By (5.36), each ζ in the intersection satisfies $\delta(\zeta) \le \delta$. $\square$

With increasing δ, the pseudozero set Z_δ of an empirical polynomial $(\bar{p}_\kappa, e_\kappa)$ expands and covers larger and larger portions of $\mathbb{C}$ so that the δ-pseudozero sets of the $(\bar{p}_\kappa, e_\kappa)$ must have a nonempty intersection for sufficiently large δ. This establishes the finiteness of $\delta(\overline{P}, E)$ for any empirical system of univariate polynomials.

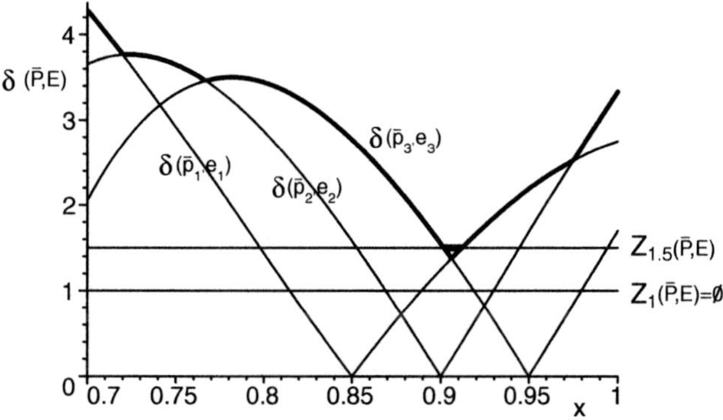

Figure 6.4.

Heuristically, if one zero of *each* $\bar{p}_\kappa$ is located very close to a point $\tilde{\zeta} \in \mathbb{C}$, this value is likely to represent a common pseudozero of $(\overline{P}, E)$; e.g., the arithmetic mean of those zeros may be chosen as a candidate for $\tilde{\zeta}$ to be tested by (6.63). We may also take the zero set $Z_{0,1}$ of one of the $\bar{p}_\kappa$ (say $\bar{p}_1$) and evaluate (6.63) for $\zeta \in Z_{0,1}$. If $\delta(\zeta) = O(1)$ for one or several zeros of $\bar{p}_1$, we have found common pseudozeros which we may further refine as will be discussed in section 6.4.4. Also, the use of well-known localization theorems for the zeros of univariate polynomials may restrict the search for a common pseudozero considerably; in particular, bounds b_κ for the moduli of the zeros of the $\bar{p}_\kappa$ generate a bound $b = \min_\kappa b_\kappa$ for $|\zeta|$ beyond which there cannot exist a common pseudozero.

From the divisor point of view (cf. section 6.2), a common pseudodivisor or valid approximate common divisor of $(\overline{P}, E)$ is a polynomial $\tilde{g}$ of positive degree d such that, for δ of $O(1)$,

$$\exists\, \tilde{p}_\kappa \in N_\delta(\bar{p}_\kappa, e_\kappa) \quad \text{with } \tilde{g} \mid \tilde{p}_\kappa, \quad \kappa = 1(1)k. \tag{6.66}$$

With a linear $\tilde{g}(x) = x - \tilde{\zeta}$, this requirement is clearly equivalent to (6.61). Thus, (6.64) in Proposition 6.19 is just as well a criterion for the existence of valid approximate divisors.

For $d > 1$, the d zeros ζ_μ, $\mu = 1(1)m$, of a valid approximate common divisor $g \in \mathcal{P}$ of $(\overline{P}, E)$ are not only valid approximate common zeros individually, but they form a set of *simultaneous* common pseudozeros of $(\overline{P}, E)$ (cf. section 5.2.1): There exists a set of $\tilde{p}_\kappa \in N_\delta(\bar{p}_\kappa, e_\kappa)$, $\kappa = 1(1)k$, such that

$$\tilde{p}_\kappa(\zeta_\mu) = 0, \quad \mu = 1(1)d, \quad \kappa = 1(1)k.$$

Vice versa, a set of d pseudozeros $\tilde{\zeta}_\mu$ of $(\overline{P}, E)$ is equivalent to a common pseudodivisor of degree d *only* if the $\tilde{\zeta}_\mu$ are *simultaneous pseudozeros* for each $(\bar{p}_\kappa, e_\kappa)$. An arbitrary combination of $d > 1$ pseudozeros of $(\overline{P}, E)$ will, generally, not generate a common pseudodivisor of $(\overline{P}, E)$. This becomes apparent when we relate the existence of common pseudozeros to the pseudozero sets of the individual $(\bar{p}_\kappa, e_\kappa)$.

Consider the case of two real empirical polynomials $(\bar{p}_1, e_1)$, $(\bar{p}_2, e_2)$, with their indetermination restricted to real deviations in the coefficients. Assume that, for the δ under consideration, there are no cluster domains for either polynomial, i.e. for each of the two polynomials each pseudozero domain contains only one zero of a $\tilde{p}_i \in N_\delta(\bar{p}_i, e_i)$. A common pseudozero must lie in the nonempty intersection of the pseudozero sets $Z_\delta^{(1)}$ and $Z_\delta^{(2)}$. Now assume that the real interval $Z_{\delta,1}^{(1)} \subset Z_\delta^{(1)}$ overlaps *on each of its ends* with an interval $Z_{\delta,i}^{(2)} \subset Z_\delta^{(2)}$, $i = 1, 2$, and that there are no other intersections of the pseudozero sets of the two polynomials; cf. Figure 6.4. In spite of the fact that there are *two* disjoint intervals of common pseudozeros, it would be erroneous to expect a common pseudodivisor $\tilde{g}$ of degree 2: The two zeros of $\tilde{g}$ would have to be zeros of the *same* $\tilde{p}_1 \in N_\delta(\bar{p}_1, e_1)$, but each such $\tilde{p}_1$ can have only one zero in $Z_{\delta,1}^{(1)}$.

A characterization of a set of $d > 1$ simultaneous common pseudozeros by means of the pseudozero sets $Z_\delta(\bar{p}_\kappa, e_\kappa)$ is not really feasible: Even for *one* individual $(\bar{p}_\kappa, e_\kappa)$, the selection of a particular ζ in one component of Z_δ restricts the choice of simultaneous zeros in other components of Z_δ a great deal; cf. section 5.2.2. An analysis of the interaction of these restrictions for several polynomials appears unmanageable. Naturally, cluster domains in one or several of the $(\bar{p}_\kappa, e_\kappa)$ would add further complications.

6.4.3 Algorithmic Determination of Approximate Common Divisors

The standard exact algorithm for the determination of common divisors of two intrinsic univariate polynomials is the Euclidean Algorithm of section 6.4.1, with its remarkably low number of arithmetic operations but with its well-known potential for numerical instability. The instability of this algorithm must be expected since it is equivalent to Gaussian elimination with a *fixed elimination sequence* in the Sylvester matrix of the two polynomials; cf. Exercise 6.4-1. The adaptation of the Euclidean Algorithm to approximate (= floating-point) computation has posed a challenge for some time; an excellent analysis of the situation and design of a stable algorithm is due to Beckermann/Labahn ([6.5]), with references to related work.

For two or more empirical polynomials, there appear two major approaches to an algorithmic determination of common pseudozeros or pseudodivisors, resp.:

(i) Determination of ζ with near-minimal $\delta(\zeta)$, cf. (6.63),

(ii) Determination of a near-kernel of $S(\bar{p}_k, \ldots, \bar{p}_1)$, cf. (6.57).

Some heuristic ideas for an approach of type (i) have been briefly discussed in the previous section. Let us now consider an algorithmic treatment of (6.64): Since the minimization extends only over the one (real or complex) variable ζ and rough bounds on domains for minimizing ζ are easily established (cf. section 6.4.2), this appears as a feasible task. However, the objective function is strongly nonlinear and the occurrence of the max function and of moduli excludes the use of analytic means in a *global* search for minima; cf. Figure 6.4. The use of (6.64) for a *local* refinement is discussed in the next section.

This changes when we replace the max-norm by the Euclidean norm in the definition (3.12) of the tolerance neighborhoods as well as in (6.63). With the expression (3.54) of the Euclidean backward error $\delta_E(\tilde{\zeta})$ of an approximate zero $\tilde{\zeta}$ of *one* empirical polynomial $(\bar{p}, e)$, the backward error of an approximate common zero ζ of the empirical system $(\overline{P}, E)$ becomes (with ..* for the conjugate complex)

$$\delta_E(\zeta) := \left[\sum_\kappa \frac{p_\kappa(\zeta)^* \, p_\kappa(\zeta)}{\sum_{\nu=0}^{n_\kappa} \varepsilon_{\kappa\nu}^2 (\zeta^*\zeta)^\nu} \right]^{\frac{1}{2}}. \tag{6.67}$$

Except possibly at $\zeta = 0$, $\delta_E(\zeta)^2$ is a differentiable real function $\delta_{2E}(\xi, \eta)$ of the two variables $\xi := \mathrm{Re}\,\zeta$ and $\eta := \mathrm{Im}\,\zeta$, whose stationary points are characterized by

$$\mathrm{grad}\ \delta_{2E}(\xi, \eta) = 0. \tag{6.68}$$

The numerators of the two components of (6.68) constitute a system of two bivariate polynomial equations of maximal degree $4 \sum_\kappa n_\kappa - 1$.

An approximate numerical solution of this system is well feasible when the n_κ are not large; cf. Chapter 8. However, the zeros of (6.68) not only include all the *relative* minima of $\delta_{2E}(\xi, \eta)$ but also all relative maxima and potential stationary points of other kinds. Therefore, $\delta_{2E}(\xi, \eta)$ or $\delta(\xi + i\eta)$ must be evaluated at all solutions of (6.68), with the exception of those excluded by other considerations. In the end, it may turn out that the backward error is too large at *all* zeros of (6.68). This necessity of performing the complete solution algorithm for the polynomial system before the nonexistence of a common pseudozero/pseudodivisor can be discovered, is the greatest drawback of this approach which has been proposed by Karmarkar/Lakshman.

Let us now turn to the approach (ii). For notational convenience, we write $\bar{S}$ for $S(\bar{p}_2, \bar{p}_1)$ or the generalized Sylvester matrix $S(\bar{p}_k, \ldots, \bar{p}_1)$ of (6.57), respectively.

We begin by a way to establish that $\neg \Pi_{cz}$ is valid for the empirical system $(\overline{P}, E)$, i.e. that $\delta(\overline{P}, E) > O(1)$; cf. (6.64). The following is a well-known result in numerical linear algebra (here $\|..\|_2$ is the Euclidean operator norm).

Proposition 6.21. Consider a regular matrix $A \in \mathbb{C}^{n \times n}$ and denote its singular values by σ_ν, $\nu = 1(1)n$, ordered in decreasing size. Then

$$A + \Delta A \quad \text{is nonsingular for any } \Delta A \text{ with } \|\Delta A\|_2 \prec \sigma_n$$

while there exists a matrix ΔA with Euclidean norm σ_n which renders $A + \Delta A$ singular. Also,

$$A + \Delta A \quad \text{has a rank } \geq r \text{ for any } \Delta A \text{ with } \|\Delta A\|_2 < \sigma_r$$

while there exists a matrix ΔA with Euclidean norm σ_r such that $\mathrm{rk}(A + \Delta A) < r$. Furthermore,

$$\frac{1}{\sqrt{n}} \|A\|_2 \leq \|A\|_1 \leq \sqrt{n} \|A\|_2. \tag{6.69}$$

Now we denote by E_S the "tolerance Sylvester matrix" of $(\overline{P}, E)$), i.e. the matrix obtained when each coefficient $\bar{\alpha}_{\kappa,\nu}$, $\kappa = 1(1)k$, in $\bar{S}$, with n columns, is replaced by its tolerance $\varepsilon_{\kappa,\nu}$. Either one computes $\|E_S\|_2$ by an s.v.d. or one may use

Proposition 6.22. If $\|E_S\|_1 < \frac{1}{\sqrt{n}} \sigma_n(\bar{S})$, then $\bar{S}$ is nonsingular for any selection of $\tilde{p}_\kappa \in N_1(\bar{p}_\kappa, e_\kappa)$, $\kappa = 1(1)k$. Analogously, if $\|E_S\|_1 < \frac{1}{\sqrt{n}} \sigma_{n-d}(\bar{S})$, then $\bar{S}$ has at most rank deficiency d for any set $\{\tilde{p}_\kappa\}$, $\tilde{p}_\kappa \in N_1(\bar{p}_\kappa, e_\kappa)$.

Proof: With (6.69), the assumption implies $\|E_S\|_2 < \sigma_n(\bar{S})$ or $< \sigma_{n-d}(\bar{S})$, respectively. $\square$

But the condition of Proposition 6.22 on the tolerance matrix E_S is far from being sharp: At first, the bounds in (6.69) are only attained for matrices of a very special structure. Furthermore, the matrix ΔA of Proposition 6.21 also has a very special structure which would require, e.g., a perturbation of the intrinsic elements zero in $\bar{S}$. For a Sylvester matrix, with at most $\sum_{\kappa=1}^{k}(n_\kappa+1)$ independent perturbations, the nearest singular *Sylvester matrix* is, generally, much farther away than indicated by its smallest singular value. Therefore, we may safely assume that

$$\delta(\overline{P}, E) \gg \frac{1}{\|E_S\|_1\sqrt{n}}\,\sigma_n(\bar{S}) \quad \text{or} \quad \frac{1}{\|E_S\|_2}\,\sigma_n(\bar{S}) \tag{6.70}$$

so that the hypotheses in Propositions 6.21 and 6.22 are really sufficient to confirm the validity of the assertion "the $(\bar{p}_\kappa, e_\kappa)$ in $(\overline{P}, E)$ have no valid common zero/divisor."

Analogously, if $\sqrt{n}\,\|E_S\|_1$ or $\|E_S\|_2$ is between σ_{n-d+1} and the next larger singular value, we may take this as a confirmation that "the $(\bar{p}_\kappa, e_\kappa)$ have at most d valid simultaneous common zeros" or "a valid common divisor of at most degree d," respectively. But this fact and the fact that we cannot claim that a valid common divisor has degree at most $d - 1$ under the above assumption does *not* imply that there actually exist valid common divisors of degree d. The "gray zone" which always exists between the validity of the positive and the negative assertion is widened in this case because Proposition 6.21 is not sharp for Sylvester matrices. Thus, the *existence* of common pseudozeros/pseudodivisors must actually be established constructively. For a chosen $d > 0$, this requires the determination of a *candidate* common divisor $\tilde{g} \in P_d$, the checking of its validity for $(\overline{P}, E)$, and—possibly—its refinement.

Assume that, from a comparison of $\|E_S\|_2$ with the singular values $\bar{\sigma}_\nu$ of the Sylvester matrix $\bar{S} := S(\bar{p}_2, \bar{p}_1)$, we know the maximal degree $d_{\max} > 0$ of a potential valid common divisor, i.e. the dimension of a near-kernel of $\bar{S}$. From the singular value decomposition

$$\bar{S}\,(V_1 \mid V_0) = (U_1 \mid U_0)\,\mathrm{diag}\,(\sigma_1..\sigma_{n-d} \mid \sigma_{n-d+1}..\sigma_n)\,, \tag{6.71}$$

we know that a near-kernel of $\bar{S}$ is spanned by the d columns of V_0. At the same time, if there exists a degree d common pseudodivisor $\tilde{g} = \sum_{\mu=0}^{d}\tilde{\gamma}_\mu x^\mu$, a near-kernel of $\bar{S}$ is spanned by the d vectors $\mathbf{x}(\zeta_\mu) := (1, \zeta_\mu, \ldots, \zeta_\mu^{n-1})^T$ of the d zeros ζ_μ of $\tilde{g}$, with

$$\Gamma\,\mathbf{x}(\mathbf{z}) := \begin{pmatrix} \tilde{\gamma}_0 & \cdots & \tilde{\gamma}_d & & 0 \\ & \ddots & & \ddots & \\ 0 & & \tilde{\gamma}_0 & \cdots & \tilde{\gamma}_d \end{pmatrix} \begin{pmatrix} 1 & \cdots & 1 \\ \zeta_1 & & \zeta_d \\ \vdots & & \vdots \\ \zeta_1^{n-1} & \cdots & \zeta_d^{n-1} \end{pmatrix} = 0;$$

cf. the proof of Theorem 6.7. The two bases of near-kernels must be related by a regular $d \times d$ matrix W such that $\mathbf{x}(\mathbf{z}) \approx V_0\,W$, which implies

$$\Gamma\,V_0 = \begin{pmatrix} \tilde{\gamma}_0 & \cdots & \tilde{\gamma}_d & & 0 \\ & \ddots & & \ddots & \\ 0 & & \tilde{\gamma}_0 & \cdots & \tilde{\gamma}_d \end{pmatrix} \begin{pmatrix} \vdots & & \vdots \\ v_1^{(0)} & \cdots & v_d^{(0)} \\ \vdots & & \vdots \end{pmatrix} \approx 0; \tag{6.72}$$

cf. (6.59). Without loss of generality, we assume $\tilde{\gamma}_d = 1$; then each of the $n - d$ rows of the product in (6.72) yields a linear system for the d coefficients $\tilde{\gamma}_\mu$, $\mu = 0(1)d - 1$, of the candidate common pseudodivisor $\tilde{g}$.

We can either solve one of these systems and check that this set of $\tilde{\gamma}_\mu$ leaves small residuals in the other ones, or, preferably, we determine the $\tilde{\gamma}_\mu$ such that their residuals over the $n - d$ systems are minimal:

$$\min_{\tilde{\gamma}_\mu} \quad \max_{\mu=1(1)d,\lambda=1(1)n-d} \left| \sum_{\nu=0}^{d-1} \tilde{\gamma}_\nu \, v_{\mu,\lambda+\nu} \right| .$$

If the candidate approximate common divisor thus obtained from the singular value decomposition of $\bar{S}$ turns out not to be valid and if it cannot be refined into a valid common pseudodivisor of degree d (see the next section), we may decrease d by one and repeat the whole procedure.

Since the computational effort for a singular value decomposition is negligible for the polynomial degrees commonly met in scientific computing, the above procedure with a potential successive refinement appears to be a natural approach to the determination of a valid common divisor of a set of empirical univariate polynomials. The appropriate *maximal degree* can be judged by Proposition 6.22. Often, there will be a jump in the singular values which clearly indicates the pseudorank of the Sylvester matrix. If the resulting value of $\delta(\overline{P}, E)$ is too large, the next lower d may be tried.

Naturally, one can also use a *pivoted triangularization* of $\bar{S}$, from right to left and with row exchanges to put the pivots on the diagonal. This either may proceed to the end without the appearance of tiny pivots which indicates that there is no valid common divisor, or there may be a clear jump in the size of the available pivots, with d columns remaining on the left-hand side. In this case, if the matrix in the left upper corner consists only of small elements, we may take the polynomial $g(x)$ defined by the coefficients in the row of the last pivot as a candidate for a common divisor, check it, and refine it if necessary as discussed further below.

The difficulty lies in the judgment of the size to be requested for a "valid pivot." Even with pivoting, the magnitude of the elements in the partially triangularized matrix $\bar{S}$ may change strongly so that a reference to the original tolerances is no longer possible. Therefore, the use of orthogonal matrices in the triangularization of $\bar{S}$, i.e. of Householder or Givens rotations, has been suggested. This keeps the Euclidean norm of the generated lower-triangular matrix equal to the Euclidean norm of $\bar{S}$ and also stabilizes the floating-point computation; cf. any text on numerical linear algebra. For orthogonal triangularization, with column exchanges for column pivots with maximal Euclidean norm, it is also known that the generated elements ρ_{ii} along the diagonal of the triangular factor are ordered in size and satisfy $\rho_{\nu\nu} \geq \sigma_\nu$; thus the information from their size can be used in an analogous manner as that from the singular values; cf. Proposition 6.22. Generally, if there is a gap in the σ_ν between $O(\varepsilon)$ and $O(1)$, a similar gap, with an equal number of $O(\varepsilon)$ values, will appear in the $\rho_{\nu\nu}$. In particular, if the largest remaining column has a Euclidean norm well below $\|E_S\|_2$, a continuation of the triangularization is no longer meaningful.

Proposition 6.23. In a triangularization of $\bar{S}$, if the upper d rows of the triangular factor vanish then the elements of the $(d + 1)$st row are the coefficients of the common divisor g.

Proof: By Theorem 6.8, the hypothesis implies that there exists a gcd of degree d. By (6.52), the coefficient vector of g is in the row space of the triangular factor of $\bar{S}$ and there exists no row representing a lower-degree polynomial. $\square$

In our case, the elements of the upper d rows are so small that they cannot reasonably be used as pivots by the preceeding analysis. This suggests that the elements in the $d + 1$st row yield the coefficients of a candidate for a valid common pseudodivisor.

6.4.4 Refinement of Approximate Common Zeros and Divisors

We consider a system $(\overline{P}, E)$ of $k \geq 2$ empirical polynomials $(\bar{p}_\kappa, e_\kappa)$ of degrees n_κ, and a candidate set of d approximate *common zeros* ζ_μ, $\mu = 1(1)d$; cf. section 6.4.2. Remember that this set must represent a set of *simultaneous* pseudozeros for each $(\bar{p}_\kappa, e_\kappa)$ if it is to correspond to a common pseudodivisor of degree d for $(\overline{P}, E)$. The determination of the associated backward error (6.63) follows standard procedures. Now assume that we find this backward error to be moderately too large (whatever this may mean in a given case) and that we want to improve the set.

For each empirical polynomial $(\bar{p}_\kappa, e_\kappa)$ in $(\overline{P}, E)$, we have a (shifted) coefficient space $\Delta\mathcal{A}_\kappa$ whose components are the variations $\Delta\alpha_{\kappa\nu}$ of the empirical coefficients $(\bar{\alpha}_{\kappa\nu}, \varepsilon_{\kappa\nu})$; furthermore, we need the joint data space $\Delta\mathcal{A} = \oplus_\kappa \Delta\mathcal{A}_\kappa$. Note that the empirical polynomials may have some intrinsic coefficients which, by agreement, do not figure as components in the $\Delta\mathcal{A}_\kappa$. We search for nearby $\tilde{p}_\kappa = \bar{p}_\kappa + \Delta p_\kappa$, $\kappa = 1(1)k$, and $\zeta_\mu = \tilde{\zeta}_\mu + \Delta\zeta_\mu$, $\mu = 1(1)d$, such that

$$\tilde{p}_\kappa(\zeta_\mu) = (\bar{p}_\kappa + \Delta p_\kappa)(\tilde{\zeta}_\mu + \Delta\zeta_\mu) = 0, \quad \kappa = 1(1)k, \ \mu = 1(1)d. \tag{6.73}$$

Since we expect small variations Δp_κ, $\Delta\zeta_\mu$, we linearize (6.73) into

$$\Delta r_{\kappa\mu} := \bar{p}_\kappa(\tilde{\zeta}_\mu) + \bar{p}'_\kappa(\tilde{\zeta}_\mu)\,\Delta\zeta_\mu + \Delta p_\kappa(\tilde{\zeta}_\mu) = 0, \quad \kappa = 1(1)k, \ \mu = 1(1)d. \tag{6.74}$$

This is a set of $k\,d$ linear equations in the $\leq \sum(n_\kappa + 1)$ variables $\Delta\alpha_{\kappa\nu}$ of $\Delta\mathcal{A}$ and the d free parameters $\Delta\zeta_\mu$. From each set of k equations $\Delta r_{\kappa\mu} = 0$, $\kappa = 1(1)k$, for a fixed μ, we may eliminate the only parameter $\Delta\zeta_\mu$; thus we obtain a total of $(k-1)\,d$ linear equations in the $\Delta\alpha_{\kappa\nu}$ which represent a linear manifold of codimension $(k-1)\,d$ in $\Delta\mathcal{A}$ whose shortest norm distance from the origin we can determine. Naturally, there must be at least $(k-1)\,d$ empirical coefficients and no fatal degeneracies.

The location on the manifold of the shortest norm distance tells us the minimizing $\Delta\alpha_{\kappa\nu}$ and, via (6.74), the associated $\Delta\zeta_\mu$. We can now evaluate the backward errors δ_κ of the $\tilde{\zeta}_\mu + \Delta\zeta_\mu$ as a simultaneous pseudozero set of each $(\bar{p}_\kappa.e_\kappa)$. If all δ_κ are O(1), we have verified the $\zeta_\mu = \tilde{\zeta}_\mu + \Delta\zeta_\mu$ as common pseudozeros of $(\overline{P}, E)$; otherwise, we might try another correction from the refined values. But, generally, in an overdetermined problem this simply means that no pseudosolution exists or that it cannot be reached from our candidate approximation: Since (6.73) with $\Delta\zeta_\mu = 0$ has been the basis for the backward error assessment of the candidate zeros $\tilde{\zeta}_\mu$, it is only through the shifts $\Delta\zeta_\mu$ that we have gained more freedom to decrease the backward error. This may well not be sufficient to overcome an original backward error $> $ O(1), but it does not exclude that a valid set of d simultaneous common pseudozeros may be reached from a different candidate set; cf. the end of this section.

Previously, we had remarked that it is difficult to use (6.64) for a global analysis of (6.61). *Locally*, the denominators in (6.64) may be considered as constant; then the minimization of the $|\Delta r_{\kappa\mu}|$ of (6.74) finds that intersection of the various tangents at each ζ_μ which has the largest modulus.

When we consider the refinement procedure from the *common divisor* point of view, we have a monic polynomial $\tilde{g} \in \mathcal{P}_d$, $d \leq \min_\kappa n_\kappa$, which we consider as an approximate common divisor of the $(\bar{p}_\kappa, e_\kappa)$. We have checked the validity of $\tilde{g}$ as a divisor individually for each $(\bar{p}_\kappa, e_\kappa)$ via its backward errors:

$$\delta_\kappa(\tilde{g}) \; := \; \min_{q_\kappa \in \mathbf{P}_{n_\kappa - d}} \| \, q_\kappa \, \tilde{g} - \bar{p}_\kappa \, \|^*_{e_\kappa} \; ; \tag{6.75}$$

cf. (6.19) in section 6.2.1. It has turned out that $\tilde{g}$ is not valid by a moderate margin, so we wish to refine it.

The refinement of a divisor of one empirical polynomial has been considered in section 6.2.3, but now we must take all $(\bar{p}_\kappa, e_\kappa)$ into account simultaneously. Assume that we have found that

$$\bar{p}_\kappa(x) \; = \; q_\kappa(x) \cdot \tilde{g}(x) + r_\kappa \, , \quad \kappa = 1(1)k \, ; \tag{6.76}$$

as in section 6.2.3 (cf. (6.24) and (6.25)), we look for corrections Δg and Δq_κ, $\kappa = 1(1)k$, such that $\bar{p}_\kappa + \Delta p_\kappa = (q_\kappa + \Delta q_\kappa)(\tilde{g} + \Delta g)$. For this purpose, we determine the polynomials $\Delta q_\kappa \in \mathcal{P}_{n_\kappa - d}$ and $\Delta g \in \mathcal{P}_{d-1}$ such that the linearized deviations

$$\Delta p_\kappa \; = \; q_\kappa \, \Delta g + \tilde{g} \, \Delta q_\kappa - r_\kappa \, \in \, \mathcal{P}_{n_\kappa} \tag{6.77}$$

become minimal in the norms of the spaces $\Delta \mathcal{A}_\kappa$:

$$\delta(\tilde{g} + \Delta g) \; \approx \; \min_{\Delta g \in \mathbf{P}_{d-1}, \Delta q_\kappa \in \mathbf{P}_{n_\kappa - d}} \; \max_\kappa \| \, \Delta p_\kappa \, \|^*_{e_\kappa} \, . \tag{6.78}$$

Since the Δp_κ are linear in Δg and the Δq_κ, this is a standard linear minimization problem. If (6.78) yields a value of $O(1)$, this may be expected also to hold after the omitted quadratic terms $\Delta g \, \Delta q_\kappa$ have been added to the Δp_κ. On the other hand, if the original $\max_\kappa \delta_\kappa(\tilde{g})$ is only marginally diminished by the linearized refinement, further refinements will generally not help. Again, this does not exclude the existence of a valid pseudodivisor of degree d with quite different coefficients.

From a more abstract point of view, we deal with the empirical data space $\Delta \mathcal{A}$ of the *overdetermined* system (6.60), with its origin at the data of the system $\overline{P}$. The data of systems of the same structure which possess an *exact* gcd of a specified degree $d \geq 1$ lie on a nonlinear manifold $\mathcal{S}_d \subset \Delta \mathcal{A}$, and we want to find the approximate shortest $\|..\|^*_E$-norm distance of $\mathcal{S}_d$ from the origin. Our problem has a solution iff the manifold $\mathcal{S}_d$ intersects with neighborhoods $N_\delta(\overline{P}, E)$ for $\delta = O(1)$. By (6.76), the system $\widehat{P} := \{\bar{p}_\kappa - r_\kappa\}$ lies on $\mathcal{S}_d$ and (6.77) represents the *tangential linear manifold* $\partial \mathcal{S}_d \subset \Delta \mathcal{A}$ of $\mathcal{S}_d$ at $\widehat{P}$. The minimization (6.78) yields the shortest norm distance of that linear manifold from the origin. If this distance is large relative to $O(1)$, the shortest distance of $\mathcal{S}_d$ from the origin will, in the vicinity of $\widehat{P}$, also be too large. This does not exclude the possibility that $\mathcal{S}_d$ attains a shorter distance to the origin of $\Delta \mathcal{A}$ in some other part.

A reduction of d means that we move to a different manifold $\mathcal{S}_{d'}$, $d' < d$, which necessarily has a smaller minimal distance from the origin; thus, there is a greater chance that it will intersect with neighborhoods $N_\delta(\overline{P}, E)$ for $\delta = O(1)$.

6.4.5 Example

We consider the two polynomials $(\bar{p}_1(x), \bar{p}_2(x)) =$

$$(7.5225 + 4.5022 \, x - 4.7449 x^2 - 9.6733 \, x^3 - 2.4490 \, x^4 + .9036 \, x^5 + x^6 ,$$

$$-5.5752 + 3.4077 \, x + 9.1337 \, x^2 + 11.3156 \, x^3 - .4096 \, x^4 - 2.3048 \, x^5 - 2.5172 \, x^6 + x^7)$$

and we assume a tolerance of $.5 \cdot 10^{-4}$ for each coefficient (except the leading 1's), with the indetermination restricted to real variations.

At first, we regard the *overdetermined system* (6.60) composed of the two empirical univariate polynomials $(\bar{p}_1, e_1)$ and $(\bar{p}_2, e_2)$. The individual zeros of the $\bar{p}_i$ are

$$\bar{p}_1 : \quad -1.180783 \pm 1.278785\,\mathrm{i}, \ -.841474 \pm .720516\,\mathrm{i}, \ .904857, \ 2.236055,$$

$$\bar{p}_2 : \quad -.841473 \pm .720512\,\mathrm{i}, \ -.589964 \pm 1.073453\,\mathrm{i}, \ .515089, \ 2.236095, \ 2.628891.$$

This lets us expect that there are common pseudozeros at ≈ 2.23607 and $\approx -.84147 \pm .72051\,\mathrm{i}$, resp., which is readily confirmed by evaluating their backward errors with respect to each of the two polynomials. Remember that—for a conjugate complex pair of zeros and real variations of a real polynomial—the pair has to be tested as a simultaneous pair of zeros; cf. section 5.2.1. Thus we also know that there is a valid linear common divisor $x - 2.23607$ and a valid quadratic common divisor $x^2 + 1.68294\,x + 1.22721$.

When we want to decide whether the three zeros can be simultaneous common pseudozeros, or whether the cubic polynomial with these 3 zeros is a valid common divisor, we have to compute the minimal norm distances from the origins of the $\Delta \mathcal{A}_\kappa$ of the linear manifolds $\mathcal{M}_\kappa$, $\kappa = 1, 2$, of codimension 3 which contain the coefficients of the neighboring polynomials with the above 3 zeros as exact zeros. The resulting backward errors are $\approx .63$ for $(\bar{p}_1, e_1)$ and $\approx .24$ for $(\bar{p}_2, e_2)$. This confirms the validity of the three simultaneous common pseudozeros and, equivalently, the existence of valid third degree common pseudodivisors, e.g.,

$$g(x) \ = \ (x - 2.23607)\,(x^2 + 1.68294\,x + 1.22721) \ \approx \ x^3 - .55313\,x^2 - 2.53597\,x - 2.74412.$$

At the same time, it is clear from the separated location of all other zeros that $d = 3$ is the highest feasible degree of a valid common pseudodivisor.

Now we approach the same problem from the *divisor* point of view, without information about the zeros. We form the 13×13 Sylvester matrix $\bar{S} := S(\bar{p}_2, \bar{p}_1)$ (cf. the preceeding section) and compute its singular values decomposition which yields the singular values (rounded)

$$31.52, \ 30.49, \ \ldots, \ 1.94, \ 1.41, \ .000012, \ .000008, \ .000002.$$

Since $\sqrt{13}\,\|E_S\|_1 \approx .002$, there is a clear indication of a valid divisor of degree 3; cf. Proposition 6.22. From the 3 singular vectors associated with the 3 tiny singular values, we obtain 10 versions of the system (6.72); a comparison of the 10 slightly differing coefficient sets $\tilde{\gamma}_\mu$ barely yields safe 3rd decimal digits. Therefore, we minimize the residuals over the 10 systems and obtain (rounded)

$$g(x) \ = \ x^3 - .55312\,x^2 - 2.53594\,x - 2.74410,$$

with a backward error $\delta(g) = \max_\kappa \delta_\kappa(g) \approx .20$; cf. (6.75). The zeros of g are ≈ 2.23606 and $-.84147 \pm .72051\,\mathrm{i}$.

With a pivoted triangularization of $\bar{S}$ by Gaussian elimination (from right to left), there occurs no increase in the size of the elements; the largest element of an intermediate matrix is about 16. When we begin with the pivot row 13, the further pivot sequence is 6,5,4,12,11,3,2,1,7 (in terms of the original row nos.), and all pivots have a modulus ≥ 1. After 10 elimination steps, the 3 leftmost elements in the remaining 3 rows 8,9,10 are (rounded)

$$\begin{pmatrix} -.00326 & -.00444 & -.00269 \\ -.00063 & -.00078 & -.00044 \\ .00819 & .01117 & .00662 \end{pmatrix},$$

while the previous pivot was 1.76620. This jump in potential pivot size and the uniform smallness of the remaining elements is a strong indication of $d = 3$. The 4 nonzero elements of the last pivot row yield (rounded) $\tilde{g}(x) = 4.8483 + 4.4812\,x + .9782\,x^2 - 1.7662\,x^3$ as candidate for a degree 3 common divisor.

When we test the normalized version of $\tilde{g}$, we obtain $\delta(\tilde{g}) \approx 27$, which is moderately too large. Therefore, we perform our refinement procedure (6.77)/(6.78). This generates a refined common divisor (rounded)

$$g(x) = x^3 - .55311\,x^2 - 2.53595\,x - 2.74413,$$

with a backward error $\delta(g) \approx .22$, so that we have found a valid degree 3 common divisor of $(\bar{p}_1, e_1)$, $(\bar{p}_2, e_2)$; its zeros agree with those of the divisor above within the 5 decimal digits shown.

A comparison of the three valid common divisors which we have found shows that they agree within two units of 10^{-5}. Thus it appears meaningful to specify 5 decimal digits of their coefficients (after the decimal point). Also, it appears that the backward error of a degree 3 common pseudodivisor cannot be pushed significantly below .2 for a tolerance level of $.5 \cdot 10^{-4}$. This means that valid common divisors of degree 3 will gradually cease to exist as the tolerance level in the two empirical polynomials becomes tighter than 10^{-5}. Compare also Exercise 6.4-2.

In an intuitive assessment of the three approaches, we may say that the computation and comparison of the zeros is very straightforward if the situation is so clear-cut as in our case where the near-common zeros have been obvious. If the common pseudozeros are ill conditioned so that they can differ substantially from the exact zeros of the individual polynomials, their selection may not be feasible even by human inspection.

The use of the singular value decomposition (s.v.d.) of the Sylvester matrix can be fully automated and offers the additional advantage of a candidate obtainable by averaging over many choices. Thus, in many cases, a further refinement may not be necessary.

In a pivoted triangularization of the Sylvester matrix, we lose that averaging facility: the candidate appears directly in the process. Also, the decision about the lower limit for a pivot to be acceptable is not so obvious. As in the case of the s.v.d., a reduction of the anticipated degree d of the candidate is possible by simply continuing the computation.

Finally, we report the reaction of Maple's gcd procedure: With a decimal floating-point precision of ≥ 8, the answer is 1, i.e. no common divisor. For values of Digits between 7 and 5, the procedure reports common pseudodivisors of degree 3 which are near the ones which we have found: For 6-digit decimal arithmetic, the result of $\mathrm{gcd}(\bar{p}_1, \bar{p}_2)$ is $x^3 - .553339\,x^2 - 2.53631\,x - 2.74441$, with backward errors (not provided by Maple) of ≈ 8 and 1 in $\bar{p}_1$ and $\bar{p}_2$, respectively. For values of Digits below 5, Maple refuses to produce an answer—and rightly so.

Exercises

1. (a) Find how the Euclidean Algorithm may be interpreted as a triangularization of $S(p_2, p_1)$ by Gaussian elimination steps (subtraction of a multiple of some row from another row). Formulate the algorithm in this form.

(b) What is the reason for the low number of $O(n^2)$ arithmetic operations in this triangularization? Why is this property lost when pivoting is used?

(c) Find in which situations the Euclidean Algorithm will generally become unstable.

2. Consider the Chebyshev polynomials T_4, T_5, T_6 (cf. (5.10)) and assume that their nonvanishing coefficients are empirical, with a uniform tolerance ε.

(a) Compute the singular values of the 10×9 generalized Sylvester matrix $S(T_6, T_5, T_4)$, cf. (6.57). For which size of ε would you expect the 3 empirical polynomials to have a valid common pseudodivisor of degree 2 ?

(b) Consider the extreme zeros of the T_κ, $\kappa = 4, 5, 6$, form a candidate pair $(-\zeta, \zeta)$ of common simultaneous pseudozeros, and compute its backward error in the (T_κ, ε). By varying ζ slightly, find the pair $(-\zeta^*, \zeta^*)$ with the (near-)minimal backward error. How small can you take ε such that $(-\zeta^*, \zeta^*)$ is a valid pair of common zeros? Compare with a).

(c) Use (6.74) in place of trial-and-error to find an optimal ζ^* from your original ζ.

(d) Use (6.78) to find $g^*(x) = x^2 - (\zeta^*)^2$ from $g(x) = x^2 - \zeta^2$.

(e) Determine the closest $\widetilde{T}_4$, $\widetilde{T}_5$, $\widetilde{T}_6$ such that they have *exact* common zeros at $\pm\zeta^*$.

Historical and Bibliographical Notes 6

The use of polynomials for the modelling of real-life situations, with their natural indetermination, has posed a novel problem: Algebraic predicates for such polynomials cannot be assigned a truth value (true or false) because this value may jump within the tolerance neighborhood of the polynomial(s); cf. paragraph 4 in section 3.2.1. Interval analysis has been suggested and used to resolve that dilemma; it may confirm a unique answer for all polynomials in a sharp polynomial interval if this is possible. Generally, this requires highly accurate computation and disregards the fact that empirical data cannot be represented by sharp intervals; cf. section 4.4.3.

It appears that our model of empirical data is particularly suitable for obtaining useful answers for predicates of empirical polynomials (cf. [6.1]): The interpretation of $\widetilde{\Pi}(\bar{p}, e) \in \mathbb{R}_+$ refers back to the specification of the tolerance e and leads to a meaningful and reliable assessment, at least for the application expert who has specified the tolerances. Also, we believe that our approach yields a particularly simple access to the practically important Propositions 6.3 and 6.5 as well as to a number of similar results (e.g., in Exercise 6.1-3). The nonlinear optimization problems in Definition 6.2 may be nontrivial, but they will usually be simplified by a priori information and the fact that 1-digit accuracy suffices.

Due to the preoccupation with exact data and exact results in computer algebra, the close relation between multiple zeros and zero clusters has received little attention there. In analysis, the relation (6.38) has been known for a long time, but the approach in section 6.3.3 has not generally become known, neither in computer algebra nor in numerical analysis.

The fact that—in spite of (6.38)—the location of a potential m-fold zero is *well conditioned within the set of m-fold zeros* has been pointed out by Kahan in his seminal paper [6.2], which was never published in a journal due to unfortunate circumstances. A very recent paper [6.3] by Zeng elaborates that insight and contains a clever algorithm for the computation of multiple pseudozeros under perturbation. Related condition numbers are found in [6.3] and [6.4].

The algorithm in [6.3] contains the numerical determination of a common pseudodivisor of p, p', ... This numerical common gcd algorithm (which works for arbitrary polynomials p_1, p_2, ...) appears to be the most efficient and reliable algorithm for that purpose; I have seen it too late to include it in the text. There is a wide literature on the subject (cf., e.g., [6.5]), and the research is successfully continuing as shown by [6.3].

For empirical polynomials with nontrivial tolerances, the refinement procedure of section 6.4.4 should always be an essential part of a complete gcd algorithm. The *condition* of a pseudo-gcd w.r.t. perturbations in the original polynomials has been analyzed in [6.4]; it permits a specification of meaningful numbers of digits in a pseudo-gcd.

References

[6.1] H.J. Stetter: Algebraic Predicates for Empirical Data, in: Computer Algebra in Scientific Computing - CASC 2001 (Eds. V.G. Ganzha, E.W. Mayr, E.V. Vorozhtsov), Springer, Berlin, 499–512.

[6.2] W. Kahan: Conserving Confluence Curbs Ill-Condition; Department of Computer Science, University of California Berkeley, Tech Rep. 6, 1972.

[6.3] Zh.G. Zeng: Computing multiple roots of inexact polynomials, Math. Comput., to appear.

[6.4] H.J. Stetter: Condition Analysis of Overdetermined Algebraic Problems, in: Computer Algebra in Scientific Computing - CASC 2000 (Eds. V.G. Ganzha, E.W. Mayr, E.V. Vorozhtsov), Springer, Berlin, 345–365.

[6.5] B. Beckermann, G. Labahn: When are two numerical polynomials relatively prime, J. Symb. Comput. **26** (1998), 677–689.

B. Beckermann, G. Labahn: A Fast and Numerically Stable Euclidean-like Algorithm for Detecting Relatively Prime Numerical Polynomials, J. Symbolic Comput. **26** (1998), 691–714.

[6.6] V.L. Kharitonov: Asymptotic Stability of an Equilibrium Position of a Family of Linear Differential Equations. Differ. Eq. **14** (1979), 1483–1485.

[6.7] N.K. Karmarkar, Y.N. Lakshman: Approximate Polynomial GCDs and Nearest Singular Poynomials, in: Proceed. ISSAC 96 (Ed. Y.N. Lakshman), ACM, New York, 35–39, 1996.

M.A. Hitz, E. Kaltofen: Efficient Algorithms for Computing the Nearest Polynomial with Constrained Roots, in: Proceed. ISSAC 98 (Ed. O.Gloor), 236–243, 1998.

M.A. Hitz, E. Kaltofen, Y.N. Lakshman: Efficient Algorithms for Computing the Nearest Polynomial with a Real Root and Related Problems, in: Proceed. ISSAC 99 (Ed. S. Dooley), ACM, New York, 205–212, 1999.

[6.8] The COCONUT Project (COntinuous CONstraints - Updating the Technology), Algorithms for Solving Nonlinear Constrained and Optimization Problems: The State of the Art (Progress Report), available at `solon.cma.univie.ac.at/~neum/glopt/coconut/`

Part III

Multivariate Polynomial Problems

Introductory Observations

In Part II of this book, we have seen that the numerical algebra of *univariate* polynomials is governed by a relatively small number of guiding principles: Algebraically, there is the simple structure of a principal ideal, with one generating polynomial which is unique except for a scalar factor, and of a quotient ring and dual space of the same dimension as the degree of that generator, with their multiplicative structure defined by one multiplication matrix. In most cases, the data of a structural description are directly or closely related to the data of the given problem. Therefore, our fundamental idea of describing polynomials with coefficients of limited accuracy by a family of neighborhoods, which embeds numerical algebra into analysis and defines backward errors, could quite directly be put into action in all cases: This made it possible to define the validity of approximate results and to refine approximate results whose accuracy was unsatisfactory. In principle, algorithms for the solution of all meaningful problems in the numerical algebra of univariate polynomials are known; their efficient implementation into software systems is either available or under development.

In the numerical algebra of *multivariate* polynomials, the situation is very different. We will, at first, consider some of the main reasons for this distinction:

In $\mathcal{P}^s$, $s > 1$, there is, generally, a wide choice of *bases* which may be used for a particular computational task related to a particular polynomial ideal in $\mathcal{P}^s$. The natural basis provided by the specified system of polynomials is rarely well suited, from the algebraic point of view as well as from algorithmic aspects. The use of Groebner bases, which is predominant in theoretical and symbolic polynomial algebra, is often not so advisable for numerical purposes; the related choice of a term order introduces a further ambiguity. In any case, the choice and generation of an algorithmically suitable basis may take a considerable computational effort and may meet with various difficulties when performed in floating-point arithmetic.

Furthermore, many of the bases which are considered in computer algebra provide a *singular* representation of the ideal in the following sense: An arbitrarily small modification of some of their coefficients makes the basis *inconsistent*. This is so because there are more basis elements than variables; therefore, the basis elements have to satisfy hidden constraints, the so-called *syzygies*. It is clear that this poses a difficult situation for the numerical use of such bases.

From our considerations in Chapter 2, we know that the *quotient ring* of a polynomial ideal is more important for computational purposes than an ideal basis. However, the variety of potential bases for the quotient ring of a 0-dimensional multivariate ideal is abundant for larger numbers of variables and higher total degrees; this is true even when we restrict ourselves to *monomial bases* only. On the other hand, the numerical feasibility of these bases differs greatly; therefore, it is important to use bases with good numerical properties.

For a full specification of a quotient ring, we need the specification of its multiplicative structure; with respect to a fixed basis, it is defined by the s matrices specifying multiplication with the s variables. While the only multiplication matrix of a univariate polynomial ideal contains the specified data immediately, the determination of multivariate multiplication matrices is essentially equivalent to the computation of an ideal basis. When they are computed in floating-point arithmetic, they will contain round-off errors. But their elements also have to satisfy hidden constraints: The matrices must commute! This commutativity will not be fully present in numerically determined multiplication matrices. We must see how it is possible to live with that discrepancy.

All this is further complicated when we consider *empirical* data, which is one of the principal objectives of this book: In place of one multivariate polynomial system P, we have to consider *families of neighborhoods* $N_\delta(\bar{P}, E) \subset \mathcal{P}^s$. Generally, we will at first proceed with the specified system $\bar{P}$ and—at the end—assess the validity of the computed approximate results by their *backward error* with respect to the empirical data. Algorithms for the *refinement* of results thus gain an increased importance.

Since the numerical analysis of systems of multivariate *linear* equations is so well developed, it is also worthwhile to ask in which respect systems of polynomial equations are so much more demanding. Here, the mere quantitative aspects come to mind first:

While a linear polynomial in s variables has $s + 1$ potential terms, this number rises to $\binom{d+s}{s}$ potential terms for a polynomial of degree d. For a polynomial of degree 4 in 6 variables, e.g., we have 210 potential terms. Naturally, in almost all practical situations, only few of these will actually be present; this shows that *sparsity* issues play an important role for multivariate polynomial systems.

Similarly, the number of solutions can be very large. For the moment, we restrict attention to the *regular* case where the set of zeros in $\mathbb{C}^s$ of a system of s polynomials in s variables (with real or complex coefficients) is not empty and consists only of isolated points. In this case, a linear system has precisely one zero, while a polynomial system whose individual equations have total degrees d_ν may have as many as $\prod_\nu d_\nu$ different zero s-tuples. For a system with 6 equations of degree 4 in 6 variables and (potentially) 1260 different coefficients, this yields the possibility of 4096 different zeros. Again, the actual number is often much lower; but its (a priori) computation is nontrivial as we shall see.

On the other hand, in scientific computations, we may be interested only in zeros within a tiny section of the $\mathbb{C}^s$ and—often—in real zeros only. It is generally not clear how such restrictions may be brought into play except in the final parts of the computation.

These are just a few observations which explain why the following two parts of this book will be less definitive in their character than the preceeding ones. It will be our main goal to exhibit the many questions and difficulties which arise in connection with multivariate numerical algebraic problems and to present analyses of the situations. Algorithmic solutions will be suggested in many cases, but they will tend to have a more preliminary nature. Also, on account of the size of the computational tasks, clever algorithmic design and efficient implementation will play a much greater role than in univariate numerical algebra. Altogether, there remain great challenges for the combined efforts of numerical analysts and computational algebraists.

Chapter 7

One Multivariate Polynomial

Individual polynomials in $\mathcal{P}^s$, $s = 2$ or 3, play an important role in all areas of *computational geometry*: Their zero sets constitute curves in the plane or surfaces in 3-space, respectively. In *Constructive Solid Geometry* (CSG), e.g., solids are represented by arithmetic predicates containing linear or polynomial expressions in the three space variables x_1, x_2, x_3, like $x_1^2 + x_2^2 + x_3^2 \leq 6.25 \ \wedge \ x_1 - x_2 - x_3 \geq 1.5$. The computational handling of such multivariate expressions generally assumes the data to be exact and attempts to retain logical consistency of the results throughout the further computations, like in determining the relative positions of lines or points with respect to other lines or solid bodies.

Polynomial expressions in any number of variables also play a role in the modelling of many nonlinear phenomena in virtually all areas of Scientific Computing. Here, quite generally, the data in the model expressions have a limited, sometimes very low, accuracy. In this chapter, we concern ourselves with individual multivariate polynomials and meaningful computational problems posed for them, under the general premises of approximate data and approximate computation introduced in Chapters 3 and 4.

7.1 Analytic Aspects

7.1.1 Intuitional Difficulties with Real and Complex Data

It is a trivial observation that the zero-set $Z[p] \subset \mathbb{C}^s$ of a multivariate polynomial $p \in \mathcal{P}_d^s$, $s \geq 2$, cannot be empty except for $p = \text{constant}$. This follows immediately by the substitution of values for all but one of the variable: The remaining univariate polynomial has zeros in $\mathbb{C}$ if it is not constant. By $Z[p]$, we will always denote the set of all *complex s-tuples* $\xi = (\xi_1, \ldots, \xi_s)$ such that $p(\xi_1, \ldots, \xi_s) = 0$; cf. Definition 1.8.

As with univariate polynomials, the *real* zero-sets of *real* multivariate polynomials can very well be empty, like for $p(x, y) = x^2 + y^2 + 1$. The analysis of reality questions is already nontrivial for univariate polynomials and it presents formidable difficulties in the multivariate case. These difficulties are essentially "orthogonal" to those which stem from our consideration of data with limited accuracy and of approximate computation; they are *not considered* in this book. Except when there is a clear note of the contrary, we always assume that variables and coefficients or other data may take complex values.

However, this introduces another difficulty: Every scientist with a mathematical training has some basic understanding of the space $\mathbb{C}$ of one complex variable and of the analysis of complex functions on $\mathbb{C}$. Furthermore, although the proper visualization of an analytic complex function of one complex variable needs four real dimensions and, generally, a Riemann surface; it can, for many purposes, also be visualized as a set of two real functions in two real variables, by a simple separation of real and imaginary parts.

This becomes fundamentally different when we deal with two or more complex variables. As we have already remarked in section 1.4, human beings cannot really visualize the $\mathbb{C}^2$! Thus, we lose our intuitive grip even in the simplest multivariate situations as soon as we must consider complex values for the components of our variables, and we must do that even for real data since the real domain is not algebraically closed. Furthermore, due to the fact that we can *see* the $\mathbb{R}^2$ and $\mathbb{R}^3$, we have a strong associative intuition even for the $\mathbb{R}^n$ with more than three dimensions (or at least we believe that we have). But without a vivid picture of the $\mathbb{C}^2$, we have no chance with a complex space of several dimensions. Also, the theory of analytic functions in two or more complex variables has a number of concepts and phenomena which are not generalizations of univariate concepts. Thus, when we formally deal with the ring $\mathcal{P}^s = \mathbb{C}[x_1, \ldots, x_s]$ of polynomials in s complex variables, with complex coefficients, our *intuition* automatically deals with real coefficients and real variables. This shortcoming may lead to faulty conclusions because we do not realize all aspects of a situation.

Example 7.1:   A simple example of this difficulty is provided by the consideration of the zero set in $\mathbb{C}^2$ of one polynomial (say with real coefficients) in two variables. To show the fundamentality of the problem, take the polynomial $p(x, y) = x^2 + y^2 - 1$. When we talk about its zero set, we must make a special mental effort to realize that—in $\mathbb{C}^2$—the real unit circle is merely a very special section of a manifold of *complex* dimension 1. Who is able to estimate quickly the y-components of the two points on that manifold with x-component $2 + 3\,\mathrm{i}$; they are $\approx \pm(3.116 - 1.926\,\mathrm{i})$! If our intuition deserts us so badly for the unit circle, what must we expect for some nontrivial polynomial in several variables. $\Box$

As a consequence of this intuitional deficiency, we will—implicitly or explicitly—consider real situations even when we formally proceed in the $\mathbb{C}^s$. In particular, almost all examples and exercises will employ real data. I apologize for this shortcoming; a systematic consideration of the $\mathbb{C}^s$ would have further confused many issues which are sufficiently intricate so that whatever intuition is available is badly needed. Also, I must confess, it would probably have surpassed my capabilities. On the other hand, I am in good company: I have yet to see an algebra book where the treatment of several complex variables goes beyond a formality.

7.1.2 Taylor Approximations

Fortunately, there is one approach in the theory of complex functions of complex variables which is not so prone to all these difficulties because it is formal rather than intuitive: It is the use of *power series*. This is particularly useful in dealing with polynomial problems because, obviously, polynomials are *finite* powers series, which eliminates a good deal of hard core analytic problems. The power series approach makes it possible to deal formally with multivariate polynomials as functions of several complex variables without consideration of many of the analytic traps lying about. Actually, since we deal with numerical polynomial *algebra*, we will neglect these traps wherever possible.

Formally, each polynomial in $\mathcal{P}^s$ of total degree d may be represented by its *Taylor expansion* (1.9) at some specified finite point $\xi \in \mathbb{C}^s$ which possesses nonvanishing terms up to degree d : Let $p(x) = \sum_{j \in J} \alpha_j x^j$, then, for any $\xi \in \mathbb{C}^s$,

$$p(x) \;=\; \sum_{j \in J} \alpha_j \, [(x - \xi) + \xi]^j \;=\; \sum_{|j| \leq d} \bar{\alpha}_j[\xi]\, (x - \xi)^j\,, \qquad (7.1)$$

where

$$\bar{\alpha}_j[\xi] \;=\; \partial_j[\xi]\,(p) \;:=\; \frac{1}{j_1! \,\cdots\, j_s!} \,\Big(\frac{\partial^{|j|}}{\partial x_1^{j_1} \,\cdots\, \partial x_s^{j_s}} \, p \,\Big)(\xi)\,; \qquad (7.2)$$

cf. Definition 2.13 and (2.37) in Chapter 2. In particular, the coefficients α_j of a polynomial specified in terms of monomials from $\mathcal{T}^s$—as is the overwhelming practice in polynomial algebra—satisfy $\alpha_j = \partial_j[0]\,(p)$.

From the algorithmic point of view, however, different representations (7.1) of the same polynomial p are not at all equivalent; e.g., a potential sparsity of some polynomial may only come to light in the expansion of the polynomial about a certain point z because a good number of the partial derivatives of p vanish there; cf. (7.2). Also, some algorithms may require the expansion of p about a particular z. It is rarely considered that the determination of the $\bar{\alpha}_j[z]$ for a polynomial of total degree d in s variables involves the evaluation of up to $\binom{d+s}{s}$ polynomials at z which may represent a major part of the total computational effort.

For empirical polynomials, the situation is even worse: Not only may the numerical computation of the $\bar{\alpha}_j$ introduce a good deal of round-off error, but the propagation of the tolerances of empirical coefficients onto the transformed coefficients presents a formidable and practically untractable problem; cf. the end of section 7.2.1.

Again we will follow the common practice and generally consider monomial representations, i.e. expansions about the origin. But we must be well aware of the fact that the necessity of proceeding differently may present considerable difficulties in the efficient implementation of numerical algorithms for the solutions of multivariate algebraic tasks in scientific computing.

Another important aspect of Taylor expansions (7.1) is the following: They provide a sequence of increasingly better approximations of a polynomial $p \in \mathcal{P}^s$ in the neighborhood of a point $\xi \in \mathbb{C}^s$: For small $\|\Delta x\|$,

$$\begin{aligned}
p(\xi + \Delta x) \;&=\; p(\xi) + O(\|\Delta x\|) \\
&=\; p(\xi) + \textstyle\sum_{\sigma=1}^{s} \bar{\alpha}_\sigma[\xi]\, \Delta x_\sigma + O(\|\Delta x\|^2) \\
&=\; p(\xi) + \textstyle\sum_{\sigma=1}^{s} \bar{\alpha}_\sigma[\xi]\, \Delta x_\sigma + \sum_{\sigma_1,\sigma_2} \bar{\alpha}_{\sigma_1 \sigma_2}[\xi]\, \Delta x_{\sigma_1} \Delta x_{\sigma_2} + O(\|\Delta x\|^3) \\
&=\; \cdots
\end{aligned}$$

$$(7.3)$$

By far the most important one of these approximations is the *linear* approximation

$$p(\xi + \Delta x) \;\approx\; p(\xi) + \sum_{\sigma=1}^{s} \partial_\sigma[\xi]\, \Delta x_\sigma \;=\; p(\xi) + p'(\xi) \cdot \begin{pmatrix} \Delta x_1 \\ \vdots \\ \Delta x_s \end{pmatrix} \;=:\; \bar{p}(\Delta x;\, \xi)\,, \qquad (7.4)$$

with the Fréchet derivative $p'(\xi) = \mathrm{grad}\, p(\xi)$ (a *row* vector); cf. section 1.2. Equation (7.4) describes the approximate effect of a small perturbation Δx of the argument on the value of p by a linear function of the perturbation.

For a univariate polynomial p, we had also considered finite expansions in powers of a fixed polynomial s of a degree $n > 1$; cf. Proposition 5.3, (5.18), and Example 5.4. These generalized Taylor series provide good approximations of p in the vicinity of *all n zeros of s* simultaneously.

There are analogous finite expansions for multivariate polynomials: In (7.1), we can interpret the set of components $(x_\sigma - \xi_\sigma)$, $\sigma = 1(1)s$, of $(x - \xi)$ as the basis elements of the ideal $\langle x_\sigma - \xi_\sigma \rangle$ with the only zero $\xi \in \mathbb{C}^s$. This suggests that we may replace the $(x_\sigma - \xi_\sigma)$ by the s basis polynomials $p_\sigma \in \mathcal{P}^s$ of the 0-dimensional ideal $\langle p_\sigma, \sigma = 1(1)s \rangle$, with m zeros $z_\mu \in \mathbb{C}^s$, $\mu = 1(1)m$. Such finite expansions

$$p(x) = d_0(x) + \sum_\nu d_{1,\nu}(x)\, p_\nu(x) + \sum_{\nu \le \nu_1} d_{2,\nu\nu_1}\, p_\nu(x)\, p_{\nu_1}(x) + \ldots$$
$$+ \sum_{\nu \le \nu_1 \le \ldots \le \nu_{k-1}} d_{k,\nu\nu_1..\nu_{k-1}}\, p_\nu(x)\, p_{\nu_1}(x) \ldots p_{\nu_{k-1}}(x) \tag{7.5}$$

will be introduced and discussed in section 8.3.3, cf. Corollary 8.26 and (8.38).

The difference between p and the linear part of (7.5) has (at least) a *double* zero at each zero z_μ of the ideal $\langle \{p_\sigma, \sigma = 1(1)s\} \rangle$; therefore, it provides a good approximation of p in the vicinity of *all* these zeros simultaneously. Like in a Taylor expansion, the inclusion of more terms enhances the approximation quality; cf. (7.3).

Example 7.2: Consider the polynomial $p(x, y) =$

$$-2.07 - 3.16\,x + 2.63\,y + .86\,x^2 - 4.09\,x\,y + 2.73\,y^2 + 1.71\,x^3 - 3.55\,x^2 y + .38\,xy^2 - 2.34\,y^3 + x^4;$$

we wish to have a lower degree polynomial which approximates p well around $(+1, 0)$ *and* $(-1, 0)$. The ideal with these zeros is generated by $p_1 = x^2 - 1$ and $p_2 = y$. The expansion (7.5) in powers of these p_σ is

$$p(x, y) = -.21 - 1.45\,x + (2.86 + 1.71\,x)\,(x^2 - 1) + (-.92 - 4.09\,x)\,y$$
$$+ (x^2 - 1)^2 - 3.55\,(x^2 - 1)\,y + (2.73 + .38\,x)\,y^2 - 2.34\,y^3 .$$

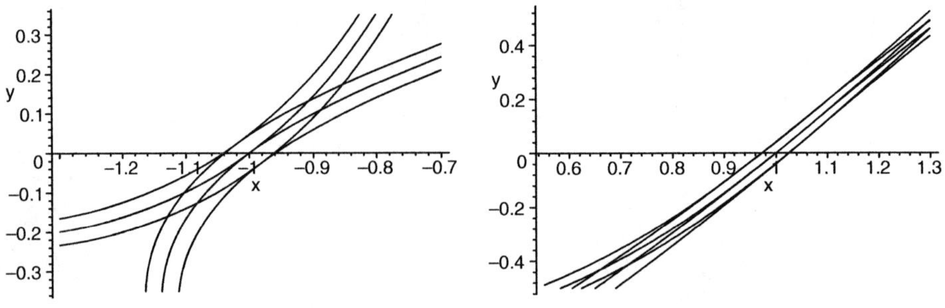

Figure 7.1.

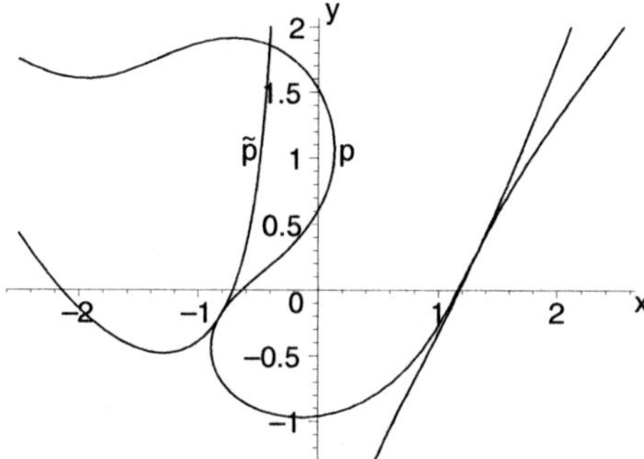

Figure 7.2.

It is immediately obvious that the 2nd line remainder has vanishing first-order derivatives at both $(+1, 0)$ and $(-1, 0)$; thus, the "linear" part $\tilde{p}(x, y)$ in the first line has the requested approximation property. When we plot level curves of both p and $\tilde{p}$ near the two points, the local approximation is obvious, cf. Figure 7.1.

On the other hand, when we plot the *manifolds* $p = 0$ and $\tilde{p} = 0$ in the x, y-plane (cf. Figure 7.2), we find that, globally, the manifolds differ strongly. The match near $(+1, 0)$ and $(-1, 0)$ is moderate because these points are not very close to either manifold. This has to be kept in mind when such approximations for a *function* p are employed to analyze the *manifold* $p = 0$. $\quad\square$

7.1.3 Nearest Points on a Manifold

A natural use of the approximations to a multivariate polynomial $p \in \mathcal{P}^s$ provided by truncated Taylor expansions (7.3) is in finding points on the manifold represented by $p(x) = 0$. When we have a point $\xi \subset \mathbb{C}^s$ such that $|p(\xi)|$ is small, i.e. when ξ is *close* to the manifold (cf. Example 7.2), we may look for the smallest modification Δx such that $p(\xi + \Delta x) = 0$. For this purpose, we may *locally linearize* p, i.e. replace it by (7.4), and find the smallest Δx with $\bar{p}(\Delta x; \xi) = 0$.

In the $\mathbb{C}^s$-space of Δx, this equation describes a linear manifold of codimension 1 and we ask for its shortest $\|..\|$-distance from the origin. We have treated exactly the same formal problem in section 3.3.1 : There, we have considered linear manifolds in the space $\Delta \mathcal{A} = \mathbb{C}^M$ of perturbations Δa of empirical data and their shortest $\|..\|^*$-distance form the origin. These two problems are duals; but formally they are identical if we observe the switch in norms and notation. To avoid confusion, we reformulate the results in terms of our present task.

Proposition 7.1. With a norm $\|..\|$ in the $\mathbb{C}^s$ of x (or Δx) and the dual norm $\|..\|^*$ in the space of the functionals a^T on $\mathbb{C}^s$ (cf. (3.6)), and with $\gamma \in \mathbb{C}$,

$$\min_{a^T x = \gamma} \|x\| = \frac{|\gamma|}{\|a^T\|^*} . \tag{7.6}$$

Proof: $\gamma = a^T x$ implies $|\gamma| \le \|a^T\|^* \|x\|$; cf. (3.6). It is well known from linear algebra that, for given a^T, equality may be attained by the choice of x. (Note that—with a somewhat modified notation—(7.6) is the dual statement of (3.52) in section 3.3.1.) $\square$

As in section 3.3.1, the minimizing x can be explicitly specified for the standard norms (3.8) used in this book:

Proposition 7.2. For specified $a^T \in \mathbb{C}^s$, $a^T \ne 0$, $\gamma \in \mathbb{C}$, the following $x \in \mathbb{C}^s$ satisfies $a^T x = \gamma$ and $\|x\| = |\gamma|/\|a^T\|^*$:

$$
x = \begin{pmatrix} x_1 \\ \vdots \\ x_s \end{pmatrix}
\quad \text{with} \quad
x_\sigma = \frac{\gamma}{(\|a^T\|^*)^2} \cdot
\begin{cases}
\frac{\alpha_\sigma^*}{|\alpha_\sigma|} \|a^T\|^*, & p = \infty, \\[2mm]
\alpha_\sigma^*, & p = 2, \\[2mm]
\alpha_{\sigma_0}^*, & \sigma = \sigma_0, \\
0, & \sigma \ne \sigma_0,
\end{cases}\ p = 1,
\tag{7.7}
$$

where, as elsewhere, $..^*$ denotes the complex conjugate and σ_0 designates $|\alpha_{\sigma_0}| = \max_\sigma |\alpha_\sigma|$; if σ_0 is not unique, any one may be chosen or the contribution may be split on several x_σ in an obvious manner.

Proof: The proposition is proved by substitution and elaboration. $\square$

Corollary 7.3. Consider $\bar{p}(\Delta x; \xi) = p(\xi) + \sum_\sigma \partial_\sigma p[\xi]\, \Delta x_\sigma$, with $p'(\xi) \ne 0$. Then

$$
\widehat{\Delta x_\sigma} = -p(\xi) \cdot
\begin{cases}
\frac{\partial_\sigma p[\xi]^*/|\partial_\sigma p[\xi]|}{\|p'(\xi)\|^*}, & p = \infty, \\[2mm]
\frac{\partial_\sigma p[\xi]^*}{(\|p'(\xi)\|^*)^2}, & p = 2, \\[2mm]
\frac{\partial_{\sigma_0} p[\xi]^*}{(\|p'(\xi)\|^*)^2}, & \sigma = \sigma_0, \\
0, & \sigma \ne \sigma_0,
\end{cases}\ p = 1,
\tag{7.8}
$$

satisfies $\bar{p}(\widehat{\Delta x}; \xi) = 0$ and $\|\widehat{\Delta x}\|$ minimal.

For *real* $p'(\xi)$, the complex conjugate is irrelevant and the expressions for $p = \infty$ and 1 may be simplified: For $p = \infty$, the numerator of the fraction is $\mathrm{sign}(\partial_\sigma p[\xi])$; likewise, for $p = 1$, we get $\mathrm{sign}(\partial_{\sigma_0} p[\xi])$ when we cancel one of the powers in the denominator. For *complex* $\partial_\sigma p[\xi] = \rho \exp(i\varphi)$, one obtains $\exp(-i\varphi)$ in place of the sign.

When ξ is sufficiently close to the manifold $p(x) = 0$ and $\|p'(\xi)\|^*$ bounded away from zero, an iterative use of linearization and choosing the closest point on the tangential linear manifold will generate a sequence which converges quadratically to a point on the manifold. Rigorous quantitative conditions for the convergence can be specified (using the well-known theory of the Newton iteration), but it may be virtually impossible to verify the assumptions in nontrivial practical situations. The observation of $|p(\xi + \widehat{\Delta\xi})| \ll |p(\xi)|$ will generally be a sufficient indication of an acceptable situation. Furthermore, it will rarely be meaningful to perform more than very few corrections; often, one correction will be sufficient.

The use of (7.8) will not work when ξ is close to a point where $p'(\xi)$ vanishes so that the quadratic terms in (7.3) are dominant. This case corresponds to the failure of Newton's method near an extremum of a univariate function. In both cases, the correction Δx obtained from a local linear approximation of the function tends to be far too large. Also like in the univariate case, if p and ξ are real, the smallness of $|p(\xi)|$ need not imply the closeness of a *real* zero of p; consider, e.g., $p(x, y) = \varepsilon + (x^2 + y^2)$ near the origin. If we proceed to a nearby point where $p'(\xi)$ *vanishes*, the problem reduces to finding a minimal-norm Δx such that $p(\xi) + \Delta x^T p''(\xi) \Delta x = 0$.

In the real case, $p''(\xi)$ is a real symmetric matrix so that it can be diagonalized, with *real* eigenvalues, by a real orthogonal transformation. This makes it possible to find a minimal-norm Δx for the usual norms; cf. Exercise 7.1-3 below. In the general case, where quantities are complex, $p''(\xi)$ is complex symmetric but *not Hermitian*. Such complex symmetric matrices A, with $A^T = A$, can be factorized in the following form (Takagi's factorization; cf., e.g., [2.15]):

$$A U = (U^T)^{-1} \Lambda = (U^{-1})^T \Lambda, \qquad \text{with unitary } U \text{ and real diagonal } \Lambda \geq 0; \qquad (7.9)$$

note that $U^T \neq U^{-1} = U^H$. Now, we can only find a lower bound for the minimal-norm Δx from $|p(\xi)| = |\Delta x^T p''(\xi) \Delta x| \leq \|\Delta x\|^* \|p''(\xi)\| \|\Delta x\|$, but it is not certain that the equality is attained and, generally, it is a difficult task to find a minimal Δx explicitly.

Example 7.3:   Consider the real polynomial[10] $p \in \mathcal{P}_5^3$

$$\begin{aligned}
p(x, y, z) \; := \; & 1 + .75x + .42y - 1.00z - 5.50x^2 - 1.42xy - 1.58y^2 - 6.63xz - .79yz \\
& - .75z^2 + 4.75x^3 + 3.75x^2y + 2.25xy^2 + .58y^3 - 1.50x^2z - .50xyz + 2.50y^2z \\
& + 1.75xz^2 + 2.08yz^2 + 2.00z^3 + 2.00x^4 - 2.00x^3y - 1.00xy^3 + 9.50x^3z \\
& - 3.00x^2yz + 3.00xy^2z - 1.00y^3z - 3.00x^2z^2 - 1.00xyz^2 - 1.00y^2z^2 + 3.00xz^3 \\
& - 1.00yz^3 - 1.00z^4 - 3.00x^5 - 1.00x^3y^2 - 1.00x^3z^2 \, .
\end{aligned}$$

$$(7.10)$$

The polynomial p is only mildly sparse: 36 of the potentially 56 terms are actually present.

In the vicinity of the origin, the approximations (7.3) are

$$\begin{aligned}
p(x, y, z) \; = \; & 1 + .75\,x + .42\,y - 1.00\,z + \mathrm{O}(\|\mathbf{x}\|^2) \\
= \; & \ldots - 5.50\,x^2 - 1.42\,xy - 1.58\,y^2 - 6.63\,xz - .79\,yz - .75\,z^2 + \mathrm{O}(\|\mathbf{x}\|^3) \, .
\end{aligned}$$

At $\xi_r = (.92, .92, .92)^T \in \mathbb{R}^3$, the linear approximation (7.4) is

$$\bar{p}(\Delta x, \Delta y, \Delta z; \, \xi_r) \approx .010510 + 4.5326\,\Delta x + 1.0635\,\Delta y + 1.7352\,\Delta z \, ;$$

at the complex value $\xi_c = (-.25 + .58\,\mathrm{i}, .75 + .58\,\mathrm{i}, 1.75 + .58\,\mathrm{i})^T$, (7.4) becomes

$$\begin{aligned}
\bar{p}(\Delta x, \Delta y, \Delta z; \, \xi_c) \; \approx \; & -.013204 - .006296\,\mathrm{i} \\
& + (10.9865 - 16.8762\,\mathrm{i})\,\Delta x - (2.9949 + 4.2136\,\mathrm{i})\,\Delta y - (12.9805 + 2.0571\,\mathrm{i})\,\Delta z \, .
\end{aligned}$$

In both cases, the smallness of $|p(\xi)|$ tells us that ξ is close to the algebraic manifold $p(\mathbf{x}) = 0$; in fact, if there are empirical coefficients with a sufficiently large tolerance in p, both ξ_r and ξ_c are valid zeros of p. We will use the linear parts of the Taylor series of $p(\xi + \Delta \xi)$ in powers of $\Delta \xi$ to approximate the closest point on the manifold by the closest point on the tangential linear manifold at $(\xi, p(\xi))$.

---

[10]We will use this polynomial also in other places. Naturally, x, y, z are the 3 components of points $\mathbf{x} \in \mathbb{C}^3$.

At first, we consider the real point $\xi_r = (.92, .92, .92)^T$. From (7.8), we obtain for the max-norm in $\mathbb{R}_3$,

$$\widehat{\Delta x} = \widehat{\Delta y} = \widehat{\Delta z} = -\frac{.010510}{4.5326 + 1.0625 + 1.7352} \approx -.0014336,$$

because all three components of $p'(\xi_r)$ have the same sign. For the Euclidean norm in $\mathbb{R}^3$, we obtain

$$\widehat{\Delta \xi} = -\frac{.010510}{4.5326^2 + 1.0625^2 + 1.7352^2} \cdot \begin{pmatrix} 4.5326 \\ 1.0625 \\ 1.7352 \end{pmatrix} \approx \begin{pmatrix} -.0019297 \\ -.0004528 \\ -.0007387 \end{pmatrix}.$$

For the 1-norm in $\mathbb{R}^3$, we have $\sigma_0 = 1$ (the first component of $p'(\xi_r)$ has maximal modulus) so that

$$\widehat{\Delta x} = \frac{.010510}{4.5326} \approx -.0023187, \qquad \widehat{\Delta y} = \widehat{\Delta z} = 0.$$

Naturally, these corrections yield points on the tangential manifold in $(\xi_r, p(\xi_r))$ and not on the manifold $p(\mathbf{x})$ itself, but the residuals upon substitution into p show that the corrected points are much closer to the manifold than ξ_r:

$$p(\xi_r + \widehat{\Delta \xi}_\infty) \approx .000022, \quad p(\xi_r + \widehat{\Delta \xi}_2) \approx .000026, \quad p(\xi_r + \widehat{\Delta \xi}_1) \approx .000031.$$

Now we consider the complex point $\xi_c = (-.25 + .58\,\mathrm{i}, .75 + .58\,\mathrm{i}, 1.75 + .58\,\mathrm{i})^T$. We determine only the nearest point on the tangential linear manifold with respect to the max-norm in $\mathbb{C}^3$ and leave the other two cases to Exercise 7.1-2. From (7.8), we obtain $\widehat{\Delta \xi} =$

$$\frac{-.013204 - .006296\,\mathrm{i}}{|10.9865 - 16.8762\,\mathrm{i}| + |-2.9949 - 4.2136\,\mathrm{i}| + |-12.9805 - 2.0571\,\mathrm{i}|} \begin{pmatrix} .54558 + .83806\,\mathrm{i} \\ -.57933 + .81509\,\mathrm{i} \\ -.98767 + .15652\,\mathrm{i} \end{pmatrix}$$

$$\approx \begin{pmatrix} .000050 + .000377\,\mathrm{i} \\ -.000332 + .000185\,\mathrm{i} \\ -.000365 - .000108\,\mathrm{i} \end{pmatrix}.$$

The residual at the corrected point is $p(\xi_c + \widehat{\Delta \xi}) \approx .000004 - .000012\,\mathrm{i}$. □

Exercises

1. (a) In Example 7.3, compute the Taylor expansion (7.1) of the polynomial (7.10) about the points ξ_r and ξ_c, respectively. If we assume p to be intrinsic, how many decimal digits must be carried in the computation in order that all coefficients $\bar{\alpha}_j[\xi]$ in the expansion are exact? How many decimal digits after the point have the exact values of the linear terms in the Taylor expansion?

(b) Generate approximate versions of the Taylor series about the above two points by taking various values for `Digits` in the computation. With the same arithmetic precision, recompute the original representation of p and compare. Perform the same forward/backward transformations with some values of ξ with very large moduli.

(c) Evaluate various exact and approximate representations of p at the same real and complex values and compare.

2. (a) In Example 7.3, perform one further correction of $\xi_r + \widehat{\Delta\xi}$ for one or several of the norms. Check the new value(s) of the residual.

(b) At the complex initial point ξ_c, determine the corrections for the 2-norm and the 1-norm in $\mathbb{C}^3$ and check the values of $|p|$ at the $\xi + \widehat{\Delta x}$.

3. For some $p \in \mathcal{P}^s$, $s > 1$, take a point $\xi \in \mathbb{C}^s$ where the gradient vector $p'(\xi)$ vanishes and consider the Taylor expansion (7.1) with remainder term $O(\|\Delta x\|^3)$ and the related quadratic approximation (cf. (7.3))

$$\bar{p}(\Delta x; \xi) \;=\; p(\xi) + \frac{1}{2}\, p''(\xi)(\Delta x, \Delta x) \;=\; p(\xi) + \frac{1}{2}\,\Delta x^T H(\xi)\Delta x\,,$$

where $H(x) = \left(\frac{\partial^2 p}{\partial x_i \partial x_j}(x) \right)$ is the Hessian matrix of p ; cf. section 1.2.

(a) Is it natural to expect that there are points $\xi \in \mathbb{C}^s$ with $p'(\xi) = 0$? Is it natural to expect such points which are zeros of p at the same time (cf. section 7.4)?

(b) Assume $p(\xi) \in \mathbb{R}$ and $H(\xi) \in \mathbb{R}^{s \times s}$ so that there exist an orthogonal matrix U and a real diagonal matrix Λ with $H(\xi)\,U = U\,\Lambda$. What is the form of $\bar{p}(U\Delta x; \xi)$? For the three standard norms in $\mathbb{R}^s$, find a real $\widehat{\Delta x}$ such that $\bar{p}(\widehat{\Delta x}; \xi) = 0$ and $\|\widehat{\Delta x}\|$ minimal, if such a $\widehat{\Delta x}$ exists. What is a necessary and sufficient condition for the existence?

(c) When $H(\xi)$ is a *complex* symmetric matrix, there exist a unitary matrix U and a nonnegative diagonal matrix Λ such that $H(\xi)\,U = (U^T)^{-1}\Lambda$; cf. (7.9). Try to obtain analogous results as in (b). What are the obstacles for proceeding like in (b)? In an example with $s = 2$, find lower bounds for the minimal-norm distances of the manifold $\bar{p}(\Delta x; \xi) = 0$ from ξ; can the lower bound be assumed?

7.2 Empirical Multivariate Polynomials

7.2.1 Valid Results for Empirical Polynomials

Almost all multivariate polynomials employed as models of real-life phenomena in scientific computing contain some coefficients with a limited accuracy. In Chapter 3, we have called such data and the polynomials in which they occur "empirical." Without further explanations, we recall some definitions and propositions of Chapter 3 in their appearance for multivariate polynomials before we analyze particular problems with empirical multivariate polynomials.

Definition 7.1. (Compare Definitions 3.3 and 3.4 and Example 3.2.) An empirical polynomial $(\bar{p}, e)$ in $s > 1$ variables defines a family of neighborhoods $N_\delta(\bar{p}, e)$, $\delta > 0$, which contain the polynomials $\tilde{p} \in \mathcal{P}^s$, $\tilde{p}(x) = \sum_{j \in J} \tilde{\alpha}_j x^j$, with

$$\| \tilde{p} - \bar{p} \|_e^* \;:=\; \| (\dots, \tfrac{|\tilde{\alpha}_j - \bar{\alpha}_j|}{\varepsilon_j}, \dots,\; j \in \tilde{J})\|^* \;\le\; \delta\,,$$
$$\tilde{\alpha}_j \;=\; \bar{\alpha}_j\,, \qquad j \in J \setminus \tilde{J}\,. \tag{7.11}$$

Here, $J \subset \mathbb{N}_0^s$ is the *support* of the *specified polynomial* $\bar{p}(x) = \sum_{j \in J} \bar{\alpha}_j\, x^j$, $\bar{\alpha}_j \in \mathbb{C}$ or $\mathbb{R}$, and $\tilde{J} \subset J$ is the *empirical support* of $(\bar{p}, e)$. $e := (\varepsilon_j > 0\,,\; j \in \tilde{J})$ is the *tolerance vector* and

$\|..\|^*$ is a *dual norm* in $\mathbb{R}^M$, $M := |\tilde{J}|$; cf. (3.6). We will generally use $\|a^T\|^* = \max_j |\alpha_j|$. Compare (3.3) for an intuitive meaning of the parameter δ. Polynomials $\tilde{p} \in N_\delta(\bar{p}, e)$, $\delta = O(1)$, are *valid instances* of the empirical polynomial $(\bar{p}, e)$. $\square$

As we have remarked in the introductory observations of Part III, multivariate polynomials have, potentially, a very large number of terms while, generally, only few are present. For an empirical multivariate polynomial $(\bar{p}, e)$, we must specify whether its valid instances $\tilde{p} \in N_\delta(\bar{p}, e)$, with $\delta = O(1)$, have to share the sparsity of the specified polynomial $\bar{p}$ fully, partially, or not at all. This is achieved through the specification of the empirical support set $\tilde{J}$: If $\tilde{J}$ contains no exponent j for a vanishing coefficient $\bar{\alpha}_j$, then, by (7.11), the sparsity of $\bar{p}$ is fully intrinsic and *all* neighboring polynomials must share it. On the other hand, if we permit a monomial x^j with a sufficiently small coefficient in $\tilde{p} \in N_\delta(\bar{p}, e)$ although that monomial is absent in $\bar{p}$, then j has to be a member of the empirical support set $\tilde{J}$ and a tolerance component ε_j has to be specified in e. In principle, one could admit a positive tolerance for all vanishing terms in $\bar{p}$, but this would rarely make sense.

Example 7.4: In $\mathcal{P}_2^2$, consider $(\bar{p}, e)$ with $\bar{p}(x, y) = \bar{\alpha}_{20}\, x^2 + \bar{\alpha}_{02}\, y^2 - 1$. If the *symmetry* with respect to the origin and the coordinate axes are intrinsic properties and not subject to a potential indetermination, then (cf. Definition 7.1) $\tilde{J} = \{20, 02\}$ and $J \setminus \tilde{J} = \{11, 10, 01, 00\}$ preserves the full sparsity of $\bar{p}$. But, in another situation, it may be meaningful to admit small linear terms $(\tilde{J} = \{20, 02, 10, 01\})$ or a small mixed product term $(\tilde{J} = \{20, 11, 02\})$. The pseudoresult sets of an algebraic problem with $(\bar{p}, e)$ and the appropriateness of the problem solution may delicately depend on that choice. $\square$

In section 3.1.3, we have further introduced the *data$\rightarrow$result mapping* F from the space $\mathcal{A}$ of the empirical data of an algebraic problem to its result space $\mathcal{Z}$; cf. Definition 3.6. F maps a data value $\tilde{a} \in \mathbb{C}^M$ into the *exact result(s)* of the algebraic problem for these data; intrinsic data are assumed fixed and considered as part of the mapping F. The F-images of the neighborhoods $N_\delta(\bar{a}, e)$ are the *pseudoresult sets* $Z_\delta(\bar{a}, e)$; cf. Definition 3.7. This concept permits the definition of *valid results* of an empirical algebraic problem: They are results in $Z_\delta(\bar{a}, e)$ with $\delta = O(1)$, i.e. *exact* results of the genuine algebraic problem, with a proper account for the indetermination in the data of the problem; cf. Definition 3.8 and Examples 3.3 and 3.4.

A numerical result $\tilde{z}$ obtained by whatever approach may be *tested for its validity*. For this purpose, we have introduced the *equivalent-data manifold* $\mathcal{M}(\tilde{z}) := \{\tilde{a} \in \mathcal{A} : F(\tilde{a}) = \tilde{z}\}$; cf. Definition 3.11. When we define a weighted norm $\|..\|_e^*$ with the reciprocals of the tolerances ε_j (cf. (3.5) and (7.11)), the minimal $\|..\|_e^*$-norm distance of $\mathcal{M}(\tilde{z})$ from the specified data point $\bar{a} \in \mathcal{A}$ equals the smallest value of δ for which $\tilde{z} \in Z_\delta(\bar{a}, e)$, or the smallest δ such that $\tilde{z}$ is the exact result of the algebraic problem for some empirical data $\tilde{a} \in N_\delta(\bar{a}, e)$. According to Proposition 3.8, this minimal distance is uniquely defined.

Definition 7.2. (Compare Definition 3.12.) In the situation under discussion,

$$\delta(\tilde{z}) := \min_{a \in \mathcal{M}(\tilde{z})} \|a - \bar{a}\|_e^* \tag{7.12}$$

is the *backward error* of the approximate result $\tilde{z}$ for the empirical algebraic problem with data $(\bar{a}, e)$. $\square$

Obviously, $\tilde{z}$ is a valid result if its backward error $\delta(\tilde{z}) = O(1)$. Thus, the approximate computation of the backward error of a computed result of an empirical algebraic problem is a crucial part of solving the problem. The vague notion of $O(1)$ in the definition and the testing of a valid result is the appropriate concept in a realistic model of empirical data: Since data tolerances are known only as orders of magnitude, an order of magnitude concept must also be applied to the judgment of the results of problems with empirical data.

Since the indetermination of an empirical polynomial is generally expressed by the indetermination of certain coefficients *in a particular representation* of the polynomial, it is clear that the representation of a polynomial plays an important role in dealing with empirical polynomials. When tolerances are prescribed for the coefficients of a certain representation of the specified polynomial $\bar{p}$, it is generally not possible to assign meaningful tolerances for the coefficients in a different representation of the same polynomial $\bar{p}$. Naturally, one can always compute strict upper bounds for the potential variation of a particular coefficient in the new representation within a fixed neighborhood $N_\delta(\bar{p}, e)$ referring to the original representation, but such bounds may grossly misrepresent the indetermination in the polyomial as we have found in section 3.3.2. Because of the importance of this issue for multivariate polynomials, we visualize the situation once more:

Consider the data space $\mathcal{A} = \mathbb{R}^M$ or $\mathbb{C}^M$ of the empirical coefficients in the original representation of $(\bar{p}, e)$; the neighborhood $N_1(\bar{p}, e)$ corresponds to the tolerance weighted norm unit ball $\{\|a - \bar{a}\|_e^* \leq 1\}$ about the data vector $\bar{a}$ of $\bar{p}$. A transformation which converts $\bar{p}$ and its neighbors to their new representation generates a transformation R of a domain A in the data space $\mathcal{A}$, with $\bar{a} \in A$, to a domain A' in the data space $\mathcal{A}'$ of the empirical coefficients in the new representation. The dimension M' of $\mathcal{A}'$ may *differ* from the dimension M of $\mathcal{A}$; moreover, the symmetries of the $\|..\|_e^*$ unit ball about $\bar{a}$ with respect to the coordinates α_j in $\mathcal{A}$ will generally be lost to a large extent. (In algebraic computations, transformations of representations are often *nonlinear*.) Therefore, if we enclose the image $R(\{\|a - \bar{a}\|_e^* \leq 1\})$ in $\mathcal{A}'$ by a tolerance weighted unit norm ball about $\bar{a}' := R(\bar{a}) \in \mathcal{A}'$, the necessary tolerances e' may have to be quite large; but $R(\{\|a - \bar{a}\|_e^* \leq 1\})$ may only fill a very small part of $\{\|a' - \bar{a}'\|_{e'}^* \leq 1\}$.

This practical impossibility of following specified tolerances through an extensive algebraic computation without a meaningless explosive expansion excludes the use of interval mathematics except in very special contexts; cf. section 4.4. The backward error analysis, as we propose it in this book, refers the *result* of the computation back to the problem formulation in the original data space, with its given, meaningful tolerance specification. Together with our generous $O(1)$ interpretation of the backward error, this permits a *realistic* treatment of empirical multivariate algebraic problems.

Example 7.5:   Consider a bivariate empirical polynomial $(\bar{p}, e)$ of degree 3, in monomial representation, with all its 10 coefficients empirical and nonvanishing. Assume that—in the course of some computation—the polynomial is to be represented with respect to a specified *exact* Groebner basis $\mathcal{G} = \{g_1, g_2, g_3\}$, with leading monomials x^2, xy, y^3, resp., and normal set $\{1, y, x, y^2\}$. Generically, a potential form for this representation is $\bar{p}(x, y) :=$

$$(\beta_{10} + \beta_{11}\, x + \beta_{12}\, y)\, g_1(x, y) + (\beta_{20} + \beta_{22}\, y)\, g_2(x, y) + \beta_{30}\, g_3(x, y) + \gamma_0 + \gamma_1\, x + \gamma_2\, y + \gamma_3\, y^2 ,$$

which has also 10 coefficients. Furthermore, it may be seen that the transformation between the vector a^T of the monomial coefficients of $\bar{p}$ and the vector bc^T of the coefficients β_j, γ_j

in its representation with respect to $\mathcal{G}$ is *linear*. Therefore, the same linear transformation translates *variations* Δa^T of the monomial coefficients into the corresponding variations Δbc^T of the Groebner representation coefficients. By an unfortunate choice of the coefficients in the Groebner basis $\mathcal{G}$ (which are subject to syzygies), some of the nonvanishing elements in the matrix of the transformation and hence its max-norm can be arbitrarily large; thus, the max-tolerances of some of the bc coefficients must be very large in this case.

Although they are correct as bounds, these tolerances are not realistic. This is seen from the fact the matrix of the inverse transformation (back from the Groebner basis representation to the monomial representation) can also have some large elements and a large max-norm in the same situation. Thus, the back-transformation will expand some of the tolerances even further and the resulting tolerances for a^T will be very much larger than the original ones; yet the polynomial $\bar{p}$ has not undergone any change in the process (if we assume exact computation). If the transformation is nonlinear, the situation may yet become much more involved. $\square$

7.2.2   Pseudozero Sets of Empirical Multivariate Polynomials

A fundamental task for empirical multivariate polynomials is checking the validity of a point $\tilde{z} \in \mathbb{C}^s$ as a (pseudo)zero. This problem has been considered at length in section 3.1 in a general context. Therefore it suffices to quote, comment, and extend the results obtained there.

We consider $(\bar{p}, e)$ with $\bar{p} \in \mathcal{P}^s$, $s > 1$, with support sets $\tilde{J} \subset J$, $|\tilde{J}| = M$, and a tolerance vector $e = (\varepsilon_j, \ j \in \tilde{J}) \in \mathbb{R}_+^M$. At first, we make no assumptions regarding reality so that all neighborhoods of empirical coefficients admit complex deviations $\Delta\alpha_j$ from $\bar{\alpha}_j$, $j \in \tilde{J}$. Also the prospective pseudozero $\tilde{z} = (\tilde{\zeta}_1, \dots, \tilde{\zeta}_s) \in \mathbb{C}^s$ is unrestricted. $\|..\|_e$ and $\|..\|_e^*$ denote a pair of e-weighted dual norms in $\mathbb{C}^s$; cf. (3.7) and (3.5). Then we have from Proposition 3.9:

Proposition 7.4. Let $\tilde{\mathbf{z}} := (\tilde{z}^j, \ j \in \tilde{J})^T$ and $\Delta a^T := (\Delta\alpha_j, \ j \in \tilde{J})$. Then the backward error of $\tilde{z}$ as a zero of $(\bar{p}, e)$ is

$$\delta(\tilde{z}) \ := \ \min_{\Delta a^T \tilde{\mathbf{z}} + \bar{p}(\tilde{z}) = 0} \| \Delta a^T \|_e^* \ = \ \frac{|\bar{p}(\tilde{z})|}{\| \tilde{\mathbf{z}} \|_e} . \tag{7.13}$$

The components of $\tilde{\mathbf{z}}$ are the evaluations of the monomials x^j, $j \in \tilde{J}$, of the M empirical terms of $(\bar{p}, e)$ at $\tilde{z}$. If and only if the evaluations of all these terms vanish, the backward error of the approximate zero $\tilde{z}$ is not defined. This may happen if terms with a positive power of some component x_σ are the only ones with empirical coefficients and if $\tilde{\zeta}_\sigma = 0$. In this case, $\tilde{z}$ cannot be interpreted as an exact zero of a polynomial in $N_\delta(\bar{p}, e)$ for any $\delta > 0$.

If we have a fully real situation, with real coefficients $\bar{a}^T$, real deviations $\Delta\alpha_j$, and real $\tilde{z}$, (7.13) holds just as well. However, a mixture of real coefficients and a complex zero requires special attention; in this case, the complex conjugate zero must be taken into account as a simultaneous zero. This has been analyzed for the case of a univariate polynomial in section 5.2.1; the same analysis applies for a multivariate polynomial.

The minimizing Δa^T for (7.13) is obtained just like the minimizing x in Corollary 7.3; with the approach of section 3.2.1, we have the following:

Proposition 7.5. Consider the s-variate empirical polynomial $(\bar{p}, e)$ and $\tilde{z} \in \mathbb{C}^s$. Then,

$$
\widehat{\Delta \alpha_j} \;=\; -\,\bar{p}(\tilde{z}) \cdot
\begin{cases}
\dfrac{(\tilde{z}^j)^* / |\tilde{z}^j|}{\|\tilde{\mathbf{z}}\|}\,, & p = 1\,, \\[2ex]
\dfrac{(\tilde{z}^j)^*}{\|\tilde{\mathbf{z}}\|^2}\,, & p = 2\,, \\[2ex]
\begin{array}{ll}
\dfrac{(\tilde{z}^{j_0})^*}{|\tilde{z}^{j_0}|^2}\,, & j = j_0, \\[1.5ex]
0\,, & j \ne j_0,
\end{array} & p = \infty\,,
\end{cases}
\tag{7.14}
$$

satisfies $\bar{p}(\tilde{z}) + \sum_{j \in \bar{j}} \widehat{\Delta \alpha_j}\, \tilde{z}^j = 0$ and $\|\widehat{\Delta a}^T\|^*$ minimal, with the same convention as in Proposition 7.2 that j_0 designates $|\tilde{z}^{j_0}| = \max_j |\tilde{z}^j|$.

Example 7.6: We return to Example 7.3 with the polynomial (7.10) and the approximate zeros ξ_r and ξ_c. Now we regard p as the specified polynomial of an empirical polynomial $(\bar{p}, e)$ by assuming that all coefficients in (7.10) which are not integer or have obvious rational values with denominators 2 or 4 have been rounded to their values in (7.10). This concerns the coefficients of the following monomials: x, y, xy, y^2, xz, yz, y^3, yz^2, and it assigns a tolerance of $.5 \cdot 10^{-2}$ to each of their coefficients. We check whether the approximate zeros $\xi_r = (.92, .92, .92)^T$ and $\xi_c = (-.25 + .58\,\mathrm{i}, .75 + .58\,\mathrm{i}, 1.75 + .58\,\mathrm{i})^T$ are valid (pseudo)zeros of this empirical polynomial, with our usual max-norm in $\mathcal{A}$.

For ξ_r, we obtain the backward error

$$
\delta(\xi_r) = \frac{|p(\xi_r)|}{.005\,(|\xi_{r1}| + |\xi_{r2}| + |\xi_{r1}\xi_{r2}| + |\xi_{r2}|^2 + |\xi_{r1}\xi_{r3}| + |\xi_{r2}\xi_{r3}| + |\xi_2|^3 + |\xi_{r2}\xi_{r3}^2|)} \approx .36\,;
$$

this qualifies ξ_r as a valid zero of $(\bar{p}, e)$. The residual $|p(\xi_r)| \approx .0105$ by itself would not have given a reliable information. By (7.14), ξ_r is an exact zero of the polynomial $p + \Delta p$, in which the specified values of the empirical coefficients have all been diminished by $p(\xi_r)/(|\xi_{r1}| + \ldots + |\xi_{r2}\xi_{r3}^2|) \approx .0018$.

This mechanism works just as well for the complex approximate zero ξ_c. We use the same expression as above with the only difference that now the absolute values refer to complex numbers. We obtain $p(\xi_c) \approx .01320 + .00630\,\mathrm{i}$ and $\delta(\xi_c) \approx .31$, again for the max norm in the coefficient space; thus, ξ_c is also a valid zero of $(\bar{p}, e)$. The corrections which must be attached to the specified coefficients in order that the corrected polynomial $p + \Delta p$ has an exact zero ξ_c have the same modulus $\approx .00155$ throughout, but they are complex and their arguments depend on the arguments of the monomials ξ^j, cf. (7.14). If we wish to have a *real* correction such that $p + \Delta p$ has ξ_c as an exact zero, we must require that ξ_c and its conjugate-complex value are *simultaneous* zeros of $p + \Delta p$; cf. section 5.2.1 for the corresponding univariate situation.

When we assume that *all* nonvanishing coefficients in (7.10) except the constant term 1 are empirical and have a tolerance of $.01$, we must accept points in $\mathbb{C}^s$ as valid zeros which are a good deal away from the specified zero manifold. For example, the point $(1., 1., 1.)^T$ leaves a residual of $.66$ upon substitution into p, but its backward error is only ≈ 1.89 which is still an $O(1)$ value. Thus, with maximal modifications of $.0189$ in the nonvanishing coefficients, we reach a polynomial with an exact zero at $(1.,1.,1.)$. $\quad\square$

Next, we consider the problem of characterizing the families of (pseudo)zero sets $Z_\delta[p]$ when p is an empirical s-variate polynomial $(\bar{p}, e)$. Here, we meet a new problem: In our introductory discussion of pseudoresult sets in sections 3.1.3, we had—implicitly or explicitly—assumed that an individual result is a point in some m-dimensional real or complex space $\mathcal{Z}$; this had always been satisfied in connection with empirical univariate polynomials. Now, our exact results are positive-dimensional manifolds and the pseudozero sets $Z_\delta[(\bar{p}, e)]$ "envelop" the algebraic manifold $Z[\bar{p}]$ with increasing "thickness" as δ increases. However, except for the fully real case in 2 dimensions (or trivial situations), it appears impossible to plot or otherwise visualize this enveloping family of sets of the algebraic manifold $Z[\bar{p}]$.

In this situation, there are some questions which we would like to be able to answer:

(1) At some point $\bar{z} \in Z[\bar{p}]$, what are lower and upper bounds for the "thickness" of a pseudozero set $Z_\delta[(\bar{p}, e)]$, or—more precisely—for the minimal distance from $\bar{z}$ of a point *not contained* in $Z_\delta[(\bar{p}, e)]$?

(2) In some specified part of $\mathbb{C}^s$, do the sets $Z_\delta[(\bar{p}, e)]$ preserve the "topological structure" of the manifold $Z[\bar{p}]$, in particular with respect to its decomposition into component manifolds? More precisely, if there are no singular points $\hat{z}$ with $\bar{p}'(\hat{z}) = 0$ in that domain, is this also true for all neighboring polynomials in $N_\delta(\bar{p}, e)$?

7.2.3 Condition of Zero Manifolds

The first question posed at the end of the previous section is about the sensitivity of the zeros of one multivariate polynomial in $\mathbb{C}^s$, or of the points of an algebraic manifold of codimension 1 in $\mathbb{C}^s$, to perturbations of the coefficients of the polynomial. In a slight generalization of the general usage of the term, we may call this the *condition of the manifold*.

A second glance on the problem tells us that we must be more specific: If the manifold moves within itself under the perturbation of the coefficients but remains invariant as a geometric object, we will not consider this as a noteworthy perturbation of the manifold. When we consider a fixed point z on the manifold, we are interested only in variations of z normal to the manifold. Generally, under a specified small perturbation in the coefficients, z will move in a direction which combines normal and parallel components. Since we can associate the parallel components with a change from z to a neighboring point on the manifold, we may restrict our attention to the normal component of the variation.

Let us, at first, apply this line of thought to the zero set of a *linear* polynomial, i.e. to a hyperplane, and restrict ourselves to the real domain for additional intuition. Consider $\bar{\ell} \in \mathcal{P}_1^s$, $\bar{\ell}(x_1, \ldots, x_s) = 1 + \bar{a}^T\mathbf{x}$, $\bar{a}^T \in \mathbb{R}^s$, $\mathbf{x} := (x_1, \ldots, x_s)^T$. (Without loss of generality, we have normalized $\alpha_0 = 1$ for notational simplicity.) With $e = (\varepsilon_1, \ldots, \varepsilon_s)$, the neighborhoods $N_\delta(\bar{\ell}, e)$ contain the linear polynomials $\tilde{\ell}(x) = 1 + (\bar{a} + \Delta a)^T\mathbf{x}$, with $\|\Delta a\|_e^* \leq \delta$. The zero set $Z[\tilde{\ell}]$ of each $\tilde{\ell} \in N_\delta((\bar{\ell}, e))$ is a linear manifold (hyperplane) "close" to the hyperplane $\bar{\ell}(x) = 0$.

We restrict our attention to a *compact part* of the $\mathbb{R}^s$, say $\overline{M} := \{|x_\sigma| \leq M_\sigma < \infty, \sigma = 1(1)s\}$. Within $\overline{M}$, the points on the hyperplanes $\tilde{\ell}(x) = 0$, $\tilde{\ell} \in N_\delta(\bar{\ell}, e)$, i.e. the points in the pseudozero set $Z_\delta[(\bar{\ell}, e)]$, fill a "sheet" which envelops the specified hyperplane $\bar{\ell}(x) = 0$. We select a point $z = (\zeta_1, \ldots, \zeta_s)$ on that hyperplane and proceed in the unique *normal* direction until we reach the boundary of the sheet. The normal direction is given by

$$n(z) = t \cdot (\operatorname{grad} \bar{\ell}(z))^T = t\,\bar{a}, \quad t \in \mathbb{R},$$

and we want

$$\Delta_\delta t(z) \;:=\; \max \; \{ \; |t| \; : \; \tilde{\ell}(z + t\,\bar{a}) = 0 \quad \text{for some } \tilde{\ell} \in N_\delta(\bar{\ell}, e) \; \} \,, \tag{7.15}$$

which indicates the local "thickness" of the sheet. From

$$\tilde{\ell}(z + t\,\bar{a}) \;=\; 1 + (\bar{a} + \Delta a)^T (z + t\,\bar{a}) \;=\; \Delta a^T z + t\,(\bar{a}^T\bar{a}) + \mathrm{O}(t\|\Delta a\|) \;=\; 0 \,,$$

we obtain, after neglecting the quadratic term as usual,

$$\Delta_\delta t(x) \;\doteq\; \frac{1}{\bar{a}^T\bar{a}} \max_{\tilde{\ell} \in N_\delta(\bar{\ell},e)} |\Delta a^T z| \;\le\; \frac{1}{\bar{a}^T\bar{a}} \max_{\|\Delta a\|_e^* \le \delta} \|\Delta a^T\|_e^* \,\|z\|_e \;=\; \frac{\delta}{\bar{a}^T\bar{a}} \sum_{\sigma=1}^{s} \varepsilon_\sigma |\zeta_\sigma| \,,$$

with our standard max norm for the indetermination in the coefficients. For the diameter $d_\delta(z)$ of the enveloping sheet at z, this implies

$$d_\delta(z) \;:=\; 2\,\Delta_\delta t(z) \,\|\bar{a}\|_2 \,(1 + \mathrm{O}(\|e\|)) \;=\; 2\,\delta\,\frac{\|z\|_e}{\|\bar{a}\|_2} \,(1 + \mathrm{O}(\|e\|)) \,; \tag{7.16}$$

the $(1 + \mathrm{O}(\|e\|))$ factor takes care of the omitted quadratic term.

Equation (7.16) is confirmed by the backward error as a zero of $(\bar{\ell}, e)$ of the extreme points $\tilde{z}_{\max} = z \pm \frac{\delta}{\bar{a}^T\bar{a}} \|z\|_e \,\bar{a}$ on the normal vector at z :

$$\delta(\tilde{z}_{\max}) \;=\; \frac{\bar{\ell}(\tilde{z}_{\max})}{\|\tilde{z}_{\max}\|_e} \;=\; (1 + \bar{a}^T (z \pm \frac{\delta}{\bar{a}^T\bar{a}} \|z\|_e \,\bar{a})) / \|\tilde{z}_{\max}\|_e \;=\; \delta\,\frac{\|z\|_e}{\|\tilde{z}_{\max}\|_e} \;=\; \delta\,(1 + \mathrm{O}(\|e\|)) \,.$$

Example 7.7: We take $\bar{\ell}(x, y) = 2.47 + 1.68\,x + 3.21\,y \in \mathcal{P}_1^2$ and $e = (.5, .5, .5) \cdot 10^{-2}$. Here, we have *not* normalized the representation of $\bar{\ell}$ and assumed all coefficients to be empirical; this implies $\|(\xi, \eta)\|_e = \varepsilon_0 + \varepsilon_1|\xi| + \varepsilon_2|\eta|$. At $(\xi, \eta) = (-2.206, 1.924) \in Z[\bar{\ell}]$, we find a "sheet diameter" $d_\delta(\xi, \eta) \approx 2\,\delta\,.005\,(1 + 2.206 + 1.924)/\sqrt{1.68^2 + 3.21^2} \approx .014\,\delta$; this indicates a well-conditioned situation. In this linear case, we can directly see that the analysis is *scaling invariant*: If we multiply $\bar{\ell}$ by some factor, the ε_j pick up the same factor and the result remains the same.

For $\delta = 1$, the two extremal points $(\xi \pm 1.68\,\Delta_1 t(\xi, \eta), \eta \pm 3.21\,\Delta_1 t(\xi, \eta))$ along the normal vector $n(\xi, \eta) = (1.68, 3.21)^T$ have a backward error of .998 and 1.002, respectively. $\square$

When we consider a *complex* linear polynomial $\bar{\ell}(x) = \alpha_0 + \sum_\sigma \alpha_\sigma x_\sigma$, with $\alpha_\sigma \in \mathbb{C}$ and x varying in $\mathbb{C}^s$, the normal direction $n(x)$ which induces the *strongest variation* in the value of $\bar{\ell}$ along $x + t\,n(x)$ is determined by the requirement

$$a^T n(x) \;=\; \|a^T\|^* \cdot \|n(x)\| \,, \qquad \text{which implies} \quad n(x) \;:=\; (\alpha_1^*, \ldots, \alpha_s^*)^T \,, \tag{7.17}$$

where $..^*$ is the conjugate-complex value. Otherwise, the above analysis remains unaltered.

In section 3.2.3, we have noticed that the diameters of pseudozero sets associated with isolated zeros z grow proportionally to the distance $\|z\|$ of z from the origin, for sufficiently large $\|z\|$. The estimate (7.16) shows that, likewise, the thickness of the pseudozero sheet $Z_\delta[(\bar{\ell}, e)]$ about the zero manifold $Z[\bar{\ell}]$ grows proportionally with the distance of the location from the origin. This also holds for the zero sets of general s-variate polynomials and suggests the following approach:

If we are interested in the perturbation sensitivity of a zero manifold in a region at a moderate distance from the origin, we may consider absolute effects as we have generally done so far. In this regime, local effects will generally dominate the condition of the manifold. At large distances from the origin, we must consider the perturbation effects at z *relative to* $\|z\|$; this relative condition often tends to a limit for growing $\|z\|$, at least along a fixed direction in $\mathbb{R}^s$ or $\mathbb{C}^s$.

In the general case of an empirical polynomial $(\bar{p}, e)$, with $\bar{p} \in \mathcal{P}^s$, we have, at some point z in the zero set $Z[p] \subset \mathbb{C}^s$ of $\bar{p}$,

$$\bar{p}(z + \Delta x) = \bar{p}'(z) \cdot \Delta x + O(\|\Delta x\|^2), \qquad \text{with } \bar{p}'(z) = (\partial_1 \bar{p}(z), \ldots, \partial_s \bar{p}(z)),$$

so that grad $\bar{p}(z) = \bar{p}'(z)$ takes the role of $\bar{a}$ in the previous analysis. Obviously, we must watch that we are sufficiently away from a potential singular zero of $\bar{p}$ with $\bar{p}'(z) = 0$; such points will be considered in the following section. Then (7.16) with (7.17) implies Proposition 7.6:

Proposition 7.6. Except in the neighborhood of a singular zero, the diameter $d_\delta(z)$ of the enveloping manifold sheet $Z_\delta[(\bar{p}, e)]$ of the manifold $Z[\bar{p}]$ satisfies

$$d_\delta(z) = 2\delta \frac{\|\mathbf{z}\|_e}{\|\bar{p}'(z)\|_2} (1 + O(\|e\|)). \tag{7.18}$$

The backward error of a point on the boundary of the δ-manifold sheet is $\delta (1 + O(\|e\|))$.

This result is the immediate generalization of (5.38) for the condition of a zero of a univariate polynomial p: For a multivariate polynomial, the local rate of change of p is represented by the gradient vector $p'(z)$. As in Example 7.5, it is also obvious that (7.18) is invariant against a simple scaling of the empirical polynomial: $(\bar{p}, e) \rightarrow (\lambda \bar{p}, \lambda e)$. Furthermore, the proportional growth of $d_\delta(z)$ with $\|\mathbf{z}\|$ is directly displayed in (7.18).

Example 7.8: We consider the empirical quadratic polynomial $(\bar{p}, e)$ with $\bar{p}(x) = \bar{\alpha}_0 + \bar{\alpha}_1 x_1^2 + \bar{\alpha}_2 x_2^2 + \bar{\alpha}_3 x_3^2$, real α_ν, intrinsic sparsity, and real variations of the coefficients only. At $z = (z_1, z_2, z_3) \in \mathbb{R}^3$ on the quadratic manifold $\bar{p}(x) = 0$, the normal vector which indicates the direction of the strongest variation of $\bar{p}$ is $\bar{p}'(z) = 2 (\bar{\alpha}_1 z_1, \bar{\alpha}_2 z_2, \bar{\alpha}_3 z_3)^T$. The vector $\mathbf{z}$ is simply $(1, z_1^2, z_2^2, z_3^2)^T$. From (7.18) we have

$$d_\delta(z) = \delta \frac{\varepsilon_0 + \varepsilon_1 z_1^2 + \varepsilon_2 z_2^2 + \varepsilon_3 z_3^2}{\sqrt{(\bar{\alpha}_1 z_1)^2 + (\bar{\alpha}_2 z_2)^2 + (\bar{\alpha}_3 z_3)^2}} (1 + O(\|e\|)).$$

When we assume that $\bar{\alpha}_1 \geq \bar{\alpha}_2 \geq \bar{\alpha}_3 > 0 > \bar{\alpha}_0$, the manifold $\bar{p}(x) = 0$ is a standard ellipsoid in 3-space. At the vertices $v = (\pm \sqrt{\frac{|\bar{\alpha}_0|}{\bar{\alpha}_1}}, 0, 0)$, e.g., the thickness of the "wall" of the pseudozero ellipsoid is

$$d_\delta(v) \approx \delta \frac{\varepsilon_0 + \varepsilon_1 |\bar{\alpha}_0|/\bar{\alpha}_1}{\sqrt{|\bar{\alpha}_0| \bar{\alpha}_1}} = \delta |v_1| \left(\frac{\varepsilon_0}{|\bar{\alpha}_0|} + \frac{\varepsilon_1}{\bar{\alpha}_1}\right).$$

This result may also be obtained by differentiation of $\bar{\alpha}_1 v_1^2 = |\bar{\alpha}_0|$: $\bar{\alpha}_1 v_1 2 dv_1 + d\alpha_1 v_1^2 = d\alpha_0$ yields the above estimate for $d_\delta(v_1) = 2 dv_1$ with $|d\alpha_i| \leq \delta \varepsilon_i$, $i = 1, 2$.   □

Exercises

1. Consider the real quadratic polynomial in 3 variables

$$\bar{p}(x, y, z) = 5.46\, x^2 - 3.97\, x\, y + 9.71\, y^2 - 1.48\, x + 5.23\, y - 4.18\, z - 2.35\,.$$

What is the sparsity structure of $\bar{p}$? What is the geometric characterization of the set $Z[\bar{p}] \subset \mathbb{R}^3$ of the real zeros of $\bar{p}$; how is it related to the sparsity of $\bar{p}$? Does the set $Z_{\mathbf{C}}[\bar{p}] \subset \mathbb{C}^3$ of the complex zeros of $\bar{p}$ reflect the same dependence on the sparsity of $\bar{p}$?

(a) For each of the nonvanishing coefficients individually, assume that it is the *only* empirical coefficient of an empirical polynomial $(\bar{p}, e)$. Which geometric aspect of the real zero manifold is affected by each of these assumptions ?

(b) Perform the same analysis for each of the vanishing coefficients in $\bar{p}$ and describe the geometric effects.

(c) Find some exact zeros (except for round-off) of $\bar{p}$ by selecting their x and y components and computing the associated z component. Perturb these zeros and check, for a specified tolerance vector e, whether they have remained valid pseudozeros of $(\bar{p}, e)$.

2. Consider a generic polynomial $p = a^T \mathbf{x} \in \mathcal{P}_3^2$ with $\mathbf{x} := (1, x_1, x_2, x_1^2, \dots, x_1 x_2^2, x_2^3)^T$.

(a) Transform $p(x_1, x_2)$ into $q(\xi_1, \xi_2) = b^T \xi := p(c_1 + \xi_1, c_2 + \xi_2)$, i.e. form the Taylor expansion of p about c. Assume that the coefficients $\alpha_{j_1 j_2}$ in a^T are empirical, with tolerances $\varepsilon_{j_1 j_2}$. Find the largest absolute deviations $\Delta\beta_{j_1 j_2}$ which can arise in the coefficients b^T of q when a^T varies in $N_1(p, e)$ and define these as tolerances $\hat{\varepsilon}_{j_1 j_2}$ of the $\beta_{j_1 j_2}$. Show that the lower triangular 10×10 matrix $C(c)$ which transforms the row vector e^T of the $\varepsilon_{j_1 j_2}$ into the row vector $\hat{e}^T$ of the $\hat{\varepsilon}_{j_1 j_2}$ by multiplication from the right has the columns $(\mathbf{x}(|c|),\ \partial_{10}[|c|]\mathbf{x},\ \partial_{01}[|c|]\mathbf{x},\ \partial_{20}[|c|]\mathbf{x},\ \dots,\ \partial_{12}[|c|]\mathbf{x},\ \partial_{03}[|c|]\mathbf{x})$; cf. (7.2) for the notation. What is the impliciation for the "tolerances" $\hat{e}$?

(b) Obviously, the back transformation from q to p implies an analogous transformation of the $\hat{e}^T$ vector to a $\tilde{e}^T$ vector of new tolerances for the a^T coefficients, with the transformation matrix $C(-c)$; thus, $\tilde{e}^T = e^T\, C(c)\, C(-c) = e^T\, C(c)^2$. Compute $C(c)^2$ and consider the implications for the components of $\tilde{e}^T$. Discuss the precise meaning of $\tilde{e}$.

3. Consider the transformations described in Example 7.4, for some polynomial $p \in \mathcal{P}_3^2$ with floating-point coefficients and an arbitrarily chosen Groebner basis of the structure indicated there. To avoid the syzygy problem, generate the g_κ as Groebner basis elements for two quadratic polynomials with rational coefficients.

(a) Find the 10×10 matrices which transform the Δa^T vector into the Δbc^T vector and back. From specified max-tolerances for the α_j compute the max-tolerances for the β_j, γ_j; then compute new tolerances for the α_j from these β_j, γ_j tolerances and compare.

(b) Try to modify the data in p and in the g_κ such that the expansion effect of the tolerances gets worse.

(c) Is it possible to choose coefficients for p and the g_κ such that the tolerances of the α_j remain unchanged after the forward and backward transformation ?

(d) Consider the problem of finding bounds for the potential variation of the bc coefficients when we assume that the coefficients in the Groebner basis are also empirical, with known tolerances.

4. Take the empirical polynomial $(\bar{p}, e)$ of the continuation of Example 7.2.

7.3 Singular Points on Algebraic Manifolds

7.3.1 Singular Zeros of Empirical Polynomials

One of our major tools in dealing with nonlinearities in polynomials is *local linearization*. Therefore, it is important to recognize situations where local linearization may fail or be likely to give an unsatisfactory answer; we have mentioned such a case after Proposition 7.4 and discussed it further in Exercise 7.1-3. This section is fully devoted to the case where the gradient vector vanishes at some point on the algebraic manifold specified by $p(x) = 0$, i.e. on the zero manifold $Z[p]$ of p. This analysis is closely connected with the 2nd question posed at the end of section 7.3.2 regarding the perturbation invariance of the geometrical structure of the zero manifolds defined by the polynomials in a tolerance neighborhood $N_\delta(\bar{p}, e)$.

Definition 7.3. For a polynomial $p \in \mathcal{P}^s$, a point $\xi \in \mathbb{C}^s$ with $p(\xi) = 0$ *and* $p'(\xi) = 0$ is called a *singular point* of the zero manifold $Z[p]$, or a *singular zero* of p. A zero of p which is not singular will occasionally be called a *regular zero* of p. $\quad\square$

Definition 7.4. For the empirical polynomial $(\bar{p}, e)$, with $\bar{p} \in \mathcal{P}^s$, a point $\xi \in \mathbb{C}^s$ is a *valid singular zero* of $(\bar{p}, e)$ if there exists a polynomial $\tilde{p} \in N_\delta(\bar{p}, e)$, $\delta = O(1)$, for which ξ is a singular zero. $\quad\square$

Let us at first consider the structure of the manifold $p(x) = 0$ at a regular zero ξ. At such points, the manifold possesses a unique *tangential manifold* which is the zero set of the linear approximation (7.4) of p at ξ.

Proposition 7.7. For $p \in \mathcal{P}^s$, consider a point $\xi \in Z[p] \subset \mathbb{C}^s$ and an arbitrary path $\bar{x}(t)$ in $Z[p]$ through $\xi = \bar{x}(0)$. If $p'(\xi) \neq 0$, the tangential vector $\bar{x}'(0)$ of the path at ξ satisfies

$$\sum_{\sigma=1}^{s} \partial_\sigma p(\xi)\, \bar{x}'_\sigma(0) \; = \; p'(\xi) \cdot \bar{x}'(0) \; = \; 0\,. \tag{7.19}$$

Proof: We have assumed $p(\bar{x}(t)) \equiv 0$; hence, at each t, $p'(\bar{x}(t)) \cdot \bar{x}'(t) = 0$ which is (7.19) for $t = 0$. $\quad\square$

Example 7.9: Consider the generic second-degree polynomial in two variables $p(x, y) = \alpha_{11}\, x^2 + \alpha_{21}\, xy + \alpha_{22}\, y^2 + \alpha_1\, x + \alpha_2\, y + \alpha_0$, $\alpha_j \in \mathbb{C}$, with

$$p'(x, y) \; = \; (\partial_x p(x, y), \partial_y p(x, y)) \; = \; (2\,\alpha_{11}\, x + \alpha_{21}\, y + \alpha_1,\; \alpha_{21}\, x + 2\,\alpha_{22}\, y + \alpha_2)\,.$$

At (ξ, η) with $p(\xi, \eta) = 0$, the tangent of the "conic section" $p(x, y) = 0$ is given by the linear polynomial

$$(2\,\alpha_{11}\, \xi + \alpha_{21}\, \eta + \alpha_1)\, (x - \xi) + (\alpha_{21}\, \xi + 2\,\alpha_{22}\, \eta + \alpha_2)\, (y - \eta) \; = \; 0\,.$$

Iff $p'(\xi, \eta) = 0$, this polynomial is not defined. We want to understand what happens in this case: Let $(\bar{x}(t), \bar{\eta}(t))$ be a path through (ξ, η) inside the manifold as in Proposition 7.7 and set $\bar{x}(t) = \xi + \Delta x(t)$, $\bar{y}(t) = \eta + \Delta y(t)$; then (cf. (7.3) and Exercise 7.1-3)

$$p(\bar{x}(t), \bar{y}(t)) \; = \; p(\xi + \Delta x, \eta + \Delta y) \; = \; \tfrac{1}{2}\, p''(\xi, \eta)\, ((\Delta x, \Delta y), (\Delta x, \Delta y)) \; =$$

$$(\Delta x, \Delta y) \begin{pmatrix} \alpha_{11} & \alpha_{21}/2 \\ \alpha_{21}/2 & \alpha_{22} \end{pmatrix} \begin{pmatrix} \Delta x \\ \Delta y \end{pmatrix} \; = \; \alpha_{11}\, \Delta x^2 + \alpha_{21}\, \Delta x \Delta y + \alpha_{22}\, \Delta y^2 \; = \; 0\,.$$

Over $\mathbb{C}$, the quadratic form in Δx, Δy always factors into two linear factors which yield

$$\Delta y(t) \; = \; \frac{1}{2} \left(-\alpha_{21} \pm \sqrt{\alpha_{21}^2 - 4\alpha_{11}\alpha_{22}} \,\right) \Delta x(t) \,. \tag{7.20}$$

Thus there are *two* possible directions for a path through (ξ, η) within the manifold, i.e. the singular zero is a point of *self-intersection* of the manifold. To possess such a self-intersection is a distinctive geometric quality of the manifold which, in this case, is a pair of straight lines in $\mathbb{C}^2$ intersecting at (ξ, η). The two lines may be real, like in $x^2 - y^2 = (x - y)(x + y)$, or complex, like in $x^2 + y^2 = (x - \mathrm{i}y)(x + \mathrm{i}y)$. If $\alpha_{21}^2 - 4\alpha_{11}\alpha_{22} = 0$, they coincide; all points on this "double line" are two-fold zeros of p.

It is well known that quadratic curves have no self-intersections except when they are degenerate; therefore it is natural to ask for the set of those coefficients in a polynomial $p \in \mathcal{P}_2^2$ for which a degeneration occurs. They must permit that the three polynomials $p(x, y)$, $\partial_x p(x, y)$, $\partial_y p(x, y)$ have a common zero (ξ, η). We may assume $\alpha_{21}^2 - 4\alpha_{11}\alpha_{22} \neq 0$ and solve the two linear equations for ξ, η in terms of the α_{ij} and substitute into $p = 0$; this yields a homogeneous polynomial of degree 3 in the 6 coefficients α_{ij}:

$$r(\alpha_{ij}) \; := \; (4\alpha_{11}\alpha_{22} - \alpha_{21}^2)\,\alpha_0 - \alpha_{22}\,\alpha_1^2 - \alpha_{11}\,\alpha_2^2 + \alpha_1\,\alpha_{21}\,\alpha_2 \,; \tag{7.21}$$

iff r vanishes for coefficients a of $p(x; a)$, there exists a singular zero of p and the manifold $Z[p]$ is degenerate. (The case $\alpha_{21}^2 - 4\alpha_{11}\alpha_{22} = 0$ is special: There may be one line consisting of singular points or two parallel lines, with their intersection at ∞.) Note that $r(a) = 0$ describes an algebraic manifold of codimension 1 in the data space $\mathcal{A} = \mathbb{C}^6$ of p.

This analysis shows why there is a problem: For an empirical polynomial $(\bar{p}, e)$, the specified coefficients $\bar{a}$ may not satisfy $r(\bar{a}) = 0$ but the neighborhood families $N_\delta(\bar{a}, e)$ may intersect with the manifold $r(a) = 0$ for $\delta = O(1)$. In this case, the family of quadratic curves associated with polynomials $\tilde{p} \in N_\delta(\bar{p}, e)$ for $\delta = O(1)$ contains nondegenerate and degenerate curves, and the nondegenerate ones may fall into different geometric patterns. This may cause an ambiguity in the interpretation of results; it may also happen that the unique degenerate pattern is the "true" pattern modelled by the empirical polynomial.

A trivial example for a situation with a valid singular zero is the empirical polynomial $(\bar{p}, e)$ with $\bar{p}(x, y) = x^2 - y^2 + .01$, $e = (0, 0, .05)$. For $\delta > .2$, the family of manifolds of the polynomials in $N_\delta(\bar{p}, e)$ contains the degenerate line pair $x - y = 0$, $x + y = 0$ as well as regular hyperbolas with the x-axis as their principal axis while $\bar{p} = 0$ represents a hyperbola with the y-axis as principal axis. Within the family of manifolds, when the constant term changes from 0 to $\pm\varepsilon$, the branches of the hyperbolas retreat from the origin at a rate $O(\sqrt{\varepsilon})$ while the tangential directions of the manifolds jump discontinuously. $\square$

In the general multivariate case, we have an empirical polynomial $(\bar{p}, e) \subset \mathcal{P}_d^s$ and wish to find out whether there are polynomials $\tilde{p} \in N_\delta(\bar{p}, e)$, $\delta = O(1)$, which possess one or several singular zeros. If this is true the family $Z_\delta[(\bar{p}, e)]$ of manifolds, i.e. the family of the manifolds $Z[\tilde{p}]$ for $\tilde{p} \in N_\delta(\bar{p}, e)$, will contain manifolds whose geometric structure differs from that of the manifold $Z[\bar{p}]$. At first, we consider the criterion for the existence of a singular zero: According to Definition 7.3, a singular zero must satisfy the *overdetermined* polynomial system of $s + 1$ polynomials in s variables $x_1, \ldots, x_s$

$$S[p] \; := \; \{\, p(x), \; \partial_1 p(x), \; \ldots, \; \partial_s p(x) \,\} \,; \tag{7.22}$$

generically, this system is inconsistent and has no solutions.

There is an extensive theory in polynomial algebra which analyzes the conditions on the coefficients of an *overdetermined system* $P = \{p_\nu \in \mathcal{P}^s, \ \nu = 0(1)s\}$ which imply that the p_ν have at least one common zero, i.e. that the ideal $\langle P \rangle$ is not the trivial ideal $\langle 1 \rangle$. This is the theory of the so-called *resultants*; like other parts of computational algebra, it was flourishing in the second part of the 19th and the beginning of the 20th century, was then pushed into oblivion through the growth of abstract algebra, and has seen a revival with the advent of computer algebra. An introduction into resultant theory would surpass the purpose of this book; we only summarize a few results which are of interest in the context of our present investigations.

A resultant for a system P of $s+1$ polynomials $p_\nu \in \mathcal{P}^s$ is a polynomial in the coefficients of the p_ν such that the resultant vanishes if and only if the system P has a common zero. Generally, it is assumed that the system P has been *homogenized* by the introduction of a dummy variable x_0 and the transformations $p_\nu \rightarrow P_\nu \in \mathcal{P}_{d_\nu}^{s+1}$

$$P_\nu(x_0, x_1, \ldots, x_s) := x_0^{d_\nu} \, p_\nu(\frac{x_1}{x_0}, \ldots, \frac{x_s}{x_0}), \quad \nu = 0(1)s, \tag{7.23}$$

where $d_\nu > 0$ is the total degree of p_ν and thus the homogeneous degree of P_ν. The coefficients of P_ν are the same as those of p_ν. The following is well known:

Theorem 7.8. There exists a unique (except for a scalar factor) polynomial *Res* in the *coefficients* $a_\nu := (\alpha_{\nu,j})$ of the P_ν, with integer coefficients, such that the P_ν have a common zero different from $(0,0,..,0)$ if and only if $Res(a) = 0$. The polynomial *Res* is homogeneous of degree $\sum_{\nu=0}^{s} d_0 .. d_{\nu-1}\, d_{\nu+1} .. d_s$ and irreducible over $\mathbb{C}^M$, with M the number of coefficients in P.

The best-known cases are $s = 1$, with arbitrary degrees d_0, d_1, and $s = n > 1$, $d_\nu = 1$. In the first case, the resultant is the determinant of the Sylvester matrix $S[P_0, P_1] := S[p_0, p_1]$ of (6.23); its vanishing is necessary and sufficient for the existence of a common zero, cf. section 6.2.2. This determinant is clearly a homogeneous polynomial of degree $d_0 + d_1$ in the coefficients, with all *its* coefficients equal to ± 1. In the second case, we have a system of $n + 1$ *linear* homogeneous polynomials in $x_0, \ldots, x_n$ and the resultant is the determinant of the matrix of this system which is homogeneous of degree $n + 1$ in the coefficients.

For all other nontrivial cases, the degree tends to become large and the number of terms excessive. For example, for $s = 2$ and $d_\nu = 2$, i.e. for a system of 3 quadratic equations in 2 variables or in 3 homogeneous variables, resp., with its $3 \cdot 6 = 18$ coefficients, the resultant is a homogeneous polynomial of degree 12 (cf. Theorem 7.8) in the $3 \cdot 6 = 18$ coefficients, with 21894 terms! But there are a number of ways to represent a resultant more economically which make it possible to compute them and to use them algorithmically.

In (7.21), we have met a special case of the resultant for $s = 2$, $d_0 = 2$, $d_1 = d_2 = 1$, in inhomogeneous form: While the generic resultant for such a system has degree $1 \cdot 1 + 2 \cdot 1 + 2 \cdot 1 = 5$ in the 12 coefficients and 21 terms, (7.21) identifies the coefficients of the linear equations with certain coefficients of the quadratic equation; thus there are only the 6 coefficients of the quadratic equation, the degree is reduced to 3 and the number of terms to 5.

Obviously, it is the generalization of this *special* situation to higher values of s and a higher degree $d_0 = d$ of the polynomial p_0 (our original polynomial p) in which we are interested now. Therefore, we will not use resultants but rather approach the potential solvability of the overdetermined system (7.22) and its neighbors directly, with tools that we have used previously.

7.3.2 Determination of Singular Zeros

In the univariate case, singular zeros ξ, with $p(\xi) = p'(\xi) = 0$, are also multiple zeros of $p \in \mathcal{P}^1$. There, the following approach to the determination of potential multiple zeros of the empirical polynomial $(\bar{p}, e)$ is natural (cf. section 6.3.4): We look for a potential $\tilde{p}$, with (near-)minimal $\|\tilde{p} - \bar{p}\|_e^*$ of $O(1)$, such that there exists a ξ with $\tilde{p}(\xi) = \tilde{p}'(\xi) = 0$. For this purpose, we find the zeros ξ_ν of $\bar{p}'(x)$ and check their backward errors as zeros of $(\bar{p}, e)$ to select the potential candidates for common pseudozeros (if any). In the neighborhood of such a ξ_ν, the value of $\bar{p}(x)$ is near-stationary; therefore the backward error $\delta(\xi_\nu)$ is practically identical to $\|\tilde{p} - \bar{p}\|_e^*$ for the closest $\tilde{p}$ with a multiple zero. For ξ_ν with $\delta(\xi_\nu) = O(1)$, we find the closest neighbor p_1 of $\bar{p}$ with $p_1(\xi_\nu) = 0$ and then take a Newton step towards an approximate zero $\xi_{\nu,1}$ of $p_1'(x)$. Because of $p_1'(\xi_{\nu,1}) \approx 0$, the step x leaves $p_1(\xi_{\nu,1}) \approx 0$; the backward error of $\xi_{\nu,1}$ as a *simultaneous zero* of $\tilde{p}$ and $\tilde{p}'$ confirms the existence of the requested $\tilde{p} \approx p_1$ in an $O(1)$ neighborhood of $\bar{p}$.

The core observation that the value of p remains nearly stationary near a zero ξ of p' translates directly to the multivariate case: For $\partial_\sigma p(\xi) = 0$, $\sigma = 1(1)s$, we have $p(\xi + \Delta x) = p(\xi) + O(\|\Delta x\|^2)$ by (7.3). Thus, we may proceed in the following way:

For the empirical polynomial $(\bar{p}, e)$, with $\bar{p} \in \mathcal{P}_d^s$, we consider the system dP of the s polynomials $\partial_\sigma \bar{p} \in \mathcal{P}_{d-1}^s$, $\sigma = 1(1)s$; for the following, we assume that dP and its neighbors $d\tilde{P}$ for $\tilde{p} \in N_\delta(\bar{p}, e)$ are 0-dimensional. We find numerical approximations ξ_ν, $\nu = 1(1)n$, for the finitely many zeros of dP; at first, we assume that there are no multiple zeros or close clusters of zeros which is equivalent to the requirement that the symmetric $s \times s$ matrices $\bar{p}''(\xi_\nu)$ are not near-singular. Then we evaluate the backward errors

$$\delta(\xi_\nu) = |\bar{p}(\xi_\nu)| / \| \left(\xi_\nu^j \right) \|_e, \qquad \nu = 1(1)n ;$$

the vector in the denominator contains those power products whose coefficients in $(\bar{p}, e)$ are empirical. If all backward errors $\delta(\xi_\nu)$ are well greater than $O(1)$, there are no polynomials $\tilde{p} \in N_\delta(\bar{p}, e)$ with $\delta = O(1)$ with a singular zero. This implies that the geometrical structure of the algebraic manifold $Z[p]$ remains invariant for all polynomials in the tolerance neighborhood of $\bar{p}$.

Now we assume that there exists some $\xi_\nu \in \mathbb{C}^s$ with $\delta(\xi_\nu) = O(1)$; we denote it by $\bar{\xi}$. From (7.14), we find $\Delta a = (\Delta\alpha_j, j \in \tilde{J})$ with $\|\Delta a\|_e^* = \delta(\bar{\xi})$ such that $p_1(\bar{\xi}) := p(\bar{\xi}; \bar{a} + \Delta a) = 0$. Then we form the polynomial system $dP_1 := \{\partial_\sigma p_1(x), \sigma = 1(1)s\}$. With our assumption on $\bar{p}''$, we may expect that the matrix $p_1''(\bar{\xi})$ whose elements differ from those of $\bar{p}''(\bar{\xi})$ by $O(\|e\|^*)$ is not near-singular; thus, we can perform the Newton refinement step

$$p_1''(\bar{\xi})\, \Delta\xi = -(p_1'(\bar{\xi}))^T \tag{7.24}$$

for $\Delta\xi$ and form $\xi_1 := \bar{\xi} + \Delta\xi$.

Now we determine the backward error of ξ_1 as a simultaneous pseudozero of the overdetermined empirical system which consists of $(\bar{p}, e)$ and its s partial derivatives (cf. section 3.3.1): In the data space $\Delta\mathcal{A}$ of the empirical coefficients of $(\bar{p}, e)$, with origin at the empirical components of $\bar{a}$, we consider the $s + 1$ linear manifolds

$$\Delta a^T \begin{pmatrix} \vdots \\ \xi_1^j \\ \vdots \end{pmatrix} + \bar{p}(\xi_1) = 0, \qquad \Delta a^T \begin{pmatrix} \vdots \\ \partial_\sigma x^j(\xi_1) \\ \vdots \end{pmatrix} + \partial_\sigma \bar{p}(\xi_1) = 0, \ \sigma = 1(1)s .$$

As above, the vectors range over the _M empirical_ terms of $(\bar{p}, e)$, and we must assume that $M \geq s + 1$ so that the intersection $\widetilde{\mathcal{M}}$ of the manifolds is not empty. The backward error of ξ_1 as a simultaneous pseudozero of $(\bar{p}, e)$ and its partial derivatives equals the shortest $\|..\|_e^*$ distance of the linear manifold $\widetilde{\mathcal{M}}$ from the origin of $\Delta\mathcal{A}$.

The following asymptotic argument shows that our procedure will succeed for sufficiently small tolerances e: Let $\bar{\varepsilon} := \|e\|^*$; then $\delta(\bar{\xi}) = O(1)$ implies $\|\Delta a\|^* = O(\bar{\varepsilon})$ so that $\bar{p}\,'(\bar{\xi}) = 0$ implies $p_1'(\bar{\xi}) = O(\bar{\varepsilon})$ and $\bar{p}''(\bar{\xi})^{-1} = O(1)$ implies $p_1''(\bar{\xi})^{-1} = O(1)$. Hence, (7.24) yields a $\Delta\xi = O(\bar{\varepsilon})$ and $p_1(\xi_1) = p_1(\bar{\xi}) + p_1'(\bar{\xi})\,\Delta\xi = O(\bar{\varepsilon}^2)$. On the other hand, the Newton refinement step (7.24) reduces $p_1(\bar{\xi}) = O(\bar{\varepsilon})$ to $p_1(\xi_1) = O(\bar{\varepsilon}^2)$. Thus, both p_1 and p_1' are $O(\bar{\varepsilon}^2)$ at ξ_1, which should make ξ_1 simultaneous pseudozero of $(\bar{p}, e)$ and of its derivatives. (A strict formalization of this argument is possible, with unwieldy quantitative assumptions about the derivatives of $\bar{p}$.)

For a ξ_ν in the "gray zone" between $\delta(\xi_\nu) = O(1)$ and $\delta(\xi_\nu) > O(1)$ which we have discussed in section 6.1.1, we may still try the above approach, with the hope for a potential success.

Example 7.10: We consider the following empirical polynomial $(\bar{p}, e)$ with

$$\bar{p}(x, y) := -.48 - 4.49\,x + 1.64\,y + 13.59\,x^2 + 9.01\,xy + 17.44\,y^2 - 4.68\,x^3 - 25.78\,x^2 y$$
$$-47.34\,xy^2 - 28.98\,y^3 + 1.20\,x^4 + 8.83\,x^3 y + 24.32\,x^2 y^2 + 29.78\,xy^3 + 13.67\,y^4\,,$$

and each coefficient rounded to two digits after the point, i.e. $\varepsilon_j = .005$ for each j, $|j| \leq 4$. A (real) plot of the manifold $V[\bar{p}]$ shows the slanted eight of Figure 7.3; but the structure of the figure near what looks like a self-intersection is ambiguous: The two loops may be separated or there may be a passage between them. A fine resolution shows that the latter situation prevails; but is this the case for all polynomials $\tilde{p} \in N_\delta(\bar{p}, e)$, $\delta = O(1)$? If there are some nearby $\tilde{p}$ with a singular zero and an exact self-intersection of $V[\tilde{p}]$, then there also exist polynomials with separated loops in the tolerance neighborhood of $\bar{p}$.

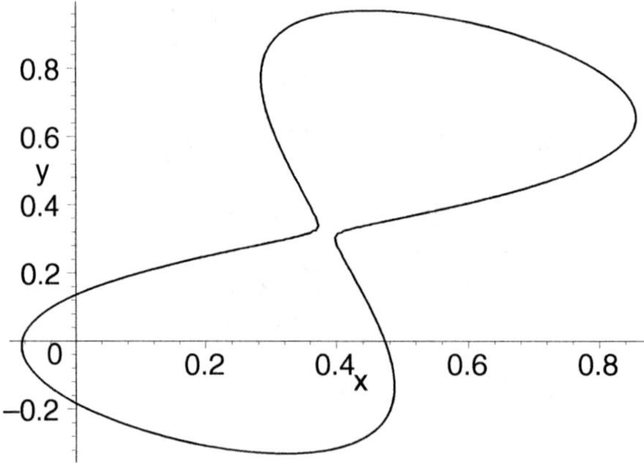

Figure 7.3.

The two polynomials $\partial_x \bar{p}(x, y)$, $\partial_y \bar{p}(x, y)$ have a total of 9 common zeros; there is actually one at the observed locus: $(\xi, \eta) \approx (.38526, .32024)$. Its backward error as a zero of $(\bar{p}, e)$ is

$\approx .31$ which is a strong indication for a valid singular zero. Its existence is confirmed by the evaluation of the backward error of (ξ, η) as a *simultaneous zero* of $\bar{p}$ and its derivatives which yields $\approx .48$. Thus the further computation along the lines of the algorithm indicated above serves only to discover the closest $\tilde{p}$ with a singular zero, and for illustration:

The modification of $\bar{p}$ into p_1 with an exact zero at (ξ, η) changes the coefficients of p by $.31 \cdot .005 = 00155$ or less which is well within the rounding domain. The Newton step for (ξ, η) to make it a good approximate common zero of the p_1 derivatives yields an increment of $\approx (.000048, .000365)$ or $(\xi_1, \eta_1) \approx (.38531, .32061)$; the backward error for this point as a simultaneous zero is reduced to $\approx .31$. This supports our observation that, since p is stationary near a singular zero of a nearby polynomial, the modification of (ξ, η) is practically irrelevant with respect to p and that a significant further reduction of the backward error is impossible.

Thus we have established the following assertions about $(\bar{p}, e)$:
- $(\bar{p}, e)$ has a valid singular zero at (ξ_1, η_1); therefore the geometric structure of the manifolds $V[\tilde{p}]$ for $\tilde{p} \in N_\delta(\bar{p}, e)$, $\delta = O(1)$, is *not invariant*.
- When the tolerance level e is reduced by a factor of 10 or more, the geometric structure of $V[\bar{p}]$, viz. the existence of a passage between the two real loops, becomes significant.

There are 6 further valid approximate singular points of $(\bar{p}, e)$ (two conjugate-complex pairs and two real ones), but none of them is of significance for the appearance of the *real* zero manifold. $\square$

7.3.3 Manifold Structure at a Singular Point

To understand and visualize the geometric structure of the algebraic manifold $Z[p]$ in the vicinity of a singular zero of $p \in \mathcal{P}_d^s$, we consider the special cases $d = 2$, $s > 2$ and $d > 2$, $s = 2$; the special case $d = s = 2$ has been considered in our introductory Example 7.9.

For the case of a quadratic polynomial in three or more variables, we can follow closely the analysis in Example 7.9 except for the use of a more compact notation, with $\mathbf{x} := (x_1, \ldots, x_s)^T$:

$$p(x) := \mathbf{x}^T A \mathbf{x} + 2 a^T \mathbf{x} + \alpha_0,$$

where A is the symmetric real or complex $s \times s$ matrix

$$A := \begin{pmatrix} \alpha_{11} & \alpha_{21}/2 & \cdots & \alpha_{s1}/2 \\ \alpha_{21}/2 & \alpha_{22} & \cdots & \alpha_{s2}/2 \\ \vdots & & \ddots & \vdots \\ \alpha_{s1}/2 & \alpha_{s2}/2 & \cdots & \alpha_{ss} \end{pmatrix},$$

and $a^T = (\alpha_1, \ldots, \alpha_s)$ is a real or complex s-vector. The s components of the gradient (row) vector $p'(x)$ are the linear polynomials:

$$p'(x) = 2 (\mathbf{x}^T A + a^T);$$

therefore, if A is regular, there is exactly one zero $\xi = -A^{-1} a$, which is a zero of p iff $r(a) = -p(\xi) = a^T A^{-1} a - \alpha_0 = 0$. Thus, with the right choice of α_0, each quadratic polynomial may become degenerate and have a singular zero.

Now we assume that $\xi = -A^{-1} a$ is a singular zero of p and analyze the manifold $V[p]$ in the neighborhood of ξ. By assumption, $p(\xi)$ and $p'(\xi)$ vanish and all derivatives of p of order greater $d = 2$ vanish; thus, by (7.3),

$$p(\xi + t\,\Delta x) \;=\; \frac{t^2}{2}\, p''(\xi)(\Delta x, \Delta x) \;=\; (t\,\Delta x)^T A\,(t\,\Delta x)\,. \qquad (7.25)$$

$p(\xi + t\,\Delta x) = 0$ is the equation of a quadratic *cone*, generated by straight lines through ξ. The geometric structure of the cone depends on the eigenstructure of A: If A is real symmetric, there exists the well-known real orthogonal decomposition $A\,U = U\,\Lambda$, with $U^T = U^{-1}$, $\Lambda = \mathrm{diag}\ \lambda_\sigma$; the λ_σ are the (real) eigenvalues of A. For a complex symmetric matrix, there is Takagi's factorization (7.9) $A\,U = (U^T)^{-1}\Lambda$, with unitary U and a *real* diagonal $\Lambda \geq 0$. We consider the real case first: With $\Delta x =: U\,\Delta y$,

$$\Delta x^T A \Delta x \;=\; \Delta y^T U^T A\, U \Delta y \;=\; \Delta y^T \Lambda\, \Delta y \;=\; \sum_{\sigma=1}^{s} \lambda_\sigma\, \Delta y_\sigma^2 \;=\; 0\,,$$

and the geometric structure of the cone depends on the number of positive, vanishing, and negative eigenvalues of A, as has been thoroughly investigated in the algebraic theory of *quadratic forms*. In $\mathbb{R}^3$, e.g., we have the cones $\lambda_1 \Delta y_1^2 + \lambda_2 \Delta y_2^2 + \lambda_3 \Delta y_3^2 = 0$; for $\lambda_1 = \lambda_2 = 1$, $\lambda_3 = -1$, this is a circular cone around the Δy_3 axis. Since the transformation between Δy and Δx is orthogonal, the geometric structure of the Δy cone describes the original manifold $V[p]$ as well. The tangential hyperplanes of the cone in its vertex ξ touch the cone along one of its generating straight lines.

In the complex case, the same substitution $\Delta x =: U\,\Delta y$ leads to $\Delta x^T A \Delta x = \Delta y^T \Lambda\, \Delta y = 0$ as previously, but now the transformation matrix U is unitary so that the reality and signs of Λ and the corresponding Δy cone structure have no intuitive meaning for the structure of $V[p]$ in $\mathbb{C}^s$. But straight lines transform into straight lines and the property of the tangential hyperplanes in the vertex of the complex cone to touch the cone along one of its generating lines remain formally valid.

As in Example 7.9, an $O(\bar{\varepsilon})$ perturbation of the degenerate quadratic polynomial leads to an $O(\sqrt{\bar{\varepsilon}})$ movement of the manifold away from the vertex ξ and to a discontinuous change in the tangential manifolds. In the real case, if none of the eigenvalues of A vanishes, the sign of $\Delta p(\xi)$ determines which of two potential geometric structures is assumed by the manifold $V[p + \Delta p]$, cf. Example 7.9. If the perturbation affects the vanishing of an eigenvalue of A, the situation is more complicated. In the complex case, an intuitive interpretation of the effects—except for the above statements—is virtually impossible.

The case $d > 2$, $s = 2$, is exemplified in Example 7.10 : For regular $p''(\xi, \eta)$, there are two isolated directions in which the manifold (= curve) $p(x, y) = 0$ passes through the singular zero (ξ, η). But now, these curves are no longer straight lines; only their tangents at (ξ, η) take one of the two distinguished directions determined by the quadratic form of the matrix $p''(\xi, \eta)$. The geometric structure of $V[p]$ away from (ξ, η) is independent of the local structure at the self-intersection. A perturbation of p leads to a switch to one of the two possible local hyperbola structures, depending on the sign of the perturbation at the singular zero.

Example 7.10, continued: At the singular zero (ξ_1, η_1) of p_1, the quadratic form of p_1'' is

(rounded)

$$\Delta x^T \begin{pmatrix} 13.51 & -16.08 \\ -16.08 & -11.19 \end{pmatrix} \Delta x = \Delta y^T \begin{pmatrix} -19.116 & 0 \\ 0 & 21.435 \end{pmatrix} \Delta y$$

for $\Delta x = \begin{pmatrix} .442 & .897 \\ .897 & -.442 \end{pmatrix} \Delta y$. Thus the quadratic form vanishes for

$$\Delta y_1 : \Delta y_2 = \pm \sqrt{\tfrac{21.435}{19.126}} \approx \pm 1.059 \quad \text{or} \quad \Delta x_1 : \Delta x_2 = 2.688 \text{ and } -.308 \,.$$

These are the directions of the tangents of the self-intersecting curve $p_1(x, y) = 0$ at the singular point (ξ_1, η_1); cf. Figure 7.4.

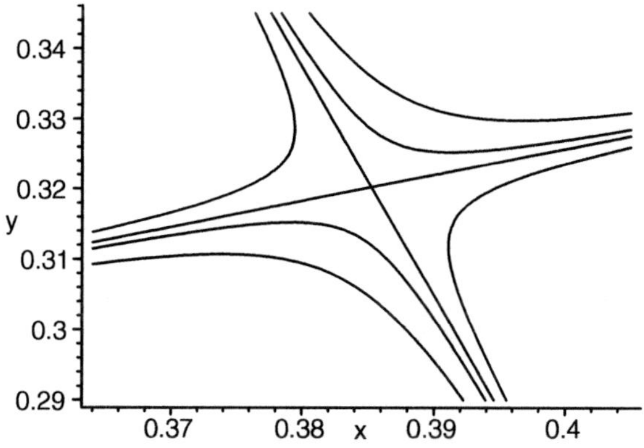

Figure 7.4.

With a generic perturbation Δp, the geometric structure of $V[p_1 + \Delta p]$ is either that of $V[\bar{p}]$ or that of two separated loops, depending on the sign of $\Delta p(\xi_1, \eta_1)$. For small Δp, the tangential direction of these curves changes smoothly but rapidly between the two directions Δx_i while the curves pass by (ξ_1, η_1); cf. Figure 7.4. $\square$

Without a further formal analysis, we conclude that the generic *local* structure of an algebraic manifold $p(x) = 0$, with $p \in \mathcal{P}_d^s$, $s > 2$, $d > 2$, at a singular point $\xi \in \mathbb{C}^s$ is modelled by an s-dimensional quadratic cone with vertex in ξ whose geometric type is determined by the eigenstructure of the matrix $p''(\xi)$. The asymptotic behavior under a perturbation corresponds to the one described above.

Exercises

1. For an algebraic manifold $\widehat{V}$ defined by $w = p(x)$ in $\mathbb{C}^{s+1}$, with $x \in \mathbb{C}^s$, $w \in \mathbb{C}$, its *stationary points* (ξ, ω) are distinguished by $\omega = p(\xi)$, $\partial_\sigma p(\xi) = 0$, $\forall \sigma$, i.e. by the occurrence of a tangential hyperplane with normal vector $(0, \ldots, 0, 1)^T$. Thus, the stationary points of $\widehat{V}$ are

singular zeros of $\hat{p}(x) := p(x) - \omega$ and vice versa. Translate our analysis of the geometric structure of $V[\hat{p})$ in the vicinity of a singular zero into an analysis of the geometric structure of the manifold $\widehat{V}$ in the vicinity of a stationary point.

2. Apply this analysis to the manifold $\widehat{V} \subset \mathbb{C}^3$ defined by $w = \bar{p}(x, y)$ with $\bar{p}$ of Example 7.10 :

(a) Characterize the geometric structure of $\widehat{V}$ in the vicinity of the stationary point $(\xi, \eta, \bar{p}(\xi, \eta))$.

(b) Find the values of $w_\mu := \bar{p}(\xi_\mu, \eta_\mu)$ at the other 4 real zeros (ξ_μ, η_μ) of $\partial_x \bar{p}(x, y) = \partial_y \bar{p}(x, y) = 0$ and analyze the singular zeros of the polynomials $p_\mu(x, y) := \bar{p}(x, y) - w_\mu$ or the stationary points $(\xi_\mu, \eta_\mu, w_\mu)$ of $\widehat{V}$, resp., for $\mu = 1, .., 4$. Consider the quadratic forms $(\Delta x, \Delta y) p_\mu''(\xi_\mu, \eta_\mu) (\Delta x, \Delta y)^T$ at these points and verify the expected behavior of $V[p_\mu]$ near (ξ_μ, η_μ) or of $\widehat{V}$ near $(\xi_\mu, \eta_\mu, w_\mu)$, resp., by plotting. (Hint: Add small positive or negative increments to w_μ to obtain satisfactory plots.)

(c) The (ξ_μ, η_μ) form two pairs with nearly equal values of w_μ. Try to generate a plot which covers the vicinity of *both* points in one pair simultaneously.

(d) Verify that, for suitably chosen values of $\bar{w}_\mu$, there exist polynomials $\tilde{p}_\mu$ in $N_\delta(\bar{p} - \bar{w}_\mu, e)$ with *two* singular zeros as it is made plausible by the plots of (c). Hint: Find the backward error of the $(\xi_{\mu_i}, \eta_{\mu_i})$, $i = 1, 2$, in one pair as *simulteaneous zeros* of $\bar{p} - \bar{w}$, $\partial_x \bar{p}$, and $\partial_y \bar{p}$; this backward error is the minimal $\|..\|_e^*$ distance of the intersection of 6 hyperplanes from the origin of the empirical data space $\Delta\mathcal{A} = \mathbb{C}^{15}$, cf. section 3.3.1.

3. Formulate the discussion about the invariance of the geometric structure of the manifolds $V[\tilde{p}]$ for $\tilde{p} \in N_\delta(\bar{p}, e)$ in terms of predicates; cf. section 6.1.1.

7.4   Numerical Factorization of a Multivariate Polynomial

7.4.1   Analysis of the Problem

In numerical polynomial algebra, each *univariate* polynomial decomposes into linear factors in $\mathbb{C}$ and in linear and quadratic factors in $\mathbb{R}$; cf. section 6.2. With *multivariate* polynomials, this is different: In the data space $\mathcal{A}$ of s-variate polynomials with a specified degree d and support $J \subset \mathbb{N}^s$, the coefficients of polynomials which possess nontrivial[11] factors occupy algebraic manifolds of dimensions generally much lower than dim $\mathcal{A}$. Therefore, exact multivariate factorizability is a *singular property* and a multivariate polynomial with (some) floating-point coefficients cannot be expected to be factorizable in the strict sense. This is the situation of a singular data$\rightarrow$result mapping which we have considered in paragraph 2 in section 3.2.1.

On the other hand, factorizability exhibits an important structural property of the zero set $Z[p]$ of a multivariate polynomial p : If $p(x) = u(x) \cdot v(x)$, then $Z[p] = Z[u] \cup Z[v]$ where, generally, $Z[u]$ and $Z[v]$ are completely independent algebraic varieties in $\mathbb{C}^s$.

Definition 7.5. If a polynomial $p \in \mathcal{P}^s$ factors into $m > 1$ (nontrivial) factors u_μ:

$$p(x) = u_1(x) \ldots u_m(x), \tag{7.26}$$

---

[11] Trivial factors are 1 and p.

the associated zero sets $Z[u_\mu]$ are *components* of the variety $Z[p]$; otherwise, p and $Z[p]$ are called *irreducible*. A component whose defining polynomial u_μ is irreducible is an *irreducible component*. $\square$

For all practical purposes, the information about the factors and components, resp., composes the information about p and its zero manifold.

In (7.26), it may happen that the u_μ are not all different, i.e. that there is a factor $[u_{\bar\mu}]^k$ with $k > 1$. In this case, the associated component is a *k-fold component* of the variety $Z[p]$. (An empirical polynomial with a pseudofactorization $[u(x)]^3$ is discussed in Example 7.15.)

If a multivariate polynomial p models some real-life situation, the factorizability of p may be a natural consequence of the model and interest may focus on the coefficients of a particular factor only. It can also happen that the factorizability of p implies a special property of the modelled situation which may or may not prevail. Thus, factorizability and the actual determination of factors play an important role in scientific computing. But, in a modelling situation, we will generally deal with *empirical polynomials* $(\bar p, e)$.

Definition 7.6. An empirical polynomial $(\bar p, e)$, $\bar p \in \mathcal{P}^s$, $s > 1$, is *(pseudo)factorizable* and has *pseudofactors* or *valid approximate factors* $\tilde u_\mu$ iff there exists a polynomial $\tilde p \in N_\delta(\bar p, e)$, $\delta = O(1)$, which is factorizable (cf. Definition 7.5) and has the exact factors $\tilde u_\mu$, $\mu = 1(1)m$, $m \geq 2$. $\square$

It is easy to verify whether a given set of polynomials $\tilde u_1, \ldots, \tilde u_m$ represents a valid factorization of some empirical polynomial $(\bar p, e)$. But it is not so clear how one should establish the *existence* of a pseudo-factorization. Even when we know that $(\bar p, e)$ has pseudofactors of a specified degree and structure, how do we determine the numerical values of their coefficients?

In the following, we restrict ourselves to the consideration of *two* factors u, v only; this is no essential restriction of generality. For a factorizable s-variate polynomial p of degree d, assume that u and v have degrees d_1 and $d - d_1$, with $d_1 \leq \frac{d}{2}$; then the coefficients α_j of p and β_j, γ_j of u, v satisfy

$$p(x) = \sum_{k=0}^{d} \sum_{|j|=k} \alpha_j x^j = \sum_{k=0}^{d_1} \sum_{|j|=k} \beta_j x^j \sum_{k=0}^{d-d_1} \sum_{|j|=k} \gamma_j x^j = u(x)\, v(x).$$

More explicitly, the β_j, γ_j have to satisfy the system

$$\sum_{|j_1|\leq d_1, |j-j_1|\leq d-d_1} \beta_{j_1}\, \gamma_{j-j_1} = \alpha_j, \qquad j \in \mathbb{N}_0^s : |j| \leq d. \tag{7.27}$$

For specified coefficients α_j, (7.27) is a system of bilinear equations for the coefficients β_j, γ_j with a very special structure:

- all nonvanishing coefficients are intrinsic and $= 1$;
- the only data are the constant right-hand terms, which may be empirical;
- each equation is sparse in a very special way.

Moreover, we observe that we can *normalize* the coefficients of p, u and v without affecting the factorization situation. The most natural normalization is

$$\alpha_0 = \beta_0 = \gamma_0 = 1, \tag{7.28}$$

which replaces the equation for $j = 0$ in (7.27); it has the further advantage of being impartial with respect to the variables. Of course, it requires $\alpha_0 \neq 0$, or rather $|\bar{\alpha}_0| \gg \varepsilon_0$, in an empirical setting. If this is not satisfied, there are many other possibilities for a normalization; cf. Exercise 7.4-3.

The system (7.27), with the normalization (7.28), has one equation for each $j \in \mathbb{N}_0^s$, $1 \leq |j| \leq d$, which gives a total of $\binom{d+s}{s} - 1$ equations; there are $\binom{d_1+s}{s} - 1$ unknowns β_j and $\binom{d_2+s}{s} - 1$ unknowns γ_j. It is easily checked that the number of equations always exceeds the number of unknowns and the more so the higher the degree d and the dimension s. Thus, (7.27) is an *overdetermined* system of polynomial equations in the β_j, γ_j.

While an intrinsic overdetermined system has a solution only for data a on some manifold $\mathcal{M}$ in its data space $\mathcal{A}$ and no solution for $a \notin \mathcal{M}$, there is the usual smooth transition between these states for overdetermined empirical systems. For empirical data $(\bar{a}, e)$, we have a family of neighborhoods $N_\delta(\bar{a}, e)$ and an approximate solution (i.e. factorization) with backward error δ exists when $\mathcal{M} \cap N_\delta$ is not empty. Thus, with no bound on δ, an approximate factorization *always* exists though its quality may be very poor. With our $O(1)$ concept for the *validity* of an approximate solution (cf. section 3.1.3), the transition between a valid and an invalid factorization is continuous, and the boundary between $\delta = O(1)$ and $\delta > O(1)$ can be adapted to the modelled situation.

In the algorithmic solution of (7.27), we will generally reach a "candidate factorization" (u_0, v_0) without having used all of the equations in the system. Naturally, we cannot expect these equations to be satisfied within their tolerances by the coefficients of (u_0, v_0). But we can employ our usual refinement procedure to reduce the backward error for those equations without generating an excessive backward error in the original subsystem. Since the data of the factorizable polynomial $\tilde{p} \in N_\delta$ of Definition 7.6 must be on $\mathcal{M}$, there is a positive lower bound $\underline{\delta} := \min_{\tilde{a} \in \mathcal{M}} \|\tilde{a} - \bar{a}\|_e$ to the achievable backward error (except for $\bar{a} \in \mathcal{M}$). If $\underline{\delta} > O(1)$, the pseudo-solution set $Z_\delta(\bar{p}, e)$ of (7.27) is empty for $\delta = O(1)$ and there does not exist a valid approximate factorization of $(\bar{p}, e)$.

Example 7.11: We consider the empirical polynomial $(\bar{p}, e)$ with

$$\bar{p}(x, y) := 1 - .82\,x + .91\,y + .15\,x^2 + 11.22\,xy - 8.71\,y^2 + 4.69\,x^3 - .65\,x^2y - 12.08\,xy^2 + 7.14\,y^3$$

and tolerances $\varepsilon_j = .005$ (except for the normalizing constant term 1). We assume that we know that there exists a real pseudofactorization, with deg $u = 1$, deg $v = 2$. The system (7.27) has 9 equations for the 7 unknown coefficients of the (normalized) factors u, v. The equations for $j = 10, 01, 20, 02, 30, 03$ form a subsystem of 6 equations for the 2 β_j and 4 of the 5 γ_j. The system has 9 solution 6-tuples, with 3 of them real. Substitution of a solution into the 11-equation yields the associated γ_{11}; the backward errors in the remaining two equations ($j = 21$ and 12) are approx. 6, 1000, and 20000 for the three real 6-tuples which leaves only one of them a candidate, viz. $\tilde{b} \approx (1.415, -1.738)$, $\tilde{c} \approx (-2.235, 2.648; 3.314, 3.588, -4.109)$.

We attach increments $\Delta b, \Delta c$ to the components of these $\tilde{b}, \tilde{c}$, substitute into (7.27), and drop the quadratic terms in the increments. Then we minimize the modulus of the residual of each equation. This yields a backward error of $\approx .67$ over the complete system and the following two valid approximate factors:

$$\begin{aligned} \tilde{u}(x, y) &= 1 + 1.4153\,x - 1.7330\,y, \\ \tilde{v}(x, y) &= 1 - 2.2353\,x + 2.6442\,y + 3.3137\,x^2 + 3.6011\,xy - 4.1228\,y^2. \quad \square \end{aligned}$$

For more complicated situations, the selection of a suitable subsystem may not be obvious. Also the number of isolated solutions of the subsystem may be very large and their determination may need a considerable computational effort; and yet all of them except one will be eliminated. When we have no a priori information about the existence of (pseudo)factors, the same approach requires that we test all solutions for all splittings of the degree d; but all of this computation may simply end in realizing that there is no pseudofactorization. Therefore, we will suggest a different approach in section 7.4.2.

The *nonexistence* of an *exact* factorization in $\mathbb{C}^s$ may be established in the following way which requires no a priori information:

We observe that two s-variate polynomials ($s > 1$) always have common zeros if we admit zeros at infinity. Therefore, when we consider the homogenized projective version $p(x, x_0)$ of $p(x)$, with the homogenizing variable x_0, the existence of a factorization $p(x) = u(x) \cdot v(x)$ of an s-variate polynomial, $s > 1$, implies that the zero set $Z[u, v] \subset P\mathbb{C}^s$ of the system $u(x, x_0) = v(x, x_0) = 0$ is not empty. For the projective version $p(x, x_0)$ of $p(x)$, we define *singular zeros* as in Definition 7.3 but we append the component $\partial_{x_0} p(x, x_0)$ to $p'(x, x_0)$; the trivial zero $x = x_0 = 0$ is disregarded as usual.

Proposition 7.9. For $p \in \mathcal{P}^s$, let $\overline{Z}[p] \subset P\mathbb{C}^s$ be the set of singular zeros of $p(x, x_0)$; cf. above. If $p(x) = u(x) \cdot v(x)$ then $Z[u, v] \subset \overline{Z}[p]$. Thus, $\overline{Z}[p] = \emptyset$ implies $Z[u, v] = \emptyset$ and the nonexistence of a factorization of p.

Proof: $p(x) = u(x) v(x)$ implies $p'(x) = v(x) u'(x) + u(x) v'(x)$ so that $x \in Z[u, v]$ implies $x \in \overline{Z}[p]$. □

A weaker form of Proposition 7.9 can be extended to empirical polynomials, with the same extension to the projective $P\mathbb{C}^s$ as above:

Corollary 7.10. If the empirical polynomial $(\bar{p}, e)$ is pseudo factorizable, then $(\bar{p}, e)$ possesses valid approximate singular zeros. Thus, if $(\bar{p}, e)$ does not possess valid approximate singular zeros, it cannot be pseudofactorizable.

Proof: Pseudofactorizability of $(\bar{p}, e)$ implies the existence of some $\tilde{p} \in N_\delta(\bar{p}, e)$, $\delta = O(1)$, with nontrivial factors $\tilde{u}, \tilde{v}$. By Proposition 7.9 we have, for this polynomial $\tilde{p}$, $Z[\tilde{u}, \tilde{v}] \subset \overline{Z}[\tilde{p}]$ so that $\overline{Z}[\tilde{p}] \neq \emptyset$ for a $\tilde{p} \in N_\delta(\bar{p}, e)$. □

Example 7.11, continued: To check the existence of valid approximate singular zeros of $(\bar{p}, e)$, we compute the 4 zeros of $(\partial_x \bar{p}, \partial_y \bar{p})$ and substitute them into $\bar{p}$. For the conjugate-complex pair $(.2658 \pm .5158\,i, .7908 \pm .4230\,i)$, the backward error in $(\bar{p}, e)$ is well below 1. This makes the pseudofactorizability of $(\bar{p}, e)$ possible.

Actually, in this simple case with $s = 2$ and $d = 3$, we know that a potential factorization must have $d_1 = 1$ and $d_2 = 2$; thus the intersection $Z[\tilde{u}, \tilde{v}]$ of the zero-sets of potential pseudofactors $\tilde{u}, \tilde{v}$ must consist of two points and the linear pseudo-factor $\tilde{u}$ is the straight line between these two points. The straight line between the two points in $\overline{Z}_\delta[(\bar{p}, e)]$ generates approximately the linear polynomial which we have previously found for $\tilde{u}$. □

For $s \geq 3$, the zero-sets $Z[u], Z[v]$ of potential factors have a dimension ≥ 2 and their intersection $Z[u, v]$ a dimension ≥ 1. The polynomial system $p'(x) = 0$, on the other hand, with s equations in s variables, continues to have a 0-dimensional zero-set if it is regular. Thus, by Corollary 7.10, the establishment of the 0-dimensionality of $Z[p']$ suffices for the establishment

of the nonfactorizability of p. As is well known (cf. also section 8.4), the 0-dimensionality of
the zero-set of a polynomial system can be directly observed from any Groebner basis of the
ideal generated by the polynomials in the system.

Unfortunately, this observation is not directly extendible to an empirical polynomial
$(\bar{p}, e) \in \mathcal{P}^s$, $s \geq 3$: Since a generic system of s polynomials in s variables is 0-dimensional, the
system $\bar{p}'(x) = 0$ obtained from $\bar{p}$ will generally not display the potential positive dimension
of a neighboring system $\tilde{p}'(x) = 0$. Actually, the existence of a positive-dimensional zero
manifold of p' for $p \in \mathcal{P}^s$ is another singular phenomenon whose numerical treatment requires
special care. We will deal with such singular systems in section 9.4.

Example 7.12: We consider the empirical polynomial $(\bar{p}, e)$, with $\bar{p} \in \mathcal{P}_5^3$ from (7.10) in Example 7.3 and $\varepsilon_j = .005$ for all nonvanishing coefficients except the normalizing constant term 1.
In section 7.4.3, we will establish that $(\bar{p}, e)$ is pseudofactorizable, with factors of degree 2 and
3. When we determine an exact Groebner basis of $\langle \partial_x \bar{p}(x, y, z), \partial_y \bar{p}(x, y, z), \partial_z \bar{p}(x, y, z) \rangle$,
we find that it is 0-dimensional, with 34 isolated zeros; this excludes the factorizability of $\bar{p}$
regarded as an exact polynomial.

Upon closer inspection of the Groebner basis, with *leading terms normalized to* 1, we
find that some of the basis elements contain very large coefficients, with moduli up to ≈ 50000.
This is a strong hint that the structure of this 0-dimensional Groebner basis of p' is not invariant
within the set of polynomials in $N_\delta(\bar{p}, e)$, $\delta = O(1)$, and that there exist valid neighboring
polynomials $\tilde{p}$ with a positive-dimensional ideal $\langle \tilde{p}' \rangle$.　　□

7.4.2　An Algorithmic Approach

Our analysis of the factorization of a multivariate polynomial has been based on a direct consideration of the product of two multivariate polynomials; cf. (7.27) and Proposition 7.9. The
standard *algebraic* approach to multivariate factorization is different: The problem is projected
onto a one-dimensional subspace, e.g., by substituting values for all but one variable. If this
univariate polynomial is factorizable—which is *not* a matter of course over the rational numbers
or some extension field—then the univariate factors are "lifted" to include the other variables,
if this is feasible. Otherwise, the polynomial is not factorizable. A more detailed description of
this *lifting technique* may be found in most books on algorithmic algebra.

For our empirical polynomials, the imitation of this algebraic procedure appears to fail
at the very beginning: In $\mathbb{C}$, a univariate polynomial is always factorizable, and, for larger
degrees, there are many ways to collect the linear factors into two polynomial factors for further
lifting. But if we are able to eliminate those combinations of univariate linear factors which have
no chance of serving as a "germ" for a multivariate pseudofactor, with a minimal effort, this
variant of the univariate approach becomes attractive: Either no univariate germ at all survives
the screening procedure, then we have established nonfactorizability; or one germ remains
(or perhaps a few), then we may attempt to extend it into a multivariate pseudofactor. In the
following sections, we will explain and elaborate this algorithmic approach in detail.

At first, we observe that a factorization (7.26) is an identity in the variables $x = (x_1, \ldots, x_s)$
and that it remains a correct relation upon substitution of any numerical values $\xi_\sigma \in \mathbb{C}$ for some
or all x_σ. In particular, if we substitute for all but one component (say x_1), we obtain a correct

factorization of the remaining univariate polynomial; for $m = 2$ this yields,

$$p(x_1, \xi_2, \ldots, \xi_s) = u(x_1, \xi_2, \ldots, \xi_s) \cdot v(x_1, \xi_2, \ldots, \xi_s). \qquad (7.29)$$

Reversely, the two univariate factors in (7.29) may be considered as univariate "germs" for the s-variate factors of p.

In particular, when we choose $\xi_2 = \ldots = \xi_s = 0$, we find that the coefficients of pure x_1-powers in a factorization of p are identical with the coefficients in a factorization of $p(x_1, 0, \ldots, 0)$. For a polynomial of degree d in x_1, there are at most $2^{d-1} - 1$ different possibilities for a factorization

$$p(x_1, 0, \ldots, 0) =: p_{10}(x_1) = u_{10}(x_1) \cdot v_{10}(x_1); \qquad (7.30)$$

if none of these can be extended ("lifted") into a factorization of p then such a factorization cannot exist. Surprisingly, the nonextendibility of a germ $u_1(x_1)$ can be discovered in a simple way:

Consider a second one of the variables (say x_2) and assume the remaining ones (if any) set to 0; then

$$p(x_1, x_2, 0, ..) =: p_{10}(x_1) + p_{11}(x_1) x_2 + p_{12}(x_1) x_2^2 + \ldots =$$

$$(u_{10}(x_1) + u_{11}(x_1) x_2 + u_{12}(x_1) x_2^2 + \ldots) \cdot (v_{10}(x_1) + v_{11}(x_1) x_2 + v_{12}(x_1) x_2^2 + \ldots);$$
$$(7.31)$$

for low degrees, certain terms vanish. For specified p_{11} from an expansion of $p(x_1, x_2, 0, ..)$ in powers of x_2 and a specified choice of u_{10}, v_{10} from (7.30), it is easy to compute the d_1 coefficients of u_{11}: Comparison of the coefficients of x_2 in (7.31) yields

$$u_{11}(x_1) v_{10}(x_1) + u_{10}(x_1) v_{11}(x_1) = p_{11}(x_1); \qquad (7.32)$$

substitution of the d_1 zeros of u_{10} yields d_1 linear equations for the coefficients of u_{11}. This assumes a generic situation in (7.32); in section 7.5.3, we will show that the determination of u_{11} from (7.32) is always possible.

Actually, we are interested only in the constant term β_{01} of $u_{11}(x_1)$ because this β_{01} also figures in a factorization of $p(0, x_2, 0, ..)$ into factors of degrees d_1, d_2:

$$p(0, x_2, 0, ..) = u_{01}(x_2) \cdot v_{01}(x_2) = (1 + \beta_{01} x_2 + \ldots) \cdot v_{01}(x_2);$$

cf. (7.30) and (7.31). Thus there are only a finite number of values which β_{01} can take *if a factorization* (7.31) *exists*: If the value computed from (7.32) does not *agree with one of them* (within a margin to be discussed later), then the selected germs $u_{10}(x_1)$, $v_{10}(x_1)$ from (7.30) cannot be extended even with respect to the variable x_2. The "target values" of β_{01} derive from the observation that, for any polynomial,

$$\prod_\nu (1 + \alpha_\nu x) = (1 + \beta_1 x + \beta_2 x^2 + \ldots + \beta_{d_1} x^{d_1}) \cdot (1 + \gamma_1 x + \ldots + \gamma_{d_2} x^{d_2})$$

implies

$$\beta_1 = \sum_{\lambda=1}^{d_1} \alpha_{\nu_\lambda}, \qquad (7.33)$$

where the sum is over some selection of d_1 of the α's. Thus the potential target values can be computed a priori for specified d_1.

If some β_{01} has passed the test and $s > 2$, the associated germ must also be checked against the remaining variables in an obvious analogous way. Assume that this further confirms the choice of the germ. Then we have to extend the germ to a full s-variate polynomial $u(x)$ of degree d_1. An analysis of the system (7.27) shows that it can be split into subsystems which permit the computation of groups of further coefficients in the two polynomial factors from *linear* equations. Details of this extension procedure will be explained in connection with examples below. Finally, all coefficients of the two factors have been determined—but not all of the equations in the overdetermined system (7.27) have been used.

In an exact factorization problem, the remaining equations will either be satisfied (within round-off), or the construction of a factorization from that germ has failed. (If this was the only germ which passed the screening this would imply nonfactorizability.) For our empirical polynomial $(\bar{p}, e)$, we cannot expect the remaining equations to be satisfied even within their tolerances because we have—unnecessarily—satisfied the other equations within round-off accuracy by determining the coefficients in floating-point arithmetic. Thus we must now perform the usual modification of the computed coefficients such that the overall backward error in the system (7.27) is minimized. Since each equation contains only the one empirical quantity α_j, the associated tolerances are directly the ε_j. Naturally, products of modifications are dropped so that the minimization problem is a standard linear one.

We now explain this algorithm by applying it at first to the simple empirical polynomial of Example 7.11. In the following section, we will use the more involved factorization of the 3-variable degree 5 polynomial (7.10) of Example 7.3 to discuss some details of our algorithmic approach.

Example 7.11, continued: Contrary to our assumption in the first part of Example 7.11, we now assume that we have no a priori knowledge about a potential pseudofactorizability; therefore we have to ascertain this when we begin to determine a factorization. We have (cf. (7.31))

$$\bar{p}_{10}(x) = 1 - .82\,x + .15\,x^2 + 4.69\,x^3 \approx (1 + 1.4153\,x)\,(1 + (-1.1177 \mp 1.4369\,\mathrm{i})\,x)\,,$$

$$\bar{p}_{01}(y) = 1 + .91\,y - 8.71\,y^2 + 7.14\,y^3 \approx (1 + 3.7449\,y)\,(1 - 1.7378\,y)\,(1 - 1.0971\,y)\,,$$

$$\bar{p}_{11}(x) = .91 + 11.22\,x - .65\,x^2\,.$$

With the choice $u_{10}(x) = 1 + 1.4153\,x$, $\xi = -.7066$, $u_{11}(x) = \beta_{01}$, $v_{10}(x) = 1 - 2.2353\,x + 3.3137\,x^2$, we have from (7.32)

$$0 = u_{11}(\xi)\,v_{10}(\xi) - p_{11}(\xi) = \beta_{01} \cdot 4.2336 + 7.3420\,,$$

or $\beta_{01} \approx -1.7342$. With $d_1 = 1$, the target values are simply the coefficients of the factors of p_{01} (cf. (7.33)), and we find that the computed β_{01} agrees with the coefficient in the 2nd factor within our assumed tolerance (cf. section 7.4.3, item 4). Thus we have found a feasible germ which, in this case, is the complete linear factor $\tilde{u}(x, y) = 1 + 1.4153\,x - 1.7342\,y$.

The conjugate-complex factors of p_{10} cannot be used for a linear potential factor if we assume that the indetermination in $(\bar{p}, e)$ does not introduce complex coefficients.

We can now use the equations (7.27) for $j = 10, 01, 20, 11, 02$ to obtain directly the coefficients of the quadratic factor $\tilde{v}(x, y) = 1 - 2.2353\,x + 2.6442\,y + 3.3136\,x^2 + 3.6012\,xy - $

$4.1244\, y^2$. Note that the equations for $j = 30, 21, 12, 03$ have not been used; but $\bar{\alpha}_{30}$ and $\bar{\alpha}_{03}$ have previously entered into p_{10} and p_{01}. The residuals of these preliminary factors $\tilde{u}, \tilde{v}$ in the remaining equations of (7.27) are $-.0003, .0003, -.0025, .0125$; thus the largest backward error is 2.5 from the y^3-term. Therefore, we may accept $\tilde{u}, \tilde{v}$ as valid factors, or we may perform a refinement step:

With $\Delta u(x, y) := \Delta\beta_{10}\, x + \Delta\beta_{01}\, y$, $\Delta v(x, y) := \Delta\gamma_{10}\, x + \ldots + \Delta\gamma_{02}\, y^2$, we consider

$$(\tilde{u} + \Delta u)(x, y) \cdot (\tilde{v} + \Delta v)(x, y) - \bar{p}(x, y),$$

drop the quadratic terms in the increments, and minimize the moduli of the x, y-coefficients. This leads to the refined factors:

$$
\begin{aligned}
u(x, y) &= 1 + 1.4157\, x - 1.7342\, y, \\
v(x, y) &= 1 - 2.2335\, x + 2.6464\, y + 3.3143\, x^2 + 3.6023\, xy - 4.1185\, y^2,
\end{aligned}
$$

with a backward error of .46 for the factorization.

When we compare these pseudofactors with those obtained in the first part of Example 7.9 for the same empirical polynomial, we find that their last two digits differ, sometimes distinctly. This shows that only the first two digits after the decimal points are firmly defined at the indetermination level which we have specified. We will consider the related question of the *condition* of a pseudofactorization in section 7.4.4. $\square$

7.4.3 Algorithmic Details

(1) Checking for nonexistence of a pseudofactorization:

Corollary 7.10 is suitable for a $\bar{p} \in \mathcal{P}_d^s$ with moderate values of s and d. The regular system $\partial_{x_\sigma}\bar{p}(x)$, $\sigma = 1(1)s$, has s^{d-1} zeros (in the projective s-space); these have to be computed and checked against $\bar{p}$.

Our alternative is the search for a "germ" which may be extended into a pseudofactor of $\bar{p}$; it has the advantage that it is a first step towards factorization if it succeeds. Without information about the degrees of potential factors, we have to test the feasibility of all selections of $\leq \lceil\frac{d}{2}\rceil$ elements of the zero set of one of the univariate polynomials $p_\sigma(x_\sigma) := \bar{p}(0, ..x_\sigma, 0..)$. In the generic case $p_\sigma \in \mathcal{P}_d$, there are $\sum_{\nu=1}^{\lceil\frac{d}{2}\rceil} \binom{d}{\nu} = \frac{1}{2}\sum_{\nu=1}^{d-1}\binom{d}{\nu} = 2^{d-1} - 1$ cases. Each one requires, potentially, a checking against several variables, but this growth with s is negligible against s^{d-1}.

(2) Selection of p_σ:

For a potential p_σ of a degree $< d$, the number of zero combinations is smaller, but then we must also consider nonstandard terms. More important appears the *condition* of the zeros of the p_σ; cf. item 4 below. If there is a zero cluster relative to the tolerance of $\bar{p}(0, ..x_\sigma, 0..)$, these zeros are very ill-conditioned; cf. section 6.3.2. If there is a valid multiple zero in the cluster, we can use it; but we must consider its potential attributions to the two factors (see item 3 below). It appears that we should select a p_σ of full degree d, with well-separated zeros, if it exists. If all p_σ have (some) ill-conditioned zeros, the factorization problem is ill conditioned; cf. section 7.4.4.

For the following, we assume (w.l.o.g.) that the selected univariate polynomial is p_1.

(3) Determination of β_{01} for a selection of zeros for u_{10} :

With the notation in (7.30)–(7.32), we want to demonstrate that the determination of β_{01} from (7.32) is always possible: Let $(\, u_{11} \ \ v_{11} \,)$ denote the row vector of the successively arranged coefficients of the univariate polynomials u_{11}, v_{11}, and $(\, p_{11} \,)$ the coefficient vector of p_{11}. Then (7.32) may be written as

$$(u_{11} \ \ v_{11} \,) \ S(v_{10}, u_{10}) \ = \ (p_{11}),$$

with $S(v_{10}, u_{10})$ the Sylvester matrix; this also holds when one of u_{10}, v_{10} is of lower than the generic degree. Thus the computation of β_{01} is well defined if u_{10} and v_{10} have no (near-) common zeros; cf. Theorem 6.7. This also displays the ill-conditioning introduced by clustered zeros.

If there is a valid multiple zero ξ of p_1, its complete attribution to one of the two factors causes no problem. However, if $p_{11}(\xi) \approx 0$, we must put at least one zero ξ into each of u_{10}, v_{10} (cf. (7.32)), which makes $S(v_{10}, u_{10})$ singular. Let ξ be a simple pseudozero of p_{11}; then we can divide (7.32) by $(1 - x/\xi)$ and obtain a system for $(\, u_{11} \ \ v_{11} \,)$ which has lost two equations. These may be recovered by a consideration of the x_2^2 terms in (7.31):

$$u_{12}(x_1)\, v_{10}(x_1) + u_{11}(x_1)\, v_{11}(x_1) + u_{10}(x_1)\, v_{12}(x_1) \ = \ p_{12} \,;$$

upon substitution of $x_1 = \xi$, this reduces to $u_{11}(\xi)\, v_{11}(\xi) = p_{12}(\xi)$. Differentiation of (7.32) and substitution of ξ yields $u_{11}(\xi)\, v_{10}'(\xi) + v_{11}(\xi)\, u_{10}'(\xi) = p_{11}'(\xi)$. When ξ is a simple zero of v_{10}, the derivative does not vanish at ξ; then we may solve these two equations for $u_{11}(\xi)$ which supplies the missing information on β_{01}. For ξ a multiple zero of v_{10}, one can extend this approach further.

Naturally, the above consideration of the Sylvester matrix does not mean that we abandon the simpler way of substituting the zeros of u_{10} into (7.32) as the method of choice to obtain β_{01} in a nondegenerate case.

(4) Target values and "β-test":

We need the β_{01} target values of (7.33) for the polynomials $p_2, \ldots, p_s$, for $d_1 = 1(1)[\tfrac{d}{2}]$; they can be computed a priori. The testing proper proceeds thus:

A set of d_1 zeros of p_1 is selected and the associated value of β_{01} is computed with respect to a particular 2nd variable called x_2; cf. item 3 above. This value is matched against the d_1 target values for p_2. If an agreement is found with one of these, it is also matched, successively against the d_1 target values for the remaining p_σ if any. A failure with any p_σ deletes the respective β_{01} from the candidate list.

The crucial question is the degree of agreement which is to be requested between β_{01} and its target values. The tolerances for the target values (7.33) derive from the condition of the zeros of p_σ since $\alpha_\nu = -1/\zeta_\nu^{(\sigma)}$. But the potential variation of β_{01} depends on the variations of the coefficients in p_{10} and p_{11} in a nontrivial way; also, this variation is nearly independent from that of the ζ_ν^σ for the target values. The strongest influence on β_{01} stems from the variation of the zeros of p_{10} which enter into u_{10}, v_{10}. When we have been able to choose a p_{10} such that these zeros are well conditioned, we may, pragmatically, respect the variation of β_{01} simply through a more generous interpretation of the tolerances of the target values. In any case, it appears better not to reject a potential u_1 at this point than to lose it forever.

(5) Extension of an accepted germ to a full factor:

At this point, we possess a u-germ of the form (cf. (7.31))

$$u_1(x_1, x_2, \ldots, x_s) \;=\; u_{10}(x_1) + u_{110..}(x_1)\, x_2 + \ldots + u_{10..1}(x_s)\, x_s \,.$$

If $d_1 = 1$, this is a candidate for the complete factor u and we can turn towards the completion of v, of which we possess the germ $v_{10}(x_1)$. In all other cases, the completion of u and v has to proceed concurrently.

For this purpose, we consider the generalized version of (7.31)

$$p(x_1, x_2, \ldots, x_s) \;=:\; p_{10..0}(x_1) + \sum_{\sigma=2}^{s} p_{1 j_\sigma}(x_1)\, x_\sigma + \sum_{\sigma_1 \sigma_2} p_{1 j_{\sigma_1 \sigma_2}}(x_1)\, x_{\sigma_1} x_{\sigma_2} + \ldots \;=$$

$$(u_1(x_1, x_2, \ldots, x_s) + \sum_{\sigma_1 \sigma_2} u_{1 j_{\sigma_1 \sigma_2}}(x_1)\, x_{\sigma_1} x_{\sigma_2} + \ldots \,)\, (v_{10}(x_1) + \sum_{\sigma} v_{1 j_\sigma}(x_1)\, x_\sigma + \ldots \,)$$

$$\tag{7.34}$$

and the resulting relations for each of the polynomial coefficients $p_{1 j_\sigma}$ analogous to (7.32). The $p_{1 j_\sigma}$ are known from an expansion of p. Their representations in terms of the u, v-coefficients—used in the right sequence—constitute *linear* systems for groups of missing coefficients. From (7.32) and analogous relations, e.g., we may find the $v_{1 j_\sigma}$ after u_{11} has been determined as shown in item 3 above.

As a further example, consider

$$p_{1 j_{\sigma_1 \sigma_2}} \;=\; u_{10}\, v_{1 j_{\sigma_1 \sigma_2}} + u_{1 \sigma_1}\, v_{1 \sigma_2} + u_{1 j_{\sigma_1 \sigma_2}}\, v_{10} \,,$$

with $u_{10}, v_{10}, u_{1\sigma_1}, v_{1\sigma_2}$ already known. Each term is a polynomial in x_1 of degree $d - 2$ like $p_{1 j_{\sigma_1 \sigma_2}}$; there are $d_1 - 1$ coefficients in $u_{1 j_{\sigma_1 \sigma_2}}$ and $d - d_1 - 1$ in $v_{1 j_{\sigma_1 \sigma_2}}$ for a total of $d - 2$ unknown coefficients to be determined from the linear equations obtained from an expansion in powers of x_1. In a generic case (all coefficient polynomials are dense), we have one more relation than unknown coefficients; thus, one of the relations must be omitted. This is to be expected since the overall system (7.27) for the coefficients is overdetermined.

In a nongeneric case, various special situations may arise; for larger values of s and/or d it is hardly possible to list all cases which can arise from sparsity and/or multiple zeros in some of the p_σ. It appears that one can determine a solution in all cases. The more disturbing aspect of these many special cases is that they require modifications of the straightforward algorithm all of which can hardly be provided in a black-box implementation.

(6) Complete resolution of the overdetermined system (7.27):

Assume that we have determined values for all coefficients in the two factors u, v by the above approach. It cannot be expected that those equations in (7.27) which have not been used are satisfied by these values, within the tolerances of the respective α_j. The necessary refinement procedure follows the usual pattern:

- substitution of incremented values for the coefficients,

- linearization with respect to the increments,

- minimization of the $|\Delta \alpha_j| / \varepsilon_j$ in terms of the $\Delta \beta_j, \Delta \gamma_j$.

If the modified coefficients satisfy (7.27) with a backward error of $O(1)$, we are finished and have found a pseudofactorization; with a moderate backward error, we may repeat the procedure, with linearization about the modified coefficients of the factors.

While, in principle, there is always a closest point to $\bar{p}$ on the manifold of factorizable p in the coefficient space, it is not certain that we can reach it by *linearized* minimization from the germ with which we have started. In any case, we cannot necessarily distinguish a divergent iteration from one which converges against a far-away minimum during their initial phases. Therefore, it appears wise to terminate the iterative search when a 2nd step has not diminished the backward error substantially. Also, a near-stationary behavior with a backward error $> O(1)$ calls for a termination. If we have not yet exhausted the list of feasible germs, the whole procedure must then repeated with another germ. There may be a good number of feasible germs but no pseudofactorization. Since we are dealing with a high-dimensional global minimization problem, this situation is to be expected.

(7) Sparsity:

Real-life multivariate polynomials are nearly always sparse, often very sparse; generally, this sparsity is intrinsic, i.e. it extends to all polynomials in the tolerance neighborhood $N(\bar{p}, e)$. If pseudofactorizations exist which preserve that sparsity exactly, they must also have a certain sparsity pattern. The consideration of that pattern from the start may strongly reduce the overall effort; occasionally, it may be necessary for the successful determination of a solution. For a random sparsity pattern, however, we cannot generally require the exact preservation by a pseudofactorization; see below.

To find a potential implied sparsity structure of the factors, we consider the subsystem of the equations in (7.27) whose right-hand sides $\bar{\alpha}_j$ vanish intrinsically. Due to its bilinear structure, choices of vanishing factor coefficients which satisfy the subsystem are generally not unique. For example, assume $d = 5, d_1 = 2$ with an absent x_2^5 in p; this requires that the x_2^2-term in u *or* the x^3-term in v must be absent. Thus, this subsystem may have many solutions. However, a good number of these will contradict the *nonvanishing* of some other $\bar{\alpha}_j$ so that they must be excluded. If we are lucky, a unique sparsity pattern for the factor polynomials remains; cf. Example 7.13 below.

Otherwise we have to enforce an *approximate sparsity*. For this purpose, we introduce small tolerances ε_j for the vanishing $\bar{\alpha}_j$ and treat them as empirical data $(0, \varepsilon_j)$. These artificial tolerances should be of the order ε^2, where ε denotes the tolerance level of the truly empirical coefficients in $(\bar{p}, e)$.

A sparsity which is easy to utilize is the *evenness* of p in one or several or all variables. In this case, it appears advisable to introduce new variables for the relevant x_σ^2's. *Oddness* in a variable, on the other hand, requires oddness in one factor and evenness in the other, which leads again to a nonunique situation.

If some tiny coefficients appear in the germ determination and if the annihilation of these values is compatible with the sparsity of p, it seems reasonable to assume that these coefficients vanish intrinsically. But if this assumption results in a failure of the germ, one must introduce increments for these coefficients after all.

(8) Reality:

Often, only real factorizations of a real polynomial p are of interest. This is an aspect which can be readily accommodated throughout our algorithmic approach, and which often reduces the computational effort markedly. In the germ determination, only real zeros of p_1 can be used for $d_1 = 1$ while the two zeros of a conjugate pair may appear *simultaneously* for $d_1 \geq 2$.

Example 7.13: We take the empirical polynomial $(\bar{p}, e) \in \mathcal{P}_5^3$ of Example 7.12; cf. (7.10). $\bar{p}$ is real and sparse: Only 35 of the 55 potential coefficients appear and we assume this sparsity to be intrinsic; also, we assume that we are only interested in a real pseudofactorization.

Pairs of potential pseudofactors may have degrees 1,4 or 2,3. We try to determine their implied sparsity patterns along the lines of item 7 above. For $d_1 = 1$, $d_2 = 4$, we submitted the 20 relations from (7.27) with vanishing $\bar{\alpha}_j$ to solve of Maple6, with all 21 occurring β_j, γ_j as unknowns, and obtained 8 different solution sets which assign 0 to a large number of the coefficients. But these assignments have to permit the nonvanishing of the three 5th degree terms in $\bar{p}$: For example, the occurrence of x^5 in $\bar{p}$ requires that neither β_{100} nor γ_{400} can vanish, and similar requirements result from the other two 5th degree terms. Each of the 8 assignments disagreed with some of these requirements. If we believe that Maple generates *all* solutions, this implies that a linear factor is incompatible with the sparsity structure of $\bar{p}$.

When we applied the same approach to the splitting in a 2nd degree and a 3rd degree factor, we obtained 18 solutions; two of these satisfied the requirements from the 5th degree terms, but only one also agreed with the occurrence of a z^4 term in $\bar{p}$. The associated sparsity pattern deletes 3 of the 9 coefficients in the quadratic factor and 11 of the 19 coefficients in the cubic factor:

$$u(x, y, z) = 1 + \beta_{100}\, x + \beta_{010}\, y + \beta_{001}\, z + \beta_{200}\, x^2 + \beta_{020}\, y^2 + \beta_{002}\, z^2\,,$$

$$v(x, y, z) = 1 + \gamma_{100}\, x + \gamma_{010}\, y + \gamma_{001}\, z + \gamma_{110}\, xy + \gamma_{101}\, xz + \gamma_{011}\, yz + \gamma_{002}\, z^2 + \gamma_{300}\, x^3\,.$$
$$(7.35)$$

This leaves only 14 of the originally 28 coefficients. These coefficients have to satisfy the 35 equations of (7.27) with nonvanishing $\bar{\alpha}_j$, approximately.

The further computations are based on the factor structure (7.35). We begin by finding the zeros of $\bar{p}(x, 0, 0)$, $\bar{p}(0, y, 0)$, $\bar{p}(0, 0, z)$, of degrees 5,3,4, resp.; cf. item 2. They are (rounded)

$$\xi = 1.0000,\ -.3333,\ -1.4013,\ .7006 \pm .4719\, i\,,$$
$$\eta = 1.7241,\ 1.6180,\ -.6180\,,\quad \zeta = 1.2808,\ -.7808,\ .7500 \pm .6614\, i\,.$$

Only the two positive zeros of p_2 are slightly ill-conditioned. In accordance with item 2, we use the zeros ξ of $p_{100} := \bar{p}(x, 0, 0)$ for the construction of a germ; cf. item 3 and (7.32). To keep the germ real, we must put either 2 of the 3 real zeros or the conjugate pair into u_{100}; cf. (7.31). This leads to four candidates for u_{100}:

$$1 + 2.0000\, x - 3.0000\, x^2\,,\quad 1 - .2864\, x - .7136\, x^2\,,\quad 1 + 3.7136\, x + 2.1409\, x^2\,,$$
$$1 - 1.9636\, x + 1.4013\, x^2\,,$$

and the corresponding 3rd degree polynomials v_{100}. From $p_{110} = .42 - 1.42\, x + 3.75\, x^2 - 2.00\, x^3$, the coefficients of the corresponding four u_{110} of generic degree 1 are obtained from (7.32) via

$$u_{110}(\xi) = p_{110}(\xi)/v_{100}(\xi)\,,\quad \text{at the 2 zeros of } u_{100}\,.$$

In our particular case with the sparsity structure (7.35), a feasible u_{110} must be *constant*, which yields a further test for the selection of the correct ξ-combination. Actually, the combination of the first two ξ values yields $u_{110} \approx 1.0024 - .0024\, x$ while the other three combinations yield strongly nonzero x-coefficients.

The target values for the β-test are found from the $v_{010}(y)$; cf. item (4) and (7.33): β_1 is the coefficient of y in u_{010} and hence $\sum(-1/\eta_v)$ over selections of 2 zeros of p_{010}. With the above 3 real zeros of p_{010}, we obtain the 3 target values $\approx 1.0380, 1.0000, -1.1980$. The agreement of $\beta_{010} \approx 1.0024$ with the 2nd target value is obvious; due to the poor condition (≈ 60) of the two close positive zeros of p_{010}, the agreement with 1.038 would also be satisfactory. In any case, we have a clear confirmation of a germ with $u_{100} = 1 + 2.0000\,x - 3.0000\,x^2$.

The corresponding procedure with p_{101} and $x_2 = z$ in (7.31) yields, for our choice of the first two ξ-values, $u_{101} \approx .4992 - .0059\,x$ and (real) target values $\approx .5000, -1.5000$ from p_{001}. This is the final confirmation of our germ. At the same time, the matched target values tell us the correct combination of η's and ζ's for $u(0, y, 0)$ and $u(0, 0, z)$ and thus the coefficients of y^2 and z^2 in u. The complete u-factor candidate is

$$\tilde{u}(x, y, z) \approx 1 + 2.0000\,x + 1.0024\,y + .4992\,z - 3.0000\,x^2 - 1.0000\,y^2 - 1.0000\,z^2 .$$

From (7.31), we can now determine the complementary coefficient polynomials in v. In an obvious manner, we find

$$v_{100}(x) \approx 1 - 1.2500\,x + 1.0000\,x^3, \quad v_{010}(y) \approx 1 - .5800\,y, \quad v_{001}(z) \approx 1 - 1.5000\,z + 1.0000\,z^2 .$$

The missing x-coefficients in v_{110} and v_{101} are found from (7.32) by substitution of a pair of zeros of v_{100}: $\gamma_{110} \approx .9991$, $\gamma_{101} \approx -3.0004$. According to our sparsity structure (7.35), this leaves only the coefficient γ_{011} of the yz-term unknown. It is found from the relation $\gamma_{011} + \beta_{010}\,\gamma_{001} + \beta_{001}\,\gamma_{010} = \alpha_{011}$ in (7.27) as ≈ 1.0036; the complete v-factor candidate is now $\tilde{v}(x, y, z) \approx$

$$1 - 1.2500\,x - .5800\,y - 1.5000\,z + .9991\,xy - 3.0004\,xz + 1.0036\,yz + 1.0000\,z^2 + 1.0000\,x^3 .$$

When we multiply our candidate factors and compare with $\bar{p}$, we find that there are no excessive residuals; the overall backward error of our pseudofactorization is only ≈ 2.4. To achieve a lower backward error (and for demonstration purposes), we go through a refinement phase.

Since the sparsity in our factor polynomials matches that of $\bar{p}$, we do not face the problem of maintaining that structure; cf. item (7). We have 6 corrections $\Delta\beta_j$ and 8 corrections $\Delta\gamma_j$ for the (linearized) minimization of the 35 residuals from $(\tilde{u} + \Delta u)\,(\tilde{v} + \Delta v) - \bar{p}$. The backward error of the corrected factorization becomes $\approx .41$, with the full 10-digit corrections. When we round the corrected coefficients to 4 digits, it increases slightly to .44. Thus,

$$\bar{p}(x, y, z) \approx (1 + 2.0005\,x + 1.0008\,y + .4990\,z - 2.9990\,x^2 - .9989\,y^2 - .9990\,z^2)$$
$$(1 - 1.2512\,x - .5827\,y - 1.5011\,z$$
$$+ 1.0000\,xy - 3.0012\,xz + 1.0010\,yz + .9996\,z^2 + .9996\,x^3)$$

is a satisfactory pseudofactorization of the empirical polynomial $(\bar{p}, e)$; cf. Figure 7.5.

A plot of the (real) manifold $V[p]$ looks rather confusing, but a comparison with $V[u]$, $V[v]$ shows that it is simply the superposition of the two component manifolds. $\square$

7.4.4 Condition of a Multivariate Factorization

Assume that we have a dense polynomial $p \in \mathcal{P}^s$, with coefficient vector $\mathbf{a}^T = (\alpha_j) \in \mathbb{C}^n$, which factors into $u \cdot v$, with coefficients $b^T = (\beta_j)$ and $c^T = (\gamma_j)$, respectively. For the present

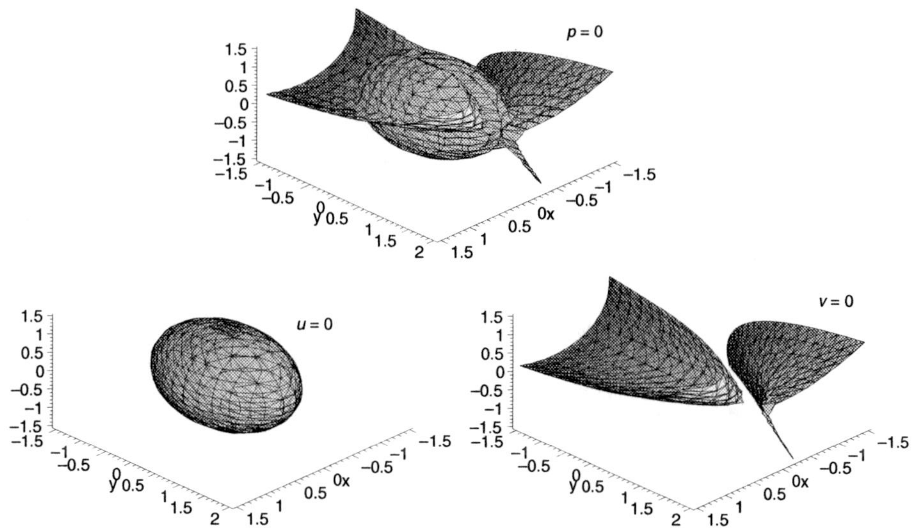

Figure 7.5.

considerations, we combine the factor coefficients into one vector $\mathbf{b}^T = (b^T \; c^T) \in \mathbb{C}^m$, $m < n$. Let $F : \mathbb{C}^m \to \mathbb{C}^n$ denote the mapping whose componentwise representation is given by the system (7.27). We want to assess the sensitivity of $\mathbf{b}^T$ to small changes of $\mathbf{a}^T$.

At first, we note that $p(\mathbf{a} + \Delta a)$ remains factorizable only if the modified coefficients remain on the manifold $\mathcal{M}$ of factorizable polynomials in the data space $\mathcal{A} = \mathbb{C}^n$. A local parametrization of $\mathcal{M}$ near $\mathbf{a}$ is given by $\Delta a^T = F(\mathbf{b}^T + \Delta b^T) - \mathbf{a}^T$; its tangential manifold in $\mathbf{a}^T$ is given by

$$\Delta a^T = \Delta b^T \, F'(\mathbf{b}^T), \qquad \text{with } F'(\mathbf{b}^T) \in \mathbb{C}^{m \times n}. \tag{7.36}$$

Let

$$F'(\mathbf{b}^T) = U \, (\Sigma \; 0) \begin{pmatrix} V_1^T \\ V_0^T \end{pmatrix} \tag{7.37}$$

be the singular value decomposition of $F'(\mathbf{b}^T)$, with the orthogonal (Hermitian) matrices $U \in \mathbb{C}^{n \times m}$, $V_1^T \in \mathbb{C}^{m \times n}$, $V_0^T \in \mathbb{C}^{(n-m) \times n}$. If $\Sigma = \operatorname{diag} \sigma_\mu$ is of full rank, V_0^T spans the subspace of those Δa^T which are not in the image of $F'(\mathbf{b})$ while V_1^T spans the tangential manifold of $\mathcal{M}$ in $\mathbf{a}^T$. For $\Delta a_1^T \in \operatorname{span} V_1^T$, the solution of $\Delta b^T \, F'(\mathbf{b}) = \Delta a_1^T$ is $\Delta b^T = \Delta a_1^T (V_1 \, \Sigma^{-1} U^T)$.

For small $\Delta a^T \notin \operatorname{span} V_1^T$, an exact factorization ceases to exist; but there are pseudo-factorizations, one of which, with a near-minimal backward error, will correspond to the exact factorization for the V_1^T-component Δa_1^T of Δa^T. With $\Delta a_1^T = \Delta a^T \, V_1 V_1^T$, we obtain

$$\Delta b^T = \Delta a^T (V_1 V_1^T)(V_1 \Sigma^{-1} U^T) = \Delta a^T (V_1 \, \Sigma^{-1} U^T),$$

and

$$\| \Delta b^T \|_{\max} \le \| V_1 \, \Sigma^{-1} U^T \|_1 \, \| \Delta a^T \|_{\max}, \tag{7.38}$$

with (7.37), is the condition estimate for which we have been looking. For an intrinsically sparse p, with a known implied factor sparsity, we may naturally restrict the coefficient vectors accordingly, which may strongly reduce the dimensionality of the matrices in the s.v.d of $F'(\mathbf{b}^T)$.

Example 7.14: We consider the situation of Example 7.11: To have the data point $\mathbf{a}^T$ on the manifold $\mathcal{M}$ for $\mathbf{b}^T$ from the coefficients of $\tilde{u}$, $\tilde{v}$, we form the exact multiple $\tilde{u}\,\tilde{v} =: p^*(\mathbf{a}^T)$. The singular value decomposition of $F'(\mathbf{b}^T)$ yields a smallest singular value of $\approx .63$ and $\|V_1 \Sigma^{-1} U^T\|_1 \approx 3$. This means that there exist factorizable polynomials in $N_1(p^*, e)$ for which some coefficients in the factors differ from those in $\tilde{u}$, $\tilde{v}$ by $O(e)$. It implies that the pseudofactors of our $(\bar{p}, e)$ are not meaningful to more than 3 decimal digits.

When we modify the coefficient vector of p^* by $\Delta a^T = (1,-1,-1,1,1,-1,-1,-1,-1)\varepsilon = \Delta a_1^T + \Delta a_2^T$, with the Δa_i^T in the subspaces span V_i^T, the pseudofactorization of $p^* + \Delta p$ equals, in linearized approximation, the exact factorization for the coefficient modification Δa_1^T. The corresponding coefficient change in the factors $\Delta b^T = \Delta a_1^T\,(V_1\,\Sigma^{-1} U^T)$ has a maximal component $\approx 3.05\,\varepsilon$. This shows that the condition estimate (7.38) is sharp for sufficiently small perturbations. $\square$

Naturally, we cannot evaluate (7.38) without the approximate knowledge of $\mathbf{b}$ and the factorization structure. But the structure of the system $F(\mathbf{b}^T) = \mathbf{a}^T$ suggests that a clustering of the zeros in the univariate polynomials $p_\sigma(x_\sigma)$ (cf. item 1 in section 7.4.3) leads to an instability in the determination of feasible germs and hence to a potential wide variety of valid pseudofactors. In this case, one should replace the clustered zeros by an approximate multiple zero, along the lines discussed in section 6.3.

Example 7.15: Consider $(\bar{p}, e)$, with $\bar{p} \in \mathcal{P}_3^3$:

$$\begin{aligned}
\bar{p} &= 1 + 4.16\,x + 4.11\,y + 3.19\,z + 5.75\,x^2 + 11.36\,xy + 5.62\,y^2 \\
&\quad + 8.85\,xz + 8.75\,yz + 3.39\,z^2 + 2.63\,x^3 + 7.82\,x^2 y + 7.74\,xy^2 \\
&\quad + 2.55\,y^3 + 6.11\,x^2 z + 12.09\,xyz + 5.98\,y^2 z + 4.70\,xz^2 + 4.65\,yz^2 + 1.20\,z^3 \,,
\end{aligned}$$

with tolerances .005 on all coefficients except 1. This polynomial has been obtained by multiplying the three linear factors

$$q_1 = 1 + 1.43\,x + 1.45\,y + 1.00\,z, \quad q_2 = 1 + 1.51\,x + 1.43\,y + 1.11\,z,$$
$$q_3 = 1 + 1.22\,x + 1.235\,y + 1.08\,z,$$

and rounding to 2 decimal digits.

Without that knowledge, one will look for a linear and a quadratic factor. Naturally, a valid pseudofactorization is obtained; but it differs substantially from any factorization $q_{i_1} \cdot q_{i_2} q_{i_3}$. Also, when one repeats the computation with alternate "leading variables," one obtains two other valid pseudofactorizations which differ from the first one by as much as several units of 10^{-1} in some coefficients! When one applies our condition analysis to these results, one finds condition factors of several hundreds which explains the extreme variation in the valid results. Obviously, the task is close to meaningless: On the one hand, one must specify the factor coefficients to at least 3 decimal digits if they are to constitute a valid pseudofactorization; on the other hand they are hardly determined to 1 decimal digit.

When we assume that we know of a factorization into 3 linear factors, we may modify our algorithmic approach to find a corresponding pseudofactorization. But again, the result differs from the original q_i and from results which are obtained with different initializations. Again, the condition factors are of O(100).

With a closer analysis of the zeros of the three univariate polynomials p_1, p_2, p_3, one finds that, with a slightly generous interpretation of O(1), all three polynomials possess a 3-cluster of zeros. This induces one to look for a potential pseudofactorization of the form $(1 + \beta_{100}x + \beta_{010}y + \beta_{001}z)^3$. This assumption leads to strongly simplified computation with a resulting linear triple factor which, after refinement, has a backward error of 3.7. What is more remarkable: The condition of that pseudofactorization is .2 ! This confirms once more our observation that replacing a zero cluster by a multiple zero can lead to enormous improvements in the condition of the related tasks; cf. section 6.3.4. It also shows that the condition of differently structured results for the same task can differ widely. $\square$

This example shows that it is not meaningful to speak of the condition of a multivariate polynomial with respect to factorization, without explicit reference to the structure of the pseudofactorization. If this structure is not known, an a priori condition analysis appears to be impossible.

Exercises

1. Consider the normalized quadratic polynomial in 2 variables

$$p(x, y; a) = 1 + \alpha_{10}x + \alpha_{01}y + \alpha_{20}x^2 + \alpha_{11}xy + \alpha_{02}y^2 .$$

(a) The manifold $\mathcal{M}$ of factorizable quadratic polynomials has codimension 1 in the data space $\mathcal{A} = \mathbb{C}^5$. Find the implicit polynomial representation M of $\mathcal{M}$ from its "parameter representation" (7.27).

(b) Let $\mathcal{S}$ be the manifold of quadratic polynomials with singular points. Find the implicit representation S of $\mathcal{S}$ from the normal form of p in the ideal $\langle \partial_x p, \partial_y p \rangle$. Convince yourself that $M \equiv S$; give an intuitive reason for this.

(c) Assume that, for some given empirical quadratic polynomial $(\bar{p}, e)$, $|S(\bar{a})| \neq 0$ but small. Find whether there exists $\tilde{p} = p(x, y; \tilde{a}) \in N_\delta(\bar{p}, e)$, $\delta = O(1)$, with $S(\tilde{a}) = 0$ or, equivalently, find min $\{\|\Delta a\|_e^* : S(\bar{a} + \Delta a) - 0\}$. Hint: Find the $\|..\|_e^*$ distance of the *linear* manifold $\overline{S}(\Delta a; \bar{a})$ from the origin in $\Delta\mathcal{A}$, where $\overline{S}(\Delta a; \bar{a})$ is the *linearization* of $S(\bar{a} + \Delta a)$ at $\bar{a}$.

2. (This example is from a lecture by E. Kaltofen.) Consider the bivariate polynomial

$$p(x, y) := 81\,x^4 + 72\,x^2y^2 + 1296 - 648\,x^2 + 16\,y^4$$
$$-288\,y^2 - 648.003\,z^4 + .002\,x^2z^2 + .001\,y^2z^2 - .007\,z^2 ;$$

the fact that

$$81\,x^4 + 72\,x^2y^2 + 1296 - 648\,x^2 + 16\,y^4 - 288\,y^2 - 648\,z^4$$
$$= (9\,x^2 + 4\,y^2 - 18\,\sqrt{2}\,z^2 - 36)(9\,x^2 + 4\,y^2 + 18\,\sqrt{2}\,z^2 - 36)$$

suggests that p is the rounded result of this product after $\sqrt{2}$ has been replaced by some approximation (different ones) in each of the two factors.

(a) Find a corresponding pseudofactorization of p under the assumption that its noninteger coefficients are empirical, with $\varepsilon = .5 \cdot 10^{-3}$.

(b) Determine the condition of the resulting pseudofactorization. Find the Δa which should lead to the greatest variation of the factors and test it computationally.

3.   Consider an empirical polynomial $(\bar{p}, e)$ with $\bar{p} \in \mathcal{P}^s$, $s \geq 2$, without a constant term but with a sufficiently large coefficient of x_1. Regard possible normalizations of $\bar{p}$ and of the candidate factors u, v. Which modifications do they imply in our factorization algorithm?

Historical and Bibliographical Notes 7

The analytic understanding of functions of several complex variables has expanded significantly in the 2nd half of the 20th century, but we will not refer to results from this research. For an introduction, cf. [7.1]. Though they are largely restricted to one variable complex analysis, the three volumes [7.2] of Henrici are a wonderful basis for the computational treatment of functions of complex variables. The inherent cross relations between multivariate polynomial algebra and algebraic geometry are nicely elaborated and explained in [2.10] and [2.11].

The reverse path from geometry to polynomial algebra is well exhibited in [7.3]; it also exposes the restrictions which appear when one deals with realistic geometric modelling in place of abstract algebraic geometry. Reference [7.3] also discusses the use of approximate computation and points to the difficulties which may arise form its careless use. This approach has been further extended in the volume [7.4]; it gives an indication of the numerous practically important but unsolved problems which prevail in numerical polynomial algebra. It also puts the exposition of our book which is exclusively based on the indetermination model of Chapter 3 and our moderate selection of topics into the right perspective. Analyses like that of the condition of an algebraic manifold (section 7.2.3) appear as fundamental tools in the design of efficient and reliable geometrical software.

Singular points are interesting from the analytic, algebraic, geometric, and numeric points of view. We have emphasized only the determination of valid singular points for empirical polynomials because this is often the key to valid answers about the local structure of a manifold; cf. Example 7.10. For further considerations from the algebraic point of view, we point again to [2.10] and [2.11]; for an analytic and/or geometric point of view, one should consult texts on multivariate analysis and differential geometry.

The factorization of a multivariate polynomial has been a challenge in polynomial algebra for a long time; however, the emphasis has been on polynomials over all kinds of coefficient fields, both finite and infinite. Efforts towards an approximate numerical factorization of a multivariate polynomial over $\mathbb{C}$, possibly with coefficients of limited accuracy, appear to have begun about 1990. A number of different independent approaches to the solution of this numerical task have been proposed:

- Our approach of section 7.4 which finds a valid approximate solution of the overdetermined polynomial system (7.27) has resulted from an Austrian–Chinese technical project; cf. [7.5].

- Sasaki and coworkers have developed and analyzed an approach based on zero-sum relations among power series root: Write p as a polynomial of degree d in one variable x, with coefficients in the other variables y, and consider the power series expansions of the zeros $x_\nu = \varphi_\nu(y)$, $\nu = 1(1)n$. A factorization of p must collect the factors $(x - \varphi_\nu(y))$ into two separate products. For further details, see [7.6].

- Corless, Watt, and coworkers have elaborated an approach based on the local construction of polynomial components: From a point x on the manifold of p, follow the manifold by numerical continuation to obtain a local parametrized representation which may be implicitized to yield the factor on whose manifold x happened to lie. For further details, see [7.7].

- Sommese, Verschelde, Wampler sample the manifold of p on lines $x(t) = x_0 + t\,y$, where $x_0, y \in \mathbb{C}^s$ are random vectors; points on different lines may then be connected by homotopy techniques. For further details, see [7.8].

An assessment of the relative merits of these approaches under varying circumstances appears premature at this time. Also, a sparsity pattern analysis as in paragraph 7 of section 7.4.3 should be further elaborated since it will be helpful in any approach. The condition analysis of the multivariate factorization problem in section 7.4.4 is from [6.4].

References

[7.1] R.M. Range: Complex Analysis: A Brief Tour into Higher Dimensions, Amer. Math. Monthly, Feb. 2003.

[7.2] P. Henrici: Applied and Computational Complex Analysis, 3 volumes, Wiley, Hoboken NJ, 1988–1993.

[7.3] Ch.M. Hoffmann: Geometric and Solid Modelling - An Introduction, Morgan Kaufmann Publ., San Mateo, CA, 1989.

[7.4] Uncertainty in Geometric Computations (Eds. J. Winkler, M. Niranjan), Kluwer Academic Publ., Boston, 2002.

[7.5] Y.Zh. Huang, H.J. Stetter, W.D. Wu, L.H. Zhi: Pseudofactors of Multivariate Polynomials, in: Proceed. ISSAC 2000 (Ed. C. Traverso), ACM, New York, 161–168, 2000.

[7.6] T. Sasaki: Approximate Multivariate Polynomial Factorization Based on Zero-sum Relations, in: Proceed. ISSAC 2001 (Ed. B. Mourrain), ACM, New York, 284–291, 2001.

and previous papers by this author, starting with

T. Sasaki, M. Suzuki, M. Kolář, M. Sasaki: Approximate Factorization of Multivariate Polynomials and Absolute Irreducibility Testing, Japan J. Indust. Appl. Math. **8** (1991), 357–375.

[7.7] R.M. Corless, M.W. Giesbrecht, M. van Hoeij, I.S. Kotsireas, S.M. Watt: Towards Factoring Bivariate Approximate Polynomials, in: Proceed. ISSAC 2001 (Ed. B. Mourrain), ACM, New York, 85–92, 2001.

[7.8] A.J. Sommese, J. Verschelde, C.W. Wampler: Numerical Decomposition of the Solution Sets of Polynomial Systems into Irreducible Components, SIAM J. Numer. Anal. **38** (2001), 2022–2046.

Chapter 8

Zero-Dimensional Systems of Multivariate Polynomials

We have previously emphasized that an indispensable prerequisite for the numerical treatment of an algebraic problem is its *embedding into analysis*. This is a straightforward matter for individual polynomials, univariate or multivariate: We can immediately see them as elements of a linear space, with the monomials of their support as basis elements; in this linear space, a topology is generated through the natural topology of the complex coefficients. Thus, as long as the structure of an algebraic task is determined by only *one* polynomial, its analytic embedding is so natural that we have often not bothered to display it explicitly. For univariate problems with several data polynomials (like g.c.d.), we have also been able to extend this approach.

For multivariate algebraic problems with several data polynomials, we have a different situation: The topology of the coefficients does not generally establish a suitable analytic embedding for the algebraic problem. In *linear* algebra, this is well known for singular or near-singular systems of linear polynomials. In polynomial algebra there are many more ways how an algebraic structure may change its character through (arbitrarily) small changes of its data. This is aggravated by the fact that many multivariate algebraic structures are commonly described in an *overdetermined* fashion: For example, the reduced Groebner basis of a polynomial ideal in s variables has, generally, more than s elements. This makes it impossible to define the "neighborhood of an ideal" naively by a neighborhood of its Groebner basis.

Fortunately, for a wide and meaningful class of multivariate polynomial systems, the *zeros* remain continuous functions of the coefficients in the system, like it is well-known for regular systems of linear equations. Therefore, in this chapter, we will consider the analog of *regular* systems of linear multivariate equations, viz. systems of multivariate polynomials with a finite, positive number of real or complex isolated zeros. Like in numerical linear algebra, we interpret *regularity* of a polynomial system to imply that it is sufficiently removed from systems which are either inconsistent or which possess a positive-dimensional zero manifold so that all sufficiently close neighboring systems are also consistent and 0-dimensional. A necessary but not sufficient prerequisite for this is that the system contains equally many polynomials as there are variables.

The considerations in this chapter will elaborate aspects of regular polynomial systems which are fundamental for their computational solution. Numerical aspects in the proper sense will be delayed to Chapter 9.

8.1 Quotient Rings and Border Bases of 0-Dimensional Ideals

Algebraically, a system P of polynomials specifies a polynomial *ideal* $\mathcal{I} = \langle P \rangle$ which defines a *quotient ring* $\mathcal{R}[\mathcal{I}]$ and a *dual space* $\mathcal{D}[\mathcal{I}]$; cf. Chapter 2. There, we have seen that—in our restricted context—all the defining relations in the diagram

$$
\begin{array}{ccc}
 & \mathcal{I} & \\
\swarrow\nearrow & & \searrow\nwarrow \\
\mathcal{R}[\mathcal{I}] \quad \underset{\leftarrow}{\rightarrow} & & \mathcal{D}[\mathcal{I}]
\end{array}
$$

may be read both ways, i.e. any member of the set $\{\mathcal{I}, \mathcal{R}[\mathcal{I}], \mathcal{D}[\mathcal{I}]\}$ defines the other two; cf. Theorems 2.7 and 2.21 etc. In sections 2.4 and 2.5, we have established how all information about the m zeros (counting multiplicities) of a 0-dimensional polynomial ideal $\mathcal{I}$ in $\mathcal{P}^s(m)$ can be determined—by numerical linear algebra techniques—from the multiplicative structure of the quotient ring $\mathcal{R}[\mathcal{I}]$, specified with respect to a suitable basis of $\mathcal{R}$ as an m-dimensional linear space over $\mathbb{C}$. A *normal set representation* of $\mathcal{I}$ specifies these data; cf. Definition 2.23.

In this section, we will continue our considerations of Chapter 2 and consider various aspects of the specification of a quotient ring in more detail. Remember that the elements proper of a quotient ring $\mathcal{R}[\mathcal{I}]$ are *residue classes* mod $\mathcal{I}$, but that—for a simpler notation and in agreement with common practice (cf. section 2.2)—we will generally formulate relations in $\mathcal{R}$ in terms of particular member polynomials of these residue classes. This implies that such relations must be interpreted as *equivalences* mod $\mathcal{I}$. A typical example is the representation of multiplication in $\mathcal{R}[\mathcal{I}]$ by the formulation (cf. (2.20))

$$
p(x) \cdot \mathbf{b}(x) \;=\; p(A)\,\mathbf{b}(x)
$$

which—as a relation in $\mathcal{P}^s$—is to be read as

$$
p(x) \cdot \mathbf{b}(x) \;\equiv\; p(A)\,\mathbf{b}(x) \bmod \mathcal{I} \qquad \text{or} \qquad p(x) \cdot \mathbf{b}(x) - p(A)\,\mathbf{b}(x) \;\in\; \mathcal{I}.
$$

This reveals the second important fact to be remembered about relations in a polynomial quotient ring $\mathcal{R}[\mathcal{I}]$: *At the zeros of* $\mathcal{I}$, relations in $\mathcal{R}$ become proper equations in $\mathbb{C}$ or $\mathbb{R}$, resp.: For $z \in Z[\mathcal{I}]$,

$$
p(z) \cdot \mathbf{b}(z) - p(A)\,\mathbf{b}(z) \;=\; 0.
$$

8.1.1 The Quotient Ring of a Specified Dual Space

Let us, at first, consider the transition from a quotient ring to the zeros of the associated polynomial ideal in the *reverse direction*: In $\mathbb{C}^s$, $s > 1$, we consider sets of m points z_μ, $\mu = 1(1)m$, which we regard as zero sets Z of 0-dimensional ideals $\mathcal{I} \subset \mathcal{P}^s$. If all z_μ are *disjoint*, we interpret them as *simple* zeros of $\mathcal{I}$ and Z defines the dual space $\mathcal{D}[\mathcal{I}]$; cf. Theorem 2.20. To a subset of $m_\mu \geq 2$ *identical* values $z_\mu = z_{\mu+1} = \ldots = z_{\mu+m_\mu-1}$ we assign the multiplicity m_μ and require that the associated $m_\mu - 1$ further basis elements (beyond evaluation at z_μ) of the dual space $\mathcal{D}[\mathcal{I}]$ be specified; cf. Definition 2.14. We will see later (cf. section 8.5) that this amounts essentially to the specification of $m_\mu - 1$ further vectors in $\mathbb{C}^s$. With this interpretation of coinciding points, the elements of $(\mathbb{C}^s)^m$ represent the ideals in $\mathcal{P}^s$ with exactly m zeros.

Definition 8.1. The set of all 0-dimensional ideals in $\mathcal{P}^s$ with exactly m zeros (counting multiplicities) will be denoted by $\mathcal{P}^s(m)$; the set $(\mathbb{C}^s)^m$, with the indicated interpretation of its elements as zero sets of ideals in $\mathcal{P}^s(m)$, will be denoted by $Z^s(m)$. $\square$

The above consideration suggests that the numerical part of the specification of $\mathcal{R}[\mathcal{I}]$ for $\mathcal{I} \in \mathcal{P}^s(m)$ should consist of either m s-tuples or s m-tuples of complex numbers. This is generally obscured by the fact that the form and the numerical data of the specification of a quotient ring depend strongly on the basis chosen for the representation. Even if we restrict our attention to *monomial* bases or *normal sets* (cf. Definition 2.19)—as we will in the following—there is a wide variety of feasible normal sets for nontrivial values of s and m: For the quotient ring of any given ideal in $\mathcal{P}^s(m)$, all or *almost all* elements from $T^s(m)$ are feasible as a basis; cf. Proposition 2.28. However, the algorithmic and numerical properties of a basis may depend strongly on the particular choice.

To avoid the technical discussion of special cases, we assume that we select only normal sets which contain *all* variables x_σ as elements; obviously, this requires $m \geq s+1$ and excludes some extremely degenerate positions of the zeros z_μ. An extension to these cases is straightforward and will be indicated in examples. We also assume that $1, x_1, \ldots, x_s$ are the *first* $s+1$ components in the normal set *vector* **b**.

For a particular zero set $Z \in Z^s(m)$, when we have selected a (feasible) normal set $\mathcal{N} \in T^s(m)$ and arranged the monomials of $\mathcal{N}$ into a basis (column) vector $\mathbf{b}(x) = (x_\mu^{j_\mu}, \mu = 1(1)m)$, we can immediately specify the multiplicative structure of the associated quotient ring $\mathcal{R}$ with respect to this basis with the aid of Theorem 2.27; cf. (2.50). Let us, at first, assume that all $z_\mu = (\zeta_{\mu,1}, \ldots, \zeta_{\mu,s}) \in Z$ are disjoint. Then (cf. also Theorem 2.23) the multiplication by x_σ in $\mathcal{R}[\mathcal{I}]$, i.e. $\mathrm{mod}\,\mathcal{I}$, is represented by the matrix

$$A_\sigma = X\,\Lambda_\sigma\,X^{-1} \in \mathbb{C}^{m\times m},$$

$$\text{with}\quad X := \begin{pmatrix} | & & | \\ \mathbf{b}(z_1) & \cdots & \mathbf{b}(z_m) \\ | & & | \end{pmatrix},\quad \Lambda_\sigma := \begin{pmatrix} \zeta_{1,\sigma} & & 0 \\ & \ddots & \\ 0 & & \zeta_{m,\sigma} \end{pmatrix}. \tag{8.1}$$

The right-hand side of (8.1) shows that all information about Z is in the $m \times m$-matrix X which is the *same* for all σ, and that even X generally contains this information in a redundant form: Except when $m = s+1$ and $\mathbf{b} = (1, x_1, ..x_s)^T$ (cf. our assumption above), the $m - s - 1$ last components of the vectors $\mathbf{b}(z_\mu)$ are immediate functions of the components no. 2 through $s+1$: For $b_\nu(x) = x^{j_\nu}$, $\nu > s+1$,

$$b_\nu(z_\mu) = z_\mu^{j_\nu} = \zeta_{\mu,1}^{j_{\nu 1}} \ldots \zeta_{\mu,s}^{j_{\nu s}} = (b_2(z_\mu))^{j_{\nu 1}} \ldots (b_{s+1}(z_\mu))^{j_{\nu s}}. \tag{8.2}$$

The fact that the complete information in (8.1) about Z, and thus about the ideal $\mathcal{I}$ with the zero set Z, coded with reference to the normal set $\mathcal{N}$, is contained in the 2nd to $(s+1)$st rows of X is in agreement with $Z \in (\mathbb{C}^s)^m$.

On the other hand, this shows that the elements of the $m \times m$-matrices A_σ whose joint eigenvectors constitute the matrix X must generally satisfy a number of *restrictions*. Trivially, the *first* row $a_1^{(\sigma)T}$ of A_σ must comply with $x_\sigma b_1(x) = x_\sigma = a_1^{(\sigma)T}\mathbf{b}$ or $a_1^{(\sigma)T} = (0..\overset{\sigma+1}{1}..0)$. Similarly, if the x_σ-multiple of some component $b_\nu(x)$ remains in $\mathcal{N}$, then the ν-th row of A_σ must be a unit vector, with the 1 in position $\bar{\nu}$ for $x_\sigma b_\nu(x) = b_{\bar{\nu}}(x)$. The remaining rows of

A_σ are nontrivial; they must comply with the relations (8.2) for the eigenvectors $\mathbf{b}(z_\mu)$ of A_σ. Generally there are more than s nontrivial rows while s of them should be sufficient, considering the number of data elements.

Moreover, it appears from (8.1) that *each one* of the A_σ carries the full information about Z in its eigenvectors which are the same for all matrices in the commuting family $\overline{A}$ generated by the A_σ; cf. Proposition 2.8 and Theorem 2.23. Therefore, the specification of *one* A_σ, say A_s, should uniquely specify all other A_σ. However, there is a fine point in Theorem 2.23: It talks about the *joint* eigenvectors of the matrices in the commuting family $\overline{A}$. And by Proposition 2.8, we can only be sure that an eigenvector of a particular A_σ is a joint eigenvector of $\overline{A}$ if the associated eigenvalue of A_σ has geometric multiplicity 1. While the *family* $\overline{A}$ is a nonderogatory commuting family by Corollary 2.26, an individual A_σ may well have a multiple eigenvalue with an eigenspace of a dimension > 1.

This is evident directly from (8.1): When $n > 1$ simple zeros have the *same* σ-component $\zeta_{\mu,\sigma}$, the associated vectors $\mathbf{b}(z_\mu)$ span an n-dimensional eigenspace of A_σ with the eigenvalue $\zeta_{\mu,\sigma}$. Naturally, when we prescribe the $\mathbf{b}(z_\mu)$ there is no harm. But when we compute the eigenvectors from A_σ, the eigenspace may appear as spanned by a different set of linearly independent eigenvectors which are not interpretable as values of $\mathbf{b}(x)$! Sometimes, it is possible to use the relations (8.2) to determine the appropriate basis vectors; cf. Example 8.1. Generally, one has to use a different matrix A_σ or a suitable linear combination of two or more A_σ. Since the projection of a set of disjoint points in $\mathbb{C}^s$ onto a 1-dimensional subspace preserves the disjointness for almost all such subspaces, almost all linear combinations of A_σ are nonderogatory.

The case of true multiple zeros is more delicate. Assume that there is one m_1-fold zero z_1 in Z, while the remaining z_μ, $\mu = m_1 + 1, .., m$, are simple. Let the associated functionals in the dual space $\mathcal{D}(Z)$ be c_μ, $\mu = 1(1)m_1$. Theorem 2.27 tells us that

$$A_\sigma \left(X_1 \mid \mathbf{b}(z_{m_1+1}) \ldots \mathbf{b}(z_m) \right) \;=\; \left(X_1 \mid \mathbf{b}(z_{m_1+1}) \ldots \mathbf{b}(z_m) \right) \begin{pmatrix} T_{\sigma 1} & & & \\ & \zeta_{m_1+1,\sigma} & & \\ & & \ddots & \end{pmatrix},$$

$$(8.3)$$

or

$$\left(c_1(x_\sigma \mathbf{b}),\; c_2(x_\sigma \mathbf{b}),\; \ldots,\; c_{m_1}(x_\sigma \mathbf{b}) \right) \;=\; \left(c_1(\mathbf{b}),\; c_2(\mathbf{b}),\; \ldots,\; c_{m_1}(\mathbf{b}) \right) T_{\sigma 1}; \qquad (8.4)$$

thus, the *joint invariant subspace* X_1 of the family $\overline{A}$ is spanned by the vectors $c_1(\mathbf{b})$, $c_2(\mathbf{b})$, $\ldots, c_{m_1}(\mathbf{b})$, and the elements in the upper triangular $m_1 \times m_1$-matrices $T_{\sigma 1}$ arise from the representation of the vectors $c_\mu(x_\sigma \mathbf{b})$, $\mu = 1(1)m_1$, in terms of the vectors in X_1: Since, for A_σ, $\zeta_{1,\sigma}$ is the eigenvalue associated with the only eigenvector $\mathbf{b}(z_1)$ in X_1, all diagonal elements of $T_{\sigma 1}$ are $\zeta_{1,\sigma}$, and—with an appropriate ordering of the c_μ—there exist unique coefficients $t_{\lambda\mu}^{(\sigma)}$ such that

$$c_\mu(x_\sigma \mathbf{b}) \;=\; \zeta_{1,\sigma}\, c_\mu(\mathbf{b}) + \sum_{\lambda=1}^{\mu-1} t_{\lambda\mu}^{(\sigma)}\, c_\lambda(\mathbf{b}), \qquad \mu = 2(1)m_1;$$

cf. section 2.3.2. Now, the information about the differential structure of the multiple zero z_1 is shared by the columns of X_1 and all of the $T_{\sigma 1}$ in a not so transparent way, and there is, generally, no single A_σ which carries the full information about the differential structure of a multiple zero. We delay a detailed analysis to section 8.5.3.

If we have the quantitative description of a quotient ring with reference to a normal set $\mathcal{N}_1$, the generation of the description with reference to another normal set $\mathcal{N}_2$ (with the same number m of elements, of course) is straightforward, cf. section 2.4.1: Denote the multiplication matrices w.r.t. the basis vectors $\mathbf{b}_i$ of $\mathcal{N}_i$ by $A_\sigma^{(i)}$, $i = 1, 2$. The elements of $\mathcal{N}_2$ are either also in $\mathcal{N}_1$ or they have a nontrivial normal form representation in terms of $\mathbf{b}_1$ which implies

$$\mathbf{b}_2(x) = M_{21}\,\mathbf{b}_1(x) \mod \mathcal{I}.$$

Thus, by Proposition 2.22,

$$A_\sigma^{(2)} = M_{21}\,A_\sigma^{(1)}\,M_{21}^{-1}, \qquad \sigma = 1(1)s. \tag{8.5}$$

If M_{21} is singular, $\mathcal{N}_2$ is not a feasible normal set for the quotient ring because the singularity of M_{21} implies that span $\mathcal{N}_2$ contains polynomials in $\mathcal{I}$.

Example 8.1: We take $s = 3$ and $m = 7$ and prescribe 7 disjoint points in $\mathbb{C}^3$ "at random," e.g., $Z = \{(2, 1, 0), (0, 4, 3), (2, 0, -1), (-1, -2, 4), (0, -1, 1), (1, i, 2+i), (1, -i, 2-i)\}$. From $\mathcal{T}^3(7)$, we choose a symmetric normal set $\mathcal{N}_1$ with $\mathbf{b}_1(x) = (1, x, y, z, xy, xz, yz)^T$ and form the vectors $\mathbf{b}_1(z_\mu)$ and the matrices X and Λ_σ of (8.1), where $\sigma = 1, 2, 3$ refers to x, y, z in this order. From $A_\sigma^{(1)} = X \Lambda_\sigma X^{-1}$, we obtain

$$A_1^{(1)} = \begin{pmatrix} 0 & 1 & 0 & 0 & 0 & 0 & 0 \\ \frac{-10}{183} & \frac{5}{3} & \frac{10}{183} & \frac{14}{183} & \frac{1}{3} & \frac{-73}{183} & \frac{-2}{61} \\ 0 & 0 & 0 & 0 & 1 & 0 & 0 \\ 0 & 0 & 0 & 0 & 0 & 1 & 0 \\ \frac{218}{61} & -2 & \frac{148}{61} & \frac{-110}{61} & 1 & \frac{42}{61} & \frac{-40}{61} \\ \frac{454}{183} & \frac{-8}{3} & \frac{644}{183} & \frac{-50}{183} & \frac{-1}{3} & \frac{130}{183} & \frac{-80}{61} \\ \frac{-1213}{183} & \frac{11}{3} & \frac{-983}{183} & \frac{527}{183} & \frac{7}{3} & \frac{-199}{183} & \frac{99}{61} \end{pmatrix},$$

$$A_2^{(1)} = \begin{pmatrix} 0 & 0 & 1 & 0 & 0 & 0 & 0 \\ 0 & 0 & 0 & 0 & 1 & 0 & 0 \\ \frac{625}{61} & -7 & \frac{534}{61} & \frac{-143}{61} & -2 & \frac{-43}{61} & \frac{-113}{61} \\ 0 & 0 & 0 & 0 & 0 & 0 & 1 \\ \frac{689}{183} & \frac{-7}{3} & \frac{409}{183} & \frac{-379}{183} & \frac{1}{3} & \frac{107}{183} & \frac{-33}{61} \\ \frac{-1213}{183} & \frac{11}{3} & \frac{-983}{183} & \frac{527}{183} & \frac{7}{3} & \frac{-199}{183} & \frac{99}{61} \\ 27 & -19 & 27 & -5 & -8 & -3 & -6 \end{pmatrix},$$

$$A_3^{(1)} = \begin{pmatrix} 0 & 0 & 0 & 1 & 0 & 0 & 0 \\ 0 & 0 & 0 & 0 & 0 & 1 & 0 \\ 0 & 0 & 0 & 0 & 0 & 0 & 1 \\ \frac{-643}{61} & 7 & \frac{-455}{61} & \frac{400}{61} & 2 & \frac{-125}{61} & \frac{151}{61} \\ \frac{-1213}{183} & \frac{11}{3} & \frac{-983}{183} & \frac{527}{183} & \frac{7}{3} & \frac{-199}{183} & \frac{99}{61} \\ \frac{-2695}{183} & \frac{35}{3} & \frac{-2795}{183} & \frac{845}{183} & \frac{10}{3} & \frac{182}{183} & \frac{315}{61} \\ \frac{703}{61} & -7 & \frac{395}{61} & \frac{-301}{61} & -2 & \frac{75}{61} & \frac{68}{61} \end{pmatrix}.$$

Each of the matrices has 4 nontrivial rows, which indicates immediately that the information in these rows must be dependent.

When we determine the eigenvectors of the $A_\sigma^{(1)}$, say by the Maple routine `Eigenvectors`, with subsequent normalization for a first component 1, we recover the vectors $\mathbf{b}_1(z_\mu)$ for $\sigma = 2$ and 3. For $\sigma = 1$, however, we have three 2-fold eigenvalues $0, 1$, and 2, because each of these values occurs as first coordinate for two different zeros; the associated 2-dimensional eigenspaces are not represented by the corresponding $\mathbf{b}_1(z_\mu)$ but by two other vectors chosen by the routine. This is immediately displayed by the appearance of a first component 0 in one vector of a pair, and by the fact that no complex components appear in the pair for the conjugate-complex zeros. Also it is easily checked that the relations (8.2) are not satisfied for these 3 pairs of eigenvectors. As a remedy, we may try to build linear combinations of the computed vectors which satisfy (8.2).

For the zeros (0,4,3), (0,-1,1), e.g., Maple gives us the eigenvectors $\bar{b}_1 = (1, 0, -\frac{7}{2}, 0, 0, 0, -\frac{15}{2})^T$, $\bar{b}_2 = (0, 0, \frac{5}{2}, 1, 0, 0, \frac{13}{2})^T$. When we determine $\gamma_1, \gamma_2 \in \mathbb{C}$ from

$$\gamma_1 \bar{b}_{1,1} + \gamma_2 \bar{b}_{2,1} = 1, \quad (\gamma_1 \bar{b}_{1,3} + \gamma_2 \bar{b}_{2,3})(\gamma_1 \bar{b}_{1,4} + \gamma_2 \bar{b}_{2,4}) = (\gamma_1 \bar{b}_{1,7} + \gamma_2 \bar{b}_{2,7}),$$

we obtain the two (γ_1, γ_2) pairs (1,1), (1,3) yielding the correct eigenvectors $(1, 0, -1, 1, 0, 0, -1)^T$ and $(1, 0, 4, 3, 0, 0, 12)^T$. For the conjugate-complex zero pair, the complex components of the eigenvectors can be recovered from the real representation because the linear combination coefficients are found from a quadratic equation.

When we wish to represent the quotient ring with respect to a basis $\mathbf{b}_2 = (1, x, y, z, x^2, xy, y^2)^T$, we need the normal forms of x^2 and y^2 in the basis $\mathbf{b}_1$. Obviously, their coefficients can be found in the 2nd row of $A_1^{(1)}$ and in the 3rd row of $A_2^{(1)}$, respectively. This yields the transformation matrix

$$M_{21} = \begin{pmatrix} 1 & 0 & 0 & 0 & 0 & 0 & 0 \\ 0 & 1 & 0 & 0 & 0 & 0 & 0 \\ 0 & 0 & 1 & 0 & 0 & 0 & 0 \\ 0 & 0 & 0 & 1 & 0 & 0 & 0 \\ \frac{-10}{183} & \frac{5}{3} & \frac{10}{183} & \frac{14}{183} & \frac{1}{3} & \frac{-73}{183} & \frac{-2}{61} \\ 0 & 0 & 0 & 0 & 1 & 0 & 0 \\ \frac{625}{61} & -7 & \frac{534}{61} & \frac{-143}{61} & -2 & \frac{-43}{61} & \frac{-113}{61} \end{pmatrix},$$

and the multiplication matrices $A_\sigma^{(2)}$ via (8.5) with respect to the basis $\mathbf{b}_2$. $\quad\square$

8.1.2 The Ideal Generated by a Normal Set Ring

In section 2.2.1, we have seen that we can recover a 0-dimensional ideal $\mathcal{I} \subset \mathcal{P}^s$ from its quotient ring $\mathcal{R}[\mathcal{I}] = \mathcal{P}^s/\mathcal{I}$ when we have a basis $\mathbf{b}(x)$ of $\mathcal{R}[\mathcal{I}]$ and generators of the commuting family $\overline{A}$ of matrices which describe the multiplicative structure of the ring with respect to $\mathbf{b}$; cf. Theorem 2.7. This raises the following question: Given an *arbitrary* commutative ring $\mathcal{R}$ on an m-dimensional vector space $V \subset \mathcal{P}^s$, can it be interpreted as the quotient ring of a polynomial ideal in $\mathcal{P}^s(m)$?

Without essential loss of generality (cf. section 2.5.1), we assume a normal set basis $\mathcal{N} \subset \mathcal{T}^s(m)$ for $\mathcal{R}$, with the associated basis *vector* $\mathbf{b}(x)$. As a ring, $\mathcal{R}$ is closed with respect to multiplication, i.e. all products of elements in $\mathcal{R}$ are in $\mathcal{R}$. But we may also interpret $\mathcal{R}$ as a subset of $\mathcal{P}^s$; then products of elements in $\mathcal{R}$ may well be outside $\mathcal{R}$. In particular, when we consider multiplication of an element $r \in \mathcal{R}$ by some x_σ, $\sigma = 1(1)s$, the product in the $\mathcal{P}^s$-sense is in $\mathcal{N} \cup B[\mathcal{N}]$; obviously, (2.58) in Definition 2.21 has been chosen to accommodate all products $x_\sigma r$ for $r \in \mathcal{R}$. Thus, the multiplicative structure of $\mathcal{R}$ specifies a *linear map* $\mathcal{A}_\mathcal{N}$ from span $\mathcal{N} \cup B[\mathcal{N}]$ into span $\mathcal{N}$ which is a *projection*, i.e. which leaves $\mathcal{N}$ invariant:

$$\mathcal{A}_\mathcal{N} \, x^j \; := \; \begin{cases} x^j & \text{for } x^j \in \mathcal{N}, \\ a_j^T \, \mathbf{b}(x) & \text{for } x^j \in B[\mathcal{N}]. \end{cases} \tag{8.6}$$

Then, for $r = c^T[r]\,\mathbf{b}(x) \in \mathcal{R}$, the result in $\mathcal{R}$ of its multiplication by x_σ is

$$x_\sigma r(x) \; = \; c^T[r]\, x_\sigma \mathbf{b}(x) \; = \; c^T[r]\, \mathcal{A}_\mathcal{N}(x_\sigma \mathbf{b}(x)) \; =: \; c^T[r]\, A_\sigma \mathbf{b}(x). \tag{8.7}$$

Obviously, these $m \times m$ matrices A_σ, $\sigma = 1(1)s$, contain trivial μ-th rows (unit vector rows) for components b_μ of $\mathbf{b}$ with $x_\sigma b_\mu \in \mathcal{N}$ and the nontrivial rows a_j^T from (8.6) for b_μ with $x_\sigma b_\mu = x^j \in B[\mathcal{N}]$. Together, the *multiplication matrices* A_σ contain the complete information about $\mathcal{A}_\mathcal{N}$, i.e. about the multiplicative structure of $\mathcal{R}$. Note that we expect $\mathcal{R}$ to specify an ideal with m zeros (with $m\,s$ components) but that there are $N = |B[\mathcal{N}]| > s$ vectors a_j^T (with $N\,m$ components) in (8.6).

Due to the projection property of $\mathcal{A}_\mathcal{N}$ and the commutativity of multiplication in $\mathcal{R}$, the A_σ must satisfy

$$\begin{array}{llll} \text{(i)} & e_j^T A_\sigma & = & e_{j'}^T & \text{for } x^{j'} = x_\sigma x^j \in \mathcal{N}, \\ \text{(ii)} & A_{\sigma_1} A_{\sigma_2} & = & A_{\sigma_2} A_{\sigma_1} & \text{for all } \sigma_1, \sigma_2 \,. \end{array} \tag{8.8}$$

By their commutativity, the A_σ define a *commuting family* $\overline{A} := \{A^j := A_1^{j_1} \ldots A_s^{j_s}, \; j \in \mathbb{N}_0^s\}$ which contains all matrix polynomials

$$p(A) \; := \; \sum_{j \in J} \alpha_j \, A^j \; \in \; \mathbb{C}^{m \times m} \qquad \text{for } p \in \mathcal{P}^s \,;$$

cf. section 2.2.2.

Theorem 8.1. Consider a commutative ring $\mathcal{R}$ with a normal set basis $\mathcal{N} \in T^s(m)$ and basis vector $\mathbf{b}(x)$ and let the multiplication in $\mathcal{R}$ be represented by the $A_\sigma \in \mathbb{C}^{m \times m}$, $\sigma = 1(1)m$, of (8.7). Then

$$\mathcal{I}[\mathcal{R}] \; := \; \{\, p \in \mathcal{P}^s \; : \; p(A) = 0 \; (\text{zero matrix})\,\} \; \subset \; \mathcal{P}^s \tag{8.9}$$

is an ideal.

Proof: Trivially, the 0-polynomial is in $\mathcal{I}[\mathcal{R}]$, and so is $p_1 + p_2$ for $p_1, p_2 \in \mathcal{I}[\mathcal{R}]$; for $q \in \mathcal{P}^s$, $p \in \mathcal{I}[\mathcal{R}]$, $q \cdot p \in \mathcal{I}[\mathcal{R}]$ holds because, by commutativity, $(q\,p)(A) = q(A)\,p(A) = 0$. $\square$

Theorem 8.1 is constructive because a basis of $\mathcal{I}[\mathcal{R}]$ can readily be specified:

Proposition 8.2. In the situation of Theorem 8.1, consider the $x^{j'} \in B[\mathcal{N}]$ and the polynomials

$$bb_{j'}(x) \; := \; x^{j'} - a_{j'}^T \, \mathbf{b}(x), \tag{8.10}$$

with the coefficient vectors $a_{j'}^T$ of (8.6). Then $\mathcal{I}[\mathcal{R}] := \langle bb_{j'}(x), \; \forall j' : x^{j'} \in B[\mathcal{N}]\rangle \,.$

Proof: Let $\mathbf{b}(x) := (x^{j_\mu}, \ \mu = 1(1)m\}$ and consider the two possible interpretations of the m-vector $A_\sigma \cdot \mathbf{b}(A)$ of $m \times m$-matrices

$$
\begin{pmatrix} \sum_{\mu=1}^{m} (a_{\sigma,1})_\mu \, A^{j_\mu} \\ \vdots \\ \sum_{\mu=1}^{m} (a_{\sigma,m})_\mu \, A^{j_\mu} \end{pmatrix} = A_\sigma \cdot \mathbf{b}(A) = \begin{pmatrix} A_\sigma \, A^{j_1} \\ \vdots \\ A_\sigma \, A^{j_m} \end{pmatrix}
$$

which follow from linear algebra and (8.7). The identity of the right-hand and left-hand expression implies that $bb_{j'}(A) = 0$ for $\forall\, x^{j'} = x_\sigma x^j \in B[\mathcal{N}]$, $x^j \in \mathcal{N}$. How each $p \in \mathcal{I}[\mathcal{R}]$ can be represented as a polynomial combination of $bb_{j'}$ is shown in section 8.2.1. $\square$

Since $|B[\mathcal{N}]| > s$ (often $\gg s$), the basis (8.10) could well be *inconsistent* so that the generated ideal would be trivial (possess no zeros).

Theorem 8.3. In the situation of Theorem 8.1, if the commuting matrix family $\overline{A}$ is nonderogatory (cf. Definition 2.9), then the border basis (8.10) is consistent and $\langle \{bb_{j'}\} \rangle \subset \mathcal{P}^s(m)$.

Proof: Compare section 2.2.2. We consider, at first, the case of m joint eigenvectors which form the regular matrix $X \in \mathbb{C}^{m \times m}$. By (2.22), the A_σ satisfy $A_\sigma = X \Lambda_\sigma X^{-1}$; but as we have begun with an arbitrary multiplicative structure, we cannot readily interpret X and the diagonal matrices Λ_σ as in Theorem 2.23. However (cf. section 2.4.1), we can interpret X as a *transformation matrix* from $\mathbf{b}$ to another basis $\mathbf{b}_0(x) = X^{-1} \mathbf{b}(x)$ of $\mathcal{R}$ with respect to which multiplication in $\mathcal{R}$ is represented by the matrices

$$
A_\sigma^{(0)} = X^{-1} A_\sigma X = \Lambda_\sigma =: \ \operatorname{diag}(\lambda_{\sigma\mu}), \quad \sigma = 1(1)s \, ;
$$

cf. Proposition 2.22. Thus, we have in $\mathcal{R}$, for $\sigma = 1(1)s$, $x_\sigma \mathbf{b}_0(x) = \Lambda_\sigma \mathbf{b}_0(x)$ or

$$
(x_\sigma - \lambda_{\sigma 1}) b_{01}(x) = \ldots = (x_\sigma - \lambda_{\sigma\mu}) b_{0\mu}(x) = \ldots = (x_\sigma - \lambda_{\sigma m}) b_{0m}(x) = 0 . \qquad (8.11)
$$

These equations in $\mathcal{R}$ must hold in $\mathbb{C}$ at the zeros $z_\mu = (\zeta_{\mu\sigma}, \ \sigma = 1(1)s)$ of $\mathcal{I}[\mathcal{R}]$ which requires that, at each z_μ, all $b_{0\sigma}(z_\mu)$ vanish except one, which we call $b_{0\mu}$. $\mathbf{b}_0(z_\mu)$ cannot vanish because this would imply $\mathbf{b}(z_\mu) = X \cdot \mathbf{b}_0(z_\mu) = 0$ while $b_1(x) \equiv 1$. With the normalization $b_{0\mu}(z_\mu) = 1$, $\mathbf{b}_0$ becomes the Lagrange basis for $Z := \{z_\mu, \ \mu = 1(1)m\}$. As in section 2.4.1, this implies that X and Λ_σ are as in section 8.1.1. Now, Theorem 2.27 tells us that, for any $p \in \mathcal{P}^s$, $p(A) = X \operatorname{diag}(p(z_1), .., p(z_m)) X^{-1}$ whence

$$
p(A) = 0 \qquad \Longrightarrow \qquad p(z_\mu) = 0, \quad \mu = 1(1)m .
$$

There cannot be another $\bar{z} \notin Z$ with $p(\bar{z}) = 0 \ \forall\, p \in \mathcal{I}[\mathcal{R}]$ because this would require an $(m+1)$st eigenvalue for the matrices A_σ.

The case of joint invariant subspaces of dimensions > 1 leads to multiple zeros whose differential structure is defined by the associated vectors in X and upper-tridiagonal blocks in the Λ_σ; cf. Proposition 2.8. As above, $p(A) = 0$ implies that all $p \in \mathcal{I}[\mathcal{R}]$ must have a multiple zero z_μ with that particular structure. Details will be explained in section 8.5. $\square$

We have shown that any commutative ring $\mathcal{R}$ with a normal set basis $\mathcal{N} \in T^s(m)$ and commuting multiplication matrices with the appropriate trivial rows generates a polynomial ideal $\mathcal{I}[\mathcal{R}] \in \mathcal{P}^s(m)$ via the border basis $\mathcal{B}_{\mathcal{N}}[\mathcal{I}[\mathcal{R}]] = \{bb_{j'}(x),\ x^{j'} \in B[\mathcal{N}]\}$ of (8.10). This establishes that the *commutativity of the multiplication matrices* is the essential criterion for the existence of a relation between a ring with a monomial basis and a 0-dimensional ideal. In a concise form, this insight has been pointed out by Mourrain ([8.1]).

For us, an important consequence is the following: Consider a 0-dimensional polynomial system $P \subset \mathcal{P}^s$, with $\langle P \rangle \in \mathcal{P}^s(m)$, with a quotient ring $\mathcal{R}[\langle P \rangle] = \mathrm{span}\,\mathcal{N}$, $\mathcal{N} \in T^s(m)$, with multiplication matrices A_σ, $\sigma = 1(1)s$. Consider matrices $\tilde{A}_\sigma \approx A_\sigma$, with the same trivial rows but slightly perturbed nontrivial rows. *If and only if* the $\tilde{A}_\sigma$ commute, they define a ring $\tilde{\mathcal{R}}$ which can be interpreted as the quotient ring of an ideal $\tilde{\mathcal{I}} \in \mathcal{P}^s(m)$; the zero set $\tilde{Z} \in (\mathbb{C}^s)^m$ is then an approximation of the zero set Z of $\langle P \rangle$. But when we compute *approximate* multiplication matrices $\tilde{A}_\sigma$ from the specification of P by approximate computation, we cannot expect them to commute exactly; thus we cannot follow our standard approach and consider the $\tilde{A}_\sigma$ as *exact* multiplication matrices of a neighboring quotient ring $\tilde{\mathcal{R}}$. The handling of this dilemma will be discussed in section 9.2.

The following result from linear algebra facilitates the checking of the commutativity of the A_σ:

Proposition 8.4. Consider a family $\overline{A}$ of matrices generated by A_σ, $\sigma = 1(1)s$. If one of the generators (say A_s) is nonderogatory, then $A_\sigma A_s = A_s A_\sigma$, $\sigma \neq s$, implies the commutativity of all matrices in $\overline{A}$.

Proof: Compare [2.15], p.139, Problem 1. $\square$

Example 8.2: For $\mathcal{N} := \{1, x, y, z\} \in T^3(4)$, consider the quotient ring $\mathcal{R}$ for the ideal from $\mathcal{P}^3(4)$ with the zero set $Z = \{(1, 2, 0),\ (-1, -2, 0),\ (2, 0, -1),\ (0, -1, -2)\}$; for $\mathbf{b}(x) = (1, x, y, z)^T$, the multiplication matrices of $\mathcal{R}$ are

$$
A_1 = \begin{pmatrix} 0 & 1 & 0 & 0 \\ 1 & 2 & -1 & 1 \\ 2 & \frac{-4}{7} & \frac{2}{7} & \frac{6}{7} \\ 0 & \frac{-8}{7} & \frac{4}{7} & \frac{-2}{7} \end{pmatrix},\quad
A_2 = \begin{pmatrix} 0 & 0 & 1 & 0 \\ 2 & \frac{-4}{7} & \frac{2}{7} & \frac{6}{7} \\ 4 & \frac{-10}{7} & \frac{5}{7} & \frac{8}{7} \\ 0 & \frac{-4}{7} & \frac{2}{7} & \frac{-8}{7} \end{pmatrix},\quad
A_3 = \begin{pmatrix} 0 & 0 & 0 & 1 \\ 0 & \frac{-8}{7} & \frac{4}{7} & \frac{-2}{7} \\ 0 & \frac{-4}{7} & \frac{2}{7} & \frac{-8}{7} \\ 0 & \frac{-4}{7} & \frac{2}{7} & \frac{-15}{7} \end{pmatrix}.
$$

The border set $B[\mathcal{N}]$ has the 6 elements x^2, xy, xz, y^2, yz, z^2; accordingly, the ideal may be defined by the 6 border basis elements $bb_1 = x^2 - (2x - y + z + 1)$, $bb_2 = xy - (-4x + 2y + 6z + 14)/7$, $bb_3 = xz - (-8x + 4y - 2z)/7$, $bb_4 = y^2 - (-10x + 5y + 8z + 28)/7$, $bb_5 = yz - (-4x + 2y - 8z)/7$, $bb_6 = z^2 - (-4x + 2y - 15z)/7$. This basis is consistent and has the joint zero set Z. (Note that A_3 has a 2-fold eigenvalue 0 and the associated eigenspace spanned by $(1, 0, 0, 0)^T$, $(0, 1, 2, 0)^T$ does not permit a determination of the two associated zeros.)

Now we round the elements in the A_σ to 5 *relative* decimal digits and compute the normalized eigenvectors of the individual $\tilde{A}_\sigma$. Each $\tilde{A}_\sigma$ yields a slightly different approximation $\tilde{Z}_\sigma$ of the original zero set Z; rounded to 5 digits after the decimal point we obtain from

$$\tilde{A}_1 : \begin{matrix} (.99985, 2.00220, .00150) \\ (-1.00002, -2.00000, .00039) \\ (2.00013, -.00018, -.99992) \\ (-.00026, -.99966, -1.99905) \end{matrix} \qquad \tilde{A}_2 : \begin{matrix} (1.00010, 1.99980, -.00010) \\ (-.99971, -1.99967, -.00051) \\ (1.99973, .00040, -.99984) \\ (.00019, -1.00013, -1.99970) \end{matrix}$$

$$\tilde{A}_3 : \begin{matrix} (.99996, 2.00001, .00001) \\ (.00000, .00000, .00000) \\ (2.00006, -.00006, -1.00006) \\ (.00002, -1.00004, -2.00004) \end{matrix}$$

(The perturbation has also split the 2-fold eigenvalue 0 of A_3, but the "singular" eigenvector $(1, 0, 0, 0)^T$ from the eigenspace representation has persevered, due to the 0-column in A_3.)　□

8.1.3　Quasi-Univariate Normal Sets

In the previous section, we have seen that the *overdetermination* which is generally present in the specification of a quotient ring $\mathcal{R}$ by a monomial basis $\mathcal{N}$ and the s multiplication matrices A_σ requires the satisfaction of the *constraints* expressed by (8.8); otherwise there exists no ideal which matches $\mathcal{R}$ (cf. Example 8.2).

In the univariate case, this problem does not arise: A quotient ring $\mathcal{R}$ for a polynomial ideal from $\mathcal{P}^1(m)$ has the unique normal set basis $\mathcal{N} = \{1, x, \ldots, x^{m-1}\}$ and the only multiplication matrix, for the basis vector $\mathbf{b}(x) = (1, x, x^2, \ldots, x^{m-1})^T$, is

$$A = \begin{pmatrix} 0 & 1 & 0 & & & \\ & \ddots & 1 & \ddots & & \\ 0 & & \ddots & \ddots & & 0 \\ & & & 0 & 1 \\ \alpha_0 & \cdots & & \cdots & \alpha_{m-1} \end{pmatrix} \in \mathbb{C}^{m \times m}.$$

Any choice of the m elements α_μ in the only nontrivial row of A defines a multiplication matrix of a proper quotient ring $\mathcal{R}$. The m eigenvalues of A (counting multiplicities) are the m zeros of the ideal $\mathcal{I}[\mathcal{R}]$. This indicates that we may be able to avoid the overdetermination in the multivariate case if we look for a specification of the multiplicative structure by *one $m \times m$-matrix with exactly s nontrivial rows*: m zeros in $\mathbb{C}^s$ represent $m\,s$ data and so do s rows of m elements.

The nontrivial rows $a_{\sigma\nu}^T$ of a multiplication matrix A_σ contain the coefficients of the normal forms of the monomials $x_\sigma\,b_\nu(x) = x^{j'} \in B[\mathcal{N}]$; these monomials constitute the *border subset* $B_\sigma[\mathcal{N}] := (x_\sigma\mathcal{N}) \setminus \mathcal{N} \subset B[\mathcal{N}]$.

Definition 8.2. A normal set $\mathcal{N} \subset \mathcal{T}^s$, with $x_1, \ldots, x_s \in \mathcal{N}$, for which one or several border subsets B_σ have exactly s elements, is called a *quasi-univariate* normal set. (For a simpler notation, the distinguished variable x_σ to which the quasi-univariate situation refers will commonly be called x_s.)　□

An element in B_s has the form $x^{k_\nu} x_s^{n_\nu}$, with a vanishing s-component in k_ν and $n_\nu \geq 1$. The requirement $1, x_1, \ldots, x_{s-1} \in \mathcal{N}$, together with $|B_s| = s$, restricts the k_ν to 0 and the unit vectors e_σ, $\sigma = 1(1)s - 1$. Thus, the nondistinguished variables $x_1, \ldots, x_{s-1}$ can occur only *disjointly and linearly* in the monomials of a quasi-univariate normal set with distinguished variable x_s. Closedness of $\mathcal{N}$ further requires that $n_\sigma \leq n_s$, $\sigma = 1(1)s - 1$, $n_s \geq 2$, where n_σ is the x_s-exponent of the multiple of x_σ in B_s. For $m \geq s$ (which we have generally assumed), quasi-univariate normal sets always exist.

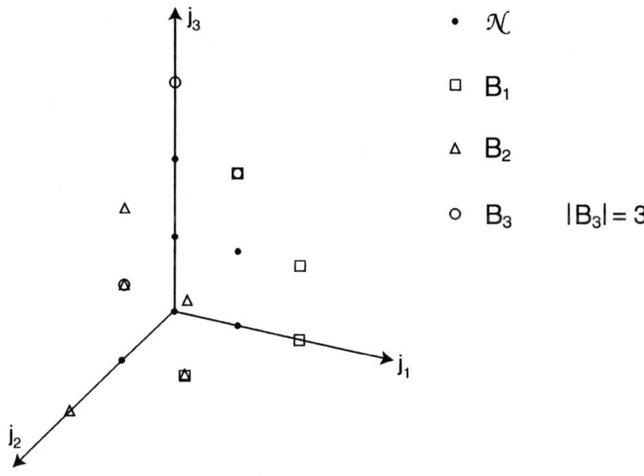

Figure 8.1.

Example 8.3: Definition 8.2 loosely says: A quasi-univariate normal set contains only products of one element from $\{1, x_1, \ldots, x_{s-1}\}$ with a power of x_s. Thus, a quasi-univariate normal set consists of s "columns" of length n_σ in the direction of x_s; cf. Figure 8.1. For $s = 1$, this reduces to $1, x, \ldots, x^{n-1}$, which explains the name.

In $T^3(7)$ (cf. Example 8.1), the following normal sets are quasi-univariate, with distinguished variable z : $(1, x, y, z, z^2, z^3, z^4)$, $(1, x, y, z, yz, z^2, z^3)$, $(1, x, y, z, yz, z^2, yz^2)$, $(1, x, y, z, xz, yz, z^2)$, and the two resulting from exchanging the roles of x and y in the 2nd and the 3rd set. The normal set $\{1, x, y, z\}$ of Example 8.2 is quasi-univariate with respect to each of the three variables. $\square$

Theorem 8.5. Consider a quasi-univariate normal set $\mathcal{N}$ from $T^s(m)$, $m \geq s + 1$, with distinguished variable x_s, and form a multiplication matrix $A_s \in \mathbb{C}^{s \times s}$ with respect to $\mathcal{N}$ by properly choosing the trivial rows and by completing the nontrivial rows with *arbitrary* complex numbers. If A_s is nonderogatory and has m linearly independent eigenvectors, it defines a commutative ring $\mathcal{R}$ with basis $\mathcal{N}$ which is the quotient ring of an ideal $\mathcal{I}[\mathcal{R}] \in \mathcal{P}^s(m)$. The m zeros of $\mathcal{I}[\mathcal{R}]$ are specified by the components of the normalized eigenvectors of A_s in the usual way.

Proof: The restriction on the eigenvectors of A_s excludes the delicate case of a multiple zero in $\mathcal{I}$; a multiple eigenvalue of A_s with an eigenspace of corresponding dimension is also excluded. Thus, after normalization, each eigenvector defines a zero $z_\mu \in \mathbb{C}^s$; also, by the structure of the basis vector $\mathbf{b}$ and the trivial rows of A_s, each eigenvector must be internally consistent, i.e. satisfy (8.2). Therefore, we can define the remaining multiplication matrices A_σ via (8.1) and obtain a *commuting family* $\overline{A}$ which generates a border basis for the ideal $\mathcal{I}[\mathcal{R}]$ with the m zeros z_μ. $\square$

Corollary 8.6. Under the assumptions of Theorem 8.5, a small perturbation of the nontrivial rows of A_s does *not* invalidate the conclusions of the Theorem. In particular, the eigenvectors remain internally consistent and the zeros are continuous functions of the perturbation.

Proof: The restrictions on A_s hold in an open domain in the data space of the elements of the nontrivial rows; therefore, they hold in a sufficiently small neighborhood of A_s. Under the assumptions of Theorem 8.5, the components of the normalized eigenvectors are continuous functions of the nontrivial elements of A_s. $\square$

Example 8.4:   We take the situation of Example 8.1, but choose the quasi-univariate normal set $\mathcal{N}_0 = \{1, x, y, z, xz, yz, z^2\}$, with the components in $\mathbf{b}_0$ in the same order. To find the transformation matrix M_{01} which yields the multiplication matrix $A_3^{(0)}$ with respect to $\mathbf{b}_0$ we have only to express the new normal set element z^2 in terms of the original basis $\mathbf{b}_1$ which is achieved by the 4th row of $A_3^{(1)}$. With

$$M_{01} = \begin{pmatrix} 1 & 0 & 0 & 0 & 0 & 0 & 0 \\ 0 & 1 & 0 & 0 & 0 & 0 & 0 \\ 0 & 0 & 1 & 0 & 0 & 0 & 0 \\ 0 & 0 & 0 & 1 & 0 & 0 & 0 \\ 0 & 0 & 0 & 0 & 0 & 1 & 0 \\ 0 & 0 & 0 & 0 & 0 & 0 & 1 \\ \frac{-643}{61} & 7 & \frac{-455}{61} & \frac{400}{61} & 2 & \frac{-125}{61} & \frac{151}{61} \end{pmatrix},$$

we obtain $A_3^{(0)} = M_{01}\, A_3^{(1)}\, M_{01}^{-1} = \begin{pmatrix} 0 & 0 & 0 & 1 & 0 & 0 & 0 \\ 0 & 0 & 0 & 0 & 1 & 0 & 0 \\ 0 & 0 & 0 & 0 & 0 & 1 & 0 \\ 0 & 0 & 0 & 0 & 0 & 0 & 1 \\ \frac{520}{183} & 0 & \frac{-520}{183} & \frac{-385}{61} & \frac{269}{61} & \frac{190}{183} & \frac{5}{3} \\ \frac{60}{61} & 0 & \frac{-60}{61} & \frac{99}{61} & \frac{-50}{61} & \frac{219}{61} & -1 \\ \frac{485}{61} & -9 & \frac{613}{61} & \frac{-191}{61} & \frac{-89}{61} & \frac{-197}{61} & 3 \end{pmatrix},$

which obviously has the right form. Since the z-component takes a different value in each of the 7 zeros, the assumptions of Theorem 8.5 are satisfied.

When we subject the elements in the lower 3 rows of $A_3^{(0)}$ to random perturbations say of maximal modulus .0003, the relevant components of the eigenvectors change only by quantities of that same order. Also, the internal consistency of the eigenvectors remains fully intact: The defects in the constraints (8.2) remain at the round-off level of the computation. $\square$

Proposition 8.7. Consider a quasi-univariate normal set and the multiplication matrix A_s for the distinguished variable x_s. If A_s satisfies the assumptions of Theorem 8.5, the nontrivial rows of the other A_σ may be determined from A_s directly without a solution of the eigenproblem.
Proof: The elements x^j in a border subset B_σ, $\sigma \neq s$, fall in one of 3 categories:

(a) x^j is also in B_s;

(b) $x^j = x_s^{\lambda_j} x^{\bar{j}}$ is "above" an element $x^{\bar{j}} \in B_s$ $\qquad$ (cf. Figure 8.1);

(c) neither (a) nor (b).

The rows $a_{x^j}^T$ for (a) and (b) can be found directly from A_s and $A_s^{\lambda_j+1}$, respectively. For the remaining nontrivial rows (case (c)), one can write down relations which relate them to an x_σ-neighbor of an element in B_s; these relations are *linear* in the rows of A_σ. They must have a unique solution since we know from Theorem 8.5 that A_σ is uniquely defined by A_s. A_s has been assumed as nonderogatory; therefore, by Proposition 8.4, the A_σ must also be *mutually commuting*. $\quad\Box$

Example 8.4, continued: For $\mathcal{N}_0$, the set B_1 consists of x^2, xy, x^2z, $x\,y\,z$, $x\,z^2$. $x\,z^2$ is in (a); the 4 others are in (c). The 4 linear equations for these nontrivial rows of A_1 are:

$$a_{x^2}^T \cdot A_3 = a_{x^2z}^T = a_{xz}^T \cdot A_1 \quad \text{and} \quad a_{xy}^T \cdot A_3 = a_{xyz}^T = a_{yz}^T \cdot A_1 \,. \quad \Box$$

The quasi-univariate situation also sheds light on the potential shortcomings of a multiplication matrix A_s whose border subset $B_s[\mathcal{N}]$ is larger than s, say e.g. $s+1$. In this case, $\mathcal{N}$ must contain one monomial nonlinear in the x_σ, $\sigma < s$, and possibly products of it with powers of x_s. Let x_1x_2 be that monomial. Then the nontrivial row in A_s for the border element $x_1x_2x_s^{k_{12}} = x_s \cdot x_1x_2x_s^{k_{12}-1}$, with $x_1x_2x_s^{k_{12}-1} \in \mathcal{N}$, is the end of a chain of trivial rows; this causes the corresponding components in the eigenvectors of A_s to be internally consistent with respect to multiplication with the respective eigenvalue: In the μ-th eigenvector, the component for $x_1x_2x_s^\kappa$ will equal $\zeta_{\mu s}$ times the component for $x_1x_2x_s^{\kappa-1}$, a consistency which holds automatically for all components $x_\sigma x_s^\kappa$.

But there *cannot* be any consistency between the values of the components $x_1x_2x_s^\kappa$ and $x_1x_s^\kappa$ or $x_2x_s^\kappa$, resp.; actually, the monomial x_1x_2 simply takes the place of a further nondistinguished variable in the quasi-univariate case whose value at the zeros is independent of the values of the other x_σ. The necessary relations with respect to factors x_1 or x_2 (or generally x_σ, $\sigma < s$) are only introduced into the eigenvectors of A_s by the fact that A_s commutes with the other A_σ.

Example 8.5: We consider Example 8.1 and round $A_3^{(1)}$ to 6 relative decimal digits. The 2nd through 4th components of the normalized eigenvectors give approximations to the zeros in Z with deviations of $O(10^{-5})$ throughout. When we compare the 5th components (for xy) with the products of the 2nd and 3rd components, the differences vary between .3 and 12 units of 10^{-5}. If we do the same with the components no. 6 and the products of no. 2 and no. 4, or with no. 7 and the products of no. 3. and no. 4, the differences are at the round-off level. This shows that the internal inconsistency of the rounded multiplication matrix is restricted to the relations between the normal set monomials x, y and xy. $\quad\Box$

The surprisingly strong result of Theorem 8.5 and Corollary 8.6 appears to suggest that the computational determination of a basis for the quotient ring associated with a 0-dimensional polynomial system should be led towards a quasi-univariate normal set. At present, it is not clear how this may generally be achieved and how a distinguished variable may be selected a priori. We will also find aspects of a normal set from a numerical point of view which do not favor quasi-univariate normal sets; for large m, the strong asymmetry in a quasi-univariate representation may result in a poor condition of the eigenproblem for $A_s^{(0)}$.

Exercises

1. In translating the representation of a quotient ring from one normal set basis $\mathcal{N}_1$ to another one $\mathcal{N}_2$, it may not always happen that all elements of $\mathcal{N}_2 \setminus \mathcal{N}_1 \cap \mathcal{N}_2$ are in the border set $B[\mathcal{N}_1]$ so that their normal forms with respect to $\mathbf{b}_1$ can be read from one of the multiplication matrices $A_\sigma^{(1)}$; cf. Example 8.1. In this case, we can form intermediate normal sets $\mathcal{N}_{1\kappa}$, with $\mathcal{N}_{10} = \mathcal{N}_1, \mathcal{N}_{1k} = \mathcal{N}_2$ such that $\mathcal{N}_{1,\kappa+1} \setminus \mathcal{N}_{1\kappa} \cap \mathcal{N}_{1,\kappa+1} \subset B[\mathcal{N}_{1\kappa}]$ and the transformation matrices for the transition from $\mathcal{N}_{1\kappa}$ to $\mathcal{N}_{1,\kappa+1}$ can be written down directly.

   (a) Use this approach with the data of Example 8.1 to find the representation of $\mathcal{R}$ with reference to the ("univariate") normal set $\widehat{\mathcal{N}} = \{1, z, z^2, z^3, z^4, z^5, z^6\}$.

   (b) From the information about the zeros, determine the only nontrivial row of $\widehat{A}_3$ directly. How can you also determine the first row of the multiplication matrices $\widehat{A}_1$ and $\widehat{A}_2$ (why are they nontrivial for $\widehat{\mathcal{N}}$)? How can you obtain the remaining rows of $\widehat{A}_1$ and $\widehat{A}_2$ from their first row and the last row of $\widehat{A}_3$?

   (c) Assume that you have been given the last row of $\widehat{A}_3$ and the first row of $\widehat{A}_1$ and $\widehat{A}_2$. Convince yourself that they contain the complete information about the zeros of $\mathcal{I}[\mathcal{R}]$. How would you find the zeros from this information?

2. (a) Consider Example 8.2 and convince yourself that its normal set $\mathcal{N} = \{1, x, y, z\}$ is quasi-univariate for *each* of the three variables. How does this agree with the fact that each of the three *perturbed* multiplication matrices $\tilde{A}_\sigma$ yields a different zero set?

   (b) For the quotient ring $\widetilde{\mathcal{R}}_2$ defined by $\mathcal{N}$ and $\tilde{A}_2$, find the associated multiplication matrices $\tilde{A}_1$ and $\tilde{A}_3$ by the procedure described in section 8.1.3 (above Example 8.4, continued). Verify that these yield the same zero set as $\tilde{A}_2$ within round-off.

8.2   Normal Set Representations of 0-Dimensional Ideals

In section 2.5.2, we had introduced our standard representation of a 0-dimensional ideal by a *normal set* and the related multiplication matrices mod $\mathcal{I}$ or—equivalently—the related border basis; cf. Definition 2.23. In this section, we consider various aspects of this representation which are important for its computational use.

For an ideal $\mathcal{I} \subset \mathcal{P}^s(m)$, we have a normal set $\mathcal{N} \in \mathcal{T}^s(m)$, with the associated monomial basis vector $\mathbf{b}(x) = (b_\mu(x), \ \mu = 1(1)m)$ of $\mathcal{R}[\mathcal{I}]$. The border set $B[\mathcal{N}]$ consists of N monomials x^j, $j \in \mathbb{N}_0^s$. By (8.6), the quantitative part of the normal set representation of $\mathcal{I}$ consists of the N row vectors $a_j^T \in \mathbb{C}^m$ which specify the normal forms of the border monomials x^j mod $\mathcal{I}$:

$$\mathrm{NF}_{\mathcal{I}}[x^j] = a_j^T \mathbf{b}(x) \qquad \forall \, x^j \in B[\mathcal{N}]. \tag{8.12}$$

These a_j^T furnish the nontrivial rows of the multiplication matrices A_σ mod $\mathcal{I}$ as well as the coefficients of the polynomials in the border basis $\mathcal{B}_\mathcal{N}[\mathcal{I}] = \{bb_j(x) := x^j - a_j^T \mathbf{b}(x)\}$.

8.2.1   Computation of Normal Forms and Border Basis Expansions

With the aid of the normal set representation of $\mathcal{I}$, we must be able to compute the normal form $\mathrm{NF}_{\mathcal{I}}[p]$ in span $\mathcal{N}$ for a specified $p \in \mathcal{P}^s$; cf. Definition 2.20.

Proposition 8.8. In the situation just described,

$$\mathrm{NF}_{\mathcal{I}}[p(x)] \;=\; \mathrm{NF}_{\mathcal{I}}[e_1^T p(x)\mathbf{b}(x)] \;=\; e_1^T p(A)\,\mathbf{b}(x) \tag{8.13}$$

so that the normal form functionals $\mathbf{c}^T$ which form the basis of the dual space $\mathcal{D}[\mathcal{R}]$ conjugate to $\mathbf{b}$ are defined by $\mathbf{c}^T[p] = e_1^T p(A)$.

Proof: Compare (2.20) in Corollary 2.6 and Proposition 2.13. $\quad\square$

For small values of m and a low total degree of p, the evaluation of (8.13) provides a quick and easy way for the computation of $\mathrm{NF}_{\mathcal{I}}[p]$. Note that, as long as $x^j \in \mathcal{N}$, with $x^j = b_\nu(x) = e_\nu^T \mathbf{b}(x)$, $\mathrm{NF}[x^j] = e_1^T A^j = e_\nu$ remains a unit vector; for $x^j \in B[\mathcal{N}]$, $\mathrm{NF}[x^j]$ equals a row in one of the A_σ. Thus the actual computation starts when the recursive evaluation of the normal forms of the monomials in $p(x) = \sum_{j\in J} \alpha_j x^j$ leaves the border set.

To deal with the "distance" of a monomial from $\mathcal{N}$, we remember the "hull sets" of a normal set $\mathcal{N}$ which are generated by an iteration of the border operation; cf. Definition 2.21, (2.59):

For a closed set $\mathcal{N} \subset T^s$, the *hull sets* $H_\ell[\mathcal{N}]$, $\ell = 0, 1, 2, \ldots$, are defined by

$$H_0[\mathcal{N}] \;:=\; \mathcal{N}\,, \quad H_{\ell+1}[\mathcal{N}] \;:=\; H_\ell[\mathcal{N}] \,\cup\, B[H_\ell[\mathcal{N}]] \;\subset\; T^s\,.$$

Obviously, we may reach any monomial x^j in $H_\ell[\mathcal{N}]$ from an appropriate monomial in $B[\mathcal{N}]$ by $\ell - 1$ successive multiplications by suitable variables so that $\mathrm{NF}[x^j]$ may be found from (8.13) by $\ell - 1$ vector-matrix multiplications.

Definition 8.3. With reference to a fixed normal set $\mathcal{N} \subset T^s$, consider the map $\ell : T^s \to \mathbb{N}_0$ defined by $\ell(x^j) := \min\{\lambda : x^j \in H_\lambda[\mathcal{N}]\}$. The $\mathcal{N}$-*index* of a polynomial $p(x) = \sum_{j\in J} \alpha_j x^j \in \mathcal{P}^s$ is $\ell(p) := \max_{j\in J} \ell(x^j)$. The terms in p whose monomials satisfy $\ell(x^j) = \ell(p)$ are the $\mathcal{N}$-*leading terms* of p. $\quad\square$

Example 8.6: By its definition (8.10), each border basis element $bb_j(x)$ has $\mathcal{N}$-index 1 and one $\mathcal{N}$-leading term x^j. A polynomial $x^k bb_j(x)$ has $\mathcal{N}$-index $|k| + 1$ and the leading term x^{k+j}. Generally, a polynomial has more than one $\mathcal{N}$-leading term; if all terms of p are in $B[H_{\ell-1}[\mathcal{N}]]$, p consists only of $\mathcal{N}$-leading terms. $\quad\square$

The number of vector-matrix multiplications necessary to find $\mathrm{NF}[p]$ from (8.13) is, of course, bounded by the sum of the $\mathcal{N}$-indices of its terms, but an intelligent evaluation will strive to use common intermediate monomials to arrive at the various monomials composing p. Thus the minimal number of necessary vector-matrix multiplications may be much smaller. Also, due to the special structure of the A_σ with many trivial and perhaps only a few nontrivial rows which may themselves be quite sparse, the cost of a vector-matrix multiplication is often much less than m^2 arithmetic operations ($m = |\mathcal{N}|$).

The more conventional way of computing normal forms is the reduction of p to a polynomial in $\mathcal{N}$ by the successive subtraction of polynomials in $\mathcal{I}$. With the border basis elements bb_j of (8.10), we have a means of reducing the $\mathcal{N}$-index of a monomial by one in one subtraction: Assume that x^j with $\ell(x^j) = \ell > 1$ is a multiple $x^k x^j$ of the leading monomial x^j of one of the border basis elements bb_j; there must be one or several such bb_j. Then, by (8.10),

$$x^j - x^k bb_j(x) \;=\; a_j^T x^k \mathbf{b}(x)\,; \tag{8.14}$$

thus, this reduction step generates a polynomial of $\mathcal{N}$-index $\ell - 1$. But this polynomial may have several $\mathcal{N}$-leading monomials so that the continuation of this reduction procedure will generally involve more and more monomials after a few steps. If it is applied to all monomials in a given polynmial p, (almost) all of the monomials in the sets $B[H_\lambda[\mathcal{N}]]$ will become involved as λ decreases; at the same time, the magnitude of these sets decreases with λ. An a priori control which minimizes the number of individual subtractions of a multiple of a border basis element is hard to conceive, taking into account the many possible shapes of a normal set in s dimensions. While the individual reduction steps are not unique, the final result $\mathrm{NF}_{\mathcal{I}}[p]$ must be unique; otherwise two different polynomials in span $\mathcal{N}$ would differ by a polynomial in $\mathcal{I}$.

For moderate values of ℓ, we must expect $\mathrm{O}(\ell\, m\, n)$ operations, where $n \sim |\, B[H_\lambda[\mathcal{N}]]\,|$ for low λ. This appears comparable to the count for an algorithm based on (8.13). Actual numbers will depend strongly on the implemented strategies for the multiple use of operations and on the specific situations at hand.

Normal form computation may also be based on

Proposition 8.9. For $\mathcal{I}$ with normal set $\mathcal{N}$, the normal form mod $\mathcal{I}$ of a polynomial $p \in \mathcal{P}^s$ is the interpolation polynomial in span $\mathcal{N}$ of the values $p(z_\mu)$ specified at the zeros $z_\mu \in Z(\mathcal{I})$. *Proof*: $p - \mathrm{NF}_{\mathcal{I}}[p] \in \mathcal{I}$ implies $p(z_\mu) = \mathrm{NF}_{\mathcal{I}}[p](z_\mu)$. $\quad\square$

Thus, if a normal form computation mod $\mathcal{I}$ follows *after* the computation of the zeros of $\mathcal{I}$ so that the zeros z_μ and the matrix X of the joint eigenvectors of the A_σ are available, the coefficients $\mathbf{c}^T(p)$ of the normal form of a polynomial p may also be computed from the linear system

$$\mathbf{c}^T(p)\, X \;=\; (\, p(z_1) \ldots p(z_m)\,)\,. \tag{8.15}$$

Multiple zeros require the usual modification; cf. section 8.5. Note that the cost of this approach does not depend on the $\mathcal{N}$-index of p but on the cost of the evaluations of p. The cost of solving (8.15) is $\mathrm{O}(m^3)$; this cost may be reduced by the use of recursive interpolation algorithms (cf. section 9.6).

Example 8.7: We consider the ideal $\mathcal{I}$ of Example 8.2, specified by the normal set $\mathcal{N} = \{1, x, y, z\}$ and the multiplication matrices A_1, A_2, A_3. We wish to find the normal form of the polynomial $p(x, y, z) = x^4 - x^2yz + 3x^2y + 2x^2z - xyz$. The hull sets are $H_\ell[\mathcal{N}] = \{\text{monomials in } x, y, z \text{ of total degree} \le \ell + 1\}$. Thus, p has $\mathcal{N}$-index 3 and the two $\mathcal{N}$-leading terms x^4 and $-x^2yz$.

When we use (8.13) to compute the normal form, we may compose $\mathbf{c}^T(p)$ from the nontrivial rows $a_{x^j}^T := \mathbf{c}^T(x^j)$ in the A_σ as

$$a_{x^2}^T A_1^2 + a_{xy}^T A_1 (-A_3 + 3\,I) + 2\,a_{xz}^T A_1 - a_{xz}^T A_2\,;$$

the only economization possible is in the computation for the second and third term of p. With the values from Example 8.2, we obtain

$$\mathrm{NF}_{\mathcal{I}}[p] \;=\; 1 + \tfrac{24}{7}\,x + \tfrac{9}{7}\,y - \tfrac{1}{7}\,z\,.$$

When we use (8.14), we may at first reduce the monomials x^4 and x^2yz by $x^2\, bb_{x^2}$ and $x^2\, bb_{yz}$, resp., where bb_{x^j} denotes the border basis element with $\mathcal{N}$-leading term x^j; this generates polynomials with monomials x^3, x^2y, x^2z, x^2. The first three terms, together with the remaining terms of p, may be reduced by x-multiples of bb_{x^2}, bb_{xy}, bb_{xz}, bb_{yz}, respectively. Now we have arrived at polynomials which contain only terms with monomials from $B[\mathcal{N}]$ or $\mathcal{N}$ itself. The final expression is, of course, the same as above.

The values of p at the zeros in Z are, in the order of the zeros in Example 8.2, 7, -5, 8, 0. The linear system of (8.15) becomes

$$(c_1(p),\ c_2(p),\ c_3(p),\ c_4(p))\begin{pmatrix} 1 & 1 & 1 & 1 \\ 1 & -1 & 2 & 0 \\ 1 & -2 & 0 & -1 \\ 0 & 0 & -1 & -2 \end{pmatrix} = (7,\ -5,\ 8,\ 0)$$

which yields directly the normal form coefficients as above. $\square$

For various purposes it is desirable to find not only the normal form $\mathrm{NF}_{\mathcal{I}}[p]$ but a full "expansion" of $p \in \mathcal{P}^s$ in terms of the border basis $\mathcal{B}_{\mathcal{N}}$:

$$p(x) = \mathrm{NF}[p] + \sum_{x^j \in B[\mathcal{N}]} q_j(x)\, bb_j(x)\,; \tag{8.16}$$

such representations exist since $p - \mathrm{NF}[p] \in \mathcal{I}$. Equation (8.16) is a generalization of the representation (5.16) in the univariate case. The sum in (8.16) cannot be unique in general because of the arbitrariness in the reduction of p to its (unique) normal form.

To determine one particular set of q_j, one has to follow a path of the reduction $p \xrightarrow{\mathcal{B}} \mathrm{NF}[p]$. If the reduction proceeds by subtraction of multiples of the bb_j's, then one simply has to accumulate these multiples over the reduction; cf. (8.14). When the normal form has been formed as $e_1^T p(A)\,\mathbf{b}$ (cf. (8.13)), we can extend this representation in the following way:

With $\mathbf{bb} := (bb_1, \ldots, bb_N)^T$, we have—as relations in $\mathcal{P}^s$—

$$x_\sigma \cdot \mathbf{b}(x) = A_\sigma\,\mathbf{b}(x) + B_\sigma\,\mathbf{bb}(x)\,, \quad \sigma = 1(1)s, \tag{8.17}$$

where $B_\sigma \in \{0, 1\}^{m \times N}$ has μ-th rows of zeros for $x_\sigma b_\mu(x) \in \mathcal{N}$ and a 1 in the j-th position for $x_\sigma b_\mu(x) = x^j \in B[\mathcal{N}]$. Equation (8.17) may now be extended recursively: Let

$$x^k \cdot \mathbf{b}(x) = A^k\,\mathbf{b}(x) + B^k(x)\,\mathbf{bb}(x)\,;$$

then

$$x_\sigma x^k \cdot \mathbf{b}(x) = A_\sigma\,[A^k\,\mathbf{b}(x) + B^k(x)\,\mathbf{bb}(x)] + B_\sigma\,x^k\,\mathbf{bb}(x)\,.$$

Thus, the $B^k(x)$ matrices have to follow the recursion

$$B^{k+e_\sigma}(x) := A_\sigma\,B^k(x) + x^k\,B_\sigma\,. \tag{8.18}$$

With this recursion, we may turn the equivalence

$$p(x) \cdot \mathbf{b}(x) = \left(\sum_k \alpha_k x^k\right) \cdot \mathbf{b}(x) \equiv \left(\sum_k \alpha_k A^k\right)\mathbf{b}(x) \quad \mathrm{mod}\ \mathcal{I}$$

into the equation in $\mathcal{P}^s$

$$p(x) \cdot \mathbf{b}(x) = \sum_k \alpha_k\,(A^k\,\mathbf{b}(x) + B^k(x)\,\mathbf{bb}(x)) \qquad \mathrm{or}$$

$$p(x) \ = \ p(x)\, e_1^T \mathbf{b}(x)$$
$$= \ \sum_k \alpha_k \left(e_1^T A^k\, \mathbf{b}(x) + e_1^T B^k(x)\, \mathbf{bb}(x) \right) \ = \ \mathrm{NF}[p] + \sum_{x^j \in B[\mathcal{N}]} q_j(x)\, bb_j(x)\,. \tag{8.19}$$

Example 8.7, continued: For the normal set, the multiplication matrices and the border basis of Example 8.2, we have the matrices

$$B_1 = \begin{pmatrix} 0 & 0 & 0 & 0 & 0 & 0 \\ 1 & 0 & 0 & 0 & 0 & 0 \\ 0 & 1 & 0 & 0 & 0 & 0 \\ 0 & 0 & 1 & 0 & 0 & 0 \end{pmatrix}, \quad B_2 = \begin{pmatrix} 0 & 0 & 0 & 0 & 0 & 0 \\ 0 & 1 & 0 & 0 & 0 & 0 \\ 0 & 0 & 0 & 1 & 0 & 0 \\ 0 & 0 & 0 & 0 & 1 & 0 \end{pmatrix},$$

$$B_3 = \begin{pmatrix} 0 & 0 & 0 & 0 & 0 & 0 \\ 0 & 0 & 1 & 0 & 0 & 0 \\ 0 & 0 & 0 & 0 & 1 & 0 \\ 0 & 0 & 0 & 0 & 0 & 1 \end{pmatrix}.$$

With an iterated application of (8.18), we obtain the first rows b_k of the B^k, $k \in \mathbb{N}_0^3$; e.g.,

$$b_{4,0,0} \ = \ (x^2 + 2x + \tfrac{31}{7},\ -x - \tfrac{12}{7},\ x + \tfrac{6}{7},\ 0,\ 0,\ 0)\,,$$
$$b_{2,1,1} \ = \ (-\tfrac{4}{7}x,\ \tfrac{2}{7}x,\ -\tfrac{8}{7}x,\ 0,\ x^2,\ 0)$$

$$\text{etc.}$$

Then (8.19) yields, with the normal form found previously,

$$p(x) \ = \ (1 + \tfrac{24}{7}x + \tfrac{9}{7}y - \tfrac{1}{7}z)$$
$$+ (1 + \tfrac{18}{7}x + x^2)\, bb_1(x) + \tfrac{12}{7}x\, bb_2(x) + (4 + \tfrac{29}{7}x)\, bb_3(x) + (-x^2 - x)\, bb_5(x)\,. \ \square$$

8.2.2 The Syzygies of a Border Basis

The row vectors $a_j^T \in \mathbb{C}^m$, $x^j \in B[\mathcal{N}]$, which define a normal set representation of $\mathcal{I}$ determine the nontrivial rows of the $m \times m$ matrices A_σ as well as the border basis elements $bb_j(x)$. Therefore, the commutativity conditions (8.8) on the A_σ must also be representable in terms of the border basis elements bb_k. Let $\mathbf{b}(x) = (b_\mu(x))^T = (x^{j_1}, \dots, x^{j_m})^T$ be the normal set vector for $\mathcal{N}$; for the present purpose, we denote the μth row of A_σ by $a_{\sigma\mu}^T$; iff $x^j = x_\sigma b_\mu(x) \in B[\mathcal{N}]$ then $a_{\sigma\mu}^T = a_j^T$ is nontrivial.

Definition 8.4. The *boundary set* $\partial\mathcal{N}$ of a closed monomial set $\mathcal{N} \subset T^s$ is defined by (cf. Definition 2.17)

$$\partial\mathcal{N} \ := \ \{x^j \in \mathcal{N} \ : \ \text{At least one positive neighbor of } x^j \text{ is in } B[\mathcal{N}]\}\,. \quad \square \tag{8.20}$$

Definition 8.5. For a closed monomial set $\mathcal{N} \subset T^s$, $|\mathcal{N}| = m$, with boundary set $\partial\mathcal{N} = \{b_\mu\}$, $|\partial\mathcal{N}| =: \bar{m} \leq m\}$, we consider the following *edges* $e_{\mu\sigma_1\sigma_2} = [x^{j_1}, x^{j_2}]$ between the N monomials of $B[\mathcal{N}]$: For each triple $(\mu, \sigma_1, \sigma_2)$, $b_\mu \in \partial\mathcal{N}$, $\sigma_1 \neq \sigma_2 \in \{1, .., s\}$,

(i) if $x_{\sigma_1} b_\mu \in B[\mathcal{N}]$, $x_{\sigma_2} b_\mu \notin B[\mathcal{N}]$,

$$e_{\mu\sigma_1\sigma_2} \ := \ [x_{\sigma_1}\, b_\mu,\ x_{\sigma_1} x_{\sigma_2}\, b_\mu]\,, \tag{8.21}$$

(ii) if $x_{\sigma_1} b_\mu$, $x_{\sigma_2} b_\mu \in B[\mathcal{N}]$,

$$e_{\mu\sigma_1\sigma_2} := [x_{\sigma_1} b_\mu, x_{\sigma_2} b_\mu], \tag{8.22}$$

(iii) if $x_{\sigma_1} b_\mu$, $x_{\sigma_2} b_\mu \notin B[\mathcal{N}]$: $e_{\mu\sigma_1\sigma_2}$ not defined.

The *border web* $BW_\mathcal{N}$ of $\mathcal{N}$ is the set of all edges $e_{\mu\sigma_1\sigma_2}$ in $B[\mathcal{N}]$. Obviously, $\bar{N} := |BW_\mathcal{N}| \le \bar{m}\,\frac{s(s-1)}{2}$. $\square$

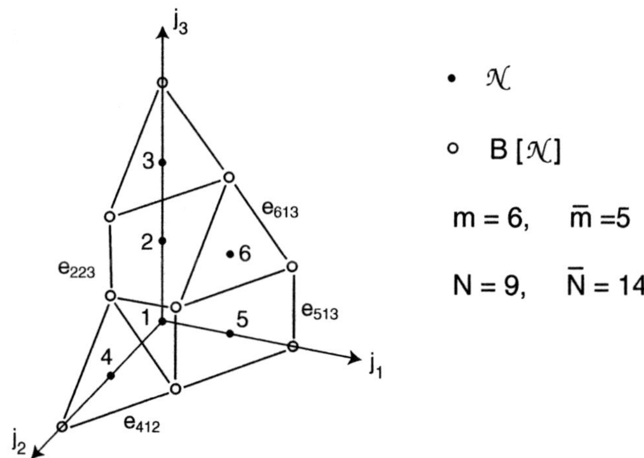

Figure 8.2.

Proposition 8.10. For a normal set $\mathcal{N} \subset T^s$, consider the set of N nontrivial row vectors $a_{\sigma\mu}^T \in \mathbb{C}^m$ which define the potential multiplication matrices A_σ, $\sigma = 1(1)s$, for a ring $\mathcal{R} = \text{span } \mathcal{N}$. The A_σ *commute* iff their trivial rows have been set properly and the following conditions hold (cf. Figure 8.2):

- for each $e_{\mu\sigma_1\sigma_2} \in BW_\mathcal{N}$ of type (i), where $a_{\sigma_2\mu}^T = e_{\bar{\mu}}^T$ is a trivial row in A_{σ_2} but $x_{\sigma_1} b_{\bar{\mu}}(x) = x_{\sigma_2} x_{\sigma_1} b_\mu(x)$ is in $B[\mathcal{N}]$:

$$a_{\sigma_1\bar{\mu}}^T = e_{\bar{\mu}}^T A_{\sigma_1} = e_\mu^T A_{\sigma_2} A_{\sigma_1} = e_\mu^T A_{\sigma_1} A_{\sigma_2} = a_{\sigma_1\mu}^T A_{\sigma_2}. \tag{8.23}$$

- for each $e_{\mu\sigma_1\sigma_2} \in BW_\mathcal{N}$ of type (ii), with both $a_{\sigma_i\mu}^T$ nontrivials rows of A_{σ_i}:

$$a_{\sigma_2\mu}^T A_{\sigma_1} = e_\mu^T A_{\sigma_2} A_{\sigma_1} = e_\mu^T A_{\sigma_1} A_{\sigma_2} = a_{\sigma_1\mu}^T A_{\sigma_2}. \tag{8.24}$$

Proof: For fixed σ_1, σ_2, the condition determined by $e_{\mu\sigma_1\sigma_2}$ implies that the μth row of $A_{\sigma_1} A_{\sigma_2} - A_{\sigma_2} A_{\sigma_1}$ vanishes. $\square$

Let $\mathcal{S}$ be the set of all equations (8.23) and (8.24). For each triple $(\mu, \sigma_1, \sigma_2)$ which occurs in an equation of $\mathcal{S}$, we consider the monomial $x_{\sigma_1} x_{\sigma_2} b_\mu$; cf. Figure 8.2. In case (i), with $x_{\sigma_2} b_\mu =: b_{\bar{\mu}}$, we consider the two border basis elements $bb_{\bar{k}}$, with leading term $x_{\sigma_1} b_{\bar{\mu}}$, and bb_k, with leading term $x_{\sigma_1} b_\mu$, and form

$$\begin{aligned}
bb_{\bar{k}}(x) - x_{\sigma_2} bb_k(x) &= (x_{\sigma_1} b_{\bar{\mu}}(x) - a_{\sigma_1\bar{\mu}}^T \mathbf{b}(x)) - (x_{\sigma_2} x_{\sigma_1} b_\mu(x) - x_{\sigma_2} a_{\sigma_1\mu}^T \mathbf{b}(x)) \\
&= -(a_{\sigma_1\bar{\mu}}^T - a_{\sigma_1\mu}^T x_{\sigma_2}) \mathbf{b}(x).
\end{aligned}$$

Some components of $x_{\sigma_2}\mathbf{b}(x)$ are in $B[\mathcal{N}]$; with the aid of border basis elements, they may be reduced to the components of $A_{\sigma_2}\mathbf{b}(x) \in \mathcal{N}$. Thus,

$$a^T_{\sigma_1\bar\mu} - a^T_{\sigma_1\mu}A_{\sigma_2} = 0 \qquad \Longleftrightarrow \qquad bb_{\bar k}(x) - x_{\sigma_2}bb_k(x) \;\xrightarrow{\;\mathcal{B}\;}\; 0 . \tag{8.25}$$

In case (ii), we consider the two border basis elements bb_{k_1}, with leading term $x_{\sigma_1}b_\mu$, and bb_{k_2}, with leading term $x_{\sigma_2}b_\mu$, and form

$$
\begin{aligned}
x_{\sigma_1}bb_{k_2}(x) - x_{\sigma_2}bb_{k_1}(x) &= (x_{\sigma_1}x_{\sigma_2}b_\mu(x) - x_{\sigma_1}a^T_{\sigma_2\mu}\mathbf{b}(x)) - (x_{\sigma_2}x_{\sigma_1}b_\mu(x) - x_{\sigma_2}a^T_{\sigma_1\mu}\mathbf{b}(x)) \\
&= -(a^T_{\sigma_2\mu}x_{\sigma_1} - a^T_{\sigma_1\mu}x_{\sigma_2})\,\mathbf{b}(x) .
\end{aligned}
$$

Again, we can reduce the right-hand side with border basis elements to components in $\mathcal{N}$ and obtain

$$a^T_{\sigma_2\mu}A_{\sigma_1} - a^T_{\sigma_1\mu}A_{\sigma_2} = 0 \qquad \Longleftrightarrow \qquad x_{\sigma_1}bb_{k_2}(x) - x_{\sigma_2}bb_{k_1}(x) \;\xrightarrow{\;\mathcal{B}\;}\; 0 . \tag{8.26}$$

Above (cf. (8.23) and (8.24)), we have found that the left-hand sides of (8.25) and (8.26) are equivalent to the commutativity of the A_σ matrices. Now, we have found that the same relations are equivalent to the fact that certain polynomial combinations of the border basis elements can be reduced, by subtraction of suitable elements from the border basis set $\mathcal{B}$, to the zero polynomial. Thus, the right-hand sides in (8.25) and (8.26), formed over all triples $(\mu, \sigma_1, \sigma_2)$ described above, are equivalent to the commutativity of the A_σ.

All these right-hand sides combine two "neighboring" border basis elements in a way which implies the cancellation of the $\mathcal{N}$-leading term of the combination.

Definition 8.6. Consider a set $\mathcal{B}$ of polynomials in $\mathcal{P}^s$ for which leading monomials are defined. Take two polynomials p_1, $p_2 \in \mathcal{B}$, with leading monomials x^{j_1}, x^{j_2}; their least common multiple is $x^k := \mathrm{l.c.m.}(x^{j_1}, x^{j_2}) = x^{k-j_1}x^{j_1} = x^{k-j_2}x^{j_2}$. The S-*polynomial*[12] of p_1, p_2 is

$$S[p_1, p_2] := l.c.(p_2)\,x^{k-j_1}p_1(x) - l.c.(p_1)\,x^{k-j_2}p_2(x) , \tag{8.27}$$

where $l.c.$ denotes the $\mathcal{N}$-leading coefficient. $\square$

Note that $S[p_2, p_1] = -S[p_1, p_2]$ so that, in most contexts, the order of the arguments does not matter.

Theorem 8.11. In the situation described at the beginning of this section, the following facts are equivalent:

- The multiplication matrices A_σ which represent the multiplicative structure of $\mathcal{R}$ with respect to the normal set basis $\mathbf{b}$ commute.

- All S-polynomials formed for neighboring elements $bb_k(x)$ of the border basis $\mathcal{B}$ may be reduced to 0 by polynomials from $\mathcal{B}$.

- The S-polynomials formed for *any* two elements of the border basis $\mathcal{B}$ may be reduced to 0 by polynomials from $\mathcal{B}$.

Proof: It remains only to show that the 2nd fact implies the 3rd one. Consider bb_k, $bb_\ell \in \mathcal{B}$ with leading terms x^k, x^ℓ, resp., and their least common multiple $x^K := \mathrm{l.c.m.}(x^k, x^\ell) = x^{K-k}x^k =$

---

[12]This terminology has been introduced by B. Buchberger in his thesis; the S in S-polynomial refers to syzygy.

$x^{K-\ell} x^\ell$. The S-polynomial of bb_k, bb_ℓ is $S[bb_k, bb_\ell] = x^{K-k} bb_k(x) - x^{K-\ell} bb_\ell(x)$. Due to the closedness of a normal set, there must be a chain of monomials $x^{k_\nu} \in B[\mathcal{N}]$, $\nu = 0, 1, \ldots, n$, such that $k_0 = k$, $k_n = \ell$, and $x^{k_{\nu+1}}$ and x^{k_ν} are "neighboring monomials" in $B[\mathcal{N}]$ which means that they satisfy one of the following two relations:

$\quad$ (i) $\exists x_\sigma :$ $\qquad x^\kappa = x_\sigma x^{\kappa'}$ $\quad$ or $\quad x^{\kappa'} = x_\sigma x^\kappa$,

$\quad$ (ii) $\exists x_{\sigma_1}, x_{\sigma_2} : x_{\sigma_2} x^\kappa = x_{\sigma_1} x^{\kappa'}$,

where the monomial in the equation is a divisor of x^K. Thus, by the chain, we may compose the relation $x^{K-k} x^k = x^{K-\ell} x^\ell$ by a sum of x^j-multiples of the relations for neighboring monomials. Accordingly, we may compose the S-polynomial of bb_k, bb_ℓ by a sum of multiples of S-polynomials of neighboring border basis elements. Since each of these may be reduced to 0 by $\mathcal{B}$, this is also true for the sum. $\quad\square$

Corollary 8.12. Consider a set $\mathcal{N} \in T^s(m)$ and a polynomial set $\mathcal{B} = \{bb_j\} \subset \mathcal{P}^s$, $|\mathcal{B}| = |B[\mathcal{N}]|$, where each $bb_j \in \mathcal{B}$ has its support in $\mathcal{N} \cup B[\mathcal{N}]$, with only one monomial $x^j \in B[\mathcal{N}]$. Iff the S-polynomials of any two polynomials in $\mathcal{B}$ may be reduced to 0 by polynomials from $\mathcal{B}$, then $\mathcal{B}$ generates an ideal $\langle \mathcal{B} \rangle \subset \mathcal{P}^s(m)$, with normal set $\mathcal{N}$.

Proof: Compare Theorems 8.1, 8.3, and 8.11. $\quad\square$

Example 8.8: We take the situation in Example 8.2, with $\mathcal{N} = \{1, x, y, z\}$, $\partial \mathcal{N} = \{x, y, z\}$, and $B[\mathcal{N}] = \{x^2, xy, xz, y^2, yz, z^2\}$ and 9 edges in $BW_\mathcal{N}$. All edges are of type (ii); the corresponding 9 commutativity relations (8.24) for the multiplication matrices are

$$a_{\sigma_2 \mu}^T A_{\sigma_1} = a_{\sigma_1 \mu}^T A_{\sigma_2}, \qquad \mu = 2(1)4, \quad \sigma_1 \neq \sigma_2 \in \{1, 2, 3\}.$$

For the triple $\mu = 2$, $\sigma_1 = 2$, $\sigma_2 = 1$, e.g., we have $a_{12}^T A_2 = a_{22}^T A_1$ or

$$(1, 2, -1, 1) \begin{pmatrix} 0 & 0 & 1 & 0 \\ 2 & \frac{-4}{7} & \frac{2}{7} & \frac{6}{7} \\ 4 & \frac{-10}{7} & \frac{5}{7} & \frac{8}{7} \\ 0 & \frac{-4}{7} & \frac{2}{7} & \frac{-8}{7} \end{pmatrix} = (2, \tfrac{-4}{7}, \tfrac{2}{7}, \tfrac{6}{7}) \begin{pmatrix} 0 & 1 & 0 & 0 \\ 1 & 2 & -1 & 1 \\ 2 & \frac{-4}{7} & \frac{2}{7} & \frac{6}{7} \\ 0 & \frac{-8}{7} & \frac{4}{7} & \frac{-2}{7} \end{pmatrix}$$

$$= (0, \tfrac{-2}{7}, \tfrac{8}{7}, \tfrac{-4}{7}) ;$$

for $\mu = 4$, $\sigma_1 = 2$, $\sigma_2 = 1$, we have $a_{14}^T A_2 = a_{24}^T A_1$ or

$$(0, \tfrac{-8}{7}, \tfrac{4}{7}, \tfrac{-2}{7}) A_2 = (0, 0, \tfrac{3}{7}, \tfrac{8}{7}) = (0, \tfrac{-4}{7}, \tfrac{2}{7}, \tfrac{-8}{7}) A_1 .$$

Correspondingly, there are 9 reduction relations (8.26) for the 6 border basis elements bb_κ of Example 8.2. For the triple $(1,2,1)$, we have

$$
\begin{aligned}
S[bb_1, bb_2] &= y\, bb_1(x, y, z) - x\, bb_2(x, y, z) \\
&= y\, (x^2 - (2x - y + z + 1)) - x\, (xy - (-4x + 2y + 6z + 14)/7) \\
&= (-4x^2 - 12xy + 6xz + 7y^2 - 7yz + 14x - 7y)/7
\end{aligned}
$$

which is reduced to 0 by subtraction of $\frac{-4}{7} bb_1 - \frac{12}{7} bb_2 + \frac{6}{7} bb_3 + bb_4 - bb_5$. Similarly, $y\, bb_3(x, y, z) - x\, bb_5(x, y, z) = (-\frac{4}{7} x^2 + \frac{10}{7} xy - \frac{8}{7} xz - \frac{4}{7} y^2 + \frac{2}{7} yz)$ reduces to 0 by subtraction of the border basis polynomials with the respective leading terms.

For non-neighboring border basis elements, e.g., bb_1 and bb_6 with leading terms x^2 and z^2, resp., we may use the chain (x^2, xz), (xz, z^2) to obtain

$$S[bb_1, bb_6] = z^2\, bb_1(x, y, z) - x^2\, bb_6(x, y, z) = z\, (z\, bb_1 - x\, bb_3) + x\, (z\, bb_3 - x\, bb_6) = 0$$

by (8.26).

If we had been presented with the 6 polynomials bb_κ of Example 8.2, without further information, we could have chosen the set $\mathcal{N}$ as $\{1, x, y, z\}$ to conform with Corollary 8.12. We could then have formed the 9 reduction relations (8.26) and verified the reduction to 0. This would have told us that the bb_κ generate an ideal $\mathcal{I}$ with 4 zeros. From the bb_κ, we could have formed the multiplication matrices A_σ of $\mathcal{R}[\mathcal{I}]$ and computed the zeros. $\quad\square$

Theorem 8.11 shows that the elements $bb_j(x)$ of a border basis $\mathcal{B}$ of a polynomial ideal $\mathcal{I} \subset \mathcal{P}^s$ satisfy a multitude of identities: Assume (for notational simplicity) that the elements $b_{\bar\mu}$ through b_m are in $\partial\mathcal{N}$ and have an x_1-neighbor in $B[\mathcal{N}]$. When we spell out the reduction $x_1\,\mathbf{b}(x) \xrightarrow{\;\mathcal{B}\;} A_1\,\mathbf{b}(x)$, we obtain—as a relation in $\mathcal{P}^s$—

$$x_1\,\mathbf{b}(x) = A_1\,\mathbf{b}(x) + (0, \ldots, 0, bb_{k_{\bar\mu}}(x), \ldots, bb_{k_m}(x))^T, \tag{8.28}$$

where bb_{k_μ} is the border basis element with leading term $x_1 b_\mu(x)$; such relations hold for all x_σ-multiples of $\mathbf{b}$. Thus, (8.26) implies, e.g.,

$$
\begin{aligned}
x_{\sigma_1} bb_{k_2}(x) - x_{\sigma_2} bb_{k_1}(x) &= (a^T_{\sigma_2\mu} x_{\sigma_1} - a_{\sigma_1\mu} x_{\sigma_2})\,\mathbf{b}(x) \\
&= (a^T_{\sigma_2\mu} A_{\sigma_1} - a_{\sigma_1\mu} A_{\sigma_2})\,\mathbf{b}(x) + a^T_{\sigma_2\mu}\,\mathbf{bb}_{\sigma_1}(x) - a^T_{\sigma_1\mu}\,\mathbf{bb}_{\sigma_2}(x),
\end{aligned}
$$

where $\mathbf{bb}_\sigma(x)$ is the m-vector of zeros and bb_j's arising from the reduction of $x_\sigma \mathbf{b}(x)$; cf. (8.28). Since the first member of the right-hand side vanishes for commuting A_σ, we have the identity in $\mathcal{P}^s$

$$x_{\sigma_1} bb_{k_2}(x) - x_{\sigma_2} bb_{k_1}(x) - a^T_{\sigma_2\mu}\,\mathbf{bb}_{\sigma_1}(x) + a^T_{\sigma_1\mu}\,\mathbf{bb}_{\sigma_2}(x) = 0, \tag{8.29}$$

which must be satisfied by the border basis elements if they are to be consistent and define a nontrivial 0-dimensional ideal. Similar identities arise from the relations (8.25).

These nontrivial syzygies (cf. Definition 2.3) of the border basis $\mathcal{B}$ of a 0-dimensional ideal $\mathcal{I}$ represent nontrivial representations of the 0-polyomial by the bb_k; they show that the bb_k are not independent. This was, of course, to be expected since they possess $N\,s$ coefficients while $\mathcal{I}$ is determined by $m\,s$ data in $\mathbb{C}^s$. Because of these syzygies, we cannot expect uniqueness in the representation (8.16) of a polynomial $p \in \mathcal{P}^s$ as we may add arbitrary multiples of a syzygy (8.29) to it.

The set of all syzygies of a border basis is a linear space with an algebraic structure (a "module") and can be generated from a basis; but we will not make use of this structure in a formal way. *Trivial* syzygies $\sum_\nu q_\nu p_\nu = 0$, with $q_\nu \in \langle\{p_\nu\}\rangle$, exist in any polynomial system $\{p_\nu\}$, e.g., $p_{\nu_2} p_{\nu_1} - p_{\nu_1} p_{\nu_2} = 0$.

Example 8.8, continued: We may rewrite the reductions, e.g., the ones displayed above, into the syzygies $y\, bb_1 - x\, bb_2 = \frac{-4}{7}\, bb_1 - \frac{12}{7}\, bb_2 + \frac{6}{7}\, bb_3 + bb_4 - bb_5$ or

$$(y+\tfrac{4}{7})\, bb_1(x, y, z) + (-x+\tfrac{12}{7})\, bb_2(x, y, z) - \tfrac{6}{7}\, bb_3(x, y, z) - bb_4(x, y, z) + bb_5(x, y, z) = 0,$$

and $y\, bb_3 - x\, bb_5 = \frac{-4}{7}\, bb_1 + \frac{10}{7}\, bb_2 - \frac{8}{7}\, bb_3 - \frac{4}{7}\, bb_4 + \frac{2}{7}\, bb_5$ or

$$\tfrac{4}{7}\, bb_1(x, y, z) - \tfrac{10}{7}\, bb_2(x, y, z) + (y+\tfrac{8}{7})\, bb_3(x, y, z) + \tfrac{4}{7}\, bb_4(x, y, z) - (x+\tfrac{2}{7})\, bb_5(x, y, z) = 0.$$

8.2.3   Admissible Data for a Normal Set Representation

At the end of section 8.1.2, we observed that *computed* multiplication matrices will generally not commute exactly. Corollary 8.12 expresses the same dilemma in terms of border bases: In a *computed* border basis, the S-polynomials will generally not reduce to 0 exactly. Let us take another look at this situation:

When we fix a (feasible) normal set $\mathcal{N} \in T^s(m)$ for the representation of an ideal $\mathcal{I} \in \mathcal{P}^s(m)$, the data of the normal set representation of $\mathcal{I}$ consist of the N row vectors $a_j^T \in \mathbb{C}^m$, $x^j \in B[\mathcal{N}]$; cf. the beginning of section 8.2. As we have seen in section 8.2.2, the a_j^T cannot take arbitrary values but must satisfy the relations (8.23) and (8.24) or (8.25) and (8.26), resp., which are quadratic polynomials in the components of the a_j^T. Thus they define an algebraic manifold in the data space $\mathcal{A} = \mathbb{C}^{Nm}$ of the a_j^T.

Definition 8.7. For a normal set representation with normal set $\mathcal{N} \in T^s(m)$, the algebraic manifold

$$\mathcal{M}_{\mathcal{N}} := \{ a_j^T \in \mathbb{C}^m \text{ satisfying the commutativity constraints in Proposition 8.10} \} \subset \mathcal{A} \tag{8.30}$$

is the *admissible-data manifold* of $\mathcal{N}$. A set of $N = |B[\mathcal{N}]|$ vectors $a_j^T \in \mathbb{C}^m$ specifies an ideal $\mathcal{I} \in \mathcal{P}^s(m)$ iff $\{a_j^T\} \in \mathcal{M}_{\mathcal{N}}$.   □

Proposition 8.13. The admissible-data manifold $\mathcal{M}_{\mathcal{N}}$ has dimension $m\,s$.

Proof: Take a zero set $Z \subset (\mathbb{C}^s)^m$ of m disjoint zeros $z_\mu \in \mathbb{C}^s$ and consider the nontrivial row vectors a_j^T in the A_σ matrices generated (for $\mathcal{N}$) by (8.1) so that the $m \times m$ matrix $\mathbf{b}(\mathbf{z})$ is the normalized joint eigenvector matrix of the commuting matrix family $\overline{A}$ generated by the A_σ. Since $\mathbf{b}(\mathbf{z})$ is regular, (8.1) defines a *bijective* mapping between a full ms-dimensional neighborhood of Z and the associated neighborhood of the nontrivial rows of the A_σ on $\mathcal{M}_{\mathcal{N}}$. Thus $\mathcal{M}_{\mathcal{N}}$ is ms-dimensional at all of its points which correspond to a set of m disjoint zeros. The situation at a point which corresponds to a zero set with some multiple zero will be analyzed in section 9.3.   □

The *codimension* of $\mathcal{M}_{\mathcal{N}}$ in $\mathcal{A} = \mathbb{C}^{Nm}$ is positive except when either s or m is 1: A univariate polynomial is its own border basis and all coefficient values are admissible. For $\mathcal{I} \in \mathcal{P}^s(1), \mathcal{N} = \{1\}$ is the only possible normal set and the s values $a_j^T \in \mathbb{C}^1$ are the components of the only zero z. In all other cases, we have $N > s$ and $\mathrm{codim}\ \mathcal{M}_{\mathcal{N}} = (N - s)\,m > 0$. However,[13] at least for $s > 2$, the set $\mathcal{S}$ of relations (8.23) and (8.24) in Proposition 8.10 which expresses the commutativity constraints defining the admissible-data manifold $\mathcal{M}_{\mathcal{N}}$, contains $\bar{N} := |B W_{\mathcal{N}}| > N - s$ equations. Thus $\mathcal{S}$ constitutes a consistent *overdetermined* representation of $\mathcal{M}_{\mathcal{N}}$. In Example 8.8, e.g., we have $N - s = 3$ but $\bar{N} = 9$.

This is an unfortunate situation for computational purposes: Assume that, from a representation $(\mathcal{N}, \{\tilde{a}_j\})$, we wish to reach a proper neighboring representation $(\mathcal{N}, \{a_j\})$, with $a_j^T = \tilde{a}_j^T + \Delta a_j^T$ and small modifications Δa_j^T; the Δa_j^T are to be found by a Newton step applied to the equations for some requested property and for the position on $\mathcal{M}_{\mathcal{N}}$. Thus we need the Jacobian of the system $\mathcal{S}$ at the $\tilde{a}_j$; because of the quadratic character of $\mathcal{S}$, the elements of this matrix contain the $\tilde{a}_j^T$. In a computational situation, their values carry round-off errors

---

[13]The nontypical situation for $s = 2$ variables is discussed in Exercise 8.2-5.

(or worse) which raise the rank of the Jacobian above its theoretical value $N - s$; this may lead to serious computational difficulties. Therefore, we must attempt to specify a subset $\mathcal{S}_0$ of $\mathcal{S}$, with $\bar{N}_0 = N - s$ relations, which defines the admissible-data manifold $\mathcal{M}_\mathcal{N}$ *without overdetermination*. Fortunately, there are some well-known rules which permit a reduction of the set $\mathcal{S}$:

Proposition 8.14. Consider a system $\mathcal{S}$ of edge conditions of type (8.25) and (8.26) on the border web $BW_\mathcal{N}$ of a normal set $\mathcal{N}$.

(a) Consider three monomials $x^{k_j} \in B[\mathcal{N}]$, $j = 1(1)3$, and let $x^{k_{j_1 j_2}}$ be the least common multiples (l.c.m.) of $x^{k_{j_1}}$ and $x^{k_{j_2}}$. Assume that $x^{k_{12}}$ and $x^{k_{23}}$ divide $x^{k_{13}}$. Then the condition on the edge $[x^{k_1}, x^{k_3}]$ follows from those on $[x^{k_1}, x^{k_2}]$ and $[x^{k_2}, x^{k_3}]$.

(b) Consider two monomials $x^{k_1}, x^{k_2} \in B[\mathcal{N}]$ which are *coprime*, i.e. l.c.m.$(x^{k_1}, x^{k_2}) = x^{k_1} x^{k_2}$. Then the S-polynomial $S[bb_{k_1}, bb_{k_2}]$ reduces to 0 with $\{bb_{k_1}, bb_{k_2}\}$.

Proof:

(a)
$$x^{k_{13}-k_1} bb_{k_1} - x^{k_{13}-k_3} bb_{k_3} =$$
$$(x^{k_{12}-k_1} bb_{k_1} - x^{k_{12}-k_2} bb_{k_2}) x^{k_{13}-k_{12}} + (x^{k_{23}-k_2} bb_{k_2} - x^{k_{23}-k_3} bb_{k_3}) x^{k_{13}-k_{23}}.$$

(b)
$$S[bb_{k_1}, bb_{k_2}] = x^{k_2} \cdot (x^{k_1} - a_{k_1}^T \mathbf{b}(x)) - x^{k_1} \cdot (x^{k_2} - a_{k_2}^T \mathbf{b}(x))$$
$$= -(x^{k_2} - a_{k_2}^T \mathbf{b}(x)) a_{k_1}^T \mathbf{b}(x) + (x^{k_1} - a_{k_1}^T \mathbf{b}(x)) a_{k_2}^T \mathbf{b}(x). \qquad \square$$

It is obvious that part (a) of Proposition 8.14 can be extended to the case where some edge $[x^{k_1}, x^{k_r}]$ closes a longer "chain" of edges $[x^{k_j}, x^{k_{j+1}}]$, $j = 1(1)r - 1$, if all l.c.m.$(x^{k_j}, x^{k_{j+1}})$ divide l.c.m.(x^{k_1}, x^{k_r}). This permits the elimination of various conditions on edges which "close a circle" in $BW_\mathcal{N}$. However, the assumption on the l.c.m. of the chained edges is rather restrictive. If this assumption is not met, the chain relation will only lead to a representation of some monomial multiple of the relation for the closing edge in terms of the relations on the edges of the chain; this happens because we must multiply the relation in the proof above by a monomial x^ℓ which makes $x^{k_{1r}} x^\ell$ a multiple of all the l.c.m.$(x^{k_j}, x^{k_{j+1}})$.

In terms of multiplication matrices, this implies only $(a_{k_1}^T A^{k_{1r}-k_1} - a_{k_r}^T A^{k_{1r}-k_r}) A^\ell = 0$, which yields the relation (8.24) for the closing edge only if A^ℓ is regular. In terms of zeros of the underlying ideal, this means that none of the zeros must have a vanishing x_λ component for λ a nonvanishing component of ℓ. But the zero sets $Z^s(m)$ which do not meet this condition comprise a low-dimensional subset of the $\mathbb{C}^{sm}$ which parametrizes the admissible-data-manifold $\mathcal{M}_\mathcal{N}$. Therefore, the commutativity relation on the closing edge is a consequence of the relations along the chain *almost everywhere* on $\mathcal{M}_\mathcal{N}$; hence it must be an *algebraic* consequence of these relations. By continuity, this algebraic consequence must also hold on the low-dimensional parts of $\mathcal{M}_\mathcal{N}$ where it cannot be derived in the above straightforward manner. This implies:

Proposition 8.15. On the border web $WB_\mathcal{N}$, conditions of $\mathcal{S}$ on the closing edge of a loop are satisfied if they are satisfied on the remaining edges of the loop.

Thus, one may delete from the web *all* edges which close circles. The remaining web possesses exactly $N - 1$ edges which connect the N monomials in $B[\mathcal{N}]$; it may, e.g., have the form of one continuous thread which proceeds from one monomial to another without any branches. But we must still remove $s - 1$ further edges to arrive at $\bar{N}_0 = N - s$. This is made possible by part (b) of Proposition 8.14 which permits the introduction of "virtual" edges between certain monomials of $BW_\mathcal{N}$ along which the relation (8.26) is automatically satisfied.

If such a virtual edge closes a circle on $BW_\mathcal{N}$, it permits the elimination of another real edge of the web. For example, we may select one "extremal" monomial (power of one variable) $x_{\sigma'}^{j_{\sigma'}}$; it is connected to the other $s-1$ extremal monomials by virtual edges. This permits the elimination of the $s-1$ edges which previously met these monomials and reduces the number of real edges in $\mathcal{S}$, with active conditions, to $N-s$. Thus we may reach a subset $\mathcal{S}_0$ of $\mathcal{S}$ with the correct number of relations and with each $x^j \in B[\mathcal{N}]$ appearing in at least one relation.

From these arguments and from the evidence in nontrivial—but admittedly not very large—examples, it appears that the relations of such a subset $\mathcal{S}_0$ characterize the admissible-data-manifold $\mathcal{M}_\mathcal{N}$ completely and without overdetermination. A rigorous proof of this fact would naturally be highly desirable.

From the practical point of view, an easy way to eliminate many unnecessary relations from $\mathcal{S}$ is given by Proposition 8.4: In a generic situation, it suffices to consider the relations which originate from the commutativity of a fixed multiplication matrix, say A_s, with the remaining matrices A_σ. This eliminates those relations (8.23) and (8.24) where neither σ_1 nor σ_2 equals s. In the remaining web, circles are easier to spot and the elimination of the connections to the extremal points k_σ, $\sigma = 1(1)s-1$, is straightforward.

Example 8.9: Consider $\mathcal{N} = \{1, x_3, x_3^2, x_2, x_1, x_1x_3\}$, with $B[J_\mathcal{N}] = \{(0,0,3), (0,1,2), (1,0,2), (1,1,1), (2,0,1), (0,1,1), (0,2,0).(1,1,0), (2,0,0)\}$, $N = 9$, $\overline{N} = 14$; cf. Figure 8.2. The consideration of Proposition 8.4, with $\sigma_3 = 1$, removes the 5 edges $e_{\mu\sigma_1\sigma_2}$ with $\sigma_1\sigma_2 = 23$ and leaves only one closed loop; cf. Figure 8.3. By Proposition 8.15, we can remove (say) e_{413} and replace the edges to the extremal points $(0,2,0)$ and $(0,0,3)$ by virtual edges. This leaves us with a subset $\mathcal{S}_0$ with $\overline{N}_0 = 9 - 3 = 6$ edges.

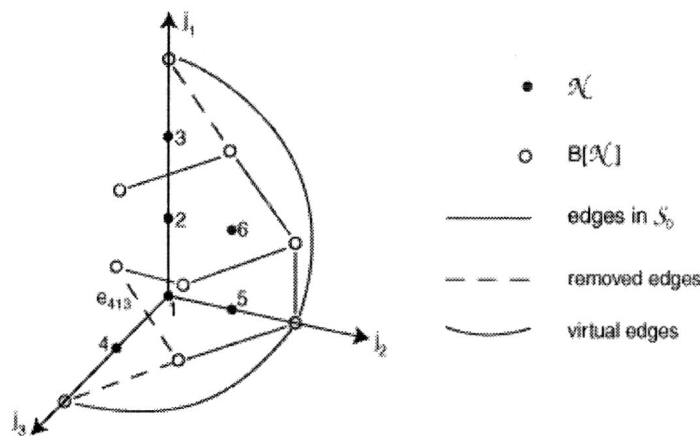

Figure 8.3.

When we form the 36×24 Jacobian matrix $J(\mathcal{S})$ of the full system $\mathcal{S}$ and assume that the $\tilde{a}_j^T$ *satisfy the consistency conditions*, we find rank $J(\mathcal{S}) = 12$. The same rank is found for the 12×24 Jacobian matrix of a minimal set $\mathcal{S}_0$ of conditions exhibited above. If the consistency conditions are violated, the rank of $J(\mathcal{S})$ increases; for generic elements in the $\widetilde{A}_\sigma$ or *for generic perturbations of the* $\tilde{a}_j^T$, it becomes 24.

Of course, with a small violation of the consistency conditions, the singular values of $J(\mathcal{S})$ may still reflect the generic rank to some extent; but this reflection may be so vague that it cannot be utilized numerically: In an experimental computation for the above normal set $\mathcal{N}$, we formed exact rational matrices $\tilde{A}_\sigma$, $\sigma = x, y, z$, for a set of 4 "random" zeros with integer components of $O(1)$. Then we formed the 36×24 Jacobian $J(\mathcal{S})$ for the exact $\tilde{a}_j^T$ and for approximations obtained by rounding them to 5 decimal digits. There were precisely 12 nonvanishing singular values (between 223 and 36) for the exact data, as predicted by our considerations. For the perturbed values, there were 24 nonvanishing singular values decreasing smoothly from 242 to .187. In the critical places from the 11th to the 14th singular value (ordered in decreasing size), we found the values $\approx 19.87, 14.92, 12.29, 11.61$ without any indication of a jump in size! Thus the only chance for obtaining a Newton step with 12 further conditions for the 24 increments would have been the immediate use of a subsystem $\mathcal{S}_0$ of only 12 equations.

The quasi-univariate feature of $\mathcal{N}$ gives rise to a further special situation: Take z as the distinguished variable and eliminate the relations generated by $A_x A_y = A_y A_x$. According to the considerations in section 8.1.3, the remaining 6 relations must be *linear* and of full rank in the rows $a_{x^2}^T, a_{xy}^T, a_{y^2}^T$; thus, the 24×12 Jacobian matrix of these relations with respect to $a_{x^2}^T, a_{xy}^T, a_{y^2}^T$ *only* must not contain any components of these rows and have rank 12 for generic $a_{xz}^T, a_{yz}^T, a_{z^2}^T$. This is also confirmed by the respective computations. This also reconfirms Proposition 8.7 which asserts that the rows $a_{xz}^T, a_{yz}^T, a_{z^2}^T$ of A_z can be chosen arbitrarily and that the other rows of A_x, A_y are then determined by $\mathcal{S}_0$. $\quad\square$

In the general case, the selection of the free parameters is not so simple. For example, the conjecture that there may always exist a subset of s vectors a_j^T whose data may be chosen arbitrarily and thus determine the generated ideal can be disproved by a simple counter-example: In $\mathcal{P}^2$, consider the normal set $\{1, y, x, y^2, xy, x^2\}$, with $B[\mathcal{N}] = \{y^3, xy^2, x^2y, x^3\}$. When we choose any two of the monomials in $B[\mathcal{N}]$ as leading monomials of two border basis polynomials, with generic coefficients, and analyze the ideals generated by them, we find that each of these ideals has *more than* 6 zeros. Thus it cannot be possible to generate an ideal in $\mathcal{P}^2(6)$ by two border basis elements for the above normal set.

Exercises

1. Consider the set $\mathcal{B}$ of the following 10 polynomials:

$$
\begin{pmatrix} bb_1 \\ bb_2 \\ bb_3 \\ bb_4 \\ bb_5 \\ bb_6 \\ bb_7 \\ bb_8 \\ bb_9 \\ bb_{10} \end{pmatrix}
=
\begin{pmatrix} 56z^2 \\ 56y^2 \\ 14x^2 \\ 16yz^2 \\ 112xz^2 \\ 112y^2z \\ 16xyz \\ 14x^2z \\ 16xy^2 \\ 4x^2y \end{pmatrix}
-
\begin{pmatrix}
28 & -62 & 14 & -2 & 28 & -13 & -1 \\
252 & 162 & 14 & 54 & -84 & 15 & 27 \\
28 & -34 & 14 & -16 & 28 & -26 & -2 \\
16 & 8 & 8 & 2 & -4 & 1 & 1 \\
-28 & -338 & -14 & 74 & 140 & -79 & 37 \\
0 & -144 & 0 & -6 & 364 & 45 & -3 \\
4 & -34 & 2 & 2 & 28 & 13 & 1 \\
14 & 121 & 7 & -4 & -28 & 2 & -2 \\
60 & 66 & 30 & 54 & -12 & 15 & -5 \\
16 & 8 & 8 & 2 & 4 & 1 & -3
\end{pmatrix}
\begin{pmatrix} 1 \\ z \\ y \\ x \\ yz \\ xz \\ xy \end{pmatrix}.
$$

(a) If $\mathcal{B}$ is the border basis of a nontrivial ideal $\mathcal{I}$, what is the associated normal set basis

b of the quotient ring $\mathcal{R}[\mathcal{I}]$? What are the multiplication matrices A_σ of $\mathcal{R}[\mathcal{I}]$ with respect to **b** ?

(b) Prove that $\mathcal{R} = \text{span } \mathbf{b}$ with a multiplicative structure defined by the A_σ is indeed the quotient ring of an ideal $\mathcal{I} \subset \mathcal{P}^s(7)$. (Compare Theorem 8.1.)

(c) Compute and reduce at least a few of the S-polynomials $S[bb_j, bb_{j'}]$ for a (partial) *direct* verification that $\mathcal{B}$ is a border basis. (Compare Corollary 8.12.)

(d) Compute the 7 zeros of $\mathcal{I}$ from the nonderogatory one of the three multiplication matrices. From the zeros, explain why the other two A_σ are derogatory. Find a simple linear combination of these two matrices which is nonderogatory and determine the zeros from it.

2. Consider the normal set $\mathcal{N}_1$ of Example 8.1.

(a) Identify the border set $B[\mathcal{N}_1]$ and its subsets $B_\sigma[\mathcal{N}_1]$, $\sigma = 1, 2, 3$; cf. Definition 8.2. Verify that $\mathcal{N}_1$ is not quasi-univariate with respect to either variable.

(b) Identify the boundary set $\partial \mathcal{N}_1$ (cf. Definition 8.4) and the monomials $b_\mu \in \mathcal{N}_1$ for which $x \cdot b_\mu \in B[\mathcal{N}_1]$ (there must be as many as there are nontrivial rows in $A_1^{(1)}$). For each of these b_μ, consider $y \cdot b_\mu$ and $z \cdot b_\mu$ and write down the respective relations (8.23) or (8.24). Check their validity for the $A_\sigma^{(1)}$ of Example 8.1.

(c) Write down all elements of the border basis $\mathcal{B}$ for the ideal of Example 8.1. Translate the relations (8.23) and (8.24) found in (b) into syzygies of $\mathcal{B}$ and verify their validity.

3. Consider the "rectangular" normal set $\mathcal{N} = \{1, x, y, xy, z, xz, yz, xyz, z^2, xz^2, yz^2, xyz^2\} \subset T^3(12)$. In answering the following questions, sketch the situation in $\mathbb{N}_0^3$.

(a) How many elements (N) has the border set $B[J_\mathcal{N}]$? How many edges ($\overline{N}$) are there in the web representing the consistency relations in $\mathcal{S}$?

(b) Compile the $\overline{N}$ consistency relations for the nontrivial rows of the multiplication matrices A_σ of a quotient ring with basis $\mathcal{N}$. Determine the Jacobian of these relations with respect to the nontrivial elements of the A_σ.

(c) How many edges are left after the deletion of all edges representing relations occurring only in $A_x A_y = A_y A_x$? Which further edges can you delete (and why) to bring the number of the remaining ones to $\overline{N}_0 = N - 3$?

(d) Write down the remaining set $\mathcal{S}_0$ of relations in terms of the row vectors a_j^T of the A_σ and in terms of syzygies for the border basis elements bb_j.

4. (a) Consider the ideal $\mathcal{I}$ with the *zero set* $J_\mathcal{N} \subset \mathbb{N}_0^3$ from the previous exercise. Verify that the normal set $\mathcal{N}$ is feasible for $\mathcal{I}$. Determine the multiplication matrices A_x, A_y, A_z of $\mathcal{R}[\mathcal{I}]$ with respect to the normal set basis $\mathcal{N}$.

(b) Evaluate the Jacobian from (b) in the previous exercise at the elements of the current A_σ. Verify that the rank of the Jacobian is $\overline{N}_0$.

(c) Select the rows from the Jacobian which correspond to the relations in $\mathcal{S}_0$ and verify that this submatrix has full rank.

5. Consider the determination of the admissible-data manifold $\mathcal{M}_\mathcal{N}$ for $s = 2$ variables.

(a) Convince yourself that, for a normal set with $|B[\mathcal{N}]| = N$, $\overline{N} = N - 1$ and that $\overline{N}_0 = N - 2$ is obtained by taking into account the virtual edge between the two extremal points of $B[J_\mathcal{N}]$.

(b) Compute the rank and the singular values of the Jacobian of the relations in S and S_0, resp., for numerical examples with exact and perturbed multiplication matrices.

8.3 Regular Systems of Polynomials

8.3.1 Complete Intersections

It would be nice if the situation at each simple zero of a polynomial system resembled that at a zero of a regular system of s *linear* equations p_ν in s variables. This situation is very clear cut: Each of the equations $p_\nu(x) = 0$ represents a hyperplane in $\mathbb{C}^s$ or $\mathbb{R}^s$ and there is the well-known *alternative*:

- if the normal vectors of the s hyperplanes span the s-space there is a unique intersection point z ("regular case");

- else there exists a d-dimensional subspace ($d > 0$) which is parallel to each hyperplane, in which case the hyperplanes either have a common linear manifold of dimension d or they have no point in common ("singular case").

Algebraically, in the regular case the ideal $\langle p_\nu, \nu = 1(1)s \rangle$ consists of all polynomials in $\mathcal{P}^s$ which vanish at z and is a maximal ideal; in the singular case, the ideal is either d-dimensional, with a linear zero manifold, or trivial. The corresponding *criterion* is well known:

Proposition 8.16. Let $p_\nu(x) = \alpha_{\nu 0} + \sum_{\sigma=1}^{s} \alpha_{\nu\sigma} x_\sigma$, $\nu = 1(1)s$, and let $A := (\alpha_{\nu\sigma}, \sigma > 0) \in \mathbb{C}^{s \times s}$, $a := (\alpha_{\nu 0}) \in \mathbb{C}^s$, $\overline{A} := (A \mid a) \in \mathbb{C}^{s \times (s+1)}$. For *nonsingular* A, the situation is regular and $z = -A^{-1}a$ is the unique solution. For *singular* A, with $\mathrm{rk}(A) = s - d =: r$ and $A z = 0$ for $z \in Z$, $\dim(Z) = d$, if $\mathrm{rk}(\overline{A}) = r$ then the zero set $Z[\langle p_\nu \rangle] = -A^+ a + Z$ where A^+ is the Moore–Penrose pseudo-inverse of A; else $Z[\langle p_\nu \rangle] = \emptyset$ and $\langle p_\nu \rangle = \langle 1 \rangle$.

From *numerical* linear algebra we know that this strict, discontinuous separation between the regular and the singular case makes sense only for *exact* coefficients $\alpha_{\nu\sigma}$ and can only be verified with *exact computation*. From a computational point of view, there is a *transition regime* where A becomes increasingly *ill-conditioned*; with *empirical* coefficients, this means that the pseudozero sets containing the valid approximations for z become larger and larger and that the naive use of approximate computation can lead to large deviations in the result. We will return to this aspect in section 9.4; at present we consider the intrinsic case.

We observe that there is a natural one-to-one correspondence between points z in $\mathbb{C}^s$ and 0-dimensional ideals $\mathcal{I}_z$ in $\mathcal{P}^s$ with z as their only (simple) zero. Moreover, for each $z = (\zeta_\sigma) \in \mathbb{C}^s$, there exist systems P of s polynomials $p_\sigma \in \mathcal{P}^s$ such that $\langle P \rangle = \mathcal{I}_z$, e.g., $p_\sigma(x) = x_\sigma - \zeta_\sigma$, $\sigma = 1(1)s$. Thus, each 0-dimensional ideal in $\mathcal{P}^s(1)$ may be *generated* by s polynomials. Does this observation generalize to $m > 1$?

Definition 8.8. A 0-dimensional ideal $\mathcal{I} \subset \mathcal{P}^s$, $s > 1$, which can be generated by s polynomials, is called a *complete intersection ideal* and its generating system a *complete intersection system*. The variety (=zero set) of such an ideal is also called a *complete intersection*.   □

More generally, an $(s - n)$-dimensional ideal in $\mathcal{P}^s$ which can be generated by n polynomials and the associated variety may also be called a complete intersection (ideal) and the generating system a complete intersection system. The essential characteristic of a complete intersection system $\{p_\nu, \nu = 1(1)n\}$ is the following: When we consider the sequence of ideals $\mathcal{I}_\nu :=$ $\langle p_1, \ldots, p_\nu \rangle$, $\nu = 1(1)n$, we must have $\dim \mathcal{I}_\nu = s - \nu$ for any numbering of the polynomials in the system; with other words, each further generating polynomial must reduce the dimension exactly by one. In this text, we will restrict the use of the term to the case $n = s$ as in Definition 8.8. Naturally, a complete intersection ideal may also have bases of a structure which requires more than s basis elements; cf. Example 8.2.

In the introduction to Chapter 8, we have required a *regular polynomial system* in s variables to consist of s polynomials and to have a 0-dimensional zero set; thus, each regular polynomial system in $\mathcal{P}^s$ is a complete intersection system and generates a complete intersection ideal. However, with a regular system we have also associated that it is sufficiently removed from a singular situation.

Any set Z of m points in $\mathbb{C}^s$ represents a complete intersection when it is interpreted as the set of *simple* zeros of a polynomial ideal in $\mathcal{P}^s$. This can be seen as follows:

Definition 8.9. A linear form $a : x \to a^T x$ on $\mathbb{C}^s$ is *separating* for a finite set $Z \subset \mathbb{C}^s$ if it takes different values for each $z \in Z$. $\square$

Example 8.9: If the points $z_\mu \in Z$ have different $\bar\sigma$-th components, the linear form $a^T x = x_{\bar\sigma}$ is separating on Z. Only in a set Z where no such $\bar\sigma$ exists do we have to take recourse to a less simple form. Take, e.g., $Z = \{(0, 0), (1, 0), (0, 1)\} \subset \mathbb{R}^2$; now, no individual component is separating, but a natural separating linear form is $a^T x = x_1 - x_2$. $\square$

Proposition 8.17. For any m-element set $Z \subset \mathbb{C}^s$, there exist infinitely many separating linear forms.

Proof: For each pair μ_1, μ_2, the relation $a^T z_{\mu_1} = a^T z_{\mu_2}$ represents an $(s - 1)$-dimensional linear subspace in the s-dimensional vector space $\mathcal{A}$ of the a^T. The $m(m-1)/2$ (finitely many) subspaces defined by the disjoint pairs of z_μ, $\mu = 1(1)m$, which have to be *avoided*, cannot fill $\mathcal{A}$. $\square$

Theorem 8.18.   For any finite set $Z = \{z_\mu, \mu = 1(1)m\} \subset \mathbb{C}^s$, there exist sets P of s polynomials $p_\sigma \in \mathcal{P}^s$, $\sigma = 1(1)s$, such that the ideal $\mathcal{I} = \langle P \rangle \subset \mathcal{P}^s$ has a simple zero at each $z_\mu \in Z$ but no other zeros.

Proof: Choose a separating linear form $a_s^T x$ for Z which exists by Proposition 8.17, with distinct values $\omega_{s\mu} := a_s^T z_\mu \in \mathbb{C}^s$, $\mu = 1(1)m$. Let $q_s(w) := \prod_{\mu=1}^m (w - \omega_{s\mu}) \in \mathcal{P}_m^1$.

Now choose $s - 1$ row vectors $a_\sigma^T \in \mathbb{C}^s$, $\sigma = 1(1)s - 1$, such that they span the $\mathbb{C}^s$ together with a_s^T. For each $\sigma = 1(1)s - 1$, let $\omega_{\sigma\mu} := a_\sigma^T z_\mu$, $\mu = 1(1)m$, and form the interpolation polynomial $q_\sigma(w) \in \mathcal{P}_{m-1}^1$ with $q_\sigma(\omega_{s\mu}) = \omega_{\sigma\mu}$, $\mu = 1(1)m$. By their construction, the s polynomials

$$p_\sigma(x) := a_\sigma^T x - q_\sigma(a_s^T x), \quad \sigma = 1(1)s - 1, \qquad p_s(x) := q_s(a_s^T x), \qquad (8.31)$$

vanish at each $z_\mu \in Z$; these can only be simple common zeros by construction of q_s. Assume that there exists $\bar z \notin Z$ with $p_\sigma(\bar z) = 0$, $\sigma = 1(1)s$. For $\sigma = s$, this implies $a_s^T \bar z = a_s^T z_{\bar\mu}$ for some $\bar\mu \in \{1, .., m\}$; hence $a_\sigma^T \bar z = q_\sigma(a_s^T z_{\bar\mu}) = a_\sigma^T z_{\bar\mu}$ for $\sigma = 1(1)s - 1$. But this implies $\bar z = z_{\bar\mu}$ since the a_σ^T span the $\mathbb{C}^s$. $\square$

Example 8.10: If we have an ideal $\mathcal{I}$ with m simple zeros whose last components differ, a complete intersection basis for $\mathcal{I}$ may have the structure

$$x_\sigma - q_\sigma(x_s)\,, \quad \sigma = 1(1)s - 1, \quad q_s(x_s)\,.$$

Note that all q_σ in this basis are *univariate* polynomials of degrees $m - 1$ and m resp. This is the Groebner basis of the ideal for a lexicographic ordering, with x_s the lowest variable; cf. section 8.4.2.

For the ideal with $Z = \{(0, 0), (1, 0), (0, 1)\} \subset \mathbb{R}^2$ in Example 8.9, with a separating linear form $a_2^T x = x_1 - x_2$, we have $q_2(w) = w^3 - w$; with $a_1^T x = x_1 + x_2$, we obtain $q_1(w) = w^2$ so that a complete intersection representation of this ideal is

$$p_1(x_1, x_2) \;=\; x_1 + x_2 - (x_1 - x_2)^2\,, \quad p_2(x_1, x_2) \;=\; (x_1 - x_2)^3 - (x_1 - x_2)\,.$$

This representation appears unnaturally complicated for the simple location of the zeros. There is the more "natural" basis $\{x_1^2 - x_1,\ x_1 x_2,\ x_2^2 - x_2\}$; but no two of its elements suffice to describe the ideal correctly.　　□

Example 8.11: For a nondegenerate set $Z = \{z_\mu,\ \mu = 1(1)4\} \subset \mathbb{C}^2$ (no three points on a straight line), there exists a one-parametric family of conic sections which pass through the 4 points since a conic section is defined by 5 of its points. Any two of the quadratic polynomials which describe the conic sections generate the ideal with Z as zero set.　　□

Since we have seen that a finite set of simple zeros always represents a complete intersection, zero constellations which are not complete intersections must contain multiple zeros. Even then, a complete intersection often prevails:

Theorem 8.19. All 0-dimensional polynomial ideals in $\mathcal{P}^s$ with no zeros of a multiplicity greater than 2 are complete intersection ideals.

Proof: Without loss of generality, assume that z_1 is a double zero, with the associated dual basis element (cf. Definition 2.14) $\sum_{\sigma=1}^{s} \gamma_{1\sigma} \partial_{x_\sigma}[z_1]$. In the choice of the separating linear form a_s^T of the proof of Theorem 8.18, we introduce the additional subspace restriction $a_s^T c_1 \neq 0$, with $c_1 := (\gamma_{1\sigma}) \in \mathbb{C}^s$; this leaves the choice of a_s^T feasible and we choose the a_σ^T as in the proof of Theorem 8.1. To the univariate polynomial $q_s(w)$ which vanishes at the $\omega_\mu = a_s^T z_\mu$ we attach a second factor $(w - \omega_1)$. For the univariate interpolation polynomials $q_\sigma(w)$ we add the requirement that $q_\sigma'(\omega_1) = a_\sigma^T c_1 / a_s^T c_1$. Now, the polynomials (8.31) satisfy $\sum_{\sigma=1}^{s} \gamma_{1\sigma} \partial_{x_\sigma} p_\sigma(z_1) = 0$, $\sigma = 1(1)s$, in addition to $p_\sigma(z_\mu) = 0$. For each further double zero, the same procedure can be employed.　　□

Example 8.10, continued: We choose the same zero set $Z = \{(0, 0), (1, 0), (0, 1)\}$ and assume that the polynomials in the ideal have a vanishing x_2-derivative at $(1,0)$. With a_σ^T as previously, we have $a_1^T c = 1$, $a_2^T c = -1$, and

$$p_1(x_1, x_2) \;=\; x_1 + x_2 - (3(x_1 - x_2) + 2(x_1 - x_2)^2 - 3(x_1 - x_2)^3)/2\,,$$
$$p_2(x_1, x_2) \;=\; (x_1 - x_2 + 1)(x_1 - x_2)(x_1 - x_2 - 1)^2\,;$$

surprisingly, this set permits a reduction to a much simpler basis of only 2 polynomials, viz. $\{x_1^2 - x_1,\ x_2^2 + (x_1 - 1)x_2\}$. In any case, the ideal with the above zero set is a complete intersection ideal.　　□

Example 8.11, continued: When we specify three points in $\mathbb{C}^2$ and a tangential direction in one of them, there is again a one-parametric family of conic sections satisfying these data. Thus, the respective ideal can be generated by two quadratic equations as previously. Note that the previous example is a special case of this situation so that the reduced representation was to be expected. $\square$

From the proof of Theorem 8.19, we expect that difficulties may arise when the dual space of an ideal in $\mathcal{P}^2$ contains more than one first-order derivative at the same zero z_μ, which is a perfectly reasonable case; cf. section 8.5. Following the construction above, we should now satisfy more than one condition for the derivatives q'_σ at the respective ω_μ. Also, in $\mathcal{P}^s$, if s first-order derivative conditions are associated with the same z_μ, this implies that *all* first-order derivatives must vanish at z_μ. The following classical counterexample shows that, in $\mathcal{P}^2$, there are simple (but very degenerate) polynomial ideals with a triple zero of this type which cannot be generated by only 2 polynomials; the analogous construction works for an $s + 1$-fold zero and s variables:

Example 8.12: In $\mathcal{P}^2$, consider $\mathcal{I} := \{p \in \mathcal{P}^2 \ : \ p(0,0) = \partial_{x_1} p(0,0) = \partial_{x_2} p(0,0) = 0\}$, with no further zero or multiplicity, and assume $\mathcal{I} = \langle p_1, p_2 \rangle$. Each $p \in \mathcal{I}$ must have a Taylor expansion

$$p(x_1, x_2) \; = \; x_1^2 \, q_1(x_1, x_2) + x_1 \, x_2 \, q_2(x_1, x_2) + x_2^2 \, q_3(x_1, x_2),$$

which implies

$$\partial_{x_1^2} p_\nu(0,0) = q_{\nu 1}(0,0), \quad \partial_{x_1 x_2} p_\nu(0,0) = q_{\nu 2}(0,0), \quad \partial_{x_2^2} p_\nu(0,0) = q_{\nu 3}(0,0), \quad \nu = 1, 2.$$

Since there exists a vector $(\gamma_1, \gamma_2, \gamma_3)$ such that $\sum_{j=1}^3 \gamma_j \, q_{\nu j}(0,0) = 0$ for $\nu = 1$ and 2, there exists a 2nd order derivative at (0,0) which vanishes for all $p \in \mathcal{I}$. Thus the assumption of a complete intersection is incompatible with a triple zero of the above kind. Note that this implies that a zero of this type cannot occur with a regular polynomial system.

When we keep the 3-fold zero at (0,0) but change the derivative conditions, we may well have a complete intersection ideal: For $\mathcal{I} := \{p \in \mathcal{P}^2 \ : \ p(0,0) = \partial_{x_1} p(0,0) = \partial_{x_1}^2 p(0,0) = 0\}$, with no further zero or multiplicity, there is the trivial basis $\{x_1^3, x_2\}$. $\square$

Obviously, the dual spaces $\mathcal{D}$ whose associated ideal $\mathcal{I}[\mathcal{D}]$ is not a complete intersection ideal constitute a very "thin" subset of all dual spaces of dimension m in s variables; furthermore, they cannot appear with regular polynomial systems.

In the normal set representation of an ideal $\mathcal{I}$ with respect to a *quasi-univariate* normal set $\mathcal{N}$, with distinguished variable x_s, consider the border basis subset $\mathcal{B}_s = \{bb_1, \ldots, bb_s\}$ of the border basis $\mathcal{B}_\mathcal{N}$ of $\mathcal{I}$ whose members bb_σ, $\sigma = 1(1)s$, have their $\mathcal{N}$-leading monomials x^{j_σ} in the border subset $B_s[\mathcal{N}]$.

Proposition 8.20. $\mathcal{B}_s$ is a complete intersection system, with $\langle \mathcal{B}_s \rangle = \langle \mathcal{B}_\mathcal{N} \rangle = \mathcal{I}$.
Proof: Compare Theorem 8.5 and Proposition 8.7. $\square$

Thus, the border basis of a quasi-univariate normal set representation contains at least one complete intersection system as a subset.

Example 8.13: In Example 8.2, the normal set $\mathcal{N} = \{1, x, y, z\}$ is quasi-univariate with respect to each variable; the assumptions of Theorem 8.5 are satisfied for x and y as distinguished variables. Hence (with the notation of Example 8.2), the following border basis subsets are complete intersection systems for $\mathcal{I}[\mathcal{R}]$: $\{bb_1, bb_2, bb_3\}$ and $\{bb_2, bb_4, bb_5\}$. $\square$

8.3.2 Continuity of Polynomial Zeros

In section 5.1.1, we have convinced ourselves that the zeros of a univariate polynomial p are continuous functions of the coefficients of p, and analytic functions in the case of simple zeros. Of course, the quantitative meaning of that assertion depends on the particular *representation* of p which we have in mind (cf. section 5.1.2), but a change between representations by different bases amounts essentially to a regular linear transformation between the coefficients. The simplicity of the situation rests strongly on the fact that there is a unique relation between a univariate polynomial p and the ideal $\langle p \rangle$; cf. section 5.1.3.

With 0-dimensional systems P of *multivariate* polynomials and the associated ideals $\langle P \rangle$, this situation is quite different because of the many potential basis representations of ideals in $\mathcal{P}^s$ which cannot be related in a simple linear way. Also many basis representations consist of more than s polynomials and are thus overdetermined; cf. section 8.2.2. Even when we restrict ourselves to complete intersection ideals and bases with s polynomials represented in terms of monomials, the supports of the basis polynomials may be quite distinct in two different bases for the same ideal. Thus, when we consider the maps from the coefficients of one basis to the individual zeros of $\mathcal{I} = \langle P \rangle$ and the analogous maps from the coefficients of another basis, it may not be obvious at all how these maps are related.

Therefore, in our analysis of the relations between an ideal $\mathcal{I} = \langle P \rangle \subset \mathcal{P}^s$ and the zero set $Z[\mathcal{I}]$, we assume throughout that a particular complete intersection system P has been specified as generating system for $\mathcal{I}$. This defines a data space $\mathcal{A}$ for the coefficients in $P = P(x; a)$, with $a \in \mathcal{A}$. With our regularity concept, each neighboring system $P(x; a + \Delta a)$, with $\|\Delta a\|$ sufficiently small, defines an ideal $\tilde{\mathcal{I}}$, with a zero set $Z[\tilde{\mathcal{I}}]$ of the same magnitude (counting multiplicities), and we can introduce the data$\rightarrow$result maps

$$ F_\mu \; : \quad \mathcal{A} \;\rightarrow\; \mathbb{C}^s \,, \qquad \text{with } F_\mu(a) = z_\mu \,, \; \mu = 1(1)m \,, $$

whose domains are neighborhoods of $a \in \mathcal{A}$, and use them as reference for our analysis of the continuity of the zeros.

Due to the linearity of polynomials in their coefficients, we have

$$ P(x; a + \Delta a) \;=\; P(x; a) + P(x; \Delta a) \;=\; \{\, p_\nu(x; a) + p_\nu(x; \Delta a) \,\} \,, $$

and due to the differentiability with respect to the variables x, we have the Taylor expansions

$$ P(x + \Delta x; a) \;=\; P(x; a) + P'(x; a)\, \Delta x + \dots \;=\; \{\, p_\nu(x; a) + p_\nu'(x; a)\, \Delta x + \dots \} \,; $$

cf. (1.9). Thus, the zeros $z_\mu + \Delta z_\mu$ of the ideal generated by a neighboring system with coefficients $a + \Delta a$, satisfy

$$ P(z_\mu + \Delta z_\mu; a + \Delta a) \;= $$

$$ P(z_\mu; a) + P'(z_\mu; a)\, \Delta z_\mu + P(z_\mu; \Delta a) + O(\|\Delta z_\mu\|^2) + O(\|\Delta a\| \|\Delta z_\mu\|) \;=\; 0 \,. \tag{8.32} $$

At a simple zero z_μ of $P(x; a)$, with a regular Jacobian $P'(z_\mu; a)$, this implies

$$ \Delta z_\mu \;=\; F_\mu(a + \Delta a) - F_\mu(a) = \tfrac{d}{da} F_\mu(a)\, \Delta a + O(\|\Delta a\|^2) $$

$$ =\; -\big(P'(z_\mu; a)\big)^{-1} P(z_\mu, \Delta a) + O(\|\Delta a\|^2) \tag{8.33} $$

$$ =\; \big(P'(z_\mu; a)\big)^{-1} \Big(-\textstyle\sum_{j \in J_\nu} \Delta \alpha_{\nu j} z_\mu^j\Big) + O(\|\Delta a\|^2) \,. $$

Proposition 8.21. In a sufficiently small neighborhood of the specified coefficients, a simple zero z_μ of the regular polynomial system $P(x; a)$ is a differentiable (and hence continuous) function of the coefficients in the system P.

From (8.33), we observe that the increments of a simple zero are really determined by the *residuals* $\Delta p_\nu(z_\mu) = \sum_{j \in J_\nu} \Delta \alpha_{\nu j} z_\mu^j$ of the original zeros in the modified polynomials. When we consider P as a map from the $\mathbb{C}^s$ of the x-arguments to the $\mathbb{C}^s$ of the $p_\nu(x)$, the implicit function theorem tells us that – for regular $P'(x)$ – there is an analytic function $\Delta P(x) \to \Delta x$ such that $P(x + \Delta x) = P(x) + \Delta P(x)$; cf., e.g., [8.4], section 10.2.

Theorem 8.22. In the neighborhood of a simple zero $z_\mu \in \mathbb{C}^s$ of a regular polynomial system $P \in (\mathcal{P}^s)^s$, there exists a bijective analytic map between approximate zeros $z_\mu + \Delta z_\mu$ and residuals $\Delta P := P(z_\mu + \Delta z_\mu)$.

Quantitatively, we have, with appropriately matching norms,

$$\|\Delta P\| \leq \|P'(z_\mu)\| \, \|\Delta z_\mu\| + O(\|\Delta z_\mu\|^2) \qquad \text{and}$$

$$\|\Delta z_\mu\| \leq \|(P'(z_\mu))^{-1}\| \, \|\Delta P\| + O(\|\Delta P\|^2) \, . \tag{8.34}$$

Proposition 8.23. The absolute condition of a simple zero $z_\mu \in \mathbb{C}^s$ of the regular polynomial system $P \in (\mathcal{P}^s)^s$ is quantified by $\|(P'(z_\mu))^{-1}\|$, where $\|..\|$ is the operator norm for the norms in the solution and the residual spaces.

Example 8.14: In $\mathcal{P}^2$, consider the zero set of the ideal generated by two nondegenerate quadratic polynomials $(\mathbf{x}:=(x,y))$

$$p_\nu(x, y) = \mathbf{x}^T A_\nu \, \mathbf{x} + a_\nu^T \mathbf{x} + \alpha_{\nu 0} \, , \quad \nu = 1, 2 \, ,$$

with regular symmetric $A_\nu \in \mathbb{C}^{2 \times 2}$, $a_\nu^T \in \mathbb{C}^2$, $\alpha_{\nu 0} \in \mathbb{C}$, i.e. the 4 real or complex intersection points of two conic sections. If the intersection angles are not very acute, small shifts in the conic sections lead to small changes in the zeros. At a zero $z_\mu = (\xi_\mu, \eta_\mu)$, we have from (8.33)

$$\Delta z_\mu = - \begin{pmatrix} 2\,(\xi_\mu, \eta_\mu)\,A_1 + a_1^T \\ 2\,(\xi_\mu, \eta_\mu)\,A_2 + a_2^T \end{pmatrix}^{-1} \begin{pmatrix} (z_\mu^T \Delta A_1 + \Delta a_1^T)\,z_\mu + \Delta \alpha_{10} \\ (z_\mu^T \Delta A_2 + \Delta a_2^T)\,z_\mu + \Delta \alpha_{20} \end{pmatrix} ,$$

which displays the effect of a change in individual coefficients on the components of the zeros. $\square$

For *near-singular* $P'(z_\mu)$, the condition of the zero z_μ may become arbitrarily bad; cf. (8.34). This reflects a situation where the gradient vectors $p_\nu'(z_\mu)$ are nearly linearly dependent or—equivalently—two or more of the manifolds $p_\nu(x) = 0$ are nearly tangential at z_μ so that the zero must react extremely sensitively to certain small perturbations in the p_ν.

Singularity of $P'(z_\mu)$ characterizes z_μ as a *multiple zero* of P. In the univariate case, at a multiple zero, differentiability of the coefficient→zero map disappears but continuity is retained as Hölder-continuity; cf. Proposition 5.1. For a multivariate complete intersection system, this remains true, but the analysis must consider the particular derivative structure of the multiple zero. We will regard this in detail in sections 8.5 and 9.3; at this point we only demonstrate the situation with a simple example:

Example 8.15: In $\mathcal{P}^2$, consider $p_1(x, y) = x^2 + y^2 - 1$, $p_2(x, y) = x^2 + y^2 + 2x - 3$, with a real double zero of $\langle p_1, p_2 \rangle$ at $(1,0)$. For perturbations of the two constant terms,

$$p_\nu(1 + \Delta x, \Delta y) = 2\nu \Delta x + (\Delta x)^2 + (\Delta y)^2 + \Delta\alpha_{\nu 0} = 0, \quad \nu = 1, 2.$$

This implies

$$\Delta x = (\Delta\alpha_{10} - \Delta\alpha_{20})/2, \quad \Delta y = \pm\sqrt{\Delta\alpha_{20} - 2\Delta\alpha_{10}}\,(1 + O(\|\Delta a\|)),$$

which displays the Hölder-continuity of the y-component while the x-component remains differentiable. $\square$

8.3.3 Expansion by a Complete Intersection System

In section 8.2.1 (cf. (8.16)), we had observed that, in the expansion

$$p(x) = \mathrm{NF}_{\mathcal{I}}[p] + \sum_{x^j \in B[\mathcal{N}]} q_j(x)\, bb_j(x)$$

of a polynomial $p \in \mathcal{P}^s$ with respect to an arbitrary border basis $\mathcal{B}_{\mathcal{N}}$ of a 0-dimensional ideal $\mathcal{I}$, the q_j are generally not unique because there exist nontrivial syzygies between the bb_j. If we consider the same type of an expansion with respect to a complete intersection system, the non-uniqueness can be removed in as much as one pleases. This is due to a fundamental property of complete intersection systems:

Proposition 8.24. A complete intersection system $P = \{p_1, \ldots, p_s\} \subset \mathcal{P}^s$ has no nontrivial syzygies, i.e.

$$\sum_{\nu=1}^{s} q_\nu(x)\, p_\nu(x) = 0 \quad \text{(zero polynomial)} \quad \text{implies} \quad q_\nu \in \langle P \rangle, \quad \nu = 1(1)s. \tag{8.35}$$

Proof: An algebraic proof has been pointed out to me by D. Cox, but it requires too many technicalities to be reproduced here. Geometrically, the plausibility of (8.35) is seen thus: Let $\widehat{V}_\sigma := \cap_{\nu \neq \sigma} V[p_\nu]$, $\sigma = 1(1)s$. For a complete intersection, $\dim \widehat{V}_\sigma = 1$ and $\dim V[p_\sigma] \cap \widehat{V}_\sigma = 0$. For some $\sigma \in \{1, .., s\}$, take $x \in \widehat{V}_\sigma$, $x \notin V[p_\sigma]$. Substitution into the sum in (8.35) shows that q_σ vanishes on $\widehat{V}_\sigma$ with the possible exception of its intersection with $V[p_\sigma]$; by continuity, it must vanish on all of $\widehat{V}_\sigma$ which implies $q_\sigma \in \langle p_\nu, \nu \neq \sigma \rangle \subset \langle P \rangle$. $\square$

When we write (8.35) in the equivalent form

$$\text{With } q_\nu \in \mathcal{R}[\langle P \rangle], \quad \sum_{\nu=1}^{s} q_\nu(x)\, p_\nu(x) \equiv 0 \quad \text{implies} \quad q_\nu = 0, \quad \nu = 1(1)s, \tag{8.36}$$

then it is a natural extension of the fundamental concept of (scalar) linear independence of a system of s linear polynomials in s variables in linear algebra:

$$\text{With } \gamma_\nu \in \mathbb{C}, \quad \sum_{\nu=1}^{s} \gamma_\nu\, p_\nu(x) \equiv 0 \quad \text{implies} \quad \gamma_\nu = 0, \, \nu = 1(1)s.$$

Thus it would be meaningful to call a system $P \subset \mathcal{P}^s$ which satisfies (8.36) *polynomially linearly independent*. But, surprisingly, a term for this property appears not to exist in polynomial algebra.

Theorem 8.25. Consider a complete intersection ideal $\mathcal{I} = \langle p_1, \ldots, p_s \rangle \subset \mathcal{P}^s$ and its quotient ring $\mathcal{R}[\mathcal{I}]$, with an arbitrary but fixed basis $\mathbf{b}$. In the expansion of $p \in \mathcal{P}^s$

$$p(x) \ = \ \mathrm{NF}_{\mathcal{I}}[p] + \sum_{\nu=1}^{s} q_\nu(x) \, p_\nu(x) \,, \tag{8.37}$$

the normal forms $\mathrm{NF}_{\mathcal{I}}[q_\nu] =: d_{1,\nu}(x) \in \mathcal{R}[\mathcal{I}]$ are unique.

Proof: Let $q_\nu(x) = d_{1,\nu}(x) + \sum_{\nu_1} q_{\nu\nu_1} p_{\nu_1}(x)$; consider another expansion (8.37) of p with coefficients $\hat{q}_\nu$ and $\hat{d}_{1,\nu}(x) := \mathrm{NF}_{\mathcal{I}}[\hat{q}_\nu]$ so that

$$\sum_{\nu=1}^{s} \left[(\hat{d}_{1,\nu}(x) - d_{1,\nu}(x)) + \sum_{\nu_1=1}^{s} (\hat{q}_{\nu\nu_1}(x) - q_{\nu\nu_1}(x)) \, p_{\nu_1}(x) \right] p_\nu(x) \ = \ 0 \,.$$

By Proposition 8.24, this requires $[\ldots] \in \mathcal{I}$ and hence $\hat{d}_{1,\nu} = d_{1,\nu}$, $\nu = 1(1)s$. $\square$

The idea of the proof of Theorem 8.25 may directly be extended to representations of the zero polynomial by an expression homogeneous in the p_ν of a degree greater than 1. Therefore, one can extend the uniqueness assertion to the normal forms of the $q_{\nu\nu_1}$ above and further. Thus we arrive at

Corollary 8.26. In the situation of Theorem 8.25, there exists a unique (finite) expansion of an arbitrary polynomial $p \in \mathcal{P}^s$ of the form

$$\begin{aligned}
p(x) \ = \ &d_0(x) + \sum_\nu d_{1,\nu}(x) \, p_\nu(x) + \sum_{\nu \le \nu_1} d_{2,\nu\nu_1} \, p_\nu(x) \, p_{\nu_1}(x) + \ldots \\
&+ \sum_{\nu \le \nu_1 \le \ldots \le \nu_{k-1}} d_{k,\nu\nu_1\ldots\nu_{k-1}} \, p_\nu(x) \, p_{\nu_1}(x) \ldots p_{\nu_{k-1}}(x) \,,
\end{aligned} \tag{8.38}$$

with all coefficients $d_{\ldots}(x) = \sum_\mu \delta_{\ldots,\mu} b_\mu(x) \in \mathcal{R}[\mathcal{I}]$.

The importance of the expansions (8.37) and (8.38) will appear when we consider empirical systems of polyomials: Since the specified indetermination in such systems refers to the particular form in which the system is given, it is important to refer other data also to this given system. Naturally, for a p of high degree, there are also intermediate forms between an expansion (8.37) to linear terms in the p_ν and the full expansion (8.38). Note that for an ideal with only *one* zero $z = (\zeta_1, ..\zeta_s)$ and the generating complete intersection system $\{x_1 - \zeta_1, \ldots, x_s - \zeta_s\}$, (8.38) becomes the Taylor expansion of p about z.

Example 8.16: The ideal considered in Examples 8.2 and 8.7 is a complete intersection ideal by Theorem 8.19; by Proposition 8.20, generating complete intersection systems are easily obtained since the normal set $\{1, x, y, z\}$ in Example 8.2 is quasi-univariate w.r.t each variable. Therefore, each of the three subsets $\mathcal{B}_1$, $\mathcal{B}_2$, $\mathcal{B}_3$ of the border basis $\mathcal{B}$ in Example 8.7 is a generating complete intersection system. In the following, we use $\mathcal{B}_1 = \{bb_1, bb_2, bb_3\}$.

With $\mathcal{B}_1$, two different expansions (8.37) of the polynomial $p = x^4 - x^2yz + 3x^2y +$

$2x^2z - xyz$ in Example 8.7 are, e.g., with $\mathrm{NF}_{\mathcal{I}}[p] = 1 + \frac{24}{7}x + \frac{9}{7}y - \frac{1}{7}z$, $p(x, y, z) =$

$$
\begin{aligned}
\mathrm{NF}_{\mathcal{I}}[p] \quad &+ \left(-\tfrac{11}{21} - \tfrac{18}{7}x + \tfrac{8}{7}y - \tfrac{11}{3}z + x^2 - 2xy - 3xz + \tfrac{2}{3}y^2 - \tfrac{2}{3}yz - z^2\right) bb_1(x, y, z) \\
&+ \left(\tfrac{16}{21} + \tfrac{14}{3}x - \tfrac{16}{21}y + \tfrac{27}{14}z + 2x^2 - \tfrac{2}{3}xy + \tfrac{25}{6}xz - \tfrac{7}{6}yz + \tfrac{7}{6}z^2\right) bb_2(x, y, z) \\
&+ \left(\tfrac{34}{21} - 2x + \tfrac{17}{14}y - \tfrac{50}{21}z + 3x^2 - \tfrac{9}{2}xy + xz + \tfrac{7}{6}y^2 - \tfrac{7}{6}yz\right) bb_3(x, y, z)
\end{aligned}
$$

and

$$
\begin{aligned}
\mathrm{NF}_{\mathcal{I}}[p] \quad &+ \left(1 + \tfrac{22}{21}x - \tfrac{11}{21}x^2 - \tfrac{2}{3}xy - xz - \tfrac{2}{3}x^2y - x^2z\right) bb_1(x, y, z) \\
&+ \left(\tfrac{52}{21}x + \tfrac{10}{7}x^2 + \tfrac{7}{6}xz + \tfrac{2}{3}x^3 + \tfrac{7}{6}x^2z\right) bb_2(x, y, z) \\
&+ \left(4 + \tfrac{37}{21}x - \tfrac{29}{21}x^2 - \tfrac{7}{6}xy + x^3 - \tfrac{7}{6}x^2y\right) bb_3(x, y, z) \, .
\end{aligned}
$$

For the coefficients q_ν of the bb_ν in *both* expressions, we find

$$
\mathrm{NF}_{\mathcal{I}}[q_1] = \tfrac{6}{7} + 4x - \tfrac{15}{7}y + \tfrac{1}{7}z \, , \quad \mathrm{NF}_{\mathcal{I}}[q_2] = \tfrac{10}{7} + \tfrac{30}{7}x - \tfrac{4}{7}y + z \, , \quad \mathrm{NF}_{\mathcal{I}}[q_3] = \tfrac{2}{7} + \tfrac{31}{7}x - 2y - \tfrac{6}{7}z \, .
$$

With a full expansion of the q_ν in terms of the bb_ν and a collection of terms, we obtain the *unique* representation (8.38) of p in terms of the ideal basis $\{bb_1, bb_2, bb_3\}$

$$
\begin{aligned}
p(x, y, z) = \quad &\left(1 + \tfrac{24}{7}x + \tfrac{9}{7}y - \tfrac{1}{7}z\right) + \left(-\tfrac{6}{7} + 4x - \tfrac{15}{7}y + \tfrac{1}{7}z\right) bb_1(x, y, z) \\
&+ \left(\tfrac{10}{7} + \tfrac{30}{7}x - \tfrac{4}{7}y + z\right) bb_2(x, y, z) + \left(\tfrac{2}{7} + \tfrac{31}{7}x - 2y - \tfrac{6}{7}z\right) bb_3(x, y, z) \\
&+ \left(bb_1(x, y, z)\right)^2 - bb_2(x, y, z)\, bb_3(x, y, z) \, . \qquad \square
\end{aligned}
$$

An expansion (8.37) or (8.38) can rarely be determined directly, except in trivial cases. Normally, one must first have a normal set for the quotient ring $\mathcal{R}[\langle p_1, \ldots, p_s \rangle]$ and an associated border basis of the complete intersection ideal; then, one may proceed as explained in section 8.2.1. In order to proceed further from (8.16) to (8.37), we must have a representation of the border basis elements bb_j in terms of the complete intersection system $P = \{p_\nu\}$:

$$
bb_j(x) = \sum_{\nu=1}^{s} v_{j\nu}(x) p_\nu(x) \quad \forall \, bb_j \in \mathcal{B} \, . \tag{8.39}
$$

If the bb_j are determined by iterated linear combination from the original system P, the determination of the coefficients $v_{j\nu}$ in (8.39) requires simply a bookkeeping in the procedure by which the border basis $\mathcal{B}$ is determined. Unfortunately, current computer algebra systems will not furnish that bookkeeping, not even for a Groebner basis computation, but it is clear that it can easily be implemented. If the complete intersection basis is a *subset* of the border basis, as in Example 8.13 above, (8.39) requires that the border basis elements not used are represented by the complete intersection basis.

From (8.16) and (8.39), the expansion (8.37) is then immediately obtained:

$$
\begin{aligned}
p(x) = \quad &\mathrm{NF}[p] + \sum_j q_j(x)\, bb_j(x) = d_0(x) + \sum_j q_j(x) \sum_\nu v_{j\nu}(x)\, p_\nu(x) \\
= \quad &d_0(x) + \sum_\nu \left(\sum_j q_j(x)\, v_{j\nu}(x)\right) p_\nu(x) \, .
\end{aligned} \tag{8.40}
$$

Thus, an implementation of (8.37) and the more detailed expansions which may be further derived from it meets no principal difficulties if a normal set and border basis algorithm with sufficient bookkeeping is available. Normal set and border basis algorithms will be further discussed in later sections; cf. also section 8.4.4 for the special case of Groebner bases.

In section 5.1.3, we have "expanded" a polynomial $p \in \mathcal{P}^1$ in powers of a specified polynomial s; cf. Proposition 5.3 and (5.18). This expansion is a generalization of the Taylor expansion of p; truncated copies of the expansion furnish higher order approximations simultaneously in the vicinity of *all* zeros of s (or $\langle s \rangle$). Equation (8.38) is the multivariate counterpart of (5.18):

Proposition 8.27. In the situation of Theorem 8.25, let the zero set $Z[\mathcal{I}] \subset \mathbb{C}^s$ consist of m simple zeros. Let $r_k \in \mathcal{P}^s$ be the remainder of the expansion (8.38) truncated after the k-th order terms. Then all derivatives of r_k of an order $\leq k$ vanish at each point $z_\mu \in Z[\mathcal{I}]$.

Proof: The proof follows immediately from the fact that each p_ν vanishes at each z_μ. $\hspace{0.5cm}\square$

Example 8.16, continued: $r_1 = bb_1^2 - bb_2\, bb_3$; the Taylor expansion of r_1 at the zero $(1,2,0)$ of $\mathcal{I}$ is, for example,

$$r_1(x, y, z) = -\tfrac{144}{49}\,(x-1)^2 + \tfrac{32}{49}\,(x-1)(y-2) - \tfrac{114}{49}\,(x-1)z + \tfrac{69}{49}\,(y-2)^2 - \tfrac{167}{49}\,(y-2)z + \tfrac{103}{49}\,z^2 \,.$$

Thus the linear part of the expansion of p at the end of Example 8.16 is a good approximation of p simultaneously at each of the 4 zeros of the ideal. $\hspace{0.5cm}\square$

8.3.4 Number of Zeros of a Complete Intersection System

A strict upper bound for the number of zeros (counting multiplicities) of a regular multivariate polynomial system has been known for a long time:

Proposition 8.28 (Bézout). For a regular system $P = \{p_\nu \in \mathcal{P}^s,\ \nu = 1(1)s\}$, $d_\nu := \deg p_\nu$, the potential number of zeros is bounded by

$$m_{\text{Bézout}} = \prod_{\nu=1}^{s} d_\nu \,. \tag{8.41}$$

It is well known (cf., e.g., [2.11]) that the bound (8.41) is assumed if all p_ν are generic and *dense*, i.e. if they contain all terms of total degree $\leq d_\nu$. But most polynomial systems are extremely *sparse*; for such systems, the actual number of zeros may be considerably smaller than $m_{\text{Bézout}}$, perhaps by an order of magnitude. That it is the presence of the high degree monomials which determines the true number of zeros is obvious from the simplest examples: Two generic quadratic equations in two variables have 4 zeros, but the two quadratic equations

$$\alpha_{11}^{(\nu)}\, x\, y + \alpha_{10}^{(\nu)}\, x + \alpha_{01}^{(\nu)}\, y + \alpha_{00}^{(\nu)} = 0, \quad \nu = 1, 2,$$

can only have two zeros because we may readily elminate the xy-term from one of the equations which leaves us with a quadratic and a linear equation.

Fortunately, there exists a bound on the number m of zeros which takes into account the sparsity structure of the p_ν. It is associated with the names of Bernstein, Khovanski, and Kushnirenko, and with terms like "Newton polytopes" and "mixed volumes"; in this text, it will be denoted as BKK-bound. Although its formal specification is rather straightforward, its concise derivation and its computation for $s > 2$ are rather complicated. Therefore, we will not formulate an explicit expression for the BKK-bound but rather explain its meaning and use. A more thorough introduction (which still avoids unnecessary mathematical technicalities) may be found in Chapter 7 of [2.11] which also contains references to the original literature about the subject.

In 2002, a procedure for the evaluation of the BKK-bound was not yet available in either Maple 7 or in Mathematica 4. However, there are some special packages which serve that purpose, e.g., PHC pack by J. Verschelde ([8.5]). In any case, in this introductory report, we assume that it is possible to retrieve the value $BKK(P)$ of the BKK-bound for a specified regular system P of polynomial equations. Naturally, the computational effort increases with the number s of variables and the degrees d_ν of the polynomials.

What is the meaning of the integer number $BKK(P)$ which is generated by a BKK-bound algorithm? Consider the system

$$P = \{p_\nu, \ \nu = 1(1)s\} \subset (\mathcal{P}^s)^s, \ \text{with } p_\nu(x) = \sum_{j \in J_\nu} \alpha_j^{(\nu)} x^j, \ \alpha_j^{(\nu)} \in \mathbb{C}, \ \nu = 1(1)s.$$

For *fixed supports* J_ν, almost all instantiations of the coefficients lead to the same number m of isolated zeros z_μ (counting multiplicities) of P; thus m is invariant and independent of the particular coefficient values for large regions of the data space $\mathcal{A}$ of P which is determined by the support $J_\nu, \nu = 1(1)s$. The BKK-bound $BKK(P)$ is *equal* to this number m of zeros which prevails for almost all coefficient values in $\mathcal{A}$ or—as it is often expressed—for "generic coefficients." Moreover, it is the *maximal* number of isolated zeros which can occur for a system with the supports J_ν.

Only for coefficient values from some lower-dimensional manifold in $\mathcal{A}$, the actual number of zeros may be smaller than $BKK(P)$, or there may exist a zero manifold. That a zero which exists for generic coefficients may *disappear* for a particular instantiation of the coefficients is well known: A generic linear system $A x = b$ has 1 zero; this zero disappears when the coefficient matrix A is singular and the right-hand side b not in the image space of A. Actually, the zero moves towards ∞ as the coefficients move towards values for which the system is inconsistent. A trivial nonlinear example is $p_1(x, y) = (x - \alpha_1)(y - \beta_1) - 1, \ p_2(x, y) = (x - \alpha_2)(y - \beta_2) + 1$. For generic α_ν, β_ν, there are 2 zeros, in agreement with $BKK(\{p_1, p_2\}) = 2$. For $\alpha_2 \to \alpha_1$, one of the zeros disappears to ∞; the other one follows as $\beta_2 \to \beta_1$ and the system has become inconsistent. Polynomial systems with "diverging zeros" will be considered in section 9.5.

All this applies with the important reservation that the BKK-bound does not always count a simple or multiple zero with one or several components 0 ! This may happen when one or several of the p_ν possess *no constant term*. That such a reservation is necessary is explained by the observation that the BKK-bound is invariant against multiplication of the p_ν by monomials: It is obvious that such multiplications will generally introduce further zeros, at 0 or with some zero components.

Fortunately, there is a simple trick to get rid of this deficiency; it was proposed by T. Y. Li: As we have seen in section 8.3.2, the zeros of a regular polynomial system are continuous functions of the coefficients of the p_ν and hence of their constant terms. This remains valid when a constant term happens to vanish as long as there is a *generic* constant term in each equation.

Thus, the BKK-bound gives the correct number of zeros, independently of their location in the finite $\mathbb{C}^s$, when we append a generic constant term to the p_ν with $\alpha_0^{(\nu)} = 0$. When this is done after multiplication of the p_ν by monomials, the BKK-bound is no longer invariant against such multiplications.

Example 8.17 (from [2.11]): In $\mathcal{P}^2$, consider the very sparse system with generic coefficients

$$
P(x, y) = \left\{
\begin{array}{rcl}
p_1(x, y) &=& \alpha_{32}^{(1)} x^3 y^2 + \alpha_{02}^{(1)} y^2 + \alpha_{10}^{(1)} x + \alpha_{00}^{(1)} \\
p_2(x, y) &=& \alpha_{14}^{(2)} x y^4 + \alpha_{30}^{(2)} x^3 + \alpha_{01}^{(2)} y
\end{array}
\right\} = 0. \tag{8.42}
$$

Upon input of the supports of p_1 and p_2, a BKK-package delivers $\mathrm{BKK}(\{p_1, p_2\}) = 18$. Since p_2 does not contain a constant term, this value may not include some zero with a vanishing component. However, it is easily seen that, for p_2, the vanishing of one zero component would imply that of the other one, which is incompatible with p_1. Accordingly, $\mathrm{BKK}(\{p_1, p_2 + \alpha_{00}^{(2)}\})$ has the same value 18.

This value is also generated for $\mathrm{BKK}(\{p_1, x\, p_2\})$ and for $\mathrm{BKK}(\{y\, p_1, x\, p_2\})$, although the actual number of zeros for these systems is 20 and 24, respectively. The correct values are now obtained from $\mathrm{BKK}(\{p_1, x\, p_2 + \alpha_{00}^{(2)}\})$ and $\mathrm{BKK}(\{y\, p_1 + \beta_{00}^{(1)}, x\, p_2 + \alpha_{00}^{(2)}\})$. The additional zeros with zero components are easily spotted: $\{p_1, x\, p_2\}$ inherits all zeros from P and it has the additional zeros $(0, \pm\sqrt{-\alpha_{00}^{(1)}/\alpha_{02}^{(1)}})$. $\{y\, p_1, x\, p_2\}$ has a further zero at $(0, 0)$; this is a 4-fold zero because the first three partial x-derivatives also vanish at $(0, 0)$. With a nonzero constant term, this zero splits into 4 isolated zeros .

Note that the Bézout numbers (8.41) for the above systems are 25, 30, and 36, resp., which are rather misleading values. $\quad\square$

For a superficial explanation of how $\mathrm{BKK}(P)$ is *defined*, we must introduce the following notion:

Definition 8.10. In $\mathbb{N}_0^s$, the grid of nonnegative integer s-tuples, consider the set of the $j \in J$, the support of $p \in \mathcal{P}^s$. The *convex hull* of this set is the *Newton polytope* of p :

$$
\mathrm{NP}(p) := \mathcal{C}\{j \in J\} \subset \mathbb{R}^s . \tag{8.43}
$$

Due to its convexity, the Newton polytope of a multivariate polynomial $p \in \mathcal{P}^s$ reflects the sparsity of p in a very special way: Only those monomials x^j whose exponent j generates a *corner* of $\mathrm{NP}(p)$ are essential for the shape of the polytope. Exponents $k \in \mathbb{N}_0^s$ which lie on a *face* or in the *interior* of $\mathrm{NP}(p)$ are irrelevant; the presence or absence of these exponents does not influence the position and shape of $\mathrm{NP}(p)$. Since we will see below that $\mathrm{BKK}(P)$ is exclusively determined by the Newton polytopes $\mathrm{NP}(p_\nu)$, $\nu = 1(1)s$, this implies that the number of zeros of an s-variate polynomial system $\{p_\nu, \nu = 1(1)s\}$ depends only on the presence of certain distinguished monomials in each p_ν while the presence or absence of the remaining terms is irrelevant for the number of zeros (with the exception of some special instantiations of the coefficients).

Example 8.17, continued: Figure 8.4 shows the Newton polytopes of the two polynomials P_1, p_2 in (8.42) and of the polynomials $y\, p_1$ and $x\, p_2$. With p_1, p_2 of (8.42), all terms contribute to the definition of $\mathrm{NP}(p_1)$ and $\mathrm{NP}(p_2)$, resp.; cf. Figure 8.4. Additional terms with the monomials y, xy, $x^2 y$ in p_1 would not affect $\mathrm{NP}(p_1)$. The addition of a constant term to p_2 does affect

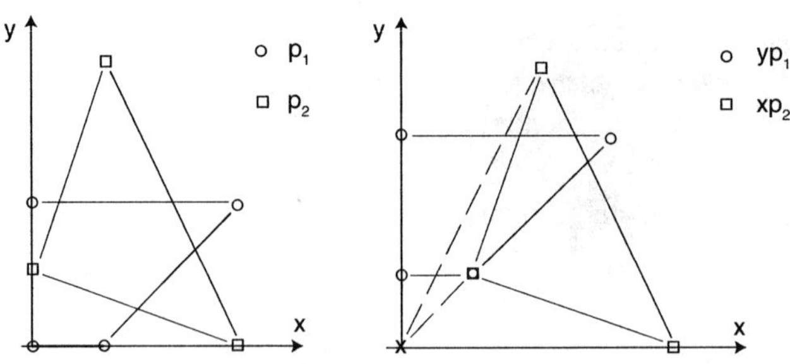

Figure 8.4.

NP(p_2), but it turns out that the BKK-bound is not affected. The addition of constant terms to $x\,p_2$ and to $y\,p_1$ affects the Newton polytopes of these polynomials, and this change is reflected by the BKK-bound as it should be; cf. above. $\square$

For a set of s polytopes P_σ, $\sigma = 1(1)s$, with corners in $\mathbb{N}_0^s$ (so-called lattice polytopes), a mapping

$$\mathrm{MV}\ :\quad P_1,\ldots,P_s\ \to\ \mathbb{Z}_0$$

to the nonnegative integers has been defined which is called the *mixed volume* of $P_1,\ldots,P_s$; it is a symmetric function of its arguments. This function defines the BKK-bound :

$$\mathrm{BKK}(\{p_1,\ldots,p_s\})\ :=\ \mathrm{MV}\,(\mathrm{NP}(p_1),\ldots,\mathrm{NP}(p_s))\,. \tag{8.44}$$

We refrain from giving a formal definition of the function MV and point the reader once more to the explanations in [2.11]. Since a mixed volume is only defined for s lattice polytopes in $\mathbb{N}_0^s$, the BKK-bound is only defined for a system of s polynomials in $\mathcal{P}^s$, i.e. for a regular system. When a 0-dimensional ideal in $\mathcal{P}^s$ is defined by more than s polynomials, the number of its zeros cannot be determined by (8.44). This is natural because BKK(P) depends only on the *support* of P and an overdetermined system is inconsistent for almost all instantiations of its coefficients.

The number $m = \mathrm{BKK}(P)$ of zeros of P equals the dimension of the quotient ring $\mathcal{R}[\langle P\rangle]$; thus, the knowledge of m puts the elements of the set $T^s(m)$ (cf. Definition 2.18) at our disposal for a basis of $\mathcal{R}[\langle P\rangle]$. There are two potential reasons why some particular normal set $\mathcal{N} \in T^s(m)$ may not be feasible for the specified system P :

(i) $\mathcal{N}$ contains the complete support of one or more of the p_ν; this would imply $p_\nu \in \mathcal{R}[\langle P\rangle]$ which is a contradiction.

(ii) For a basis $\mathbf{c}^T$ of the dual space $\mathcal{D}[\langle P\rangle]$, the matrix $\mathbf{c}^T(\mathbf{b})$ is singular; cf. Proposition 2.12.

Condition (i) may easily be checked; it may exclude various sets $\mathcal{N} \in T^s(m)$ for *all* systems with a particular support structure. Condition (ii), on the other hand, cannot be checked a priori when we do not know the zeros of P. Also, $\mathbf{c}^T(\mathbf{b})$ will generally not be singular for all P of a given support structure but only when their coefficients lie on some manifold in the data space. Therefore, an algorithm for the determination of a normal set basis vector $\mathbf{b}(x)$ for $\mathcal{R}[\langle P\rangle]$ can comply with (i) but must rely on intermediate numerical results for compliance with (ii). This is also true for the exceptional case that $BKK(P)$ is not the correct dimension of $\mathcal{R}[\langle P \rangle]$; cf. the remarks above.

Example 8.18: Consider the three systems $\{p_1, p_2\}$, $\{p_1, x \, p_2\}$, $\{y \, p_1, x \, p_2\}$ of Example 8.17, with $m = 18, 20, 24$, respectively. When we consider "nice" normal sets for (8.42), we may try the 16 monomials $x^{j_1} y^{j_2}$, $0 \le j_1, j_2 \le 3$, and two further ones, say x^5, y^5. But this $\mathcal{N}$ would contain the full support of p_1 and is therefore not admissible.

In another attempt, we could choose the 15 monomials $x^{j_1} y^{j_2}$, $0 \le j_1 + j_2 \le 4$, and look for 3 further ones: Among the monomials of total degree 5, we have to avoid $x^3 y^2$ and $x \, y^4$, but we could take y^5, $x^2 y^3$, $x^4 y$ and obtain a satisfactory normal set.

Composing a normal set for $\{p_1, x \, p_2\}$ in the same fashion, we can now include all degree 5 monomials except $x^3 y^2$ which provides the requested 20 basis elements. For $\{y \, p_1, x \, p_2\}$, we can begin with the 21 monomials of total degree ≤ 5 and avoid $x^3 y^3$, $x^2 y^4$ in choosing 3 further monomials. $\quad\square$

Example 8.19: Consider a generic dense system in $(\mathcal{P}^s)^s$, with all polynomials of total degree d. Here the BKK-bound agrees with the Bézout bound d^s. The "hypercube" normal set $\mathcal{N} = \{x^j, \, \|j\|_{\max} \le d - 1\}$ is in $T^s(d^s)$ and not excluded by (i). $\quad\square$

The compliance with (i) generally excludes only a small part of the wide variety of sets in $T^s(m)$ as potential normal sets for some specified P.

Exercises

1. Consider the ideal $\mathcal{I} \in \mathcal{P}^3(8)$ whose zero set Z consists of the 8 corners of the unit cube in $\mathbb{R}^3$.

(a) Find a simple separating functional for Z. By the construction in the proof of Theorem 8.18, design a 3-element basis P for $\mathcal{I}$. What are the degrees of the 3 polynomials in P and hence the Bézout number of P.

(b) Obviously, there are several pairs of planes in $\mathbb{R}^3$ which contain all points of Z. By an appropriate selection of three such pairs, compose a basis for $\mathcal{I}$ consisting of 3 quadratic equations.

(c) Choose some polynomial in $\mathcal{P}_4^3$ and form its expansion (8.38) in terms of the quadratic complete intersection system found in (b).

2. Design pairs (p_1, p_2) of quadratic polynomials in $\mathcal{P}^2$ such that their common zeros have various condition properties with respect to changes in the coefficients.

(a) Use the manifolds $p_\nu = 0$ in the x, y-plane to find pairs whose 4 zeros are well-conditioned. Check by forming the singular values of P' at the zeros. Find a pair with 4 well-conditioned complex zeros.

(b) In the same fashion, find pairs with ill-conditioned zeros and verify via P'. Can you make all 4 zeros very ill conditioned? Find changes of the coefficients which display the ill-conditioning fully.

3. With a software package for the computation of the BKK-bound, find experimentally how the generic presence of certain terms in a specified polynomial system in $(\mathcal{P}^s)^s$, $s = 2$ and 3, influences the number of zeros of that system. Compare with the Bézout number for the same system.

8.4   Groebner Bases

In virtually all texts on constructive polynomial algebra, Groebner bases play a central role, both as the standard representation of polynomial ideals and as a tool for performing various tasks with polynomial ideals; cf., e.g., [2.10] and many others. In section 2.5.3, we have indicated why we have not followed that path in this book; the reader should refer to these explanations now.

On the other hand, powerful software for the determination of Groebner bases for polynomial systems with *rational* coefficients is available in Maple and Mathematica, and in practically all more specialized computer algebra systems. Software for general border bases is—at this time—still restricted to local developments. Therefore, Groebner bases often present the only available access to some computational task and it is important to have a clear view of their characterization, their potential, and their shortcomings. This section is devoted to this objective.

8.4.1   Term Order and Order-Based Reduction

A term order specifies a strict sequential order in the infinite set T^s of all monomials or terms x^j in s variables.

Definition 8.11. A linear order $\prec$ in the set T^s, $s \geq 1$, which is compatible with multiplication so that

$$x^{j_1} \prec x^{j_2} \implies x^j x^{j_1} \prec x^j x^{j_2}, \quad \forall j \in \mathbb{N}^s, \tag{8.45}$$

and for which $1 = x^0$ is the first ("lowest") element, is called a *term order*. Naturally, $x^{j_2} \succ x^{j_1}$ is synonymous with $x^{j_1} \prec x^{j_2}$. $\square$

In T^1, $1 \prec x \prec x^2 \prec x^3 \prec \ldots$ is the only possible term order. For $s > 1$, there are a few widely used choices:

Definition 8.12. Assume that we have specified a linear order between the components x_σ of $x = (x_1, \ldots, x_s)$ and numbered them such that $1 \prec x_s \prec x_{s-1} \prec \ldots \prec x_2 \prec x_1$; further assume that we write the elements in T^s as $x_1^{j_1} \ldots x_s^{j_s}$:

(1) the *lexicographic* order $\prec_{lex}$ is defined by

$$x^j \prec_{lex} x^k \text{ if either } j_1 < k_1 \text{ or } j_\sigma = k_\sigma \text{ for } \sigma = 1(1)\bar{s} - 1 \text{ and } j_{\bar{s}} < k_{\bar{s}} ;$$

(2) the *graded lexicographic* orders $\prec_{glex}$ and $\prec_{grevlex}$ are defined by

$$x^j \prec_{g(rev)lex} x^k \text{ if either (total) } \deg(x^j) < \deg(x^k) \text{ or the total degrees are equal and}$$
$x^j \prec_{(rev)lex} x^k$; here, $x^j \prec_{revlex} x^k$ if either $j_s > k_s$ or $j_\sigma = k_\sigma$ for $\sigma = s(-1)\bar{s} + 1$ and $j_{\bar{s}} > k_{\bar{s}}$. $\square$

For equal total degrees, the graded lexicographic orders can also be characterized by

$$x^j \quad \genfrac{}{}{0pt}{}{\prec_{glex}}{\prec_{grevlex}} \quad x^k \quad \text{if the} \quad \genfrac{}{}{0pt}{}{\text{"leftmost"}}{\text{"rightmost"}} \quad \text{nonvanishing component of } k - j \text{ is} \quad \genfrac{}{}{0pt}{}{\text{positive}}{\text{negative}} \quad .$$

Maple implements the *lex* order, with the notation `plex` ("pure lexicographic"), and the *grevlex* order, with `tdeg` ("total degree") and as a *default*. This is also the order which will most often be used in this book. Note that the reverse lexicographic order without a superimposed degree ordering would not satisfy $1 \prec x_\sigma$. There exist abstract definitions of a term order which admit further possibilities, but they will be of no concern in our context.

Example 8.20: For the lexicographic order, we have, e.g., $x_1^3 x_2 x_3^3 \succ_{lex} x_1^2 x_2^4 x_3^2$, or $x_1^3 x_2 x_3^2 \succ_{lex} x_1^3 x_3^3$, etc.

For the *grevlex* order, we have, e.g., $x_1^3 x_2 x_3^3 \prec_{grevlex} x_1^2 x_2^4 x_3^2$ but $x_1^3 x_2 x_3^2 \succ_{grevlex} x_1^3 x_3^3$ though for a different reason than in the *lex* order. □

A term order defines a linear order in the terms of a multivariate polynomial by the order of their monomials. In particular, there is now a well-defined *leading term* in each polynomial in $\mathcal{P}^s$:

Definition 8.13. Consider $p(x) = \sum_{j \in J} \alpha_{j_1 .. j_s} x_1^{j_1} \ldots x_s^{j_s} \in \mathcal{P}^s$. With respect to some specified term order $\prec$, the term $\alpha_k x^k$ with $x^j \prec x^k$ for all $j \in J$, $j \neq k$, is the $\prec$-*leading term* (l.t.) of p, with the $\prec$-*leading coefficient* (l.c.) α_k and the $\prec$-*leading monomial* (l.m.) x^k. If it is evident or irrelevant which order is referred to, we will also say *order-leading term, etc.* or simply leading term, etc. □

Example 8.20, continued: For a univariate polynomial, the leading term is always the one with the highest power of the variable. In $p(x) = 2 x_1^3 x_2 x_3^3 - 3 x_1^2 x_2^4 x_3^2 \in \mathcal{P}^3$, for *grevlex* order, the leading term is the second one, with a leading coefficient -3; for *lex* order, the leading term is the first one, with leading coefficient 2 . □

The definition of leading monomials permits the introduction of an order-based *reduction procedure* for multivariate polynomials: In $\mathcal{P}^1$, when we divide $p(x)$ of degree n by $s(x)$ of degree $m \leq n$, we form (cf. (5.41) in section 5.3)

$$r_1(x) := p(x) - \frac{l.t.(p)}{l.t.(s)} \cdot s(x) \tag{8.46}$$

and know that $\deg r_1 < \deg p$. We continue this procedure with the successive remainders r_λ until $\deg r_\ell < \deg s$; then r_ℓ is the unique *remainder* of the division and the accumulated factors of s in (8.46) compose the *quotient q*. The procedure must terminate because the degree of the remainder decreases in each step. In terms of the ideal $\langle s \rangle$ and its quotient ring spanned by $\{1, x, \ldots, x^{m-1}\}$, this procedure can also be interpreted as the reduction of p to $\mathrm{NF}_{\langle s \rangle}[p] = r_\ell$.

In $\mathcal{P}^n$, $n > 1$, with a specified term order, assume that the order-l.m. of the divisor polynomial s divides the order-l.m. of the dividend polynomial p. Then we can perform the same reduction step (8.46); again we can be sure that $\mathrm{l.m.}(r_1) \prec \mathrm{l.m.}(p)$, which is now not simply a consequence of a diminished degree. Naturally, our initial assumption may cease to hold after one or few steps.

But when $\mathrm{l.m.}(s)$ not or no longer divides $\mathrm{l.m.}(r_\ell)$, it may still divide other terms in r_ℓ. If we reduce such terms in the same fashion, in the sequence of their term order, we will arrive at a polynomial r none of whose terms are divisible by $\mathrm{l.m.}(s)$. This r is the result of the *reduction of p by s* with respect to the term order $\prec$, which is often denoted by

$$p \xrightarrow{s} r \,;$$

if the term order is important, it should be denoted explicitly.

Proposition 8.29. For a specified term order, the result r of the reduction $p \overset{s}{\to} r$ in $\mathcal{P}^n$ is uniquely defined. A polynomial p is irreducible by s, with $\mathrm{l.m.}(s) = x^k$, iff none of the monomials x^j in p satisfy $j \geq k$, where $\geq$ is the componentwise partial order in $\mathbb{N}_0^s$.

Proof: Assume that p could also be reduced to r'; then r and r' must differ by a multiple of s : $r' = r + q \cdot s$. Clearly, $q\,s$ contains terms which are divisible by $\mathrm{l.m.}(s)$, e.g., the term $\mathrm{l.t.}(q)\cdot \mathrm{l.t.}(s)$ which cannot cancel in $q\,s$. Thus, r' cannot be the result of the reduction of p by s. The observation about the irreducibility is obvious. $\square$

If $x^k = \mathrm{l.m.}(s)$, all other monomials x^j in s must satisfy $x^j \prec x^k$. For the `tdeg` order $\prec_{grevlex}$, this means that either their total degree is smaller than $|k|$, or—if it is equal—$x^j \prec_{revlex} x^k$. Note that each exponent $k \in \mathbb{N}^s$ separates the grid hyperplane $|j| := \sum_{\sigma=1}^s j_\sigma = |k|$ into three disjoint sets: $\{x^j \prec_{revlex} x^k\}$, $\{x^j \succ_{revlex} x^k\}$, $\{x^k\}$; cf. Figure 8.5.

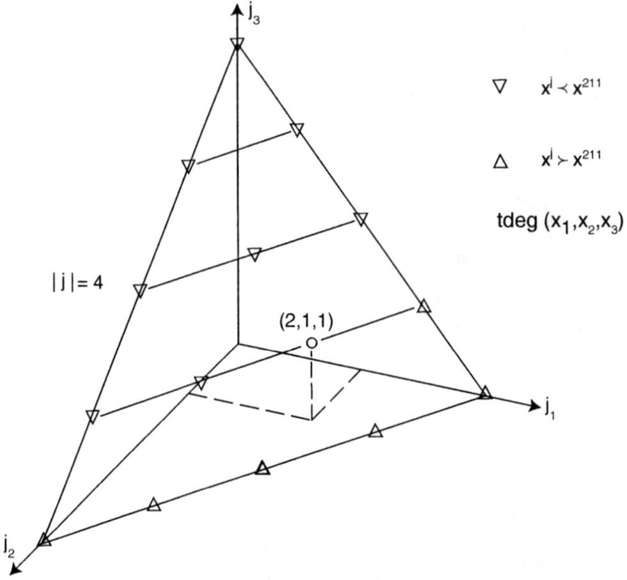

Figure 8.5.

Now consider a set $S = \{s_1, \ldots, s_n\} \subset (\mathcal{P}^s)^n$, $n > 1$, of divisor polynomials and define order-reduction by S as successive order-reduction by any one of the s_ν. The potential for reduction is now strongly increased, but it turns out that the result of the reduction need no longer be unique: It may happen that there exist polynomials $\bar{s}(x) = \sum_{\nu=1}^n q_\nu\, s_\nu \neq 0$ which are irreducible by any one of the s_ν. Then $r(x)$ and $r'(x) = r(x) + \bar{s}(x)$ may both be S-reduced forms of p. A simple example (from [2.10]) is the following:

Example 8.21: In $\mathcal{P}^2$, let $p(x, y) = x^2y + xy^2 + y^2$, $S = \{s_1, s_2\}$ with $s_1(x, y) = xy - 1$, $s_2(x, y) = y^2 - 1$. With `tdeg` order,

$$p = x^2y + xy^2 + y^2 \begin{cases} \overset{s_1}{\longrightarrow} & x + y + y^2 & \overset{s_2}{\longrightarrow} & x + y + 1 & =: r, \\[2mm] \overset{s_2}{\longrightarrow} & x^2y + x + 1 & \overset{s_1}{\longrightarrow} & 2x + 1 & =: r'. \end{cases}$$

Obviously, both r and r' are irreducible by S, and so is $\bar{s} = r' - r = x - y$. $\quad\square$

On the other hand, the normal form computation discussed in section 8.2.1 is a reduction very much like our present reduction procedure for the divisor set $\mathcal{B} = \{bb_j\}$, with their $\mathcal{N}$-leading monomials x^j not defined by a term order but by $x^j \in B[\mathcal{N}]$. There, the result of the reduction, viz. $\mathrm{NF}_{\langle\mathcal{B}\rangle}[p]$, has been unique. We had also observed that the elements bb_j of a border basis $\mathcal{B}$ must satisfy a large number of constraints; cf. section 8.2.2. This suggests that—in the present setting with a term order—the polynomials $s_\nu \in S$ also have to satisfy *constraints* if the reduction by S is to be unique.

8.4.2 Groebner Bases

Can we have a normal set $\mathcal{N}$ and associated border basis elements whose $\mathcal{N}$-leading monomials x^k are also their order-leading monomials with respect to some specified term order $\prec$? At first sight, this appears to require that $x^j \prec x^k$ for each $x^j \in \mathcal{N}$ and $x^k \in B[\mathcal{N}]$.

Definition 8.14. The *generic normal set* $\mathcal{N}_\prec$ of m elements in s variables for a specified term order $\prec$ is the set of the m $\prec$-lowest elements in $\mathcal{T}^s$. $\quad\square$

Example 8.22: Consider $s = 3$ and $m = 7$. For $\prec_{grevlex}$, we have $\mathcal{N}_\prec = \{1, x_3, x_2, x_1, x_3^2, x_2 x_3, x_1 x_3\}$ while *glex* generates the slightly different generic normal set $\{1, x_3, x_2, x_1, x_3^2, x_2 x_3, x_2^2\}$. For $\prec_{lex}$, we obtain the univariate generic normal set $\{1, x_3, x_3^2, x_3^3, x_3^4, x_3^5, x_3^6\}$. Generic lexicographic normal sets are always univariate which exhibits why the lexicographic order is inferior to a graded order in many respects.

The *symmetric* normal set $\mathcal{N}_{symm} = \{1, x_3, x_2, x_1, x_2 x_3, x_1 x_3, x_1 x_2\}$ (cf. Example 8.1) cannot be generic for *any* term order since this would require $x_2 x_3 \prec x_3^2$ and $x_2 x_3 \prec x_2^2$, but neither $x_2 \prec x_3$ nor $x_3 \prec x_2$ can support both relations if $\prec$ is compatible with multiplication. $\square$

Definition 8.15. The border basis $\mathcal{B}_\mathcal{N} = \{bb_k(x)\}$ for the normal set $\mathcal{N}$ is *order-compatible* for a specified term order $\prec$ iff

$$\mathcal{N}\text{-leading monomial of } bb_k = \prec\text{-leading monomial of } bb_k \quad \forall\, bb_k \in \mathcal{B}_\mathcal{N}. \quad\square \qquad (8.47)$$

For the generic normal set $\mathcal{N}_\prec \in \mathcal{T}^s(m)$ of some term order, a border basis is automatically order compatible; cf. Definition 8.14. For a nongeneric normal set $\mathcal{N} \in \mathcal{T}^s(m)$, compatibility with a specified term order can be obtained iff those monomials from $\mathcal{N}$ which succeed the $\mathcal{N}$-leading monomial of some bb_k *do not occur* in bb_k.

Example 8.22, continued: Consider the symmetric normal set $\mathcal{N}_{symm}$ and *grevlex* order. Of the monomials in $B[\mathcal{N}_{symm}]$, x_3^2 precedes $x_2 x_3$, $x_1 x_3$, $x_1 x_2$ and x_2^2 precedes $x_1 x_2$. Thus, in order to satisfy (8.47), bb_{002} must have no further quadratic terms at all and bb_{020} must not contain a term with $x_1 x_2$. (We will see later that this corresponds to certain degeneracies in the zero set $Z[P]$ of the system P.) $\quad\square$

In principle, *each* normal set in $\mathcal{T}^s(m)$ can support an order-compatible border basis for a specified term order, if the corresponding restrictions on the occurrence of coefficients in the bb_k are met. These restrictions imposed by order compatibility are independent of the intrinsic restrictions which we have derived in section 8.2.3 and supplementary to them. Thus, the specification and use of a term order *restricts* the liberty in representing a polynomial ideal for computational purposes. Actually, for a specified term order, the restriction imposed by (8.47) admits only *one uniquely defined*[14] border basis for the ideal of a regular polynomial system:

Theorem 8.30. For a 0-dimensional polynomial system $P \subset (\mathcal{P}^s)^s$ and a specified term order $\prec$, there exists a unique $\prec$-compatible border basis $\mathcal{B}[\langle P \rangle]$.

Proof: Since the unique existence of a Groebner basis (cf. Definition 8.16 below) has been proved in different ways and is a well-known fact, we give only an intuitive survey of a proof.

Assume at first that the generic normal set C for the specified term order is a feasible monomial basis for $\mathcal{R}[\langle P \rangle]$. As we have just observed, the border basis $\mathcal{B}_{\mathcal{N}_\prec}[\langle P \rangle]$ is automatically order-compatible. There cannot exist an order-compatible border basis for a different normal set: This normal set would have to include (at least) one of the leading monomials of $\mathcal{B}_{\mathcal{N}_\prec}[\langle P \rangle]$, say x^k, and a monomial x^j of $\mathcal{N}_\prec$ would now function as leading monomial. But this requires that x^j was present in the border basis element bb_k which implies that x^k is present in the new border basis element bb_j, which contradicts the term order.

Now assume that $\mathcal{N}_\prec$ is not a feasible normal set for $\mathcal{R}[\langle P \rangle]$. We form new normal sets by successively replacing a monomial of highest term order in the edge of $\mathcal{N}_\prec$ by a lowest order monomial in the previous border set (respecting the closedness of the result, of course). Since almost all sets in $\mathcal{T}^s(m)$ are feasible normal sets for a specified system P, we must reach a feasible normal set after finitely many steps. Assume that the replacement of *one* element from $\partial \mathcal{N}_\prec$ (the highest one in term order) by the lowest order monomial in $B[\mathcal{N}_\prec]$ is sufficient. This means that the violation of the feasibility of $\mathcal{N}_\prec$ has been caused only by the removed monomial $x^{k_m} = b_m(x) \in \mathcal{N}_\prec$ which implies

$$b_m(z_\mu) \;\in\; \mathrm{span}\,\{b_1(z_\mu), \ldots, b_{m-1}(z_\mu)\}, \qquad \text{for all } z_\mu \in Z[P],$$

and the existence of coefficients $\beta_{m\mu}$ such that

$$\mathrm{NF}_{\langle P \rangle}[b_m(x)] \;=\; \sum_{\mu=1}^{m-1} \beta_{m\mu}\, b_\mu(x)\,.$$

Thus, in the border basis for the modified normal set $\mathcal{N}$, the border basis element with the $\mathcal{N}$-leading monomial b_m (which is now in $B[\mathcal{N}]$) contains only monomials $b_\mu \prec b_m$ but *not* the monomial $b_{m+1} \succ b_m$ incorporated into $\mathcal{N}$ in place of b_m. Thus the $\mathcal{N}$-border basis is order compatible. Reversely, this argument shows how the vanishing of the coefficient of b_{m+1} in the border basis element with the $\mathcal{N}$-leading monomial b_m implies that the previous normal set cannot have been feasible; this establishes uniqueness.   □

Definition 8.16. For a specified term order $\prec$, the unique $\prec$-compatible border basis $\mathcal{B}[\langle P \rangle]$ is called the *border Groebner basis* $\mathcal{G}_\prec[\langle P \rangle]$ of $\langle P \rangle$. (The subscript on $\mathcal{G}$ will be omitted if the order is evident.)   □

Example 8.23: Consider the two quadratic polynomials in 2 variables which describe two

---

[14]Here and in the following, *uniqueness* always refers to some *normalized* form of the polynomials, e.g., with the coefficient 1 in the $\mathcal{N}$-leading term.

axiparallel ellipses:

$$P := \begin{cases} p_1(x, y) &= x^2 + 4y^2 - 4\,; \\ p_2(x, y) &= 9x^2 + y^2 - 2y - 8\,. \end{cases}$$

Obviously, $m = 4$, and the generic normal set for `tdeg` (with $x \succ y$) is $\mathcal{N}_\prec = \{1, y, x, y^2\}$. But the 4 zeros of P form two pairs with equal y-components (and two pairs with equal x-components), which makes $\mathcal{N}_\prec$ infeasible as may easily be verified. When we replace y^2 by the lowest monomial xy in $B[\mathcal{N}_\prec]$, we obtain the normal set $\mathcal{N} = \{1, y, x, xy\}$, with $B[\mathcal{N}] = \{y^2, x^2, xy^2, x^2y\}$. It is easily checked that the feasibility of this normal set is not impaired by the symmetries of the zeros. By the argument in the proof of Theorem 8.30, the border basis $\mathcal{B}_{\mathcal{N}}[P]$ should be order compatible.

Solving for x^2, y^2 in P yields

$$bb_{20}(x, y) = x^2 - \tfrac{8}{35}y - \tfrac{5}{4}\,, \qquad bb_{02}(x, y) = y^2 + \tfrac{2}{35}y - \tfrac{5}{4}\,;$$

substitution of bb_{02} into $y\, bb_{20}$ and $x\, bb_{02}$ yield

$$bb_{21}(x, y) = x^2 y - \tfrac{6061}{4900}y - \tfrac{2}{7}\,, \qquad bb_{12}(x, y) = xy^2 + \tfrac{2}{35}xy - \tfrac{5}{4}x\,.$$

The critical border basis element is b_{02} with the $\mathcal{N}$-leading monomial y^2; it must not contain the monomial $xy \succ y^2$. Since this is the case, the above $\mathcal{B}_{\mathcal{N}}[\langle P \rangle]$ is indeed order compatible and thus the border Groebner basis $\mathcal{G}_\prec[\langle P \rangle]$. $\quad\square$

The order-compatibility of a border Groebner basis has the following important consequence:

Theorem 8.31. Let $\mathcal{N}$ be the normal set associated with the border Groebner basis $\mathcal{G}[\langle P \rangle]$ for some specified term order. Consider the corner set $C[\mathcal{N}] \subset B[\mathcal{N}]$; cf. Definition 2.21, (2.57). The elements of $\mathcal{G}[\langle P \rangle]$ with leading monomials in $C[\mathcal{N}]$ constitute a complete basis of $\langle P \rangle$.
Proof: Denote the corner subset of $\mathcal{G}[\langle P \rangle]$ by $C[\langle P \rangle]$. We show that the remaining elements of $\mathcal{G}[\langle P \rangle]$ are uniquely determined by the elements in $C[\langle P \rangle]$. By the definition of $C[\mathcal{N}]$, each element in $B[\mathcal{N}] \setminus C[\mathcal{N}]$ is a monomial multiple of some element in $C[\mathcal{N}]$. Since the corner basis elements cb_k are order compatible, this is also true, by (8.45) for all polynomials which are monomial multiples of them.

Now we consider the monomials in $B[\mathcal{N}] \setminus C[\mathcal{N}]$ in increasing term order and determine their associated border basis elements by taking appropriate multiples of elements in $C[\langle P \rangle]$. Due to the order-compatibility of the elements in $C[\langle P \rangle]$ and the consideration of the border basis elements in increasing term order of their leading monomials, the multiples will only contain monomials from $\mathcal{N}$ or leading monomials of basis elements in $C[\langle P \rangle]$ or leading monomials of border basis elements already processed and represented. Thus, all border basis elements not in $C[\langle P \rangle]$ can be recursively determined. $\quad\square$

The fact that the "corner basis" $C[\langle P \rangle]$ may be extended into a full border basis implies that the multiplication matrices of $\mathcal{R}[\langle P \rangle]$ may also be determined from the polynomials in $C[\langle P \rangle]$. Thus, the corner basis contains the full information necessary for the computation of the zero set of the underlying ideal.

Definition 8.17. The corner basis subset $C_\prec[\langle P \rangle]$ of the border Groebner basis $G_\prec[\langle P \rangle]$ is called the *reduced Groebner basis* of P. $\quad\square$

In section 8.4.1, we had observed that order-based reduction by a polynomial set S is, generally, not unique. According to section 8.2.1, uniqueness prevails for a set which is an order-compatible border basis. By reversing the argument in the proof of Theorem 8.31, we see that the non-corner elements of a Groebner basis may be reduced to 0 by the corner elements. Thus, order-based reduction by a reduced Groebner basis is unique.

Example 8.23, continued: The corner set $C[\mathcal{N}]$ is $\{y^2, x^2\}$ and the corner basis $C_{grevlex}[\langle P \rangle]$ is $\{bb_{02}, bb_{20}\}$. The remaining two border basis elements were formed as $bb_{12} := x\, bb_{02}$ and $bb_{21} := y\, bb_{20} + \frac{8}{35}\, bb_{02}$, which displays their reduction to 0 by $\{bb_{02}, bb_{20}\}$. $\quad\square$

Example 8.24: Consider the polynomial system

$$
P := \begin{cases}
p_1(x, y) &=& x^2 + \alpha_{02}^{(1)} y^2 + \alpha_{11}^{(1)} x\, y + \alpha_{10}^{(1)} x + \alpha_{01}^{(1)} y + \alpha_{00}^{(1)}, \\
p_2(x, y) &=& y^3 + \alpha_{02}^{(2)} y^2 + \alpha_{11}^{(2)} x\, y + \alpha_{10}^{(2)} x + \alpha_{01}^{(2)} y + \alpha_{00}^{(2)}.
\end{cases}
$$

Because of the absence of xy^2 terms, we may conjecture that P is the corner subset of a `tdeg`-compatible border basis of $\langle P \rangle$ for the normal set $\{1, y, x, y^2, xy, xy^2\}$. To confirm that conjecture for generic coefficients, our present knowledge requires that we extend P to a full border basis and check the commutativity of the multiplication matrices. This is an awkward job, but it leads to a positive result. Thus, P is the reduced Groebner basis $C[\langle P \rangle]$. $\quad\square$

In the algebraic literature, the term Groebner basis denotes any basis of $\langle P \rangle$ which is a superset of the *reduced* Groebner basis $C_\prec[\langle P \rangle]$; at the same time, the general term is often directly associated with the reduced Groebner basis. Also, most computer algebra software generates the reduced Groebner basis $C_\prec[\langle P \rangle]$, e.g., the procedure `gbasis` in Maple. While we will follow this convention, the border Groebner basis $G_\prec[\langle P \rangle]$ is also of central importance for us.

The property of border Groebner bases formulated in Theorem 8.31 does not *characterize* Groebner bases; there are polynomial systems P and normal sets $\mathcal{N}$ where the associated border basis $B_\mathcal{N}[\langle P \rangle]$ is not order-compatible for any term order but where the subset $C_\mathcal{N}[\langle P \rangle] \subset B_\mathcal{N}[\langle P \rangle]$ of the basis elements with leading monomials in $C[\mathcal{N}]$ constitutes a basis for $\langle P \rangle$. But this is not true in general: It is easy to construct examples where the corner subset of a border basis does not generate the same ideal. In this sense, Groebner bases are distinguished border bases.

This "shortcoming" of arbitrary border bases is irrelevant in most computational contexts: In the algorithmic determination of normal sets and the associated border bases from some polynomial system P, the corner basis elements do not occur separately from the remaining border basis elements and the full border basis is generally determined anyway; cf. section 10.1.2. In the multiplication matrices which dominate computational tasks for the ideal $\langle P \rangle$, there is no distinction between corner and other border monomials of a normal set; generally, they require the full border basis for their determination.

The uniqueness (modulo normalization) of the reduced Groebner basis $C_\prec[\mathcal{I}]$ for a specified term order $\prec$ (cf. Theorems 8.30 and 8.31) permits an easy decision about the identity of the ideals generated by two polynomial systems P_1, $P_2 \subset \mathcal{P}^s$: Iff $C_\prec[\langle P_1 \rangle] \equiv C_\prec[\langle P_2 \rangle]$, for an arbitrary fixed term order, the two polynomial systems generate the same polynomial ideal. Naturally, this is also a decision procedure for the identity of their zero sets.

8.4.3 Direct Characterization of Reduced Groebner Bases

According to Definition 8.17, the considerations of the previous section and Corollary 8.12, a reduced Groebner basis $\mathcal{C} = \{g_1, \ldots, g_k\} \subset \mathcal{P}^s$, $k \geq s$, must satisfy the following requirements:

(1) There exists a term order $\prec$ such the set L of the $\prec$-leading monomials of the g_κ constitutes the corner set $C[\mathcal{N}]$ of a closed finite set $\mathcal{N} \subset T^s$;

(2) All S-polynomials $S[g_{\kappa_1}, g_{\kappa_2}]$, $g_{\kappa_1} \neq g_{\kappa_2} \in \mathcal{C}$, can be reduced to 0 by $\mathcal{C}$;

(3) Extend $\mathcal{C}$ to a border basis $\mathcal{G}_\mathcal{N}$: All S-polynomials $S(\bar{g}_{\kappa_1}, \bar{g}_{\kappa_2})$, $\bar{g}_{\kappa_1} \neq \bar{g}_{\kappa_2} \in \mathcal{G}_\mathcal{N}$, can be reduced to 0 by $\mathcal{C}$.

Proposition 8.32. $L = \{x^{j_\kappa}, \ \kappa = 1(1)k\} \subset T^s$ is the corner set of a finite closed subset of T^s iff

(i) L contains a "pure power" $x_\sigma^{m_\sigma}$, $m_\sigma \in \mathbb{N}$, of each variable x_σ, $\sigma = 1(1)s$;

(ii) none of the $x^{j_\kappa} \in L$ divides another monomial in L.

Proof: Let $\mathcal{I}_L := \langle L \rangle$ be the *monomial ideal* generated by L, i.e. the set of all polynomial multiples of the x^{j_κ}. The set $\mathcal{N}_L := T^s \setminus \mathcal{I}_L$ is the only candidate for an $\mathcal{N}$ such that $C[\mathcal{N}] = L$; cf. section 2.5.1.

If there is no element $x_\sigma^{m_\sigma}$ in L for some σ, then $\mathcal{I}_L$ cannot contain any power x_σ^ℓ, $\ell \in \mathbb{N}$, and $T^s \setminus \mathcal{I}_L$ is an infinite set. With (i), on the other hand, all elements x^j, $j \geq (m_1, \ldots, m_s)$ are in $\mathcal{I}_L$ and $T^s \setminus \mathcal{I}_L$ is finite.

Let $x^{j_{\kappa'}} | x^{j_\kappa}$ so that $x^{j_\kappa} = x^j x^{j_{\kappa'}}$. Since all $x^j x^{j_{\kappa'}} \in \mathcal{I}_L$, x^{j_κ} has a negative neighbor $\notin \mathcal{N}$ so that $x^{j_\kappa} \notin C[\mathcal{N}]$; cf. Definitions 2.17 and 2.21. With (ii), on the other hand, none of the x^{j_κ} can have a negative neighbor in $\mathcal{I}_L$. $\square$

The insistence on a *finite* normal set stems from our current restriction to 0-dimensional ideals. Groebner bases may also be defined for positive-dimensional ideals; then condition (i) is irrelevant.

When we have a set L satisfying condition (ii), the monomial ideal $\mathcal{I}_L$ contains all monomials in the union of the closed positive orthants with vertices at the $x^{j_\kappa} \in L$; cf. Figure 8.6. Condition (ii) is often expressed by the intuitive phrase "the x^{j_κ} must form a *staircase*" and $\mathcal{N}_L$ is called "the set under the staircase."

A verification of requirement 2 appears to be demanding since there are $(k-1)k/2$ combinations of k corner basis elements g_κ. However, the rules of Proposition 8.14 apply immediately to the web of the monomials of the corner set $C[\mathcal{N}]$ where only edges of type (ii) occur so that the number of necessary reductions is diminished. Clearly, when the Groebner basis computation does not assume *a priori knowledge* about the existence of a *finite* normal set and its dimension, the extension in Proposition 8.15 *cannot* be applied.

Proposition 8.33. The above requirement (3) is automatically satisfied if requirements (1) and (2) are satisfied.

Proof: The additional basis elements $\bar{g}_{\bar{\kappa}}$ are formed by multiplying the appropriate g_κ by the appropriate monomial and reducing the nonleading monomials $\notin \mathcal{N}$ of the product with $\mathcal{C}$. Let (w.l.o.g.)

$$\bar{g}_i(x) = x^{j_i} g_i(x) + \sum_\kappa \beta_{i\kappa} g_\kappa(x), \quad i = 1, 2,$$

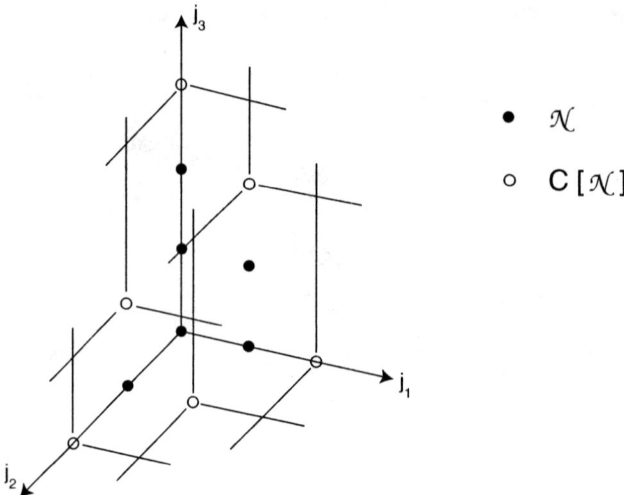

Figure 8.6.

so that the $\bar{g}_i$ have the leading monomials $x^{(j_i + \bar{j}_i)}$, and let j_{12} be the l.c.m. of these leading monomials. Then

$$S[\bar{g}_1, \bar{g}_2] \;=\; x^{j_{12}-j_1-\bar{j}_1}\,\bar{g}_1(x) - x^{j_{12}-j_2-\bar{j}_2}\,\bar{g}_2(x) \;=\; x^{j_{12}-j_1}\,g_1(x) - x^{j_{12}-j_2}\,g_2(x) + \text{multiples of } g_\kappa \,.$$

Since l.c.m.(x^{j_1}, x^{j_2}) divides j_{12}, the first two terms are a multiple of $S[g_1, g_2]$ whose reduction to 0 has been assumed. $\square$

Thus we have proved

Theorem 8.34. A set of $k \geq s$ polynomials $g_\kappa \in \mathcal{P}^s$ is the reduced Groebner basis $\mathcal{C}_{\prec}[\mathcal{I}]$ of a 0-dimensional ideal $\mathcal{I} \subset \mathcal{P}^s$ with respect to the term order $\prec$ iff

(1) the $\prec$-leading monomials x^{j_κ} of the g_κ form a set L which satisfies the assumptions of Proposition 8.32;

(2) all S-polynomials of the g_κ reduce to 0 with $\mathcal{C}_{\prec}$.

Example 8.24, continued: With our newly gained insights, we can immediately recognize that the polynomial system P is the reduced Groebner basis of the ideal $\langle P \rangle$, with respect to the term order tdeg, $x \succ y$: For this term order, the set L of leading monomials of P is $\{x^2, y^3\}$; L satisfies the conditions of Proposition 8.32 for the 6-element normal set $\mathcal{N} = \{1, y, x, y^2, xy, xy^2\}$. Note that $\mathcal{N}$ is not identical with the support of P; it is precisely the fact that xy^2 is in $\mathcal{N}$ but not in the support of P which permits the interpretation. The only S-polynomial of P reduces to zero with P for arbitrary choices of the coefficients according to rule (a) of Proposition 8.14. This also establishes that $\langle P \rangle$ has 6 zeros. $\square$

In virtually all texts on computational polynomial algebra, the contents of Theorem 8.34 (or some equivalent formulations) serve as the *Definition* of a reduced Groebner basis. Moreover, any set of polynomials which satisfies (2), with leading monomials defined by a *term order* (cf.

Definition 8.13), is usually called a Groebner basis. Thus, our definitions of a Groebner basis and of a reduced Groebner basis describe the same concepts but from a different end:

According to our observations in Chapter 2 and in section 8.1, computational polynomial algebra happens essentially in the quotient ring of a polynomial ideal rather than in the ideal itself. Therefore, our central focus is on monomial bases for quotient rings and the associated multiplication matrices; this makes it natural to consider bases for the ideal which carry the same symbolic and numeric information, viz. *border bases* (cf. section 2.5). All this proceeds without any consideration of a term order which is not a central concept from the algebraic point of view. When a term order is introduced, one particular border basis is singled out for each specific order, which is the Groebner (border) basis. For this particular border basis, a reduction to a corner basis is always possible, which is the reduced Groebner basis.

When one follows the usual approach which focuses on reduced Groebner bases, it is not so obvious that the concept of an ideal basis which admits the full set of potential monomial bases of the quotient ring, is not that of a corner basis but a border basis. As is easily established by counter examples, corner bases do not generally exist for an ideal $\mathcal{I} \subset \mathcal{P}^s$ with all normal sets which are feasible as bases for the quotient ring $\mathcal{R}[\mathcal{I}]$.

Example 8.25: Consider an ideal $\mathcal{I} \in \mathcal{P}^2(6)$, with $\mathcal{N} = \{1, y, x, y^2, xy, xy^2\}$ as a feasible normal set, and the associated border basis $\mathcal{B}_{\mathcal{N}}[\mathcal{I}]$. Because $\mathcal{N}$ is quasi-univariate with respect to y, we know that the two polynomials from $\mathcal{B}_{\mathcal{N}}[\mathcal{I}]$ with $\mathcal{N}$-leading monomials y^3, xy^3 constitute a complete intersection basis of $\mathcal{I}$; cf. Proposition 8.20. Assume that the corner subset $\mathcal{C}_{\mathcal{N}}$ of $\mathcal{B}_{\mathcal{N}}$, i.e. the two polynomials with $\mathcal{N}$-leading monomials x^2 and y^3, would constitute a basis for $\mathcal{I}$. We may assume that the support of both of these polynomials contains all 6 monomials in $\mathcal{N}$, because we can choose the coefficients of the quasi-univariate basis arbitrarily; cf. Theorem 8.5. When we now form the BKK-number of $\mathcal{C}$ (cf. section 8.3.4), we obtain $\mathrm{BKK}(\mathcal{C}_{\mathcal{N}}) = 7$. Thus, there are polynomial systems in $\mathcal{P}^2(6)$ for which $\mathcal{N}$ is a feasible normal set but for which there does not exist an $\mathcal{N}$-corner basis. □

8.4.4 Discontinuous Dependence of Groebner Bases on P

While the uniqueness of the reduced Groebner basis $\mathcal{C}_{\prec}[\langle P \rangle]$ for $P \in \mathcal{P}^s$ is a desirable property, it carries a fundamental drawback with it: Groebner bases cannot be *continuous functions* of P uniformly. We explain this vague statement at first with an intuitive example:

Example 8.26: Consider the family of polynomial systems

$$P_\varepsilon = \begin{cases} p_1(x, y; \varepsilon) &= x^2 + \varepsilon\, x\, y + y^2 - 1\,, \\ p_2(x, y) &= y^3 - 3\, x^2\, y\,. \end{cases}$$

For small $|\varepsilon|$, $p_1 = 0$ describes a slightly distorted unit circle while $p_2 = 0$ consists of three straight lines through the origin under $0^0, 60^0, 120^0$; cf. Figure 8.7. It is obvious that the 6 zeros of P_ε depend smoothly on ε as ε varies in a neighborhood of 0. In this sense (cf. also section 8.3.2), the ideal $\langle P \rangle$ varies smoothly for small $|\varepsilon|$.

For `tdeg` order with $x \succ y$, the reduced Groebner basis $\mathcal{C}[\langle P_\varepsilon \rangle]$, with indeterminate ε, consists of the 3 polynomials

$$g_1 = p_1(x, y; \varepsilon)\,, \quad g_2 = xy^2 + \frac{4}{3\varepsilon}y^3 - \frac{1}{\varepsilon}y\,, \quad g_3 = y^4 + \frac{9\varepsilon}{16 - 3\varepsilon^2}xy - \frac{12}{16 - 3\varepsilon^2}y^2\,;$$

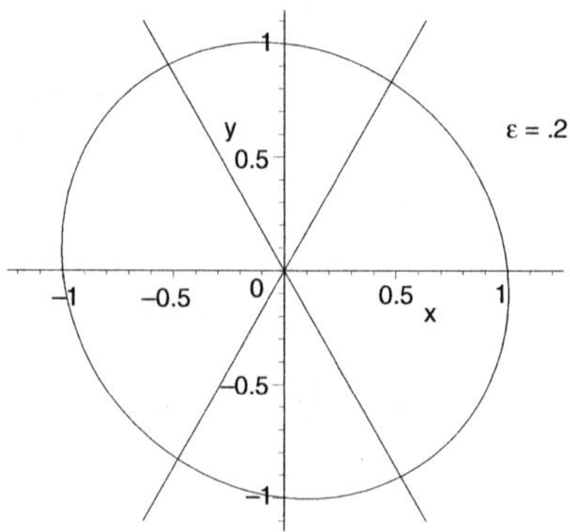

Figure 8.7.

the associated normal set is $\mathcal{N}_\varepsilon = \{1, y, x, y^2, xy, y^3\}$ with $C[\mathcal{N}_\varepsilon] = \{x^2, xy^2, y^4\}$. The multiplication matrices of $\mathcal{R}[\langle P_\varepsilon \rangle]$ with respect to the basis vector $\mathbf{b} = (1, y, x, y^2, xy, y^3)^T$ are

$$
A_x = \begin{pmatrix}
0 & 0 & 1 & 0 & 0 & 0 \\
0 & 0 & 0 & 0 & 1 & 0 \\
1 & 0 & 0 & -1 & -\varepsilon & 0 \\
0 & \frac{1}{\varepsilon} & 0 & 0 & 0 & \frac{-4}{3\varepsilon} \\
0 & 0 & 0 & 0 & 0 & \frac{1}{3} \\
0 & 0 & 0 & \frac{-3\varepsilon}{16-3\varepsilon^2} & \frac{12}{16-3\varepsilon^2} & 0
\end{pmatrix}, \quad
A_y = \begin{pmatrix}
0 & 1 & 0 & 0 & 0 & 0 \\
0 & 0 & 0 & 1 & 0 & 0 \\
0 & 0 & 0 & 0 & 1 & 0 \\
0 & 0 & 0 & 0 & 0 & 1 \\
0 & \frac{1}{\varepsilon} & 0 & 0 & 0 & \frac{-4}{3\varepsilon} \\
0 & 0 & 0 & \frac{12}{16-3\varepsilon^2} & \frac{-9\varepsilon}{16-3\varepsilon^2} & 0
\end{pmatrix}.
$$

It is immediately obvious that this representation of the ideal $\langle P_\varepsilon \rangle$ can only be valid for $\varepsilon \neq 0$. As $|\varepsilon|$ becomes small, the basis polynomial g_2 diverges if we leave its leading coefficient normalized; otherwise, its leading coefficient tends to 0. *Independently of this normalization*, the matrices A_x and A_y diverge and the condition number of their eigenproblems grows like $O(1/\varepsilon)$! Thus, while the representation remains mathematically correct for $\varepsilon \neq 0$, its computational use for numerically specified, tiny ε is not advisable.

For $\varepsilon = 0$ and the same term order `tdeg`, the reduced Groebner basis $C[\langle P_0 \rangle]$ becomes

$$
g_1 = p_1(x, y; 0) = x^2 + y^2 - 1, \quad g_2 = y^3 - \frac{3}{4} y,
$$

with $\mathcal{N}_0 = \{1, y, x, y^2, xy, xy^2\}$ and $C[\mathcal{N}_0] = \{x^2, y^3\}$. The multiplication matrices of $\mathcal{R}[\langle P_0 \rangle]$

with respect to the monomial basis $\mathcal{N}_0$ are

$$
A_x^{(0)} = \begin{pmatrix}
0 & 0 & 1 & 0 & 0 & 0 \\
0 & 0 & 0 & 0 & 1 & 0 \\
1 & 0 & 0 & -1 & 0 & 0 \\
0 & 0 & 0 & 0 & 0 & 1 \\
0 & \frac{1}{4} & 0 & 0 & 0 & 0 \\
0 & 0 & 0 & \frac{1}{4} & 0 & 0
\end{pmatrix}, \quad
A_y^{(0)} = \begin{pmatrix}
0 & 1 & 0 & 0 & 0 & 0 \\
0 & 0 & 0 & 1 & 0 & 0 \\
0 & 0 & 0 & 0 & 1 & 0 \\
0 & \frac{3}{4} & 0 & 0 & 0 & 0 \\
0 & 0 & 0 & 0 & 0 & 1 \\
0 & 0 & 0 & 0 & \frac{3}{4} & 0
\end{pmatrix}.
$$

There is no immediate way to understand the transition of $C[\langle P_\varepsilon \rangle]$, $\mathcal{N}_\varepsilon$, A_x, A_y to $C[\langle P_0 \rangle]$, $\mathcal{N}_0$, $A_x^{(0)}$, $A_y^{(0)}$ in the smooth fashion in which it occurs in the system P_ε and in its zeros. $\quad\square$

In Exercise 2.5-3 b, we have observed that, for $s \geq 2$, $m \geq 2$, there does not exist a normal set which is uniformly feasible for all ideals in $\mathcal{P}^s(m)$. Thus, when we "parametrize" the ideals in $\mathcal{P}^s(m)$ by the components of their zeros, the map from a point in this parameter space to the *unique* normal set $\mathcal{N}_C \in T^s(m)$ which supports the Groebner basis for that particular zero set must be *discontinuous* because the image space $T^s(m)$ is a discrete set.

For fixed s, m and a specified term order $\prec$, consider the generic normal set $\mathcal{N}_g \in T^s(m)$; cf. Definition 8.14. In the parameter space $\mathbb{C}^{s \times m}$, the relation (2.53) defines the singularity manifold $S_{\mathcal{N}_g}$ of codimension 1 which contains the zero locations for which $\mathcal{N}_g$ is not feasible. When the zero set moves onto $S_{\mathcal{N}_g}$, the image $\mathcal{N}_C$ *jumps* from $\mathcal{N}_g$ to another normal set $\mathcal{N}_1$ which is distinguished by having one element in its corner set $C[\mathcal{N}_1]$ which precedes one element in $\mathcal{N}_1$. But $\mathcal{N}_1$ has its own singularity manifold $S_{\mathcal{N}_1}$ which (excepting some trivial situations) intersects with $S_{\mathcal{N}_g}$ in a manifold of codimension 2. When the zero set moves onto this manifold—either from within $S_{\mathcal{N}_g}$ or from outside—the Groebner normal set $\mathcal{N}_C$ jumps to a different element $\mathcal{N}_2$ of $T^s(m)$, with two order violations in its corner set, etc.

This mechanism generates a tree of manifolds in $\mathbb{C}^{s \times m}$ of higher and higher codimensions dg; each of these singularity manifolds is embedded in those which preceed it in the tree. The value of dg denotes, in an informal sense, the "degree of degeneracy" of the associated zero sets. In spite of this recursive embedding, a zero set of a high degree of degeneracy can be approached by a path consisting of generic zero sets; thus $\mathcal{N}_C$ can jump over arbitrarily many intermediate values of dg.

Definition 8.18. A singularity of a normal set representation of a polynomial ideal of the kind described above is called a *representation singularity*.

Example 8.26, continued: Due to the symmetry of pairs of zeros of P_ε with respect to the origin, the `tdeg` Groebner normal set $\mathcal{N}_\varepsilon$ is not generic for that term order but has $dg = 1$: The corner element x^2 precedes the normal set element y^3. At $\varepsilon = 0$, the zero set becomes also symmetric to the x- and y-axes; this zero location lies on the singularity manifold $S_{\mathcal{N}_\varepsilon}$ of $\mathcal{N}_\varepsilon$ and induces the jump of the Groebner normal set to $\mathcal{N}_0$, with $dg = 2$: In $C[\mathcal{N}_0]$, the two corner monomials x^2, y^3 precede the normal set monomial xy^2. The zero positions at the corners of the unit hexagon could also have been reached in a continuous way from an unsymmetric generic position of the zeros, inducing an immediate jump from $\mathcal{N}_g$ to $\mathcal{N}_0$.

When we choose the term order `tdeg(y,x)`, it turns out that the transition from P_ε to P_0 does not affect the normal set $\mathcal{N}_\varepsilon' = \{1, x, y, x^2, xy, x^3\}$ of the associated Groebner bases; here, the zero location of P_0 does not lie on the singularity manifold of $\mathcal{N}_\varepsilon'$. For both lexicographic term orders `plex(x,y)` and `plex(y,x)`, on the other hand, the degeneration degree of the normal set jumps in the transition from P_ε to P_0 and we have analogous discontinuity phenomena as described above. $\square$

This discontinuity dilemma becomes a source of serious difficulties when we use coefficients of limited accuracy and approximate computation; it *cannot be avoided* for Groebner bases. Its origin is the insistence on order-compatibility in the selection of a basis for $\langle P \rangle$ which is implicit in the Groebner basis concept. (It is similar to the fact that we must have a singularity in the representation of a closed 2-manifold in 3-space when we insist on $z = \varphi(x, y)$.) Without this restriction, we can always choose a normal set $\mathcal{N} \in T^s(m)$ whose singularity manifold $S_\mathcal{N}$ in $\mathbb{C}^{s \times m}$ is sufficiently removed from the location of the zero set $Z[\langle P \rangle]$ so that it can be safely used not only for P but also for all systems $\tilde{P}$ in a suitable neighborhood of P, due to the continuous behavior of the zeros.

The fact that a zero set $Z[\mathcal{I}]$ is close to the singularity manifold $S_{\mathcal{N}_\prec}$ of the unique normal set $\mathcal{N}_\prec$ associated with the Groebner basis $\mathcal{G}_\prec[\mathcal{I}]$ is displayed by the coefficients of $\mathcal{G}_\prec[\mathcal{I}]$ or—equivalently—by the elements of the multiplication matrices A_σ for $\mathcal{R}[\mathcal{I}]$ for the monomial basis $\mathcal{N}_\prec$: The closeness of $S_{\mathcal{N}_\prec}$ lets some elements in the A_σ *diverge* towards ∞. Depending on the normalization employed for $\mathcal{G}_\prec[\mathcal{I}]$, this means that either some coefficients in the Groebner basis have extremely large moduli (for l.c.=1) or that the modulus of some leading coefficients is excessively small; cf. the polynomial g_2 in $\mathcal{C}[\langle P_\varepsilon \rangle]$ in Example 8.26 above.

More generally, consider a regular system $P_0 \in (\mathcal{P}^s)^s$ which is *degenerate* in the following sense: For a specified term order, the Groebner basis $\mathcal{G}_\prec[\langle P_0 \rangle]$ employs a normal set $\mathcal{N}_0$ different from the generic normal set $\mathcal{N}_g$ for $\prec$; this implies that the zero set $Z[P_0]$ lies on the singularity manifold $S_{\mathcal{N}_g}$. When we modify the coefficients of P_0 (keeping the convex hull of the supports J_ν of the polynomials in P_0 invariant; cf. section 8.4.3) so that the zero set $Z[\tilde{P}]$ moves away from $S_{\mathcal{N}_g}$, the border basis $\mathcal{B}_{\mathcal{N}_0}[\langle \tilde{P} \rangle]$ with the *original* normal set $\mathcal{N}_0$ cannot remain a Groebner basis: Order-incompatible monomials from $\mathcal{N}_0$ will appear, with small coefficients, in one or several of the elements of $\mathcal{B}_{\mathcal{N}_0}$ some of whose $\mathcal{N}_0$-leading monomials are $\prec$-lower than some elements in $\mathcal{N}_0$, due to the assumed degeneracy.

Definition 8.19. In the situation just described, the border basis $\mathcal{B}_{\mathcal{N}_0}[\langle \tilde{P} \rangle]$ which is the continuous extensions of $\mathcal{G}_\prec[\langle P_0 \rangle]$ to neighboring polynomial systems $\tilde{P}$ of P_0, is called an *extended Groebner basis*. $\square$

A fuller discussion of this concept appears in section 10.1.2. The term "extended Groebner basis" has been introduced by the author in his paper [8.6] where it was used for the extension of the *reduced* Groebner basis, i.e. for the *corner subset* of $\mathcal{B}_{\mathcal{N}_0}[\langle \tilde{P} \rangle]$ in the above context. There it has been proved that—for systems $\tilde{P}$ sufficiently close to the degenerate system P_0—this extended reduced Groebner basis is indeed a *basis* of $\langle \tilde{P} \rangle$; it determines the remaining elements of $\mathcal{B}_{\mathcal{N}_0}[\langle \tilde{P} \rangle]$ and thus the multiplication matrices of $\mathcal{R}[\langle \tilde{P} \rangle]$ with respect to the monomial basis $\mathcal{N}_0$.

Example 8.26, continued: The *extended* reduced Groebner basis $\mathcal{C}_{\mathcal{N}_0}[\langle P_\varepsilon \rangle]$ is easily found as

$$cb_1(x, y) \; = \; x^2 + \varepsilon\, x\, y + y^2 - 1, \qquad cb_2(x, y) \; = \; y^3 - \frac{3}{4}\, y + \frac{3}{4}\, \varepsilon\, x\, y^2 \, ;$$

it is well-defined for $|\varepsilon| < 4/\sqrt{3} \approx 2.3$; the multiplication matrices of $\mathcal{R}[\langle P_\varepsilon \rangle]$ with respect to the normal set vector $\mathbf{b}_0 = (1, y, x, y^2, xy, xy^2)^T$ which also display the remaining elements of the extended Groebner basis $\mathcal{B}_{\mathcal{N}_0}[\langle P_\varepsilon \rangle]$ are

$$
A_x = \begin{pmatrix}
0 & 0 & 1 & 0 & 0 & 0 \\
0 & 0 & 0 & 0 & 1 & 0 \\
1 & 0 & 0 & -1 & -\varepsilon & 0 \\
0 & 0 & 0 & 0 & 0 & 1 \\
0 & \frac{1}{4} & 0 & 0 & 0 & -\frac{\varepsilon}{4} \\
0 & 0 & 0 & \frac{4}{16-3\varepsilon^2} & \frac{-3\varepsilon}{16-3\varepsilon^2} & 0
\end{pmatrix}, \quad
A_y = \begin{pmatrix}
0 & 1 & 0 & 0 & 0 & 0 \\
0 & 0 & 0 & 1 & 0 & 0 \\
0 & 0 & 0 & 0 & 1 & 0 \\
0 & \frac{3}{4} & 0 & 0 & 0 & -\frac{3\varepsilon}{4} \\
0 & 0 & 0 & 0 & 0 & 1 \\
0 & 0 & 0 & \frac{-3\varepsilon}{16-3\varepsilon^2} & \frac{12}{16-3\varepsilon^2} & 0
\end{pmatrix}.
$$

These matrices and $\mathcal{B}_{\mathcal{N}_0}[\langle P_\varepsilon \rangle]$ turn smoothly into the $A_x^{(0)}$, $A_y^{(0)}$ matrices and $\mathcal{G}[\langle P_0 \rangle]$ for $|\varepsilon| \to 0$. This shows that $\mathcal{B}_{\mathcal{N}_0}[\langle P_\varepsilon \rangle]$ rather than $\mathcal{G}[\langle P_\varepsilon \rangle]$ or its subset $\mathcal{C}[\langle P_\varepsilon \rangle]$ is the appropriate representation of the ideal $\langle P_\varepsilon \rangle$ for small values of $|\varepsilon|$. $\square$

Extended Groebner bases present a possibility to overcome the discontinuous behavior of classical Groebner bases at the singularity manifolds of their normal sets. Since these discontinuities stem from the specification of a term order and the insistence on order-compatibility, the better approach appears to be a *term-order-free* determination of a monomial basis for $\mathcal{R}[\langle P \rangle]$ and of the associated multiplication matrices, with a careful observation of *numerical stability*. This will be one of our objectives in Chapter 10.

In the consideration of perturbations of the degenerate polynomial system P_0, we had restricted the *support* of the perturbations to the convex hull of the supports J_ν of the respective polynomials p_ν in P_0. Otherwise, we must consider the larger supports $\tilde{J}_\nu$ of $\tilde{P}$ as implicitly holding also for P_0; then the degeneracy in P_0 occurs explicitly as the coefficients of some monomials which define the Newton polygon of some $\tilde{p}_\nu$ tend to 0. If this makes BKK(P_0) smaller than BKK($\tilde{P}$), the corresponding number of zeros of $\tilde{P}$ must "disappear" to ∞ as the coefficients tend to 0; for $\tilde{P}$ close to P_0, these zeros will have very large moduli. In this case, which violates our regularity assumptions, a continuous extension from a border basis of $\langle P_0 \rangle$ to one of $\langle \tilde{P} \rangle$ cannot exist because of the distinct dimensions of the normal sets. Such situations will be considered in section 9.5.

Example 8.27:   In Example 8.26, when we replace the polynomial p_1 in P_ε by $\tilde{p}_1(x, y; \varepsilon) = p_1(x, y; 0) - \varepsilon\, xy^2$, this does not affect P_0; but now BKK($\tilde{P}_\varepsilon$) $= 8 >$ BKK(P_0) $= 6$. When we numerically compute the zeros of $\tilde{P}_\varepsilon$ for $\varepsilon = 10^{-3}$, we obtain the two zeros at $(\pm 1, 0)$ and 4 further zeros which lie within $O(10^{-3})$ of the zeros of P_0, but there are two further zeros at approx. $(1333, -2308)$ and $(1334, 2310)$. $\square$

Exercises

1. (a) For $s = 2$, there is no difference between $\prec_{glex}$ and $\prec_{grevlex}$; but this is no longer true for $s \geq 3$. Find the difference in the orderings of the quadratic and cubic monomials for $s = 3$ and 4.

(b) For `tdeg`, visualize the generic normal sets in $T^3(m)$, for $m = 5(1)20$ (cf. Definition 8.14). Form some of the adjacent normal sets with low degrees of degeneracy.

2. (a) Consider the `tdeg`-based reduction of a monomial $x^j \in T^2$ by two dense polynomials $s_i \in \mathcal{P}^2$, $i = 1, 2$, with coprime leading monomials x^{j_i} both of which divide x^j. Find the potential support of the remainders r_i of the reductions $x^j \xrightarrow{s_i} r_i$, $i = 1, 2$.

(b) Determine the potential supports of the remainders r_{12} and r_{21} defined by $r_1 \xrightarrow{s_2} r_{12}$ and $r_2 \xrightarrow{s_1} r_{21}$. Construct an example where the two supports differ.

(c) Assume that s_1, s_2 are two elements of a Groebner basis for `tdeg`. Verify that now the two supports in (b) coincide.

3. Use a suitable computer algebra system to determine the reduced Groebner bases $C_{\prec}[\langle P \rangle]$ for various freely chosen polynomial systems P in 2, 3, and more variables.

(a) Verify the satisfaction of the criteria in Theorem 8.34.

(b) Try to design systems near the singularity manifold of the normal set associated with $C_{\prec}[\langle P \rangle]$ by observing the magnitudes of the coefficients for l.c. $= 1$. Try to identify the causes of the degeneration.

4. By the procedure introduced in section 2.5.2, transform some of the Groebner bases found in Exercise 3 into border bases for other normal sets $\mathcal{N}$ which you choose.

(a) Check whether the corner subset of $\mathcal{B}_{\mathcal{N}}$ defines the same ideal (hint: determine its Groebner basis).

(b) For the near-singular Groebner bases found in Exercise 3 b , find normal sets $\mathcal{N}$ such that the $\mathcal{N}$-border basis does not contain large coefficients for l.c. $= 1$.

8.5   Multiple Zeros of Intrinsic Polynomial Systems

For univariate polynomials, we have seen in section 6.3 that a cluster of zeros is best understood and analyzed by considering it as the effect of appropriate perturbations on a multiple zero. To follow that same approach in the multivariate case, we must at first understand the structure and the properties of multiple zeros of polynomial ideals $\mathcal{I} \subset \mathcal{P}^s$ in $s > 1$ dimensions, a task which we have repeatedly delayed in Chapter 8. The key to that understanding is the structure of the subspace $\mathcal{D}_0$ of the dual space $\mathcal{D}[\mathcal{I}]$ which corresponds to the multiple zero $z_0 \in \mathbb{C}^s$ and of the associated ideals $\mathcal{I}_0 := \mathcal{I}[\mathcal{D}_0]$ and quotient rings $\mathcal{R}_0 := \mathcal{R}[\mathcal{D}_0]$.

8.5.1   Dual Space of a Multiple Zero

By our considerations in section 2.3.2, an m-fold zero z_0 of $\mathcal{I}$ must contribute m functionals $c_{01}, \ldots, c_{0m}$ to a basis of $\mathcal{D}[\mathcal{I}]$. These functionals define a dual space $\mathcal{D}_0$ which characterizes the ideal $\mathcal{I}_0 \supset \mathcal{I}$ of all polynomials with an m-fold zero of that particular structure at z_0. In the univariate case, the $c_{0\mu}$ are simply $\partial_{\mu-1}[z_0]$; cf. section 6.3.

With s variables, we have a much larger variety of potential differential functionals at our hand: Not only are these the $\binom{s+d-1}{d}$ functionals $\partial_j[z_0]$ of order $d = |j|$, $j \in \mathbb{N}_0^s$, of (2.37) in Definition 2.13, we have also to consider linear combinations of such functionals of equal or different orders. On the other hand, it is to be expected that there must be some restraints for combining such functionals into a basis of $\mathcal{D}_0$. The formal answer to that problem is provided by Definition 2.15 and Theorem 2.21 in section 2.3.2:

Definition 8.20. A zero z_0 of an 0-dimensional ideal $\mathcal{I} \subset \mathcal{P}^s$ is an *m-fold zero* of $\mathcal{I}$ if there exists a *closed* set of *m linearly independent* differentiation functionals $c_{0\mu} = \sum_j \beta_{\mu j} \partial_j[z_0]$ in the dual space $\mathcal{D}[\mathcal{I}]$. The m-dimensional dual space $\mathcal{D}_0 := \mathrm{span}\,(c_{0\mu},\ \mu = 1(1)m)$ defines the *multiplicity structure* of z_0; $\mathcal{I}_0 := \mathcal{I}[\mathcal{D}_0] \supset \mathcal{I}$ is the principal ideal of all polynomials which have an m-fold zero of the structure $\mathcal{D}_0$ at z_0. $\quad\square$

Definition 8.19 implies that—for $s \geq 2$—an s-variate m-fold zero z_0 is not fully characterized by its multipicity m but that there are as many qualitatively different versions of an m-fold zero as there are different m-dimensional dual spaces $\mathcal{D}_0$. To visualize the extent of this variability, we consider the case $s = 3$ with $m = 2$ and 3; the location $z_0 \in \mathbb{C}^3$ of the multiple zero is fixed and will not be denoted. Note that the fact that z_0 is a *zero* implies that $c_1 = \partial_0[z_0]$ must be a basis functional in any $\mathcal{D}_0$.

For $m = 2$, the other basis functional c_2 must be a differentiation functional at z_0 which forms a closed space jointly with c_1, i.e. $c_1(p) = c_2(p) = 0$ must imply $c_2(q\,p) = 0$ for each $q \in \mathcal{P}^3$; cf. Definition 2.15. It is easily seen that any functional

$$c_2 := \beta_{100}\,\partial_{100} + \beta_{010}\,\partial_{010} + \beta_{001}\,\partial_{001}\,, \quad \text{with } b^T = (\beta_{100}, \beta_{010}, \beta_{001}) \neq 0\,, \tag{8.48}$$

satisfies this requirement: With e_σ the σ-th unit vector,

$$\begin{aligned}
c_2(qp) &= \textstyle\sum_{\sigma=1}^{3} \beta_{e_\sigma}\,\partial_{e_\sigma}[z_0](qp) = q(z_0) \sum_\sigma \beta_{e_\sigma}\,\partial_{e_\sigma}[z_0](p) + p(z_0) \sum_\sigma \beta_{e_\sigma}\,\partial_{e_\sigma}[z_0](q) \\
&= c_1(q)\,c_2(p) + c_2(q)\,c_1(p) = 0\,.
\end{aligned}$$

It is also clear that there are no other candidates because a second derivative would violate the closedness condition. Thus c_2 is a *directional first derivative* which appears as a natural generalization of the univariate situation. But even a double zero needs the specification of a (normalized) vector $b \in \mathbb{C}^3$ to characterize its multiplicity structure.

Equation (8.48) has an immediate intuitive interpretation: At a 2-fold zero $z_0 \in \mathbb{C}^3$ of $P = \{p_1, p_2, p_3\} \subset \mathcal{P}^3$, the tangential hyperplanes of the 3 manifolds $p_\nu = 0$ at z_0 intersect in the common *line* $x(t) = (\zeta_{01} + \beta_{100}t, \zeta_{02} + \beta_{010}t, \zeta_{03} + \beta_{001}t)$ so that this line is simultaneously tangential to all 3 manifolds in z_0.

Now take $m = 3$ and assume that a particular c_2 as in (8.48) has already been found as a basis element of $\mathcal{D}_0$. What are candidates for a functional c_3 such that $\mathcal{D}_0 = \mathrm{span}\,(c_1, c_2, c_3)$ is the dual space of a primary ideal $\mathcal{I}_0$ with a zero at z_0? Two possible choices come to mind:

(i) Take another directional derivative $c_3^{(1)} := \sum_\sigma \gamma_{e_\sigma}\,\partial_{e_\sigma}$, with coefficients c^T linearly independent from b^T.

(ii) Take a second derivative in the direction specified by b:

$$c_3^{(2)} := \Big(\sum_\sigma \beta_{e_\sigma}\,\partial_{e_\sigma}\Big)\Big(\sum_\sigma \beta_{e_\sigma}\,\partial_{e_\sigma}\Big)[z_0] = 2\,\Big(\sum_{\sigma_1 \geq \sigma_2} \beta_{e_{\sigma_1}}\,\beta_{e_{\sigma_2}}\,\partial_{e_{\sigma_1}+e_{\sigma_2}}\Big)[z_0]\,.$$

Naturally, we may then also take a linear combination $c_3 := \lambda_1\,c_3^{(1)} + \lambda_2\,c_3^{(2)}$ which constitutes the general case. This yields a wide range of possibilities for the multiplicity structure of a triple zero; they may be parametrized by $\beta_{100} : \beta_{010} : \beta_{001}$ for c_2, $\gamma_{100} : \gamma_{010} : \gamma_{001}$ for $c_3^{(1)}$, and $\lambda_1 : \lambda_2$ for c_3. To go beyond $m = 3$, we clearly need a systematic formalization, particularly for $s > 3$.

Example 8.28: Consider the system $P \subset \mathcal{P}^3$ consisting of

$$\begin{aligned}
p_1(x, y, z) &= 3x^2 - y^2 + 2yz - z^2 - 8x - 8y + 5z - 5\,, \\
p_2(x, y, z) &= x^3 - 6x^2 - 6xy - 4y^2 + z^2 + 3x + 7y - 7z + 15\,, \\
p_3(x, y, z) &= z^3 + 4x^2 + 2xy - 3z^2 - 13x - 5y + 6z + 5\,.
\end{aligned}$$

An analysis along the lines of section 8.1 exhibits a triple zero $z_0 = (2, -1, 1)$ (plus 15 further simple zeros). Since a shift of the origin does not change the differential functionals, we move the origin to z_0. The Taylor-expansion of P about $(2,-1,1)$ yields, with $\xi := x - 2$, $\eta := y + 1$, $\zeta := z - 1$,

$$
\begin{aligned}
p_1(\xi, \eta, \zeta) &:= 4\xi - 4\eta + \zeta + 3\xi^2 - \eta^2 + 2\eta\zeta - \zeta^2, \\
p_2(\xi, \eta, \zeta) &:= -3\xi + 3\eta - 5\zeta - 6\xi\eta - 4\eta^2 + \zeta 2 + \xi^3, \\
p_3(\xi, \eta, \zeta) &:= \xi - \eta + 3\zeta + 4\xi^2 + 2\xi\eta + \zeta^3.
\end{aligned}
$$

The vanishing at 0 of the first order directional derivative $c_2 = \partial_\xi + \partial_\eta$ for all 3 polynomials is obvious. With some trial and error, one finds that $c_3 = ((\partial_\xi + \partial_\eta)^2 - 2\,\partial_\zeta)[(0, 0, 0)]$ also vanishes for all 3 polynomials (remember that $\partial_{\xi^2} = \frac{1}{2}\frac{\partial^2}{\partial\xi^2}$). Thus, the multiplicity structure of the triple zero z_0 of the original system is specified by the dual space $\mathcal{D}_0 = \operatorname{span}(\partial_0, \partial_{x+y}, \partial_{x+y}^2 - 2\,\partial_z)[z_0]$. $\quad\square$

For a formal treatment, we must at first understand the restrictions imposed by the required closedness of a basis of $\mathcal{D}_0$. The following formula is well known in multivariate analysis and often quoted as Leibniz' rule; cf. (1.8):

Proposition 8.35. Consider a differentiation functional ∂_j (2.37) with $j \in \mathbb{N}_0^s$. For $p, q \in \mathcal{P}^s$,

$$
\partial_j(q\, p) = \sum_{0 \le k \le j} \partial_k(q)\, \partial_{j-k}(p), \tag{8.49}
$$

where the sum runs over all $k \in \mathbb{N}_0^s$, with $k \le j$ componentwise. Note that the factors in (2.37) imply that no numerical factors appear in (8.49).

Definition 8.21. The *anti-differentiation operators* s_σ, $\sigma = 1(1)s$, are defined by

$$
s_\sigma\, \partial_j[z] := \begin{cases} \partial_{j-e_\sigma}[z] & \text{if } j_\sigma > 0, \\ 0\text{-functional} & \text{if } j_\sigma = 0, \end{cases} \quad \text{and} \quad s_\sigma\Big(\sum_j \gamma_j\, \partial_j[z_0]\Big) := \sum_j \gamma_j\, s_\sigma\, \partial_j[z_0]. \quad\square
$$
$$\tag{8.50}$$

Example 8.29: In $\mathcal{P}^3$, $s_1\, \partial_{210} = \partial_{110}$, $s_2\, \partial_{210} = \partial_{200}$, $s_3\, \partial_{210} = 0$; $s_2\,(2\,\partial_{210} - \partial_{021} + 3\,\partial_{102}) = 2\,\partial_{200} - \partial_{011}$. $\quad\square$

Theorem 8.36. In $\mathcal{P}^s$, consider a linear space $\mathcal{D}(z_0)$ of differentiation functionals c, with evaluation at z_0. $\mathcal{D}(z_0)$ is closed iff

$$
c \in \mathcal{D}(z_0) \implies s_\sigma\, c \in \mathcal{D}(z_0), \quad \sigma = 1(1)s. \tag{8.51}
$$

Proof: Closedness of $\mathcal{D}$ requires that $l \in \mathcal{D} \Rightarrow (l \cdot q) \in \mathcal{D}\ \forall q \in \mathcal{P}^s$; cf. Definition 2.15. By (8.49), all derivative evaluations of p which occur in an evaluation of $\partial_j(q\, p)$ are of the form $\partial_{j-k}(p) = s^k\, \partial_j(p) := s_{\sigma_1}^{k_1} \dots s_{\sigma_s}^{k_s}\, \partial_j(p)$, $k \le j$. If $\partial_j(q\, p)$ is to vanish for *arbitrary* $q \in \mathcal{P}^s$ and $\partial_j(p) = 0$ then *all* $s^k\, \partial_j(p)$, $k \le j$, must vanish and vice versa. Linearity of the s_σ extends this to linear combinations of ∂_j's. $\quad\square$

Let us now derive an *algorithmic approach* for the determination of the dual space of a multiple zero $z_0 \in \mathbb{C}^s$ of a polynomial system $P \in \mathcal{P}^s$, assuming that we know the position of z_0. It appears natural to proceed incrementally from ∂_0 and to look for further candidate functionals c_μ, with free parameters. Then we can attempt to determine the parameters such that span $(\mathcal{D}_0 \cup c_\mu)$ remains closed and $c_\mu(p_\nu)$ vanishes for the polynomials in the given system P. If this is possible, a new basis functional for $\mathcal{D}_0$ has been found. If it is not possible, we *save* the candidate c_μ, with its parameters partially chosen such that closedness is attained, for use in linear combinations with other candidates. If, at some point, none of the candidate functionals annihilates the polynomials in P, we are finished and $\mathcal{D}_0$ is complete. Naturally, this happens after finitely many steps.

For an intuitive development of such an algorithm, we assume at first that there exists a "monomial basis" $\mathbf{c}^T = \{\partial_{j_1}, \partial_{j_2}, \ldots, \partial_{j_m}\}$ of plain derivatives for $\mathcal{D}_0$. Then $\mathcal{D}_0$ may be viewed as a vector space $V \subset \mathbb{C}^s$ with basis $\{j_1, j_2, \ldots, j_m\}$. The operator s_σ moves $j \in V$ to its negative σ-neighbor (cf. Definition 2.17) or annihilates it if the σ-component of j is 0. Closedness appears as the direct analog of the closedness of a set of monomials $\mathcal{N} = \{x^{j_1}, \ldots, x^{j_m}\}$: Each negative neighbor of an $x^{j_\mu} \in \mathcal{N}$ must be in $\mathcal{N}$ or outside the first orthant. We define the degree $|j_\mu| = |(j_{\mu 1}, \ldots, j_{\mu s})| := \sum_\sigma |j_{\mu\sigma}|$ and the "total degree" $|d|$ of $d = \sum_\mu \gamma_\mu \partial_{j_\mu}$ by $\max_\mu |j_\mu|$.

Now we construct a monomial basis $\mathbf{c}^T$ incrementally by total degree, assuming that $\mathcal{D}_0$ admits such a basis. $\mathbf{c}^T$ must contain $\partial_{0,\ldots,0}$, the only element of degree 0; this permits the ∂_{e_σ}, $\sigma = 1(1)s$, to be considered as candidates for further basis elements because they are consistent with closedness. A candidate is accepted if it actually annihilates the polynomials $p_\nu \in P$. Assume that—after a potential renumbering of components—the ∂_{e_σ} for $\sigma = 1(1)s_0$, $1 \leq s_0 \leq s$, pass this acceptance test. If $s_0 < s$, further candidates ∂_{j_μ} with a higher degree must have vanishing components $j_{\mu,s_0+1}, \ldots, j_{\mu,s}$ to comply with closedness.

Now we form "quadratic" candidates $\partial_{e_{\sigma_1}+e_{\sigma_2}}$, $\sigma_1, \sigma_2 \in \{1, .., s_0\}$, all of which satisfy closedness. If they are all inconsistent with P so that none of them is accepted, we are finished. But we are also finished if the accepted ∂_{j_μ} with $|j_\mu| = 2$ do not permit a further closed extension of the current basis $\mathbf{c}^T$ which requires the existence of ∂_j with $|\partial_j| = 3$ with all negative neighbors in $\mathbf{c}^T$. Otherwise, we continue with the existing "cubic" candidate(s) in the same fashion.

It is obvious that our assumption about $\mathcal{D}_0$ is restrictive and will not be satisfied in general. There are two principal ways in which we must extend the approach: Assume, at first, as previously that there are $s_0 < s$ plain derivatives ∂_{e_σ} which vanish for the $p_\nu \in P$ and no other first degree basis elements. Then, trivially, not only the $\partial_{e_{\sigma_1}+e_{\sigma_2}}$, $\sigma_1, \sigma_2 \in \{1, .., s_0\}$, retain closedness but also any linear combination of them. And we can add a linear combination of the *discarded* ∂_{e_σ}, $\sigma \in \{s_0 + 1, .., s\}$ to such a quadratic term and retain closedness. This gives us *one* candidate with a sizeable number of *free parameters* as a candidate which retains closedness. We can now require that this parametrized differentiation functional annihilates the p_ν and solve for the parameters which achieve that. Each linearly independent solution yields a basis element for $\mathcal{D}_0$. If there exists no solution for the parameters there are no basis elements beyond the linear ones.

How do we continue from existing quadratic basis functionals $d_1^{(2)}, \ldots, d_k^{(2)}$, each of which contains 2nd derivatives only with respect to $\sigma \in \{1, .., s_0\}$. A potential "cubic" functional $d^{(3)}$ must have a 3rd derivative part which is reduced to the 2nd derivative part of one of the $d_k^{(2)}$ (or to a linear combination of them) by s_σ, $\sigma = 1(1)s_0$. Generally, this requires $k = s_0$ different $d_k^{(2)}$. If $k < s_0$, $d^{(3)} = \sum_{|j|=3} \gamma_j^{(3)} \partial_j$ must be reduced to the *same* linear combination $\sum_\kappa \beta_\kappa d_\kappa^{(2)}$

by two *different* s_{σ_1}, s_{σ_2}, which requires that, for all $|j| = 2$, the ratio $\gamma_{j+e_{\sigma_1}}^{(3)} : \gamma_{j+e_{\sigma_2}}^{(3)}$ has the same fixed value; this reduces the number of free parameters considerably, and even more so if the results of more than two s_σ are to coincide. With this understanding, one can set up the equations necessary for closedness and annihilation of the p_ν and try to solve them. A potential further continuation beyond degree 3 has to follow the same principles.

Example 8.30: Consider the following system $P = \{p_1, p_2, p_3\} \subset \mathcal{P}^3$ with a multiple zero of unknown multiplicity at $z_0 = 0$:

$$
\begin{aligned}
p_1(x_1, x_2, x_3) &= x_1^2 - 4x_1x_2 + 4x_2^2 + x_3, \\
p_2(x_1, x_2, x_3) &= x_1^2 + x_2^2 + x_3^2 - 2x_3, \\
p_3(x_1, x_2, x_3) &= x_1^2 x_2 + x_1 x_2^2 + x_1 x_2 x_3.
\end{aligned}
$$

It can immediately be seen that ∂_{100} and ∂_{010} are the basis functionals c_2 and c_3 so that $s_0 = 2$; furthermore, ∂_{001} is naturally consistent with closedness but not with P. Above, we have seen that such functionals can be added to higher degree candidates; thus, the general quadratic functional which retains closedness is

$$
d^{(2)} = \gamma_{200}^{(2)} \partial_{200} + \gamma_{110}^{(2)} \partial_{110} + \gamma_{020}^{(2)} \partial_{020} + \gamma_{001}^{(2)} \partial_{001}.
$$

Consistency with P requires $d^{(2)} p_\nu = 0$, $\nu = 1(1)3$, or

$$
\gamma_{200}^{(2)} - 4\gamma_{110}^{(2)} + 4\gamma_{020}^{(2)} + \gamma_{001}^{(2)} = 0, \qquad \gamma_{200}^{(2)} + \gamma_{020}^{(2)} - 2\gamma_{001}^{(2)} = 0,
$$

while the annihilation of p_3 is trivial. This system has a 2-dimensional solution space so that there are two linearly independent $d^{(2)}$ functionals. As basis functionals c_4, c_5 we take

$$
c_4 = d_1^{(2)} = 4\partial_{200} - 3\partial_{110} - 4\partial_{020}, \qquad c_5 = d_2^{(2)} = 2\partial_{200} + 3\partial_{110} + 2\partial_{020} + 2\partial_{001}.
$$

Since there are $2 = s_0$ $d^{(2)}$-functionals, we can set up the cubic candidate without restrictions: We "shift" (differentiate) $d^{(2)}$ in the x_1 and x_2 directions and obtain

$$
d^{(3)} = \gamma_{300}^{(3)} \partial_{300} + \gamma_{210}^{(3)} \partial_{210} + \gamma_{120}^{(3)} \partial_{120} + \gamma_{030}^{(3)} \partial_{030} + \gamma_{101}^{(3)} \partial_{101} + \gamma_{011}^{(3)} \partial_{011}.
$$

In this particular case, the introduction of functionals which have satisfied closedness but not consistency is futile: $d^{(2)}$, with free parameters, trivially annihilates p_3, and consistency with p_1, p_2 turns it into a linear combination of c_4, c_5 as we have seen. Closedness requires

$$
s_1 d^{(3)} = \beta_{11} c_4 + \beta_{12} c_5, \qquad s_2 d^{(3)} = \beta_{21} c_4 + \beta_{22} c_5, \qquad s_3 d^{(3)} = \beta_{31} c_2 + \beta_{32} c_3. \tag{8.52}
$$

These are $4+4+2 = 10$ homogeneous equations for the 12 parameters $\gamma_{300}^{(3)}, \ldots, \gamma_{011}^{(3)}, \beta_{11}, \ldots, \beta_{32}$. $d^{(3)}$ annihilates p_1 and p_2 trivially, and $d^{(3)} p_3 = 0$ gives only one further homogeneous equation. The 11×12 homogeneous system has full rank so that there is exactly one nontrivial solution for a c_6: With smallest integer coefficients, we obtain

$$
c_6 = 51\partial_{300} + 9\partial_{210} - 9\partial_{120} - 11\partial_{030} + 21\partial_{101} - \partial_{011}.
$$

With $1 < s_0$ cubic functionals, a continuation would require a constant proportionality of the consecutive $\gamma_j^{(3)}$ which clearly is not there. Thus we are finished: The multiplicity of $z_0 = 0$ for the system P is 6 and its multiplicity structure is given by $\mathcal{D}_0 = \text{span } \mathbf{c}^T$, with $\mathbf{c}^T = (c_1, \ldots, c_6)$ as computed above.

In a review of the result, we observe: c_4 and c_5 are "natural," their coefficients complete the $(x_1^2, x_1 x_2, x_2^2, x_3)$-coefficient vectors $(1, -4, 4, 1)$ and $(1, 0, 1, -2)$ of p_1 and p_2 to a basis of the $\mathbb{C}^4$. c_6, on the other hand, is essentially determined by closedness conditions; consistency with p_3 is only incorporated through $\gamma_{210}^{(3)} = -\gamma_{120}^{(3)}$. c_6 could hardly have been found without explicit use of the system (8.52). Yet this particular form of c_6, together with $c_1, \ldots, c_5$, determines the details of the splitting of the 6-fold zero upon a perturbation of P, as we shall see in section 9.3. Also, P has a total of 12 zeros so that not even the multiplicity $m = 6$ could have easily been found without an algorithmic analysis of the above kind. $\square$

There remains one last shortcoming of our algorithmic procedure: Generally, the s_0 first order differentials $c_2, \ldots, c_{s_0+1}$ in a basis of $\mathcal{D}_0$ will not be plain ∂_{e_σ} but s_0 linearly independent combinations of such derivatives. This may easily be repaired: A linear transformation of the variables which takes the vectors of the linear combinations into different unit vectors reduces the situation to the one which we have discussed. The appropriate transformation is found thus:

If there are s_0 linearly independent combinations of first derivatives at z_0 which vanish, the Jacobian $P'(z_0)$ has deficiency s_0 and there are s_0 column vectors $r_\tau = (\rho_{\tau 1}, \ldots, \rho_{\tau s})^T \in \mathbb{C}^s$ such that $P'(z_0) r_\sigma = 0$. When we complete these columns into a regular $s \times s$-matrix R and substitute $x = z_0 + R(y - z_0)$ in P to form $\widehat{P}(y)$, then the Jacobian of $\widehat{P}$ at z_0 will have s_0 leading vanishing columns which implies that the $\hat{p}_\nu(y)$ have vanishing first derivatives with respect to $y_1, \ldots, y_{s_0}$ at z_0. The columns r_τ may be found by Gaussian elimination in $P'(z_0)$.

Example 8.31: We return to our initial Example 8.28 and take the system in its shifted form; for notational clarity, we rename the variables ξ, η, ζ as x_1, x_2, x_3. The Jacobian

$$P'(0) = \begin{pmatrix} 4 & -4 & 1 \\ -3 & 3 & -5 \\ 1 & -1 & 3 \end{pmatrix} \quad \text{is annihilated by } r_1 = \begin{pmatrix} 1 \\ 1 \\ 0 \end{pmatrix}.$$

We complete this column by $r_2 = (1, -1, 0)^T$, $r_3 = (0, 0, 1)^T$ and form $P(Ry) =: \widehat{P}(y)$:

$$\hat{p}_1(y_1, y_2, y_3) = 8 y_2 + y_3 + 2 y_1^2 + 8 y_1 y_2 + 2 y_2^2 + 2 y_1 y_3 - 2 y_2 y_3 - y_3^2,$$
$$\hat{p}_2(y_1, y_2, y_3) = -6 y_2 - 5 y_3 - 10 y_1^2 + 8 y_1 y_2 + 2 y_2^2 + y_3^2 + y_1^3 + 3 y_1^2 y_2 + 3 y_1 y_2^2 + y_2^3,$$
$$\hat{p}_3(y_1, y_2, y_3) = 2 y_2 + 3 y_3 + 6 y_1^2 + 8 y_1 y_2 + 2 y_2^2 + y_3^3.$$

Now we have the plain derivative ∂_{100} for c_2 and $s_0 = 1$. Thus, a quadratic functional consistent with closedness can only have the form $\partial_{200} + \gamma_{010} \partial_{010} + \gamma_{001} \partial_{001}$. Consistency with $\widehat{P}$ yields 3 inhomogeneous equations for $\gamma_{010}, \gamma_{001}$, with a solution $\gamma_{010} = 0$, $\gamma_{001} = -2$, so that $c_3 = \partial_{200} - 2 \partial_{001}$.

A cubic functional consistent with closedness is $\partial_{300} - 2 \partial_{101} + \gamma_{010}^{(3)} \partial_{010} + \gamma_{001}^{(3)} \partial_{001}$. But now the 3 inhomogeneous equations for the two parameters are inconsistent so that we are finished. A return to the variables before the transformation turns c_2 into $\partial_{100} + \partial_{010}$ and c_3 into $\partial_{200} + \partial_{110} + \partial_{020} - 2 \partial_{001}$. $\square$

The practical difficulty in the application of the described procedure to a nontrivial polynomial system P lies in the fact that, generally, the multiple zero z_0 will only be known approximately; thus, $P'(z_0)$ will only be *close* to a matrix of deficiency s_0. This means that even for intrinsic systems, the determination of the differentiability structure of a multiple zero may have to follow the same lines as for an empirical system which will be discussed in section 9.3.

The algorithmic determination of a basis for the dual space $\mathcal{D}_0(z_0)$ associated with a multiple zero of a complete intersection polynomial system has been described in [2.6]; its first (and supposedly only) implementation has been achieved by my student G.Thallinger; cf. [8.3]. There, a term order has been used as incremental guideline; the above presentation shows that a term order is really not necessary.

8.5.2 Normal Set Representation for a Multiple Zero

From the m-dimensional dual space $\mathcal{D}_0 = \text{span } \mathbf{c}^T = \text{span } (c_1, \ldots, c_n)$ describing the multiplicity structure of an m-fold zero $z_0 \in \mathbb{C}^s$ of $P \subset \mathcal{P}^s$, we want to determine the associated quotient ring $\mathcal{R}_0 = \mathcal{R}[\mathcal{D}_0]$ and primary ideal $\mathcal{I}_0 = \mathcal{I}[\mathcal{D}_0]$. We proceed in a standard way; cf. sections 2.3.2 and 8.1.1.

We select a suitable normal set $\mathcal{N}_0 = \{b_1, \ldots, b_m\}$ from $T^s(m)$ which yields a regular matrix $\mathbf{c}^T(\mathbf{b})$. From section 2.3.2, we note that $x_\sigma \, \mathbf{b}(x) \equiv A_\sigma \, \mathbf{b}(x) \bmod \mathcal{I}_0$ implies $\mathbf{c}^T(x_\sigma \mathbf{b}(x)) = A_\sigma \, \mathbf{c}^T(\mathbf{b}(x))$ so that

$$A_\sigma = \mathbf{c}^T(x_\sigma \mathbf{b}(x)) \cdot (\mathbf{c}^T(\mathbf{b}(x)))^{-1}, \qquad \sigma = 1(1)s. \tag{8.53}$$

Thus, we gain a normal set representation of $\mathcal{R}_0$ and $\mathcal{I}_0$. Generally, all matrices involved will be extremely sparse; cf. the examples below.

For the analysis of the zero cluster originating from a perturbation of the multiple zero, it will turn out to be advantageous to have a representation of $\mathcal{I}_0$ by a complete intersection system, i.e. by a basis of only s elements. If we select the normal set $\mathcal{N}_0$ considerately, we may be able to obtain such a basis as a subset of the full border basis $\mathcal{B}[\mathcal{I}_0]$:

Assume, e.g., that $\mathbf{c}^T$ does not contain differentiations with respect to $x_{s_0+1}, \ldots, x_s$; then we need not introduce these variables into the normal set $\mathcal{N}_0$. If we then choose $\mathcal{N}_0$ as quasi-univariate in the "active" variables $x_1, \ldots, x_{s_0}$ (cf. section 8.1.3), we can try to take the s_0 elements from a B_σ subset of the border basis (where x_σ is a distinguished variable) and complete the basis of $\mathcal{I}_0$ by $x_{s_0+1}, \ldots, x_s$.

Even when we take a valid quasi-univariate normal set in all variables, it is not clear that we can select a complete intersection from the border basis because $\mathcal{I}_0$ may not be a complete intersection! Remember that the classic example of a 0-dimensional ideal which is *not* a complete intersection is the principal ideal of a triple zero in two variables, with $\mathcal{D}_0 = \text{span } (\partial_{20}, \partial_{11}, \partial_{02})$; cf. Example 8.12 in section 8.3.1. But there we had also found that such zeros cannot occur in regular systems.

Example 8.32: We take the system $\widehat{P}(y)$ of Example 8.31, with $\mathcal{D}_0 = \text{span } (\partial_{000}, \partial_{100}, \partial_{200} - 2\partial_{001})$. No differentiation with respect to y_2 appears in the multiplicity structure of the triple zero at 0; therefore, we take $\mathcal{N}_0 = \text{span } (1, y_1, y_3)$. This yields $\mathbf{c}^T(\mathbf{b}) = \begin{pmatrix} 1 & 0 & 0 \\ 0 & 1 & 0 \\ 0 & 0 & -2 \end{pmatrix}$

and $\mathbf{c}^T(y_1\mathbf{b}) = \begin{pmatrix} 0 & 1 & 0 \\ 0 & 0 & 1 \\ 0 & 0 & 0 \end{pmatrix}$, $\mathbf{c}^T(y_2\mathbf{b}) = \begin{pmatrix} 0 & 0 & 0 \\ 0 & 0 & 0 \\ 0 & 0 & 0 \end{pmatrix}$, $\mathbf{c}^T(y_3\mathbf{b}) = \begin{pmatrix} 0 & 0 & -2 \\ 0 & 0 & 0 \\ 0 & 0 & 0 \end{pmatrix}$.

A_1, A_2, A_3 are the same matrices with their last columns multiplied by $-\frac{1}{2}$; cf. (8.53). The border basis $\mathcal{B}_{\mathcal{N}_0}[\mathcal{I}_0]$ which can be read from the A_σ, is $\{2\,y_1^2 + y_3,\ y_1 y_3,\ y_3^2,\ y_2,\ y_1 y_2,\ y_2 y_3\}$. Because A_1 is nonderogatory, the subset $\{2\,y_1^2 + y_3,\ y_1 y_3,\ y_2\}$ generates the full basis; it is a complete intersection basis for $\mathcal{I}_0$. $\square$

Example 8.33: When we consider the 6-fold zero $z_0 = 0$ of the system $P(x)$ of Example 8.30, with its rather nontrivial multiplicity structure $\mathcal{D}_0$, we have to be more considerate in the choice of the normal set $\mathcal{N}_0$ to reach a regular matrix $\mathbf{c}^T(\mathbf{b})$: For each component c_μ of $\mathbf{c}^T$, the normal set vector $\mathbf{b}(x) = (b_1(x), \ldots, b_m(x))^T$ must contain at least one component which is not annihilated by c_μ. Thus, on account of c_1, c_2, and c_3, we must have 1, x_1, x_2 in $\mathcal{N}_0$. When we further attempt to bypass the variable x_3, the functionals c_4, c_5 suggest the inclusion of x_1^2 and $x_1 x_2$. Finally, to keep the normal set quasi-univariate in x_1, x_2, we take x_1^3 as $b_6(x)$. Now we have $\mathbf{b} = (1, x_1, x_2, x_1^2, x_1 x_2, x_1^3)^T$ which yields

$$
\mathbf{c}^T(\mathbf{b}) =
\begin{pmatrix}
1 & 0 & 0 & 0 & 0 & 0 \\
0 & 1 & 0 & 0 & 0 & 0 \\
0 & 0 & 1 & 0 & 0 & 0 \\
0 & 0 & 0 & 4 & 2 & 0 \\
0 & 0 & 0 & -3 & 3 & 0 \\
0 & 0 & 0 & 0 & 0 & 51
\end{pmatrix}
\quad \text{and} \quad
(\mathbf{c}^T(\mathbf{b}))^{-1} =
\begin{pmatrix}
1 & 0 & 0 & 0 & 0 & 0 \\
0 & 1 & 0 & 0 & 0 & 0 \\
0 & 0 & 1 & 0 & 0 & 0 \\
0 & 0 & 0 & \frac{1}{6} & \frac{1}{9} & 0 \\
0 & 0 & 0 & \frac{1}{6} & \frac{-2}{9} & 0 \\
0 & 0 & 0 & 0 & 0 & \frac{1}{51}
\end{pmatrix}.
$$

We must now form the matrices $\mathbf{c}^T(x_\sigma \mathbf{b})$, $\sigma = 1(1)3$, and multiply them by $(\mathbf{c}^T(\mathbf{b}))^{-1}$ to obtain the multiplication matrices A_σ for $\mathcal{R}_0 = \mathrm{span}\ \mathbf{b}(x)$. This yields

$$
A_1 =
\begin{pmatrix}
0 & 1 & 0 & 0 & 0 & 0 \\
0 & 0 & 0 & 1 & 0 & 0 \\
0 & 0 & 0 & 0 & 1 & 0 \\
0 & 0 & 0 & 0 & 0 & 1 \\
0 & 0 & 0 & 0 & 0 & \frac{3}{17} \\
0 & 0 & 0 & 0 & 0 & 0
\end{pmatrix}
\quad \text{and} \quad
A_2 =
\begin{pmatrix}
0 & 0 & 1 & 0 & 0 & 0 \\
0 & 0 & 0 & 0 & 1 & 0 \\
0 & 0 & 0 & \frac{-1}{3} & \frac{8}{9} & 0 \\
0 & 0 & 0 & 0 & 0 & \frac{3}{17} \\
0 & 0 & 0 & 0 & 0 & \frac{-3}{17} \\
0 & 0 & 0 & 0 & 0 & 0
\end{pmatrix}
$$

for the multiplication matrices in the x_1, x_2-subspace, with a total of 5 different nontrivial rows from the normal forms of x_2^2, $x_1 x_2^2$, $x_1^2 x_2$, $x_1^3 x_2$, x_1^4; this defines the 5 border basis elements in this subspace. The multiplication matrix A_1 for the distinguished variable x_1 of $\mathcal{N}_0$ is derogatory, but it turns out that the border basis elements with $\mathcal{N}_0$-leading monomials x_2^2 and $x_1^2 x_2$ define the other 3 ones.

From p_1 and the normal form of x_2^2, we obtain the normal form of $x_3 \notin \mathcal{N}_0$ as $\frac{1}{3} x_1^2 + \frac{4}{9} x_1 x_2$. With A_1, A_2, this defines all other rows in A_3. Thus we have a complete intersection basis for $\mathcal{I}_0$ of the form

$$
b_1 = x_2^2 + \tfrac{1}{3} x_1^2 - \tfrac{8}{9} x_1 x_2, \qquad
b_2 = x_1^2 x_2 - \tfrac{3}{17} x_1^3, \qquad
b_3 = x_3 - \tfrac{1}{3} x_1^2 - \tfrac{4}{9} x_1 x_2.
$$

For each polynomial combination p of these 3 polynomials, $\mathbf{c}^T[p]$ vanishes. $\square$

8.5.3 From Multiplication Matrices to Dual Space

For a fixed normal set basis $\mathcal{N}_0$ of the quotient ring $\mathcal{R}_0$ of a multiple zero $z_0 \in \mathbb{C}^s$, the dual space $\mathcal{D}_0$ determines the multiplication matrices $A_{0\sigma}$. Thus we should also be able to determine a basis $\mathbf{c}^T$ of $\mathcal{D}_0$ from $\mathcal{N}_0$ and the $A_{0\sigma}$, $\sigma = 1(1)s$. This task appears when we have found a normal set representation $(\mathcal{N}, \mathcal{B}_{\mathcal{N}})$ for the ideal $\langle P \rangle$ of the polynomial system $P \subset \mathcal{P}^s$ and the eigenanalysis of the A_σ exhibits a *joint invariant subspace* of a dimension $m_0 > 1$; cf. section 2.4.3. Note that $\mathcal{N} \in T^s(m)$ is a normal set of $\langle P \rangle$ and not of the ideal $\mathcal{I}_0$ associated with z_0. In this case, we have (cf. (2.50)), for $\sigma = 1(1)s$,

$$
A_\sigma\, X_0 = A_\sigma \begin{pmatrix} | & & | \\ x_{01} & .. & x_{0m_0} \\ | & & | \end{pmatrix} = \begin{pmatrix} | & & | \\ x_{01} & .. & x_{0m_0} \\ | & & | \end{pmatrix} \begin{pmatrix} \zeta_{0\sigma} & & \cdots \\ & \ddots & \vdots \\ 0 & & \zeta_{0\sigma} \end{pmatrix} = X_0\, T_{0\sigma},
$$

$$(8.54)$$

where $x_{01} = \mathbf{b}(z_0) \in \mathbb{C}^m$ is the only joint eigenvector in the joint invariant subspace span X_0 of the commuting family $\overline{A}$ of the A_σ, and the $T_{0\sigma} = (t^{(\sigma)}_{\nu\mu}) \in \mathbb{C}^{m_0 \times m_0}$ are upper triangular, with a diagonal of $\zeta_{0\sigma}$, the σ-component of z_0.

Equation (8.54) characterizes z_0 as an m_0-fold zero of $\langle P \rangle$ whose *multiplicity structure* we want to find from the $x_{0\mu} \in \mathbb{C}^m$, $\mu = 2(1)m_0$, and the matrices $T_{0\sigma}$, $\sigma = 1(1)s$. This means that we want to find a basis $\mathbf{c}_0^T = (c_1, \ldots, c_{m_0})$, with $c_\mu = \sum_j \gamma_{\mu j} \partial_j[z_0]$, of a dual space $\mathcal{D}_0$ such that

$$
x_{0\mu} = c_\mu(\mathbf{b}(x)), \qquad \mu = 1(1)m_0, \tag{8.55}
$$

after the first component of x_{01} has been normalized to 1 as usual.

At first, we note that the upper triangularity of the $T_{0\sigma}$ requires that the vectors $x_{0\mu}$ have been arranged in a particular order; from linear algebra, we know that this is always possible. When we interpret the $x_{0\mu}$ as $c_\mu(\mathbf{b})$, this is equivalent to the fact that each leading subset $(c_1, \ldots, c_{\bar\mu})$ of $\mathbf{c}_0^T$ is *closed*:

Proposition 8.37. In $(\mathcal{P}^s)^*$, let $c_\mu = \sum \gamma_{\mu j} \partial_j[z_0]$, $\mu = 1(1)m_0$, and let $\mathbf{b}(x)$ be the normal set vector of $\mathcal{N} \in T^s(m)$. Each leading subset of $\mathbf{c}_0^T = (c_1, \ldots, c_{m_0})$, $m_0 \leq m$, is closed iff there exist upper triangular matrices $T_\sigma \in \mathbb{C}^{m_0 \times m_0}$, with diag $T_\sigma = (\zeta_{0\sigma} \ldots \zeta_{0\sigma})$, such that

$$
\mathbf{c}_0^T(x_\sigma \mathbf{b}(x)) = \mathbf{c}_0^T(\mathbf{b}(x))\, T_\sigma, \qquad \sigma = 1(1)s. \tag{8.56}
$$

Proof: By Proposition 8.35 and (8.54),

$$
c_\mu(x_\sigma \mathbf{b}) = \partial_0[z_0]x_\sigma\, c_\mu(\mathbf{b}) + s_\sigma c_\mu(\mathbf{b}) = \zeta_{0\sigma}\, c_\mu(\mathbf{b}) + \sum_\nu t^{(\sigma)}_{\nu\mu}\, c_\nu(\mathbf{b}).
$$

If each leading subset of $\mathbf{c}^T$ is closed, then $s_\sigma c_\mu$ is in the subspace span $(c_1, \ldots, c_{\mu-1})$ for all σ (cf. Theorem 8.36) and $t^{(\sigma)}_{\nu\mu} = 0$ for $\nu \geq \mu$. Vice versa, the triangularity of the T_σ implies $s_\sigma c_\mu \in$ span $(c_1, \ldots, c_{\mu-1})$ for all σ, since $\mathbf{b}$ is a basis of $\mathcal{R}$. $\square$

Thus, when we interpret the elements $t^{(\sigma)}_{\nu\mu}$ of the $T_{0\sigma}$ in (8.54) in terms of the coefficients $\gamma_{\mu j}$ of the dual space basis elements $c_\mu = \sum_j \gamma_{\mu j} \partial_j[z_0]$, we can assume that the c_μ are ordered in a closedness-consistent way, with $c_1 = \partial_0[z_0]$. Next, there must be at least one pure first-order differential $c_2 = \sum_\sigma \gamma_{2\sigma} \partial_{e_\sigma}[z_0]$; with $c_2(x_\sigma \mathbf{b}) = \zeta_{0\sigma} c_2(\mathbf{b}) + \gamma_{2\sigma} c_1(\mathbf{b})$, this yields immediately $\gamma_{2\sigma} = t^{(\sigma)}_{12}$ for $\sigma = 1(1)s$. If there are further pure first-order differentials $c_\mu = \sum_\sigma \gamma_{\mu\sigma} \partial_{e_\sigma}[z_0]$, $\mu = 3, \ldots$, we have correspondingly $c_\mu(x_\sigma \mathbf{b}) = \zeta_{0\sigma} c_\mu(\mathbf{b}) + \gamma_{\mu\sigma} \mathbf{b}(z_0)$, which implies $t^{(\sigma)}_{1\mu} = \gamma_{\mu\sigma}$, $t^{(\sigma)}_{\nu\mu} = 0$, $\nu = 2(1)\mu - 1$. Thus, the vanishing of all $t^{(\sigma)}_{\nu\mu}$, $\nu = 2(1)\mu - 1$, for some μ-th column of T_σ signals the presence of a pure first-order basis element c_μ in $\mathbf{c}_0^T$, with coefficients given by the $t_{1\mu}^{(\sigma)}$. Note that an ordering of the c_μ like $\partial_0,\ \partial_{100},\ \partial_{200},\ \partial_{010},\ \ldots$ does not violate the closedness assumption.

Let $M_1 = \{2, \ldots\}$ be the set of subscripts μ which refer to a pure first-order c_μ. Then the first appearance of nonvanishing elements $t_{\nu\mu}^{(\sigma)}$ in rows $\nu \in M_1$ signals that the respective c_μ is a second-order differential $c_\mu = \sum_\sigma \gamma_{\mu e_\sigma} \partial_{e_\sigma}[z_0] + \sum_{|j|=2} \gamma_{\mu j}\partial_j[z_0]$. From $c_\mu(x_\sigma \mathbf{b}) = \zeta_{0\sigma} c_\mu(\mathbf{b}) + s_\sigma c_\mu(\mathbf{b})$, with

$$s_\sigma c_\mu(\mathbf{b}) = \gamma_{\mu e_\sigma} \mathbf{b}(z_0) + \sum_{|j|=2} \gamma_{\mu j}\, \partial_{j-e_\sigma}[z_0] = \sum_{\nu<\mu} t_{\nu\mu}^{(\sigma)}\, c_\nu(\mathbf{b}),$$

we find that the $t_{1\mu}^{(\sigma)}$ again display the $\gamma_{\mu e_\sigma}$ while, by (8.50),

$$\text{2nd order part of } c_\mu = \sum_{\nu\in M_1} t_{\nu\mu}^{(\sigma)}\, \partial_{e_\sigma} c_\nu(\mathbf{b}) + \text{ differentials } \partial_j \text{ with } s_\sigma\, \partial_j = 0, \quad \sigma = 1(1)s.$$

$$(8.57)$$

The relations (8.57) must have a unique solution since $s_\sigma c_\mu \in \text{span}(c_1, \ldots, c_{\mu-1})$. Similarly, all further columns with vanishing elements except in row 1 and rows $\nu \in M_1$ refer to second-order differentials whose coefficients may be found from the nonvanishing $t_{\nu\mu}^{(\sigma)}$ via (8.57).

When we put all these μ into a set M_2, the first appearance of a column μ with a non-vanishing element in a row $\nu \in M_2$ of one of the T_σ signals that c_μ is a third-order differential whose coefficients may be determined in an analogous fashion. The continuation of the procedure is now obvious.

Example 8.34: We take the system $P(x, y, z)$ of Example 8.28 in its original form, before the shifting of the triple zero to the origin. A feasible normal set $\mathcal{N} \in T^3(18)$ for $\langle P \rangle$ is (in the sequence of the components in the normal set vector) $\{1, x, y, z, x^2, xy, y^2, xz, yz, \ldots\}$, the associated multiplication matrix A_x has an invariant subspace X_0 of dimension 3, with eigenvalue 2:

$$
\begin{pmatrix}
0 & 1 & 0 & 0 & 0 & 0 & 0 & 0 & 0 & \ldots \\
0 & 0 & 0 & 0 & 1 & 0 & 0 & 0 & 0 & \ldots \\
0 & 0 & 0 & 0 & 0 & 1 & 0 & 0 & 0 & \ldots \\
0 & 0 & 0 & 0 & 0 & 0 & 0 & 1 & 0 & \ldots \\
-10 & 5 & 1 & 2 & 3 & 6 & 5 & 0 & -2 & \ldots \\
 & & & & \ldots
\end{pmatrix}
\begin{pmatrix}
1 & 0 & 0 \\
2 & 1 & 0 \\
-1 & 1 & 0 \\
1 & 0 & -2 \\
4 & 4 & 1 \\
-2 & 1 & 1 \\
1 & -2 & 1 \\
2 & 1 & -4 \\
-1 & 1 & 2 \\
\vdots & \vdots & \vdots
\end{pmatrix}
$$

$$
= \begin{pmatrix}
| & | & | \\
x_{01} & x_{02} & x_{03} \\
| & | & |
\end{pmatrix}
\begin{pmatrix}
2 & 1 & 0 \\
 & 2 & 1 \\
 & & 2
\end{pmatrix},
$$

where the dots in the first 5 rows of A_x are short for 9 further 0 elements. It turns out that this subspace X_0 is a joint invariant subspace of A_x, A_y, A_z, with

$$T_{01} = \begin{pmatrix} 2 & 1 & 0 \\ & 2 & 1 \\ & & 2 \end{pmatrix}, \quad T_{02} = \begin{pmatrix} -1 & 1 & 0 \\ & -1 & 1 \\ & & -1 \end{pmatrix}, \quad T_{03} = \begin{pmatrix} 1 & 0 & -2 \\ & 1 & 0 \\ & & 1 \end{pmatrix}.$$

From the eigenvector $x_{01} = (1, 2, -1, 1, \ldots)^T$ or from the diagonal elements of the $T_{0\sigma}$, we have the position of the triple zero at $z_0 = (2,-1,1)$. The 2nd column of the $T_{0\sigma}$ displays $c_2 = \partial_{100} + \partial_{010}$. Since $2 \in M_1$ and the 3rd columns of T_{01} and T_{02} have a nonvanishing element in row 2, c_3 must be a 2nd order differential. From (8.57), we have

$$c_3 = \partial_{100} c_2 + \ldots = \partial_{010} c_2 + \ldots = -2 \partial_{001} c_1 + \ldots$$

which implies $c_3 = \partial_{200} + \partial_{110} + \partial_{020} - 2 \partial_{001}$, as we had obtained it analytically. All evaluations of the above partials are at z_0, of course. $\square$

Example 8.35: We take the system $P(x_1, x_2, x_3)$ of Example 8.30. A Groebner basis algorithm for the term order `tdeg(x3,x2,x1)` produces the normal set $\{1, x_1, x_2, x_3, x_1^2, x_1 x_2, x_1 x_3,$ $x_2 x_3, x_1^3, x_1^2 x_2, x_1^2 x_3, x_1^4\} \in T^3(12)$ and associated multiplication matrices $A_1, A_2, A_3 \in \mathbb{C}^{12 \times 12}$. These matrices have a joint invariant subspace of dimension 6 which is spanned by the vectors $x_{01} = (1, 0, .., 0)^T$, $x_{02} = (0, 1, 0, .., 0)^T$, $x_{03} = (0, 0, 1, 0, .., 0)^T$, $x_{04} = (0, 0, 0, 0, 4, -3, 0, .., 0)^T$, $x_{05} = (0, 0, 0, 2, 2, 3, 0, .., 0)^T$, $c_{06} = (0, .., 0, 21, -1, 51, 9, 0, 0)^T$, with a decomposition (8.54) with the matrices

$$\begin{pmatrix} 0 & 1 & 0 & 0 & 0 & 0 \\ & 0 & 0 & 4 & 2 & 0 \\ & & 0 & -3 & 3 & 0 \\ & & & 0 & 0 & \frac{15}{2} \\ & & & & 0 & \frac{21}{2} \\ & & & & & 0 \end{pmatrix}, \quad \begin{pmatrix} 0 & 0 & 1 & 0 & 0 & 0 \\ & 0 & 0 & -3 & 3 & 0 \\ & & 0 & -4 & 2 & 0 \\ & & & 0 & 0 & \frac{5}{2} \\ & & & & 0 & \frac{-1}{2} \\ & & & & & 0 \end{pmatrix}, \quad \begin{pmatrix} 0 & 0 & 0 & 0 & 2 & 0 \\ & 0 & 0 & 0 & 0 & 21 \\ & & 0 & 0 & 0 & -1 \\ & & & 0 & 0 & 0 \\ & & & & 0 & 0 \\ & & & & & 0 \end{pmatrix},$$

as T_{01}, T_{02}, T_{03}. The 0 diagonals appear because the multiple zero is at $(0,0,0)$. Furthermore, we see that none of the A_σ is nonderogatory: Besides the only joint eigenvector x_{01}, there are the additional eigenvectors x_{03} for A_1, x_{02} for A_2, and x_{02}, x_{03}, x_{04} for A_3; but none of these is a joint eigenvector.

The functionals $c_2 = \partial_{100}$ and $c_3 = \partial_{010}$ are immediately read from T_{01} and T_{02}; thus $M_1 = \{2, 3\}$. The 5th column of T_{03} might indicate another first-order differential, but the 5th columns of the other $T_{0\sigma}$ have nonvanishing elements in rows 2, 3 $\in M_1$. Thus we have, for $\mu = 4$ and 5, and for $\sigma = 1, 2, 3$,

$$c_\mu = t_{1\mu}^{(\sigma)} \partial_{e_\sigma} + t_{2\mu}^{(\sigma)} \partial_{e_\sigma} c_2 + t_{3\mu}^{(\sigma)} \partial_{e_\sigma} c_3 + \ldots,$$

which yields

$$c_4 = \partial_{100}(4 c_2 - 3 c_3) + \ldots = \partial_{010}(-3 c_2 - 4 c_3) + \ldots = 4 \partial_{200} - 3 \partial_{110} - 4 \partial_{020},$$
$$c_5 = \partial_{100}(2 c_2 + 3 c_3) + .. = \partial_{010}(3 c_2 + 2 c_3) + .. = 2 \partial_{001} + ..$$
$$= 2 \partial_{200} + 3 \partial_{110} + 2 \partial_{020} + 2 \partial_{001}.$$

There are no further 2nd order differentials and the nonvanishing elements in rows 4, 5 $\in M_2$ signal a third order functional c_6. From the generalization of (8.57), we have

$$c_6 = \partial_{100}\left(\tfrac{15}{2}c_4 + \tfrac{21}{2}c_5\right) + \ldots = \partial_{010}\left(\tfrac{5}{2}c_4 + \tfrac{-1}{2}c_5\right) + \ldots = \partial_{001}\left(21\,c_2 - c_3\right) + \ldots\,.$$

This yields a consistent representation of c_6 as

$$c_6 = 51\,\partial_{300} + 9\,\partial_{210} - 9\,\partial_{120} - 11\,\partial_{030} + 21\,\partial_{101} - \partial_{011}\,.$$

Thus we have found the same expressions as previously for the basis functionals of the dual space $\mathcal{D}_0$ which defines the multiplicity structure of the 6-fold zero at $(0,0,0)$. $\square$

Exercises

1. For $s = 2$ and 3, interpret the vanishing of various differential functionals geometrically in terms of the manifolds $p_\nu = 0$, $\nu = 1(1)s$; cf. our interpretation of (8.48).

2. Consider the polynomial system P specified by

$$\begin{aligned}
p_1 &= 2x_1^2 + 3x_1x_2 + 3x_2^2 + 15x_1 + 21x_2 + 5x_3 + 35\,,\\
p_2 &= x_1^2 - 6x_1x_3 - 4x_3^2 + 5x_1 + 7x_2 - 6x_3 + 25\,,\\
p_3 &= x_2^2 - 3x_2x_3 - 3x_3^2 - 3x_1 + 13x_2 - 4x_3 + 22\,.
\end{aligned}$$

(a) Form the Jacobian P' of P and check the overdetermined system $\{p_1, p_2, p_3, \det(P')\}$ $=: \bar{P}$ for a potential common zero, i.e. a multiple zero z_0 of P. (Compute a Groebner basis of $\bar{P}$.) Shift the origin of the $\mathbb{C}^3$ to z_0 so that the transformed system has no constant terms.

(b) Find the multiplicity and the multiplicity structure of the multiple zero 0 of the transformed system. (Use the approach of section 8.5.1.)

(c) Find a normal set and a border basis for the primary ideal of the multiple zero, in the transformed and the original coordinates.

3. For some polynomial ideal $\mathcal{I} \subset \mathcal{P}^3$, let the family $\overline{A}$ of multiplication matrices with respect to the normal set $\{1, x, y, z, \ldots\}$ have a joint invariant subspace X_0 of dimension 4. When its basis vectors x_1, x_2, x_3, x_4, in suitable order, are chosen as the first 4 vectors in the matrix $X = \mathbf{c}^T(\mathbf{b})$, the following upper triangular matrices are obtained from (cf. Example 8.34)

$$A_\sigma \begin{pmatrix} | & | & | & | \\ x_1 & x_2 & x_3 & x_4 \\ | & | & | & | \end{pmatrix} = \begin{pmatrix} | & | & | & | \\ x_1 & x_2 & x_3 & x_4 \\ | & | & | & | \end{pmatrix} T_{0\sigma}\,, \qquad \sigma = x, y, z\,:$$

$$T_{0x} = \begin{pmatrix} 5 & 1 & 0 & -1 \\ & 5 & 0 & 2 \\ & & 5 & -3 \\ & & & 5 \end{pmatrix}, \quad T_{0y} = \begin{pmatrix} 2 & 3 & 1 & 2 \\ & 2 & 0 & 3 \\ & & 2 & -9 \\ & & & 2 \end{pmatrix}, \quad T_{0z} = \begin{pmatrix} -3 & 0 & -1 & 1 \\ & -3 & 0 & 3 \\ & & -3 & 0 \\ & & & -3 \end{pmatrix}.$$

(a) What is the location and the multiplicity structure of the associated 4-fold zero z_0 of $\mathcal{I}$?

(b) Determine the multiplication matrices A_{0x}, A_{0y}, A_{0z} and the border basis $\mathcal{B}_0$ of the ideal $\mathcal{I}_0$ of z_0 with respect to the normal set $\mathcal{N}_0 = \{1, x, y, z\}$.

(c) Convince yourself that none of the multiplication matrices of $\mathcal{I}_0$ is nonderogatory and that none of the 3-element border basis subsets derived from the quasi-univariate structure of $\mathcal{N}_0$ generates $\mathcal{I}_0$ (but a positive-dimensional ideal). To establish $\mathcal{I}_0$ as a complete intersection, find another 3-element subset of $\mathcal{B}_0$ which generates $\mathcal{I}_0$.

4. In $\mathcal{P}^3$, consider the differentiation functionals (with evaluation at a fixed $z_0 \in \mathbb{C}^3$)

$$c_1 = \partial_0, \quad c_2 = \partial_{100} + \partial_{010}, \quad c_3 = \partial_{100} + \partial_{001}, \quad c_4 = \partial_{200} + \partial_{110} + \partial_{020} + 3\,\partial_{100},$$
$$c_5 = \partial_{300} + \partial_{210} + \partial_{120} + \partial_{030} + 6\,\partial_{200} + 3\,\partial_{110} - \partial_{020} + \partial_{011} + \partial_{002} - 2\,\partial_{010} + \partial_{001}.$$

Let $\mathcal{D}_0 = \operatorname{span}(c_1, \ldots, c_5)$, with $z_0 = (1, -2, 3)$.

(a) Verify that each subset $(c_1, \ldots, c_\mu)$, $\mu = 2, 3, 4, 5$, is closed.

(b) Find a feasible normal set basis $\mathcal{N}_0 \in \mathcal{T}^3(5)$ for $\mathcal{R}_0 = \mathcal{R}[\mathcal{D}_0]$ (i.e. that $\mathbf{c}^T(\mathbf{b})$ is regular). Verify that all normal sets with $1, x_1, x_2, x_3$, and some quadratic monomial are feasible. Verify that all these normal sets are quasi-univariate.

(c) For a chosen $\mathcal{N}_0$ and associated normal set vector $\mathbf{b}$, determine the multiplication matrices A_{01}, A_{02}, A_{03} of $\mathcal{R}_0$ via (8.53) and the associated border basis $\mathcal{B}_0$ of $\mathcal{I}_0 = \mathcal{I}[\mathcal{D}_0]$. Check whether there exist 3-element subsets $\mathcal{B}_c$ of $\mathcal{B}_0$ (complete intersection bases) such that $\langle \mathcal{B}_c \rangle = \langle \mathcal{B}_0 \rangle = \mathcal{I}_0$.

(d) Form polynomial combinations of the polynomials in $\mathcal{B}_0$ and check that they are annihilated by the functionals of $\mathcal{D}_0$.

(e) Compute the associated upper-triangular matrices $T_{0\sigma}$, $\sigma = 1(1)3$, of (8.54)

    (i) by representing the columns of $\mathbf{c}^T(x_\sigma \mathbf{b})$ in terms of the basis $\mathbf{c}^T(\mathbf{b})$,

    (ii) from the interpretation of the elements of the $T_{0\sigma}$ in terms of the c_μ as in section 8.5.3.

Historical and Bibliographical Notes 8

The central role of the S-polynomial criterion in the work of Buchberger has dominated the attitude of computer algebraists towards basis representations of polynomial ideals for decades. There appear to have been no serious attempts to develop alternate approaches and criteria, at least for the important special case of 0-dimensional polynomial ideals where the finite dimension of the quotient ring and the dual space designates those as natural representations. Not even the gradual recognition of the Central Theorem (cf. section 2.4) as the proper tool for the determination of the zeros of a 0-dimensional polynomial system P during the 1990s changed that situation, at first: Normal sets and multiplication matrices for $\mathcal{R}[\langle P \rangle]$ had to be found via Groebner bases for $\langle P \rangle$.

Eventually, the related but largely independent efforts of B. Mourrain and the author of this book have initiated a change: In numerous conference presentations, I have emphasized the desirability of a direct determination of a basis for $\mathcal{R}[\langle P \rangle]$ and of its multiplicative structure. The formal basis for the novel approach was laid by Mourrain's paper [8.1] and further work. Simultaneously, there began first attempts towards an algorithmic realization, e.g., in [8.2].

Various open questions remained: For the (generally) overdetermined representation of $\langle P \rangle$ by a reduced Groebner basis, the S-polynomial criterion (Theorem 8.34) guarantees completeness and consistency in a minimal way. For a highly overdetermined normal set representation $(\mathcal{N}, \mathcal{B}_\mathcal{N})$ of $\langle P \rangle$, the commutativity of the multiplication matrices A_σ constitutes a sufficient but highly *redundant* set of conditions. Are there minimal sets of conditions which *guarantee* commutativity and hence consistency? Preliminary answers to this and other natural questions have been given in sections 8.1 and 8.2.

While the regularity of a *linear* multivariate system is considered as its most important property, the analogous regularity property of polynomial systems—often denoted by the term "complete intersection"—has received far less attention. From the numerical point of view, it is crucial in almost every aspect, as will be seen in Chapter 9. Also, the remarkable fact that the number of zeros of a 0-dimensional regular system can be precisely determined from its sparsity pattern by *symbolic* computation has not been widely utilized in polynomial algebra so far. Actually, the knowledge of the BKK-number of a regular polynomial system makes the use of term order obsolete in many respects (cf., e.g., Chapter 10).

One of the reasons for the introduction of dual spaces in [2.6] was the analysis of the structure of multiple zeros of polynomial systems. The content of section 8.5 is largely from [8.3]. In spite of their many facets, multiple zeros have attracted little attention in polynomial algebra so far.

In view of the immense literature on the subject of Groebner bases (cf., e.g., [2.10]–[2.13]), further notes on this subject appear unnecessary.

References

[8.1] B. Mourrain: A New Criterion for Normal Form Algorithms, in: Applied Algebra, Algebraic Algorithms and Error-Correcting Codes, Lecture Notes in Comput. Science Vol. 1719, Springer, Berlin, 1999, 430–443.

[8.2] B. Mourrain, Ph. Trébuchet: Solving Projective Complete Intersections Faster, in: Proceed. ISSAC 2000 (Ed. C. Traverso), ACM, New York, 231–238, 2000.

B. Mourrain, Ph. Trébuchet: Normal Form Computation for Zero-Dimensional Ideals, in: Proceed. ISSAC 2002 (Ed. T. Mora), ACM, New York, 2002.

[8.3] H.J. Stetter: Analysis of Zero Clusters in Multivariate Polynomial Systems, in: Proceed. ISSAC 1996 (Ed. Y.N. Lakshman), ACM, New York, 127–135, 1996.

G. Thallinger: Analysis of Zero Clusters in Multivariate Polynomial Systems, Diploma Thesis, Tech. Univ. Vienna, 1996.

[8.4] J. Dieudonné: Foundations of Modern Analysis, Academic Press, New York, 1968 (7th Ed.).

[8.5] J. Verschelde: Algorithm 795: PHC-pack: A General-Purpose Solver for Polynomial Systems by Homotopy Continuation, Trans. Math. Software **25** (1999), 251–276.

[8.6] H.J. Stetter: Stabilization of Polynomial Systems Solving with Groebner Bases, in: Proceed. ISSAC 1997 (Ed. W. Kuechlin), ACM, New York, 117–124, 1997.

Chapter 9

Systems of Empirical Multivariate Polynomials

Empirical polynomials, i.e. polynomials with some coefficients of limited accuracy, have been introduced in Chapter 3; in the multivariate setting, they have been further considered in section 7.2. Everything said there now refers to the individual empirical polynomials $(\bar{p}_\nu, e_\nu)$ of a system $(\overline{P}, E)$ of such polynomials. For easy reference, we reformulate the definitions thus obtained:

Definition 9.1. (Compare Definitions 3.3–3.5 and Definition 7.1) A *system* $(\overline{P}, E) = \{(\bar{p}_\nu, e_\nu),$ $\nu = 1(1)n\}$ *of empirical polynomials* in $s > 1$ variables defines a family of neighborhoods $N_\delta(\overline{P}, E)$ of the system $\overline{P} = \{\bar{p}_\nu, \nu = 1(1)n\} \in (\mathcal{P}^s)^n$:

$$N_\delta(\overline{P}, E) := \{\widetilde{P} \in (\mathcal{P}^s)^n : \widetilde{P} = \{\tilde{p}_\nu, \nu = 1(1)n\}, \text{ with } \tilde{p}_\nu \in N_\delta(\bar{p}_\nu, e_\nu) \,\forall\, \nu\}, \qquad (9.1)$$

where $N_\delta(\bar{p}_\nu, e_\nu)$ is defined by (7.11). $\qquad \square$

Note that we employ *one common* validity parameter δ for the n empirical polynomials in the system $(\overline{P}, E)$. This makes it desirable that the individual tolerance vectors e_ν are *compatible* in the following sense: For a fixed $\delta > 0$, the neighborhoods $N_\delta(p_\nu, e_\nu)$, $\nu = 1(1)n$, include instances of polynomials of (approximately) the same validity.

Remember that an s-variate empirical polynomial $(\bar{p}_\nu, e_\nu)$ has an *empirical support* $\widetilde{J}_\nu \subset \mathbb{N}_0^s$ which contains the subscript vectors j of those M_ν coefficients $(\bar{\alpha}_{\nu j}, \varepsilon_{\nu j})$ of $(\bar{p}_\nu, e_\nu)$ which are empirical; cf. Definition 7.1. The neighborhoods $N_\delta(\bar{p}_\nu, e_\nu)$ are defined in terms of weighted norms $\|..\|_{e_\nu}^*$ in the *data spaces* $\mathcal{A}_\nu$ of these coefficients, or the increment data spaces $\Delta\mathcal{A}_\nu$ with origins at $\bar{a}_\nu := (\ldots, \bar{\alpha}_{\nu j}, \ldots)$; cf. (5.32), and the remarks in section 5.2.1 on the one hand, and (7.11) and the remarks in section 7.2.1 on the other hand.

For the empirical system $(\overline{P}, E)$, we have the combined data space $\mathcal{A} := \prod_\nu \mathcal{A}_\nu$, or the increment data space $\Delta\mathcal{A} := \prod_\nu \Delta\mathcal{A}_\nu$, with origin at $\bar{a} := (\bar{a}_\nu, \nu = 1(1)n)$. In accordance with (9.1), we define the norm in $\Delta\mathcal{A}$ as

$$\|\Delta a\|_E^* := \max_\nu \|\Delta a_\nu\|_{e\nu}^*. \qquad (9.2)$$

For a *system* of empirical multivariate polynomials even more than for a single such polynomial, it is important to realize that—in our context—a neighborhood of a polynomial system always refers to *one particular representation* of that system and that it is generally not meaningful to transform this neighborhood into a new one when the representation is transformed; cf. the last part of section 7.2.1 and Example 7.5. For empirical systems, there is the further difficulty that the transformed system may have *more equations* than the original regular system; in this case, the variation of the coefficients of the new representation must be restricted to a submanifold of the new data space. We will have to deal with such problems in particular contexts, but one should definitely not consider the transformation of neighborhoods of polynomial systems as a general tool.

In section 3.2, we have introduced the *data→result mapping* F from a domain A in the space $\mathcal{A}$ of the empirical data to some result space $\mathcal{Z}$ (cf. Definition 3.6) and the *pseudoresult sets* $Z_\delta \subset \mathcal{Z}$ which are the images of the $N_\delta(\bar{a}, e)$ neighborhoods under F (cf. Definition 3.7). We have also introduced the *equivalent-data manifold* $\mathcal{M}(\tilde{z}) := \{\tilde{a} \in \mathcal{A} : F(\tilde{a}) = \tilde{z}\}$ (cf. Definition 3.11) which permits the definition of the *backward error* of an approximate result $\tilde{z} \in \mathcal{Z}$ as the smallest δ such that $\tilde{z}$ is the exact result of the algebraic problem for some data $\tilde{a} \in N_\delta(\bar{a}, e)$ (cf. Definition 3.12 and Definition 7.2). $\tilde{z}$ is a *valid* approximate result if its backward error is $\leq O(1)$, i.e. if it is the exact result for valid data $\tilde{a}$ (cf. Definition 3.8). All this can be immediately applied to systems of polynomials: With (9.1) and (9.2) and the above conventions, and with the backward errors $\delta_\nu(\tilde{z})$ for the individual polynomials p_ν, the *backward error* of the approximate result $\tilde{z}$ for $(\overline{P}, E)$ is

$$\delta(\tilde{z}) := \max_{\nu} \delta_\nu(\tilde{z}). \quad \Box \tag{9.3}$$

It is true that it may often be possible and meaningful to assess the backward errors $\delta_\nu(z)$ individually. But, formally at least, we will use *one* backward error (9.3) for the result of a computational task with an empirical system.

9.1 Regular Systems of Empirical Polynomials

In the introductory remarks of Chapter 8, we have specified that, until later, we deal with *regular* systems of s polynomials in s variables ($s > 1$) only. There, our intuitive definition of regularity included a reference to the embedding of the system. For an empirical polynomial system, this embedding is formally specified:

Definition 9.2. An empirical polynomial system $(\overline{P}, E) = \{(\bar{p}_\nu, e_\nu), \nu = 1(1)s\} \subset (\mathcal{P}^s)^s$ is *regular* if all systems $\widetilde{P} \in N_\delta(\overline{P}, E)$, $\delta = O(1)$, are consistent and 0-dimensional, with the same number m of zeros (counting multiplicities). $\quad \Box$

The regularity of an empirical polynomial system is often known from the context in which it has arisen; otherwise, there is no a priori test for regularity. The necessary condition $n = s$ has been incorporated into Definition 9.2. For regular systems of empirical multivariate polynomials, we can transcribe most of the concepts which we have introduced for one univariate empirical polynomial in section 5.2.

9.1.1 Backward Error of Polynomial Zeros

From the numerical point of view, the most important task in the context of a polynomial system is the computation of approximations for some or all of its zeros. In an empirical system, zeros have a natural indetermination which permits the use of approximate computation. Therefore, it is of principal importance to check the *validity* of an approximate zero, independently of the way in which it has been found. This is done by the computation of its backward error; cf. the remarks in connection with (9.3). With the notation explained at the beginning of section 7.2.2, we have

Proposition 9.1. For $\tilde{z} \in \mathbb{C}^s$, let $\tilde{\mathbf{z}}_\nu \in \mathbb{C}^{M_\nu}$ be the column vector $(\tilde{z}^j,\ j \in \tilde{J}_\nu)$. Then the backward error of $\tilde{z}$ as an approximate zero of the empirical system $(\overline{P}, E)$ is

$$\delta(\tilde{z}) \ := \ \max_\nu \ \frac{|\bar{p}_\nu(\tilde{z})|}{\|\tilde{\mathbf{z}}_\nu\|_{e_\nu}} \ . \tag{9.4}$$

If one or several of the $\tilde{\mathbf{z}}_\nu$ vanish, the backward error is not defined.

Proof: Compare (9.3) and Proposition 7.4.   □

The surprising fact that there is an explicit expression for the backward error of an approximate zero of any multivariate polynomial system is simply a consequence of the *linearity* of polynomials in their coefficients, as we have remarked on various occasions; the respective equivalent-data manifolds are hyperplanes in the data spaces ΔA_ν whose minimal distance from the origin is explicitly known. In the exceptional case mentioned in Proposition 9.1, $\tilde{z}$ cannot be interpreted as an exact zero of a polynomial system in $N_\delta(\overline{P}, E)$ for any δ; cf. the remark below Proposition 7.4.

With our convention to use only *one* tolerance parameter δ for an empirical system (cf. (9.1), the backward error on an approximate zero will generally stem from one particular polynomial in $(\overline{P}, E)$: This polynomial of "worst fit" requires the largest modification in its empirical coefficients to accommodate $\tilde{z}$ as an exact zero. For some of the other polynomials in the system, the necessary modifications may be much smaller, in terms of their respective $\|..\|_{e_\nu}^*$ norms. It will rarely be meaningful to utilize the full extent of $\delta(\tilde{z})$ also for the modifications of these polynomials; rather one should use the individual minimal-norm modifications from (7.14) for each polynomial $\bar{p}_\nu$.

Example 9.1: Consider the following regular system $\overline{P}$ in 3 variables (called x, y, z):

$$\begin{aligned}
\bar{p}_1(x, y, z) &= 1.853\,x^3 - 5.192\,xy^2 + 2.397\,x^2z + 4.862\,y^2z - 5.227\,yz^2 \\
&\quad +.864\,x^2 - 2.077\,y^2 + 4.312\,yz + 5.113\,x - 4.728\,z\,, \\
\bar{p}_2(x, y, z) &= 5.338\,xy - 3.286\,xz + 3.117\,z^2 + 4.319\,y - 5.223\,, \\
\bar{p}_3(x, y, z) &= 2.451\,x^2 - .973\,y^2 + 4.286\,yz + 6.210\,y + 4.771\,z - 8.642\,.
\end{aligned} \tag{9.5}$$

Assume that all coefficients are empirical and have a tolerance of .001.

Consider the approximate zero $\tilde{u} = (\tilde{\xi}, \tilde{\eta}, \tilde{\zeta}) = (.751, -.112, 1.855)$. Its (max-norm) backward errors for the individual equations

$$\delta_\nu(\tilde{u}) \ = \ |p_\nu(\tilde{u})|/(.001 \sum_{j \in J_\nu} |\tilde{u}^j|)\,, \quad \nu = 1(1)3\,,$$

are $\delta_1 \approx 1.16$, $\delta_2 \approx 1.29$, $\delta_3 \approx 2.03$, which implies a system backward error of 2.03; cf. (9.4). According to our lenient concept of backward error assessment, we should accept this approximate zero as valid for the empirical system $(\overline{P}, E)$; cf. (3.3).   □

As in the univariate case, the polynomial system with an exact *complex* zero $\tilde{z}$ which is closest to a *real* system $\bar{P}$ has complex coefficients and may not be admissible if the variation of the coefficients is restricted to the real domain. In this case, one has to look for the closest system which has both $\tilde{z}$ and its conjugate-complex value $\tilde{z}^*$ as simultaneous zeros. More generally, if we want to assess the *simultaneous* validity of *several* approximate zeros $\tilde{z}_\mu$, $\mu = 1(1)m$, we have to proceed as in section 5.2.2 and form the corresponding equivalent-data manifolds $\mathcal{M}_\nu(\tilde{z}_1, \ldots, \tilde{z}_m)$ in the data spaces $\Delta\mathcal{A}_\nu$ of the individual polynomials $(\bar{p}_\nu, e_\nu)$:

$$\mathcal{M}_\nu(\tilde{z}_1, \ldots, \tilde{z}_m) := \{\Delta\alpha_{\nu j} \in \Delta\mathcal{A}_\nu : \sum_{j \in \tilde{J}_\nu} \Delta\alpha_{\nu j} \tilde{z}_\mu^j + \bar{p}_\nu(\tilde{z}_\mu) = 0, \quad \mu = 1(1)m\}; \quad (9.6)$$

cf. (5.39). The $\mathcal{M}_\nu$ have codimension m (except in some degenerate situations).

The backward error of the $\tilde{z}_\mu$, $\mu = 1(1)m$, as *simultaneous* approximate zeros of the empirical system $(\bar{P}, E)$ is now defined as (cf. (7.12) and (9.4))

$$\delta(\tilde{z}_1, \ldots, \tilde{z}_m) = \max_\nu \delta_\nu(\tilde{z}_1, \ldots, \tilde{z}_m) = \max_\nu \min_{\Delta a_\nu \in \mathcal{M}_\nu(\tilde{z}_1,\ldots,\tilde{z}_m)} \|\Delta a_\nu\|_{e_\nu}^* . \quad (9.7)$$

The determination of the individual δ_ν is a standard minimization problem as previously, but there is no longer a closed form solution. Naturally, there must be sufficiently many empirical coefficients in *each* of the $(\bar{p}_\nu, e_\nu)$ to make the minimization feasible, i.e. we must have $M_\nu \geq m$ for all ν.

In the case of a complex approximate zero $\tilde{w} \in \mathbb{C}^s$ of a real empirical system, i.e. with the tolerance neighborhoods of all coefficients restricted to the real domain, it is advantageous to replace the defining equations

$$\sum_{j \in \tilde{J}_\nu} \Delta\alpha_{\nu j} \tilde{w}^j + \bar{p}_\nu(\tilde{w}) = \sum_{j \in \tilde{J}_\nu} \Delta\alpha_{\nu j} (\tilde{w}^*)^j + \bar{p}_\nu(\tilde{w}^*) = 0$$

in (9.6) by

$$\sum_{j \in \tilde{J}_\nu} \Delta\alpha_{\nu j} \operatorname{Re}(\tilde{w}^j) + \operatorname{Re}(\bar{p}_\nu(\tilde{w})) = 0, \qquad \sum_{j \in \tilde{J}_\nu} \Delta\alpha_{\nu j} \operatorname{Im}(\tilde{w}^j) + \operatorname{Im}(\bar{p}_\nu(\tilde{w})) = 0, \quad (9.8)$$

in order to obtain a fully real minimization problem.

Example 9.1, continued: Consider the complex approximate zero $\tilde{w} = (-2.295 - .002\,\mathrm{i}, -1.240 + .755\,\mathrm{i}, -2.185 - .985\,\mathrm{i})$. When we form the backward errors $\delta_\nu(\tilde{w})$ for this individual complex zero, we obtain, from (7.13), $\delta_1 \approx .16$, $\delta_2 \approx .36$, $\delta_3 \approx .15$; these are all well below 1 but their realization requires a complex modification of the $\bar{p}_\nu$ to make $\tilde{w}$ an exact zero.

If we employ (9.8), we obtain $\delta_1 \approx .23$, $\delta_2 \approx .66$, $\delta_3 \approx .65$; this shows that there is a valid *real* neighboring system $\tilde{P}$ which has $\tilde{w}$ as an exact zero.

In view of the small imaginary part in the nearly real $\tilde{\xi}$-component above, one may ask whether this imaginary part may be spurious. The backward errors of $\tilde{w}$ with the imaginary part of $\tilde{\xi}$ omitted turn out as 5.31, .85, 7.48. Thus one will probably hesitate to accept the modified zero but rather look for a near-by approximate zero with a real ξ-component but a smaller backward error.

As to be expected from the mild sparsity of (9.5), there are $\mathrm{BKK}(\overline{P}) = m_{\mathrm{B\acute{e}zout}}(\overline{P}) = 12$ zeros. Since the numbers M_ν of empirical coefficients in the $(\bar{p}_\nu, e_\nu)$ are only $10, 5, 6$, resp., it would not be possible to find a neighboring system *of the same sparsity* with exact zeros at, say, 6 specified locations. Specifications $\tilde{w}_\mu$ for 5 zeros would generally define a *unique* $\tilde{p}_2$ since the correction $\tilde{p}_2 - \bar{p}_2$ must be the *interpolation polynomial* of the negative residuals of $\bar{p}$ at the zeros:

$$(\tilde{p}_2 - \bar{p}_2)(\tilde{w}_\mu) \;=\; -\,\bar{p}_2(\tilde{w}_\mu)\,, \quad \mu = 1(1)5\,;$$

this linear system for the 5 $\Delta\alpha_{2j}$, $j \in \tilde{J}_2$, will generally have a unique solution. These $\Delta\alpha_{2j}$ would supposedly determine the backward error in this case because of the remaining minimization potential in the other two polynomials. $\square$

The assessment of approximate *multiple* zeros is much more delicate in several variables than in one variable; we delay its treatment to section 9.3.

9.1.2 Pseudozero Domains for Multivariate Empirical Systems

In section 3.1.3, we have introduced the family of δ-pseudoresult sets Z_δ, $\delta > 0$, of an empirical algebraic problem; cf. Definition 3.7 and (3.16). For zeros of univariate polynomials, the introduction of δ-pseudoresult sets Z_δ has naturally led to the δ-pseudozero domains of Definition 5.3; cf. (5.37) and (5.39). For empirical polynomial systems in s variables, we have analogously:

Definition 9.3. For a regular empirical polynomial system $(\overline{P}, E)$, $\overline{P} \in (\mathcal{P}^s)^s$,

$$Z_\delta(\overline{P}, E) \;:=\; \{\tilde{z} \in \mathbb{C}^s : \exists\, \tilde{P} \in N_\delta(\overline{P}, E) \text{ with } \tilde{P}(\tilde{z}) = 0\} \subset \mathbb{C}^s\,, \quad 0 < \delta < \bar{\delta}, \qquad (9.9)$$

is the family of $(\delta\text{-})$*pseudozero sets* of $(\overline{P}, E)$. (The cut-off bound $\bar{\delta} > 1$ is to avoid the loss of regularity if necessary.) $\square$

Due to the continuity of polynomial zeros (cf. section 8.3.2), $\lim_{\delta\to 0} Z_\delta(\overline{P}, E) = Z(\overline{P})$, the zero set of the specified system $\overline{P}$, which consists of individual points $\bar{z}_\mu$, $\mu = 1(1)m$, in the regular case. This indicates that—at least for small values of δ—$Z_\delta(\overline{P}, E)$ must consist of m disjoint domains. With increasing δ, some of these domains may meet and merge.

Definition 9.4. A connected subset of $Z_\delta(\overline{P}, E)$ is a $(\delta\text{-})$*pseudozero domain* of $(\overline{P}, E)$. $\square$

For sufficiently small δ, pseudozero domains $Z_{\delta,\mu}$ which issue from a *simple* zero $\bar{z}_\mu$ of $\overline{P}$ cannot contain another zero of $\overline{P}$. In this case, they contain exactly one simple zero $\tilde{z}_\mu$ of each $\tilde{P} \in N_\delta(\overline{P}, E)$, like in the univariate case; cf. Proposition 5.7.

Definition 9.5. For a fixed δ of $O(1)$, a pseudozero domain $Z_{\delta,\mu}(\overline{P}, E)$ has *multiplicity m_μ* if it contains $m_\mu > 1$ zeros of $\overline{P}$ (counting multiplicities); in this case, it is called an *m-cluster domain* (for tolerance level δ); cf. Definition 6.6. $\square$

Proposition 9.2. For a fixed value of δ, if a pseudozero domain $Z_{\delta,\mu}$ contains m_μ zeros of $\overline{P}$, it also contains m_μ zeros (counting multiplicities) of each $\tilde{P} \in N_\delta(\overline{P}, E)$.
Proof: The proof of the analogous Proposition 6.12 in the univariate case may be transcribed in an obvious manner. Details of the splitting mechanism of potential multiple zeros will be analyzed in section 9.3.4. $\square$

Obviously, the multiplicity of a domain $Z_{\delta,\mu}(\overline{P}, E)$ is an increasing function of δ. A pseudozero domain $Z_{\delta,\mu}$ which envelops an m_μ-fold zero of $\overline{P}$ has multiplicity $m_\mu > 1$ for arbitrarily small values of δ. Otherwise, a multiplicity > 1 can only appear for larger values of δ.

In the univariate case, it is possible to plot the pseudozero domain family of an individual zero in the complex domain; cf. section 5.2.2. Such a plot gives some intuitive impression of the indetermination of the zero due to the indetermination in the empirical coefficients. But it is practically impossible to visualize pseudozero domains in two complex variables, let alone in more than two; cf. section 7.1.1. Therefore, pseudozero domains for multivariate systems may be a useful conceptual tool; their practical importance is negligible.

We may, however, immediately apply our general considerations in section 3.2.2 about the quantitative assessment of the condition of the result of an algebraic problem to zeros of a polynomial system. Obviously, we have the situation of (3.35), with P taking the place of G and some other notational changes. As a function of the empirical coefficients a in P, a zero $z(a)$ must satisfy (cf. (3.36)), with $\bar{\mathbf{a}} := (\bar{a}_\nu, \ \nu = 1(1)s)$,

$$P(z(a); a) \equiv 0 \qquad \text{for } a \in N_\delta(\bar{\mathbf{a}}, E). \tag{9.10}$$

This implies (cf. (3.37))

$$\frac{\partial P}{\partial x}(z(a); a) \cdot z'(a) + \frac{\partial P}{\partial a}(z(a); a) = 0. \tag{9.11}$$

If the associated pseudozero set has multiplicity 1 for $\delta < \bar{\delta}$, the Jacobian matrix $\frac{\partial P}{\partial x}(z(a); a)$ is regular for $a \in N_\delta(\bar{\mathbf{a}}, E)$ and we have

$$z'(a) = -\left(\frac{\partial P}{\partial x}(z(a); a)\right)^{-1} \cdot \frac{\partial P}{\partial a}(z(a); a) \qquad \text{for } a \in N_{\bar{\delta}}(\bar{\mathbf{a}}, E). \tag{9.12}$$

When we denote the inverse $s \times s$-matrix in (9.12) by $K(a) = (K_{\sigma\nu})$, the expression for the derivative of the individual components z_σ of a simple zero with respect to a particular empirical coefficient $\alpha_{\nu j}$ in p_ν becomes

$$\frac{\partial z_\sigma}{\partial \alpha_{\nu j}}(a) = K_{\sigma\nu}\, z(a)^j, \tag{9.13}$$

since the (ν, j)-elements of $\frac{\partial P}{\partial a}(z(a); a)$ are the evaluations of the monomials x^j in p_ν at $z(a)$.

We may now employ these expressions for the sensitivity of simple zeros with respect to perturbations in the coefficients of a polynomial system to derive various kinds of condition estimates. With the linearized approach explained in detail in section 3.2.3, we may further derive quantitative estimates of the potential variation of the components of a simple zero caused by the indetermination of the coefficients of an empirical system; cf. Proposition 3.7 and (3.43). In all this, it is important not to forget the purpose of such estimates: We wish to have an indication of the *number of meaningful digits* in the components of an approximate zero of an empirical polynomial system. Thus it is fully sufficient to perform a rough evaluation of the *order of magnitude* of the respective expressions.

Example 9.2: We take the system (9.5) and analyze the sensitivity of the real zero $\tilde{u}$ and the complex zero $\tilde{w}$ which have been considered in Example 9.1. They are not the exact zeros of

$\overline{P}$, but this does not matter except if $\frac{\partial \overline{P}}{\partial x}$ should be near-singular there. We obtain (approx.) .17 and .45, resp., for $\|(\frac{\partial \overline{P}}{\partial x})^{-1}\|_\infty$ at the two zeros. This tells us that the sensitivity of both zeros with respect to the indetermination in the coefficients is low and that it should be meaningful to display these zeros to 3 decimal digits, as we have done it in Example 9.1.

Actually, from (9.13) we obtain the linearized relation

$$\Delta z_\sigma \approx \sum_\nu K_{\sigma\nu} \sum_{j \in \tilde{J}_\nu} z^j \, \Delta\alpha_{\nu j} \quad \Rightarrow \quad |\Delta z_\sigma| \le \sum_\nu |K_{\sigma\nu}| \left(\sum_{j \in \tilde{J}_\nu} |z^j| \right) \max_j |\Delta\alpha_{\nu j}| ;$$

cf. section 3.2.3. Its application to (9.5) and the two zeros $\tilde{u}$ and $\tilde{w}$, resp., yields, for $\max_j |\Delta\alpha_{\nu j}| = .001$, indetermination bounds of the order .001 for the components of $\tilde{u}$ and a little larger ($\le .003$) for the components of $\tilde{w}$. This confirms our consideration above.

For a numerical test, when we add .001 to each coefficient in (9.5) and compute more accurate approximations for the two zeros of the unmodified and the modified system $\overline{P}$, we obtain the following approximate changes in the components of $\tilde{u}$ and $\tilde{w}$:

$$(.00026, -.00013, .00093), \quad (-.00119 + .00006\,i, -.00032 - .00105\,i, -.00130 + .00034\,i).$$

Naturally, this modification of $\overline{P}$ is not the one which generates the largest changes in the zeros, but its effect agrees with our consideration. $\quad \square$

9.1.3 Feasible Normal Sets for Regular Empirical Systems

In section 2.5.1, we have observed that each element of $T^s(m)$, the set of all closed sets of m monomials in s variables, is a candidate for a monomial basis of the quotient ring $\mathcal{R}[\langle P \rangle]$ of a 0-dimensional polynomial ideal $\langle P \rangle \in \mathcal{P}^s(m)$; in order to qualify as a normal set $\mathcal{N}[\langle P \rangle]$ for the system P, the monomials must satisfy (2.53) in Proposition 2.28 (or its generalization (2.54) for multiple zeros), i.e. they must actually span $\mathcal{R}[\langle P \rangle]$ (cf. Definition 2.20). This means that an arbitrary element $\mathcal{N} \in T^s(m)$ is admissible as a normal set for all ideals $\mathcal{I} \in \mathcal{P}^s(m)$ except those from a "singular" set $S_\mathcal{N}$; in the $(\mathbb{C}^s)^m$ of all m-tuples of zeros, $S_\mathcal{N}$ is an algebraic manifold of codimension 1, represented by the polynomial equation $s_\mathcal{N} = 0$ of (2.53) in the components of the zeros; cf. section 8.4.4.

For a regular empirical system $(\overline{P}, E)$ with m zeros (cf. Definition 9.2), we must employ a normal set which is admissible for *all* systems $\tilde{P} \in N_\delta(\overline{P}, E)$, $\delta = O(1)$, if we wish to avoid principal complications. The existence of such normal sets may appear questionable at first: Although each candidate normal set $\mathcal{N}$ from T^s has a singular manifold $S_\mathcal{N}$ of codimension 1 in $(\mathbb{C}^s)^m$, it could happen that the $S_\mathcal{N}$ for all $\mathcal{N} \in T^s(m)$ intersect in one or several points of $(\mathbb{C}^s)^m$, or that there exist zero constellations in $(\mathbb{C}^s)^m$ which are arbitrarily close to *each* $S_\mathcal{N}$. The first situation is excluded by the fact that there exists at least one normal set for *each* constellation of m zeros (e.g., the one for a Groebner basis of the associated ideal; cf. section 8.4.2), and the second situation cannot occur because there are only finitely many singular manifolds (of increasing codimension) for each pair s, m.

Definition 9.6. For a regular empirical system $(\overline{P}, E)$ with m zeros, a normal set $\mathcal{N} \in T^s$ is *feasible* if it is an admissible normal set for all $\tilde{P} \in N_\delta(\overline{P}, E)$, $\delta = O(1)$. $\quad \square$

For $(\overline{P}, E)$ with a generic specified system $\overline{P}$ and sufficiently small tolerances, *all* $\mathcal{N} \in T^s(m)$ are feasible: A generic point avoids a finite number of manifolds of codimension 1. Critical situations may arise when $\overline{P}$ is degenerate in one of many possible ways (cf. section 8.4.4), or very close to such a degeneration. In many applications, such a situation will be highly probable, or even certain because of the underlying model.

Among feasible normal sets, we would like to select one whose singular manifold is far from the location of the pseudozeros of $(\overline{P}, E)$. Generally, these locations are not known and can only be found with the help of the normal set, and $\mathcal{N}$ is determined within the procedure which finds the multiplicative structure of the quotient ring $\mathcal{R}[\langle P \rangle]$. Thus, the algorithmic selection of a "sufficiently feasible" normal set becomes one of the central problems in the numerical solution of polynomial systems. We have discussed this problem already in section 8.4.4 and found the use of extended Groebner bases as a potential remedy.

Example 9.3: The system P_ε of Example 8.26, with $n = s = 2$, $m = 6$, which describes the intersections of three straight lines through the origin with a near-circular ellipse with center at the origin, has the admissible normal set $\mathcal{N}_\varepsilon = \{1, y, x, y^2, xy, y^3\}$ for $\varepsilon \neq 0$. For $\varepsilon = 0$, the ellipse is a circle and the symmetry of the intersections puts them on the singular manifold $S_{\mathcal{N}_\varepsilon}$. If we now consider the empirical system $(\overline{P}, E)$ with $\overline{P} = P_\varepsilon$, with a small numerical value for ε and a tolerance $\geq \varepsilon$ for the coefficient of xy in $p_1(x, y; \varepsilon)$ in the system, $\mathcal{N}_\varepsilon$ is not a feasible normal set for $(\overline{P}, E)$.

In the continuation of Example 8.26, on the other hand, we have found that the normal set $\mathcal{N}_0 = \{1, y, x, y^2, xy, xy^2\}$ is an admissible normal set for all $|\varepsilon| < 4/\sqrt{3} \approx 2.3$ (for these values of ε, 2 zeros move to infinity); thus is it is definitely a feasible normal set in the above situation. In the *algorithmic* determination of a normal set for $(\overline{P}, E)$, it is therefore important to reach $\mathcal{N}_0$ and not $\mathcal{N}_\varepsilon$. We will pursue this further in section 10.2.2 $\square$

The joint normal set $\mathcal{N}$ for the multitude of systems $\widetilde{P}$ in an empirical polynomial system $(\overline{P}, E)$ provides the common reference frame which makes a joint *algebraic* consideration of the systems $\widetilde{P}$ possible. This explains why we have emphasized the role of the quotient ring and its monomial basis $\mathcal{N}$ in dealing with a 0-dimensional polynomial ideal in Chapter 8. The fact that $\mathcal{N} = \{x^j\}$ is determined by discrete, integer data (the set of exponent vectors j) provides a rigid basis for the handling of the indetermination in the $(\overline{P}, E)$. The expansions (8.37) and (8.38) of an arbitrary $p \in \mathcal{P}^s$ in terms of the specified polynomials $\bar{p}_\nu$ of a regular empirical system $(\overline{P}, E)$ which rely on a fixed normal set $\mathcal{N}$ play a major role in this context.

9.1.4 Sets of Ideals of System Neighborhoods

Consider a neighborhood $N_\delta(\overline{P}, E) \in \mathcal{P}^s$, with fixed $\delta > 0$; cf. Definition 9.1. For a regular empirical system (cf. Definition 9.2), each $\widetilde{P} \in N_\delta(\overline{P}, E)$ defines a 0-dimensional ideal $\widetilde{\mathcal{I}} := \langle \widetilde{P} \rangle \subset \mathcal{P}^s$. We refrain from calling the set of these $\widetilde{\mathcal{I}}$ a neighborhood of ideals because this would indicate the existence of a metric for ideals. We simply denote this set of ideals by

$$\langle N_\delta(\overline{P}, E) \rangle := \{ \langle \widetilde{P} \rangle : \widetilde{P} \in N_\delta(\overline{P}, E) \} \tag{9.14}$$

and we note that an empirical system $(\overline{P}, E)$ defines a family of such sets. The notation of (9.14) also shows that the "closeness" of the ideals in such a set is only defined through the closeness of the polynomial systems in the empirical system $(\overline{P}, E)$, with the particular representation of $\overline{P}$.

The following task plays a fundamental role: Given some $p \in \mathcal{P}^s$, determine its "membership" in $\langle N_\delta(\overline{P}, E)\rangle$, i.e. find whether there exists some $\widetilde{P} \in N_\delta(\overline{P}, E)$ such that $p \in \langle \widetilde{P}\rangle$. Moreover, we want to find the smallest δ such that this is the case. This task will be solved later in this section.

Naturally, each ideal $\widetilde{\mathcal{I}} \in \langle N_\delta(\overline{P}, E)\rangle$ has an associated quotient ring $\widetilde{\mathcal{R}} := \mathcal{P}^s/\widetilde{\mathcal{I}} = \mathcal{R}[\langle \widetilde{P}\rangle]$, with $\widetilde{P} \in N_\delta(\overline{P}, E)$. Analogously to (9.14), we denote the set of these quotient rings by

$$\mathcal{R}[N_\delta(\overline{P}, E)] := \{\, \mathcal{R}[\langle \widetilde{P}\rangle] \,:\, \widetilde{P} \in N_\delta(\overline{P}, E)\,\} \tag{9.15}$$

and note that an empirical system defines a family of such sets.

The essential aspect is that we may assume that all these polynomial rings employ *one and the same feasible normal set* $\mathcal{N}$ as a common monomial basis. Then they differ only in the nontrivial rows of their multiplication matrices with respect to this basis. Each of the elements in these rows lies in a neighborhood of the value for $\mathcal{R}[\langle \overline{P}\rangle]$, but they are also interconnected by the commutativity conditions, as we have seen in section 8.2.3. Thus, given the multiplication matrices $\widetilde{A}_\sigma$, $\sigma = 1(1)s$, of some polynomial ring $\widetilde{\mathcal{R}}$ with basis $\mathcal{N}$, the procedure for determining whether $\widetilde{\mathcal{R}}$ is in $\mathcal{R}[N_\delta(\overline{P}, E)]$ must include a commutativity check. This will also be considered further in section 9.2.

Finally, each ideal $\widetilde{\mathcal{I}} \in \langle N_\delta(\overline{P}, E)\rangle$ has an associated dual space $\widetilde{\mathcal{D}} := \mathcal{D}[\langle \widetilde{P}\rangle]$, with $\widetilde{P} \in N_\delta(\overline{P}, E)$, and we define

$$\mathcal{D}[N_\delta(\overline{P}, E)] := \{\, \mathcal{D}[\langle \widetilde{P}\rangle] \,:\, \widetilde{P} \in N_\delta(\overline{P}, E)\,\}. \tag{9.16}$$

A natural basis of a dual space $\widetilde{\mathcal{D}} = \mathcal{D}[\langle \widetilde{P}\rangle]$ is given by the evaluation functionals at the zeros of the system $\widetilde{P}$, with the extensions introduced in sections 2.3.2 and 8.5.1 for the case of multiple zeros; note that, for $\delta = O(1)$, these zeros must be valid *simultaneous* zeros of the empirical system $(\overline{P}, E)$. The conjugate basis to the normal set basis $\mathcal{N}$ of the associated quotient ring is given by the map from $\mathcal{P}^s$ to the normal set coefficients. While $\mathcal{N}$ is fixed for all $\widetilde{\mathcal{D}} \in \mathcal{D}[N_\delta(\overline{P}, E)]$, this conjugate basis reflects the variations in the $\widetilde{P}$ because it refers to the normal forms mod $\widetilde{\mathcal{I}}$.

Example 9.4: Consider the empirical system $(\overline{P}, E)$ of two quadratic polynomials in 2 variables, with specified polynomials

$$\bar{p}_1 := x^2 + \frac{1}{4} y^2 + 1.576\, x - 2.324\, y - 3.069\,, \quad \bar{p}_2 := x^2 - 3\, xy + 4\, y^2 + 1.234\, x - 1.963\, y - 2.354\,,$$

with intrinsic quadratic terms and with tolerances of .001 on the coefficients of the linear and constant terms. Let $\bar{\mathcal{I}} := \langle (\bar{p}_1, \bar{p}_2)\rangle$ and take the feasible normal set $\mathcal{N} = \{1, y, x, xy\}$ as basis of $\overline{\mathcal{R}} := \mathcal{R}[\bar{\mathcal{I}}]$. The multiplication matrices $\bar{A}_x$, $\bar{A}_y$ of $\overline{\mathcal{R}}$ are (rounded to 4 decimal digits)

$$\begin{pmatrix} 0 & 0 & 1 & 0 \\ 0 & 0 & 0 & 1 \\ 3.1167 & 2.3481 & -1.5988 & -.2000 \\ -.4350 & 2.4550 & .2426 & .2608 \end{pmatrix}, \quad \begin{pmatrix} 0 & 1 & 0 & 0 \\ -.1907 & -.0963 & .0912 & .8000 \\ 0 & 0 & 0 & 1 \\ -.0637 & 2.1781 & -.1424 & .0942 \end{pmatrix},$$

so that the associated border basis $\overline{B}$ of $\overline{\mathcal{I}}$ is (rounded)

$$bb_1 = x^2 + .2000\,xy + 1.5988\,x - 2.3481\,y - 3.1167\,,$$
$$bb_2 = x^2 y - .2608\,xy - .2426\,x - 2.4550\,y + .4350\,,$$
$$bb_3 = xy^2 - .0942\,xy + .1424\,x - 2.1781\,y + .0637\,,$$
$$bb_4 = y^2 - .8000\,xy - .0912\,x + -0963\,y + .1907\,.$$

The zeros of $\overline{P}$ which determine the dual space $\overline{\mathcal{D}} := \mathcal{D}[\overline{\mathcal{I}}]$ are (rounded to 4 decimal digits)

$$(-2.5674, -.2201),\ \ (-1.6638, -1.1222),\ \ (1.1966, .1083),\ \ (1.6966, 1.2319)\,.$$

For $\delta = 1$, the ideal set $\langle N_\delta(\overline{P}, E)\rangle$ contains, e.g., the ideal $\langle(\tilde{p}_1, \tilde{p}_2)\rangle$, with

$$\tilde{p}_1 := x^2 + \frac{1}{4}\,y^2 + 1.577\,x - 2.323\,y - 3.068\,, \qquad \tilde{p}_2 := x^2 - 3\,xy + 4\,y^2 + 1.233\,x - 1.962\,y - 2.353\,.$$

The associated quotient ring $\widetilde{\mathcal{R}}$ has the multiplication matrices $\tilde{A}_x$, $\tilde{A}_y$ (rounded)

$$\begin{pmatrix} 0 & 0 & 1 & 0 \\ 0 & 0 & 0 & 1 \\ 3.1157 & 2.3471 & -1.5998 & -.2000 \\ -.4351 & 2.4540 & .2438 & .2592 \end{pmatrix}, \quad \begin{pmatrix} 0 & 1 & 0 & 0 \\ -.1907 & -.0963 & .0917 & .8000 \\ 0 & 0 & 0 & 1 \\ -.0622 & 2.1785 & -.1424 & .0927 \end{pmatrix};$$

these coefficients also appear in the border basis $\widetilde{B}$ which has the same structure as $\overline{B}$. The zeros of $\widetilde{P}$ are (rounded)

$$(-2.5674, -.2209),\ \ (-1.6649, -1.1222),\ \ (1.1956, .1076),\ \ (1.6958, 1.2319)\,.$$

The data of these and other "neighboring" ideals, quotient rings and dual spaces reflect the closeness of the elements in the sets $\langle N_\delta(\overline{P}, E)\rangle$, $\mathcal{R}[N_\delta(\overline{P}, E)]$, and $\mathcal{D}[N_\delta(\overline{P}, E)]$. $\square$

The essential lesson from this section is the following: What an empirical polynomial system defines is *not* an ideal (quotient ring, dual space) with "vague" data but a *set of* ideals (quotient rings, dual spaces). Each *member* of these sets is completely standard in the sense of algebra and may thus be treated computationally like an ordinary algebraic object.

The introduction of the set $\langle N_\delta(\bar{P}, E)\rangle$ of ideals associated with an empirical polynomial system is meaningful only if we can solve the task specified below (9.14): For $p \in P^s$, what is the smallest $\delta \geq 0$ such that $p \in \langle \widetilde{P}\rangle \in \langle N_\delta(\bar{P}, E)\rangle$.

We assume that we are able to represent p in terms of the specified regular system $\overline{P}$ which is a complete intersection system; cf. (8.37) and (8.38) in section 8.3.3. This generally requires that we can compute normal forms mod $\langle\overline{P}\rangle$ in terms of a fixed normal set basis $\mathcal{N} = \{x^{j_\mu},\ \mu = 1(1)m\}$ of $\mathcal{R}[\langle\overline{P}\rangle]$ and the associated border basis $\mathcal{B}_\mathcal{N}[\langle\overline{P}\rangle]$, a task considered in section 8.2.1. So let (cf. (8.38))

$$p(x) = d_0(x) + \sum_{\nu=1}^{s} d_{1\nu}(x)\,\bar{p}_\nu(x) + \sum_{\nu \leq \nu_1} q_{\nu\nu_1}(x)\,\bar{p}_\nu(x)\,\bar{p}_{\nu_1}(x)\,, \tag{9.17}$$

with

$$d_0 =: \sum_{\mu=1}^{m} \delta_{0\mu}\, x^{j_\mu} \in \mathcal{R}[\langle \overline{P} \rangle]\,, \quad d_{1\nu} =: \sum_{\mu} \delta_{1\nu\mu}\, x^{j_\mu} \in \mathcal{R}[\langle \overline{P} \rangle]\,, \quad q_{\nu\nu_1} \in \mathcal{P}^s\,.$$

By Corollary 8.26, the coefficients $\delta_{0\mu}$, $\delta_{1\nu\mu} \in \mathbb{C}$ are uniquely determined, and so are the normal forms of the polynomials $q_{\nu\nu_1}$.

If $d_0 \neq 0$, we modify the empirical coefficients in the $\bar{p}_\nu$ with the goal of making the normal form of p mod $\langle \overline{P} + \Delta P \rangle$ vanish; naturally, this will also change the $d_{1\nu}$ and $q_{\nu\nu_1}$:

$$p(x) = 0 + \sum_{\nu=1}^{s}(d_{1\nu} + \Delta d_{1\nu})\,(\bar{p}_\nu + \Delta p_\nu) + \sum_{\nu \leq \nu_1}(q_{\nu\nu_1} + \Delta q_{\nu\nu_1})\,(\bar{p}_\nu + \Delta p_\nu)(\bar{p}_{\nu_1} + \Delta p_{\nu_1})\,, \quad (9.18)$$

with

$$\Delta p_\nu = \sum_{j \in \tilde{J}_\nu} \Delta\alpha_{\nu j}\, x^j\,, \quad \Delta d_{1\nu} = \sum_{\mu=1}^{m} \Delta\delta_{1\nu\mu}\, x^{j_\mu}\,, \quad \Delta q_{\nu\nu_1} \in \mathcal{P}^s\,.$$

As usual, we assume that d_0 has been small enough so that the necessary modifications in (9.17) are sufficiently small that we can neglect the quadratic terms in the modifications, at least in a first computational phase. Equations (9.17) and (9.18) imply

$$\begin{aligned} d_0(x) = \quad & \textstyle\sum_{\nu=1}^{s} d_{1\nu}(x)\,\Delta p_\nu(x) \\ & + \textstyle\sum_{\nu} \Delta d_{1\nu}\, \bar{p}_\nu + \sum_{\nu \leq \nu_1}(q_{\nu\nu_1}\,(\bar{p}_\nu \Delta p_{\nu_1} + \bar{p}_{\nu_1} \Delta p_\nu) + \Delta q_{\nu\nu_1}\, \bar{p}_\nu \bar{p}_{\nu_1}) + O(\|\Delta..\|^2)\,. \end{aligned}$$

Since the left-hand side is in $\mathcal{R}[\langle \overline{P} \rangle]$, it must equal the normal form of the right-hand side mod $\langle \overline{P} \rangle$ which leaves (after neglection of the quadratic terms)

$$\sum_{\mu} \delta_{0\mu}\, x^{j_\mu} = d_0(x) = \mathrm{NF}_{\langle \overline{P} \rangle}\Big[\sum_{\nu} d_{1\nu}(x)\,\Delta p_\nu(x)\Big] = \sum_{\nu=1}^{s}\sum_{j \in \tilde{J}_\nu} \Delta\alpha_{\nu j} \sum_{\mu=1}^{m} \delta_{1\nu\mu}\,\mathrm{NF}[x^{j+j_\mu}]\,.$$

$$(9.19)$$

These are m linear equations in the $M := \sum_\nu M_\nu$ unknown modifications $\Delta\alpha_{\nu j}$ of the empirical coefficients in $(\overline{P}, E)$. If $M \geq m$, we may solve (9.19) for the $\Delta\alpha_{\nu j}$ and (for $M > m$) simultaneously minimize $\delta := \max_\nu \|\Delta a_\nu\|^*_{e_\nu}$. Except for a potential effect of the quadratic terms, this is the smallest δ such that $p \in \langle \widetilde{P} \rangle \in \langle N_\delta(\bar{P}, E) \rangle$.

Definition 9.7. Given an empirical system $(\overline{P}, E)$ and a polynomial p in $\mathcal{P}^s$, the smallest δ such that there exists a $\widetilde{P} \in N_\delta(\bar{P}, E)$ with $p \in \langle \widetilde{P} \rangle$ is the *backward error* of p as a member in $\langle N_\delta(\bar{P}, E) \rangle$. $\square$

By (9.18), the only contribution of the quadratic terms to (9.19) would be $\mathrm{NF}[\sum_{\nu \leq \nu_1} q_{\nu\nu_1} \Delta p_\nu \Delta p_{\nu_1}]$. If the Δp_ν obtained from (9.19) are tiny, this contribution should be negligible; otherwise, it may be estimated and bounded. Generally, the size of δ obtained from the above linearized minimization will be a sufficiently accurate value of the backward error of p as a member of $\langle N_\delta(\bar{P}, E) \rangle$.

In exceptional situations, the linear equations (9.19) can be inconsistent; then there exists no $\widetilde{P}$ near $\overline{P}$ with $p \in \langle \widetilde{P} \rangle$. The other case where our approach must generally fail is when

$M < m$ so that there is not sufficient indetermination in $(\overline{P}, E)$ to permit a full adaptation to p. In these cases, we can only modify $\overline{P}$ such that the normal form coefficients with respect to a $\langle\widetilde{P}\rangle$ are minimized. The interpretation of the result of this analysis will depend on the situation.

Example 9.5: Consider the empirical system $(\overline{P}, E)$ of Example 9.4 and the cubic polynomial

$$p(x, y) := 1.793\,x^3 + 4.663\,x^2 y - 3.761\,xy^2 + .865\,y^3$$
$$+1.779\,x^2 - 7.471\,xy + 2.748\,y^2 - 9.048\,x - .909\,y + 5.572\,.$$

The expansion (9.17) of p with respect to the specified system $\overline{P}$ for the normal set $\{1, x, y, xy\}$ is (rounded)

$$p(x, y) \approx -.0024 - .0015\,y + .0038\,x + .0010\,xy$$
$$+ (-3.1985 + 2.8160\,x + 1.4697\,y)\,\bar{p}_1(x, y) + (1.8019 - 1.0230\,x + .1244\,y)\,\bar{p}_2(x, y)\,.$$

We form the linear system (9.19) for the 6 coefficients $\Delta\alpha_{\nu,j}$ of the corrections of the empirical coefficients in the $\bar{p}_\nu$

$$-.0024 - .0015\,y + .0038\,x + .0010\,xy =$$
$$\mathrm{NF}_{\langle N_\delta(\overline{P}, E)\rangle}\,[\,(-3.1985 + 2.8160\,x + 1.4697\,y)\,(\Delta\alpha_{1,0} + \Delta\alpha_{1,10}x + \Delta\alpha_{1,01}y)$$
$$+ (1.8019 - 1.0230\,x + .1244\,y)\,(\Delta\alpha_{2,0} + \Delta\alpha_{2,10}x + \Delta\alpha_{2,01}y)\,]\,;$$

the formation of the normal form requires the reduction of the x^2 and y^2-terms. The solution of these 4 linear equations combined with the minimization of the moduli of the $\Delta\alpha_{\nu,j}$ leads to corrections of the linear terms in the $\bar{p}_\nu$ of maximum modulus $\approx .0013$ or to a backward error of 1.3 for p as a member in $\langle N_\delta(\overline{P}, E)\rangle$. Thus, p may be considered a valid member in the set of ideals associated with the empirical system $(\overline{P}, E)$.

When we append the corrections found above to the system $\overline{P}$ and compute the expansion of p with respect to the resulting modified system $\widetilde{P} \in N_{1.3}(\overline{P}, E)$, all normal form coefficients are below 10^{-4} in modulus and we obtain (rounded) $p(x, y) \approx$

$$(-3.1980 + 2.8160\,x + 1.4697\,y)\,(x^2 + .25\,y^2 + 1.5763\,x - 2.3242\,y - 3.0681)$$
$$+(1.8020 - 1.0230\,x + .1244\,y)\,(x^2 - 3\,xy + 4\,y^2 + 1.2353\,x - 1.9643\,y - 2.3527)\,.$$

This confirms that our linear correction procedure is sufficient.

Let us now assume that the only empirical coefficients in $(\overline{P}, E)$ are the constant terms. Then there exists no neighboring system $\widetilde{P}$ such that $p \in \langle\widetilde{P}\rangle$. Also, we can only minimize the *linear* coefficients in the normal form of p because the coefficient of xy does not depend on the constant terms in the p_ν. When we perform that minimization, we can reduce the normal form of p to

$$\mathrm{NF}_{\langle\widetilde{P}\rangle}[p] \approx -.0015 - .0015\,x + .0015\,y + .0010\,xy\,,$$

with corrections $\Delta\alpha_{1,0} \approx .00014$, $\Delta\alpha_{2,0} \approx -.00188$; these are just barely valid with our tolerances .001. $\square$

Exercises

1. (a) Find approximate values of some of the further real and complex zeros of $(\overline{P}, E)$ in Example 9.1 and determine their backward errors, perhaps after rounding them to less accurate values.

(b) Analyze the absolute and relative sensitivity of these zeros with respect to the indetermination of the coefficients of $(\overline{P}, E)$.

2. (a) For $s = 2$, $m = 3$, $T^2(3)$ consists of $\{1, x, x^2\}$, $\{1, x, y\}$, $\{1, y, y^2\}$. For each of these candidate normal sets, determine the singular manifold $S_{\mathcal{N}}$ and interpret it geometrically in terms of the zero locations. Consider also the various possibilities of confluent zeros (a double and a simple zero or a triple zero).

(b) Convince yourself that there is at least one admissible normal set for each zero location (= dual space specification). Are there zero locations which are close to each of the three singular manifolds ?

3. Consider the complete intersection system $\overline{P} \subset \mathcal{P}^3$ with

$$\begin{aligned}
\bar{p}_1(x, y, z) &:= \quad xz + 2.864 - 1.198\,z - 5.762\,y + 2.793\,x - .683\,z^2\,, \\
\bar{p}_2(x, y, z) &:= \quad yz - 3.427 - 5.781\,z + 3.384\,y - .956\,x + 2.528\,z^2\,, \\
\bar{p}_3(x, y, z) &:= \quad z^3 - .971 + 4.764\,z - 6.351\,y + 4.663\,x - 5.228\,z^2\,.
\end{aligned}$$

(a) Interpret $\overline{P}$ as part of a border basis; what is the associated normal set $\mathcal{N}$? With respect to which variable is $\mathcal{N}$ a quasi-univariate normal set? Which multiplication matrix is fully specified by $\overline{P}$? Determine the remaining border basis elements and multiplication matrices of this normal set representation of $\langle \overline{P} \rangle$; cf. section 8.1.3.

(b) For which term order is $\mathcal{N}$ a generic normal set so that the border basis of (a) is the Groebner basis? What is the reduced Groebner basis for this term order?

(c) Assume that $\overline{P}$ is the specified system of an empirical system $(\overline{P}, E)$, with tolerances $.5 \cdot 10^{-3}$ on all noninteger coefficients. Find (by systematic experimentation) how strongly the remaining border basis elements may vary when P varies within $N_1(\overline{P}, E)$.

(d) What does that imply for the meaningful accuracy with which the elements of the Groebner basis of $\overline{P}$ (other than those in $\overline{P}$) may be computed when it is known that $\overline{P}$ has been obtained by rounding to 3 decimal digits?

9.2 Approximate Representations of Polynomial Ideals

9.2.1 Approximate Normal Set Representations

It is one of the goals of this book to promote the use of floating-point arithmetic and other approximations in algebraic computations. For empirical polynomials, the use of approximate computation is natural because the inherent indetermination in the data prevents the existence of "exact results" in the sense of classical algebra. Generally, the use of approximate computation greatly diminishes the computational effort for the determination of numerical results of algebraic tasks.

The determination of a normal set representation for the ideal $\langle P \rangle$ generated by a 0-dimensional polynomial system $P \subset \mathcal{P}^s$ is an algebraic task of central importance. In all current computer algebra systems, it is routinely executed in rational (integer) arithmetic, which is possible because all numerical operations in the algorithm are rational ones. But, generally, this involves an enormous growth in the numerators and denominators of the intermediate and final data. When (some of) the original coefficients have been floating-point numbers which were *interpreted* as rational numbers, the situation may become prohibitive. This case generally prevails when P is the specified system $\overline{P}$ of an empirical system. But even for systems with integer coefficients, the number of digits in the exact coefficients of the basis is often so large that it appears unreasonable to *use* the exact coefficients in further computations, e.g., in the computation of zeros. If only approximations of the coefficients are ever used, why should one not make use of approximate computation for their determination.

In the computation of a normal set representation, what is the "approximate result" which we expect to obtain? Apparently, we expect to obtain a normal set $\mathcal{N} \in T^s(m)$ which is a correct feasible normal set for $\langle P \rangle$, and numerical row vectors $\tilde{a}_j^T \in \mathbb{C}^m$ (i.e. nontrivial rows of multiplication matrices or border basis coefficients, resp., cf. Definition 2.23) which may differ slightly from the exact row vectors a^T in the $\mathcal{N}$-normal set representation of $\langle P \rangle$. We expect to interpret these approximate $\tilde{a}_j^T$ as the exact vectors for a problem with slightly different data and assess this data perturbation relative to the potential indetermination in the specified data. However, we are in a situation which has been discussed in section 8.2.3: The set of N row vectors $a_j^T \in \mathbb{C}^m$ which specifies a normal set representation of an ideal $\mathcal{I} \in \mathcal{P}^s(m)$ must lie on the *admissible-data manifold* $\mathcal{M}_\mathcal{N} \subset \mathbb{C}^{Nm}$ (cf. Definition 8.6); otherwise it cannot be interpreted as data of a normal set representation *at all*.

If the computed $\tilde{a}^T$ approximate the exact a^T but do not lie on $\mathcal{M}_\mathcal{N}$, the multiplication matrices $\tilde{A}_\sigma$ formed with the $\tilde{a}^T$ *do not commute* and the polynomials $\tilde{bb}_j$ in the $\mathcal{N}$-border basis $\widetilde{\mathcal{B}}_\mathcal{N}$ are *inconsistent*: They have no common zeros and $\langle \widetilde{\mathcal{B}}_\mathcal{N} \rangle = \langle 1 \rangle$. This fact explains why many algebraists have viewed all attempts to determine ideal bases by approximate computation with a high degree of skepticism. On the other hand, we may take the following pragmatic view which has a long and successful tradition in Applied Mathematics[15]: We realize that an "approximate normal set representation" is not strictly a representation of any nontrivial ideal but that its data are close to the data of the exact representation of an ideal we are looking for. Therefore, we can derive *valid information* about this ideal from the approximate representation if we proceed cleverly.

For example, we may select *one* of the multiplication matrices, say A_1, and consider only those components of the normalized eigenvectors of A_1 which correspond to $x_1, .., x_s$ in the normal set. If A_1 is nonderogatory, these components provide approximations $\tilde{z}_\mu$ of all zeros z_μ of P in the usual sense; cf. Example 8.2.

Example 9.6: We start with a system of two polynomials P in $\mathbb{Q}[x, y]$, which we have constructed from their 6 real rational zeros. The exact coefficients of P have numerators and denominators with between 10 and 15 digits. Rounded to 5 decimal digits, the system is

$$\bar{p}_1(x, y) = y^3 + .48423\, xy^2 + .05784\, xy - .09135\, y^2 - .25145\, x - 1.21464\, y + .45580,$$
$$\bar{p}_2(x, y) = xy^3 - .51904\, xy^2 - .86818\, xy + 2.36473\, y^2 + .73810\, x + .76642\, y - 2.30780.$$

We may think of $\bar{P} = \{\bar{p}_1, \bar{p}_2\}$ as the specified system of an empirical system $(\bar{P}, E)$, with

---

[15]C. F. Gauss has demonstrated how to obtain excellent approximate results from the *inconsistent, overdetermined* linear systems which arise from the use of surplus measurements in surveying.

tolerances of $.5 \cdot 10^{-5}$ on all except the leading coefficients. All zeros of P are reasonably well-conditioned so that the zeros of $\bar{P}$ differ by $O(10^{-5})$ from those of P.

When we compute the Groebner basis of P, for $\texttt{tdeg(x,y)}$, we obtain excessively long numerators and denominators; when we convert $\bar{P}$ to an integer system by multiplication with 10^5 and compute its Groebner basis, the coefficients become even more unwieldly. Rounded to 5 decimal digits, this Groebner basis $\widetilde{\mathcal{G}}$ looks like

$$\tilde{g}_1(x, y) = y^3 - .18428\,x^2 + .27064\,xy + .17425\,y^2 + .39572\,x - 1.31238\,y + .17877\,,$$

$$\tilde{g}_2(x, y) = xy^2 + .38056\,x^2 - .43946\,xy - .54850\,y^2 - 1.33649\,x + .20184\,y + .57210\,,$$

$$\tilde{g}_3(x, y) = x^2y - .34518x^2 - .74861xy - 5.31753y^2 - 1.08966x - 2.44432y + 6.20218\,,$$

$$\tilde{g}_4(x, y) = x^3 - 1.22257x^2 - .34229xy + 2.95924y^2 - 3.59317x + 1.32856y - 1.26231\,.$$

Due to the rounding, this system $\widetilde{\mathcal{G}}$ of 4 polynomials in 2 variables must be *inconsistent*; strictly speaking, it does not possess common zeros. When we form the two multiplication matrices $\tilde{A}_x$ and $\tilde{A}_y$ from the $\tilde{g}_\nu$ and test their commutativity, we obtain a residual matrix with elements of $O(10^{-5})$ in the lower 3 rows; also the normalized eigenvectors of $\tilde{A}_x$ and $\tilde{A}_y$ differ by about $O(10^{-5})$. But because the $\tilde{g}_\nu$ are very close to the elements of the exact Groebner basis of $\langle \bar{P} \rangle$, all these quantities are *close to the exact quantities* for $\bar{P}$: For example, when we consider the 2nd and 3rd components of the normalized eigenvectors of $\tilde{A}_x$, we find that they reproduce the components of the exact zeros of $\bar{P}$ within $O(10^{-5})$. $\square$

The system of the $\tilde{g}_\nu$ above is an example of what we intuitively mean by an approximate normal set representation. (Here, the $\mathcal{N}$-border basis and the Groebner basis coincide.) Therefore, we propose the following:

Definition 9.8. An *approximate normal set representation* for an ideal $\mathcal{I} \in \mathcal{P}^s(m)$ consists of a normal set $\mathcal{N} \in T^s(m)$ which is feasible for $\mathcal{I}$ and of N row vectors (nontrivial rows of multiplication matrices, coefficient vectors of border basis polynomials) $\tilde{a}_j^T \in \mathbb{C}^m$, $N = |B[\mathcal{N}]|$, with $\tilde{a}_j^T \approx a_j^T$, where the a_j^T are the respective exact vectors for $\mathcal{I}$. $\square$

The meaning of $\approx$ may strongly depend on the particular situation. In any case, the crucial requirement in the above definition is the feasibility of the normal set: If the normal set $\mathcal{N}$ of the approximate representation cannot be used for all systems in a neighborhood of the specified system (cf. section 9.1.3), we cannot use continuity arguments to establish that quantities found from the approximate representation will approximate the respective quantities of the specified system. This will also become clear in the discussion of an algorithm for the computation of an approximate normal set representation in section 10.2.

As Example 9.6 has shown, an approximate normal set representation is, generally, well satisfactory for the determination of approximate zeros of an intrinsic polynomial system, or of pseudozeros of an empirical system $(\bar{P}, E)$, respectively. In any case, we may assess the quality of approximate zeros from an approximate normal set representation by applying one Newton step (for an intrinsic system) or by evaluating the backward error (for an empirical system); these are low-cost operations compared to the determination of the approximate zeros.

For some other purposes, an approximate normal set representation may not be fully satisfactory, e.g., for the computation of normal forms of higher degree polynomials: Due to the missing full commutativity of the approximate multiplication matrices or, equivalently, the slight remainders in the reduction of the S-polynomials of the approximate border basis elements, the resulting normal forms will depend on the path which has been followed in the reduction of the polynomials. Thus, it may appear necessary at times to *refine* a computed approximate normal set representation towards a *proper* representation of an ideal.

9.2.2 Refinement of an Approximate Normal Set Representation

Assume that we have an approximate normal set representation, i.e. a normal set $\mathcal{N} \in \mathcal{T}^s(m)$ and $N = |B[\mathcal{N}]|$ row vectors $\tilde{a}_j^T \in \mathbb{C}^m$; the set $\{\tilde{a}_j^T, \ j \in B[J_\mathcal{N}]\}$ does not lie on the admissible-data manifold $\mathcal{M}_\mathcal{N}$ but is close to it in a suitable sense (see below). We wish to modify the set of the $\tilde{a}_j^T$ such that it lies on $\mathcal{M}_\mathcal{N}$, except for round-off.

There are two principal ways to deal with this task:

(1) Take a minimizing Newton step towards the consistency of the set $\{\tilde{a}_j^T\}$;

(2) Transform the representation to one on a quasi-univariate normal set; from the consistent multiplication matrix with respect to the distinguished variable, recompute the representation for the original normal set.

The first approach uses the commutativity constraints for the multiplication matrices; cf. section 8.2.3. The approximate row vectors $\tilde{a}_j^T$ provide us with approximate multiplication matrices $\widetilde{A}_\sigma, \ \sigma = 1(1)s$, which are not fully commuting; they yield small commutativity residual matrices

$$R_{\sigma_1 \sigma_2} := \widetilde{A}_{\sigma_1} \widetilde{A}_{\sigma_2} - \widetilde{A}_{\sigma_2} \widetilde{A}_{\sigma_1} \in \mathbb{C}^{m \times m}, \quad \sigma_1 \neq \sigma_2 \in \{1, \ldots, s\}. \tag{9.20}$$

We want to attach corrections ΔA_σ to the $\widetilde{A}_\sigma$ such that the $\widetilde{A}_\sigma + \Delta A_\sigma$ form a commuting family, or—more realistically—have commutativity residuals at round-off level in our chosen floating-point environment. Note that we cannot computationally *verify* commutatitvity beyond round-off level.

In section 8.2.3, we have analyzed the set S of relations which the N nontrivial row vectors a_j^T of the multiplication matrices A_σ of a proper normal form representation have to satisfy so that $\{a_j^T\} \in \mathcal{M}_\mathcal{N}$; we have found that—at least for $s > 2$—the number $\overline{N}$ of these relations is often much larger than N. This discrepancy perseveres even when we take Proposition 8.4 into account and consider (9.20) only for combinations $(\bar{\sigma}, \sigma)$ with an appropriate fixed $\bar{\sigma}$. Thus, the quadratic system in the Δa_j^T

$$(\widetilde{A}_{\sigma_1} + \Delta A_{\sigma_1})(\widetilde{A}_{\sigma_2} + \Delta A_{\sigma_2}) - (\widetilde{A}_{\sigma_2} + \Delta A_{\sigma_2})(\widetilde{A}_{\sigma_1} + \Delta A_{\sigma_1}) = 0, \quad \text{for appropriate } \sigma_1, \sigma_2,$$

or rather its Newton linearization

$$R_{\sigma_1 \sigma_2} + \widetilde{A}_{\sigma_1} \Delta A_{\sigma_2} - \widetilde{A}_{\sigma_2} \Delta A_{\sigma_1} + \Delta A_{\sigma_1} \widetilde{A}_{\sigma_2} - \Delta A_{\sigma_2} \widetilde{A}_{\sigma_1} = 0, \tag{9.21}$$

has generally more equations than unknown Δa_j^T components. In section 8.2.3, we have seen that the matrix of the system (9.21), i.e. the Jacobian of the quadratic system in the Δa_j^T, has only rank $(N - s)\,m$ *if and only if* the rows of the $\widetilde{A}_\sigma$ are in $\mathcal{M}_\mathcal{N}$. But in the situation under consideration, this is not the case. For generic $\widetilde{A}_\sigma$, (9.21) has full rank and is thus an *overdetermined, inconsistent* linear system.

But we have also been able to specify subsets $\mathcal{S}_0$ of exactly $N-s$ relations from $\mathcal{S}$, which are sufficient for $\{a_j^T\} \in \mathcal{M}_{\mathcal{N}}$ and whose Jacobian is of rank $(N-s)\,m$ on $\mathcal{M}_{\mathcal{N}}$. In section 8.2.3, we have concluded that the use of these relations for $\{\tilde{a}_j^T\}$ nearly in $\mathcal{M}_{\mathcal{N}}$ will retain the regularity of the system and yield reasonable approximations for the quantities to be determined. The use of such a minimal subset $\mathcal{S}_0$ will also be necessary in the case $s = 2$ where we have $\overline{N} = N-1$ which is smaller than N but still greater than $N-s$; cf. Exercise 8.2-5. Actually, with $N-s$ vector equations, we now have an *underdetermined* system and may prescribe further conditions on the Δa_j^T. A reasonable choice in the present situation is to determine the modifications with minimal norm; this means that we aim for the data on $\mathcal{M}_{\mathcal{N}}$ which are closest to $\{\tilde{a}_j^T\}$. In the spirit of our text, we will use a maximum norm minimization; however, other norms may also be appropriate in this context.

This procedure will generally lead to modified multiplication matrices $\tilde{A}_\sigma + \Delta A_\sigma$ and row vectors $\tilde{a}_j^T + \Delta a_j^T$ for which the consistency conditions (8.23)–(8.24) and (8.25)–(8.26) are satisfied within round-off (or within squares of the corrections). Thus, the matrices and polynomials with these elements will behave like genuine multiplication matrices and border basis elements except for minute deviations; they will thus provide a proper normal set representation of a neighboring system $\tilde{P}$ of P or $\overline{P}$, respectively. Only if the approximate representation is so far from $\mathcal{M}_{\mathcal{N}}$ that the omission of the quadratic terms in (9.21) has a significant impact and the linearized approach may fail.

The deviation of this $\tilde{P}$ from P or $\overline{P}$, say in terms of zero positions, is not diminished in this refinement procedure; while the refined A_σ now produce practically identical zeros from their eigenvectors, these zeros will generally differ from those of P or $\overline{P}$ by the same order of magnitude as the individual zeros from the previous approximate $\tilde{A}_\sigma$.

Example 9.7: In the situation of Example 9.6, the approximate multiplication matrices defined by $\tilde{\mathcal{G}}$, for the normal set $\mathcal{N} = \{1, y, x, y^2, xy, x^2\}$ of $\tilde{\mathcal{G}}$, are

$$
\tilde{A}_x = \begin{pmatrix}
0 & 0 & 1 & 0 & 0 & 0 \\
0 & 0 & 0 & 0 & 1 & 0 \\
0 & 0 & 0 & 0 & 0 & 1 \\
-.57210 & -.20184 & 1.33649 & .54850 & .43946 & -.38056 \\
-6.20218 & 2.44432 & 1.08966 & 5.31753 & .74861 & .34518 \\
1.26231 & -1.32856 & 3.59317 & -2.95924 & .34229 & 1.22258
\end{pmatrix},
$$

$$
\tilde{A}_y = \begin{pmatrix}
0 & 1 & 0 & 0 & 0 & 0 \\
0 & 0 & 0 & 1 & 0 & 0 \\
0 & 0 & 0 & 0 & 1 & 0 \\
-.17877 & 1.31238 & -.39572 & -.17425 & -.270639 & .18428 \\
-.57210 & -.20184 & 1.33649 & .54850 & .43946 & -.38056 \\
-6.20218 & 2.44432 & 1.08966 & 5.31753 & .74861 & .34518
\end{pmatrix}.
$$

The first 3 rows of the matrix R_{xy} of (9.20) vanish automatically; the 18 elements in the lower 3 rows are of $O(10^{-5})$, the largest element has modulus $\approx 2.8 \cdot 10^{-5}$.

Here, (9.21) constitutes a linear system of $3 \cdot 6 = 18$ equations for the $4 \cdot 6 = 24$ correction components in the Δa_j^T, $j = 03, 12, 21, 30$. From section 8.2.3, we know that this system is a perturbation of a rank 12 system and hence inconsistent but that we may omit one of the conditions (8.24) in favor of the virtual relation between 03 and 30 which is automatically satisfied. We omit the relation corresponding to the edge from 12 to 21 and, with the $2 \cdot 6 = 12$ remaining equations for the 24 components of the Δa_j^T as constraints, we minimize their moduli.

The resulting correction components are all below $.3 \cdot 10^{-5}$; the commutation residuals (9.20) of the corrected multiplication matrices are now of $O(10^{-9})$ or less. This means that, except for "round-off," we have obtained a genuine Groebner basis. To permit a comparison with the approximate basis in Example 9.6, we display it rounded to 6 digits:

$$y^3 - .184281\, x^2 + .270637\, xy + .174249\, y^2 + .395719\, x - 1.312377\, y + .178772\,,$$

$$xy^2 + .380565\, x^2 - .439471\, xy - .548476\, y^2 - 1.336477\, x + .201853\, y + .572074\,,$$

$$x^2 y - .345205\, x^2 - .748649\, xy - 5.317555\, y^2 - 1.089621\, x - 2.444281\, y + 6.202176\,,$$

$$x^3 - 1.222582\, x^2 - .342329\, xy + 2.959279\, y^2 - 3.593209\, x + 1.328599\, y - 1.262349\,.$$

This Groebner basis $\mathcal{G}$ represents the polynomial system $\widetilde{P}$ whose normal set representation is (near-)closest to the approximate normal set representation $\widetilde{\mathcal{G}}$ of Example 9.6, in the sense of our approach. For the refined multiplication matrices, the normalized eigenvectors now agree up to at most a few digits of 10^{-9} in the crucial 2nd and 3rd components. Thus, within round-off, these are the zeros of $\widetilde{P}$. The agreement of these approximate zeros with the (rational) zeros of the original system has not been affected, it is still of $O(10^{-5})$, due to the original difference between $\widetilde{P}$ and P. $\quad\square$

Our second approach utilizes the special property of quasi-univariate normal sets to permit a syzygy-free representation of a 0-dimensional ideal by a complete intersection system; cf. section 8.1.3 and Proposition 8.20. We select a feasible quasi-univariate normal set $\mathcal{N}_0 \in T^s(m)$ which coincides with the normal set $\mathcal{N}$ of our approximate normal set representation in as many monomials as possible; let x_s be the distinguished variable of $\mathcal{N}_0$; cf. Definition 8.2. Now we determine a multiplication matrix A_{0s} for $\mathcal{N}_0$ from the information of the $\widetilde{A}_\sigma$ of our approximate normal set representation.

For this purpose, we consider the border subset B_s of $\mathcal{N}_0$ (cf. Definition 8.2); to define A_{0s} we must have normal forms of the monomials in B_s with respect to $\mathcal{N}_0$. We obtain these normal forms from our approximate representation, *disregarding its deficiency*. Let $\mathbf{b}$ and $\mathbf{b}_0$ be the normal set vectors of $\mathcal{N}$ and $\mathcal{N}_0$, resp., and consider an $x^j \in B_s$. To transform the normal form $\mathrm{NF}_{\mathcal{N}}[x^j] = d_j^T\, \mathbf{b}$ into a normal form $\mathrm{NF}_{\mathcal{N}_0}[x^j] = d_{0j}^T\, \mathbf{b}_0$ with respect to $\mathcal{N}_0$ we must replace, in $d_j^T\, \mathbf{b}$, those monomials x^k which are in $\mathcal{N}$ but not in $\mathcal{N}_0$ by their normal forms with respect to $\mathcal{N}_0$; cf. Figure 9.1.

Let $\check{\mathcal{N}} := \mathcal{N} \cap \mathcal{N}_0$, $\hat{\mathcal{N}} := \mathcal{N} \setminus \check{\mathcal{N}}$, $\hat{\mathcal{N}}_0 := \mathcal{N}_0 \setminus \check{\mathcal{N}}$. $\hat{\mathcal{N}}$ and $\hat{\mathcal{N}}_0$ have the same number $\hat{m}$ of monomials. The monomials $\hat{x}^{j_0}$ in $\hat{\mathcal{N}}_0$ have normal forms with respect to $\mathcal{N}$ which can be found from the $\widetilde{A}_\sigma$; if an $\hat{x}^{j_0}$ is not in $B[\mathcal{N}]$, its normal form may depend slightly on the reduction path but we simply take one particular copy. Since $\mathcal{N}_0$ has been assumed to be feasible for the underlying system—and hence for all sufficiently close neighboring systems (cf. Definition 9.6)—it must be possible to solve these relations between the monomials in $\hat{\mathcal{N}}_0$ and those in $\hat{\mathcal{N}}$ for the latter ones; this yields representations of the $\hat{x}^j \in \hat{\mathcal{N}}$ in span $\mathcal{N}_0$. The substitution of these representations into the $\mathcal{N}$-normal forms of the $x^j \in B_s$ produces their representations in span $\mathcal{N}_0$ and thus the s nontrivial rows for the matrix A_{0s}.

By the quasi-univariate feature of $\mathcal{N}_0$, A_{0s} is a *proper* multiplication matrix. If it is nonderogatory and possesses a full eigenvector system, it defines the dual space $\mathcal{D}_0$ of a 0-dimensional ideal $\mathcal{I}_0 \subset \mathcal{P}^s(m)$ and the associated quotient ring $\mathcal{R}_0$. By Proposition 8.7, it

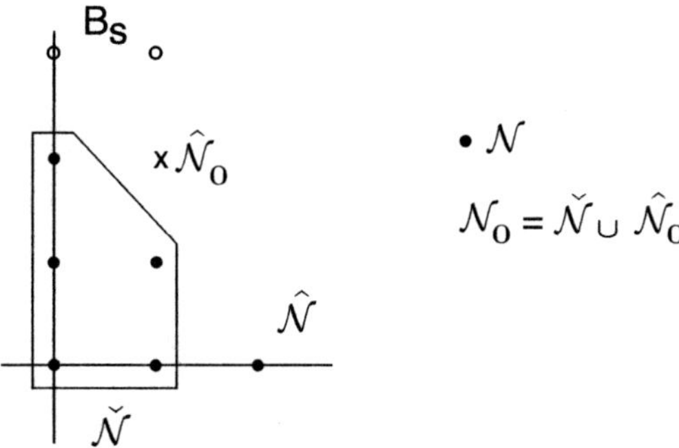

Figure 9.1.

permits the determination of the other multiplication matrices $A_{0\sigma}$ of $\mathcal{R}_0$ with respect to $\mathcal{N}_0$ from a system of linear equations such that the $A_{0\sigma}$, $\sigma = 1(1)s$, form a *commuting family*. We can now transform this *exact* normal set representation of $\mathcal{I}_0$ to one on the normal set $\mathcal{N}$. Except for round-off errors in the computation, its defining row vectors lie on the admissible-data manifold $\mathcal{M}_\mathcal{N}$ because they have been obtained from data on $\mathcal{M}_{\mathcal{N}_0}$.

It is not obvious in which sense the proper normal set representation thus obtained from the initial approximate normal set representation is "close" to it. If the residuals of the approximate representation in the consistency restraints are small, the zero set of $\mathcal{I}_0$ should be very close to the zero sets defined by the various $\tilde{A}_\sigma$. Note that no linearization or minimization is involved in this second approach, but there may be some ambiguity from non-unique normal forms in the transition to A_{0s}.

Example 9.8: Again we start with the approximate normal form representation $\widetilde{\mathcal{G}}$ of Example 9.6; cf. also Example 9.7. As quasi-univariate normal set, we choose $\mathcal{N}_0 = \{1, y, x, y^2, xy, xy^2\}$, with the distinguished variable y and $B_y = \{y^3, xy^3\}$; cf. Figure 9.1. It differs from $\mathcal{N}$ by one monomial only; thus, the sets $\hat{\mathcal{N}}$ and $\hat{\mathcal{N}}_0$ consist only of one monomial each, viz. x^2 and xy^2, respectively.

To transform the representation from $\mathcal{N}$ to $\mathcal{N}_0$, we must invert the approximate representation of xy^2 in $\mathcal{N}$; since $xy^2 \in B[\mathcal{N}]$, it is immediately given by the 4th row of $\tilde{A}_x$ (or 5th row of $\tilde{A}_y$):

$$xy^2 = -.57210 - .20184\,y + 1.33649\,x + .54850\,y^2 + .43946\,xy - .38056\,x^2\,,$$

which inverts to (rounded to 5 digits)

$$x^2 = -1.50331 - .53038\,y + 3.51190\,x + 1.44130\,y^2 + 1.15477\,xy - 2.62771\,xy^2\,.$$

Now we can rewrite the approximate $\mathcal{N}$-normal forms of the two monomials $y^3, xy^3 \in B_y$ into representations in $\mathcal{N}_0$: The normal form coefficients of y^3 are in the 4th row of $\tilde{A}_y$ and yield the $\mathcal{N}_0$-representation (rounded to 5 digits)

$$y^3 \approx -.45580 + 1.21464\,y + .25145\,x + .09135\,y^2 - .05784\,xy - .48423\,xy^2\,;$$

those of xy^3 may be obtained either from $\tilde{a}_{y^3}^T \widetilde{A}_x$ or from $\tilde{a}_{xy^2}^T \widetilde{A}_y$:

$$
\begin{aligned}
x y^3 &\approx 2.01086 - .87118\,y - .04441\,x - 2.08004\,y^2 + 1.09628\,xy - .19753\,x^2 \\
&\approx 2.30781 - .76642\,y - .73811\,x - 2.36473\,y^2 + .86818\,xy + .51905\,xy^2, \quad \text{or} \\[6pt]
&\approx 2.01083 - .87117\,y - .04440\,x - 2.08001\,y^2 + 1.09628\,xy - .19753\,x^2 \\
&\approx 2.30777 - .76641\,y - .73809\,x - 2.36470\,y^2 + .86818\,xy + .51904\,xy^2.
\end{aligned}
$$

The slight differences display the dependence of the normal form on the reduction path in an approximate normal form representation.

From Theorem 8.5, we know that the matrix

$$
A_{0y} =
\begin{pmatrix}
0 & 1 & 0 & 0 & 0 & 0 \\
0 & 0 & 0 & 1 & 0 & 0 \\
0 & 0 & 0 & 0 & 1 & 0 \\
.. & .. & a_{y^3}^T & & .. & .. \\
0 & 0 & 0 & 0 & 0 & 1 \\
.. & .. & a_{xy^3}^T & & .. & ..
\end{pmatrix}
$$

is a proper multiplication matrix for any choice of the nontrivial rows, with the exception of a few singular situations, and that it fully determines an associated 0-dimensional ideal (quotient ring, dual space). In particular it determines the associated multiplication matrix A_{0x} from a *linear* system in the elements of the nontrivial rows of A_{0x}; this system consists of the 3 nontrivial relations in $A_{0x} A_{0y} - A_{0y} A_{0x} = 0$ (cf. Proposition 8.7). Because of the trivial rows in A_{0x}, the system is inhomogeneous; it constitutes a minimal system S_0 in the sense of section 8.2.3 and is therefore of full rank. When we take the first version of a_{xy^3}, we obtain (rounded to 5 digits)

$$
A_{0x} =
\begin{pmatrix}
0 & 0 & 1 & 0 & 0 & 0 \\
0 & 0 & 0 & 0 & 1 & 0 \\
-1.50302 & -.53048 & 3.51171 & 1.44098 & 1.15460 & -2.62762 \\
0 & 0 & 0 & 0 & 0 & 1 \\
-6.72084 & 2.26112 & 2.30182 & 5.81478 & 1.14713 & -.90703 \\
-4.74364 & 1.03721 & 2.13164 & 4.93721 & 1.17804 & -2.13938
\end{pmatrix}.
$$

Now we have a full $\mathcal{N}_0$-normal set representation of an ideal $\mathcal{I}_0$ from whose approximate (Groebner basis) representation $\widetilde{\mathcal{G}}$ we had started our computation.

We can now rewrite A_{0x} and A_{0y} as multiplication matrices of $\mathcal{R}[\mathcal{I}_0]$ with respect to the normal set $\mathcal{N}$. For this purpose, we must use the 3rd row of A_{0x} which represents x^2 in terms of xy^2 and invert it to obtain xy^2 in terms of x^2. Substitution into the $\mathcal{N}_0$-representations of y^3, xy^2, x^2y, x^3 yields (within round-off) the exact Groebner basis of $\mathcal{I}_0$. This yields the following $g_{0\kappa}$ (rounded to 6 digits)

$$
\begin{aligned}
&y^3 - .184286\,x^2 + .270615\,xy + .174201\,y^2 + .395705\,x - 1.312402\,y + .178815, \\
&xy^2 + .380573\,x^2 - .439410\,xy - .548398\,y^2 - 1.336459\,x + .201886\,y + .572008, \\
&x^2y - .345192\,x^2 - .748566\,xy - 5.317364\,y^2 - 1.089606\,x - 2.444235\,y + 6.202007, \\
&x^3 - 1.222487\,x^2 - .341689\,xy + 2.960615\,y^2 - 3.592598\,x + 1.329081\,y - 1.263836.
\end{aligned}
$$

This Groebner basis deviates from the approximate Groebner basis in Example 9.6 more substantially than the one obtained in Example 9.7, particularly in the 4th basis polynomial. However, its commutativity residuals are, on the whole, smaller than those in Example 9.7; in this sense, it is a "better" exact Groebner basis. Clearly, from an approximate normal set representation *not on* $\mathcal{M}_\mathcal{N}$, we may reach many different representations on $\mathcal{M}_\mathcal{N}$; their properties cannot readily be predicted. $\quad\square$

The above two examples have shown that it is possible, with a small amount of floating-point computation, to refine an approximate normal set representation (e.g., an approximate Groebner basis) of a 0-dimensional ideal into a nearby exact (within round-off) representation. Since the ideal behind the approximate representation is unknown to the refinement procedure, the relation between the original ideal and the computed one cannot readily be described.

Example 9.6, continued: When we now revisit Example 9.6, we realize that $\tilde{p}_1$, $\tilde{p}_2$ may immediately be interpreted as a specification of the two nontrivial rows of a multiplication matrix A_{0y} with respect to the quasi-univariate normal set $\mathcal{N}_0$ of Example 9.8 so that the zeros of $\widetilde{P}$ are directly determined by the eigenvectors of this matrix A_{0y}.

When we proceed like in Example 9.8 and compute A_{0x} for this A_{0y}, we obtain the *exact* normal set representation of $\langle \tilde{p}_1, \tilde{p}_2 \rangle$. This can then be converted into the exact normal set representation (Groebner basis) for the normal set $\mathcal{N}$ which, naturally, agrees with the one obtained by rational computation within round-off. $\quad\square$

9.2.3 Refinement Towards the Exact Representation

In the previous section, we have seen how one may modify an approximate normal set representation of an ideal $\mathcal{I} = \langle P \rangle$ into a proper normal set representation for an ideal $\tilde{\mathcal{I}}$ generated by a neighboring system $\widetilde{P}$. Both of our approaches did not use information about the original system P but simply projected the approximate normal set representation onto the admissible-data manifold $\mathcal{M}_\mathcal{N}$ in a practicable way. But amongst the potential images of such a projection, there must also be the *exact* normal set representation of $\langle P \rangle$; can we perform a projection which—within linearization and round-off—leads to that particular image? The positive answer to this question given in the following implies that we can computationally transform an approximate normal set representation of $\langle P \rangle$, say an approximate Groebner basis, into the exact representation. Naturally, the approximate representation must employ a normal set $\mathcal{N}$ which is feasible for $\langle P \rangle$ and be sufficiently good so that a linearized approach succeeds.

In the previous section, for a suitable subset $\mathcal{S}_0$ of the consistency conditions, the linear system (9.21) for the N correction vectors Δa_j^T has only rank $(N-s)\, m$; cf. section 8.2.3. Thus we may append $s\, m$ further conditions if these are linear and independent of (9.21). When we use this freedom for a Newton step towards the annihilation of the m normal form coefficients of the $p_\nu \in P$, $\nu = 1(1)s$, we should obtain a full rank linear system for the Δa_j^T such that the refined representation essentially lies on $\mathcal{M}_\mathcal{N}$ *and* represents $\langle P \rangle$. We have not done this in section 9.2.2 to keep the discussion transparent and sufficiently general for other uses of that approach. Naturally, we must be aware that the normal forms of the p_ν with respect to an *approximate* representation of $\langle P \rangle$ are not sharply defined.

The approximate border basis representation $(\mathcal{N}, \{\tilde{a}_j^T\})$ of $\langle P \rangle$ defines the N $\mathcal{N}$-border basis polynomials $\widetilde{bb}_j$. With these, we compute the $\mathcal{N}$-border basis expansions (cf. section

8.2.1) of the polynomials p_ν in P

$$p_\nu = \tilde{d}_{\nu 0}^T \mathbf{b} + \sum_j (\tilde{d}_{\nu j}^T \mathbf{b})\, \widetilde{bb}_j + \text{ terms quadratic in the } \widetilde{bb}_j\,, \quad \nu = 1(1)s\,; \qquad (9.22)$$

the row vectors $\tilde{d}_{\nu 0}^T$ and $\tilde{d}_{\nu j}^T$ are in $\mathbb{C}^m$. Even for an exact border basis $\mathcal{B}$, only the normal form coefficient vectors $\tilde{d}_{\nu 0}^T$ are uniquely determined because of the syzygies of $\mathcal{B}$. In the expansion (9.22), also the $\tilde{d}_{\nu 0}^T$ depend slightly on how they have been obtained; cf. section 9.2.1. For the modified multiplication matrices $A_\sigma = \tilde{A}_\sigma + \Delta A_\sigma$ and border basis polynomials $bb_j = \widetilde{bb}_j + \Delta bb_j$, the p_ν must have vanishing normal forms:

$$p_\nu = \sum_j ((\tilde{d}_{\nu j}^T + \Delta d_{\nu j}^T)\mathbf{b})\,(\widetilde{bb}_j + \Delta bb_j) + \text{ terms quadratic in the modified } \widetilde{bb}_j\,, \quad \nu = 1(1)s\,.$$

A comparison of the two expansions yields

$$\tilde{d}_{\nu 0}^T \mathbf{b} = \sum_j (\tilde{d}_{\nu j}^T \mathbf{b})\, \Delta bb_j + \text{ terms with } \widetilde{bb}_j + \text{ terms quadratic in corrections}\,.$$

Since the left-hand side is a "normal form" with respect to the $\widetilde{bb}_j$, we also reduce the right-hand side with respect to the $\widetilde{bb}_j$, and we omit the quadratic terms in the corrections. This yields, with $\Delta bb_j =: -\Delta a_j^T \mathbf{b}$ and $\tilde{d}_{\nu 0}^T =: (\tilde{\delta}_{\nu 0 \mu}\,, \mu = 1(1)m)$, $\tilde{d}_{\nu j}^T =: (\tilde{\delta}_{\nu j\mu}\,, \mu = 1(1)m)$,

$$\sum_\mu \tilde{\delta}_{\nu\mu}\, b_\mu = -\sum_{\mu,\mu'} \sum_j \tilde{\delta}_{\nu j\mu}\, \Delta\alpha_{j\mu'}\, \mathrm{NF}_{\{\widetilde{bb}_j\}}[b_\mu b_{\mu'}]\,, \quad \nu = 1(1)s\,, \qquad (9.23)$$

where the polynomials $\mathrm{NF}[b_{0\mu} b_{0\mu'}]$ are in span $\mathcal{N}$. These are sm equations for the Nm correction components of the nontrivial row vectors in the A_σ or of the coefficients of the bb_j, respectively. They must be appended to the $(N-s)\,m$ equations for the correction components which place the corrected quantities onto the admissible-data manifold $\mathcal{M}_\mathcal{N}$; cf. section 9.2.2. Altogether we obtain a square linear system in the components of the Δa_j^T. With the corrections from this system, the corrected multiplication matrices $A_\sigma := \tilde{A}_\sigma + \Delta A_\sigma$ provide the exact $\mathcal{N}$-representation for $\langle P \rangle$ (except for linearization and round-off effects).

Example 9.9: We resort once more to our Example 9.6, with

$$p_1 = y^3 + .48423\,xy^2 + .05784\,xy - .09135\,y^2 - .25145\,x - 1.21464\,y + .45580\,,$$
$$p_2 = xy^3 - .51904\,xy^2 - .86818\,xy + 2.36473\,y^2 + .73810\,x + .76642\,y - 2.30780\,,$$

and with the approximate Groebner basis $\widetilde{G} = \{\tilde{g}_1, \tilde{g}_2, \tilde{g}_3, \tilde{g}_4\}$ displayed there; $\widetilde{G}$ is an approximate $\mathcal{N}$-border basis for $\mathcal{N} = \{1, y, x, y^2, xy, x^2\}$, it has syzygy errors of $O(10^{-5})$. While the computation of a normal form for p_1 with $\widetilde{G}$ remains unambiguous, the reduction of xy^3 in p_2 may employ either $x\,\tilde{g}_1$ or $y\,\tilde{g}_2$; the resulting normal forms differ by terms of $O(10^{-5})$.

We explain the evaluation of (9.23) for p_2: Its expansion (9.22), with xy^3 reduced by $x\,\tilde{g}_1$, has the form

$$p_2 = \mathrm{NF}_{\widetilde{G}}[p_2] + x\,\tilde{g}_1 - .69329\,\tilde{g}_2 - .27064\,\tilde{g}_3 + .18428\,\tilde{g}_4\,,$$

where $\mathrm{NF}[p_2] = \tilde{d}_2^T \mathbf{b}$ has coefficients of $O(10^{-5})$. When we write the corrected basis elements as $g_j = \tilde{g}_j - \Delta a_j^T \mathbf{b}$ (cf. above), we obtain the correction equation

$$\tilde{d}_2^T \mathbf{b} = -\Delta a_1^T \, x \, \mathbf{b} + .69329 \, \Delta a_2^T \, \mathbf{b} + .27064 \, \Delta a_3^T \, \mathbf{b} + .18428 \, \Delta a_4 \, \mathbf{b}$$

in which we have to replace $x\mathbf{b}$ by $\tilde{A}_x \mathbf{b}$. This yields the equation (9.23) for $v = 2$. The corresponding equation for $v = 1$ is obtained more directly because (9.22) does not contain monomial factors with the $\tilde{g}_j$.

For the linearized relations which put the corrected row vectors $a_j^T = \tilde{a}_j^T + \Delta a_j^T$ onto the admissible-data manifold $\mathcal{M}_{\mathcal{N}}$, we take the same two of three possible relations which we have used in Example 9.7; remember that this avoids overdetermination and inconsistency. This gives us a linear system of 4 vector relations for the Δa_j^T, with 6 components each. The resulting correction components are $O(10^{-6})$ as they must be because $\tilde{G}$ had been obtained from the exact Groebner basis by rounding to 5 decimal digits.

The coefficients of the corrected approximate Groebner basis elements differ from those of the exact ones by $O(10^{-8})$ to $O(10^{-10})$. When we reduce the p_v with the corrected basis elements, we obtain normal forms of $O(10^{-9})$; this does not depend on the reduction path taken for p_2. Also the commutativity residual matrix for the corrected multiplication matrices has elements of $O(10^{-9})$ or less. Thus, for most purposes, the corrected elements may be considered as exact. $\square$

Though of a simple structure, our example has shown the feasibility of the procedure explained above; thus it is possible to refine a reasonably approximate normal set representation of the ideal generated by a regular system into the exact representation of the ideal. Remaining linearization errors can be eliminated by another run through the procedure, round-off errors by the choice of a higher precision. In general, the computational effort for this refinement is far below that for the computation of the approximate representation.

Exercises

Take a system P of three dense quadratic polynomials p_v, $v = 1(1)3$, in three variables, with integer coefficients, and determine the Groebner basis $\mathcal{G}[P]$ of $\langle P \rangle \subset \mathcal{P}^3(8)$ for $\mathrm{tdeg}(x_1, x_2, x_3)$; the associated normal set is $\mathcal{N} = \{1, x_3, x_2, x_1, x_3^2, x_2x_3, x_1x_3, x_3^3\}$ (except for very specially chosen coefficients; why?!). Normalize the 6 elements g_κ of $\mathcal{G}$ for leading coefficient 1 and round the rational coefficients to 5 decimal digits. With these $\tilde{g}_\kappa$ in place of the exact g_κ, determine the 5 remaining elements of an $\mathcal{N}$-border basis and round them to 5 decimal digits; now you have an approximate border basis $\tilde{\mathcal{B}}_{\mathcal{N}} = \{\tilde{bb}_j\}$ for $\langle P \rangle$. With the coefficients of these $\tilde{bb}_j$, compose the approximate multiplication matrices $\tilde{A}_\sigma$, $\sigma = 1(1)3$. This approximate normal set representation $(\mathcal{N}, \{\tilde{a}_j^T\})$ of $\langle P \rangle$ is to be used in the following exercises.

1. (a) Use the quasi-univariate feature of $\mathcal{N}$ to compute the zero set of P; cf. Example 9.6 continued Compute the approximate zero sets $\tilde{Z}_\sigma$ defined by the three approximate multiplication matrices $\tilde{A}_\sigma$ and compare.

(b) Assign tolerances to the coefficients in P and compute the backward error of each of the zero sets $\tilde{Z}_\sigma$. (Note that you must form the *simultaneous* backward error of all zeros in the zero set; how many empirical coefficients in the p_v do you need for that, at least?) Comment the "validity" of the approximate border basis $\tilde{\mathcal{B}}_{\mathcal{N}}$.

2. At first, we want to refine $(\mathcal{N}, \{\tilde{a}_j^T\})$ into a proper representation, with the $a_j^T = \tilde{a}_j^T + \Delta a_j^T \in \mathcal{M}_{\mathcal{N}}$, without reference to the p_ν. The three commutativity residual matrices $R_{\sigma_1 \sigma_2}$ of (9.20) yield a total of $\overline{N} = 17$ linear relations (9.21) for the 11 row vectors $\Delta a_j^T \in \mathbb{C}^8$.

(a) Try to solve all 17 relations simultaneously to verify their inconsistency. Then minimize the moduli of their componentwise residuals and interpret the result. Test the linearization effect by substituting the refined $A_\sigma = \tilde{A}_\sigma + \Delta A_\sigma$ into (9.20).

(b) Take (9.21) with a fixed σ_1 and delete remaining circles from the "web" of edges between the points of $J_{\mathcal{N}}$; cf. section 8.2.3. Now, we can set the remaining 10 relations as equality constraints for the minimization of the moduli of the components of the Δa_j^T. Interpret the result and test (9.20) for the refined multiplication matrices.

(c) Delete two further commutativity relations by considering the "virtual edges" between the extremal points of $J_{\mathcal{N}}$ and proceed like in (b). Compare the results.

(d) Perform (b) and (c) also for the two other choices of σ_1 in (b). Compare the results of the three computations.

3. Now we want to adapt the refined representation to $\langle P \rangle$. Convince yourself that, in this particular situation, the computation of the normal forms of the p_ν with respect to $\tilde{\mathcal{B}}_{\mathcal{N}}$ is trivial and that their nonvanishing is only due to the rounding of three of the g_j.

(a) Append the system of the 8 commutativity constraints for the Δa_j^T of 2 (c) above by the equations for the vanishing of the normal form coefficients for the p_ν and solve. Compare the refined approximate Groebner basis with the exact Groebner basis $\mathcal{G}$.

(b) Convince yourself that, in this particular situation, (a) decomposes into the correction of the three g_j mentioned above and the adaptation of the remaining bb_j to this correction.

9.3 Multiple Zeros and Zero Clusters

In Section 8.5, we have considered multiple zeros of multivariate polynomial systems P; we have characterized such zeros by their *multiplicity structure* defined by the dual space $\mathcal{D}_0$ of the primary ideal $\mathcal{I}_0$ of the multiple zero; cf. Definition 8.18. We have been able to determine this structure from the data of the associated joint invariant subspace of the multiplication matrices of $\mathcal{R}[\mathcal{I}_0]$. We have, however, not discussed the fact that an m-fold zero, $m > 1$, is a *singular phenomenon*: It disappears under generic perturbations of the system P and decomposes into a cluster of m isolated zeros whose location is extremely ill-conditioned. This correlates with the analogous properties of invariant subspaces of linear maps whose numerical determination is a highly sensitive affair, even for exact data; cf., e.g., [3.11]. Thus, even when the multiplication matrices of an intrinsic polynomial system with integer coefficients have been found by exact rational computation, the determination and characterization of an existing multiple zero with irrational coordinates may present serious difficulties.

As with other singular phenomena in polynomial algebra, a full understanding of multiple zeros can only be achieved by a suitable analytic embedding of the situation. Therefore, like in the univariate case, we will immediately consider regular systems of *empirical* multivariate polynomials; intrinsic sytems may be considered as empirical systems with very small tolerances.

The definitions of a cluster domain and of a valid zero cluster may directly be adapted from the univariate situation (cf. Definitions 6.6–6.7 and 9.5):

Definition 9.9. For each $\tilde{P} \in N_\delta(\bar{P}, E)$, the m-tuple $(z_1, \ldots, z_m)$ of zeros of $\tilde{P}$ in an m-cluster domain of the empirical system $(\bar{P}, E)$ is a *valid m-cluster of zeros* of $(\bar{P}, E)$ at tolerance level δ. $\square$

Definition 9.9 is meaningful because of Proposition 9.2. The m zeros in a valid m-cluster must be *simultaneous* pseudozeros of each of the $\bar{p}_\nu$ of P, which is a strong requirement. The implications for their geometrical arrangement will be discussed in section 9.3.4.

As in the univariate case, the immediate computation of the locations of the zeros in a cluster requires a particular effort because it is an extremely ill conditioned task; cf. Proposition 6.11. Therefore, it is imperative to consider an m-cluster of zeros as a perturbed m-fold zero and to use the information which can be derived from that.

In section 6.3, in our treatment of multiple zeros and zero clusters of empirical univariate polynomials $(\bar{p}, e)$, we have not stressed the algebraic aspects because of the simple algebraic structure of the univariate situation. In algebraic terms, our analysis of a univariate m-cluster of zeros has been like this (cf. section 6.3.3):

(i) The location $z_0 \in \mathbb{C}$ and multiplicity m of a potential m-fold zero are derived from the eigenanalysis of the multiplicative structure of $\mathcal{R}[\langle p \rangle]$; this yields the basis polynomial $s_0 = (x - z_0)^m$ for the associated primary ideal.

(ii) If $\langle s_0 \rangle$ contains a polynomial $\tilde{p} \in N_\delta(\bar{p}, e)$, [16] $\delta = O(1)$, z_0 is a valid m-fold zero of $(\bar{p}, e)$; otherwise:

(iii) Correct s_0 into a "cluster polynomial" s such that there exists a neighboring $\tilde{p} \in \langle s \rangle$.

(iv) The zeros of s compose a valid zero cluster of $(\bar{p}, e)$; the coefficients of s are well-conditioned functions of the coefficients of p.

None of these steps generalizes easily to the case of a regular empirical system $(\bar{P}, E)$, $\bar{P} \in \mathcal{P}^s$. In the following, we will consider this generalization.

9.3.1 Approximate Dual Space for a Zero Cluster

Assume that we have determined a normal set representation $(\mathcal{N}, \mathcal{B}_\mathcal{N})$ for a specified regular system $P = \{p_\nu, \nu = 1(1)s\} \subset \mathcal{P}^s$; here, P stands either for the system $\bar{P}$ of $(\bar{P}, E)$ or for an intrinsic system. If the eigenanalysis of the multiplication matrices A_σ for $\mathcal{N}$ exhibit a set of $m > 1$ very close zeros $z_\mu \in \mathbb{C}^s$, $\mu = 1(1)m$, we may conjecture that they form a cluster and look for an m-fold zero z_0 of a neighboring system $\tilde{P}$ through whose perturbation the cluster may have originated.

In the analogous univariate situation, we simply chose $z_0 = \frac{1}{m} \sum_\mu z_\mu$ and $s_0(x) = (x - z_0)^m$ as basis of the primary ideal $\mathcal{I}_0$ of an m-fold zero at z_0; cf. step (i) above. In the multivariate situation, it is not immediately obvious how we should choose z_0, and in view of the great variety of potential ideals with an m-fold zero at z_0 (cf. section 8.5.1), it is not clear at all how we should select a basis for $\mathcal{I}_0$.

Let us, at first, consider the choice of z_0. In section 6.3.3, our suggested choice for the univariate case was based on intuition rather than a formal argument which could have proceeded along the following lines: Let $p \in P$ be the specified polynomial with the m-cluster of zeros z_μ, $\mu = 1(1)m$, and $p - \varepsilon\, p_1 =: p_0 = (x - z_0)^m q(x)$ possess an m-fold zero at z_0. By

---

[16] That is, if s_0 is a valid divisor of $(\bar{p}, e)$.

Proposition 6.10, we have $|z_\mu - z_0| = O(\varepsilon^{\frac{1}{m}})$ so that the individual z_μ's will not provide good estimates for z_0. But by Proposition 6.13 we have, for $p = p_0 + \varepsilon\, p_1$,

$$\sum_{\mu=1}^{m} (z_\mu - z_0) = \sum_{\mu=1}^{m} z_\mu - m\, z_0 = O(\varepsilon)$$

so that, for small ε, the arithmetic mean of the z_μ should provide a reasonable estimate for z_0. Similarly, one finds that $p^{(m-1)}(x) = (x - z_0)\, q_m(x) + \varepsilon\, p_1^{(m-1)}(x)$, with q_m bounded away from 0 near the cluster, must have a zero $\bar{z}_0$ with $|\bar{z}_0 - z_0| = O(\varepsilon)$; this was our alternative suggestion in section 6.3.3.

The first one of these arguments permits a generalization to multivariate multiple zeros:

Proposition 9.3. Consider a regular polynomial system $P_0 = \{p_{0\nu}\} \subset \mathcal{P}^s$ with an m-fold zero at $z_0 \in \mathbb{C}^s$. Consider a perturbed system $P = \{p_\nu\}$, with $p_\nu = p_{0\nu} + \varepsilon\, p_{1\nu}$, $\nu = 1(1)s$, where the Newton polytope (cf. Definition 9.10) of each p_ν is contained in that of $p_{0\nu}$. For sufficiently small $|\varepsilon|$, the m zeros z_μ, $\mu = 1(1)m$, of P near z_0 satisfy

$$\frac{1}{m} \sum_{\mu=1}^{m} z_\mu = z_0\, (1 + O(|\varepsilon|)). \tag{9.24}$$

Proof: Consider a normal set representation $(\mathcal{N}, \mathcal{B}_\mathcal{N})$ of $\langle P_0 \rangle$ and the expansions of the $p_{0\nu}$ in the elements bb_j of $\mathcal{B}_\mathcal{N}$ (cf. (9.22)):

$$p_{0\nu} = \sum_j (d_{0\nu j}^T\, \mathbf{b})\, bb_j + \text{terms quadratic in the } bb_j, \quad \nu = 1(1)s\,;$$

for the p_ν (which are not in $\langle P_0 \rangle$) we obtain

$$p_{0\nu} + \varepsilon\, p_{1\nu} = \varepsilon\, (d_{1\nu}\, \mathbf{b}) + \sum_j ((d_{0\nu j}^T + \varepsilon\, d_{1\nu j}^T)\, \mathbf{b})\, bb_j + \ldots, \quad \nu = 1(1)s\,.$$

When we adapt the bb_j to $\langle P \rangle$, they acquire corrections Δbb_j of $O(\varepsilon)$ which cancel the $O(\varepsilon)$ normal forms of the perturbations $\varepsilon\, p_{1\nu}$; cf. section 9.2.3. Thus, the nontrivial rows of the multiplication matrices $A_{0\sigma}$ of $\langle P_0 \rangle$ are changed by the coefficients of these Δbb_j into $A_\sigma = A_{0\sigma} + \varepsilon\, A_{1\sigma}$, $\sigma = 1(1)s$.

The eigenvalues of the $A_{0\sigma}$ and A_σ, resp., are the σ-components $\zeta_{0\sigma\mu}$ and $\zeta_{\sigma\mu}$ of the zeros $z_{0\mu}$ and z_μ of P_0 and P, respectively. Thus, the characteristic polynomials $w_{0\sigma}$ of the $A_{0\sigma}$ each have an m-fold zero $\zeta_{0\sigma 0}$, the σ-component of z_0. The characteristic polynomials w_σ of the A_σ differ from the $w_{0\sigma}$ by ε-perturbations; these decompose the m-fold zeros into m isolated zeros $\zeta_{\sigma\mu}$, $\mu = 1(1)m$, for which Proposition 6.13 holds. $\square$

As in the univariate case, (9.24) implies that the arithmetic mean of m zeros which are supposed to form an m-cluster should be a valid zero of $(\bar{P}, E)$. If this is not the case, the set of m zeros is more likely to consist of several individual clusters. Meaningful quantitative criteria are not available for the multivariate situation.

Now we must try to find the multiplicity structure of an m-fold zero with which we may expect to validate our estimated z_0 as an m-fold zero of a neighboring system $\tilde{P}$. In section

8.5.1, we have solved that task for a system with an exact m-fold zero at a specified location. We are now supposed to solve the same task for perturbed specifications. It is clear that we cannot expect to obtain more than estimates of the numerical data in the specifications of the dual space $\mathcal{D}_0$ for which we are looking, but we would hope to predict the correct *structure* of $\mathcal{D}_0$, e.g., how many linearly independent first-order derivatives vanish at the multiple zero.

Vanishing derivatives at some point exhibit themselves by vanishing coefficients in the Taylor expansion about that point; therefore, we expand the p_ν about z_0. Since z_0 is not a multiple zero of P, we cannot expect any strictly vanishing coefficients in these expansions, but we should see the multiplicity structure reflected in the magnitudes of the coefficients. However, as we have seen in Examples 8.28 and 8.31, even vanishing first derivatives may show up only after a rearrangement of the local coordinate system: We had to take the eigenvector(s) of the eigenvalue 0 of the Jacobian as local basis vectors. Analogously, we should find one or several tiny eigenvalues for the Jacobian of P at z_0; we may then take their eigenvectors as basis vectors and complete them into a full basis. In the new coordinates, the vanishing first derivatives with respect to one or more variables display the first-order derivative functionals for the dual space to be constructed.

If the number m of dual space functionals is not yet complete, we must look for second-order derivatives which nearly vanish for the transformed p_ν. As in section 8.3.2, we must keep the closedness conditions in mind, but—in contrast to the exact situation—we cannot expect any equations for the coefficients in the differentials to be satisfied exactly, which creates a delicate situation. Hopefully, the existence of the m-fold zero in a nearby polynomial system will lead to distinct levels of magnitudes which permit a reasonable judgment. It is hard to see how this analysis is to be performed, for larger values of s and m, on a purely algorithmic basis. Therefore, we show and discuss only a sufficiently nontrivial example.

Example 9.10: We consider a regular empirical system from $\mathcal{P}^3$ which consists of two dense quadratic equations and one linear equation; the noninteger coefficients have 4 decimal digits, with an assumed tolerance of $.5 \times 10^{-4}$. A computation of the four zeros exhibits the following 3-cluster (rounded to 5 digits)

$$(1.50100, .46804, .30063), \quad (1.50073 \pm .00049\,\mathrm{i}, .46067 \pm .00420\,\mathrm{i}, .31056 \pm .00574\,\mathrm{i}).$$

We want to confirm that this 3-cluster stems form an exact 3-fold zero of a system within the tolerance neighborhood.

The arithmetic mean of the 3 cluster zeros is $z_0 = (1.50082, .46312, .30725)$. We shift the origin to z_0 and obtain (rounded to 4 digits)

$$\begin{aligned}
p_1(x_1, x_2, x_3) &= .000011 - .7852\,x_1 - 2.4396\,x_2 - 1.7732\,x_3 + x_1^2 - 5.5870\,x_1 x_2 \\
&\quad + 1.7348\,x_1 x_3 + 7.8036\,x_2^2 - 4.8461\,x_2 x_3 + .7524\,x_3^2, \\
p_2(x_1, x_2, x_3) &= .000041 + .7853\,x_1 + 2.4395\,x_2 + 1.7730\,x_3 - 5.0799\,x_1^2 + 3.5514\,x_1 x_2 \\
&\quad + 19.5391\,x_1 x_3 - 7.2807\,x_2^2 - 7.6514\,x_2 x_3 - 10.9038\,x_3^2, \\
p_3(x_1, x_2, x_3) &= -1.0662\,x_1 + 10.7947\,x_2 + 7.9795\,x_3.
\end{aligned}$$

Quite obviously, the Jacobian of P at 0 is very nearly singular and there is one tiny eigenvalue. Thus we expect one first order derivative in the dual space $\mathcal{D}_0$ for the assumed 3-fold zero. For an easier search for a second order derivative, we transform by the eigenvector matrix R of the

Jacobian: The substitution $x = R\, y$ yields (rounded)

$$\hat{p}_1(y_1, y_2, y_3) = .000011 + .4163\, y_1 + .000002\, y_2 - 2.4270\, y_3 + 2.2973\, y_1^2 + 7.2080\, y_1 y_2$$
$$+ 1.2116\, y_1 y_3 + 5.6538\, y_2^2 + 1.9007\, y_2 y_3 + .1598\, y_3^2\,,$$
$$\hat{p}_2(y_1, y_2, y_3) = .000041 - .4164\, y1 - .000066\, y_2 + 2.4269\, y_3 - 11.7219\, y_1^2$$
$$- 16.8280\, y_1 y_2 - 15.9552\, y_1 y_3 - 5.6543\, y_2^2 - 8.1291\, y_2 y_3 - 7.5776\, y_3^2\,,$$
$$\hat{p}_3(y_1, y_2, y_3) = .5656\, y_1 + .000089\, y_2 + 9.8042\, y_3\,.$$

In the new coordinate system, the obvious first order basis differential is ∂_{010}. The only second order differential consistent with closedness (cf. section 8.5.1) must be of the form $\partial_{020} + c_1\, \partial_{100} + c_3\, \partial_{001}$ evaluated at 0; application to $\hat{P}$ yields 3 equations for c_1, c_3 whose residuals are minimized for $c_1 \approx -10.162$, $c_3 \approx .5862$. This completes the basis for our tentative dual space $\mathcal{D}_0$.

Since the tolerances in our empirical system refer to the original coordinate system, we must backtransform this basis. The shift of the origin affects only the evaluation point; for the transformation of first order differentials, we use the following formalism: We collect the partial differentiation operators $\frac{\partial}{\partial x_\sigma}$ and $\frac{\partial}{\partial y_\sigma}$, resp., into row vectors ∂_x and ∂_y, with the understanding that $\partial_x(\mathbf{x}) = I = \partial_y(\mathbf{y})$ for the coordinate column vectors $\mathbf{x} = (x_1, \ldots, x_s)^T$ and $\mathbf{y}$, respectively. We can now form linear combinations of first order differentials as $\partial_x(..)\, c$, with a column vector c, and apply them to linear functions $a^T \mathbf{x}$:

$$(\partial_x c)(a^T \mathbf{x}) = a^T\, \partial_x(\mathbf{x})\, c = a^T\, I\, c = a^T c. \tag{9.25}$$

For $\mathbf{x} = R\,\mathbf{y}$, $\mathbf{y} = R^{-1}\mathbf{x}$, we obtain, with $\partial_y = \partial_x\,(dx/dy) = \partial_x\, R$, the following transformation rule for first order differentials

$$\partial_y\, \hat{c} = \partial_x\, R\, \hat{c} = \partial_x\, c \quad \text{with } c := R\,\hat{c}. \tag{9.26}$$

This formalism may be extended to higher order differentials. In our present situation, (9.26) yields the following basis for $\mathcal{D}_0$ in the original coordinate system (evaluations at z_0, coefficients rounded):

$$c_{01} = \partial_0\,, \qquad c_{02} = .0236\, \partial_{100} - .5927\, \partial_{010} + .8050\, \partial_{001}\,,$$
$$c_{03} = .00056\, \partial_{200} - .0140\, \partial_{110} + .0190\, \partial_{101} + .3513\, \partial_{020} - .4772\, \partial_{011} + .6481\, \partial_{002}$$
$$+ 5.3187\, \partial_{100} + 5.3198\, \partial_{010} - 6.4861\, \partial_{001}\,.$$

Application of these three functionals to the original three polynomials gives the following residuals (rounded):

$$p_1:\ .000011,\ -.000002,\ .000167 \qquad p_2:\ .000041,\ -.000066,\ .000167$$
$$p_3:\ 0,\ .000089,\ .000167\,.$$

While this looks promising, we must verify that there is actually a $\tilde{P}$ in the tolerance neighborhood for which these residuals vanish. $\quad\square$

9.3.2 Further Refinement

When we have found a tentative dual space $\mathcal{D}_0 = \text{span}\,(c_{01}, \ldots, c_{0m})$ for our conjectured m-fold zero, the search for the nearest system $\tilde{P} = (\tilde{p}_1, \ldots, \tilde{p}_s)$ with

$$c_{0\mu}(\tilde{p}_\nu) = 0 \quad \text{for } \mu = 1(1)m, \; \nu = 1(1)s, \tag{9.27}$$

is straightforward: We attach a correction $\Delta\alpha_{\nu j} = \varepsilon_{\nu j}\,\delta$ to each empirical coefficient and require (9.27) for $\tilde{p}_\nu = \bar{p}_\nu + \Delta p_\nu$. If there are more than m empirical coefficients in $(\bar{p}_\nu, e_\nu)$, we may minimize δ. If there are fewer than m empirical coefficients in some $(\bar{p}_\nu, e_\nu)$, a strict satisfaction of (9.27) will generally not be possible for that ν, with any choice of δ.

Note that the requirements (9.27) for the individual $\tilde{p}_\nu$ are *unrelated* except through the common parameter δ. The minimization of δ for the individual $\tilde{p}_\nu$ will yield different δ_ν, but by our general assumptions (cf. section 9.1.1), it is $\delta = \max_\nu \delta_\nu$ which counts as the *backward error* of an m-fold zero with multiplicity structure $\mathcal{D}_0$. Thus, if (9.27) cannot be satisfied at all for one ν, this implies $\delta_\nu = \infty$ and there is no neighboring system which has the requested m-fold zero. Otherwise, z_0 is a *valid m-fold zero with multipicity structure $\mathcal{D}_0$* of $(\bar{P}, E)$ iff $\delta = O(1)$.

If δ is moderately too large, we may try to reduce the backward error by a refinement of the assumed location z_0 and multiplicity structure $\mathcal{D}_0$ of the m-fold zero. Here, we will not change the dual space *structure* but the coefficients in the linear combinations. If there is an indication that we have found a faulty structure by our heuristic procedure, we must return there.

For the correction of z_0 and the coefficients $\gamma_{\mu j}$ in $c_\mu = \sum_j \gamma_{\mu j}\partial_j$, one would think of a Newton step for the equations $c_\mu(p_\nu) = 0$, $\mu = 1(1)m$, $\nu = 1(1)s$. With

$$
\begin{aligned}
c_\mu + \Delta c_\mu &= \sum_j (\gamma_{\mu j} + \Delta\gamma_{\mu j})\,\partial_j[z_0 + \Delta z_0] \\
&= c_\mu + \sum_j \Delta\gamma_{\mu j}\partial_j[z_0] + \sum_j \gamma_{\mu j}\sum_\sigma \Delta z_{0\sigma}\partial_{j+e_\sigma}[z_0] + \text{quadratic terms},
\end{aligned}
$$

one obtains the following set of equations for the $\Delta\gamma_{\mu j}$ and $\Delta z_{0\sigma}$:

$$\sum_{j\in J_\mu} \left(\Delta\gamma_{\mu j}\partial_j + \gamma_{\mu j}\sum_{\sigma=1}^{s}\Delta z_{0\sigma}\partial_{j+e_\sigma}\right) p_\nu(z_0) = -c_\mu(p_\nu), \quad \mu = 1(1)m, \; \nu = 1(1)s, \tag{9.28}$$

where the J_μ are the sets of j's occurring in c_μ. However, not all of the $\Delta\gamma_{\mu j}$ can be considered as independent: Just as the c_μ, the $c_\mu + \Delta c_\mu$ must satisfy the closedness constraints (8.51). Thus, a good deal of the $\Delta\gamma_{\mu j}$ for $\mu > 2$ are actually functions of the corrections in previous c_μ. Also, these functions are often nonlinear and must be linearized.

But (9.28) is a treacherous approach in any case: Let us consider the same approach for a univariate polynomial with an m-cluster about $z_0 \in \mathbb{C}$. The only possible $\mathcal{D}_0$ is span $(\partial_0, \partial_1, \ldots, \partial_{m-1})$ and the system analogous to (9.28) is

$$\Delta z_0\, \partial_\mu\, p(z_0) = -\partial_{\mu-1}\, p(z_0), \quad \mu = 1(1)m.$$

But in an m-cluster about z_0, all derivatives of p very nearly vanish at z_0 up to the $(m-1)$st one. Thus, only the equation for $\mu = m$ is suitable for a determination of Δz_0.

A direct transfer of this argument to the multivariate case works only for $\mu = 1$: When we consider (9.28), for $\nu = 1(1)s$, as a linear system for the $\Delta z_{0\sigma}$, the matrix is the Jacobian of P whose near-vanishing was the indication of a near-multiple zero. For the further differentials, the more complicated situation permits a formal analysis only when all higher order differentials are "powers" of the only first order differential, which is a "quasi-univariate" case. In any case, it is to be expected that (9.28) has to be handled with great care; on the other hand, the univariate case indicates that an appropriate subsystem should be suitable for the determination of the corrections.

Example 9.11: We continue our Example 9.10 with the tentative dual space $\mathcal{D}_0$ determined there. The explicit expressions for the original p_ν before the shift of the origin have not been displayed; it is sufficient to know that they are dense and that they have empirical coefficients throughout (except for the x^2 term in p_1). In each polynomial, we have more than m empirical coefficients, but this is only barely true for the linear polynomial p_3. The backward errors for a 3-fold zero of the determined structure at the determined z_0 are, for the individual polynomials, .27, .68, 3.55, so that the overall backward error is 3.55. This shows that there are polynomials $\tilde{p}_1$ and $\tilde{p}_2$ with an exact 3-fold zero as specified which round to our specified polynomials p_1, p_2, but the nearest such polynomial $\tilde{p}_3$ differs from p_3 by 2 units in the 4th decimal digit. In many situations, one will be satisfied with this result; here we attempt a correction of $\mathcal{D}_0$, for the purpose of demonstration.

For $\mu = 1$ and the quadratic p_1, p_2, (9.28) is a plain Newton step which is not a reasonable procedure near a multiple zero, so we omit the cases $\mu = 1$, $\nu = 1, 2$. For the linear p_3, on the other hand, (9.28) expresses a continued vanishing of the residual after a shift to $z_0 + \Delta z_0$: $\alpha_{3,100}\Delta z_{01} + \alpha_{3,010}\Delta z_{02} + \alpha_{3,001}\Delta z_{03} = 0$. Clearly, we want to satisfy this relation.

For the first order differential c_2, we may use (9.28) just as specified for all three p_ν; for p_3, the shift by Δz_0 is without effect since the second derivatives vanish. We obtain three equations for the $\Delta\gamma_{2j}$, $j = 100, 010, 001$, and the $\Delta z_{0\sigma}$.

When we consider a correction in the second order differential c_3, we must realize that the closedness constraints must remain intact for the corrected differential. This means that the second order part of $c_3 + \Delta c_3$ is fully determined by $c_2 + \Delta c_2$ and that only the supplementary first order part can contain further independent $\Delta\gamma_{3j}$. Furthermore, when we write the second order part of $c_3 + \Delta c_3$ in terms of the coefficients in $c_2 + \Delta c_2$, we obtain also quadratic terms in the $\Delta\gamma_{2j}$ which we drop. On the other hand, the shift by Δz_0 affects only the result of the first order part of c_3, and everything is even simpler for p_3. Finally, we have three linear equations for the $\Delta\gamma_{2j}$ and $\Delta\gamma_{3j}$, $j = 100, 010, 001$, and the $\Delta z_{0\sigma}$.

Altogether, this gives us 7 inhomogeneous linear equations for 9 correction parameters; we use the remaining flexibility to minimize the corrections in the coefficients of the differentials. The resulting values for the $\Delta\gamma_{\mu j}$ are all $\approx .00002$ in modulus while the correction of z_0 is only in the 6th decimal digit of the components. This confirms that the arithmetic mean of the cluster zeros is a very good estimate for the position of a potential valid multiple zero.

Since we have employed linearization, we must now check the actual residuals of the $(c_\mu + \Delta c_\mu)(p_\nu)$. For p_1 and p_2, for which we had not included a condition on the value at $z_0 + \Delta z_0$, we obtain residuals at $z_0 + \Delta z_0$ of .000007 and .000045, respectively. These can be trivially cancelled by a corresponding correction in the constant terms of p_1 and p_2 which is within their tolerance and does not affect the residuals for the differentials. The only further residual which is not $O(10^{-7})$ or less is $(c_3 + \Delta c_3)(p_2) \approx -.000034$. This can be cancelled by a change in the linear coefficients of p_2 which avoids significant contributions to the other two residuals of p_2; it remains well below the tolerance limit.

This informal analysis of the backward error of the corrected dual space shows clearly that it lies below 1. Thus we have confirmed the assertion that our empirical system of Example 9.10 has a valid 3-fold zero at z_0, whose multiplicity structure we have also determined. $\quad\square$

9.3.3 Cluster Ideals

For a zero cluster of a *univariate* polynomial p, the associated *cluster polynomial* has the cluster zeros as its only zeros; cf. section 6.3.3. The cluster polynomial of an m-cluster about $z_0 \in \mathbb{C}$ is a perturbation of $(x - z_0)^m$:

$$s(x) = (x - z_0)^m + \sum_{\mu=0}^{m-1} \hat{\sigma}_\mu \, (x - z_0)^\mu, \quad \text{with small } |\hat{\sigma}_\mu|;$$

cf. (6.40). While the locations of the individual zeros z_μ in a cluster are very volatile and depend on the coefficients of p in a highly ill-conditioned way, the coefficients $\hat{\sigma}_\mu$ of s are well-conditioned functions of the coefficients of p. Also they describe the potential cluster configurations in a stable manner; cf. Proposition 6.13. We will try to generalize this approach to zero clusters of multivariate polynomials.

It is clear that a *multivariate m-cluster* cannot be specified by one polynomial but only by an *ideal* $\mathcal{I}_c \subset \mathcal{P}^s$ or by a basis $\mathcal{B}_c$ of $\mathcal{I}_c$, respectively. To represent $\mathcal{I}_c$ as a perturbation of the primary ideal $\mathcal{I}_0$ of an m-fold zero, $\mathcal{B}_c$ should consist of the basis polynomials of $\mathcal{I}_0$ appended by yet-to-be-determined expansions about z_0 with small coefficients:

$$\mathcal{B}_c = \begin{cases} cb_1(x) &=& bb_{01}(x) + \sum_\mu \hat{\gamma}_{1\mu}\,(x - z_0)^{j_\mu}, \\ &\cdots& \\ cb_k(x) &=& bb_{0k}(x) + \sum_\mu \hat{\gamma}_{k\mu}\,(x - z_0)^{j_\mu}, \end{cases} \quad \text{with small } |\hat{\gamma}_{\kappa\mu}|, \qquad (9.29)$$

where $\langle \mathcal{B}_0 \rangle := \langle bb_{01}, \ldots, bb_{0k} \rangle$ has an m-fold zero at z_0 as its only zero.

For the initial specification of $\mathcal{I}_0$, we can use the procedure proposed in section 9.3.1 to find a tentative specification of $\mathcal{D}_0 = \mathcal{D}[\mathcal{I}_0]$ and turn this into a suitable normal set representation of $\mathcal{I}_0$. In some situations, $\mathcal{D}_0$ or $\mathcal{I}_0$ may actually be known a priori. But $\mathcal{B}_0$ is *not uniquely* defined by $\mathcal{D}_0$: $\mathcal{I}_0$ can have many *different bases* $\mathcal{B}_0$, with varying numbers k of basis elements. Furthermore, it is not clear *which monomials* $(x - z_0)^{j_\mu}$ should be admitted in the perturbation. To match the $s\,m$ components of the zeros in the cluster, we would like to represent it by a basis of exactly s polynomials $bb_{0\kappa}$ and admit m perturbation terms per basis polynomial to have a total of $s\,m$ parameters .

Without loss of generality, we may assume that we have shifted the origin to the location of the expected multiple zero z_0; in the shifted variables, the cb_κ of (9.29) then take the form

$$cb_\kappa(x) = bb_{0\kappa}(x) + \sum_\mu \gamma_{\kappa\mu}\, x^{j_\mu}; \qquad (9.30)$$

cf. section 6.3.3 for the univariate case and Example 9.10. A potential realization of the above design could consist in a normal set representation $(\mathcal{N}, \mathcal{B}_\mathcal{N})$ for $\mathcal{I}_0$ which contains a complete intersection $(bb_{01}, \ldots, bb_{0s})$ as a subset; the perturbation monomials would then naturally be furnished by the normal set $\mathcal{N}$. A more generally applicable approach uses a standard normal set representation $(\mathcal{N}, \mathcal{B}_{\mathcal{N}})$, with $N = |B[\mathcal{N}]|$ elements. This permits N potential perturbation polynomials from span $\mathcal{N}$, but the cb_κ must satisfy the $N - s$ independent syzygies of section 8.2.3, linearized with respect to the perturbations. This gives us once more $s\,m$ free parameters for the description of the m cluster in $\mathbb{C}^s$.

In the univariate case, the perturbation coefficients have been determined such that the cluster polynomial s becomes a divisor of the full polynomial p or a valid divisor of $(\bar{p}, e)$, or equivalently, such that the ideal $\langle s \rangle$ *contains* p or some $\tilde{p} \in N_\delta(\bar{p}, e)$, $\delta = O(1)$. In analogy, we must strive to determine the perturbation coefficients $\hat{\gamma}_{\kappa\mu}$ in (9.29) such that the s polynomials p_ν of the full system P or some nearby system $\tilde{P} \in N_\delta(\overline{P}, E)$ are in $\langle \mathcal{B}_c \rangle$. Algorithmically, this is the same task as the refinement of an approximate ideal representation in section 9.2.3.

Definition 9.10. An ideal $\mathcal{I}_c$ with a basis of the form (9.29)–(9.30), where the primary ideal $\langle bb_{01}, \ldots, bb_{0k} \rangle$ has an m-fold zero and the monomials x^j are from the associated normal set, is called an *m-cluster ideal*. It is a valid m-cluster ideal *for the empirical system* $(\overline{P}, E)$ if there exists a system $\widetilde{P} \in N_\delta(\overline{P}, E)$, $\delta = O(1)$, with $\widetilde{P} \subset \mathcal{I}_c$. $\qquad\square$

Let us briefly recall the approach of section 9.2.3 for our current situation: Assume that we have a tentative dual space $\mathcal{D}_0$ for an associated m-fold zero at the origin. This implies that we can construct a normal set representation $(\mathcal{N}, \mathcal{B}_0)$, with a total of $N = |B[\mathcal{N}] > s$ elements in $\mathcal{B}_0 = \{bb_{0j}\}$, for the ideal $\mathcal{I}_0 = \mathcal{I}[\mathcal{D}_0]$ and the associated multiplication matrices $A_{0\sigma}$, $\sigma = 1(1)s$; cf. section 9.2.1.

Assume, at first, that we can select s elements $bb_{0\sigma}$, $\sigma = 1(1)s$, which form a basis of $\mathcal{I}_0$; let $\overline{\mathcal{B}}_0$ be the subset of these elements. Then we may compute the expansions (9.22) of the p_ν in terms of $\overline{\mathcal{B}}_0$; since $\overline{\mathcal{B}}_0$ is a minimal basis without nontrivial syzygies, these expansions are unique. Now, just as in section 9.2.3, we set out to cancel the part in span $\mathcal{N}$ by adapting the $bb_{0\sigma}$; as usual, we drop terms which are quadratic in the corrections, which leads to the equation set (9.23). With our basis of only s elements, there are only $s\,m$ correction coefficients, which generally permits a straightforward solution of the linear system. With these corrections attached to our $bb_{0\sigma}$ (cf. (9.29)), we have a basis $\overline{\mathcal{B}}_c = \{cb_\sigma, \sigma = 1(1)s\}$ of a cluster ideal $\mathcal{I}_c$. Except for linearization and round-off effects, we have $P \subset \mathcal{I}_c$. The generalization of this procedure to a basis with $N > s$ elements is shown in Example 9.12.

If P is intrinsic and its remainders with respect to the ideal of the computed cb_σ are not sufficiently small, we may wish to repeat the adaptation procedure. Naturally, if the choice of $\mathcal{D}_0$ was structurally mistaken, no convergence can be expected and the procedure must fail. If P is the specified system of the empirical system (P, E), we can compute the backward error δ of $\overline{\mathcal{B}}_c$, i.e. the minimal tolerance-weighted distance in the data space of (P, E) of the manifold representing the polynomials in $\langle \overline{\mathcal{B}}_c \rangle$ from the data of P. If $\delta = O(1)$, we have found the representation of the exact cluster ideal of a polynomial system $\widetilde{P}$ in the tolerance neighborhood of (P, E); this means that $\langle \overline{\mathcal{B}}_c \rangle$ is a valid cluster ideal of (P, E) and that its zeros compose a valid zero cluster of (P, E); cf. Definitions 9.10 and 9.9.

Example 9.12: We return to Examples 9.10 and 9.11 but assume that the system P is intrinsic so that the cluster cannot be interpreted as a valid 3-fold zero. We want to find a representation of the cluster ideal $\mathcal{I}_c$ of the three clustered zeros. As a starting point, we use the basis $\mathbf{c}_0 =$

(c_{01}, c_{02}, c_{03}) of the dual space $\mathcal{D}_0$ for a 3-fold zero at the arithmetic mean z_0 of the cluster which we have determined in Example 9.10. At first, we must find a representation for $\mathcal{I}_0 = \mathcal{I}[\mathcal{D}_0]$:

With $m = 3$, we can only have two of the variables in the normal set; since x_1 has the smallest coefficients in c_{02} and c_{03}, we choose $\mathcal{N} = \{1, x_2, x_3\}$, with $\mathbf{b}_0 = (1, x_2, x_3)^T$. This gives 6 elements for the border basis $\mathcal{B}_0$, but the quasi-univariate form of $\mathcal{N}$ (in x_2, x_3) will permit us to select a basis of $\mathcal{I}_0$ with just 3 elements. At first, we form the multiplication matrices $A_{0\sigma} = \mathbf{c}_0(x_\sigma \mathbf{b}_0)\, (\mathbf{c}_0(\mathbf{b}_0))^{-1}$ (rounded):

$$
A_{01} = \begin{pmatrix} -5.4864 & 10.1229 & 7.4828 \\ -3.2182 & 6.1632 & 3.4465 \\ -2.1709 & 3.1452 & 3.8257 \end{pmatrix}, \quad
A_{02} = \begin{pmatrix} 0 & 1 & 0 \\ -.6595 & 1.5718 & .4753 \\ .4621 & -.5696 & -.1825 \end{pmatrix},
$$

$$
A_{03} = \begin{pmatrix} 0 & 0 & 1 \\ .4621 & -.5696 & -.1825 \\ -.9153 & 1.1908 & 1.4913 \end{pmatrix}.
$$

The six different nontrivial rows also yield the coefficients of the six elements bb_{0j} of $\mathcal{B}_0$; at first, we use only $bb_{0,002}, bb_{0,011}, bb_{0,100}$ which form a complete intersection basis. The p_ν (which we have never explicitly displayed) have the following expansions (9.22) in $\overline{\mathcal{B}}_0$:

$$
\begin{aligned}
p_1 &= -.00018 + .00027\, x_2 + .00020\, x_3 + (61.4951 - 45.1115 x_2)\, bb_{0,002} \\
&\quad + (29.4362 + 45.1115 x_3)\, bb_{0,011} + (-12.7053 + 14.6589 x_2 + 16.7005 x_3)\, bb_{0,100} + bb_{0,100}^2, \\
p_2 &= .00053 - .00067\, x_2 - .00057\, x_3 + (-73.7646 + 413.0589 x_2)\, bb_{0,002} \\
&\quad + (298.3885 - 413.0589 x_3)\, bb_{0,011} + (64.1264 - 99.2956 x_2 - 56.4850 x_3)\, bb_{0,100} \\
&\quad - 5.0799\, bb_{0,100}^2, \\
p_3 &= -.00115 + .00162\, x_2 + .00130\, x_3 - 1.0662\, bb_{0,100}.
\end{aligned}
$$

With $cb_j := bb_{0,j} + \Delta cb_j$, $\Delta cb_j := \hat{\gamma}_{j1} + \hat{\gamma}_{j2}\, (x_2 - \zeta_{02}) + \hat{\gamma}_{j3}\, (x_3 - \zeta_{03})$ and after the reduction of products not in $\mathcal{N}$, we obtain nine equations (9.23) for the nine correction coefficients; these give us (approximately) the border basis $\mathcal{B}_c$ of the cluster ideal $\mathcal{I}_c$.

The magnitude of the correction coefficients is $O(10^{-4})$ for cb_{002} and cb_{011} and $O(10^{-3})$ for cb_{100}. But although we have omitted quadratic terms in the corrections, the three zeros of our computed $\mathcal{B}_c$ agree with the cluster zeros of the original system P up to a few units of 10^{-8} or less. In view of the fact that—for a 3-cluster—perturbations of order ε in P or $\mathcal{B}_c$ may change the zeros by $O(\varepsilon^{1/3})$, this is an extremely good agreement and confirms that $\langle P \rangle \subset \langle \mathcal{B}_c \rangle$ holds to a high degree of accuracy.

The determination of the expansion (9.22) (which we have not discussed) is much simpler if we use *all* elements of $\mathcal{B}_c$ whose leading terms coincide with the monomials in the p_ν in a natural way, but then we have to introduce linearized syzygies for the Δcb_j: According to section 8.2.3, we have to consider the edges of the border web $W B_{\mathcal{N}}$ which relate to the nonvanishing rows in $A_{\sigma_1} A_{\sigma_2} - A_{\sigma_2} A_{\sigma_1}$; cf. the relations (8.23) and (8.24). Of these $\overline{N}$ edges, we may delete all those which belong to relations with $\sigma_1 \neq \bar{\sigma}, \sigma_2 \neq \bar{\sigma}$ and all those which close a loop. This leaves precisely $|B[J_{\mathcal{N}}]| - s$ edges or relations. In our situation (cf. Figure 9.2), we have $\overline{N} = 8$. When we choose $\bar{\sigma} = 3$, we can delete the edges $e_{112}, e_{212}, e_{312}$, which leaves no loops; then we can replace e_{223} and e_{113} by virtual edges. This leaves the 3 edges $e_{323}, e_{313}, e_{213}$, which

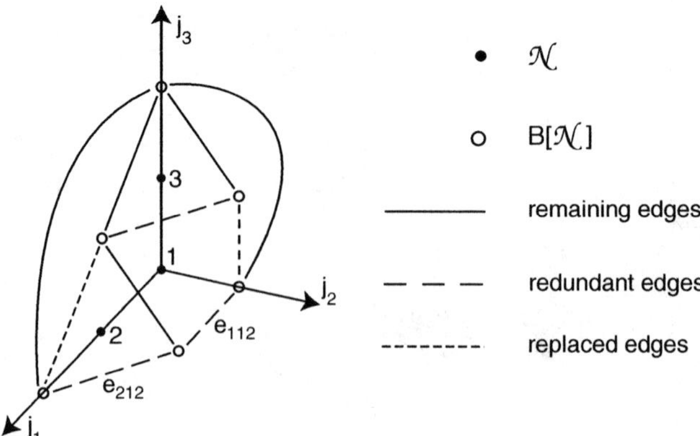

Figure 9.2.

correspond to the rows a_{111}^T, a_{012}^T, a_{102}^T in the commutativity relations. Accordingly, we take the 2nd and 3rd row in $A_1 A_3 - A_3 A_1$ and the 3rd row in $A_2 A_3 - A_3 A_2$ but replace the A_σ by $A_{0\sigma} + \Delta A_\sigma$. This yields, after dropping the quadratic terms in the corrections, 9 equations like

$$[(A_{01} + \Delta A_1)(A_{03} + \Delta A_3) - (A_{03} + \Delta A_3)(A_{01} + \Delta A_1)]_{2,1} \; = \; 0 \qquad \Rightarrow$$

$$1.6497 \, \Delta\alpha_{011,1} + .4621 \, \Delta\alpha_{110,2} + 3.4465 \, \Delta\alpha_{002,1} - .9153 \, \Delta\alpha_{110,3} - .4621 \, \Delta\alpha_{100,1}$$

$$+.5696 \, \Delta\alpha_{110,1} + 3.2182 \, \Delta\alpha_{011,2} + .1825 \, \Delta\alpha_{101,1} + 2.1709 \, \Delta\alpha_{011,3} \; = \; 0\,.$$

Note that the $\Delta\alpha_{j\mu}$ in these equations correspond to a representation of the $\Delta c b_j$ above as $\Delta c b_j := \Delta\alpha_{j1} + \Delta\alpha_{j2}\, x_2 + \Delta\alpha_{j3}\, x_3$ so that $\Delta\alpha_{j\mu}\, x^{j_\mu} = \hat{\gamma}_{j\mu}\, (x - z_0)^{j_\mu}$. These 9 syzygy equations are appended to the 9 equations which cancel the normal form coefficients of the p_ν with respect to $\mathcal{B}_c$ which now contains the 18 correction coefficients for the 6 basis elements. Within round-off, this linear system yields the same corrections for the $c b_{100}$, $c b_{011}$, $c b_{002}$ as found previously and corresponding corrections for the remaining cb's. $\square$

From examples, at least, it appears that both approaches can be turned into complete algorithms with comparable ease. There remains again the question to which extent the shift of the origin is to be used in the computation; cf. the analogous dichotomy in sections 6.3.3 and 6.3.4.

9.3.4 Asymptotic Analysis of Zero Clusters

For univariate zero clusters, the cluster polynomial provides qualitative and quantitative information about the location of the individual zeros in the cluster; cf. section 6.3.3. If the perturbation of a polynomial with an exact m-fold zero z_0 tends to zero, the cluster zeros tend to z_0 in a specifiable way; cf. (6.38) and Proposition 6.13. We would hope that analogous information may be obtained about a multivariate zero cluster from representations of the cluster ideal.

We consider the following situation: The system $P_0 \subset \mathcal{P}^s$ has an m-fold zero at z_0 which splits into an m-cluster for $P = P_0 + \varepsilon\, P_1$; we wish to determine the *asymptotic behavior* of the cluster zeros for $\varepsilon \to 0$. For small ε, these asymptotic expressions should also provide good approximations of the actual locations of the zeros.

At first, we determine the derivatives of the coefficients in (9.29)/(9.30) with respect to changes ΔP of the coefficients in P. If we apply this procedure at P_0 (i.e. for $\varepsilon = 0$), with $\Delta P = \varepsilon\, P_1$, we obtain the asymptotic values $\hat{\gamma}^0_{\kappa\mu}$ or $\gamma^0_{\kappa\mu}$ of the $\hat{\gamma}_{\kappa\mu}$ or $\gamma_{\kappa\mu}$, resp., for $\varepsilon \to 0$.

Definition 9.11. In the above situation, the polynomial system

$$\mathcal{B}^0_c(P_1) := \{cb^0_\kappa(x) = bb_{0\kappa}(x) + \varepsilon \sum_\mu \hat{\gamma}^0_{\kappa\mu}\,(x - z_0)^{j_\mu}\,,\ \kappa = 1(1)k\} \tag{9.31}$$

is an *asymptotic cluster basis* and the ideal $\langle \mathcal{B}^0_c(P_1)\rangle$ the *asymptotic cluster ideal* of P_0 generated by a perturbation $\varepsilon\, P_1$. $\quad\square$

For simplicity, we assume that we have a representation of $\mathcal{I}_c$ by a complete intersection basis $\overline{\mathcal{B}}_c$ so that there are exactly s basis polynomials cb_σ in (9.29); we start from the expansion (9.22) of the system P with respect to $\overline{\mathcal{B}}_c$: For $v = 1(1)s$,

$$\begin{aligned}
p_v &= \textstyle\sum_\sigma d_{v\sigma}(x)\, cb_\sigma + \text{ terms quadratic in the } cb_\sigma\,, \\
&= \textstyle\sum_\sigma d_{v\sigma}(x)\, (bb_{0\sigma}(x) + \sum_\mu \hat{\gamma}_{\sigma\mu}\,(x - z_0)^{j_\mu}) + \dots\,.
\end{aligned} \tag{9.32}$$

The $\hat{\gamma}_{\sigma\mu}$ are uniquely determined by the p_v; thus, in a sufficiently small neighborhood of P, (9.32) defines each coefficient $\hat{\gamma}_{\sigma\mu}$ and also the functions $d_{v\sigma}(x)$ as functions of the individual coefficients in the p_v. Assume that we are away from a singular situation where this map is not smooth and differentiate (9.32) with respect to one particular coefficient $\alpha_{v'j}$ in P; then we have, for that v',

$$x^j = \sum_\sigma [(\tfrac{\partial}{\partial \alpha_{v'j}} d_{v'\sigma}(x))\, cb_\sigma(x) + d_{v'\sigma}(x) \sum_\mu (\tfrac{\partial}{\partial \alpha_{v'j}} \hat{\gamma}_{\sigma\mu})\,(x - z_0)^{j_\mu} + \dots]; \tag{9.33}$$

for the remaining v, the left-hand side vanishes. Now we apply to (9.33) the m basis functionals c_λ of the dual space $\mathcal{D}_c$; by their definition, these annihilate the polynomials in $\mathcal{B}_c$ like $\tfrac{\partial}{\partial \alpha_{vj}} d_{v\sigma}(x)\, cb_\sigma(x)$ or the derivatives of the quadratic terms in the cb_σ. Thus we have

$$\sum_{\sigma,\mu} c_\lambda\,(d_{v\sigma}(x)\,(x - z_0)^{j_\mu})\,(\tfrac{\partial}{\partial \alpha_{v'j}} \hat{\gamma}_{\sigma\mu}) = \begin{cases} c_\lambda(x^j) & \text{for } v = v', \\[4pt] 0 & \text{for } v \neq v'. \end{cases} \tag{9.34}$$

For $v = 1(1)s$, $\lambda = 1(1)m$, these are $m\,s$ equations for the derivatives of the ms $\hat{\gamma}_{\sigma\mu}$. These derivatives describe the reaction of the basis $\mathcal{B}_0$ to changes of particular coefficients in P_0, some of which may actually vanish in P_0. This makes it possible to predict in which way a multiple zero of a multivariate system will split upon a certain perturbation of its coefficients.

Example 9.13: To keep the presentation transparent, we consider only two quadratic equations: The following system has an exact 3-fold zero at the origin:

$$P_0 = \{p_{01}(x_1, x_2) = x_1^2 + x_1 x_2 - 2\, x_2^2 + 2\, x_2,\ p_{02}(x_1, x_2) = 5\, x_1^2 + 2\, x_2^2 + 10\, x_2\}\,.$$

The multiplicity structure of z_0 is given by $\mathcal{D}_0 = \mathrm{span}\,(c_{01} = \partial_0,\ c_{02} = \partial_{10},\ c_{03} = 2\,\partial_{20} - \partial_{01})$; a suitable representation for $\mathcal{I}_0$ consists of the normal set $\mathcal{N} = \{1, x_1, x_2\}$ and the border basis subset $\overline{B}_0 = \{bb_{01} = x_1^2 + 2x_2,\ bb_{02} = x_1 x_2\}$. The expansion (9.32) of P_0 becomes

$$p_{01} = (1 - x_2)\, bb_{01} + (1 + x_1)\, bb_{02}, \qquad pp_{02} = (5 + x_2)\, bb_{01} - x_1\, bb_{02}.$$

Upon a perturbation $\varepsilon\, P_1 = \varepsilon\, (p_1, p_2)$ of P_0, the 3-fold zero at the origin will generally split and the polynomials cb_1, cb_2 which generate the ideal $\mathcal{I}_c$ of the 3-cluster will have nonvanishing coefficients $\gamma_{\sigma\mu}$ with the normal set monomials $1, x_1, x_2$. We compute the rates $\varepsilon\, \gamma_{\sigma\mu}^0$ at which these coefficients appear.

For a perturbation $\varepsilon\, x^j$ in p_{01}, the system (9.34) has the form

$$c_{0\lambda}\, [(1 - x_2)\, \gamma_{11}^0 + (1 - x_2)\, x_1\, \gamma_{12}^0 + (1 - x_2)\, x_2\, \gamma_{13}^0$$
$$+ (1 + x_1)\, \gamma_{21}^0 + (1 + x_1)\, x_1\, \gamma_{22}^0 + (1 + x_1)\, x_2\, \gamma_{23}^0] = c_{0\lambda}(x^j),$$
$$c_{0\lambda}\, [(5 + x_2)\, \gamma_{11}^0 + (5 + x_2),\, x_1\, \gamma_{12}^0 + (5 + x_2)\, x_2\, \gamma_{13}^0$$
$$- x_1\, \gamma_{21}^0 - x_1\, x_1\, \gamma_{22}^0 - x_1\, x_2\, \gamma_{23}^0] = 0,$$

for $\lambda = 1, 2, 3$. We note at first that all three $c_{0\lambda}(x^j)$ vanish for $j = 11$ and 02 so that the system becomes homogeneous; this implies that *no splitting* of the zero at 0 occurs when the coefficients of $x_1 x_2$ and/or of x_2^2 are perturbed (the 4th zero changes, of course). For $j = 00, 10, 01, 20$, exactly one of the $c_{0\lambda}(x^j)$ is nonzero; individual perturbations of each kind generate quite distinguished effects: For $j = 00$, e.g., the system (9.34) becomes

$$\gamma_{11}^0 + \gamma_{21}^0 = 1, \qquad \gamma_{12}^0 + \gamma_{21}^0 + \gamma_{22}^0 = 0, \qquad \gamma_{11}^0 - \gamma_{13}^0 + 2\gamma_{22}^0 - \gamma_{23}^0 = 0,$$
$$5\gamma_{11}^0 = 0, \qquad 5\gamma_{12}^0 - \gamma_{21}^0 = 0, \qquad -\gamma_{11}^0 - 5\gamma_{13}^0 - 2\gamma_{22}^0 = 0;$$

it yields $\gamma_{11}^0 = 0$, $\gamma_{12}^0 = .2$, $\gamma_{13}^0 = .48$, $\gamma_{21}^0 = 1$, $\gamma_{22}^0 = -1.2$, $\gamma_{23}^0 = -2.88$. Thus we have obtained the asymptotic cluster basis

$$\mathcal{B}_c^0((1, 0)) = \{x_1^2 + 2x_2 + \varepsilon\,(.2\,x_1 + .48\,x_2),\ x_1 x_2 + \varepsilon\,(1 - 1.2\,x_1 - 2.88\,x_2)\}$$

for the asymptotic cluster ideal of P_0 generated by a perturbation of the constant term in p_{01} only.

To show the validity of our approach, we take $\varepsilon = 10^{-6}$ and compute the zeros of $\{p_{01} + \varepsilon,\ p_{02}\}$ and of $\mathcal{B}_c^0((1, 0))$; we obtain identical sets

$$(.0125366,\ -.0000786),\qquad (-.0062670 \pm .0109662\,i,\ .0000405 \pm .0000687\,i)$$

of cluster zeros, up to the digits shown. Note that, apparently, the x_1-components behave like $O(\varepsilon^{1/3})$ while the x_2-components behave like $O(\varepsilon^{2/3})$. $\square$

We can now derive the asymptotic behavior of the cluster zeros from $\mathcal{B}_c^0(P_1)$: We set

$$x_\sigma(\varepsilon) \sim \varphi_\sigma\, \varepsilon^{\chi_\sigma} \quad \text{for } \varepsilon \to 0, \qquad \sigma = 1(1)s, \tag{9.35}$$

and substitute into the basis polynomials cb_κ^0 of (9.31):

$$cb_\kappa^0(x(\varepsilon)) = bb_{0\kappa}(x(\varepsilon)) + \varepsilon \sum_\mu \gamma_{\kappa\mu}^0\, x(\varepsilon)^{j_\mu} = 0, \qquad \kappa = 1(1)k.$$

Since the χ_σ are the *lowest* exponents in an asymptotic expansion of the $x_\sigma(\varepsilon)$, we must find the smallest χ_σ which permit the vanishing of the lowest order terms in the cb_κ^0; cf. the continuation of Example 9.13 below. With these exponents, we can then, generally, solve the equations for the lowest order terms and obtain the φ_σ. This provides us with the full asymptotic information about the splitting of the m-fold zero.

Example 9.13, continued: We substitute (9.35) into our asymptotic basis for $P_1 = (1, 0)$:

$$\varphi_1^2\,\varepsilon^{2\,\chi_1} + 2\,\varphi_2\,\varepsilon^{\chi_2} + .2\,\varphi_1\,\varepsilon^{1+\chi_1} + .48\,\varphi_2\,\varepsilon^{1+\chi_2} \;=\; 0,$$

$$\varphi_1\varphi_2\,\varepsilon^{\chi_1+\chi_2} + \varepsilon - 1.2\,\varphi_1\,\varepsilon^{1+\chi_1} + 2.88\,\varphi_2\,\varepsilon^{1+\chi_2} \;=\; 0.$$

The first equation requires $\chi_1 > 0$, $\chi_2 > 0$; thus ε^1 is the lowest order in the second equation and we have $\chi_1 + \chi_2 = 1$ and $\chi_1 < 1$, $\chi_2 < 1$. This makes $2\,\chi_1 = \chi_2$ the lowest exponent in the first equation. Together, this yields $\chi_1 = \frac{1}{3}$, $\chi_2 = \frac{2}{3}$, which confirms our previous observation about the asymptotic behavior of the zeros.

Now, we obtain from the first two terms of each equation $\varphi_1^2 + 2\,\varphi_2 = 0$, $\varphi_1\varphi_2 + 1 = 0$, which yields $\varphi_1 = 2^{\frac{1}{3}}$ and

$$x_1(\varepsilon) \;\sim\; 2^{\frac{1}{3}}\,\varepsilon^{\frac{1}{3}}, \quad x_2(\varepsilon) \;\sim\; -x_1(\varepsilon)^2/2,$$

with the 3 complex values of $2^{\frac{1}{3}}$. Quantitatively, $\varepsilon = 10^{-6}$ yields

$$x_1(10^{-6}) \approx \begin{cases} +.0125992 \\ -.0062996 \pm .0109112\,\mathrm{i} \end{cases}, \quad x_2(10^{-6}) \approx \begin{cases} -.0000794 \\ +.0000397 \pm .0000687\,\mathrm{i} \end{cases}.$$

These values agree quite well with the exact locations of the cluster zeros of $(p_{01}+10^{-6}, p_{02})$. $\square$

This example also shows that we cannot generally expect the same asymptotic order for all components of the cluster zeros. If the χ_σ differ, this means that the cluster initially develops *tangentially* to some lower dimensional subspace defined by the components with the lowest order exponents.

This subspace need not be a coordinate subspace; it is essentially determined by the first order differentials in $\mathcal{D}_0$. Assume, e.g., that $z_0 = 0$ and that $c_{02} = \sum_{,\sigma} \gamma_{1\sigma}\,\partial_{e_\sigma}$ is the only first order differential in $\mathcal{D}_0$. Then we may rotate the coordinate system such that $y_1 = \gamma_{11}x_1 + \ldots + \gamma_{1s}x_s$ while the other $y_\sigma = \sum_{\sigma'} \gamma_{\sigma\sigma'}x_{\sigma'}$ satisfy $\sum_{\sigma'} \gamma_{1\sigma'}\gamma_{\sigma\sigma'} = 0$; in the new coordinate system, the first order differential is now ∂_{e_1}. A basis $\mathcal{B}_0$ of $\mathcal{I}[\mathcal{D}_0]$ can therefore not have a linear term with y_1 while the other y_σ may well occur linearly. In the search for the lowest exponent, χ_1 will appear with an integer factor > 1; this will generally imply $\chi_1 < \chi_\sigma$, $\sigma \neq 1$. Thus the y_1-component of the cluster zeros will be the *dominant* component, i.e. the zeros will approach z_0 *tangentially to the* y_1-axis for $\varepsilon \to 0$.

Naturally, there may be several first order differentials in $\mathcal{D}_0$. The above considerations apply as long as there are *fewer* than s first order differentials; cf. Example 9.14. If there are s first order differentials, *all* first order derivatives vanish at the multiple zero and there are no distinguished directions.

Example 9.14: We take the quadratic system in 3 variables

$$P_0 = \begin{cases} p_{01} &=& x_1 + 2\,x_2 + 3\,x_3 - 2\,x_1^2 + 27\,x_1x_2 - 12\,x_1x_3 + 12\,x_2^2 + 31\,x_2x_3 - 6\,x_3^2, \\ p_{02} &=& -2\,x_1 - 4\,x_2 - 6\,x_3 + 8\,x_1^2 + 5\,x_1x_2 - 4\,x_1x_3 - 21\,x_2^2 - 25\,x_2x_3 - 40\,x_3^2, \\ p_{03} &=& 3\,x_1 + 6\,x_2 + 9\,x_3 - 2\,x_1^2 - 31\,x_1x_2 + 8\,x_1x_3 - 33\,x_2^2 - 25\,x_2x_3 + 14\,x_3^2; \end{cases}$$

the linear terms in P_0 are proportional to $x_1 + 2x_2 + 3x_3$, which indicates that there are two linearly independent first order differentials at the multiple zero $z_0 = 0$. We rotate the coordinate system so that $y_1 = x_1 + 2x_2 + 3x_3$ while the other two y-coordinates are orthogonal to y_1, e.g., $y_2 = x_1 + x_2 - x_3$, $y_3 = x_1 - 2x_2 + x_3$. In the new variables, P_0 becomes

$$P_0(y) = \begin{cases} p_{01}(y) &= y_1 + y_1^2 + 2y_1y_2 - 3y_1y_3 + 2y_2^2 - y_2y_3 - 3y_3^2, \\ p_{02}(y) &= -2y_1 - 3y_1^2 + 5y_1y_2 + 2y_1y_3 + y_2^2 + 4y_2y_3 - y_3^2, \\ p_{03}(y) &= 3y_1 - y_1^2 - 5y_1y_2 + 4y_1y_3 - 3y_2^2 + 2y_2y_3 + y_3^2. \end{cases}$$

Now, the first order differentials in $\mathcal{D}_0$ are ∂_{e_2} and ∂_{e_3}. But (cf. Example 8.12) there must also be a 2nd order basis differential which is found as $55\,\partial_{020} + 13\,\partial_{011} + 43\,\partial_{002} + 32\,\partial_{100}$; thus, we have a 4-fold zero at 0. A feasible normal set for a basis $\mathcal{B}_0$ of $\mathcal{I}_0$ is $\{1, y_1, y_2, y_3\}$, which leads to the border basis

$$y_1^2, \quad y_1y_2, \quad y_1y_3, \quad y_2^2 - \tfrac{55}{32}y_1, \quad y_2y_3 - \tfrac{13}{32}y_1, \quad y_3^2 - \tfrac{43}{32}y_1.$$

The last three polynomials form a complete intersection which generates $\mathcal{I}_0$, we denote them by $bb_{01}, bb_{02}, bb_{03}$. With them, we have

$$\begin{aligned} p_{01}(y) = \;& [\,(1098 - 172\,y_1 + 824\,y_3)\,bb_{01} + (-549 + 104\,y_1 - 824\,y_2 + 568\,y_3)\,bb_{02} \\ & + (-1647 - 220\,y_1 - 568\,y_2)\,bb_{003}\,]\,/\,549 + \text{ quadr. terms in the } bb_{0\kappa}\,; \end{aligned}$$

there are the analogous expansions for p_{02} and p_{03}.

We perturb the $bb_{0\kappa}$ into some asymptotic cluster basis $\mathcal{B}_c^0$, e.g.,

$$cb_1^0 = bb_{01} + \varepsilon, \quad cb_2^0 = bb_{02} + 2\varepsilon, \quad cb_3^0 = bb_{03} + 3\varepsilon\,;$$

cf. (9.31). In this case, the asymptotic analysis is straightforward; we obtain $\chi_1 = \varepsilon$, $\chi_2 = \chi_3 = \varepsilon^{1/2}$. The resulting system

$$\varphi_2^2 - \tfrac{55}{32}\varphi_1 + 1 = \varphi_2\varphi_3 - \tfrac{13}{32}\varphi_1 + 2 = \varphi_3^2 - \tfrac{43}{32}\varphi_1 + 3 = 0$$

yields the four solutions for the $y_\sigma(\varepsilon)$ (rounded)

$$(2.4626\,\varepsilon,\ \pm1.7979\,\varepsilon^{\frac{1}{2}},\ \mp.5560\,\varepsilon^{\frac{1}{2}}),\quad (-.1894\,\varepsilon,\ \pm1.1513\,i\,\varepsilon^{\frac{1}{2}},\ \pm1.8040\,i\,\varepsilon^{\frac{1}{2}}).$$

This shows that, for the above perturbation, the explosive initial $O(\varepsilon^{\frac{1}{2}})$ spread of the cluster zeros occurs only within the y_2, y_3-subspace. Note that there is no order $\varepsilon^{\frac{1}{4}}$ which one might have expected for a 4-fold zero.

When we substitute the chosen perturbation into the linear terms in the expansion of the $p_{0\nu}$ and solve the perturbed system $P(y)$, with $\varepsilon = 10^{-4}$, we obtain values for the cluster zeros which coincide with the asymptotic values in 6 or more decimal digits. When we rotate the perturbed system back to the original x-coordinates, the cluster zeros of the perturbed system $P(x)$ show an $O(\varepsilon^{\frac{1}{2}})$ behavior in *all* components. Now the smooth $O(\varepsilon)$ behavior is hidden in the distance from the subspace spanned by the y_2, y_3 coordinate axes; it may be recovered by forming $x_1 + 2x_2 + 3x_3$ for the cluster zeros. $\square$

Exercises

1. Consider the system $P(x, y, z) \subset \mathcal{P}^3$ of Examples 8.28, 8.31, and 8.32 and form the perturbed system

$$P_\varepsilon(x, y, z) \;=\; P(x, y, z) + \varepsilon\,(x - 2y + 1,\; y - 2z,\; z - 2x - 1),$$

with $\varepsilon = 10^{-5}$, which has a 3-cluster of zeros about $(2,-1,1)$.

    (a) Use the approach of section 9.3.1 to find, from the values of the clustered zeros of P_ε, an approximate dual space $\widetilde{\mathcal{D}}_0$ for the 3-fold zero "behind" the 3-cluster.

    (b) Find a system $\widetilde{P}$ near P which has a 3-fold zero of the multiplicity structure $\widetilde{\mathcal{D}}_0$. Compare with P and P_ε.

2. (a) In the situation of Example 9.13, form the systems (9.34) for all other values of $j \in \{00, 10, 01, 20\}$, with a perturbed term with x^j in either p_{01} or p_{02}, and solve them. Then compute the cluster zeros of P and of $\overline{B}_c$, resp., for each case and check against the zero sets.

    (b) Analyze the different behavior of the zeros in the cluster for the eight different perturbation cases by determining the values of the parameters in (9.35). Compare with the actual zero locations.

    (c) Check that a perturbation of the $x_1 x_2$ or x_2^2 terms leaves the triple zero intact.

3. Consider the system $P \subset \mathcal{P}^3$ of Examples 8.30 and 8.33.

    (a) By the analysis in section 9.3.4, find the asymptotic splitting pattern of the 6-fold zero at 0 for some chosen perturbation(s) $\varepsilon\,P_1$ of P.

    (b) Are there perturbations which do not split the 6-fold zero ?

9.4 Singular Systems of Empirical Polynomials

For a regular system $(\bar{P}, E)$ of s empirical polynomials in s variables with m zeros, we have stipulated that there are no systems $\widetilde{P}$ in its tolerance neighborhood $N_\delta(\bar{P}, E)$, $\delta = O(1)$, which have either a *positive-dimensional* zero set or *fewer than m zeros*; cf. Definition 9.2. This agrees with the standard notion of a regular linear system, with coefficients of limited accuracy, in numerical linear algebra.

A system of linear equations is commonly called *singular* if the coefficient matrix of its linear terms is rank-deficient. It is well known that this implies that the system has either a positive-dimensional zero manifold or no zeros at all, i.e. fewer than the standard one zero. In numerical linear algebra, a system with coefficients of limited accuracy is *near-singular* if there are singular systems within its tolerance neighborhood. Thus, the natural counterpart of Definition 9.2 is

Definition 9.12. A polynomial system $P \in (\mathcal{P}^s)^s$ is called *singular* if it has either a positive-dimensional zero set or fewer zeros (counting multiplicities) than its BKK-number requires; cf. section 8.3.4. An *empirical* polynomial system $(\overline{P}, E) \subset (\mathcal{P}^s)^s$ is *singular* if there exist singular systems $\widetilde{P} \in N_\delta(\overline{P}, E)$, $\delta = O(1)$. $\quad\square$

The term "singular" is commonly used in mathematics for phenomena which disappear under generic perturbations of the situation. This holds for the present context: A singular matrix becomes regular under a generic perturbation so that the singularity of a linear system vanishes under almost all perturbations of its coefficients. Likewise, an intrinsic singular polynomial system becomes regular under generic perturbations. This fact appears to contradict the continuous dependence of zeros on the coefficients of a multivariate system: While the disappearance of a zero towards infinity may well be seen as a continuous process, it is—at first sight—difficult to understand the spontaneous appearance or disappearance of a zero *manifold* upon arbitrarily small changes of coefficients as a continuous phenomenon. We shall see how our neighborhood concept provides a natural continuous explanation of this situation. The situation where zeros are "lost" will be discussed separately in section 9.5.

Before we turn to the discussion of singular polynomial systems, we review the well-understood situation with singular linear systems in a way which helps in the analysis of singular polynomial systems.

9.4.1   Singular Systems of Linear Polynomials

Consider the singular linear system (cf. any text on numerical linear algebra)[17]

$$P_0 = A_0 \mathbf{x} + b_0, \qquad A_0 \in \mathbb{R}^{s \times s}, \ b_0 \in \mathbb{R}^s, \qquad \text{with } \mathbf{x} := (x_1, \dots, x_s)^T; \qquad (9.36)$$

the matrix A_0 is assumed to have a rank $r < s$, with singular-value decomposition

$$A_0 \, (U_1 \ U_2) = (V_1 \ V_2) \begin{pmatrix} \Sigma_0 & 0 \\ 0 & 0 \end{pmatrix}, \qquad (9.37)$$

with orthogonal $s \times s$ matrices $(U_1 \ U_2)$ and $(V_1 \ V_2)$, $U_1, \ V_1 \in \mathbb{R}^{s \times r}$, $U_2, \ V_2 \in \mathbb{R}^{s \times (s-r)}$, $\Sigma_0 = \text{diag}(\sigma_1, \dots, \sigma_r) \in \mathbb{R}^{r \times r}$, $\sigma_\rho > 0$, $\rho = 1(1)r$. If b_0 is in the span of V_1, a parametrized representation of the $(s-r)$-dimensional zero manifold M_0 of $\langle P_0 \rangle$ is given by

$$z_0(w) = -U_1 \Sigma_0^{-1} V_1^T b_0 + U_2 \, w, \qquad w \in \mathbb{R}^{s-r} \text{ arbitrary}; \qquad (9.38)$$

$U_1 \Sigma^{-1} V_1^T =: A_0^+$ is the (Moore–Penrose) *pseudo-inverse* of A_0. For b_0 not in the span of V_1, the system P_0 has no zero.

With $\overline{U}_i := \text{span } U_i$, $i = 1, 2$, the s.v.d. (9.37) yields decompositions $\overline{U}_1 \oplus \overline{U}_2$ of the $\mathbb{R}^s$ and $U_1 U_1^T + U_2 U_2^T$ of the identity in $\mathbb{R}^s$. The first term in (9.38) is the unique projection of the zeros on M_0 onto $\overline{U}_1$, the second term the projection onto $\overline{U}_2$. The distinction of these two projections plays a fundamental role in the analysis of singular systems.

We will now analyze the *transition* between the singular system P_0 and neighboring nonsingular systems from various points of view. For an arbitrary fixed matrix A_1, we consider the linear system

$$P(\varepsilon) := P_0 + \varepsilon \, (A_1 \mathbf{x} + b_1), \qquad \varepsilon \in \mathbb{R}. \qquad (9.39)$$

Due to

$$(A_0 + \varepsilon \, A_1) \, (U_1 \ U_2) = (V_1 \ V_2) \begin{pmatrix} \Sigma_0 + \varepsilon \, V_1^T A_1 U_1 & \varepsilon \, V_1^T A_1 U_2 \\ \varepsilon \, V_2^T A_1 U_1 & \varepsilon \, V_2^T A_1 U_2 \end{pmatrix}, \qquad (9.40)$$

---

[17]In this section, we switch to a *real* context for a simpler formal representation.

$A_0 + \varepsilon A_1$ is nonsingular for all sufficiently small $|\varepsilon|$ iff $V_2^T A_1 U_2$ is regular. In this case, which we will assume in the following, $P(\varepsilon)$ has one (simple) zero $z(\varepsilon)$ which depends smoothly on ε, with a limit value $z_0 := \lim_{\varepsilon \to 0} z(\varepsilon)$. The system $P_0 = \lim_{\varepsilon \to 0} P(\varepsilon)$, on the other hand, has the $(s-r)$-dimensional zero manifold $M_0 = \{-A_0^+ b_0 + U_2 w, \ w \in \mathbb{R}^{s-r}\}$ of (9.38)! How are these facts compatible? A first explanation is given by:

Proposition 9.4. For any perturbation (A_1, b_1) with regular $V_2^T A_1 U_2$, there exists a unique $z_0 := \lim_{\varepsilon \to 0} z(\varepsilon) \in M_0$. Reversely, for any specified $\bar{z}_0 \in M_0$, there exist (A_1, b_1) such that $\lim_{\varepsilon \to 0} z(\varepsilon) = \bar{z}_0$.

Proof: We try to determine $z_0, z_1 \in \mathbb{R}^s$ such that

$$P(\varepsilon)\,(z_0 + \varepsilon\,z_1 + \mathrm{O}(\varepsilon^2)) \;=\; (A_0\,z_0 + b_0) + \varepsilon\,(A_0\,z_1 + A_1\,z_0 + b_1) + \mathrm{O}(\varepsilon^2) \overset{\varepsilon}{\equiv} 0, \quad (9.41)$$

which requires $z_0 = -A_0^+ b_0 + U_2 w \in M_0$ and $A_1 z_0 + b_1 \in \mathrm{span}\, A_0 = \mathrm{span}\, V_1$ or (cf. (9.38))

$$V_2^T\,(A_1\,z_0(w_0) + b_1) \;=\; V_2^T A_1 U_2\, w_0 + V_2^T\,(-A_1 A_0^+ b_0 + b_1) \;=\; 0;$$

this defines a unique w_0 for regular $V_2^T A_1 U_2$, i.e. a unique position of z_0 on M_0. Reversely, for a specified $\bar{z}_0 \in M_0$, any A_1, b_1 with $A_1 \bar{z}_0 + b_1 = 0$ satisfies $P(\varepsilon)\bar{z}_0 = 0$ for all ε and hence for $\varepsilon \to 0$. $\square$

Proposition 9.4 establishes the zero manifold M_0 as the set of *all possible limits* of zeros z_ε of $P(\varepsilon)$ from a neighborhood of P_0. A particular limit z_0 on M_0 is reached for a particular $P(\varepsilon)$. From the proof of Proposition 9.4, we have for $z(\varepsilon) =: z_0 + \varepsilon\,z_1 + \mathrm{O}(\varepsilon^2)$:

$$U_1^T z_0 = -\Sigma_0^{-1} V_1^T b_0\,, \qquad U_2^T z_0 = -(V_2^T A_1 U_2)^{-1} V_2^T\,(A_1 U_1 U_1^T z_0 + b_1)\,,$$
$$U_1^T z_1 = -\Sigma_0^{-1}, V_1^T\,(A_1 z_0 + b_1)\,. \tag{9.42}$$

Thus, the $\mathrm{O}(\varepsilon)$ quantities A_1, b_1 simultaneously determine the $\mathrm{O}(1)$ projection of $z_0(\varepsilon)$ onto M_0 and the $\mathrm{O}(\varepsilon)$ part of the projection of $z_0(\varepsilon)$ onto the subspace $\overline{U}_1 \perp M_0$. This indicates that the $\overline{U}_1$ component is a well-conditioned function of $P(\varepsilon)$ while the $\overline{U}_2$ component is ill-conditioned. We will now investigate this observation further.

For this purpose, we consider the effect of perturbations of the system $P(\varepsilon)$ on its zero $z(\varepsilon)$ for a fixed $\varepsilon > 0$. We perturb $P(\varepsilon)$ into

$$\hat{P}(\varepsilon) \;:=\; (A_0 + \varepsilon\,A_1 + \Delta A)\,\mathbf{x} + (b_0 + \varepsilon\,b_1 + \Delta b) \;=\; 0\,, \tag{9.43}$$

with zero $\hat{z}(\varepsilon)$. $\hat{\varepsilon} := \max(\|\Delta A\|, \|\Delta b\|)$ is assumed small, and we retain the reference to the subspace system $\overline{U}_1, \overline{U}_2$. We assume $\|A_1\|, \|b_1\| = \mathrm{O}(1)$, to permit relations between ε and $\hat{\varepsilon}$, and $A_1 z_0 + b_0 \neq 0$ for a nontrivial situation.

In (9.42), we substitute $\varepsilon A_1 + \Delta A$, $\varepsilon b_1 + \Delta b$ in place of $\varepsilon A_1, \varepsilon b_1$, and set $\hat{z}(\varepsilon) =: \hat{z}_0 + \hat{z}_1 \varepsilon + \mathrm{O}(\varepsilon^2)$. Obviously, the perturbation has no influence at all on $U_1^T z_0$, while

$$\varepsilon\,U_1^T \hat{z}_1 \;=\; -\Sigma_0^{-1} V_1^T\,(\varepsilon\,A_1 + \Delta A)\,z_0 + (b_0 + \varepsilon\,b_1 + \Delta b)\,;$$

this implies that the effect of the perturbation orthogonally to M_0 is $\mathrm{O}(\hat{\varepsilon})$ independently of ε. For the ill-conditioned $\overline{U}_2$-component $U_2^T \hat{z}_0$, on the other hand, which determines the projection of $\hat{z}(\varepsilon)$ onto M_0, we have

$$U_2^T \hat{z}_0 \;=\; -(V_2^T (A_1 + \tfrac{1}{\varepsilon}\Delta A)U_2)^{-1} V_2^T\,((A_1 + \tfrac{1}{\varepsilon}\Delta A)U_1 U_1^T z_0 + (b_1 + \tfrac{1}{\varepsilon}\Delta b))\,,$$

and the effect of the $O(\hat{\varepsilon})$ perturbation becomes $O(\frac{\hat{\varepsilon}}{\varepsilon})$! As long as $\hat{\varepsilon} \ll \varepsilon$, there is no qualitative change, but the $\overline{U}_2$ components (parallel to M_0) of the indetermination in $\hat{z}(\varepsilon)$ will extend further and further for increasing $\frac{\hat{\varepsilon}}{\varepsilon}$. Finally, when $\hat{\varepsilon} \approx \varepsilon$, simultaneously with the appearance of singular systems in the perturbation neighborhood of $P(\varepsilon)$, the pseudozero set of $P(\varepsilon)$ extends to infinity and includes the complete manifold M_0; cf. (9.42).

Thus, the embedding of $P(\varepsilon)$ into an *empirical* system $(P(\varepsilon), E)$, with $\|E\| = \hat{\varepsilon}$, establishes the transition between the regular systems $P(\varepsilon)$, $\varepsilon \neq 0$, with their isolated zeros $z(\varepsilon)$, and the singular system $P(0)$ with its zero manifold M_0 as a continuous phenomenon: The set $Z_\delta(P(\varepsilon), E)$ of the zeros of the linear systems $\tilde{P} \in N_\delta(P(\varepsilon), E)$, $\delta = O(1)$, extends further and further parallel to the zero manifold M_0 of $P(0)$ as $\hat{\varepsilon} = \|E\|$ approaches ε while its extension orthogonal to M_0 remains $O(\hat{\varepsilon})$. When $N_\delta(P(\varepsilon), E)$, $\delta = O(1)$, includes singular systems for $\hat{\varepsilon} \approx \varepsilon$, $Z_\delta(P(\varepsilon), E)$ extends to infinity and envelops the manifold M_0 completely; cf. Figure 9.3.

Example 9.15: We take a simple transparent situation, with $s = 2$, $r = 1$:

$$P(\varepsilon) = \left[\begin{pmatrix} 1 & -1 \\ 1 & -1 \end{pmatrix} + \varepsilon \begin{pmatrix} 1 & 1 \\ -1 & -1 \end{pmatrix} \right] \mathbf{x} + \left[\begin{pmatrix} -1 \\ -1 \end{pmatrix} + \varepsilon \begin{pmatrix} -1 \\ 0 \end{pmatrix} \right],$$

$$\text{with } \begin{pmatrix} 1 & -1 \\ 1 & -1 \end{pmatrix} \frac{1}{\sqrt{2}} \begin{pmatrix} 1 & 1 \\ -1 & 1 \end{pmatrix} = \frac{1}{\sqrt{2}} \begin{pmatrix} 1 & 1 \\ 1 & -1 \end{pmatrix} \begin{pmatrix} 2 & 0 \\ 0 & 0 \end{pmatrix}.$$

From (9.42), we have

$$z(\varepsilon) = \sqrt{2} \left((\frac{1}{2} + \frac{\varepsilon}{4}) U_1 + \frac{1}{4} U_2 \right) = \begin{pmatrix} \frac{3}{4} + \frac{\varepsilon}{4} \\ -\frac{1}{4} - \frac{\varepsilon}{4} \end{pmatrix};$$

there are no further terms in the asymptotic expansion of $z(\varepsilon)$. For $\varepsilon \to 0$, $z(\varepsilon)$ tends to $z_0 = (\frac{3}{4}, -\frac{1}{4})$ on the zero manifold $M_0 = \frac{1}{\sqrt{2}} U_1 + U_2 w$, $w \in \mathbb{R}$, of $P(0)$; cf. Figure 9.3.

Now we attach potential perturbations Δa_{ij}, Δb_i of maximal modulus $\hat{\varepsilon}$ to the coefficients in $P(\varepsilon)$, i.e. we consider the empirical system $(P(\varepsilon), E)$ with $E = \hat{\varepsilon} \begin{pmatrix} 1 & 1 & | & 1 \\ 1 & 1 & | & 1 \end{pmatrix}$. We regard the maximal deviations d_1 and d_2 of the $\overline{U}_1$- and $\overline{U}_2$-components of the pseudozeros $\tilde{z}(\varepsilon)$ of $\tilde{P} \in N_1(P(\varepsilon), E)$ from $z(\varepsilon)$; we obtain the d_i by requiring the backward error of $z(\varepsilon) + d_i U_i$ to be ≤ 1 (cf. section 9.1.1).

For $i = 1$, we obtain $|d_1| \leq \sqrt{2}\,(1 + \frac{\varepsilon}{4})\,\frac{\hat{\varepsilon}}{1-\hat{\varepsilon}}$; thus the extension of $Z_1(P(\varepsilon), E)$ orthogonal to M_0 is $O(\hat{\varepsilon})$ and does not vary significantly when $\varepsilon \to 0$; cf. Figure 9.3.

For $i = 2$, the backward error along the parallel to M_0 through $z(\varepsilon)$ depends on the signs of the x_i: As long as $z(\varepsilon) + d_2 U_2$ is in the quadrant of $z(\varepsilon)$, $|d_2| = \sqrt{2}\,(1 + \frac{\varepsilon}{4})\,\frac{\hat{\varepsilon}}{\varepsilon} = O(\frac{\hat{\varepsilon}}{\varepsilon})$. For $\hat{\varepsilon} \overset{>}{\approx} \frac{\varepsilon}{4}$, $z(\varepsilon) + d_2 U_2$ is in the first quadrant and $d_2 = \frac{3}{4}\sqrt{2}\,\frac{\hat{\varepsilon}}{\varepsilon - \hat{\varepsilon}}$; now the extension of Z_1 increases inversely to $\varepsilon - \hat{\varepsilon}$. For $\hat{\varepsilon} \geq \varepsilon$, $N_1(P(\varepsilon), E)$ contains singular systems and the extension of Z_1 in the U_2-direction is *infinite* while its extension in the U_1-direction remains $O(\hat{\varepsilon})$; cf. Figure 9.3.

Numerically, with $\varepsilon = .01$, we have

$$P(\varepsilon) = \bar{A}\,\mathbf{x} + \bar{b} = \begin{pmatrix} 1.01 & -.99 \\ .99 & -1.01 \end{pmatrix} \begin{pmatrix} x_1 \\ x_2 \end{pmatrix} - \begin{pmatrix} 1.01 \\ 1.00 \end{pmatrix}, \quad \text{with } z(\varepsilon) = \begin{pmatrix} .7525 \\ -.2525 \end{pmatrix}.$$

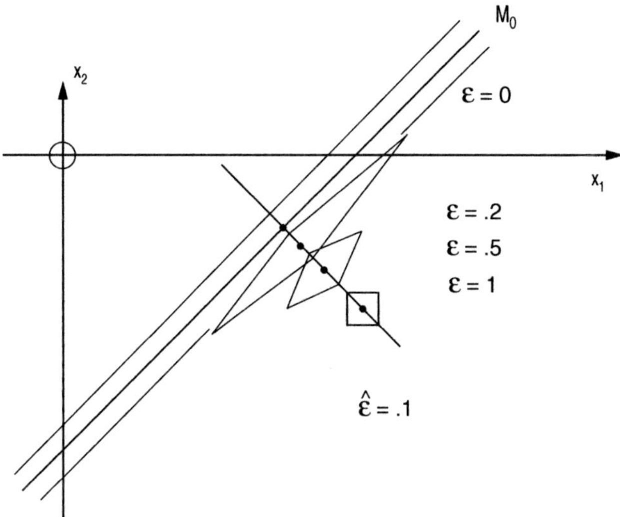

Figure 9.3.

When we assume that $P(\varepsilon)$ has been rounded to two decimal digits ($\hat{\varepsilon} = .5 \cdot 10^{-2}$), we find that valid pseudozeros $\tilde{z}$ have (rounded) $\tilde{\zeta}_1 - \tilde{\zeta}_2 \in [.995, 1.015]$, but $\tilde{\zeta}_1 + \tilde{\zeta}_2 \in [-.50, 2.00]$! Thus, only the distance of the $\tilde{z}$ from M_0 is well defined, and it is not meaningful to specify $\tilde{z}$ in terms of its x_1, x_2-components. $\quad\square$

In a real-life situation, there is, generally, no explicit parameter ε but an empirical linear system $((\bar{A}, \bar{b}), (E, e))$ with a near-singular matrix $\bar{A}$. From linear algebra, we know that $\min \sigma_i(\bar{A}) \leq \|E\|_2\, \delta$ is the criterion for $N_\delta(\bar{A}, E)$ to contain singular matrices. If the decreasingly ordered singular values of $\bar{A}$ satisfy $\sigma_i \leq \|E\|_2\, \delta$ for $i = (\bar{r}+1)(1)s$, $\bar{r}$ is the lowest rank of a matrix in $N_\delta(\bar{A}, E)$; it is attained, e.g., by

$$\tilde{A} = (\tilde{V}_1\ \tilde{V}_2) \begin{pmatrix} \tilde{\Sigma}_1 & 0 \\ 0 & 0 \end{pmatrix} (\tilde{U}_1\ \tilde{U}_2)^T = \tilde{V}_1 \tilde{\Sigma}_1 \tilde{U}_1^T. \tag{9.44}$$

Iff $\|\tilde{V}_2^T \bar{b}\|_2 \leq \|e\|_2\, \delta$, there exists $\tilde{b} = \tilde{V}_1(\tilde{V}_1^T \bar{b}) \in N_\delta(\bar{b}, e)$ so that the system $\tilde{A}\,x + \tilde{b}$ has a zero manifold. This implies that the empirical system $((\bar{A}, \bar{b}), (E, e))$ has this manifold as a valid solution. In many cases, this will be the appropriate solution for the system. The case $\|\tilde{V}_2^T \bar{b}\|_2 > O(1)\,\|e\|_2$ will be treated in section 9.5.1.

Let us finally consider the solution of $P(\varepsilon) = 0$, with a consistent singular part $A_0 x - b_0$, from the algebraic point of view. With $\mathcal{N} = \{1\}$, the computation of a border basis is equivalent with a diagonalization of $A_0 + \varepsilon A_1$. With row and column interchanges which avoid a division by an $O(\varepsilon)$ term, we can, at first, proceed to a system (rows and columns have been renumbered)

$$
\begin{aligned}
\alpha_{11}\,x_1 & & + & \alpha_{1,r+1}\,x_{r+1} + \ldots & + & \beta_1\,, \\
& \alpha_{22}\,x_2 & + & \alpha_{2,r+1}\,x_{r+1} + \ldots & + & \beta_2\,, \\
& \ddots & & \ldots & & \\
& \alpha_{rr}\,x_r \ \ + & & \alpha_{r,r+1}\,x_{r+1} + \ldots & + & \beta_r\,, \\
& & & \varepsilon\,\alpha_{r+1,r+1}\,x_{r+1} + \ldots & + & \varepsilon\,\beta_{r+1}\,, \\
& & & \ldots & & \\
& & & \varepsilon\,\alpha_{s,r+1}\,x_{r+1} + \ldots & + & \varepsilon\,\beta_s\,,
\end{aligned}
\tag{9.45}
$$

where the $\alpha_{\rho\rho}$, $\rho = 1(1)r$, are nonzero and $O(1)$ in ε (the dependence of the α_{ij} on ε is not denoted). Actually, (9.45) is a *comprehensive Groebner basis*[18] of $P(\varepsilon)$. There are two distinct ways of proceeding further:

We may divide the lower $s - r$ polynomials by ε and eliminate further to obtain a border basis system

$$\alpha_{\sigma\sigma}(\varepsilon)\,x_\sigma + \beta_\sigma(\varepsilon)\,, \qquad \sigma = 1(1)s\,,$$

with $\zeta_\sigma(\varepsilon) = -\beta_\sigma(\varepsilon)/\alpha_{\sigma\sigma}(\varepsilon)$ and $\zeta_{0\sigma} = -\beta_\sigma(0)/\alpha_{\sigma\sigma}(0)$, $\sigma = 1(1)s$.

Or we may set $\varepsilon = 0$ everywhere in (9.45) and obtain a basis for the manifold M_0.

The choice depends on the type of analysis which we intend. Note that (9.45) contains both options; this is the significance of a comprehensive Groebner basis.

Example 9.15, continued: Elimination in $(A_0 + \varepsilon A_1)\,\mathbf{x} + (b_0 + \varepsilon b_1)$ yields the comprehensive basis

$$(1 + \varepsilon)\,x_1 - (1 - \varepsilon)\,x_2 - (1 + \varepsilon)\,, \qquad 4\varepsilon\,x_2 + \varepsilon\,(1 + \varepsilon)\,.$$

The first option leads to $(1 + \varepsilon)\,x_1 = (1 + \varepsilon)\,(\tfrac{3}{4} + \tfrac{\varepsilon}{4})$ and $z(\varepsilon)$ as previously. The second option yields the representation $x_1 - x_2 - 1 = 0$ of M_0. $\square$

9.4.2   Singular Polynomial Systems; Simple d-Points

Like singular linear systems, singular polynomial systems P_0 with a d-dimensional zero manifold M_0, $d > 0$ (cf. Definition 9.12), constitute a *consistent overdetermined* representation of M_0 and a *singular* situation. In a *discrete* algebraic world, there is nothing special about them; in fact, a variety in $\mathbb{C}^s$ which consists of a manifold *and* of isolated points cannot be defined by fewer than s polynomials. In a *continuous* algebraic world, however, we must realize that almost all arbitrarily small perturbations of coefficients of P_0 make $\langle P_0 \rangle$ 0-dimensional. This fact, and the associated spontaneous appearance and disappearance of a manifold, constitute the singularity of P_0. As in the linear case, this affects the isolated zeros near M_0 of neighboring 0-dimensional systems $\tilde{P}$.

Naturally, singular polynomial systems may have many aspects not related to their singularity. For example, they may have isolated simple and multiple zeros away from M_0 which behave perfectly normally and which we will not consider further. For simplicity, we will also assume that there is only one positive-dimensional zero manifold.

---

[18]Comprehensive Groebner Bases have been introduced and discussed by V. Weispfenning in [2.13]; in this book, we do not consider them explicitly.

As in section 9.4.1, we consider at first the behavior of zeros of near-singular systems

$$P(\varepsilon)(x) := P_0(x) + \varepsilon\, P_1(x) \in (\mathcal{P}^s)^s\,, \qquad \varepsilon \in \mathbb{C}\,; \tag{9.46}$$

we have now returned to our standard context $x \in \mathbb{C}^s$. With a fixed system P_1, we have a well-defined limit process $\varepsilon \to 0$; yet the choice of P_1 permits a large degree of generality. However, we restrict the support of each $p_{1\nu} \in P_1$ to the convex hull of the support of the corresponding $p_{0\nu} \in P_0$ to avoid the appearance of additional zeros; this restriction will be generally assumed in the following without mention.

At first, we assume that $P(\varepsilon)$ is nonsingular for all sufficiently small $|\varepsilon| > 0$; the explicit condition for this will appear in (9.53) below. From our experience with singular linear systems, we expect that some m_0 of the m isolated zeros $z_\mu(\varepsilon)$ of $P(\varepsilon)$ approach specific points $z_{0\mu}$ on the d-dimensional zero manifold M_0 of P_0 while the remaining $m - m_0$ isolated zeros of $P(\varepsilon)$ approach the corresponding isolated zeros of P_0; the case $m_0 = m$ is possible. While this turns out to be true, the details are far more complicated than in the linear case.

As in Proposition 9.4, we assume

$$z_\mu(\varepsilon) \;=\; z_{0\mu} + \varepsilon\, z_{1\mu} + \mathrm{O}(\varepsilon^2)\,, \qquad \mu = 1(1)m_0\,, \tag{9.47}$$

and substitute into $P(\varepsilon)$:

$$P(\varepsilon)(z_\mu(\varepsilon)) \;=\; P_0(z_{0\mu}) + \varepsilon\, (P_0'(z_{0\mu})\, z_{1\mu} + P_1(z_{0\mu})) + \mathrm{O}(\varepsilon^2)\,. \tag{9.48}$$

For $z_0 \in M_0$, the $s \times s$ Jacobian $P_0'(z_0)$ has rank $r := s - d$, or smaller at isolated special points. Thus, the $z_{0\mu}$ in (9.47) must satisfy the condition

$$P_1(z_{0\mu}) \;\in\; \mathrm{range}\ P_0'(z_{0\mu})\,. \tag{9.49}$$

When we, at first, ignore potential points with $\mathrm{rk}\ P_0' < r$, this range condition may be represented by d polynomial equations in $z_{0\mu}$: Take a system of d linearly independent row vectors which span the space orthogonal to the image of $P_0'(z_{0\mu})$; their scalar products with $P_1(z_{0\mu})$ must vanish. Together with P_0, they define $m_0 \geq 0$ isolated points on M_0.

Definition 9.13. For a specified near-singular polynomial system (9.46), the points $z_\mu \in M_0$ (if any) which satisfy $\mathrm{rk}\ P_0'(z_{0\mu}) = r$ and (9.49) are the *simple d-points*[19] of (9.46).   □

Like in the linear case, *all* points on M_0 may be simple d-points for a suitably chosen P_1; but there may well be isolated zeros of $P(\varepsilon)$ which do *not* converge to a point on M_0 for $\varepsilon \to 0$.

At each fixed simple d-point $z_{0\mu}$, we now introduce local coordinates $U_{1\mu}$, $U_{2\mu}$ (cf. (9.37)):

$$P_0'(z_{0\mu})\begin{pmatrix} U_{1\mu} & U_{2\mu} \end{pmatrix} \;=\; \begin{pmatrix} V_{1\mu} & V_{2\mu} \end{pmatrix}\begin{pmatrix} \Sigma_{0\mu} & 0 \\ 0 & 0 \end{pmatrix}\,, \qquad P_1(z_{0\mu}) \;=\; V_{1\mu}\,(V_{1\mu}^T P_1(z_{0\mu}))\,, \tag{9.50}$$

and decompose $z_{1\mu}$ of (9.47) into its components

$$z_{1\mu} \;=\; z_{11\mu} + z_{12\mu} \;=\; U_{1\mu}\,(U_{1\mu}^T z_{1\mu}) + U_{2\mu}\,(U_{2\mu}^T z_{1\mu})\,. \tag{9.51}$$

---

[19] d-points is short for "distinguished points."

Substitution into (9.48) yields, with (9.50),

$$z_{11\mu} \;=\; -\,U_{1\mu}\,\Sigma_{0\mu}^{-1}\,V_{1\mu}^T\,P_1(z_{0\mu})\,; \tag{9.52}$$

cf. (9.42). The component $z_{12\mu}$ remains undetermined on the $O(\varepsilon)$ level of (9.48) as in the linear case; we can obtain it from the $O(\varepsilon^2)$ terms which imply the range condition

$$V_{2\mu}^T\,[\tfrac{1}{2}\,P_0''(z_{0\mu})\,(z_{1\mu}, z_{1\mu}) + P_1'(z_{0\mu})\,z_{1\mu}] \;=$$
$$V_{2\mu}^T\,[\tfrac{1}{2}\,P_0''(z_{0\mu})\,((z_{11\mu}, z_{11\mu}) + 2\,(z_{11\mu}, z_{12\mu}) + (z_{12\mu}, z_{12\mu})) + P_1'(z_{0\mu})\,(z_{11\mu} + z_{12\mu})] \;=\; 0.$$

The quadratic term in $z_{12\mu}$ vanishes because (9.50) implies $V_{2\mu}^T\,P_0''(z_{0\mu})\,(U_{2\mu}, U_{2\mu}) = 0$. Thus $z_{12\mu} = U_{2\mu}\,(U_{2\mu}^T z_{12\mu})$ is determined by the linear system

$$V_{2\mu}^T\,[P_0''(z_{0\mu})\,(z_{11\mu}, U_{2\mu}) + P_1'(z_{0\mu})\,U_{2\mu}]\,(U_{2\mu}^T z_{12\mu})$$
$$= -\,V_{2\mu}^T\,[\tfrac{1}{2}\,P_0''(z_{0\mu})\,(z_{11\mu}, z_{11\mu}) + P_1'(z_{0\mu})\,z_{11\mu}]. \tag{9.53}$$

Obviously, the necessary regularity of $V_{2\mu}^T\,[P_0''(z_{0\mu})\,(z_{11\mu}, U_{2\mu}) + P_1'(z_{0\mu})\,U_{2\mu}]$ corresponds to the regularity of $V_2^T A_1 U_2$ in the linear case (with $P_0'' = 0$, $P_1' = A_1$). But now this condition cannot be tested a priori; we will not consider the phenomena which may appear when it is not satisfied.

Example 9.16: We consider, in $\mathcal{P}^3$,

$$P(\varepsilon)(x) \;=\; P_0(x) + \varepsilon\, P_1(x) \;=\; \left\{ \begin{array}{l} x_1 x_2 + x_3 \\ x_1 x_3 + x_2^2 - x_3 \\ x_2 + x_3 \end{array} \right. \;+\; \varepsilon \left\{ \begin{array}{l} 1 \\ 1 \\ 1 \end{array} \right. , \tag{9.54}$$

with the 3 simple zeros $z_1(\varepsilon) = (2, 0, -\varepsilon)$, $z_{2,3}(\varepsilon) = (1, \pm i\sqrt{\varepsilon}, \mp i\sqrt{\varepsilon} - \varepsilon)$ for $\varepsilon \neq 0$. $P_0(x) = 0$ and $\det P_0'(x) = 2\,x_2^2 - (x_2 + x_3)(x_1 - 1) = 0$ determine the x_1-axis as a 1-dimensional zero manifold $M_0 := \{(\xi_1, 0, 0),\ \xi_1 \in \mathbb{C}\}$ of $\langle P_0 \rangle$.

$$\text{range } P_0'(\xi_1, 0, 0) \;=\; \text{range} \begin{pmatrix} 0 & \xi_1 & 1 \\ 0 & 0 & \xi_1 - 1 \\ 0 & 1 & 1 \end{pmatrix} \;=\; \text{span}\left\{ \begin{pmatrix} \xi_1 \\ 0 \\ 1 \end{pmatrix}, \begin{pmatrix} 1 \\ \xi_1 - 1 \\ 1 \end{pmatrix} \right\},$$

so that (9.49) requires $(1, 1, -\xi_1) \cdot P_1(\xi_1, 0, 0) = 0$ or $\xi_1 = 2$. Thus, $z_{01} = (2, 0, 0)$ is the only simple d-point of $P(\varepsilon)$ and $\lim_{\varepsilon \to 0} z_1(\varepsilon) = z_{01}$. From the explicit expression for $z_1(\varepsilon)$, we have $z_{111} = (0, 0, -1)$, $z_{112} = 0$ in (9.52), and there are no further terms in the expansion (9.47) of $z_1(\varepsilon)$.

Obviously, the point $z_{02} = (1, 0, 0)$ on M_0 acts as a common limit point for the other two zeros $z_2(\varepsilon)$, $z_3(\varepsilon)$ of $P(\varepsilon)$. At z_{02}, rk $P_0' = 1 < 3 - d$; such *multiple d-points* will be considered in section 9.4.4. $\square$

Let us now consider $P(\varepsilon)$ from an algebraic point of view: For $|\varepsilon| \neq 0$, $P(\varepsilon)$ is nonsingular; it has a border basis $\mathcal{B}_{\mathcal{N}}(\varepsilon)$ for a feasible normal set $\mathcal{N}$ with $|\mathcal{N}| = m$ elements. When we treat ε like an indeterminate in the computation of $\mathcal{B}_{\mathcal{N}}(\varepsilon)$, the border basis elements are polynomials in ε and $\mathcal{B}_{\mathcal{N}}(\varepsilon)$ is a basis of $\langle P(\varepsilon) \rangle$ for *all values of ε*, including $\varepsilon = 0$. In analogy to the concept of "comprehensive Groebner bases" (cf. [2.13]), we call such a $\mathcal{B}_{\mathcal{N}}(\varepsilon)$ a *comprehensive border basis* of $P(\varepsilon)$.

Since $P(0) = P_0$ is positive-dimensional, $\mathcal{B}_{\mathcal{N}}(\varepsilon)$ must contain elements which vanish for $\varepsilon = 0$, i.e. which are divisible by a power of ε. This leads to the same two options which we have found for (9.39) in section 9.4.1:

Option 1 : We divide these elements by the respective powers of ε. Since the zeros $z_\mu(\varepsilon)$, $\varepsilon \neq 0$, are not affected by this normalization of $\mathcal{B}_{\mathcal{N}}(\varepsilon)$, $\mathcal{N}$ must have remained a feasible normal set, and $\mathcal{B}_{\mathcal{N}}(\varepsilon)$ is turned into a regular border basis $\overline{\mathcal{B}}_{\mathcal{N}}(\varepsilon)$, perhaps after further reduction. $\overline{\mathcal{B}}_{\mathcal{N}}(\varepsilon)$ permits the transition to $\overline{\mathcal{B}}_{\mathcal{N}}(0)$ without a structural change; the zeros $z_\mu(\varepsilon)$ go to the zeros $z_\mu(0)$ of $\overline{\mathcal{B}}_{\mathcal{N}}(0)$.

Option 2 : We set $\varepsilon = 0$ in the comprehensive basis $\mathcal{B}_{\mathcal{N}}(\varepsilon)$; this *deletes* the elements which have been normalized in Option 1. $\mathcal{B}_0 := \mathcal{B}_{\mathcal{N}}(0)$ is now a basis of $\langle P_0 \rangle$, but it is no longer a border basis because some monomials of $B[\mathcal{N}]$ no longer appear as $\mathcal{N}$-leading monomials of elements of $\mathcal{B}_0$.

Further isolated zeros of $P(\varepsilon)$ which remain away from M_0 are unaffected; their limits for $\varepsilon \to 0$ appear in $\langle \overline{\mathcal{B}}_{\mathcal{N}}(0) \rangle$ as well as in $\langle \mathcal{B}_0 \rangle$.

Example 9.16, continued: For $\mathcal{N} = \{1, x_1, x_3\}$, the comprehensive border basis $\mathcal{B}_{\mathcal{N}}(\varepsilon)$ of $P(\varepsilon)$ is

$$
\begin{aligned}
bb_1(x; \varepsilon) &= & & \varepsilon \cdot (x_1^2 - 3x_1 + 2), \\
bb_2(x; \varepsilon) &= & x_1 x_3 - x_3 \ + & \ \varepsilon \cdot (x_1 - 1), \\
bb_3(x; \varepsilon) &= & x_3^2 \ + & \ \varepsilon \cdot (2x_3 - x_1 + 2) + \varepsilon^2, \\
bb_4(x; \varepsilon) &= & x_2 + x_3 \ + & \ \varepsilon \cdot 1, \\
bb_5(x; \varepsilon) &= & x_2 x_1 + x_3 \ + & \ \varepsilon \cdot 1, \\
bb_6(x; \varepsilon) &= & x_2 x_3 \ + & \ \varepsilon \cdot (-x_3 + x_1 - 2) - \varepsilon^2.
\end{aligned}
$$

For $\varepsilon \neq 0$, $\mathcal{B}_{\mathcal{N}}(\varepsilon)$, or its complete intersection subset $\{bb_2(x; \varepsilon), bb_3(x; \varepsilon), bb_4(x; \varepsilon)\}$, define the 3 zeros $z_\mu(\varepsilon)$ of $P(\varepsilon)$. The border basis element bb_1 which has a divisor ε indicates the near-singularity of $P(\varepsilon)$.

When we choose Option 1, we replace bb_1 by $\overline{bb}_1 := \frac{1}{\varepsilon} bb_1$ and obtain $\mathcal{B}_{\mathcal{N}}(\varepsilon)$. Its limit $\overline{\mathcal{B}}_{\mathcal{N}}(0)$ has the 3 zeros $z_1(0) = (2, 0, 0)$, $z_2(0) = z_3(0) = (1, 0, 0)$.

With Option 2, we set $\varepsilon = 0$ in $\mathcal{B}_{\mathcal{N}}(\varepsilon)$ and obtain $\mathcal{B}_0 = \{bb_2(x; 0), \ldots, bb_6(x; 0)\}$. As a basis of the 1-dimensional ideal $\langle P_0 \rangle$ with the zero manifold M_0, it has the infinite normal set $\mathcal{N}_0 = \{1, x_3, x_1, x_1^2, \ldots\}$. The meaning of the seemingly superfluous normal set element x_3 will become clear in section 9.4.4. $\square$

In section 9.4.1, we had explicitly derived how the pseudozero sets of an empirical near-singular system of linear polynomials grow in the directions parallel to the anticipated zero manifold M_0 while they remain well-behaved in the directions orthogonal to M_0. For genuine polynomial systems, such a detailed analysis is not possible in general terms. We will give an intuitive explanation for the analogous phenomena which occur with near-singular and singular empirical systems of polynomials and which visualize the transition from isolated zeros to a zero manifold as a continuous process.

As in section 9.4.1, we consider empirical systems $(P(\varepsilon), E)$, i.e. sets of systems

$$
\hat{P}(\varepsilon) := P_0(x) + \varepsilon P_1(x) + \Delta P(x), \tag{9.55}
$$

where we assume that, for a normal set $\mathcal{N}$ which is feasible for all $\varepsilon \neq 0$, the support of ΔP is in $\mathcal{N}$; for $\hat{P}(\varepsilon) \in N_\delta(P(\varepsilon), E)$, the moduli of the coefficients of ΔP are bounded by $E\,\delta$, with $\|E\| = \hat{\varepsilon}$ in a suitable norm for E.

To obtain a qualitative view of the behavior of the pseudozero sets, we analyze the asymptotic differential sensitivity of the zeros of $P(\varepsilon)$ to changes in its coefficients; cf. similar approaches in section 9.3.4. We substitute $z_\mu(\varepsilon)$ into $P(\varepsilon) + \Delta a_{\nu j}\, x^j$, differentiate with respect to the modification $\Delta a_{\nu j}$ of the coefficient of x^j in $p_{0\nu}$, and then set $\Delta a_{\nu j} = 0$. With (9.50), we obtain:

$$(P_0'(z_\mu(\varepsilon)) + \varepsilon\, P_1'(z_\mu(\varepsilon)))\,(U_1\, U_1^T\, \tfrac{\partial}{\partial \Delta a_{\nu j}} z_\mu(\varepsilon) + U_2\, U_2^T\, \tfrac{\partial}{\partial \Delta a_{\nu j}} z_\mu(\varepsilon)) + z_\mu^{(\nu)}(\varepsilon)^j = 0,$$

where $z_\mu^{(\nu)}(\varepsilon)^j$ is an s-vector with the monomial x^j evaluated at $z_\mu(\varepsilon)$ as ν-th component and zeros otherwise. If P_0' behaves smoothly when its argument $z_\mu(\varepsilon)$ moves away from the manifold in an orthogonal direction, we may assume that, for sufficiently small $|\varepsilon|$, the deomposition of $(P_0'(z_\mu(\varepsilon)) + \varepsilon\, P_1'(z_\mu(\varepsilon)))$ with respect to the coordinate systems of (9.50) is like

$$(P_0'(z_\mu(\varepsilon)) + \varepsilon\, P_1'(z_\mu(\varepsilon)))\,(U_1\, U_2) = (V_1\, V_2) \begin{pmatrix} \Sigma_0 + O(\varepsilon) & O(\varepsilon) \\ O(\varepsilon) & O(\varepsilon) \end{pmatrix},$$

with a *regular* right-hand matrix $\Sigma(\varepsilon)$. From $\Sigma(\varepsilon)^{-1} = \begin{pmatrix} \Sigma_0^{-1} + O(\varepsilon) & O(1) \\ O(1) & O(\frac{1}{\varepsilon}) \end{pmatrix}$ and

$$\begin{pmatrix} U_1^T \\ U_2^T \end{pmatrix} \tfrac{\partial}{\partial \Delta a_{\nu j}} z_\mu(\varepsilon) = -\Sigma(\varepsilon)^{-1} \begin{pmatrix} V_1^T \\ V_2^T \end{pmatrix} z_\mu^{(\nu)}(\varepsilon)^j, \tag{9.56}$$

we see that the variations of the U_2-components of $z_\mu(\varepsilon)$ caused by a transition from $P(\varepsilon)$ to a $\tilde{P} \in N_\delta(P(\varepsilon), E)$, with $\delta = O(1)$, are generally $O(\hat{\varepsilon}/\varepsilon)$ while those of the U_1-components are $O(\hat{\varepsilon})$. Since span U_1 and span U_2 are orthogonal and parallel, resp., to the manifold M_0 at $z_\mu(0)$, this agrees with our analysis for linear equations.

Example 9.16, continued: For our $P(\varepsilon)$ of Example 9.16, we consider the sensitivity of $z_1(\varepsilon) = (2, 0, -\varepsilon)$ in the empirical system $(P(\varepsilon), E)$. Since M_0 is the x_1-axis, it is not necessary to resort to an s.v.d. but we may immediately form

$$[P'(\varepsilon)(2, 0, -\varepsilon)]^{-1} = \begin{pmatrix} 0 & 2 & 1 \\ -\varepsilon & 0 & 1 \\ 0 & 1 & 1 \end{pmatrix}^{-1} = \begin{pmatrix} -1/\varepsilon & -1/\varepsilon & 2/\varepsilon \\ 1 & 0 & -1 \\ -1 & 0 & 2 \end{pmatrix}$$

which establishes the $O(\hat{\varepsilon}/\varepsilon)$ sensitivity of the x_1-component of $z_1(\varepsilon)$ to many kinds of perturbations, and the $O(\hat{\varepsilon})$ sensitivity of the other two components.

For $\varepsilon = .01$,

$$[P_0'(2, 0, -.01) + .01\, P_1'(2, 0, -.01)]^{-1} = \begin{pmatrix} -100 & -100 & 200 \\ 1 & 0 & -1 \\ -1 & 0 & 2 \end{pmatrix}.$$

Therefore, when we add $\hat{\varepsilon} \cdot x^j$ to $p_{0\nu}$ and compute the modified $\tilde{z}_1(.01)$, the weighted difference $(\tilde{z}_1(.01) - z_1(.01))/(-\hat{\varepsilon} \cdot z_1(.01)^j)$ should approximately reproduce the ν-th column of the above matrix, at least in the sign and order of magnitude of the components.

With $\hat{\varepsilon} = 0.005$, we have tested this for various choices of x^j. The following "matrices" have been obtained in this manner for the indicated monomials x^j (rounded):

$$
j = 0,0,0: \quad
\begin{array}{rrr}
-100 & -100 & 131 \\
.67 & 0 & -2.88 \\
-.67 & 0 & 3.88
\end{array}
\qquad\qquad
j = 1,0,0: \quad
\begin{array}{rrr}
-201 & -200 & 74 \\
.67 & 0 & -2.41 \\
-.67 & 0 & 3.04
\end{array}
$$

$$
j = 0,0,1: \quad
\begin{array}{rrr}
-100 & -100 & 199 \\
1.01 & 0 & -.98 \\
-1.01 & 0 & 1.97
\end{array}
\qquad\qquad
j = 1,0,0: \quad
\begin{array}{rrr}
-100 & -99.5 & 200 \\
1.02 & 0 & -.97 \\
-1.02 & 0 & 1.96
\end{array}
$$

Also, since the x_2-component of $z_1(\varepsilon)$ is zero, (9.56) predicts that a small perturbation $\hat{\varepsilon}\, x_2$ in any of the three polynomials should not have a (1st order) effect on $z_1(\varepsilon)$. And indeed, $z_1(.01)$ remains completely unaltered for such perturbations. $\square$

In a real-life situation, there will generally be no explicit parameter ε but an empirical polynomial system $(\bar{P}, E)$ with $\bar{P}$ near-singular. Moreover, the near-singularity of $\bar{P}$ will often not be known a priori. It may have become evident by the excessive sensitivity of some well-isolated zeros of $\bar{P}$ which would otherwise only be expected for clustered zeros, or by the near-singularity of the Jacobian $\bar{P}'$ at these zeros.

Let $\bar{z}$ be such a zero of $\bar{P}$ and assume that the evaluation of $\bar{P}'(\bar{z})$ reveals a near-singular matrix as discussed in the linear case. To establish the existence of a singular system in $N_\delta(\bar{P}, E)$, we may use the following approach: We try to modify the empirical coefficients $(\bar{\alpha}_j, \varepsilon_j)$ in $(\bar{P}, E)$ and the zero $\bar{z}$ such that

$$
\bar{P}(\bar{a} + \Delta a)(\bar{z} + \Delta z) = 0 \quad \text{and} \quad \det \bar{P}'(\bar{a} + \Delta a)(\bar{z} + \Delta z) = 0, \tag{9.57}
$$

with a minimal $\|\Delta a\|_E$. Naturally, we linearize (9.57) to obtain a linear equation or minimization problem.

If our assumptions have been correct, viz. $\bar{P}(\bar{a})(\bar{z}) = 0$ and $\det \bar{P}'(\bar{a})(\bar{z})$ tiny, the linearization of (9.57) should yield modifications Δa, Δz such that (9.57) holds with high accuracy. This implies that the passing of a zero manifold through $\bar{z} + \Delta z$ has been found for the coefficients $\bar{a} + \Delta a$. This result may be further confirmed by applying the same procedure to a different $\bar{z}$ with tiny $\det \bar{P}'(\bar{z})$. An example will be given in the following section.

9.4.3 A Nontrivial Example

The following example was proposed by B. Mourrain (cf. [3.13]); systems of this sort appear in molecular chemistry. For $a \in \mathbb{R}$, consider the system

$$
P(x; a) = \begin{cases}
p_1(x_1, x_2, x_3; a) &= (x_2^2 + 4x_2x_3 + x_3^2)/2 + a\,(x_2^2 x_3^2 - 1), \\
p_2(x_1, x_2, x_3; a) &= (x_3^2 + 4x_3x_1 + x_1^2)/2 + a\,(x_3^2 x_1^2 - 1), \\
p_3(x_1, x_2, x_3; a) &= (x_1^2 + 4x_1x_2 + x_2^2)/2 + a\,(x_1^2 x_2^2 - 1).
\end{cases} \tag{9.58}
$$

For $a \neq 0$, $\langle P(x; a) \rangle$ has 16 isolated zeros. Due to the high symmetry in P, the zeros can be explicitly expressed in terms of iterated squareroots: Let $w_1(a) > 0$, $w_2(a) = -1/w_1(a) < 0$ be the two zeros of

$$
q(w) := w^2 + \frac{3}{a}\,w - 1, \qquad\qquad \text{and}
$$

$$v_1(a) := \sqrt{w_1}, \quad v_2(a) := \sqrt{-w_2}, \quad v_3(a) := \frac{w_1 - 2a}{2a\,w_1 + 1}/v_1, \quad v_4(a) := -\frac{w_2 - 2a}{2a\,w_2 + 1}/v_2.$$

Then $P(x; a)$ has the zeros

$$\pm(v_1, v_1, v_1), \quad \pm(v_1, v_1, v_3), \quad \pm(v_1, v_3, v_1), \quad \pm(v_3, v_1, v_1),$$
$$\pm\mathrm{i}(v_2, v_2, v_2), \quad \pm\mathrm{i}(v_2, v_2, v_4), \quad \pm\mathrm{i}(v_2, v_4, v_2), \quad \pm\mathrm{i}(v_4, v_2, v_2).$$

All these zeros are well defined and distinct for $a \neq 0$. Correspondingly, a Groebner basis of $P(x; a)$ yields a 16 element normal set and none of its leading coefficients vanish for a real value of $a \neq 0$. Yet it turns out that the system is singular, with a one-dimensional zero manifold M_0, for $a = \pm a_0$, with $a_0 := \frac{1}{2}\sqrt{3} \approx .8660$. $P(x; a)$ also becomes singular for $a = \pm\frac{3}{2}\mathrm{i}$; but we do not consider this case.

With a standard GB-code, say `gbasis` of Maple, this cannot be discovered, since it removes common factors $16\,a^4 + 24\,a^2 - 27 = (4\,a^2 - 3)(4\,a^2 + 9)$ from some basis elements so that the exceptional values of a cannot be seen. Such a computation implicitly corresponds to Option 1 explained in the previous two sections. Only when we compute a basis of $\langle P(x; a_0)\rangle$ directly, the 1-dimensionality of the ideal is displayed. However, since $P(x; a_0)$ also has 4 isolated zeros not on M_0, its basis representation is awkward, with 4 elements x^j in the normal set which do *not* satisfy $\exists x_\sigma : x_\sigma^k x^j \in \mathcal{N}\ \forall k$; cf. section 11.1. Therefore, we eliminate these zeros by appending to $P(x; a_0)$ the polynomial $\det P'(x; a_0)$ which vanishes on M_0 but not at these isolated zeros. From the Groebner basis `gbasis({p1, p2, p3, det P'}, tdeg(x1, x2, x3))`

$$\{x_1 x_2 + x_3 x_1 + x_2 x_3 + \sqrt{3},\ 3\,x_1^2\,(x_2 + x_3) + \sqrt{3}\,(4\,x_1 + x_2 + x_3),\ p_3(x_1, x_2, x_3; a_0)\},$$

we may derive a 2-branch parameter representation of M_0 in terms of x_1 since p_3 does not contain x_3 and the second basis element is linear in x_3 while the first one is redundant for a representation of M_0. This parameter representation $(x_1, x_2(x_1), x_3(x_1))$ has been used in Figure 9.4, which shows the projection of M_0 onto the $x_1 x_2$-plane and a space curve representation.

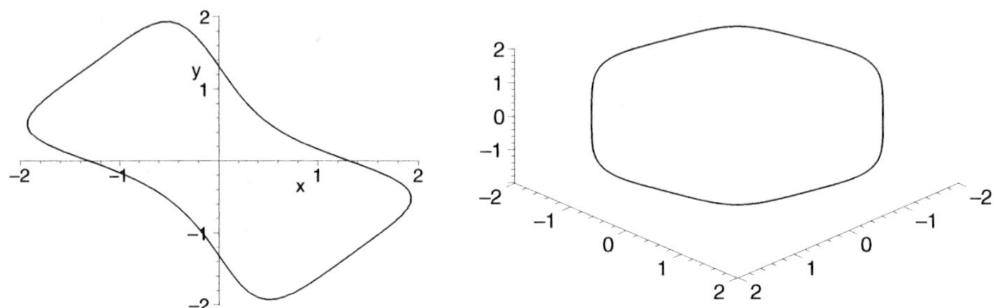

Figure 9.4.

For $a \approx a_0$, we may write $P(x; a)$ as a near-singular system (cf. (9.46))

$$P(\varepsilon) := P(x; a_0 + \varepsilon) = P(x; a_0) + \varepsilon \begin{cases} (x_2^2 x_3^2 - 1) \\ (x_3^2 x_1^2 - 1) \\ (x_1^2 x_2^2 - 1) \end{cases}.$$

For $\varepsilon \to 0$, this system has 12 simple d-points; these are the 12 zeros with only 2 equal components above; the 4 zeros with 3 equal components do not lie on the singular manifold M_0. Figure 9.4 shows the *real* parts of M_0 with the 6 real d-points and the 2 remaining real isolated zeros for $a = a_0$.

At the d-points $z_{0\mu}$, $P'(z_{0\mu}; a_0)$ has the form $\begin{pmatrix} 0 & 0 & -\beta \\ 0 & 0 & -\beta \\ \beta & \beta & 0 \end{pmatrix}$ or some permutation of it, with all β elements nonzero and equal. Thus, the tangential directions at the $z_{0\mu}$ are given by vectors like $(1, -1, 0)$ and the orthogonal directions by vectors like $(\gamma_1, \gamma_1, \gamma_2)$, with arbitrary γ_1, γ_2 (not both 0).

Let us now consider the near-singular system $\bar{P} = P(x; \bar{a})$ with $\bar{a} = .86605$, which corresponds to a value $\approx .0000246$ of ε. The values of the v_i in our explicit representation of the zeros are

$$v_1 \approx .517644 , \quad v_3 \approx -1.931852 , \quad v_2 \approx 1.931828 , \quad v_4 \approx -.517638 .$$

We consider the real zero $\bar{z} = (v_1, v_1, v_3)$ of $\bar{P}$. At $\bar{z}$,

$$\bar{P}'(\bar{z}) \approx \begin{pmatrix} 0 & .000143 & -1.79319 \\ .000143 & 0 & -1.79319 \\ 1.79319 & 1.79319 & 0 \end{pmatrix} , \quad \det \bar{P}'(\bar{z}) \approx -.000917 ;$$

thus we expect a manifold close by. Therefore we form the linearization of the system (9.57) which becomes, with $P_1 := (x_2^2 x_3^2 - 1, x_3^2 x_1^2 - 1, x_1^2 x_2^2 - 1)$,

$$P'(\bar{z}; \bar{a}) \cdot \Delta z + \Delta a \, P_1(\bar{z}) = 0 , \quad \det P'(\bar{z}; \bar{a}) + (\det P')'(\bar{z}; \bar{a}) \cdot \Delta z + \Delta a \, \frac{\partial}{\partial a} \det P'(\bar{z}; \bar{a}) = 0 ,$$

where the prime denotes differentiation with respect to x. Numerically, this yields

$$\begin{pmatrix} 0 & .000143 & -1.79319 & .0000246 \\ .000143 & 0 & -1.79319 & .0000246 \\ 1.79319 & 1.79319 & 0 & -.928200 \\ 24.003018 & 24.003018 & 9.147337 & 24.848569 \end{pmatrix} \begin{pmatrix} \Delta z_1 \\ \Delta z_2 \\ \Delta z_3 \\ \Delta a \end{pmatrix} = \begin{pmatrix} 0 \\ 0 \\ 0 \\ -.0009167 \end{pmatrix} .$$

Since we have only one parameter, no minimization is needed and we obtain directly (rounded)

$$\Delta z_1 = -.0000063655, \quad \Delta z_2 = -.0000063655, \quad \Delta z_3 = .000000008, \quad \Delta a = -.0000245950.$$

The value of $\bar{a} + \Delta a$ differs from $a_0 = \sqrt{3}/2$ only by 10^{-9}.

For a value of $a \approx a_0$ like $\bar{a}$, the sensitivity of the zeros near M_0 to small perturbations of $P(x; a)$ should be very high; more precisely, their distance from M_0 should remain well-conditioned while their position parallel to the local tangent of M_0 is ill conditioned. For a numerical test, we perturb $P(x; \bar{a})$ into $\tilde{P}(x; \bar{a})$ by adding $.00001\,(x_2, x_3, x_1)^T$ and compute the zero $\tilde{z}$ which corresponds to $\bar{z} \approx (.51764, .51764, -1.93185)$. We obtain $\tilde{z} \approx (.60848, .43306, -1.91599)$ which displays the strong sensitivity in the tangential direction $(1, -1, 0)$ of M_0; the appreciable change in the third component appears unexpected at first. However, near $\tilde{z}$, the tangential direction of M_0 has changed and contains a substantial component in the x_3-direction; in fact, the *distances* from M_0 of $\tilde{z}$ and $\bar{z}$ differ only by $4 \cdot 10^{-6}$! For comparison, we also consider the behavior of the zero $(.51764, .51764, .51764)$ of $\bar{P}$ which is away from M_0: After the perturbation, it has only moved by $1.5 \cdot 10^{-6}$.

9.4.4 Multiple d-Points

In section 9.4.2, we have excluded the case of a $z \in M_0$ with rk $P_0'(z) < s - d$ (cf. Definition 9.13); such points may or may not exist for a given singular system P_0.

Definition 9.14. For a specified singular system $P_0 \in (\mathcal{P}^s)^s$ with a d-dimensional zero manifold M_0, points $z^\dagger \in M_0$ with

$$r := \text{rk} \left(P_0'(z^\dagger) \right) < s - d \tag{9.59}$$

are *multiple d-points* of P_0. □

As the terminology suggests, multiple d-points are multiple zeros of P_0 which happen to lie on the singular manifold M_0 of P_0. Associated with such points are a number of interesting phenomena whose formal analysis requires the introduction of tools beyond the scope of this book. Therefore, we treat multiple zeros only in an intuitive way; for technical details, we refer to [9.1].

First of all, we realize that—contrary to simple d-points—the existence and the location of multiple d-points on M_0 are independent of the consideration of a perturbation of P_0. Also, it depends not directly on the geometric structure of the singular manifold M_0; in spite of the formal similarity between Definitions 7.3 and 9.14, multiple d-points need not be singular points of the manifold M_0: In section 7.3, a manifold of dimension $s - 1$ has been defined by one polynomial, and analogously a manifold of dimension d would have been defined by a (regular sequence of) $s - d$ polynomials whereas here M_0 is defined by a system of s polynomials. Thus, multiple d-points are *only* introduced through a particular *overdetermined* description of M_0 by a system P_0 of s polynomials.

While simple d-points cannot figure in the zero count of the singular system P_0, multiple d-points contribute to the dimensions of $\mathcal{R}[\langle P_0 \rangle]$ and $\mathcal{D}[\langle P_0 \rangle]$ as displayed by a normal set of $\langle P_0 \rangle$; besides being elements of the singular manifold, they also constitute isolated zeros of $\langle P_0 \rangle$.

Definition 9.15. A multiple d-point $z^\dagger$ which contributes $m - 1$ elements to a normal set $\mathcal{N}_0$ of $\langle P_0 \rangle$ has *multiplicity m*. □

This definition is supported by the fact that an m-fold d-point splits into a *cluster of m isolated zeros* upon a generic perturbation of P_0. Obviously, the fact that $z^\dagger$ lies on the zero manifold M_0 contributes one unit to the multiplicity.

Example 9.17: Consider P_0 of (9.54) in Example 9.16. A quick analysis shows that P_0 can *only* vanish on the x_1-axis which is the singular manifold M_0. Ordinarily, this manifold would be represented in $\mathcal{P}^3$ by $\langle x_2, x_3 \rangle$, with normal set $\{1, x_1, x_1^2, x_1^3, \ldots\}$.

For $\langle P_0 \rangle$, however, we have found the normal set $\mathcal{N}_0 = \{1, x_3, x_1, x_1^2, \ldots\}$ and the multiple d-point $(1, 0, 0)$, with rk $P'(1, 0, 0) = 1$; cf. section 9.4.2. Thus, $z^\dagger = (1, 0, 0)$ is a 2-fold d-point; under the perturbation P_1 of (9.54), e.g., it splits into the 2-cluster $(1, \pm i\sqrt{\varepsilon}, \mp i\sqrt{\varepsilon} - \varepsilon)$. □

For a singular system $P_0 \in (\mathcal{P}^s)^s$ with an m-fold d-point $z^\dagger$, the behavior of the m zeros near $z^\dagger$ of a near-singular system $P(\varepsilon) = P_0 + \varepsilon P_1$ for $\varepsilon \to 0$ can only be understood through an analysis of the *dual space* $\mathcal{D}_0$ of the primary ideal $\mathcal{I}_0$ of $z^\dagger$.

Definition 9.16. Consider a manifold $M_0 \subset \mathbb{C}^s$. At some $z_0 \in M_0$, a differential functional

$\partial_j[z_0]$ is called *internal* if it vanishes for all polynomials whose zero set includes M_0. All other differential functionals $\partial_j[z_0]$ are called *external*. $\square$

At any point z_0 on M_0 of dimension d, there exist $s - d$ linearly independent vectors $t_\delta \in \mathbb{C}^s$, $\delta = 1(1)d$, such that $x = z_0 + (t_1, ..t_d)\, w$, $w \in \mathbb{C}^d$, is the tangential manifold of M_0 at z_0; the vectors t_δ are the eigenvectors of $P'(z_0)$ for the eigenvalue 0. The associated first order differential functionals are internal and so are the higher order differentials arising by further differentiation along M_0. The dual space spanned by these functionals has infinite dimension and contains derivatives of arbitrary order.

If $r := \operatorname{rk} P_0'(z_0) < s - d$, there must exist $s - d - r$ further vectors $t_\delta^\dagger$ with $P'(z_0)\, t_\delta^\dagger = 0$, which are *not* parallel to the tangential manifold of M_0 at z_0. The associated external first order differentials vanish on $\langle P_0 \rangle$ but not on the primary ideal of M_0. Possibly, there are also some higher order external differentials at such a $z^\dagger$ which vanish on $\langle P_0 \rangle$.

The dual space $\mathcal{D}_0[z^\dagger]$ is spanned by these external differentials and all internal differentials. In [9.1], it has been shown that, at an m-fold d-point $z^\dagger$, the *smallest closed subspace* $\mathcal{D}_{ext}[z^\dagger]$ of $\mathcal{D}_0[z^\dagger]$ which contains no internal differentials except ∂_0 has dimension m. The ideal $\mathcal{I}_{ext}[z^\dagger]$ defined by this dual space characterizes the m-fold d-point $z^\dagger$ as an m-fold zero. For perturbations of P_0 from span $\mathcal{N}_{ext} =: \mathcal{R}_{ext}[z^\dagger]$, $\mathcal{I}_{ext}[z^\dagger]$ can be extended to an approximate (asymptotically correct) basis for the ideal of the m-cluster which issues from $z^\dagger$.

Example 9.17, continued: At $z^\dagger = (1, 0, 0)$, $P_0' = \begin{pmatrix} 0 & 1 & 1 \\ 0 & 0 & 0 \\ 0 & 1 & 1 \end{pmatrix}$ so that there is the eigenvector $(0, 1, -1)^T$ in addition to $(1, 0, 0)^T$. The associated external differential $\partial_{010} - \partial_{001}$ does not vanish on $\langle x_2, x_3 \rangle$ but for all $p \in \langle P_0 \rangle$. Thus, $\mathcal{D}_0[(1, 0, 0)] = \operatorname{span}\{\partial_0, \partial_{010} - \partial_{001}, \partial_{100}, \partial_{200}, \ldots\}$ and $\mathcal{D}_{ext}[(1, 0, 0)] = \operatorname{span}\{\partial_0, \partial_{010} - \partial_{001}\}$; the presence of ∂_0 (necessary for closedness) yields the dimension 2. With the subset $\mathcal{N}^\dagger = \{1, x_3\}$ of the normal set $\mathcal{N}_0$, we obtain the border basis $\mathcal{B}_{\mathcal{N}^\dagger}$ for the ideal $\mathcal{I}_{ext} := \mathcal{I}[\mathcal{D}_{ext}]$ as

$$x_1 - 1, \ x_2 + x_3, \ x_1 x_3 - x_3, \ x_2 x_3, \ x_3^2\, ;$$

the first, second, and last element form a complete intersection and the Groebner basis of $\mathcal{I}_{ext}[(1, 0, 0)]$. We may use it to find the dynamics of the 2-cluster which issues from $(1,0,0)$ upon a perturbation of P_0.

For the perturbation in (9.54), we have

$$\begin{aligned} p_1(x; \varepsilon) &= \varepsilon - x_3\,(x_1 - 1) + (x_2 + x_3) + (x_1 - 1)(x_2 + x_3)\,, \\ p_2(x; \varepsilon) &= \varepsilon + x_3\,(x_1 - 1) - 2\,x_3\,(x_2 + x_3) + x_3^2 + (x_2 + x_3)^2\,, \\ p_3(x; \varepsilon) &= \varepsilon + (x_2 + x_3)\,. \end{aligned}$$

As in section 9.3.2, we modify the basis elements g_κ by $\varepsilon\,(\gamma_{\kappa 1} + \gamma_{\kappa 2} x_3)$ from $\mathcal{R}_{ext}$ and drop all elements in $\mathcal{I}_{ext}$ and of $O(\varepsilon^2)$. This yields

$$\begin{aligned} 1 &= -x_3\,(\gamma_{11} + \gamma_{12} x_3) + 1 \cdot (\gamma_{21} + \gamma_{22} x_3)\,, \\ 1 &= x_3\,(\gamma_{11} + \gamma_{12} x_3) - 2\,x_3\,(\gamma_{21} + \gamma_{22} x_3) + 1 \cdot (\gamma_{31} + \gamma_{32} x_3)\,, \\ 1 &= 1 \cdot (\gamma_{21} + \gamma_{22} x_3)\,, \end{aligned}$$

and $\gamma_{11} = \gamma_{22} = 0$, $\gamma_{21} = \gamma_{31} = 1$, $\gamma_{32} = 2$, while γ_{12} may be dropped above because it multiplies $x_3^2 \in \mathcal{I}_{ext}$. Thus we have the following Groebner basis for the 2-cluster of $P(x; \varepsilon)$:

$$x_1 - 1, \quad x_2 + x_3 + \varepsilon, \quad x_3^2 + 2\,\varepsilon\,x_3 + \varepsilon,$$

and the asymptotic behavior

$$z_1(\varepsilon) = 1, \quad z_2(\varepsilon) = \pm\sqrt{-\varepsilon + \varepsilon^2}, \quad z_3(\varepsilon) = -\varepsilon \mp \sqrt{-\varepsilon + \varepsilon^2},$$

which is correct in the $O(\sqrt{\varepsilon})$ and $O(\varepsilon)$ terms. $\quad\square$

Unfortunately, the above approach works only in special situations and for special perturbations. In [9.1], it has been shown that—for a more generally applicable approach—one has to replace the dual space $\mathcal{D}_{ext}$ of above by a parametrized dual space $\overline{\mathcal{D}}_{ext}$ (the "general closed hull") in which indeterminate multiples of certain internal differentials are added to functionals in $\mathcal{D}_{ext}$. The associated ideal $\overline{\mathcal{I}}_{ext}$ (with these parameters) has to satisfy $\overline{\mathcal{I}}_{ext} \cap \mathcal{I}_{M_0} = \mathcal{I}_0$, where $\mathcal{I}_{M_0}$ is the ideal with zero set M_0. The parameters are fixed in the process of adapting the ideal to the specified perturbation. We give only an example:

Example 9.17, continued: For P_0 from Example 9.16, with M_0 the x_1-axis, we take the perturbation $P_1(x) = (x_1, x_1, 0)^T$. In replacing P_1 by $NF_{\mathcal{I}^\dagger}[P_1]$, we would replace x_1 by 1, but the effects of these two perturbations are grossly different.

For simplicity, we realize that the unperturbed p_{03} implies $x_3 = -x_2$ and reduce the problem to a two-dimensional situation with

$$p_{01} = x_1 x_2 - x_2, \quad p_{02} = -x_1 x_2 + x_2^2 + x_2,$$

with $\mathcal{I}_0 = \langle x_1 x_2 - x_2, x_2^2 \rangle$, $\mathcal{I}_{ext} = \langle x_1 - 1, x_2^2 \rangle$. In this case, the general closed hull of $\mathcal{D}_{ext} = \mathrm{span}\,\{\partial_0, \partial_{01}\}$ becomes $\overline{\mathcal{D}}_{ext} = \mathrm{span}\,\{\partial_0, \partial_{01} + \alpha\,\partial_{10}\}$ so that $\overline{\mathcal{I}}_{ext} = \langle x_1 - 1 - \alpha\,x_2, x_2^2 \rangle$. When we use this ideal, with the normal set $\mathcal{N}^\dagger = \{1, x_2\}$, for the determination of the cluster dynamics with the perturbation $P_1 = (x_1, x_1)^T$, we obtain

$$p_1(x; \varepsilon) = x_1 x_2 - x_2 + \varepsilon\,x_1 = x_2\,(x_1 - 1 - \alpha\,x_2) + \alpha\,x_2^2 + \varepsilon\,(1 + \alpha\,x_2),$$
$$p_2(x; \varepsilon) = -x_1 x_2 + x_2^2 + x_2 + \varepsilon\,x_1 = -x_2\,(x_1 - 1 - \alpha\,x_2) + (1 - \alpha)\,x_2^2 + \varepsilon\,(1 + \alpha\,x_2).$$

With the modifications $\varepsilon\,(\gamma_{i1} + \gamma_{i2}\,x_2)$, $i = 1, 2$, in the basis elements $x_1 - 1 - \alpha\,x_2$ and x_2^2, resp., to compensate the perturbation, we obtain the following conditions for the 5 parameters γ_{ij} and α:

$$x_2\,(\gamma_{11} + \gamma_{12}x_2) + \alpha\,(\gamma_{21} + \gamma_{22}x_2) = 1 + \alpha\,x_2,$$
$$-x_2\,(\gamma_{11} + \gamma_{12}x_2) + (1 - \alpha)\,(\gamma_{21} + \gamma_{22}x_2) = 1 + \alpha\,x_2,$$

which yields

$$\alpha = \frac{1}{2}, \quad \gamma_{21} = 2, \quad \gamma_{22} = 1, \quad \gamma_{11} = 0;$$

γ_{12} remains undetermined but, in the above equations, it multiplies $x_2^2 \in \overline{\mathcal{I}}_{ext}$ and drops out of the normal form.

Thus, the cluster ideal for the chosen perturbation is $\langle x_1 - 1 - \frac{1}{2}\,x_2, x_2^2 + \varepsilon\,x_2 + 2\,\varepsilon \rangle$ and the two cluster zeros are $(1 - (\varepsilon \pm \sqrt{\varepsilon^2 - 8\,\varepsilon})/4, -(\varepsilon \pm \sqrt{\varepsilon^2 - 8\,\varepsilon})/2)$. Ordinarily, this result would be asymptotically correct up to $O(\varepsilon)$, but in this simple case it is exact. $\quad\square$

Exercises

1. Consider the following polynomial system in $\mathcal{P}^3$, with a parameter c,

$$P(c) := \begin{cases} p_1(x_1, x_2, x_3; c) &=& x_1^2 + x_1 x_2 - x_1 x_3 - x_1 - x_2 + x_3\,, \\ p_2(x_1, x_2, x_3; c) &=& x_1 x_2 + c x_2^2 - x_2 x_3 - x_1 - c x_2 + x_3\,, \\ p_3(x_1, x_2, x_3; c) &=& x_1 x_3 + x_2 x_3 - x_3^2 - x_1 - x_2 + x_3\,. \end{cases} \qquad (9.60)$$

Convince yourself that the system is singular for any value $c \in \mathbb{C}$.

(a) For $c \neq 0, 1$, find the two 1-dimensional zero manifolds M_0 and M_1 of $P(c)$ (the subscript denotes the value of x_2 on the manifold) and the isolated zero z_1. From the normal set of a Groebner basis of P, we expect a 2-fold d-point in addition to the isolated zero. Find that d-point $z^\dagger$.

(b) Note that M_0 and M_1 do not depend on c. Yet for $c = 1$, the two 1-dimensional manifolds are swallowed by the 2-dimensional zero manifold M_{01} $x_3 = x_1 + x_2$ which occurs only for that value of c. Can you interpret the singular appearance of M_{10} as a continuous event for $c \to 1$?

(c) For $c = 0$, the two zero manifolds M_0, M_1 remain isolated, but there appears a third 1-dimensional zero manifold M_2 which contains both z_1 and $z^\dagger$. $z^\dagger$ is now a genuine 2-fold zero because it is the intersection of two zero components. Is it a multiple d-point for the zero manifold $M_0 \cup M_2$?

2. Now we take $c = \frac{1}{2}$ in (9.60) and analyze further.

(a) For the zero manifold M_0, find the internal dual space $\mathcal{D}_0$ and the ideal $\mathcal{I}_{M_0}$. For the multiple d-point $z^\dagger \in M_0$, determine $\mathcal{D}_{ext}$ and a (Groebner) basis of $\mathcal{I}_{ext}$, with normal set $\mathcal{N}^\dagger = \{1, x_2\}$. The general closed hull $\overline{\mathcal{D}}_{ext}$ is obtained by adding an indeterminate multiple of the first order differential in $\mathcal{D}_0$ to the external differential in $\mathcal{D}_{ext}$. Find a basis of the associated ideal $\overline{\mathcal{I}}_{ext}$. Show that $\overline{\mathcal{D}}_{ext} \cap \mathcal{I}_{M_0} = \mathcal{I}_0 = \langle P(\frac{1}{2}) \rangle$; cf. section 9.4.4.

(b) Consider the near singular system $\hat{P}(x; \varepsilon) = P(\frac{1}{2}) + \varepsilon\, (1, 1, -1)^T$. How many isolated zeros are there in $\hat{P}(\varepsilon)$ for $\varepsilon \neq 0$? The extra zero beyond the modified z_1 and the two cluster zeros issuing from $z^\dagger$ must have departed from a simple d-point on one of the singular manifolds. Determine that d-point z_0 and the $O(\varepsilon)$ terms in its asymptotic expansion.

(c) Consider the 2-cluster of $\hat{P}(\varepsilon)$: At first, expand $\hat{P}(\varepsilon)$ in terms of the basis of $\mathcal{I}_{ext}$ and try to modify the basis—as in section 9.4.4—such that $\hat{P}(\varepsilon)$ is in the modified ideal. Why does the approach fail? Now, repeat the same procedure with the basis of $\overline{\mathcal{I}}_{ext}$. For a suitably chosen factor α, we may now find a basis of the cluster ideal. For some small value of ε, find the zeros of the cluster ideal and compare their values with values obtained by some other means.

(d) Observe the sensitivity of the 4 zeros of $\hat{P}(10^{-3})$ to various small perturbations of the polynomials.

3. (a) Consider (9.60) for $c = 1$. For the 2-dimensional zero manifold M_{01}, find $\mathcal{D}_0$ and $\mathcal{I}_{M_{01}}$ as before. Is there a multiple d-point on M_{01}? Why is $z^\dagger$ from part 1) no longer a multiple d-point?

(b) Consider the singular system of 2(b) above with $c=1$. How many isolated zeros are there for $\varepsilon \neq 0$? What is the simple d-point to which the second isolated zero of $\hat{P}(\varepsilon)$ converges for $\varepsilon \to 0$? What has happened to the two zeros of $\hat{P}(\varepsilon)$ for $c \neq 1$ which formed a cluster converging to $z^\dagger$? Find out by solving $\hat{P}(\varepsilon)$ for a small value of ε and values of c which approach 1.

(c) Observe the sensitivity of the 2 zeros of $\hat{P}(10^{-3})$ to various small perturbations of the polynomials.

4. Finally consider (9.60) for $c = 0$ and proceed like in parts 2 and 3. Observe the similarities and differences. Which of the 4 zeros of $\hat{P}(\varepsilon)$ "expands" into the additional zero manifold M_2 as c tends to zero?

9.5 Singular Polynomial Systems with Diverging Zeros

In univariate polynomials of a given degree, zeros diverge to ∞ iff the coefficient(s) of the leading term(s) go to zero; cf. section 5.2.3. This phenomenon also appears in 0-dimensional polynomial systems when we replace "leading term(s)" by "term(s) which affect the BKK-number" (cf. section 8.3.4): If the coefficient of such a term goes to zero and if the BKK-count of the system decreases when that term is not present, the continuous dependence of the zeros on the coefficients requires that the appropriate number of zeros diverge to ∞ as the coefficient converges to zero. By (8.43)/(8.44), it is necessary but not sufficient that such terms span the support of the respective polynomial. As in the univariate case, this vanishing of one or more zeros to ∞ is not a genuine singularity: A generic perturbation of *low-order terms* does *not* recover the diverged zeros.

In polynomial systems, on the other hand, divergence of zeros can also happen as a truly singular behavior: When a 0-dimensional system $P_0 \in (\mathcal{P}^s)^s$ has *fewer* zeros than its BKK-number requires (cf. Definition 9.12), almost all neighboring systems P have the appropriate number of zeros, but one or several of these zeros have a very large modulus and their locations are very ill-conditioned. We will now consider these "BKK-deficient systems" which we had excluded in section 9.4. Again, we will look at linear systems first.

9.5.1 Inconsistent Linear Systems

Singular linear systems (9.36)–(9.37) with less than their one zero are *inconsistent*. This happens when $b_0 \notin \mathrm{span}\, V_1$ or $V_2^T b_0 \neq 0$; throughout this section, we assume that this is the case. We proceed immediately to neighboring systems (9.39) and denote the block matrices $V_i^T A_1 U_j$ in (9.40) by Σ_{ij}; furthermore, we set $b_0 + \varepsilon\, b_1 =: V_1(b_{01} + \varepsilon b_{11}) + V_2(b_{02} + \varepsilon b_{12})$.

Proposition 9.5. For small $|\varepsilon| \neq 0$, the near-singular systems (9.39) with regular Σ_{22} and $b_{02} \neq 0$ have a unique zero $z(\varepsilon)$ which satisfies, for $\varepsilon \to 0$,

$$u_1(\varepsilon) := U_1^T z(\varepsilon) = -\Sigma_0^{-1}(b_{01} - \Sigma_{12}\Sigma_{22}^{-1} b_{02}) + O(\varepsilon),$$

$$u_2(\varepsilon) := U_2^T z(\varepsilon) = -\tfrac{1}{\varepsilon}\Sigma_{22}^{-1} b_{02} - \Sigma_{22}^{-1}(b_{12} - \Sigma_{21}\Sigma_0^{-1}(b_{01} - \Sigma_{12}\Sigma_{22}^{-1} b_{02})) + O(\varepsilon).$$

$$\tag{9.61}$$

Proof: With the above notation and with $\mathbf{x} =: U_1 u_1 + U_2 u_2$, we obtain from (9.36)–(9.37)

$$(\Sigma_0 + \varepsilon\,\Sigma_{11})\,u_1 + \varepsilon\,\Sigma_{12}\,u_2 \;=\; -(b_{01} + \varepsilon\,b_{11})\,,$$
$$\varepsilon\,\Sigma_{21}\,u_1 + \varepsilon\,\Sigma_{22}\,u_2 \;=\; -(b_{02} + \varepsilon\,b_{12})\,.$$

We set $\varepsilon\,u_2 =: \bar{u}_2$; with the assumptions on Σ_{22} and b_{02}, this yields $\bar{u}_2(\varepsilon) = -\Sigma_{22}^{-1}\,b_{02} + O(\varepsilon)$ and $u_1(\varepsilon)$ as in (9.61). When we now introduce the O(1) part of u_1 into the second equation above, we find the O(ε) part of $\bar{u}_2$, i.e. the O(1) part of u_2. $\quad\square$

From (9.61), we see that $z(\varepsilon) = U_1 u_1(\varepsilon) + U_2 u_2(\varepsilon)$ moves to infinity alongside the manifold $\overline{M}_0 := \{\,U_1^T \mathbf{x} = u_1(0)\,\}$, approaching it as $\varepsilon \to 0$, with the projection $U_2^T z(\varepsilon) = u_2(\varepsilon) = O(1/\varepsilon)$.

What happens with the *pseudozero sets* of an empirical linear system $(P(\varepsilon), E)$ when ε tends to 0; cf. (9.43) and Example 9.15 ? We assume $E = O(\hat{\varepsilon})$ and the other assumptions of section 9.4.1 w.r.t. (9.43). In an obvious manner, we denote the contributions of a perturbation in (9.61) by Δ ... As long as $\hat{\varepsilon}$ is sufficiently smaller than $|\varepsilon|$, $\Sigma_{22} + \frac{1}{\varepsilon}\,\Delta\Sigma_{22}$ remains regular and we have

$$(\Sigma_{22} + \tfrac{1}{\varepsilon}\,\Delta\Sigma_{22})^{-1} \;=\; \Sigma_{22}^{-1}\,(1 + O(\tfrac{\hat{\varepsilon}}{\varepsilon}))\,.$$

This adds $O(\frac{\hat{\varepsilon}}{\varepsilon})$ and $O(\hat{\varepsilon})$ terms to $U_1^T z(\varepsilon)$ but $O(\frac{\hat{\varepsilon}}{\varepsilon^2})$ and $O(\frac{\hat{\varepsilon}}{\varepsilon})$ terms to $U_2^T z(\varepsilon)$. Thus, with $\varepsilon \to 0$, the extension of the pseudozero set in the $\overline{U}_2$-directions grows at the same explosive rate as the $\overline{U}_2$-components of $z(\varepsilon)$. At the same time, the extension of the pseudozero set in the subspace $\overline{U}_1$ also grows, but its width remains $O(\varepsilon)$ relative to its length. In this sense, like in section 9.4.1, the pseudozero set remains "narrow" along the manifold $\overline{M}_0$.

For $\hat{\varepsilon}$ of the same order as ε, the matrix $\Sigma_{22} + \frac{1}{\varepsilon}\,\Delta\Sigma_{22}$ may become singular for particular perturbations ΔA of $O(\hat{\varepsilon})$, i.e. $N_\delta(P(\varepsilon), E)$, $\delta = O(1)$, may contain systems which are inconsistent. This means that the pseudozero set reaches ∞ in the $\overline{U}_2$-direction; one may show further that its distance from 0 remains $O(1/\hat{\varepsilon})$. For $\varepsilon = 0$, the pseudozero set of the empirical system $(P(0), E)$ with *inconsistent* $P(0)$ reaches symmetrically from ∞ towards zero in the subspace $\overline{U}_2$ but keeps an $O(\frac{1}{\hat{\varepsilon}})$ distance from 0.

Example 9.18: We take the near-singular system of Example 9.15 but change b_0 to $(-1, +1)^T$. We have the same s.v.d. as in Example 9.15; the new constant terms give $b_{01} = 0$, $b_{02} = -\sqrt{2}$, $b_{11} = b_{12} = -\frac{1}{\sqrt{2}}$. With $\Sigma_0 = 2$, $\Sigma_{12} = \Sigma_{21} = 0$, $\Sigma_{22} = 2$, substitution into (9.61) yields $u_1 = O(\varepsilon)$, $u_2 = \frac{1}{\varepsilon\sqrt{2}} + \frac{1}{2\sqrt{2}} + O(\varepsilon)$. This means that $z(\varepsilon) = U_1 u_1 + U_2 u_2$ asymptotically behaves like $z_1(\varepsilon) = z_2(\varepsilon) = \frac{1}{2\varepsilon} + \frac{1}{4} + O(\varepsilon)$.

When we admit indeterminations $\Delta\alpha_{ij}$, $\Delta\beta_i$ bounded by $\hat{\varepsilon}$ in P_0, the $\Delta\Sigma_{ij}$ may vary within $2\hat{\varepsilon}$, and the Δb_i within $\hat{\varepsilon}$. For $\hat{\varepsilon} = .005$, e.g., the pseudozero set $Z_1(P_0, E)$ extends, in $\mathbf{x}$ coordinates, from $\pm\infty$ to $\pm(99.5, 99.5)$ along the line $x_1 - x_2 = 0$ and widens like $[.99, 1.01]\,\|x\|$ with increasing $\|x\|$. This matches well with computational evidence. $\quad\square$

When we now consider the comprehensive Groebner basis (9.45) as in section 9.4.1, the inconsistent case differs from the consistent one in the missing ε factors with the β_μ, $\mu = r + 1(1)s$. As long as the matrix of the $\alpha_{\mu\nu}$ in the lower right corner is regular, we have the same two options as with (9.45):

We may divide the lower $s - r$ polynomials by ε, which works but generates $O(1/\varepsilon)$ inhomogeneities. Elimination in this modified system yields a border basis for $P(\varepsilon)$ and the $O(1/\varepsilon)$ zero $z(\varepsilon)$; a further transition to $\varepsilon = 0$ is naturally not possible now.

We may set $\varepsilon = 0$ in (9.45); this exposes the inconsistency of P_0 since the lower inhomogeneities do not vanish now.

Example 9.18, continued: For the near-inconsistent linear system $P(\varepsilon)$ of Example 9.18, elimination yields the comprehensive basis

$$(1 + \varepsilon)\,x_1 - (1 - \varepsilon)\,x_2 - (1 + \varepsilon)\,, \qquad 4\,\varepsilon\,x_2 - (2 + \varepsilon - \varepsilon^2)\,.$$

The first option changes the second polynomial into $4\,x_2 - \frac{2}{\varepsilon} - 1 + \varepsilon$; it leads to $z(\varepsilon) = (\frac{1}{2\varepsilon} + \frac{1}{4} + O(\varepsilon),\ \frac{1}{2\varepsilon} + \frac{1}{4} + O(\varepsilon))$ which we have found previously. The second option generates the basis $\{x_1 - x_2 - 1,\ 2\}$ for $\langle P_0 \rangle$ which displays the inconsistency. $\square$

The real-life situation, without explicit parameters, is analogous to that in section 9.4.1. By Definition 9.12, the empirical linear system $((\bar{A}, \bar{b}), (E, e))$ is singular if $\min \sigma_i(\bar{A}) \le O(1)\,\|E\|_2$, and we may consider the matrix $\tilde{A} = \tilde{V}_1 \tilde{\Sigma}_1 \tilde{U}_1^T$ of (9.44). Now we must assume that $\|\tilde{V}_2^T \bar{b}\|_2 > O(1)\,\|e\|_2$; thus there is no $\tilde{b}$ with $\tilde{V}_2^T \tilde{b} = 0$ in the tolerance neighborhood of $\bar{b}$ and the system $\tilde{A}\,x + b$ has no zero for a $b \in N_\delta(\bar{b}, e)$, $\delta = O(1)$.

It is true that there will be other matrices $\hat{A} \in N_\delta(\bar{A}, E)$ and vectors $\hat{b} \in N_\delta(\bar{b}, e)$, $\delta = O(1)$ for which $\hat{A}\,x + \hat{b}$ has a zero, with a very large modulus; this is in the nature of singular systems. But generally, the fact that ∞ is a valid zero of the empirical linear system should be a serious warning or, sometimes, an important result.

9.5.2   BKK-Deficient Polynomial Systems

A polynomial counterpart of inconsistent linear systems are systems with fewer zeros than their BKK-number indicates. We restrict our analysis of this situation to a simple intuitive 2-variable example where the variety of the one polynomial is a hyperbola while the variety of the other polynomial contains an asymptote of that hyperbola as a component:

Example 9.19: Consider $P_0 \in (\mathcal{P}^2)^2$, with

$$p_{01}(x_1, x_2) = x_1 x_2 - x_1 - 1\,, \qquad p_{02}(x_1, x_2) = (x_2 - 1)\,(4\,x_1^2 + 4\,x_2^2 - 25)\,;$$

cf. Figure 9.5. The hyperbola has 4 intersections with the circle of p_{02}; generically, it should also intersect with the straight line component of p_{02} for a 5th zero which is expected from $m = \mathrm{BKK}(P_0) = 5$. But this straight line happens to be an asymptote of p_{01} so that the 5th zero is missing. $\square$

Definition 9.17. A 0-dimensional system $P_0 \subset \mathcal{P}^s$ with m_0 zeros (counting multiplicities) is called *BKK-deficient* if $m_0 < m := \mathrm{BKK}(P_0)$; $m - m_0$ is its *BKK-deficiency*. $\square$

BKK-deficiency is a *singular* phenomenon: Since any generic complete intersection system P has $\mathrm{BKK}(P)$ zeros, a generic perturbation of P_0 which affects only terms within the closed hulls of the supports J_ν of the $p_\nu \in P$ will raise m_0 to m.

Example 9.19, continued: Consider $P = \{p_1 = p_{01} + \varepsilon\,x_1,\ p_2 = p_{02} + \varepsilon\,x_1\}$. Now, the asymptote $x_2 = 1 - \varepsilon$ of p_1 intersects the asymptotic branch $x_2 \to 1$ of p_2 at $z_5(\varepsilon) = (O(\frac{1}{\varepsilon}),\ 1 + O(\varepsilon))$; cf. Figure 9.6. As $\varepsilon \to 0$, $z_5(\varepsilon)$ disappears to ∞ along $x_2 = 1$. $\square$

Generally, a BKK-deficient system P_0 is not inconsistent as in the linear case because $m_0 > 0$ zeros remain. When such a system is assumed as intrinsic and its exact normal set

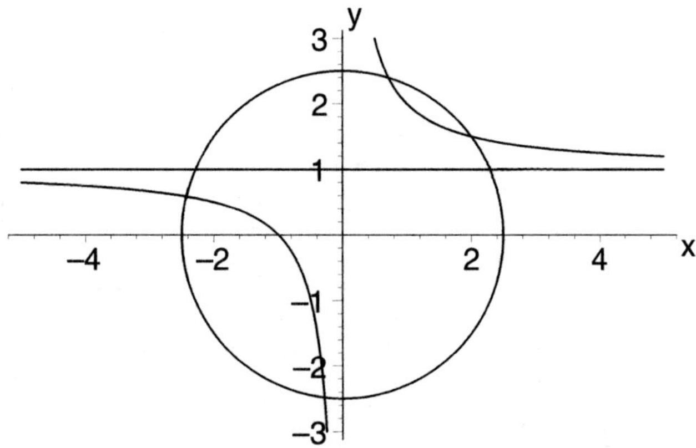

Figure 9.5.

representation is computed, nothing extraordinary appears and the normal set $\mathcal{N}_0$ has $m_0 < m$ elements. However, somewhere in the basis computation, a reduction occurs which is not possible for neighboring systems.

As previously in the context of singular systems, we consider a parametrized neighborhood of the BKK-deficient singular system P_0. Let

$$P(x;\varepsilon) \ := \ P_0(x) + \varepsilon\, P_1(x) \ \in \ (\mathcal{P}^s)^s ,$$

with a fixed perturbation P_1; we assume that the BKK-numbers of P_0 and $P(\varepsilon)$, $\varepsilon \neq 0$, agree and that $P(\varepsilon)$ has $m = \mathrm{BKK}(P_0)$ zeros for almost all feasible choices of P_1. For these neighboring systems $P(\varepsilon)$, the normal set attains its full m elements, but at least one of the border basis elements has a *tiny* $\mathcal{N}$-leading coefficient, or some huge coefficients if the leading coefficient is normalized. Here appears the difference from the situation in section 9.4.2, where *all* coefficients of such a basis polynomial were tiny: in the BKK-deficient case, the basis polynomial with a tiny leading term must also have some $O(1)$ coefficients, as it happens for inconsistent linear systems in section 9.5.1. If the tiny coefficients are annihilated, the normal set reduces to $\mathcal{N}_0$ and the border basis $\mathcal{B}_0$ of P_0 reappears.

Example 9.19, continued: In P_0, p_{01} does not admit a zero with 2nd component 1; hence p_{02} can only vanish when its second factor $4\,x_1^2 + 4\,x_2^2 - 25$ vanishes. The remaining basis elements for $\mathcal{N}_0 = \{1, x_2, x_1, x_2^2\}$ are found in a straightforward way:

$$
\begin{aligned}
bb_{01}(x_1, x_2) &= x_1 x_2 - x_1 - 1 = p_{01} , \\
bb_{02}(x_1, x_2) &= 4\,x_1^2 + 4\,x_2^2 - 25 = p_{02}/(x_2 - 1) , \ (!) \\
bb_{03}(x_1, x_2) &= x_1 x_2^2 - x_1 - x_2 - 1 = (x_2 + 1)\,p_{01} , \\
bb_{04}(x_1, x_2) &= 4\,x_2^3 - 4\,x_2^2 + 4\,x_1 - 25\,x_2 + 25 = p_{02} - 4\,x_1\,p_{01} .
\end{aligned}
$$

Now we consider the neighboring system $P(\varepsilon) := P_0 + \varepsilon\,(x_1, x_1)^T$. The replacement of the first factor in $p_2(\varepsilon) = (x_2 - 1)\,(4\,x_1^2 + 4\,x_2^2 - 25) + \varepsilon\,x_1$ by $(1 - \varepsilon x_1)/x_1$ (from $p_1(\varepsilon)$) and some reductions yield a basis polynomial

$$bb_2(x_1, x_2) \;=\; \varepsilon\,(-4\,x_1^3 + 4\,x_1^2 + 21\,x_1 - 4\,x_2 - 4 + \mathrm{O}(\varepsilon)) + (4\,x_1^2 + 4\,x_2^2 - 25)\,,$$

while the remaining $bb_{0\nu}$ have their leading terms unaltered. Thus, the normal set $\mathcal{N}$ for $P(\varepsilon)$ must pick up the additional element x_1^2 because there is a new leading term $-4\,\varepsilon\,x_1^3$ which cannot be reduced; this also requires a polynomial $bb_5 = x_1^2 x_2 - (1 - \varepsilon)\,x_1^2 - x_1 = x_1\,p_1$ in the border basis $\mathcal{B}(\varepsilon)$.

When we now choose Option 1 (cf. sections 9.4.2 and 9.5.1), i.e. normalization of bb_2, we obtain a polynomial with an $\frac{1}{\varepsilon}\,(4\,x_1^2 + 4\,x_2^2 - 25)$ part. For the 4 zeros of $P(\varepsilon)$ which approach the circle like $\mathrm{O}(\varepsilon)$, this is not disturbing, but for the 5th zero it requires an x_1-component of $\mathrm{O}(\frac{1}{\varepsilon})$ and leads to divergence towards ∞.

With Option 2, i.e. substitution of $\varepsilon = 0$, we return to the normal set representation of P_0. $\quad\square$

Let $\bar{z}(\varepsilon) \in \mathbb{C}^s$ be a zero of $P(\varepsilon)$ which diverges to ∞ for $\varepsilon \to 0$ and assume that there is no other one. We want to find an asymptotic expansion for $\bar{z}(\varepsilon)$ in powers of ε. Since this is an expansion about the point $\bar{z}(0)$ at infinity, it is natural to use a homogenized version $P(x, t; \varepsilon)$ of our system and consider the expansions

$$\bar{z}(\varepsilon) \;=\; z_0 + \varepsilon\,z_1 + \varepsilon^2\,z_2 + \dots\,, \qquad \bar{t}(\varepsilon) \;=\; t_0 + \varepsilon\,t_1 + \varepsilon^2\,t_2 + \dots\,; \tag{9.62}$$

its coefficients can be computed recursively by substitution of (9.62) into the $p_\nu(x, t; \varepsilon) = p_{0\nu}(x, t) + \varepsilon\,p_{1\nu}(x, t)$. The expansion (9.62) may not apply in special cases (e.g., the simultaneous divergence of two or more zeros); we will not analyze this further.

Example 9.19, continued: Let us derive the asymptotic expansion of $\bar{z}(\varepsilon) = z_5(\varepsilon)$; for better readability, we use (x_ν, y_ν) for the components of the z_ν in (9.62). Since we know that $x \to \infty$, $y \to 1$ for $\varepsilon \to 0$, we begin with

$$\bar{x}(\varepsilon) = x_0 + \varepsilon\,x_1 + \mathrm{O}(\varepsilon^2)\,, \quad \bar{y}(\varepsilon) = \varepsilon + \varepsilon^2\,y_2 + \mathrm{O}(\varepsilon^3)\,, \quad \bar{t}(\varepsilon) = \varepsilon + \varepsilon^2\,t_2 + \mathrm{O}(\varepsilon^3)\,.$$

Substitution into $P(x, t; \varepsilon) = (xy - xt - t^2 + \varepsilon\,xt,\ (y - t)(4x^2 + 4y^2 - 25t^2) + \varepsilon\,xt^2)$ yields

$$p_1(\bar{z}(\varepsilon), \bar{t}(\varepsilon); \varepsilon) = (x_0\,y_2 - 1 - x_0\,t_2 + x_0)\,\varepsilon^2 + \mathrm{O}(\varepsilon^3)\,, \quad p_2(\bar{z}(\varepsilon), \bar{t}(\varepsilon); \varepsilon) = 4\,(y_2 - t_2)\,x_0^2 + \mathrm{O}(\varepsilon^3)\,;$$

this implies $x_0 = 1$ and $y_2 = t_2$. Since it turns out that the t_ν remain undetermined for $\nu \geq 2$, we choose them as 0. With $\bar{x} = 1 + \varepsilon\,x_1 + \varepsilon^2\,x_2 + \dots$, $\bar{y} = \varepsilon + \varepsilon^3\,y_3 + \dots$, $\bar{t} = \varepsilon$, we obtain

$$p_1(\bar{z}(\varepsilon), \bar{t}(\varepsilon); \varepsilon) \;=\; (y_3 + x_1)\,\varepsilon^3 + \mathrm{O}(\varepsilon^4)\,, \quad p_2(\bar{z}(\varepsilon), \bar{t}(\varepsilon); \varepsilon) \;=\; (4\,y_3 + 1)\,\varepsilon^3 + \mathrm{O}(\varepsilon^4)\,,$$

and $x_1 = \frac{1}{4}$, $y_3 = -\frac{1}{4}$. In this fashion, we may continue. After dehomogenization, we have, e.g.,

$$x(\varepsilon) = \frac{1}{\varepsilon}\,(1 + \frac{1}{4}\,\varepsilon + \frac{21}{16}\,\varepsilon^3 + \frac{21}{32}\,\varepsilon^4 + \mathrm{O}(\varepsilon^5))\,, \quad y(\varepsilon) = 1 - \frac{1}{4}\,\varepsilon^2 + \frac{1}{16}\,\varepsilon^3 - \frac{85}{64}\,\varepsilon^4 + \mathrm{O}(\varepsilon^5)\,. \quad \square$$

The leading nonvanishing coefficients in the components of (9.62) depend only on the polynomials $p_{0\nu}$ of the singular system P_0. They define a manifold $\overline{M}_0$ along which the diverging zero $\bar{z}(\varepsilon)$ of $P(\varepsilon)$ moves and which it approaches asymptotically.

Example 9.19, continued: The leading coefficients $x_0 = y_1 = t_1 = 1$ define the subspace $\overline{M}_0 = (\xi, 1)$ which dominates the asymptotic behavior for any feasible P_1. $\quad\square$

Let $P(\varepsilon)$ be the specified system of an *empirical* system $(P(\varepsilon), E)$, with $\|E\| = \hat{\varepsilon}$, and assume that P_0 has BKK-deficiency 1; then, for small $|\varepsilon|$, the systems $\tilde{P} \in N_\delta(P(\varepsilon), E)$, $\delta = O(1)$, have one zero $\tilde{\tilde{z}}$ with a very large modulus, or they are also BKK-deficient. For fixed small $\hat{\varepsilon}$, the extension of the pseudozero sets $Z_\delta(P(\varepsilon), E)$ orthogonal to $\overline{M}_0$ remains moderate while their extension parallel to $\overline{M}_0$ grows with $|\bar{z}(\varepsilon)|$ as ε decreases; cf. the analogous considerations in section 9.5.1. For $\varepsilon \approx \hat{\varepsilon}$, the pseudozero sets extend to infinity along $\overline{M}_0$. Details depend on the particular case. This implies that the components of divergent zeros parallel to $\overline{M}_0$ are much worse conditioned than those orthogonal to $\overline{M}_0$.

Example 9.19, continued: When we perturb $P(.01)$ by $\Delta P = (.001\, y, .001\, y)^T$, the x-component of $\tilde{z}_5$ changes by $\approx .1 = O(\frac{\hat{\varepsilon}}{\varepsilon})$ or $O(\hat{\varepsilon})$ *relatively* while the y-component remains stable. This shows the sensitivity of the component parallel to $\overline{M}_0$. $\quad\square$

In a real-life situation with an empirical polynomial system $(\bar{P}, E)$, with $\bar{P}$ nearly BKK-deficient, one will observe one or several huge zeros $\hat{z}$ of $\bar{P}$. The Jacobian $\bar{P}'(\hat{z})$ will be near-singular, but due to the huge $|\hat{z}|$, it may also possess some very large elements and the determinant may not be small. One will want to decide if there is a system $\tilde{P}$ in $N_\delta(\bar{P}, E)$, $\delta = O(1)$, for which the zero is at ∞. Somehow, one wishes to form the backward error for a zero at ∞, but setting the homogenizing variable to zero in a homogenized version of $\bar{P}$ may delete the necessary information. It is not clear how one should generally proceed.

It appears that the further algebraic, analytic, and numerical investigation of BKK-deficient polynomial systems, particularly of those with a multiple zero at ∞ which splits into a cluster of zeros with very large moduli upon perturbation, should be an interesting and meaningful research project; cf. section 6.3.5.

Exercises

1. In $\mathcal{P}^2$, consider $p(x, y) = 3 + 4x - 5y - 2x^2 + xy - .5y^2 + x^2y - .95xy^2$ and $q(x, y) = \alpha_0 + \alpha_1 x + \alpha_2 y$.

 (a) What is the number of zeros of $\langle p, q \rangle$

 (i) for indeterminate α_j,

 (ii) for $\alpha_1 = 0$

 (iii) for $\alpha_2 = 0$ (with the remaining α_j indeterminate) ?

 (b) Find feasible normal sets for $\langle p, q \rangle$ for the 3 cases. Find the associated border basis (=GB) representations.

 (c) In the 3 cases, choose the (remaining) α_j such that the system $\{p, q\}$ is BKK-deficient. Check by solving. What are the respective manifolds $\overline{M}_0$?

 (d) Convince yourself that a generic perturbation (within the limits of the respective case) of q recovers the "lost" zeros.

2. Consider the situation in Exercise 1 geometrically :

 (a) Plot the manifold $M_p = \{p = 0\}$ and find the 3 asymptotes.

 (b) Explain the situation in (c) and (d) geometrically.

3. Consider the empirical system $(\bar{P}, E)$ with $\bar{P} = \{\bar{p}, \bar{q}\}$, $\bar{p} = p$ of Exercise 1, $\bar{q}(x, y) = 1 + 2.43\, x - 2.35\, y$. In $(\bar{p}, e_p)$, only the coefficient of xy^2 is empirical, in $(\bar{q}, e_q)$ the coefficients of x and y, all of them with $\varepsilon = .01$.

    (a) Solve the system $\{\bar{p}, \bar{q}\}$. Why would you expect a BKK-deficient system in $N_\delta(\bar{P}, E)$?

    (b) Find $\tilde{P} \in N_\delta(\bar{P}, E)$, $\delta = O(1)$, with only two zeros,

- by systematically pushing the large zero to ∞,

- by analysis and computation.

    (c) In the course of your experimentation, find a rough outline of the pseudozero domain $(\delta = 1)$ with the diverging zero.

9.6 Multivariate Interpolation

Multivariate interpolation, in particular by polynomials, is an important research area in its own right, with a considerable literature and software. In the context of this book, like in the univariate case, we only point out some algebraic aspects of the subject related to other topics in this book, in particular the ambiguity of the interpolation basis which is often not stressed. The important algorithmic aspects which depend strongly on special situations are not considered.

9.6.1 Principal Approach

Intuitively, the task of interpolation in $\mathcal{P}^s$ is the following (cf. section 5.4):

Given: a set of n evaluation functionals $c_\nu : \mathcal{P}^s \to \mathbb{C}$, $\nu = 1(1)n$;
           n associated values $w_\nu \in \mathbb{C}$, $\nu = 1(1)n$.

Find: $r \in \mathcal{P}^s$ such that

$$c_\nu(r) \;=\; w_\nu, \quad \nu = 1(1)n. \tag{9.63}$$

Example 9.20: $s = 2$, $n = 5$, $c_1(p) = p(0, 0)$, $c_2(p) = \frac{\partial}{\partial x_1} p(0, 0)$, $c_3(p) = \frac{\partial}{\partial x_2} p(0, 0)$, $c_4(p) = p(1, 0)$, $c_5(p) = p(0, 1)$. For $w^T = (1, -1, 2, 1, 2)$, a natural solution of (9.63) is

$$r(x) \;=\; 1 - x_1 + 2\,x_2 + x_1^2 - x_2^2\,;$$

but we may also add any scalar or polynomial multiples of

$$bb_1 := x_1^3 - x_1^2 \quad bb_2 := x_1^3 - x_2^2 \quad bb_3 := x_1\,x_2$$

to r without invalidating (9.63) since $c_\nu(q \cdot bb_j) = 0$ for any $q \in \mathcal{P}^2$, $\nu = 1(1)5$, $j = 1, 2, 3$. Obviously, r is not at all uniquely defined, not even when we restrict its total degree to 2, where we may still add arbitrary scalar multiples of $x_1 x_2$. $\square$

    A more formal definition of multivariate interpolation which accounts for these observations is the following:

Definition 9.18. An interpolation task in $\mathcal{P}^s$ is specified by
    (i) a *closed* set $\mathbf{c}_0^T$ of n *evaluation*[20] *functionals* $c_\nu \in (\mathcal{P}^s)^*$, $\nu = 1(1)n$;

---

[20]This is the classical case; more general functionals could be considered.

(ii) a vector of *values* $w^T = (w_1, \ldots, w_n) \in \mathbb{C}^n$.

The solution of this task is the *set* $[r]$ of all $r \in \mathcal{P}^s$ for which (9.63) holds. $\quad\square$

By Theorem 2.21, a closed set $\mathbf{c}^T$ of functionals $c_\nu \in (\mathcal{P}^s)^*$ defines a *dual space* $\mathcal{D} =$ span $\mathbf{c}^T \subset (\mathcal{P}^s)^*$ such that

$$\mathcal{I}[\mathcal{D}] := \{\, p \in \mathcal{P}^s \,:\, c(p) = 0, \ \forall c \in \mathcal{D} \,\} \subset \mathcal{P}^s$$

is an ideal. This implies that the set $[r]$ of interpolants is a *residue class* mod $\mathcal{I}[\mathcal{D}]$ and thus a member of the quotient ring $\mathcal{P}^s / \mathcal{I}[\mathcal{D}] =: \mathcal{R}[\mathcal{D}]$; cf. section 2.3.2.

With each residue class $[r]$ in an n-dimensional quotient ring $\mathcal{R} = \mathcal{P}^s/\mathcal{I}$, we can associate a *representative* $r \in \mathcal{P}^s$ by choosing a *basis* $\mathbf{b}$ of representatives $(b_\nu \in \mathcal{P}^s,\ \nu = 1(1)n)$ such that $\mathcal{R} = $ span $\mathbf{b}$ mod $\mathcal{I}$. Thus, we can make the solution of an interpolation task unique by requiring that $r \in $ span $\mathbf{b}$, where $\mathbf{b}(x)$ is a basis of $\mathcal{R}[\mathcal{D}]$; cf. section 2.2. Generally, the elements b_ν are monomials and $\mathbf{b} \in \mathcal{T}^s(n)$ is a normal set $\mathcal{N}$. As there exist a large number of feasible normal set bases for a given quotient ring in $\mathcal{P}^s$, there are many possible choices for the linear space span $\mathbf{b}$ within which we may require the interpolants for a given set $\mathbf{c}_0^T$ to lie. The actual selection will generally depend on the context and on the purposes for which the interpolant is to be used. In applications, the underlying model may require the choice of a particular interpolation basis. Only in the rare cases where $n = \binom{d+s}{s}$ and where the normal set $\{x^j : \deg(x^j) \le d\} \in \mathcal{T}^s$ is feasible, the explicit specification of a basis may be replaced by the requirement $\deg(r) \le d$. In all other cases, there exist polynomials $r' \in [r]_{\mathcal{I}[\mathcal{D}]},\ r' \ne r$, with $\deg(r') = \deg(r)$; cf. Example 9.20.

Proposition 9.6. For a specified interpolation task (cf. Definition 9.18), let $\mathcal{N} \in \mathcal{T}^s(n)$ be a feasible basis of $\mathcal{R}[\text{span } \mathbf{c}_0^T]$, with $\mathbf{b} := (b_\nu \in \mathcal{N})$ and $M_0 := \mathbf{c}_0^T(\mathbf{b}) \in \mathbb{C}^{n \times n}$. Then, for any choice of $w^T \in \mathbb{C}^n$, the unique interpolant $r \in $ span $\mathbf{b}$ which satisfies (9.63) is

$$r(x) = w^T M_0^{-1} \mathbf{b}(x). \tag{9.64}$$

Proof: $\mathbf{c}_0^T(w^T M_0^{-1} \mathbf{b}) = w^T M_0^{-1} \mathbf{c}_0^T(\mathbf{b}) = w^T$. Any other $r' \in [r]_{\mathcal{I}[\mathcal{D}]}$ must satisfy $r' - r \in \mathcal{I}[\mathcal{D}]$ and hence cannot be in span $\mathbf{b}$. $\quad\square$

Note that the set $\mathbf{c}^T$ spanned by the n interpolation functionals $c_\nu \in (\mathcal{P}^s)^*$ must be *closed* if it is to define an interpolation task which is solvable for arbitrary values $w \in \mathbb{C}^n$ by (9.64). Thus, the closedness of $\mathbf{c}^T$ combines the various special conditions which one may find for the feasibility of particular multivariate interpolation tasks.

Since the multiplication matrices of $\mathcal{R}[\mathcal{D}]$ with respect to the basis $\mathbf{b}$ are given by (cf. section 8.1.1)

$$A_\sigma = \mathbf{c}_0^T(x_\sigma \mathbf{b})\,(\mathbf{c}_0^T(\mathbf{b}))^{-1},$$

the border basis $\mathcal{B}_\mathcal{N}$ of the ideal $\mathcal{I}[\mathcal{D}]$ associated with the interpolation task may be obtained immediately: For $x^j = x_\sigma b_\nu \in B[\mathcal{N}]$, we have

$$bb_j(x) := x^j - a_{\sigma\nu}^T \mathbf{b}(x), \tag{9.65}$$

where $a_{\sigma\nu}^T$ is the ν-th row of A_σ. Note that $a_{\sigma\nu}^T \mathbf{b}(x)$ is the interpolant in span $\mathbf{b}$ for the values $w_\nu = c_\nu(x^j)$. Thus the elements of $\mathcal{B}_\mathcal{N}$ are obtained by interpolating the monomials $x^j \in B[\mathcal{N}]$.

Example 9.20, continued: For $\mathcal{N} = \{1, x_1, x_2, x_1^2, x_2^2\}$, with the components of $\mathbf{b}$ in that order, we have

$$
M_0 = \mathbf{c}_0^T(\mathbf{b}) = \begin{pmatrix} 1 & 0 & 0 & 1 & 1 \\ 0 & 1 & 0 & 1 & 0 \\ 0 & 0 & 1 & 0 & 1 \\ 0 & 0 & 0 & 1 & 0 \\ 0 & 0 & 0 & 0 & 1 \end{pmatrix}, \quad M_0^{-1} = \begin{pmatrix} 1 & 0 & 0 & -1 & -1 \\ 0 & 1 & 0 & -1 & 0 \\ 0 & 0 & 1 & 0 & -1 \\ 0 & 0 & 0 & 1 & 0 \\ 0 & 0 & 0 & 0 & 1 \end{pmatrix},
$$

and $w^T M_0^{-1} = (1, -1, 2, 1, -1)$. Furthermore, the interpolation of the values $w_j^T = \mathbf{c}_0^T(x^j)$ for the five border monomials $x_1^3, x_2^3, x_1 x_2, x_1^2 x_2, x_1 x_2^2$ yields

$$
\mathcal{B}_{\mathcal{N}} = \{x_1^3 - x_1^2, \ x_2^3 - x_2^2, \ x_1 x_2, \ x_1^2 x_2, \ x_1 x_2^2\}.
$$

Since the values of x_i^3 under $\mathcal{D}$ are equal to those of x_i^2, $i = 1, 2$, and $x_i^2 \in \mathcal{N}$, the form of the only two nontrivial border basis elements is immediate. $\quad\square$

9.6.2 Special Situations

For interpolation knots in general position, the approach sketched in the previous section gives a straightforward and reasonably efficient way for the determination of the interpolant. The recursive approach with divided differences which is favored in univariate interpolation (cf. section 5.4) cannot be well generalized to multivariate interpolation in this case.

For interpolation knots which form a *regular grid*, the situation is different: There exist a variety of algorithms which determine the interpolant for particular interpolation tasks of this sort and for associated choices of the interpolation basis. We indicate only the natural approach for the case of a rectangular grid in 2 dimensions and refer to the specialized literature for other situations.

Let function values $w_{\mu\nu} \in \mathbb{C}$ be specified at the $(m+1)(n+1)$ points (ξ_μ, η_ν), $\mu = 0(1)m$, $\nu = 0(1)n$. Then it is natural to determine the coefficients $\rho_{\mu\nu}$ of the interpolant

$$
r(x, y) = \sum_{\mu=0}^{m} \sum_{\nu=0}^{n} \rho_{\mu\nu} x^\mu y^\nu \tag{9.66}
$$

by successive univariate interpolation: For fixed $x = \xi_\mu$, we can determine the $m+1$ interpolants

$$
r_\mu(y) = \sum_{\nu=0}^{n} \rho_\nu(\xi_\mu) y^\nu, \quad \text{with } r_\mu(\eta_\nu) = w_{\mu\nu},
$$

by univariate interpolation in y and then interpolate the values $\rho_\nu(\xi_\mu)$ of the coefficients by univariate interpolation in x:

$$
\rho_\nu(x) = \sum_{\mu=0}^{m} \rho_{\mu\nu} x^\mu, \quad \text{with the values of the } \rho_\nu(\xi_\mu) \text{ from the } r_\mu(y).
$$

This yields the coefficients $\rho_{\mu\nu}$ of (9.66). Often, the ξ_μ and/or η_ν will be equidistant which reduces the effort for the univariate interpolations further.

Example 9.21: Consider the 20 interpolation knots (ξ_μ, η_ν), $\xi_\mu = 0(1)3$, $\eta_\nu = 0(1)4$, with the values

$\nu =$	0	1	2	3	4
0	$-.99$	-2.03	-3.38	-5.17	-7.44
$\mu = \quad$ 1	$-.52$	-1.01	-1.70	-2.64	-3.67
2	2.02	4.00	6.83	10.37	14.82
3	6.51	12.97	22.09	33.84	48.12

Interpolation of the 4 sets $\{w_{\mu\nu}, \; \nu = 0(1)4\}$, $\mu = 0(1)3$, by degree 4 polynomials in y yields the 4 interpolants (rounded)

$$
\begin{aligned}
r_0(y) &= -.9900 - .9508\,y - .0488\,y^2 - .0442\,y^3 + .0038\,y^4\,, \\
r_1(y) &= -.5200 - .4592\,y + .0213\,y^2 - .0608\,y^3 + .0088\,y^4\,, \\
r_2(y) &= 2.0200 + 1.4233\,y + .6508\,y^2 - .1083\,y^3 + .0142\,y^4\,, \\
r_3(y) &= 6.5100 + 5.1375\,y + 1.3129\,y^2 + .0125\,y^3 - .0029\,y^4\,.
\end{aligned}
$$

Interpolation of the 5 coefficient sets $\{\rho_\nu(\xi_\mu), \; \mu = 0(1)3\}$, $\nu = 0(1)4$, by degree 3 polynomials in x gives the 2-dimensional interpolant (rounded) $r(x, y) =$

$$
\begin{aligned}
&-.9900 - .6050\,x + 1.0950\,x^2 - .0200\,x^3 - .9508\,y - .0568\,xy + .4750\,x^2y + .0735\,x^3y \\
&-.0488\,y^2 - .3855\,xy^2 + .5433\,x^2y^2 - .0878\,x^3y^2 - .0442\,y^3 + .0651\,xy^3 - .1150\,x^2y^3 \\
&+.0332\,x^3y^3 + .0038\,y^4 - .0028\,xy^4 + .0117\,x^2y^4 - .0038\,x^3y^4\,. \qquad \Box
\end{aligned}
$$

9.6.3 Smoothing Interpolation

In many situations, the data of the interpolation have a limited accuracy so that exact interpolation, with as many basis functions as interpolation knots, is not sensible. Instead, one uses an interpolation basis $\widehat{\mathbf{b}}$ with fewer elements and replaces (9.63) by

$$
\| \mathbf{c}^T (\hat{r}^T \widehat{\mathbf{b}}) - w^T \| = \min . \tag{9.67}
$$

Furthermore, one may add bounds or minimization requests on other functionals of the interpolant, e.g., some derivative(s) of r.

Traditionally, since the pioneering work of Gauss on least squares, the square of the Euclidean norm is used in (9.67), perhaps with weights. Since $r = \hat{r}^T\widehat{\mathbf{b}}$ is linear in its coefficients and $\mathbf{c}^T$ is linear, this leads to a smooth quadratic functional in the coefficients. Differentiation with respect to each coefficient yields a symmetric linear system in the coefficients (the "normal equations") which is *positive definite* except when there is a rank deficiency; this can happen only when the columns of $\mathbf{c}^T(\widehat{\mathbf{b}})$ are linearly dependent. Numerically, it is more advantageous to solve the minimization problem (9.67) with the Euclidean norm directly with the aid of a QR-decomposition of the matrix in (9.67).

From the point of view explained in section 3.1 and employed throughout this book, the use of a weighted maximum norm in (9.67) may also be considered. Then one obtains a linear minimization problem as many times before. When we know or assume tolerances on the components of w^T, the value of the appropriately weighted maximum norm of $\mathbf{c}^T(r^T\widehat{\mathbf{b}}) - w^T$ is the *backward error* of the approximate interpolant $\hat{r}^T\widehat{\mathbf{b}}$.

In any case, the particular circumstances of the task will dominate the selection of the reduced interpolation basis $\widehat{\mathbf{b}}$. In particular, the underlying model may require the interpolant to lie in a particular linear space span $\widehat{\mathbf{b}}$. For details and particularly important special cases, we refer once more to the literature on the subject.

Example 9.22: Consider the situation and the data of Example 9.21, but assume empirical data, with w^T as specified values and a uniform tolerance of .05 on the components; the (ξ_μ, η_ν) are assumed as exact. We attempt an approximate smoothing interpolation with a biquadratic polynomial $\hat{r}(x, y) = \sum_{\mu, \nu=0}^{2} \hat{\rho}_{\mu\nu} x^\mu y^\nu$ which has only 9 parameters in place of the 20 parameters of the interpolant r in Example 9.21.

The minimization of (9.67) with the (unweighted) maximum norm yields a maximal pointwise deviation of .032 which is below the specified tolerance; thus we may consider the resulting polynomial (rounded) $\hat{r}(x, y) =$

$$-1.0217-.4500x+.9833x^2-.7833y-.4517xy+.8183x^2y-.2033y^2-.0900xy^2+.1967x^2y^2$$

as a valid continuous approximation of the specified data.

Actually, the data in Example 9.21 have been generated by taking the values of the biquadratic polynomial

$$p(x, y) = -1 - .5x + x^2 - .8y - .4xy + .8x^2y - .2y^2 - .1xy^2 + .2x^2y^2$$

at the gridpoints and perturbing them by random values from the set $-.04(.01).04$. Therefore, we can interpret the formation of $\hat{r}$ as the attempt to recover the function p from its perturbed values. The *smoothing* effect of this procedure can be seen from comparing the *derivative* values of both p and $\hat{r}$ at the gridpoints: The difference of the derivative values is nowhere greater than .05 and below .01 at most gridpoints. In comparison, the derivative values of the true interpolant r in Example 9.21 show deviations as large as .25. □

Exercises

1. In $\mathbb{R}^2$, specify a set of (say 10) interpolation knots in general position within the unit square. Take values from a nonpolynomial analytic function f as data w^T. Introduce a fixed grid G of points in the unit square at which values of f and of an interpolant are to be compared.

(a) Choose various interpolation bases, compute the interpolant, and compare its values with those of f on G. Try to correlate the quality of the fit to properties of the interpolation basis.

(b) Take bases with fewer elements than there are interpolation knots. Try to determine the best fit which you can achieve with a fixed number of elements. How does this best fit deteriorate with a decreasing number of basis elements?

(c) For identical bases, compare the fits obtained with a least squares and a maximum norm smoothing interpolant.

Historical and Bibliographical Notes 9

Seen with the eyes of an algebraist, the use of limited accuracy data and approximate computation in dealing with polynomial systems must indeed have been highly suspicious: It is really difficult to conceive of a meaningful topological embedding of polynomial ideals. While this objection may be overcome by the consideration of quotient rings and dual spaces, another aspect appears even more formidable: Since polynomial ideals *and* quotient rings are generally described in an overdetermined fashion, generically perturbed parameters in such a description do not represent an ideal or quotient ring at all! In retrospect, I believe that this fact may have been the most serious underlying obstacle for the use of algebraic tools in the numerical treatment of polynomial systems and, reversely, for the use of approximation in polynomial algebra.

With our analysis of the overdetermination mechanism in sections 8.1 and 8.2 and of the related effects of data indeterminations in section 9.2, we hope to have provided a firm basis for numerical polynomial algebra which positions it as a rather nontrivial but "ordinary" area of numerical analysis: The fact that a—strictly speaking—inconsistent approximate representation of a polynomial system may be numerically refined into an arbitrarily accurate one puts it on the same level as many other constructive approaches in applied mathematics.

The intimate relation between a zero cluster of a polynomial system and its "generating" multiple zero appears not to have received much attention so far although it is crucial in a continuous view of multiplicity, even more so than in the univariate case. The developments in section 9.3 grew out of the work with my student G. Thallinger in [8.3]. The analysis of the asymptotic behavior of clustered zeros of a polynomial system should be of interest in various applications.

Although the generalization from singular systems of linear equations to polynomial systems is so immediate, the static view of algebra has apparently disguised the appropriate asymptotic view of singular polynomial systems. The highly nontrivial and interesting mechanisms which govern the appearance of "singular manifolds"—as by a conjurer's trick—should invite numerical analysts and computer algebraists for further research. The associated potential ill-conditioning of zeros may sometimes be of great practical relevance. The preliminary analysis in section 9.4 is another result of my cooperation with G. Thallinger; cf. [9.1].

Our affine view of the BKK-approach in section 8.3.4 reveals the well-known "deficiencies" of this view as a singular phenomenon which is nothing but the natural generalization of the potential inconsistency of a square linear system. Section 9.5 provides only a first glance on this subject and will, hopefully, stimulate further research.

Multivariate polynomial interpolation, on the other hand, has developed over a long time and in many ways. The most interesting analytic foundation has been provided by de Boor and Ron in [9.2]; an algebraic approach of a very general scope may be found, e.g., in [9.3]. The direct relation of the subject to dual spaces and quotient rings has called for at least a superficial account in this text.

References

[9.1] H.J. Stetter, G.H. Thallinger: Singular Systems of Polynomials, in: Proceed. ISSAC 1998 (Ed. O. Gloor), ACM, New York, 9–16, 1998.[21]

G.H. Thallinger: Zero Behavior in Perturbed Systems of Polynomial Equations, Ph.D. Thesis, Tech. Univ. Vienna, 1998.

---

[21]An error in the printed version was corrected by the authors at the conference.

[9.2] C. de Boor, A. Ron: On Multivariate Polynomial Interpolation, Constr. Approx. **6** (1990), 287–302.

C. de Boor, A. Ron: Computational Aspects of Polynomial Interpolation in Several Variables, Computer Science Department, University of Wisconsin, Tech. Rep. #924, 1990.

[9.3] H.M. Moeller, Th. Sauer: H-Bases for Polynomial Interpolation and System Solving, Adv. Comput. Math. **12** (2000), 335–362.

Numerical Basis Computation

At the time of the writing of this book, there is still only one approach to the numerical computation of a basis for the ideal defined by a polynomial system, for which software is commonly available: It is the computation of a *Groebner basis*, for a *specified term order*, by *rational computation*. This fact is presumably the main reason why polynomial algebra has not been widely used in Scientific Computing so far; cf. also section 1.5. In the approach to polynomial algebra taken in this book, Groebner bases have played a minor role; therefore, we will only briefly describe some principal aspects of their computation in section 10.1.

We will then turn our attention to *regular systems* of polynomials and *normal set representations* for their ideals. In describing a natural approach to their computation in section 10.2, we are aware of the fact that some aspects of such algorithms are not yet fully understood; we hope that our presentation will stimulate further research on this topic. Finally, in section 10.3, we will consider the truly numerical aspects of basis computation: The use of *floating-point arithmetic* and the application to *empirical* systems.

10.1 Algorithmic Computation of Groebner Bases

10.1.1 Principles of Groebner Basis Algorithms

A basic algorithm for the computation of the reduced Groebner basis of a specified system of polynomials, for a specified term order, was developed by Buchberger in the context of his thesis [2.2]. Since then, because the computation of the reduced Groebner basis of some rather innocent looking polynomial systems in not so many variables can take extremely long and the required intermediate storage can be huge, the design of better and better algorithms for that purpose and of more and more efficient implementations of these algorithms has been a continuing challenge; there is a vast literature on the subject. It has become usual to assess the performance of a new version or implementation of a GB-algorithm by its performance on some parametrized set of sample problems: It is deemed superior when it is able to complete the computation for problems for which the computation could not be completed previously. For a list of such sample problems, see, e.g., [10.4]. In this text, we only exhibit the common basic structure of these algorithms and describe a few of the ideas which have made significant improvements possible.

When we regard the characterization of reduced Groebner bases in Theorem 8.34, it appears natural that any GB-algorithm must consist of steps which modify a current basis $G = \{ g_\kappa \}$ of the ideal such that it satisfies the criteria listed there to a fuller extent:

(1) No leading monomial of a g_κ divides the l.m. of another $g_{\kappa'}$;

(2) the S-polynomials of all pairs of g_κ's reduce to 0 with G.

There are two basic operations which modify G without changing $\langle G \rangle$:

(i) Reduction: Form $g'_\kappa(x) := g_\kappa(x) - \sum_{\kappa' \neq \kappa} \gamma_{\kappa\kappa'}(x) g_{\kappa'}(x)$ such that no monomial in g'_κ is a multiple of a l.m. of a $g_{\kappa'}$; replace g_κ by g'_κ in G.

(ii) S-polynomial formation: Form $g_{\kappa_1\kappa_2}(x) := S[g_{\kappa 1}, g_{\kappa 2}]$ (cf. (8.27)) and append it to G.

Thus, a GB-algorithm is essentially composed of the following two procedures which act on a set of polynomials:

$G' := \mathtt{autoreduce}(G)$: Uses iterated reduction to generate $G' := \{ g'_\kappa \}$ such that the l.m. of no one element divides a monomial of another element.

$G' := \mathtt{Spol}(G)$: Selects a subset $\hat{G} \subset G$, forms the S-polynomials of all pairs $\hat{g}_{\kappa_1}, \hat{g}_{\kappa_2} \in \hat{G}$ for which it is not known that $S[\hat{g}_{\kappa 1}, \hat{g}_{\kappa 2}] \xrightarrow{G} 0$ and reduces them by G; nonzero remainders are appended to G.

Algorithm 10.1.

$$
\begin{aligned}
G &:= \mathtt{autoreduce}(P) \\
\mathbf{do}\quad G' &:= G, \quad G'' := \mathtt{Spol}(G'), \quad G := \mathtt{autoreduce}(G'') \quad \mathbf{od} \\
&\mathbf{until}\quad G = G' \\
\mathcal{C}_\prec[\langle P \rangle] &:= G
\end{aligned}
$$

By Theorem 8.32, we obtain the reduced Groebner basis, i.e. the $\prec$-compatible corner basis $\mathcal{C}_\prec$, of $\langle P \rangle$, iff the algorithm ends.

Example 10.1: Consider the quadratic system P in 2 variables $\{x^2 + 3xy - 5y^2 - 6x + 3y + 4,\ 3x^2 - 2xy - 4y^2 - 7x - 2y + 1\}$. We apply Algorithm 10.1 with $\mathtt{tdeg(x,y)}$:

$\mathtt{autoreduce}(P)$ solves P for the two highest monomials x^2, xy and yields $G = \{x^2 - 2y^2 - 3x + 1,\ xy - y^2 - x + y + 1\}$. In the first loop, only one S-polynomial can be formed: $S[g_1, g_2] = y\,g_1 - x\,g_2 = xy^2 - 2y^3 + x^2 - 4xy - x + y$. Reduction yields $g_3 = y^3 + 2y^2 + x - 3y - 2$ which is appended to form G'' which becomes the new G because there is no possibility for further autoreduction.

In the next run through the loop, $S[g_1, g_3]$ need not be formed because of Proposition 8.15. $S[g_2, g_3] = y^2 g_2 - x g_3 = -y^4 - 3xy^2 + y^3 - x^2 + 3xy + y^2 + 2x$, and subtraction of $(-g_1 - (3y - 1) g_2 - y g_3)$ reduces it to 0. Thus nothing is changed and we have obtained the Groebner basis $\{x^2 - 2y^2 - 3x + 1,\ xy - y^2 - x + y + 1,\ y^3 + 2y^2 + x - 3y - 2\}$. $\square$

Example 10.2: We consider the very sparse system (8.42) of Example 8.17, again with the term order $\mathtt{tdeg(x,y)}$. Initialization gives $g_\kappa = p_\kappa$, $\kappa = 1, 2$.

During the first loop, the only S-polynomial is $S[g_1, g_2] = g_3 = \alpha_{30}^{(2)} x^5 - \alpha_{02}^{(1)} y^4 + \alpha_{01}^{(2)} yx^2 - \alpha_{10}^{(1)} y^2x - \alpha_{00}^{(1)} y^2$, with l.m. x^5, because no reduction is possible.

In the next loop, $g_4 := S[g_1, g_3] = \alpha_{30}^{(2)} x^2 g_1 - y^2 g_3 = \alpha_{02}^{(1)} y^6 - \alpha_{01}^{(2)} x^2 y^3 + \alpha_{10}^{(1)} xy^4 + \alpha_{30}^{(2)} \alpha_{02}^{(1)} x^2 y^2 + \alpha_{00}^{(1)} y^4 + \alpha_{30}^{(2)} \alpha_{10}^{(1)} x^3 + \alpha_{30}^{(2)} \alpha_{00}^{(1)} x^2$, with l.m. y^6. No further S-polynomials are to be formed since part (a) of Proposition 8.15 takes care of $S[g_2, g_3]$, and no reduction is possible.

In the following loop, $S[g_2, g_4] = \alpha_{02}^{(1)} y^2 g_2 - x g_4$ turns out to be reducible to 0 so that we are finished. The normal set extended by the l.m. of the g_κ, $\kappa = 1(1)4$, is not generic for $\texttt{tdeg(x,y)}$; it contains the two monomials $x^2 y^3$, $x^4 y$ which are higher in term order than the leading monomials of g_2, g_1, resp., but no terms with these monomials occur. In this sense, P has a degree 2 of degeneracy; cf. section 8.4.4. $\square$

In Example 10.2, we have multiplied the right-hand side of the S-polynomial definition (8.27) by the product of the leading coefficients to avoid denominators; this is always possible. Similarly, one may always avoid denominators in the autoreduction step. Naturally, this is only meaningful when the coefficients of the specified polynomial system P consist of integers and parameters; in this case, one may obtain a Groebner basis whose coefficients are polynomials in the parameters, with integer coefficients. The procedure $\texttt{gbasis}$ of Maple delivers such Groebner bases.

The above two examples are deceptive: Generally, one cannot assume that a "hand" computation of the reduced Groebner basis succeeds so readily as above, even with simple systems in 2 variables. And the attempt to compute a Groebner basis for a system with parametric coefficients very often leads to expressions which are beyond reading—if the computation ends within a reasonable time at all.

This brings us to the question why the Algorithm 10.1 must come to an end—though perhaps after an excessive running time. The proof of this fact is a major achievement of Buchberger's thesis. The central idea is the following: The S-polynomial of a pair of basis polynomials must either reduce to 0 or introduce a new polynomial with a leading monomial which is irreducible by the current set L of leading monomials. Its inclusion into L makes the ideal $\langle L \rangle$ *larger*. Thus, the successive ideals $\langle L \rangle$ form an *Ascending Chain of Ideals*, and there is a fundamental theorem (cf., e.g., [2.10], section 2.5) which states that such a chain must become *stationary*. Therefore, at some point of the algorithm, no further S-polynomials can appear which are not reducible to 0 and the algorithm comes to an end.

The very crude skeleton of Algorithm 10.1 and of its constituent procedures permits a wide variety of potential implementations: In early versions of the algorithm, only one pair was selected for $\hat{G}$ in $\texttt{Spol}$; various strategies were devised for the selection of that pair. Nowadays, the *simultaneous* formation of S-polynomials and their reductions appears more attractive:

In principle, each such operation may be interpreted as a (scalar) linear combination of coefficient vectors of elements in a linear space spanned by a sufficiently large set of monomials in $\mathcal{P}^s$. When the correct linear space has been determined, the coefficient vectors of the monomial multiples of polynomials in G employed in the current loop may be arranged as rows of a matrix, then elimination algorithms from computational linear algebra may be employed to obtain the representations of the elements of the resulting reduced polynomial set.

An algorithm which follows this strategic approach and implements it in a clever way has been designed by J.-Ch. Faugère: Around 2000, his F4 code represented the most advanced piece of software for the computation of Groebner bases. It set a number of new records in performance on Groebner basis computations; cf., e.g., [10.1]. A central feature of F4 is the following: The current basis for the linear space and of the necessary monomial multiples are determined in a purely symbolic process which need not consider the actual coefficient values. These appear only in the matrix formed after this "symbolic reduction" which is then simplified with state-of-the-art linear algebra procedures. Also the set of available "reductors" is administrated in a clever way to avoid duplication of computational work.

10.1.2 Avoiding Ill-Conditioned Representations

In section 8.4.4, we have realized that the normal set $\mathcal{N}$ which is uniquely associated with the border Groebner basis $\mathcal{G}_\prec[\langle P \rangle] = \mathcal{B}_\mathcal{N}[\langle P \rangle]$ and the reduced Groebner basis $\mathcal{C}_\prec[\langle P \rangle]$ for a specified term order $\prec$ may happen to be very close to infeasibility: In the space $\mathbb{C}^{s \times m}$ of the zero sets of polynomial systems from $\mathcal{P}^s(m)$, the zero set of P may be very close to the singularity manifold $S_\mathcal{N}$ consisting of those zero sets for which $\mathcal{N}$ is infeasible. In this case, $\mathcal{G}_\prec$ is an extremely ill-conditioned representation of $\langle P \rangle$. Even when we deal with intrinsic polynomials and use exact computation, this can lead to very undesirable situations. In section 8.4.4, we had also remarked that this singularity arises only through the representation and disappears if a more suitable normal set is used.

Example 10.3: Compare Examples 4.2 and 4.4. Consider the following two quadratic polynomials in $\mathcal{P}^2$:

$$P = \begin{cases} p_1(x, y) = 3\,(\alpha_1\,x - \alpha_2\,y)^2 + (\tfrac{1}{3} + \alpha_2\,x + \alpha_1\,y)^2 - 4\,, \\ p_2(x, y) = (\tfrac{1}{3} + \alpha_1\,x - \alpha_2\,y)^2 + 4\,(\alpha_2\,x + \alpha_1\,y)^2 - 4\,, \end{cases}$$

where $\alpha_1 = \cos(\varphi)$, $\alpha_2 = \sin(\varphi)$, and $\varphi \approx .11806$ has been chosen such that two of the 4 zeros of P have identical x-coordinates; cf. Figure 4.1 in section 4.2.3.

In the computation of the $\texttt{plex(y,x)}$ Groebner basis, the initial autoreduction yields polynomials g_1, g_2, with leading monomials y^2 and xy, resp., and further terms with $y, x^2, x, 1$. The formation and reduction of the S-polynomial $S[g_1, g_2]$ by g_1, g_2 requires the following matrix, where the positions with nonvanishing coefficients have been marked:

	xy^2	y^2	x^2y	xy	y	x^3	x^2	x	1
$x\,g_1$	$\times$			$\times$		$\times$	$\times$	$\times$	
$-y\,g_2$	$\times$	$\times$	$\times$	$\times$	$\times$				
g_1		$\times$			$\times$		$\times$	$\times$	$\times$
$x\,g_2$				$\times$	$\times$		$\times$	$\times$	$\times$
g_2				$\times$	$\times$		$\times$	$\times$	$\times$

From this symbolic reduction, we expect a polynomial g_3 with l.m. y and terms with $x^3, x^2, x, 1$; but, for the special coefficients above, the coefficient of y is annihilated in the elimination. Thus, g_3 is a univariate 3rd degree polynomial in x, which reflects the fact that there are only 3 separate x-locations in the zero set. We have a nongeneric $\texttt{plex(y,x)}$ Groebner basis $\{g_1, g_2, g_3\}$ with normal set $\mathcal{N} = \{1, x, x^2, y\}$. For $\texttt{tdeg(y,x)}$, the same basis is obtained, but here it is generic.

Now we consider a neighboring system $\bar{P}$ where the trigonometric functions have been approximated by *rational numbers* (from Maple) to 8 decimal digits. In the computation of the $\texttt{plex(y,x)}$ Groebner basis in *rational* arithmetic, the coefficient of y in g_3 is now not exactly annihilated; with Maple's integer implementation, we obtain

$$\bar{g}_3 = .36839.. \, 10^{45}\, y - .58277.. \, 10^{52}\, x^3 - .53328.. \, 10^{52}\, x^2 + .66825.. \, 10^{52}\, x + .56025.. \, 10^{52}\,.$$

Therefore, the GB-algorithm takes $\bar{g}_3$ with l.m. y into the basis and reduces g_2 with it to obtain a univariate 4th degree polynomial

$$\bar{g}_4 = .11144..\,10^{61}\,x^4 - .15005..\,10^{60}\,x^3 - .23483..\,10^{61}\,x^2 + .27006..\,10^{60}\,x + .11246..\,10^{61}$$

for the *exact* GB-basis $\{\bar{g}_3, \bar{g}_4\}$ of $\bar{P}$, with the generic normal set $\bar{\mathcal{N}} = \{1, x, x^2, x^3\}$.

The computation of the zeros from that basis appears easy: $\bar{g}_4$ should yield 4 x-components and $\bar{g}_3$ the associated y-components. But a closer look reveals the hitch: g_4 must have two zeros which very nearly coincide, yet the associated y-values differ by 1.45! This can only be possible by the cancellation of 8 leading digits in g_3, assuming that the x-values are sufficiently accurate. But these values come from a 2-cluster and are thus very ill-conditioned themselves!

When we use 10-digit decimal arithmetic to compute the zeros *from the exact integer basis* of $\bar{P}$, we obtain (rounded)

$$(-1.2044, -.7981), \quad (-.7604, 1.0614), \quad (1.0497 \pm .000007\,\mathrm{i}, -.0787 \pm 443.64\,\mathrm{i})\,.$$

The x-components of the first two zeros are correct to 10 digits (to within 1 unit), but the associated y-components have lost 8 of these and retain only two correct digits. The other two zeros are clearly nonsense, displaying the extreme ill-condition of the basis representation.

With 20-digit arithmetic, we obtain the y-components of the first two zeros to 12 digits, as to be expected. Of the other two zeros, the x-components now have 12 ($\approx 20/2$) correct digits; but that suffices only for 3-4 correct digits in the y-components. To obtain 10 correct digits for them, we need at least 18 correct digits of the x-component which requires about 30-digits in its computation! Yet all zeros of $\bar{P}$ are very well-conditioned as shown by Figure 4.1, and we have used the *exact integer* GB of the system $\bar{P}$ with *rational* coefficients.

In this transparent example, it is clear what should have been done: The tiny y-term in $\bar{g}_3$ (tiny *relative* to the level of the other coefficients) should *not* have been used as leading term of a basis polynomial. Instead, the next full-size term in $\bar{g}_3$ (with x^3) should have been *defined* as $\mathcal{N}$-leading term of an *extended* Groebner basis $\mathcal{G}_{\mathcal{N}} = \{\bar{g}_1, \bar{g}_2, \bar{g}_3\}$ with the normal set $\mathcal{N} = \{1, x, x^2, y\}$. Naturally, if we would now use the multiplication matrix A_x to find the zeros, we would reintroduce some of the ill-conditioning because this matrix has two clustered eigenvalues. But with a little further computation, we can determine the missing element with $\mathcal{N}$-leading term $x^2 y$ of the full border basis $\mathcal{B}_{\mathcal{N}}$ and use the well-conditioned matrix A_y for the zeros.

From a numerical point of view, a more reasonable approach is the following: *Drop* the y-term in $\bar{g}_3$ and solve the cubic polynomial for 3 x-components. Then compute associated y-components from $\bar{g}_2$ (linear in y) for the two negative x-components and from $\bar{g}_1$ (quadratic in y) for the one positive x-component. This yields 7–8 correct digits for both components of all 4 zeros. Further digits may be readily obtained by a Newton step in the specified system $\bar{P}$. $\square$

Naturally, for a term order with $x \succ y$, the above dilemma would not have happened. But often, term orders must be chosen without sufficient insight into the location of the zeros; it can also happen that the Groebner bases for all standard term orders are ill-conditioned. In real-life problems, symmetric positions of some zeros are frequent; if the associated polynomial systems are slightly perturbed for whatever reason, this may lead to situations as in the above example.

Analogous effects may be expected when a polynomial with a tiny leading monomial is used for elimination or S-polynomial formation somewhere during the basis computation; but this may imply that terms with a certain monomial x^j cannot be eliminated from other polynomials. If that monomial can be made a member of the normal set (as in the example above), everything is fine; the elimination is not necessary and another monomial replaces x^j as a l.m. However, if that x^j is not adjacent to the normal set which would otherwise arise, we would also have to move its negative neighbors into the normal set to keep it closed. The alternative in this case is to *delay* the treatment of the x^j-terms; a divisor of x^j may become a l.m. later and permit a reduction of the delayed terms.

All this sounds very heuristic and one may wonder whether such stabilizing measures cannot ruin the algorithm. My student A. Kondratyev has been able to put the approach into a rigid algebraic framework by the introduction of an auxiliary indeterminate ϵ which becomes attached to terms with a coefficient below a specified threshold. ϵ is an infinitesimal variable in the sense that it *precedes* 1 in any term order; at the same time, it is prevented from accumulating higher powers by an identity (say) $\epsilon^3 = 0$. This theory permits the proof of assertions about the behavior of a GB-basis algorithm which is prevented from using small pivots in reduction and S-polynomials at the cost of violations of the original term order by small elements. Details may be found in [10.3]; cf. also section 10.3.3.

10.1.3 Groebner Basis Computation for Floating-Point Systems

Consider a regular polynomial system P, with decimal fractions for (some of) its coefficients. The commonly available software systems for GB-computation either do not accept such a system, or they may react in an unpredictable way. Maple 7, e.g., appears to give a standard GB-computation in floating-point arithmetic a try; this leads to 3 possible consequences: A correct approximate result is obtained for very simple systems and very short decimals, or a result "[*float number*]" is shown which is equivalent to saying $\langle P \rangle = \langle 1 \rangle$, or an "unending" computation is entered.

Of course, Maple permits a simple way around this problem: E.g., let P:={p1,p2,p3}, and trd some term order expression; then, with(Groebner),

```
gp := gbasis(convert(P,rational),trd):
for n to nops(gp) do
   gpf[n] := convert(gp[n]/leadmon(gp[n],trd)[1],float) od;
```

will generate a Digits decimals correct approximate Groebner basis for P. The normalization is necessary because the coefficients in Maple's unnormalized basis polynomials will convert to floating-point numbers with large decimal exponents indicating the length of the integers in gp which are better not displayed. Nonetheless, one can use gp to obtain (for Maple 7 plus) the normal set $\mathcal{N}$ of the Groebner basis from SetBasis(gp,trd)[1] and the associated multiplication matrices from, e.g.,

```
ax := convert(MulMatrix(x,SetBasis(gp,trd),gp,trd),float);
```

the nontrivial rows a_j^T of the multiplication matrices yield the full border basis $\mathcal{B}_\mathcal{N}$ through

$$bb_{j_\nu}(x) := x^{j_\nu} - a_{j_\nu}^T \mathbf{b}(x), \quad \nu = 1(1)N.$$

Naturally, for large problems, or for small problems with long decimals, the computation may get drowned in the swell of the integers. Anyway, it appears unreasonable to perform a computation in integer arithmetic only to round the result when it is obtained. Therefore, the development and the efficient and safe implementation of true floating-point GB-algorithms is still a central research topic for numerical polynomial algebra. Kondratyev's Ph.D. thesis [10.3] represents a seminal first result in this direction.

Exercises

1. Specify various nontrivial polynomial systems, with integer coefficients, with various numbers of variables and degrees, and use your GB-software (e.g., in Maple) to compute an exact Groebner basis and multiplication matrices for various term orders. By observing

 - the computation time,

 - the coefficient swell,

 - the number of GB elements and the size of $m = |\mathcal{N}|$,

try to gather experience of what is an easy and a hard task.

2. Specify various polynomial systems with decimal coefficients and use the approach in section 10.1.3 to compute an approximate Groebner basis and multiplication matrices A_σ. Check the commutativity of the A_σ to confirm that the rounded exact border basis has been obtained.

3. Design polynomial systems with (some) decimal coefficients with a representation singularity for a specified term order: Introduce a parameter t so that some of the leading coefficients of the GB-basis elements become polynomials in t. Determine $\hat{t}$ such that a leading coefficient vanishes.

Observe the coefficient growth in the normalized Groebner basis when t has numerical values approaching $\hat{t}$. Change the term order to realize that this is only a representation singularity.

10.2   Algorithmic Computation of Normal Set Representations

In the remaining parts of Chapter 10, we assume that we deal with a regular polynomial system $P \subset (\mathcal{P}^s)^s$: According to our central theorem of polynomial systems solving (Theorem 2.27), when we have determined a feasible normal set basis **b** of the quotient ring $\mathcal{R}[\langle P \rangle] = \mathcal{P}^s/\langle P \rangle$ and the elements of at least one suitable multiplication matrix A_σ for $\mathcal{R}[\langle P \rangle]$ with respect to **b**, we have reduced the task of numerically computing all zeros of $\langle P \rangle$ to the eigenanalysis of A_σ; cf. section 2.4 and various parts of Chapter 8. Hence, from the point of view of numerical analysis, we may consider the system P as "solved" when a normal set representation of $\langle P \rangle$ has been found. Therefore the design and analysis of algorithms for the determination of a suitable *normal set* $\mathcal{N}$ and the associated *border basis* $\mathcal{B}_\mathcal{N}$ for P is a fundamental task in computational algebra. In this book, it is not our aim to develop and explain such algorithms in detail; we will rather try to convey a principal understanding of their structure and point out aspects which need further investigation.

A fundamental distinction between the very general algorithmic scheme 10.1 for the computation of Groebner bases and our approach consists in the fact that we assume the knowledge of $m = |\mathcal{N}| = \mathrm{BKK}(P)$. In the case of dense systems, this knowledge is trivially provided by

(8.41), for sparse systems, m may be obtained from available algorithms; cf. section 8.3.4. It is not clear whether the cost of these (purely symbolic) BKK-algorithms is generally balanced by the reduction in the cost of the basis computation. In any case, it appears that a safe and reliable *numerical* basis computation must take advantage of the knowledge of the dimension of the quotient ring $\mathcal{R}[\langle P \rangle]$.

10.2.1 An Intuitive Approach

Consider a system $P = \{p_\nu(x)\}$, $\nu = 1(1)s$, with $p_\nu \in \mathcal{P}^s$, with supports $J_\nu \subset \mathbb{N}_0^s$. Assume that we have a closed set $\overline{J}$ in $\mathbb{N}_0^s$, with $|\overline{J}| = m + s$, with the following properties:

(i) $\overline{J} \supset J := \mathcal{Cl}(\cup_{\sigma=1}^s J_\sigma)$, the closure[22] of the joint support of P;

(ii) $\overline{J}$ may be split into disjoint sets $\mathcal{N}$ and B_0, with $\mathcal{N}$ closed, $|\mathcal{N}| = m$, $B_0 \subset \partial\overline{J}$, $|B_0| = s$, $B_0 \cap J_\nu \neq \emptyset$ for $\nu = 1(1)s$; cf., e.g., Figure 10.1.

Then

$$p_\nu(x) = \sum_{x^{j_\sigma} \in B_0} \rho_{\nu\sigma} x^{j_\sigma} + \sum_{j \in \mathcal{N}} \alpha_{\nu j} x^j , \quad \nu = 1(1)s . \tag{10.1}$$

Except if the matrix $(\rho_{\nu\sigma})$ should be singular, we may "solve" for the x^{j_σ}; the resulting polynomials

$$bb_\sigma(x) := x^{j_\sigma} - \sum_{j \in \mathcal{N}} \beta_{\sigma j} x^j , \quad \sigma = 1(1)s , \tag{10.2}$$

form a complete intersection subset of a border basis $\mathcal{B}$ for $\langle P \rangle$, with normal set $\mathcal{N}$. Hence, the remaining border basis elements must be computable from (10.2).

Example 10.4: Consider two generic quadratic equations in 2 variables: $p_\nu(x) = \sum_{j \in J} \alpha_j^{(\nu)} x^j$, $J = \{(j_1, j_2) \in \mathbb{N}_0^2, \ 0 \le j_1 + j_2 \le 2\}$. With $\overline{J} = J$ and the splitting $\mathcal{N} = \{1, x_1, x_2, x_1 x_2\}$, $B_0 = \{x_1^2, x_2^2\}$, our assumptions are satisfied. If $\begin{pmatrix} \alpha_{20}^{(1)} & \alpha_{02}^{(1)} \\ \alpha_{20}^{(2)} & \alpha_{02}^{(2)} \end{pmatrix}$ is regular, we obtain a border basis subset (10.2) of the form

$$bb_\sigma(x) = x_\sigma^2 - \sum_{j \in \mathcal{N}} \beta_{\sigma j} x^j , \quad \sigma = 1, 2 .$$

To complete the border basis, we must find the normal forms of $x_1^2 x_2$ and $x_1 x_2^2$. We form

$$
\begin{aligned}
x_2 \, bb_1(x) &= x_1^2 x_2 - \sum_{j \in \mathcal{N}} \beta_{1j} x^{j+(0,1)} \\
&= x_1^2 x_2 - \beta_{1,(1,1)} x_1 x_2^2 - [\beta_{1,(0,0)} x_2 + \beta_{1,(1,0)} x_1 x_2 + \beta_{1,(0,1)} \sum_{j \in \mathcal{N}} \beta_{2j} x^j] , \\
x_1 \, bb_2(x) &= x_1 x_2^2 - \sum_{j \in \mathcal{N}} \beta_{2j} x^{j+(1,0)} \\
&= -\beta_{2,(1,1)} x_1^2 x_2 + x_1 x_2^2 - [\beta_{2,(0,0)} x_1 + \beta_{2,(0,1)} x_1 x_2 + \beta_{2,(1,0)} \sum_{j \in \mathcal{N}} \beta_{1j} x^j] .
\end{aligned}
$$

Except if $\begin{pmatrix} 1 & -\beta_{1,(1,1)} \\ -\beta_{2,(1,1)} & 1 \end{pmatrix}$ is singular, this yields the requested normal forms. $\square$

Example 10.5: Consider the system (8.42) of Example 8.17; cf. also Example 10.2. For this extremely sparse system $\{x^3 y^2 + \alpha_{02} y^2 + \alpha_{10} x + \alpha_{00}, \ xy^4 + \alpha_{30} x^3 + \alpha_{01} y\} \in \mathcal{P}_5^2$, we have $m = \mathrm{BKK}(P){=}18$. Here, the closure of $J_1 \cup J_2$ has only 16 elements (cf. Figure 10.1); thus we must append 4 more elements to obtain $\overline{J} = \mathcal{N} \cup \{x^3 y^2, xy^4\}$, a natural (but not unique) choice

---

[22] In the sense of "closed set in $\mathbb{N}^s$."

is shown in Figure 10.1. We have $|B[\mathcal{N}]| = 9$; the border basis polynomials bb_1, bb_2 with $\mathcal{N}$-leading terms $x^3 y^2$ and $x y^4$ are simply p_1 and p_2, resp., the determination of the remaining 7 elements of $\mathcal{B}_\mathcal{N}$ will be explained below. $\square$

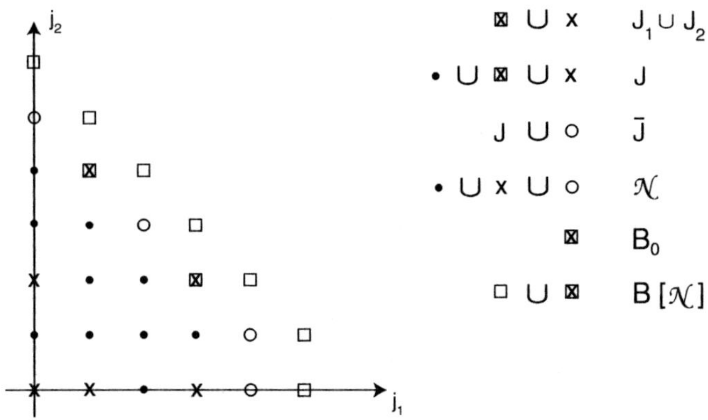

Figure 10.1.

Unfortunately, there is not always a set $\overline{J}$ as we have assumed it above; simple counterexamples are easy to find (cf. Exercise 10.1-3). But the following argument holds for all regular systems $P \subset \mathcal{P}^s$: Let $\mathcal{N}$, with $|\mathcal{N}| = m = \mathrm{BKK}(P)$ be some feasible normal set for $\langle P \rangle$. Since $\{p_v\}$ is a complete intersection basis of $\langle P \rangle$, the $bb_\kappa(x) \in \mathcal{B}_\mathcal{N}[\langle P \rangle]$ permit a representation

$$bb_\kappa(x) = x^{j_\kappa} - \sum_{j \in \mathcal{N}} \beta_{\kappa j} x^j = \sum_{v=1}^{s} \rho_{\kappa v}(x)\, p_v(x), \quad \kappa = 1(1)k,\ k = |B[\mathcal{N}]|; \qquad (10.3)$$

cf. Theorem 8.25. Let $J'_{\kappa v}$ be the supports of the polynomials $\rho_{\kappa v} \in \mathcal{P}^s$ and consider $\overline{J}'_v := \bigcup_\kappa J'_{\kappa v}$, $v = 1(1)s$. The representations (10.3) may not be unique (except for $\mathrm{NF}_\mathcal{N}[\rho_{\kappa v}]$), but we may may assume that we have chosen a representation with the smallest values of $|\overline{J}'_v| =: r_v$. Each p_v is thus multiplied by r_v different monomials x^j in (10.3), and the total number of multiples of p_v's necessary to represent all the bb_κ is $r := \sum_v r_v$. On the other hand, we can consider the supports $J''_{\kappa v}$ of the individual products $\rho_{\kappa v} p_v$ and $\overline{J}'' := \bigcup_{\kappa, v} J''_{\kappa v}$, $|\overline{J}''| =: R$.

Now, each bb_κ in (10.3) may be regarded as a *scalar* linear combination of a selection of the r polynomials $x^j p_v(x)$ which eleminates all terms in $\sum \rho_{\kappa v} p_v$ except x^{j_κ} and those with monomials in $\mathcal{N}$. We may consider the R monomials in the polynomials $x^j p_v(x)$, $j \in J'_v$, $v = 1(1)s$, as *indeterminates* t_λ, $\lambda \in \overline{J}''$, and the r equations $x^j p_v(x) = 0$ as a *linear* system in the t_λ. Equation (10.3) tells us that this linear system permits a unique solution for the $R - m$ monomials $t_\lambda \notin \mathcal{N}$ in terms of the m monomials $t_\lambda \in \mathcal{N}$. There must be at least $R - m$ equations in the system to make this possible, i.e. we must have $r + m \geq R$. Naturally, $R - m$ is generally greater than $k = |B[\mathcal{N}]|$.

This interpretation of (10.3) shows that the following holds: By the formation of sufficiently many *suitable* monomial multiples of the p_v, it must be possible to reach a situation where the number r of multiples plus m equals or exceeds the total number R of monomials involved. With the interpretation of these multiples as a linear system in the monomials (as above), standard elimination procedures from linear algebra must generate the coefficients of a border basis or (equivalently) the elements of the multiplication matrices of $\mathcal{R}[\langle P \rangle]$ with respect to the normal set basis $\mathcal{N}$.

Example 10.4, continued: When we multiply each p_ν with x_1, x_2 (and 1), we have a system of $r = 6$ polynomials $\{x_1 p_1, x_1 p_2, x_2 p_2, x_2 p_2, p_1, p_2\}$ with a total of 10 monomials so that $r + m = R$. The matrix of the associated linear system is

$$
\begin{array}{c}
\\ \\
x_1 p_1 \\
x_1 p_2 \\
x_2 p_1 \\
x_2 p_2 \\
p_1 \\
p_2
\end{array}
\begin{array}{cccccccccc}
x_1^3 & x_1^2 x_2 & x_1 x_2^2 & x_2^3 & x_1^2 & x_2^2 & x_1 x_2 & x_1 & x_2 & 1 \\
\\
\alpha_{20}^{(1)} & \alpha_{11}^{(1)} & \alpha_{02}^{(1)} & & \alpha_{10}^{(1)} & & \alpha_{01}^{(1)} & \alpha_{00}^{(1)} & & \\
\alpha_{20}^{(2)} & \alpha_{11}^{(2)} & \alpha_{02}^{(2)} & & \alpha_{10}^{(2)} & & \alpha_{01}^{(2)} & \alpha_{00}^{(2)} & & \\
& \alpha_{20}^{(1)} & \alpha_{11}^{(1)} & \alpha_{02}^{(1)} & & \alpha_{01}^{(1)} & \alpha_{10}^{(1)} & & \alpha_{00}^{(1)} & \\
& \alpha_{20}^{(2)} & \alpha_{11}^{(2)} & \alpha_{02}^{(2)} & & \alpha_{01}^{(2)} & \alpha_{10}^{(2)} & & \alpha_{00}^{(2)} & \\
& & & & \alpha_{20}^{(1)} & \alpha_{02}^{(1)} & \alpha_{11}^{(1)} & \alpha_{10}^{(1)} & \alpha_{01}^{(1)} & \alpha_{00}^{(1)} \\
& & & & \alpha_{20}^{(2)} & \alpha_{02}^{(2)} & \alpha_{02}^{(2)} & \alpha_{10}^{(2)} & \alpha_{01}^{(2)} & \alpha_{00}^{(2)}
\end{array}
$$

Generically, this system admits any one of the normal sets $\{1, x_1, x_2, x_1 x_2\}$, $\{1, x_1, x_2, x_1^2\}$, $\{1, x_1, x_2, x_2^2\}$ by "solving" for the remaining monomials in terms of those in $\mathcal{N}$. Since $k = 4$ in each case, there are always two extra leading monomials, not needed for the border basis or the multiplication matrices. $\quad\square$

Example 10.5, continued: Compare Figure 10.1. Multiplication by x and y, resp., of p_1, p_2 each generates the border basis elements with $\mathcal{N}$-leading terms $x^4 y^2$, $x^3 y^3$ and $x^2 y^4$, $x y^5$, respectively. The monomials x^5, $x^5 y$ and y^6, resp., can only be generated from p_2 by multiplication with x^2, $x^2 y$ and from p_1 by multiplication with y^4, respectively. But at the same time, we must employ a multiple of the other polynomial to cancel the $\mathcal{N}$-leading monomial and, possibly, other newly generated monomials outside $\mathcal{N}$:

$x^2 p_2$ contains x^5 but has the $\mathcal{N}$-leading monomial $x^3 y^4$; therefore we must combine it with $y^2 p_1$ which has the same $\mathcal{N}$-leading monomial $x^3 y^4$. Analogously, we must combine $x^2 y\, p_2$ with $y^3 p_1$ to obtain $x^5 y$, and $y^4 p_1$ with $x^2 y^2 p_2$ and with $x^2 p_1$ to obtain y^6. Thus, for p_1, we have $J_1' = \{1, x, x^2, y, y^2, y^3, y^4\}$ and for p_2, $J_2' = \{1, y, x, x^2, x^2 y, x^2 y^2\}$ with $\overline{J}_2' = \{1, x, x^2, y, xy, x^2 y, y^2, xy^2, x^2 y^2\}$. Thus, $r_1 = 7$, $r_2 = 9$ for $r = 16$ and, as is easily found, $R = 34 = 16 + 18$.

From the 16 "linear" equations $x^{j_{\nu 1}} y^{j_{\nu 2}} p_\nu(x, y) = 0$, $x^{j_{\nu 1}} y^{j_{\nu 2}} \in \overline{J}_\nu'$, with the monomials as indeterminates, we obtain the 9 border monomials (plus 7 further monomials outside $B[\mathcal{N}]$) in terms of the monomials in $\mathcal{N}$, i.e. the nontrivial elements of the two 18×18 multiplication matrices of $\langle P \rangle$ for $\mathcal{N}$.

The whole procedure becomes much simpler when it is carried out in a *recursive* fashion: Let $bb_j := x^j + \ldots, x^j \in B[\mathcal{N}]$. Then b_{32}, b_{14} are p_1, p_2, and $bb_{42}, bb_{33}, bb_{24}, bb_{15}$ are x or y times the bb_{32} or bb_{14}. x^5 for bb_{50} is obtained from a linear combination of $x\, bb_{24}$ with $y\, bb_{33}$ (for the removal of a surplus monomial); bb_{51} is $y\, bb_{50}$. y^6 for bb_{06} requires $y\, bb_{51}$, $x\, bb_{42}$ and bb_{14}. Thus, only 13 polynomials are actually formed and only 2 linear systems of 2 and 3 equations, resp., are solved. $\quad\square$

All algorithms for the determination of a computationally useful basis of $\langle P \rangle$ or (equivalently) for the determination of a normal set and multiplication matrices for $\mathcal{R}[\langle p \rangle]$ may be interpreted as a variant of this principal approach; this includes all kinds of GB-algorithms. Like in the last paragraph of Example 10.5 above, the complete matrix for the monomials is never formed, but it could be written down *after* the completion of the algorithm for a particular system if one wanted to see it. The central difficulty is the selection of those multiples of the p_ν which have to appear in (10.3) or an analogous representation. This selection cannot be performed a priori but only *recursively*; the various algorithmic approaches are distinguished by how they do this recursive selection and how they use the polynomials formed intermediately.

The fact that—a posteriori—the accumulated operations of these algorithms may be interpreted as a standard elimination process in a huge matrix makes it evident that, computationally, only *rational* operations on the original coefficients in the polynomial system P are employed. This explains why it is possible to run these algorithms in *rational arithmetic* if the coefficients of P are specified as rational numbers; in this case, the exact values of the elements of the multiplication matrices, or of the coefficients of the border basis elements, are obtained as rational numbers though they may have excessively large numerators and denominators. On the other hand, it suggests immediately that it must be possible to run these algorithms in floating-point arithmetic if some suitable modifications are made and precautions taken.

10.2.2 Determination of a Normal Set for a Complete Intersection

An important aspect of a GB-algorithm based on the scheme of Algorithm 10.1 is the following: It does not assume any a priori information about the system P, and it works for an arbitrary dimension of $\langle P \rangle$. The existence of a *finite* normal set is not assumed; thus, generally, the reducibility to 0 of a number of S-polynomials must be *explicitly verified* to confirm condition 2 of Theorem 8.34. Numerically, this is a critical task; cf. section 10.3.1.

On the other hand, if $P \in (\mathcal{P}^s)^s$ is a complete intersection system, we *know* that there are closed monomial sets $\mathcal{N} \in \mathcal{T}^s(m)$, $m = \mathrm{BKK}(P)$, such that $\mathcal{R}[\langle P \rangle]$ can be represented as span $\mathcal{N}$; with such a feasible normal set, the computationally appropriate representation of $\langle P \rangle$ is the normal set representation $(\mathcal{N}, \mathcal{B}_\mathcal{N})$, with the border basis $\mathcal{B}_\mathcal{N}$. The zero set of P can be directly obtained from $(\mathcal{N}, \mathcal{B}_\mathcal{N})$. Therefore, it appears natural to try to determine a feasible normal set for $P = \{p_\nu, \ \nu = 1(1)s\}$ a priori.

Let us consider which properties a candidate $\mathcal{N}^{(0)}$ for a suitable normal set should satisfy, besides the natural conditions $|\mathcal{N}^{(0)}| = m$ and $p_\nu \notin$ span $\mathcal{N}$, $\nu = 1(1)s$. As the initial stage of the computational determination of the $\mathcal{N}^{(0)}$-border basis $\mathcal{B}_{\mathcal{N}^{(0)}}$, we will always perform an autoreduction (like in Algorithm 10.1) which transforms P into a system P_0 with $\langle P_0 \rangle = \langle P \rangle$. We would wish that, like in section 10.2.1, the s polynomials $p_{0\nu}$ of P_0 are directly elements of $\mathcal{B}_{\mathcal{N}^{(0)}}$. Let $p_{0\nu} =: x^{j_\nu} - \mathrm{tail}(p_{0\nu})$ where none of the x^{j_ν} divides a monomial in $p_{0\nu'}$, $\nu' \neq \nu$; then we require

$$ x^{j_\nu} \in B[\mathcal{N}^{(0)}], \qquad \mathrm{tail}(p_{0\nu}) \in \text{span } \mathcal{N}^{(0)}, \qquad \nu = 1(1)s . \tag{10.4} $$

If the closure of the union of the supports of the tail polynomials is smaller than m, further monomials must be appended to $\mathcal{N}^{(0)}$ to reach the necessary size.

These requirements rarely distinguish a monomial set from $\mathcal{T}^s(m)$ uniquely. The remaining liberty may be used to obtain a normal set with some additional properties. For example, one may choose $\mathcal{N}^{(0)}$ such that $\mathcal{N}^{(0)}$-leading becomes $\prec$-leading for a particular term order; then the associated border basis may become the Groebner basis of P for that term order. Without a term order, one may rather wish to avoid high degrees in the normal set.

Example 10.6: Consider a system P in 3 variables of a dense cubic polynomial p_1 and two dense quadratic polynomials p_2, p_3. We have $m = 12$; thus we must choose a closed set $\mathcal{N}$ of 12 monomials in $\mathbb{N}_0^3$ such that one monomial $x^{j^{(v)}}$ of each p_{0v} is outside $\mathcal{N}$ (and the $x^{j^{(v)}}$ are disjoint). A natural choice is $\mathcal{N} = \{x_1^{j_1} x_2^{j_2} x_3^{j_3}$, with $j_1, j_2 \leq 1,\ j_3 \leq 2\}$, and $x^{j^{(1)}} = x_1^2$, $x^{j^{(2)}} = x_2^2$, $x^{j^{(3)}} = x_3^3$; cf. Figure 10.2-1.

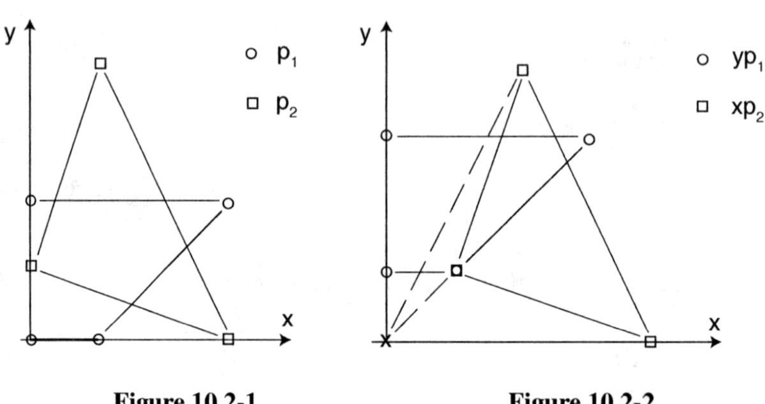

Figure 10.2-1   **Figure 10.2-2**

In the autoreduction phase, the x_2^2-term of p_2 and the x_1^2-term of p_3 are removed and the remaining 3rd degree monomials of p_1 outside $\mathcal{N}$ are removed by reduction with p_{02} and p_{03}; now the p_{0v} and $\mathcal{N}$ satisfy (10.4). The union of the supports of the tails has only 11 elements; but we had attached the further monomial $x_1 x_2 x_3^2$ to $\mathcal{N}$ from the beginning. Thus $\mathcal{N}$ is a valid candidate; it will turn out that it is feasible and supports a border basis.

The strong symmetry of this normal set prevents it from supporting a Groebner basis for any term order. In an attempt to comply with, say, `tdeg(x1,x2,x3)` we can choose the two $\prec$-highest monomials in p_2, p_3 for the autoreduction which makes x_1^2 and $x_1 x_2$ the leading monomials of p_{02} and p_{03}; this permits the removal of their multiples from p_1 and makes x_2^3 the $\prec$-leading monomial of p_{01}; cf. Figure 10.2-2. Now, the union of the supports of the tails has precisely 12 elements, which determine our candidate normal set $\mathcal{N}^{(0)}$. In the construction process of the border basis, it will turn out that $\mathcal{N}^{(0)}$ is not feasible but has to be modified in one position.

For both autoreduced systems P_0, it is obvious that $\langle P \rangle = \langle P_0 \rangle$: In both cases, p_2, p_3 are scalar linear combinations of p_{02}, p_{03}, and $p_1 = p_{01} + q_2 p_{02} + q_3 p_{03}$. Thus, each of the 12 zeros of P_0 is a zero of P which cannot have more than 12 zeros.   □

Obviously, one should begin the autoreduction with the polynomial(s) of lowest degree because they represent the strongest restriction on the choice of the normal set. Without a term order, the candidate normal set can generally be enclosed within an s-dimensional rectangle determined by monomials of highest degree from the p_v. If $\mathrm{BKK}(P) < \prod_v d_v$, the autoreduction has to be watched more carefully so that the right monomials are removed.

Concerning the requirement $\langle P \rangle = \langle P_0 \rangle$, we have

Proposition 10.1. Consider a transformation from the polynomial set $\{p_\nu, \ \nu = 1(1)s\}$ to the set $\{p_{0\nu}, \ \nu = 1(1)s\}$ of the form

$$q_1 := p_1, \quad q_2 := p_2 - c_{21}q_1, \quad \ldots, \quad q_s := p_s - c_{s1}q_1 - .. - c_{s,s-1}q_{s-1},$$

$$p_{01} := q_1 - c_{12}q_2 - .. - c_{1s}q_s, \quad p_{02} := q_2 - c_{23}q_3 - .. - c_{2s}q_s, \quad \ldots, \quad p_{0s} := q_s,$$
$$\tag{10.5}$$

where the $c_{\sigma\nu}$ are polynomials. Then the p_ν are polynomial combinations of the $p_{0\nu}$ and vice versa.

Proof: Equation (10.5) may be written as

$$
\begin{pmatrix} p_{01} \\ p_{02} \\ \vdots \\ p_{0s} \end{pmatrix}
=
\begin{pmatrix} 1 & c_{12} & .. & c_{1s} \\ 0 & 1 & .. & c_{2s} \\ & & 1 & .. \\ 0 & & & 1 \end{pmatrix}
\begin{pmatrix} 1 & & & 0 \\ c_{21} & 1 & & \\ & .. & 1 & 0 \\ c_{s1} & .. & .. & 1 \end{pmatrix}^{-1}
\begin{pmatrix} p_1 \\ p_2 \\ \vdots \\ p \end{pmatrix}.
$$

Elaboration of the matrix product and its inverse confirms the assertion. $\quad\square$

Corollary 10.2. An autoreduction of the form (10.5) implies $\langle\{p_\nu\}\rangle = \langle\{p_{0\nu}\}\rangle$.

Commonly, autoreduction procedures may be written in the form (10.5); cf. our examples.

10.2.3 Basis Computation with Specified Normal Set

Consider a complete intersection system $P \in (\mathcal{P}^s)^s$, with $\mathrm{BKK}(P) = m$, and assume that we have chosen a tentative normal set $\mathcal{N} \subset T^s(m)$ and performed an associated autoreduction of P into P_0 which satisfies (10.4) and (10.5). Assume at first that $\mathcal{N}$, with normal set vector $\mathbf{b}$, is feasible for $\langle P_0 \rangle = \langle P \rangle$, i.e. that span $\mathcal{N} = \mathcal{R}[\langle P_0 \rangle]$; cf. section 2.5.1. The determination of the border basis $\mathcal{B}_\mathcal{N}[\langle P_0 \rangle]$ may now be viewed as the following task:

$$\text{Find } \mathrm{NF}_{\langle P_0 \rangle}[x^j] \in \text{span } \mathcal{N} \qquad \text{for each } x^j \in B[\mathcal{N}]. \tag{10.6}$$

By Theorem 8.25, there exist polynomials $q_{j\nu} \in \mathcal{P}^s$ such that

$$x^j = a_j^T \mathbf{b}(x) + \sum_{\nu=1}^{s} q_{j\nu}(x)\, p_{0\nu}(x), \tag{10.7}$$

with unique vectors $a_j^T \in \mathbb{C}^m$. Assume that we have been able to find a representation (10.7) for each $x^j \in B[\mathcal{N}]$ and consider the set

$$\mathcal{B}_\mathcal{N} := \{\, bb_j(x) := x^j - a_j^T \mathbf{b}(x), \ x^j \in B[\mathcal{N}]\,\}.$$

Theorem 10.3. In the situation just described,

$$\langle \mathcal{B}_\mathcal{N} \rangle = \langle P_0 \rangle = \langle P \rangle.$$

Proof: Each zero of P_0 is a zero of $\langle \mathcal{B}_\mathcal{N} \rangle$ by (10.7) which implies $\mathcal{B}_\mathcal{N} \subset \langle P_0 \rangle$; on the other hand, (10.4) implies $P_0 \subset \langle \mathcal{B}_\mathcal{N} \rangle$. $\quad\square$

Since $\mathcal{B}_\mathcal{N}$ is the $\mathcal{N}$-border basis of $\langle P_0 \rangle$, the multiplication matrices A_σ of $\mathcal{R}[\langle P_0 \rangle]$ w.r.t. $\mathbf{b}$ specified by the a_j^T commute and define the zero structure of $\langle P_0 \rangle = \langle P \rangle$. Thus the completion of the task (10.6) "solves" the system P.

The normal forms of the border set monomials x^j or—equivalently—the border basis polynomials bb_j may be determined recursively along the border web $BW_{\mathcal{N}}$, with the $p_{0\nu} = x^{j^{(\nu)}} - a_{j^{(\nu)}}^T \mathbf{b}$ as initial elements, in the following fashion (cf. sections 8.2.2 and 8.2.3):

Consider an edge $[x^j, x^{j+e_\sigma}]$ of type (i) in $BW_{\mathcal{N}}$ and assume that $\mathrm{NF}[x^j] = a_j^T \mathbf{b}$ is known; then the right-hand side of (8.25) implies

$$(x^{j+e_\sigma} - \mathrm{NF}[x^{j+e_\sigma}]) - x_\sigma (x^j - \mathrm{NF}[x^j]) = -\mathrm{NF}[x^{j+e_\sigma}] + \mathrm{NF}[a_j^T x_\sigma \mathbf{b}] = 0.$$

$x_\sigma \mathbf{b}$ must contain some monomials $x^{j_\mu} \in B[\mathcal{N}]$, so this becomes an equation

$$\sum_\mu \alpha_\mu \, \mathrm{NF}[x^{j_\mu}] = \bar{a}_j \, \mathbf{b}. \tag{10.8}$$

Analogously, an edge $[x^{j+e_{\sigma_1}}, x^{j+e_{\sigma_2}}]$ of type (ii) implies

$$x_{\sigma_2} (x^{j+e_{\sigma_1}} - \mathrm{NF}[x^{j+e_{\sigma_1}}]) - x_{\sigma_1} (x^{j+e_{\sigma_2}} - \mathrm{NF}[x^{j+e_{\sigma_2}}]) = -\mathrm{NF}[a_{j+e_{\sigma_1}} x_{\sigma_2} \mathbf{b} - a_{j+e_{\sigma_2}}^T x_{\sigma_1} \mathbf{b}] = 0$$

and yields a relation (10.8).

Formally, over the set $\mathcal{S}$ of all $\bar{N}$ edges in $BW_{\mathcal{N}}$, this generates a system of $\bar{N}$ linear equations, with numerical coefficients, in the N quantities $\mathrm{NF}[x^j]$, $x^j \in B[\mathcal{N}]$, with right-hand sides in span $\mathcal{N}$. This includes the s "initial" equations $\mathrm{NF}[x^{j^{(\nu)}}] = a_{j^{(\nu)}}^T \mathbf{b}$ from the $p_{0\nu}$, $\nu = 1(1)s$, with their leading monomials $x^{j^{(\nu)}} \in B[\mathcal{N}]$. From section 8.2.3, we know that the $\bar{N} \times (N-s)$-submatrix of the system, without the s columns for the $x^{j^{(\nu)}}$, has rank $N - s$; cf. Proposition 8.13 and the subsequent discussion. Thus, in principle, we can solve this system for the remaining $(N - s)$ $\mathrm{NF}[x^j]$'s and complete the task (10.6).

From a computational point of view, we do not want to form more than $N - s$ equations beyond the initial s ones from the $p_{0\nu}$. Moreover, from the local structure of $BW_{\mathcal{N}}$, we would hope that we find a group of N_1 equations (10.8) which contains only N_1 unknown $\mathrm{NF}[x^j]$'s and which can be solved for these quantities. This may permit the selection of another N_2 equations for N_2 unknown $\mathrm{NF}[x^j]$'s, etc. If we can find all $\mathrm{NF}[x^j]$'s in this recursive fashion, we have also managed to use only the minimal number of $N - s$ equations from the large overdetermined system.

This *recursive* solution procedure is also necessary because, for (10.8), we have assumed that we *know* the NF's of x^j, or $x^{j+e_{\sigma_1}}, x^{j+e_{\sigma_1}}$, respectively. This is true in the first set of edges which can be chosen as issuing from the l.m. $x^{j^{(\nu)}}$ of the $p_{0\nu}$. After that, we must have further computed normal forms ready to generate new equations (10.8) with known right-hand sides.

To proceed in the recursion, one can also consider a set of $N_\lambda' < N_\lambda$ equations for a group of N_λ normal forms. One can then solve for N_λ' of them in terms of the remaining $N_\lambda - N_\lambda'$ normal forms and polynomials in span $\mathcal{N}$. The remaining quantities may become a part of another block of quantities for which equations can be selected:

Assume that we have obtained a "partial solution" for some $\mathrm{NF}[x^j]$ of the form

$$\mathrm{NF}[x^j] = \beta_{jk_1} \mathrm{NF}[x^{k_1}] + \beta_{jk_2} \mathrm{NF}[x^{k_2}] + \hat{a}_j^T \mathbf{b}.$$

For the edge $[x^j, x^{j+e_\sigma}]$, we can then formulate the equation

$$\mathrm{NF}[x^{j+e_\sigma}] - \mathrm{NF}[x_\sigma x^j] = \mathrm{NF}[x^{j+e_\sigma}] - \beta_{jk_1} \mathrm{NF}[x^{k_1+e_\sigma}] - \beta_{jk_2} \mathrm{NF}[x^{k_2+e_\sigma}] - \mathrm{NF}[\hat{a}_j^T x_\sigma \mathbf{b}] = 0,$$

and analogously for type (ii) edges.

But how do we find suitable subsystems of the original large system without forming the redundant equations. It appears that the minimal sets $\mathcal{S}_0$ of edges introduced in section 8.2.3 provide subsystems of $N - s$ equations of rank $N - s$ which permit the determination of all $\mathrm{NF}[x^j]$. Also the splitting into blocks appears to occur in a natural way if the edges in $\mathcal{S}_0$ are treated in a stagewise fashion as they issue from the leading monomials $x^{j^{(\nu)}}$ of the $p_{0\nu}$. More insight into this situation would be highly desirable so that algorithms along this pattern could be based on a firm basis.

So far, we have assumed that we have chosen a candidate normal set which is feasible, which is confirmed when it permits a solution of task (10.6). What happens if our normal set $\mathcal{N}^{(0)}$ (we had dropped the superscript in the previous analysis) is not feasible? Since we have the correct number m of elements, infeasibility means that there is some monomial in $\mathcal{N}$ which should not be there, in exchange for a monomial not in $\mathcal{N}$. During the blockwise solution of the linear system for the $\mathrm{NF}[x^j]$, this can become evident in the following way:

The matrix of a block of N_λ equations for N_λ unknown $\mathrm{NF}[x^j]$ may turn out as rank-deficient; in this case, elimination in the equations of the block leaves one equation *without* an $\mathrm{NF}[x^j]$ term. If the right-hand side also vanishes, we have simply included an equation which is dependent on the others, e.g., by using the relation of an edge which closes a circle of other edges which were used; cf. Proposition 8.15. Otherwise we have obtained an equation in span $\mathcal{N}$ which is in $\langle P_0 \rangle$. This indicates the necessity of an exchange between $\mathcal{N}$ and $B[\mathcal{N}]$. The monomials to be exchanged may not be uniquely defined by the situation; one will attempt to choose them such that the least overall modification arises. Compare, e.g., Example 10.7 below. In numerical computation, one will rather experience a *near-singularity*. Now, the relative sizes of quantities will have to be taken into account; we will return to this in section 10.3.

Example 10.6, continued: In the situation of Example 10.6, with normal set $\mathcal{N}$, we want to consider a set of $N - s = 13$ edges which should yield independent relations (10.8) according to section 8.2.3; cf. Figure 10.3. Due to the smaller supports of p_{02} and p_{03}, we begin with the edges $[x^{200}, x^{201}]$ and $[x^{020}, x^{021}]$ where $a_{200}^T x_3 \mathbf{b}$ and $a_{020}^T x_3 \mathbf{b}$ remain fully in span $\mathcal{N}$ so that we obtain $\mathrm{NF}[x^{201}]$ and $\mathrm{NF}[x^{021}]$ explicitly, without computation.

The edges $[x^{200}, x^{210}]$ and $[x^{020}, x^{120}]$ introduce $x_2 \mathbf{b}$ and $x_1 \mathbf{b}$. With $\mathrm{NF}[x^{201}]$ and $\mathrm{NF}[x^{021}]$ known, we obtain two equations for $\mathrm{NF}[x^{210}]$ and $\mathrm{NF}[x^{120}]$ which may be solved immediately.

When we move further along edges in the x_3-direction, the shifted support of p_{02}, p_{03} meets border monomials adjacent to x^{003} of p_{01}. We may now combine the relations from the 6 edges $[x^{201}, x^{202}]$, $[x^{021}, x^{022}]$, $[x^{210}, x^{211}]$, $[x^{120}, x^{121}]$, $[x^{003}, x^{103}]$, $[x^{003}, x^{013}]$, to obtain a block of 6 linear equations for the normal forms of x^{202}, x^{022}, x^{211}, x^{121}, x^{103}, x^{013}; their right-hand sides make use of the previously computed normal forms of the partners in the edges.

The remaining marked edges in Figure 10.3 yield a block of 3 equations for the normal forms of the remaining 3 border basis monomials. The employed edges form a subset $\mathcal{S}_0$ if the birder web $BW_\mathcal{N}$ which (with the virtual edges) spans the complete border set $B[\mathcal{N}]$. Thus we have completed the task (10.6) successfully, which proves that we have obtained the border basis $\mathcal{B}_\mathcal{N}$ for the ideal $\langle P \rangle$. From the coefficients of the border basis monomials, we can form the A_σ matrices and find the 12 zeros of P from their joint eigenvectors.

Note that we have exclusively used syzygy relations of type (i) in this computational determination of a basis for $\langle P \rangle$, i.e. not a single S-polynomial proper has been formed in the

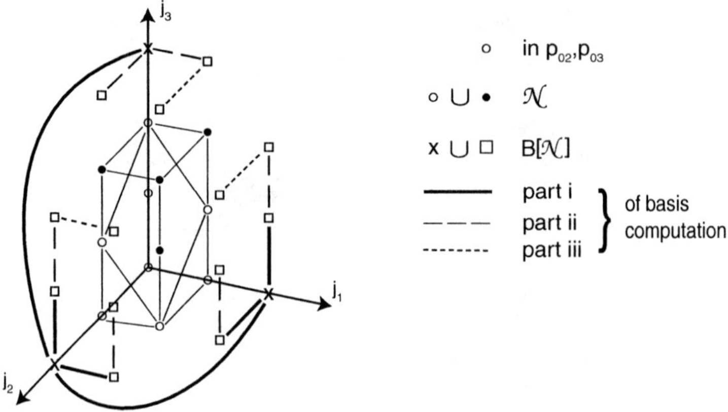

Figure 10.3.

complete basis computation.

Let us now use the candidate normal set $\mathcal{N}^{(0)}$ of Example 10.6; it appears again reasonable to begin with the x_3-shifts of x^{200} and x^{110}, the leading monomials of p_{02} and p_{03}. As before, this yields $\mathrm{NF}[x^{201}]$ and $\mathrm{NF}[x^{111}]$ without computation. Also, when we shift x^{110} by x_2, the only new monomial outside $\mathcal{N}$ in $a_{110}^T x_2 \mathbf{b}$ is x^{030} whose normal form is known from p_{03} so that we obtain $\mathrm{NF}[x^{120}]$ directly; cf. Figure 10.2-2.

When we make one further use of the two quadratic polynomials and form the S-polynomial associated with the type (ii) edge $[x^{200}, x^{110}]$, we have to form the x_1-shift of p_{02} and the x_2-shift of p_{01}. This takes a number of monomials out of $\mathcal{N}^{(0)}$, but all of them coincide with monomials in $B[\mathcal{N}^{(0)}]$ whose normal forms we already know. Thus, we have generated a nonvanishing polynomial in span $\mathcal{N}$ with a vanishing normal form, which is a contradiction.

To relieve the situation, we must remove one element from $\mathcal{N}^{(0)}$; the most natural candidate is x^{021}. It had previously figured in the normal forms of x^{201}, x^{111}, x^{120}; therefore, we must recompute the normal forms of all 4 monomials from the 4 relations used so far, which gives a unique result. The removal of x^{021} from $\mathcal{N}^{(0)}$ also removes x^{121} and x^{031} from the border set $B[\mathcal{N}^{(1)}]$ of the new normal set $\mathcal{N}^{(1)}$. It remains to append a new element to $\mathcal{N}^{(1)}$ to bring it back to 12 elements. Figure 10.4 shows the situation after the removal of x^{021}.

We can now consider the equations (10.8) from the following 5 edges (cf. Figure 10.4): $[x^{030}, x^{120}]$, $[x^{030}, x^{021}]$, $[x^{201}, x^{202}]$, $[x^{111}, x^{112}]$, $[x^{021}, x^{022}]$. They imply all the 6 remaining border set elements of the *old* normal set $\mathcal{N}^{(0)}$, but we must insert one of these elements into the normal set $\mathcal{N}^{(1)}$. Since we have aimed at compliance with the term order `tdeg(x1,x2,x3)` in the choice of $\mathcal{N}^{(0)}$, we should now move the $\prec$-lowest element of $B[\mathcal{N}^{(0)}]$ into $\mathcal{N}^{(1)}$, i.e. x^{004}; this creates 3 new monomials in $B[\mathcal{N}^{(1)}]$: x^{104}, x^{014}, x^{005}; cf. Figure 10.4. A block of 3 equations involving the normal forms of these monomials is easily found, which completes the computation. We have also generated a proper border web subset $\mathcal{S}_0^{(1)}$, connecting the elements of $B[\mathcal{N}^{(1)}]$.

A comparison reveals that $\mathcal{N}^{(1)}$ is indeed the normal set of the Groebner basis $\mathcal{G}_\prec$; thus

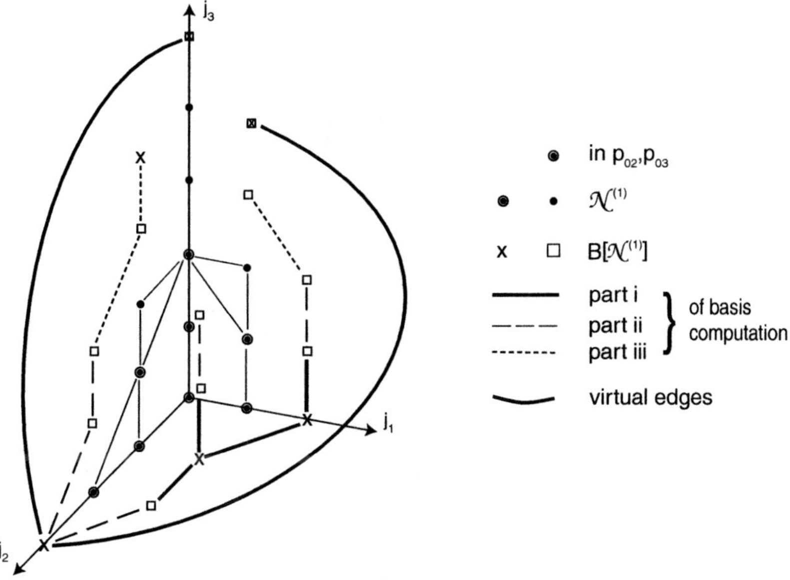

Figure 10.4.

we have actually computed the Groebner basis for this term order. A GB-algorithm can be led along the same computations which have to be performed in our normal set controlled approach; in a GB-algorithm, however, a reduction-to-0 test for several S-polynomials must be performed at the end. This test is *not necessary* in our approach since the a priori information $m = 12$ has been used. In the classical GB approach, $m = 12$ appears as a *result* of the algorithm. $\square$

It appears that the following holds:

Conjecture. For a normal set which is feasible for the regular system P_0 and satisfies (10.4), the recursive computation of the $\mathrm{NF}_{\langle P_0 \rangle}[x^j] \in \mathrm{span}\,\mathcal{N}$, for all $x^j \in B[\mathcal{N}]$, from *linear* equations of the type (10.8) is always possible.

Even if this is generally true, it requires the a priori knowledge of a feasible normal set (or a candidate very close to a feasible one) *and* the selection of appropriate subsystems for the recursion. More insight into this (essentially combinatorial) problem would be highly desirable.

If P is a (near-)BKK-deficient system with fewer zeros than BKK(P), we will have to *reduce* the normal set at some point in the basis computation. We will only show an example for this case which has been discussed in section 9.5.2.

Example 10.7: Consider the system P of Example 9.19 with

$$p_1 = x_1 x_2 - x_1 - 1\,, \quad p_2 = 4 x_1^2 x_2 + 4 x_2^3 - 4 x_1^2 - 4 x_2^2 - 25 x_2 + 25\,.$$

Initial autoreduction yields $p_{01} = p_1$ and $p_{02} = 4 x_2^3 - 4 x_2^2 + 4 x_1 - 25 x_2 + 25$. Together with BKK($P$) = 5 this makes $\mathcal{N} = \{1, x_1, x_2, x_1^2, x_2^2\}$ the only choice.

At first, we form the $\mathcal{N}$-border basis elements with leading monomials $x_1^2 x_2$ and $x_1 x_2^2$ by shifts of p_{01} and reduction:

$$bb_3 = x_1^2 x_2 - x_1^2 - x_1, \qquad bb_4 = x_1 x_2^2 - x_1 - x_2 - 1.$$

Since there is no direct route to a polynomial with $\mathcal{N}$-leading monomial x_1^3, we form $S[\bar{p}_2, bb_4]$ which reduces to a polynomial $4x_1^2 + 4x_2^2 - 25$ *inside* span $\mathcal{N}$! Thus we *must* delete x_1^2 from the normal set which causes no problems with the original autoreduction but reduces bb_3 (which is no longer necessary) to p_{02}. Instead, the reduced S-polynomial becomes the new basis element which confirms $\mathcal{N}_0 = \{1, x_1, x_2, x_2^2\}$ as the correct normal set and the system P_0 as BKK-deficient; cf. section 9.5.2. $\square$

Exercises

(1) Consider once more the system P of Example 10.6. To find a complying normal set and autoreduction, we must at first choose two different monomials from T_2^2 as $\mathcal{N}$-leading monomials x^{j_2}, x^{j_3} of p_{02}, p_{03}; this eliminates many monomials from p_1 but leaves a number of options for x^{j_1}.

(a) Except for symmetries, there are 4 different choices for the pairs (x^{j_2}, x^{j_3}). For each of these, consider the remaining monomials of p_1 and the choices for x^{j_1}; some of these may still require the choice of a monomial to bring $|\mathcal{N}|$ to 12. Try to get a view of this large variety of potential normal sets for $\langle P \rangle$.

(b) For several of these, find the border set $B[\mathcal{N}]$ and the border web $BW_{\mathcal{N}}$; compare the values of $N = |B[\mathcal{N}]|$. Try to find an initial selection of edges which yields a block of N_1 equations for the normal forms of only N_1 border monomials (beyond the x^{j_v}); cf. the continuation of Example 10.6 in section 10.2.3. Which properties of $\mathcal{N}$ are helpful or not helpful for that task?

(c) For one choice other than a symmetric copy of the two cases in Exampe 10.6, determine the outline of the computation of the complete border basis.

(d) In retrospect, why would you consider the choice of $\mathcal{N}$ in Example 10.6 the most favorable from a computational point of view?

2. Consider the system P (`cyclic5`) specified by

$$p_1 = x_1 + x_2 + x_3 + x_4 + x_5, \qquad p_2 = x_1 x_2 + x_2 x_3 + x_3 x_4 + x_4 x_5 + x_5 x_1,$$

$$p_3 = x_1 x_2 x_3 + x_2 x_3 x_4 + x_3 x_4 x_5 + x_4 x_5 x_1 + x_5 x_1 x_2,$$

$$p_4 = x_1 x_2 x_3 x_4 + x_2 x_3 x_4 x_5 + x_3 x_4 x_5 x_1 + x_4 x_5 x_1 x_2 + x_5 x_1 x_2 x_3, \qquad p_5 = x_1 x_2 x_3 x_4 x_5 - 1.$$

It is well known that $\mathrm{BKK}(P) = 70 < m_{B\acute{e}zout} = 120$. We want to find a suitable normal set for the computation of a basis of $\langle P \rangle$.

(a) The task becomes accessible when we eliminate x_1 from P and choose x_2^2 as $\mathcal{N}$-leading monomial of p_{02}. Thus $\mathcal{N}$ can only live in the two 3-dimensional sub"manifolds" $x_2 = 0$, $x_2 = 1$, of the $\mathbb{N}_0^4$ of the exponents of the (x_2, x_3, x_4, x_5)-monomials. Now we proceed according to the term order `tdeg(x2,x3,x4,x5)`: We use $p_{02} = x_2^2 + \ldots$ to eliminate monomials from p_3, p_4, p_5, and choose the highest monomial in the new p_3 as x^{j_3}; with this p_{03}, we eliminate further in p_4, p_5 and choose x^{j_4} in the same fashion, and finally arrive at p_{05}. Now, we have a relatively simple autoreduced system $P_0 = \{p_{0v}\}$.

(b) We form the union of the supports of the tails of the $p_{0\nu}$ and complete it into a closed set. How many monomials have you obtained? Even with some natural restriction, there are too many possibilities left to place the remaining normal set monomials.

(c) Compute the `tdeg(x1,x2,x3,x4x,5)` Groebner basis by software and compare its normal set.

10.3 Numerical Aspects of Basis Computation

In real-life polynomial systems, we must expect some data to be empirical and we want to use floating-point arithmetic in the basis computation. In this section, we will consider some aspects of this situation.

10.3.1 Two Fundamental Difficulties

For a regular system $P = \{p_\nu, \; \nu = 1(1)s\}$, we consider the computation of some border basis of $\langle P \rangle$ from the p_ν, either with a GB-algorithm for a specified term order or with the use of a feasible normal set. As we have seen, the computational path of such algorithms can always be represented, explicitly or implicitly, in terms of successive operations on a sequence of polynomials, beginning with the p_ν:

- multiplication of a polynomial by a monomial,

- scalar linear combination of two or more polynomials.

The first operation is *symbolic* and must be implemented as such in any numerical algorithm, exact or approximate. In the second operation, the coefficients of the linear combination are chosen to effect the cancellation of a term in the resulting polynomial, often the leading term (whatever that means). The deletion of *that* term is also done symbolically, like in numerical linear algebra; the remaining coefficients in the resulting polynomial are *computed*. In particular, it may happen, that other coefficients of the resulting polynomial are also nearly annihilated. This near-cancellation arises as a *numerical* result, hence it combines algebraic effects with perturbations from previous computational errors and the current computation. Two principal cases are of interest:

(1) The value of an *individual* coefficient becomes "tiny," i.e. of the order of the round-off level in that operation, by a severe cancellation of leading digits so that this value carries very little (or even no) algebraic information. If this coefficient is only used as a subordinate data value in the subsequent numerical operations, the small absolute error in its value may cause further errors in other values, but generally, this will not cause great harm.

However, it may happen that this coefficient should ordinarily be used as a pivot in a subsequent elimination. In section 10.1.2, we have seen that the use of a (relatively) small pivot is dangerous even when its value is exact because it may lead to excessive ill-conditioning in the computed representation, cf. Example 10.3. If its accuracy is restricted to a few digits, its use may propagate that low relative accuracy to large parts of the subsequent computation, with potentially catastrophic consequences.

(2) *All* coefficients in the resulting polynomial are tiny in the sense just discussed, though possibly not uniformly so. This suggests that the polynomial has actually been reduced to 0 by

the elimination step just performed though the size of some coefficient(s) raises doubts about whether they are indeed of round-off level. Such a situation is exceedingly dangerous:

If the reduction to 0 is correct, the computed polynomial with tiny coefficients is *spurious*, i.e. a numerical artifact. Hence, its presence changes the ideal whose basis is to be computed. If the spurious polynomial is inconsistent with $\langle P \rangle$, which is generally the case, the further computation will generate the ideal $\langle 1 \rangle$ without any zeros; otherwise, a few zeros may be lost. These potential effects may become even worse when some coefficients of the spurious polynomial are cancelled and others retained, as it may happen when a fixed threshold is used for the cancellation.

These considerations suggest that, in both cases, one should be quite generous in calling coefficients or a set of coefficients tiny, with the respective consequences. For the case of an individual tiny coefficient, we have discussed possible consequences in section 10.1.2, including the immediate cancellation of that coefficient which can generally be interpreted, a posteriori, as a perturbation of the initial coefficients in the p_ν. With an apparently tiny polynomial, the only meaningful action is its complete cancellation. What may happen when it is dropped although it would not have been reduced to 0 in an exact computation?

Then, we have lost a polynomial which should have been in $\langle P \rangle$. In a GB-algorithm, the computation may lead to the basis of a smaller ideal, with additional *spurious zeros* in the zero set or even a zero manifold. A comparison with the size of BKK(P) should reveal their presence; their larger backward error will permit their identification. In the computation of a border basis with a specified normal set, a conflict will arise in the computation.

In a nontrivial basis computation, it may be difficult to discern these events from other numerical effects although principal difficulties in numerical basis computations can always be retraced to the above situations. We discuss two simple examples to show the occurrence of both situations.

Example 10.8: We consider the approximate system $\widetilde{P}$ of Example 10.3, but now we perform the operations for the determination of a `plex(y,x)` basis in 10-digit decimal floating-point arithmetic.

There is no swell of digits now; but, again, the coefficient of y in $\tilde{g}_3$ does not vanish but remains $O(10^{-7})$ relative to the other coefficients in the polynomial. It is formed by cancellation of 7–8 leading digits so that only 3 meaningful digits remain, 2 of which are correct. When $\tilde{g}_3$ is accepted as basis polynomial, its use in the reduction of $\tilde{g}_2$ generates a $\tilde{g}_4$ with only 2–3 meaningful digits, which makes it useless for the determination of the two densely clustered zeros; cf. Example 10.3. As we have seen there, one needs a working precision beyond 30 digits to obtain reasonably correct values for the zeros of $\widetilde{P}$.

On the other hand, when we refuse to use the small l.m. for a reduction and switch to the normal set $\{1, x, x^2, y\}$ by declaring x^3 as the $\mathcal{N}$-leading monomial in $\tilde{g}_3$, we have an approximate GB (extended GB for `plex(y,x)`, ordinary reduced GB for `tdeg(y,x)`) $\{\tilde{g}_1, \tilde{g}_2, \tilde{g}_3\}$ whose coefficients have 7–8 correct decimal digits. With the use of A_y, this permits a computation of all zeros to about 7 digits, like in Example 10.3. $\square$

Example 10.9: Consider the two quadratic polynomials in $\mathcal{P}^2$:

$$P = \begin{cases} p_1(x, y) = 2.34567\,xy - 3.45678\,x + 4.56789\,y - 5.67890\,, \\ p_2(x, y) = 4.32109\,xy + 3.21098\,x - 2.10987\,y - 1.09876\,. \end{cases}$$

To show a situation as in (2) above, we employ the following nonstandard way to find a basis for $\langle P \rangle$: We multiply both p_1 and p_2 with x, y, xy, which generates a total of 8 polynomials with the joint support $\overline{J}'' = \{1, y, x, y^2, xy, x^2, xy^2, x^2y, x^2y^2\}$. As in section 10.2.1, we write the coefficients into a 8×9 matrix B and eliminate.

When we do this symbolically for *indeterminate* coefficients, we find that B has rank 7 which is compatible with $m = \mathrm{BKK}(P) = 2$. In the elimination process (from high degree monomials to low degree ones), the last two rows, with 3 elements in each, become multiples of each other so that an elimination of the leading element of the 8th row by the 7th row annihilates the 8th row completely. The basis, with leading monomials x and y^2, may then be read from the 6th and 7th row of the upper triangular factor.

When the elimination is performed for the numerical coefficients, in 10-digit floating-point arithmetic, the same last elimination step leaves the last two elements in row 8 nonzero, at a level 10^{-8}, with 2 meaningful but *no correct* digits! If we were to take these elements at their face value, we would have a 1-element normal set $\mathcal{N} = \{1\}$ and the only zero, from the last two rows, does not satisfy $P = 0$. When the 8th row is dropped, one obtains a satisfactory result. $\quad\square$

Reduction to zero is always a critical operation in a numerical basis computation; it should therefore be algorithmically avoided in as far as possible. In the example above, the formation of so many polynomials leading to a numerical 8×9 matrix for a solution in terms of $m = 2$ parameters is clearly devious. It shows why the a priori determination of $\mathrm{BKK}(P)$ (which is a symbolic and not a numerical computation) is of principal importance for avoiding the formation of overdetermined linear systems during the basis computation. This is also a goal in exact basis algorithms because it costs computational effort which is essentially wasted. There are two ways to decrease the number of necessary reductions to zero in a GB-algorithm:

- a clever use of the criteria for reducibility to 0, like the ones in Proposition 8.14;

- a clever strategy in the formation and reduction of S-polynomials.

Tricks to this end have been implemented in all of the newer versions of GB-algorithms; they may immediately be used in floating-point adaptations of these algorithms.

In a normal set controlled, term order free basis computation, *no explicit reductions to zero are necessary.* From the point of view of *numerical* basis computation, this may be the most valuable asset of this approach.

10.3.2 Pivoting

In section 10.2.1, we have observed that—a-posteriori—the computation of a normal set representation $(\mathcal{N}, \mathcal{B}_{\mathcal{N}})$ of $\langle \{p_1, .., p_s\} \rangle \subset \mathcal{P}^s$ from the p_ν can be regarded as an elimination process in a large matrix B. The columns of that $r \times R$ matrix correspond to the exponent set $\overline{J}'' \subset \mathbb{N}_0^s$ of the monomials which appear at some point in the elimination, the rows represent the polyomials $x^j p_\nu$, $j \in \overline{J}'_\nu$, which are employed; cf. Example 10.4 continued. The actual computation corresponds to eliminations in a sequence of smaller matrices whose rows are the multiples of polynomials which have been formed in the treatment of previous matrices. But the *recursive formation of linear combinations* of rows in a matrix, with the aim of a systematic generation of 0 entries, is at the bottom of any basis representation, just like in the direct solution of a multivariate system of linear equations.

For linear systems of s equations in s variables, the matrix B is $s \times (s + 1)$; the last column contains the constant terms while the other columns correspond to $x_1, \ldots, x_s$ so that there is no qualitative distinction and no dependence between them. Therefore, row and column permutations in the left-hand $s \times s$ matrix are possible without restrictions. This fact is used in the floating-point implementations of elimination algorithms to keep the procedure numerically stable (cf. section 4.2):

With *partial pivoting*, the columns are processed from left to right, but the row whose multiple is used to cancel elements in a particular column is chosen such that its element in that column is of *maximal modulus* among the eligible elements. It is well known that this simple trick greatly enhances the numerical stability of an LU-decomposition because it keeps the elements in the permuted L-matrix ≤ 1 in modulus; cf. any text in numerical linear algebra.

In the basis computation for a polynomial system, there is a principal difference from the linear case: The columns are not equivalent and independent. The "pivot" element whose multiples are used to cancel other elements in linear row combinations must be an $\mathcal{N}$-leading or the $\prec$-leading monomial of the polynomial represented by the row. This is necessary to avoid the introduction of terms of a higher $\mathcal{N}$-index (cf. Definition 8.3) or term order, resp., through the elimination. Thus, if we wish to be able to process the columns from left to right, we must have ordered the associated monomials properly:

When a term order is used, it is natural to arrange the monomials by descending term order from left to right. But this also assumes that the right-most monomials constitute the associated normal set; otherwise we must expect column interchanges to be necessary. If we *know* and use a feasible normal set, we will arrange all its monomials on the right end and proceed to the left by increasing $\mathcal{N}$-index.

With such an arrangement of the columns in the current $r_\lambda \times R_\lambda$ matrix B_λ of a sequence of such matrices, we may now use partial pivoting, just as in the linear case, in the numerical transformation of the r_λ leftmost columns into a permuted unit matrix, with a marked containment of round-off propagation effects in critical cases.

Example 10.10: For demonstration purposes, consider

$$p_1(x, y) = .02467\, x^3 - .02053\, x^2 y + 2.82741\, y^3 + .68701\, x^2 + 3.51842\, xy^2 + xy + \ldots\,,$$
$$p_2(x, y) = -4.58163\, x^2 + 3.83952\, xy + 2.44073\, y^2 + x + y + \ldots\,,$$

and assume that $\mathcal{N} = \{1, y, x, y^2, xy, xy^2\}$ is a feasible normal set which we wish to use. (This is the normal set for `tdeg (x,y)`.) Then the initial autoreduction phase will employ the matrix

	x^3	$x^2 y$	y^3	x^2	xy^2	xy	$\ldots$
p_1	.02467	$-$.02053	2.82741	.68701	3.51842	1.	$\ldots$
$x\,p_2$	$-$4.58163	3.83952		1.	2.44073	1.	$\ldots$
$y\,p_2$		$-$4.58163	2.44073		3.83952	1.	$\ldots$
p_2				$-$4.58163		3.83952	$\ldots$

In the first 3 columns, no pivoting and partial pivoting yield the pivot positions $\begin{pmatrix} 1 & & \\ & 1 & \\ & & 1 \end{pmatrix}$

and $\begin{pmatrix} & 1 & \\ 1 & & \\ & & 1 \end{pmatrix}$, respectively. In the unpivoted computation, due to two successive small pivots of $O(10^{-2})$, there arise intermediate quantities of $O(10^4)$ which are eventually reduced to $O(1)$ by cancellation of leading digits. In the pivoted computation, all intermediate values remain $O(1)$.

After the autoreduction, we have generated the border basis elements with $\mathcal{N}$-leading monomials y^3 and x^2y. With 10-digit floating-point arithmetic, we expect a loss of 4 digits in the unpivoted computation for some coefficients, which is born out: For example, the coefficient of the xy-term of the x^2y polynomial is (rounded) .2516986 in the unpivoted run and .2516957 in the pivoted one; this value is actually correct within units of 10^{-10}. If such a loss of accuracy occurs at the very beginning of a longer computation, the consequences can be dramatic. $\quad\square$

Assume that it is possible to process all intermediate matrices until the arrival at a full border basis (or GB) with pivoted eliminations and that the generated elements remain on the same order of magnitude as the original data. Then, a backward error analysis of the same nature as it is used in linear algebra shows that one may expect to obtain an approximate border basis in the sense of section 9.2.1. There we have shown how such an approximate representation may be used directly to determine an approximate zero set, and how its small inconsistencies may be removed if necessary. In particular, the potential appearance in the basis polynomials of small spurious nonzero coefficients with *nonleading* monomials causes no problems. From the point of view of multiplication matrices, it amounts simply to a small perturbation of the nontrivial rows.

In realistic situations, it may naturally happen that, in spite of pivoting, the levels of magnitude in the data may fluctuate strongly and that a cancellation of leading digits may occur— as it may also happen in realistic systems of linear equations. This raises the question which level of precision should be used for a basis computation when there is no a priori information about the well-behavedness of the computation. Nontrivial computations will generally be done in the native *binary* floating-point arithmetic of the processor where the choice is essentially between the standard 64-bit "double precision" and some higher precision available through software. In most cases, a preliminary run in double precision will be the right choice, particularly if the implementation provides a posteriori information about a potential loss of accuracy and the potential necessity of a rerun with a higher precision.

10.3.3 Basis Computation with Empirical Data

In many practical situations, one will need to determine an approximate normal set representation for a regular system $(\bar{P}, E)$ of *empirical* polynomials. Such systems have been considered in section 9.1, and approximate representations in section 9.2. By Definition 9.6, a normal set $\mathcal{N}$ is feasible for $(\bar{P}, E)$ if it's an admissible normal set for all systems $\tilde{P} \in N_\delta(\bar{P}, E)$, $\delta = O(1)$, i.e. if it is a *common* basis for all members of the set $\mathcal{R}[N_\delta(\bar{P}, E)]$ of quotient rings; cf. (9.15).

By section 8.2.3, the data a_j^T of an exact normal set representation $(\mathcal{N}, \mathcal{B}_\mathcal{N})$ must lie on the admissible-data-manifold $\mathcal{M}_\mathcal{N}$ of dimension $s\,m$ in the data space of the $N\,m$ coefficients of the a_j^T, $N = |B[\mathcal{N}]| > s$; therefore we cannot require that a computed approximate representation $(\mathcal{N}, \widetilde{\mathcal{B}}_\mathcal{N})$ is the exact representation of some $\langle\tilde{P}\rangle$, with $\tilde{P} \in N_\delta(\bar{P}, E)$, $\delta = O(1)$. All we can expect is that the computed border basis coefficient vectors $\tilde{a}_j^T \in \mathbb{C}^m$ of $\widetilde{\mathcal{B}}_\mathcal{N}$ satisfy $\tilde{a}_j^T \approx \bar{a}_j^T$ in some natural sense; cf. Definition 9.8. In section 9.2, we have found that this will generally suffice to permit the computation of approximate zeros of $\bar{P}$ which may be refined if necessary, and the refinement of the $\tilde{a}_j^T$ towards the $\bar{a}_j^T$ if necessary.

What we would like to have are thresholds for the elimination pivots to be used in the basis computation which would guarantee sufficiently accurate basis coefficients. It appears impossible to derive strict *and* realistic thresholds to be used for that purpose. Heuristically, if there is only a relative accuracy of 10^{-r} in (some of) the data, an elimination pivot with a modulus of an order close to 10^{-r} below the current level of the data size in the computation should definitely be avoided by a modification of the underlying normal set, i.e by a violation of the term order in a GB-algorithm; cf. section 10.1.2.

As in many similar situations in Scientific Computing, the simplest strategy is the following:

- Compute a sufficiently accurate approximate representation for $\langle \bar{P} \rangle$ (refine it if necesary);

- determine the order of the result indetermination due to the data indetermination a posteriori; cf. section 3.2.3.

An example for these considerations will be found in the following section.

10.3.4   A Numerical Example

We must choose a rather simple example so that we may document details of the computation; therefore, we take an empirical system $(\bar{P}, E)$ with $\bar{P} \in (\mathcal{P}_2^3)^3$:

$$
\begin{aligned}
\bar{p}_1 &= 4.831\,x^2 + 4.597\,xy + .417\,y^2 + 1.688\,xz + .351\,yz - 1.428\,z^2 \\
&\qquad\qquad + 3.344\,x + .640\,y + 3.308\,z + 2.728\,, \\
\bar{p}_2 &= 4.036\,x^2 + 3.655\,xy + 2.988\,y^2 + 2.190\,xz + 1.473\,yz + 1.960\,z^2 \\
&\qquad\qquad + 4.270\,x + 3.572\,y + .853\,z + .239\,, \\
\bar{p}_3 &= 4.229\,x^2 + 1.950\,xy + 2.988\,y^2 + 1.298\,xz + 4.860\,yz + 1.249\,z^2 \\
&\qquad\qquad + 3.056\,x + 1.267\,y + 2.887\,z + 3.853\,;
\end{aligned}
$$

$$\tag{10.9}$$

we assume that all coefficients are empirical, with a tolerance of .0005. The coefficients have been chosen such that the Groebner basis of $\tilde{P}$ for `tdeg(x,y,z)` jumps inside the neighborhood $N_1(\bar{P}, E)$, e.g., when the coefficient of z^2 in $\tilde{p}_1$ takes the value ≈ 1.4275545. Thus, the normal set $\mathcal{N}_{\prec}$ associated with $\mathcal{G}_{\prec}[\bar{P}]$ for this term order is *not* a feasible normal set for $(\bar{P}, E)$ and the border basis $\mathcal{G}_{\prec}$ does not provide an approximate normal set representation of $(\bar{P}, E)$. This should become evident in the numerical computation of $\mathcal{G}_{\prec}$ and an alternate extended Groebner basis should be computed.

We consider 3 different approaches to the computation of a representation of $\langle N_\delta[(\bar{P}, E)] \rangle$ (cf. (9.14)):

(1) a numerical Groebner basis computation, for `tdeg(x,y,z)`;

(2) a numerical border basis computation, with a normal set candidate complying with the term order `tdeg(x,y,z)`;

(3) a numerical border basis computation, with a natural symmetric normal set.

In each case, we determine a representation for the system $\bar{P}$. The computation is performed in 10-digit floating-point arithmetic; we will find that a satisfactory result is obtained even with fewer digits. In the presentation, all numbers are rounded (a posteriori) to 5 digits.

(1) According to Algorithm 10.1, we begin with an autoreduction of $\bar{P}$. We "solve" $\bar{P}$ for the three highest monomials x^2, xy, y^2, with pivoted elimination. This yields the system P_0 with

$$
\begin{aligned}
p_{01}(x, y, z) &= x^2 - .08547\,xz + 1.81917\,yz - .44649\,z^2 \\
&\quad +.10253\,x - .86043\,y + 1.52653\,z + 2.19827\,, \\
p_{02}(x, y, z) &= xy + .43222\,xz - 1.85948\,yz + .06732\,z^2 \\
&\quad +.57405\,x + .95081\,y - .82514\,z - 1.64911\,, \\
p_{03}(x, y, z) &= y^2 + .27330\,xz + .26530\,yz + 1.00600\,z^2 \\
&\quad +.50301\,x + 1.02132\,y - .65585\,z - .74556\,;
\end{aligned}
$$

we have chosen to normalize for the leading monomials. In a Groebner basis computation, we do *not* assume an a priori information about the dimension m of the quotient ring; thus we know only that the normal set must contain the support monomials $1, z, y, x, z^2, yz, xz$ (cf. Figure 10.5). The actual normal set is generated during the computation.

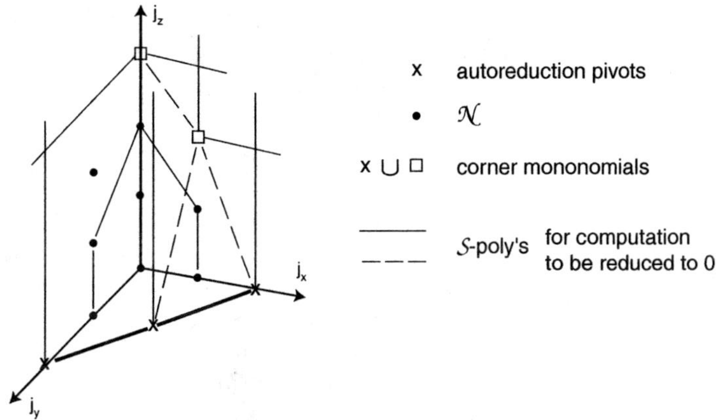

Figure 10.5.

Two S-polynomials ($S[p_{01}, p_{02}]$ and $S[p_{02}, p_{03}]$) can be formed while $S[p_{01}, p_{03}]$ reduces to zero by Proposition 8.14. The S-polynomials introduce various 3rd degree monomials; for the subsequent reduction, we need the "reductors" $z\,p_{01}$, $z\,p_{02}$, $z\,p_{03}$, p_{01}, p_{02}, p_{03} which also introduce z^3; cf. Figure 10.5. The elimination matrix, with the monomials arranged in decreasing order, takes the form

	x^2z	xyz	y^2z	xz^2	yz^2	z^3	x^2	xy	y^2		
$S[p_{01}, p_{02}]$	×	×	×	×	×		×	×	×	|	...
$S[p_{02}, p_{03}]$	×	×	×	×	×		×	×	×	|	...
$z\,p_{01}$	×			×	×	×				|	...
$z\,p_{02}$		×		×	×	×				|	...
$z\,p_{03}$			×	×	×	×				|	...
p_{01}							×			|	...
p_{02}								×		|	...
p_{03}									×	|	...

With pivoted triangularization, which leaves rows 4 to 5 at their place, we proceed until row 5 which becomes

$$0 \ \ 0 \ \ 0 \ \ 0 \ -.00012 \ \ 1.00683 \ -.09594 \ \ .49991 \ -.79403 \ \ldots$$

This means that we obtain a Groebner basis element with a leading coefficient .00012 with yz^2. With our tolerance of .0005 and all intermediate quantities in the elimination of $O(1)$, this coefficient is clearly not distinguishable from 0 within the tolerance neighborhood of $\bar{P}$ and cannot be used as a leading monomial of a basis polynomial. The next lower monomial in term order is z^3, with a coefficient of $O(1)$; therefore, we may simply exchange the roles of yz^2 and z^3 in the polynomial represented by row 5: We introduce z^3 as a corner monomial and relegate yz^2 into the normal set. The redefined Groebner basis polynomial with leading monomial z^3 contains a violation of the term order by a small element, which is typical for an extended GB; cf. section 10.1.2.

At this point of the computation, we have generated two new corner polynomials (cf. Figure 10.5):

$$
\begin{aligned}
g_4 \ = \ & xz^2 - 2.30640\,yz^2 - .58685\,xz + .98310\,yz - .83060\,z^2 \\
& \qquad\qquad -1.58942\,x - .07105\,y - 2.60755\,z - .26179 \,, \\
g_5 \ = \ & z^3 - .00012\,yz^2 + .61338\,xz + .62999\,yz - 1.32867\,z^2 \\
& \qquad\qquad -.21147\,x - 1.53674\,y - .59573\,z + .53053 \,;
\end{aligned}
$$

note that there is no term order violation in g_4. Together with the original basis polynomials p_{01}, p_{02}, p_{03}, they define a normal set $\mathcal{N}$ of 8 elements, which is the correct number m. But in the mechanism of a GB-algorithm, completion is only reached when all S-polynomials have been reduced to 0. The two new polynomials introduce a total of 7 potential S-polynomials: Each of them with the 3 old ones, and one between them. The combinations of the leading monomials z^3 with x^2, xy, y^2 and of xz^2 with y^2 are taken care of by Proposition 8.14; the remaining three S-polynomials must be formed and reduced to 0. Consider, e.g., $S[g_4, g_5] = z\,g_4 - x\,g_5$ and use all 5 basis polynomials and their multiples for the reduction; this leads to a polynomial in span $\mathcal{N}$ all of whose coefficients are $O(10^{-9})$ or less, which is round-off level. The same happens for the other two S-polynomials. Thus we have found and confirmed the set $\{p_{01}, p_{02}, p_{03}, g_4, g_5\}$ as an extended Groebner basis of $\bar{P}$ for $\mathtt{tdeg\,(x,y,z)}$.

Since the remaining coefficients in these normalized basis polynomials are all $O(1)$, and since we have not met any large elements in the elimination, it is safe to say that $\mathcal{N}$ will be a feasible normal set for the whole tolerance neighborhood of $(\bar{P}, E)$ and the computed basis an approximate extended Groebner basis for $\langle N_\delta(\bar{P}, E)\rangle$, $\delta = O(1)$.

(2) Assume that we want to determine a border basis for P, for a normal set which should be likely to coincide with that for a $\mathtt{tdeg\,(x,y,z)}$ Groebner basis. To obtain a candidate $\mathcal{N}^{(0)}$ for such a normal set, we proceed as in section 10.2.2: We autoreduce P for the 3 highest monomials in the term order and satisfy (10.4). Now we use the information $m = 8$; thus we must append one more monomial to our candidate and we choose the lowest monomial not yet in $\mathcal{N}^{(0)}$, i.e. z^3. This gives us $\mathcal{N}^{(0)} = \{1, z, y, x, z^2, zy, zx, z^3\}$; cf. Figure 10.6.

This time we begin by forming the syzygy relations defined by the edges of the border web $BW_{\mathcal{N}^{(0)}}$ which issue from the leading monomials of P_0 which is identical to $\{p_{01}, p_{02}, p_{03}\}$ found for the Groebner basis approach. The 5 edges marked in Figure 10.4 give us relations

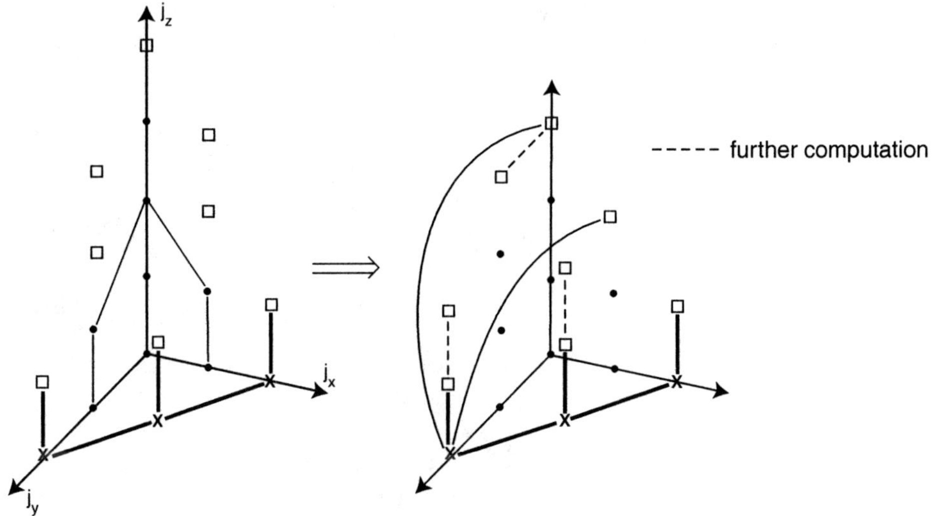

Figure 10.6.

for the normal forms of the border set monomials x^2z, xyz, y^2z, xz^2, yz^2 according to the procedure in section 10.2.3:

	x^2z	xyz	NF y^2z	xz^2	of yz^2	x^2	xy	y^2	$\mid$	$\mathcal{N}^{(0)}$
$[x^2, xy]$	$\times$	$\times$	$\times$	$\times$	$\times$	$\times$	$\times$	$\times$	$\mid$	$\ldots$
$[xy, y^2]$	$\times$	$\times$	$\times$	$\times$	$\times$	$\times$	$\times$	$\times$	$\mid$	$\ldots$
$[x^2, x^2z]$	$\times$			$\times$	$\times$				$\mid$	$\ldots$
$[xy, xyz]$		$\times$		$\times$	$\times$				$\mid$	$\ldots$
$[y^2, y^2z]$			$\times$	$\times$	$\times$				$\mid$	$\ldots$

This is the same matrix as in part (1), with the same coefficients, except that the z^3 column is missing because $z^3 \in \mathcal{N}^{(0)}$. Therefore the pivoted triangularization leads to the identical tiny pivot .00012 in the (5,5) element.

This time, we must turn to the normal set for an exchange, and, naturally, we switch roles between y^2z and z^3. This changes the border set and border web considerably: Besides yz^2, we loose the border monomials z^4, xz^3, and besides z^3, we acquire the border monomials xyz^2 and y^2z^2; cf. Figure 10.4. Since NF$[z^3]$ is obtained from the previous matrix after the exchange of roles, it remains to find the normal forms of yz^3, xyz^2, y^2z^2. These monomials are reached directly by shifts from border set monomials whose normal form we know at this point (cf. Figure 10.6) which makes the computation of their normal forms straightforward.

In the normal set controlled procedure, we are finished when we have generated a complete border basis, which is now the case. The minimal edge configuration $\mathcal{S}_0^{(1)}$ consists of the $8 = 11 - 3$ marked edges and the two virtual edges $[x^2, z^3]$, $[y^2, xz^2]$. By the same considerations as at the end of part (1), we may conclude that we have reached an approximate normal set representation $(\mathcal{N}^{(1)}, \mathcal{B}_{\mathcal{N}^{(1)}})$ for the empirical polynomial system $(\bar{P}, E)$.

(3) Without a restriction by a term order, the natural normal set to use for $\bar{P}$ is clearly $\mathcal{N} = \{1, z, y, x, yz, xz, xy, xyz\}$, i.e. the corners of the "unit cube" in $\mathbb{N}_0^3$, cf. Figure 10.7. To obtain a compatible autoreduced system P_0, we must now solve for x^2, y^2, z^2 which yields

$$
\begin{aligned}
p_{01} &= x^2 + 6.63197\,xy + 2.78103\,xz - 10.51284\,yz \\
&\qquad\qquad +3.90964\,x + 5.44530\,y - 3.94579\,z - 8.73858\,, \\
p_{02} &= y^2 - 14.94269\,xy - 6.18530\,xz + 28.05092\,yz \\
&\qquad\qquad -8.07490x - 13.18632\,y + 11.67399\,z + 23.89660\,, \\
p_{03} &= z^2 + 14.85361\,xy + 6.42010\,xz - 27.61996\,yz \\
&\qquad\qquad +8.52677\,x + 14.12293\,y - 12.25633\,z - 24.49525\,;
\end{aligned}
$$

the coefficients of $O(10)$ indicate that our choice is slightly less well conditioned than the previous one. But the computation is very straightforward:

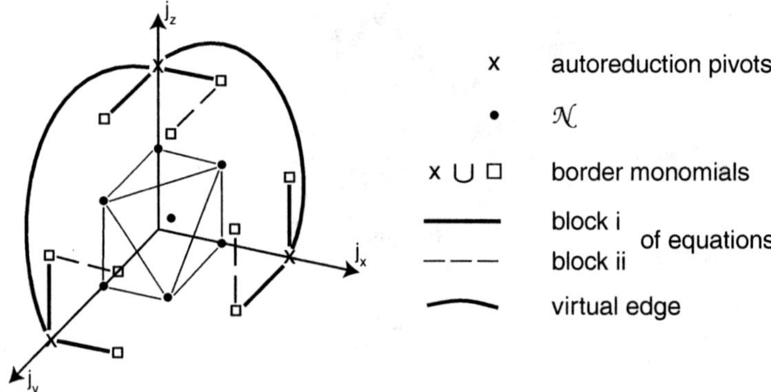

Figure 10.7.

We form the type (i) edges from all three "corners" and obtain 6 equations for the normal forms of x^2y, x^2z, y^2x, y^2z, z^2x, z^2y with a matrix

	x^2y	xy^2	x^2z	y^2z	xz^2	yz^2	x^2	y^2	z^2	$\mid$ $\mathcal{N}^{(0)}$
$[x^2, x^2y]$	×		×	×				×		$\mid$ …
$[x^2, x^2z]$		×		×	×				×	$\mid$ …
$[y^2, y^2x]$	×	×	×				×			$\mid$ …
$[y^2, y^2z]$				×	×	×			×	$\mid$ …
$[z^2, z^2x]$	×	×			×		×			$\mid$ …
$[z^2, z^2y]$			×	×		×		×		$\mid$ …

Pivoted elimination and substitution of the known normal forms of the corner monomials yield a set of six border basis polynomials. There is no instability problem at all in the elimination, all pivots are ≥ 1, some of them of $O(10)$.

There remain the three 4th degree border set monomials x^2yz, y^2xz, z^2xy, for which we can continue one edge each issuing from the corners; Figure 10.7. With the resolution of these 3 equations we are finished. The coefficients of the final basis $(\mathcal{N}, \mathcal{B}_{\mathcal{N}})$ are, besides those of the $p_{0\nu}$,

$$a_{012} = [-10.60716, -5.91100, 6.32294, 4.53652, -11.50098, 3.71530, 6.48648, 1.15930],$$
$$a_{102} = [-3.85694, -.84553, 2.92385, 4.97016, 3.82333, -4.56796, 2.62309, 2.67382],$$
$$a_{021} = [-12.27303, -6.39778, 5.59152, 3.02980, -14.89002, 2.45378, 7.67370, -1.03846],$$
$$a_{201} = [-4.13213, -2.69052, 2.61472, .22474, -4.66719, -.81635, 2.28731, -1.88037],$$
$$a_{120} = [.61376, -.39886, -.91693, -2.34924, .58281, -1.56841, -.94930, -2.44126],$$
$$a_{210} = [-2.97074, -1.04526, 1.79891, 3.46170, -3.71013, 2.51705, 1.69107, 2.49221].$$

We see that the size of the coefficients has not further increased beyond that in the $p_{0\nu}$. From Figure 10.7, we see that not a single S-polynomial has been formed, let alone been reduced to 0. This shows again that, with the a priori information about the dimension m of $\mathcal{R}[\langle P \rangle]$, S-polynomials are *not* an indispensible part of basis computation for polynomial ideals, as it is often assumed. The same argument as at the end of part (2) indicates that the normal set $\mathcal{N}$ used in part (3) is feasible for the *empirical* polynomial system $(\bar{P}, E)$.

From approaches (2) and (3), we can immediately form the multiplication matrices A_x, A_y, A_z with respect to the normal sets $\mathcal{N}^{(1)}$ and $\mathcal{N}$ respectively. In approach (1), the border basis polynomials which are not in the corner basis are also formed in the course of the reductions; they have only to be distinguished. Each of the matrices is nonderogatory, so one matrix is sufficient for the computation of all 8 zeros in each approach. The zeros are all complex (pairs of conjugate complex zeros) and in general position.

Which accuracy have we achieved in the numerical computation in 10-digit decimal floating-point arithmetic? An immediate qualitative answer is obtained by a check of the commutativity of the multiplication matrices. In approach (3), where there has been some increase in coefficient size and consequent loss of accuracy, the three matrices $A_x A_y - A_y A_x$ etc. have elements of $O(10^{-7})$ or less. This indicates that generally 7 digits after the decimal point are likely to be correct for $\bar{P}$ in the basis polynomials or the multiplication matrices, respectively. Note that a coefficient > 10 has only 8 digits after the point and that it will render the last two digits of its combination with a value below 1 meaningless. This shows that probably a large part in the accuracy loss has occurred in the initial autoreduction. But in view of the 3 valid digits in the *empirical* polynomial $(\bar{P}, E)$, a computational accuracy of 6 or 7 digits should be ample.

The accuracy in the basis computed in approach (2) is better by about one order of magnitude. This is a feature of this example and has nothing to do with the shape of the underlying normal sets. On the other hand, it shows that the collision of the singularity manifold $S_{\mathcal{N}^{(0)}}$ with the data of the empirical polynomial system (cf. section 9.1.3) is not harmful at all for its representation with respect to a different normal set. Both $S_{\mathcal{N}^{(0)}}$ and $S_{\mathcal{N}^{(1)}}$ have the same degree one of degeneracy for `tdeg(x,y,z)` which is enforced by the fact that all three polynomials are quadratic.

The question for the indetermination of the border basis coefficients as a consequence of the indetermination of the coefficients in $(\bar{P}, E)$ is beyond an easy answer. At first sight it appears that, by declaring *all* coefficients as empirical, with a relatively large tolerance, we have made the problem so indeterminate that little substance is left. But some analysis reveals that all zeros of the system are *extremely well conditioned*: By the approach in section 3.2.3, we find from

(3.44) that none of the zeros can change by more than $.0005\,\delta$ within $N_\delta(\bar{P}, E)$ as long as the linear estimate holds, i.e. for $\delta = O(1)$. On the other hand, we can determine the multiplication matrices from the zeros by (8.53); hence, we can also estimate their indetermination due to the indetermination of the zeros.

Generally, the detour from the original system to the basis via the zeros will lead to an overestimation of the indetermination because largest effects will occur for different situations in the two legs of the trip. But one will generally get an indication of the meaningful accuracy in the specification of a basis which is to fit a polynomial system with coefficients of limited accuracy. Let (cf. (8.53))

$$A_\sigma \;=\; \mathbf{c}_0^T(x_\sigma\,\mathbf{b})\,(\mathbf{c}_0^T(\mathbf{b}))^{-1} \;=:\; B_\sigma(\mathbf{z})\,B(\mathbf{z})^{-1}\,;$$

then, with a perturbation $\Delta\mathbf{z}$ in the zero set, we have

$$A_\sigma + \Delta A_\sigma \;=\; B_\sigma(\mathbf{z}+\Delta\mathbf{z})\,B(\mathbf{z}+\Delta\mathbf{z})^{-1} \;=\; (B_\sigma(\mathbf{z})+\Delta B_\sigma)\,(B(\mathbf{z})+\Delta B)^{-1}\,.$$

ΔB and ΔB_σ are easily expressed in terms of $\Delta\mathbf{z}$ and $(B+\Delta B)^{-1} = B^{-1} - B^{-1}\Delta B\,B^{-1}$, in linear approximation. This permits an estimate of

$$\Delta A_\sigma \;\approx\; \Delta B_\sigma\,B^{-1} - A_\sigma\,\Delta B\,B^{-1}\,,$$

and thus an estimate of the indetermination of the border basis coefficients from an assumed indetermination in the zeros which in turn may be estimated by the indetermination of the system coefficients. A crude evaluation in the case of our system (10.9) indicates an expected loss of 1 digit so that the border basis polynomials should actually only be specified with 2 digits after the point.

However, the border basis will rarely be a final result by itself but rather a means to solve other tasks and compute other quantities (like the zeros). Therefore, the right strategy is to compute the basis for the specified system $\bar{P}$ with a good accuracy and to do a backward error and an indetermination analysis only on the final results, interpreting them as exact results for a system from the tolerance neighborhood. While this is not possible for the border basis, as we have seen, it is generally possible for final results and leads to meaningful insights.

In our example, if we apply this to the zeros, we obtain the following insight: From a comparison of the zero sets obtained from the individual not fully commuting multiplication matrices as well as from their residuals in the original $\bar{P}$, we find that we have obtained the zeros of $\bar{P}$ to 7 digits and that their backward error is thus negligible. According to the indetermination analysis above, we must expect a potential small variation of the zeros in the 4th digit. Therefore, a meaningful zero set for the empirical system $(\bar{P}, E)$ should have either 4 or 5 digits after the decimal point:

$$(-.23422 \pm 1.10330\,i,\; -1.07323 \mp .20146\,i,\; 1.61311 \mp .52737\,i)\,,$$
$$(.19782 \pm .55559\,i,\; -1.65751 \pm .68416\,i,\; 1.36917 \pm .48644\,i)\,,$$
$$(-.15521 \pm .58136\,i,\; .33565 \mp .46568\,i,\; -.56345 \mp .29589\,i)\,,$$
$$(-.88505 \pm .06183\,i,\; .42107 \mp .62348\,i,\; -.77920 \mp .43123\,i)\,.$$

Exercises

1. Consider the example in section 10.3.4.

(a) Compute an approximate `tdeg(x,y,z)` Groebner basis of $\bar{P}$ by the procedure in section 10.1.3. How is the closeness to the representation singularity displayed? Experimentally, change a coefficient of $\bar{P}$ such that the representation becomes more singular; try to make it "jump" within the tolerance of $(\bar{P}, E)$.

(b) With the computed approximate Groebner basis of $\bar{P}$, compute (in floating-point) various S-polynomials and reduce them to zero. Observe and explain the size of the residuals.

(c) In part (2) of section 10.3.4, continue the computation without a switch in the normal set. Compare your result with the result of (a).

2. A huge increase in modulus size of the coefficients in the transition from the original system P to the (normalized) autoreduced system P_0 for a specified normal set $\mathcal{N}$ (satisfying (10.4)) indicates the near-infeasibility of $\mathcal{N}$.

(a) Consider this situation for two dense quadratic equations and the normal set $\mathcal{N} = \{1, y, x, xy\}$. For which relation between the coefficients of P is $\mathcal{N}$ infeasible? Which is the constellation of the zeros in this case?

(b) Perform the numerical basis computation with $\mathcal{N}$ for a quadratic system for which $\mathcal{N}$ is very nearly infeasible. Observe the loss of accuracy. Does pivoting have an influence?

(c) Do (a) and (b) for a different 4-element normal set.

Historical and Bibliographical Notes 10

A strong stimulus for the use of floating-point arithmetic in the computation of Groebner bases came from the potentially huge time and storage requirements of the classical Algorithm 10.1. On the one hand, this led to more and more refined implementations like in [10.1], but as long as the goal is an *exact rational* representation of the Groebner basis for a *fixed prespecified* term order, the potentially excessive size of the *result* alone must keep the computational effort potentially large. On the other hand, the naive use of floating-point arithmetic in a classical or refined GB-algorithm has led to spectacular successes as well as to equally spectacular failures, as it had to happen for an inherently numerically instable algorithmic approach. Also the interesting attempt [10.2] of a GB-algorithm in *interval arithmetic* could essentially only confirm this behavior.

Actually, my own interest in polynomial algebra stemmed initially from this apparent challenge of stabilizing an instable algorithm. While my first attempt of gaining some understanding of the underlying commutative algebra led to the unexpected insight of [2.8] and motivated my further engagement in the subject, I realized rather soon that a stabilization of the Algorithm 10.1 must require a potential *deviation from the prescribed term order* during the execution of the algorithm. This led to the concepts in [8.6], but I was not able to develop a firm algebraic basis for this approach. Only recently, this gap has been filled by A. Kondratyev ([10.3]).

Meanwhile, my own interest has shifted to approaches for 0-dimensional systems which utilize the a priori information about the dimension m of the quotient ring and do *not employ any term order*. It appears that this alternative is more satisfactory from a principal point of view and also permits more direct implementations because a normal set is present from the beginning: Its initial choice is either successful in permitting a numerically stable computation, or it has to be modified in a stepwise fashion during the computation. And—most important—no reductions to 0 have to be executed, which eliminates these numerically "impossible" operations. I hope that the considerations in sections 10.2 and 10.3 and the related work in [8.2] will aid professional implementations of this approach.

Thus, in this round-about manner, I am satisfied to have achieved some completion of a project on which I began work more than 15 years ago.

References

[10.1] J.-Ch. Faugère: A New Efficient Algorithm for Computing Gröbner Bases (F_4), J. Pure Appl. Algebra **139** (1999), 61–88.

[10.2] Ch. Jäger, D. Ratz: A Combined Method for Enclosing All Solutions of Nonlinear Systems of Polynomial Equations, Reliabl. Comput. **1** (1995), 41–64.

[10.3] A. Kondratyev: Numerical Computation of Groebner Bases, Ph.D. Thesis, Univ. Linz, 2003.

[10.4] J.-Ch. Faugère: Benchmarks for Polynomial Solvers, available from `fgbrs.lip6.fr/jcf/Benchs`.

Part IV

Positive-Dimensional Polynomial Systems

Introductory Remarks

When a polynomial system $P \subset \mathcal{P}^s$, $s \geq 2$, has positive-dimensional zero manifolds, one way of representing these manifolds is by *parametrization*:

$$\zeta_1 = z_1(t_1, .., t_d), \quad \ldots, \quad \zeta_s = z_s(t_1, .., t_d), \qquad (t_1, .., t_d) \in D \subset \mathbb{C}^d,$$

$$\text{with} \quad P(z_1(t_1, .., t_d), \ldots, z_s(t_1, .., t_d)) \overset{t}{\equiv} 0.$$

It is well known that such parametrizations do not always exist. If they exist, the $z_\sigma(t)$ are generally rational functions of t :

$$z_\sigma(t_1, .., t_d) = \frac{nz_\sigma(t_1, .., t_d)}{dz_\sigma(t_1, .., t_d)}, \quad \sigma = 1(1)s,$$

there may also exist parametrizations by polynomials $pz_\sigma(t)$.

In the context of this book, we do not wish to enter into the algorithmic theory of these parametrizations. Instead, we attempt to extend the approach which we have used for 0-dimensional zero sets: We study how it may be possible to derive parametric representations of zero manifolds of the system P from representations of the *multiplicative structure of the quotient ring* $\mathcal{R}[\langle P \rangle]$. This approach has originated in the 1990s in collaboration with the group of Wu Wenda at the Mathematics Mechanization Research Center of the Chinese Academy of Sciences; some accounts of it have been presented at workshops and conferences and appeared in conference proceedings but not in a major journal. Describing this incomplete work here—with the explicit agreement of my friend Wu—may serve to expose it to a wider public; thus, the loose ends may be picked up by others and continued. In particular, it may be clarified whether our ideas can be turned into a general algorithmic theory for the computational determination of representations of zero manifolds of polynomial systems, or whether they only constitute interesting observations which work in sufficiently restricted situations. Also, it may be possible to understand how our approach is related to other methods for determining parametrizations of zero manifolds.

Polynomial systems in $\mathcal{P}^s$ with zero sets which contain positive-dimensional manifolds either have fewer than s polynomials or they are singular systems in the sense of Definition 9.12; in the latter case, a generic perturbation will extinguish the manifolds. Similarly, a system of $s - d$ polynomials with a zero set component of a dimension greater than d is singular in the sense that the dimension reduces to d under perturbation. In the present context, because of the preliminary character of this approach, we consider only *intrinsic* polynomial systems and (at least principally) *exact* computation so that we may disregard these distinctions. A truly numerical algebra of positive-dimensional polynomial systems may, hopefully, get some stimuli from the ideas described in this short final part of my book.

Chapter 11

Matrix Eigenproblems for Positive-Dimensional Systems

It is natural to ask if and how the Central Theorem of Polynomial Systems Solving (Theorem 2.27) can be extended to systems with positive-dimensional zero sets. It was Wu Wenda (cf. [11.1]) who realized that an extension may be based on so-called *singular matrix eigenproblems* for rectangular matrices which have been introduced and analyzed by Kronecker at the end of the 19th century; a well-readable account of this subject is found in [11.3]. With this tool, the case of one-dimensional ideals, at least, permits a rather straightforward generalization of the ideas in Chapter 2 although the details are quite different. A potential generalization to higher dimensions will be indicated.

11.1 Multiplicative Structure of ∞-Dimensional Quotient Rings

11.1.1 Quotient Rings and Normal Sets of Positive-Dimensional Ideals

The quotient ring $\mathcal{R}[\mathcal{I}] \subset \mathcal{P}^s$ of a polynomial ideal $\mathcal{I} \subset \mathcal{P}^s$ consists of the residue classes mod $\mathcal{I}$. A residue class $[\bar{q}]_\mathcal{I}$ may be characterized by the fact that all $q \in [\bar{q}]_\mathcal{I}$ assume the *same values* on the zero set $Z[\mathcal{I}]$; more precisely, they yield the same values $c_\mu(q) \in \mathbb{C}$ for the functionals c_μ from a basis $\mathbf{c}^T$ of the dual space $\mathcal{D}[\mathcal{I}]$. For a 0-dimensional ideal, with $Z[\mathcal{I}]$ consisting of isolated points and $\mathcal{D}[\mathcal{I}]$ spanned by the evaluation functionals of q (and perhaps certain derivatives of q) on $Z[\mathcal{I}]$, $\mathcal{R}[\mathcal{I}]$ is a finite-dimensional vector space spanned (e.g.) by a normal set of monomials; cf. section 2.5.

When $\mathcal{I}$ is positive-dimensional so that its zero set contains at least one manifold, $\mathcal{R}[\mathcal{I}]$ and $\mathcal{D}[\mathcal{I}]$ must be *infinite-dimensional*: For any $N < \infty$, there exist polynomials $q \in \mathcal{P}^s$ whose values on a manifold M cannot be defined by N complex numbers. Take, e.g., $M = \{x_2 = \ldots = x_s = 0\}$; then the values $q(x_1, 0, \ldots, 0) =: q_1(x_1)$, with $\deg q_1 = n_1$, are defined by the $n_1 + 1$ numbers $q_1(0), q_1'(0), \ldots, q_1^{(n_1)}(0)$. For any specified finite N, we can trivially choose q with $n_1 \geq N$. Obviously, the values of the polynomials in each particular $[\bar{q}]_\mathcal{I}$ can be characterized by $\bar{n} < \infty$ numbers, but $\bar{n}$ is not bounded over all $[\bar{q}]_\mathcal{I} \in \mathcal{R}[\mathcal{I}]$.

Accordingly, a vector space basis of $\mathcal{R}[\mathcal{I}]$ must consist of infinitely many elements. When we now—as in section 2.2.1—switch from residue classes to representatives as elements of the quotient ring $\mathcal{R}[\mathcal{I}]$ and choose these representatives from span $\mathcal{N}$, with $\mathcal{N} \subset T^s$ a closed set of monomials, this *normal set $\mathcal{N}$* must also be *infinite*. As previously, $\mathcal{N}$ is not at all unique.

A Groebner basis $\mathcal{G}[\mathcal{I}]$ of a positive-dimensional ideal $\mathcal{I} \subset \mathcal{P}^s$ is characterized by the fact that there exist one or several variables x_σ such that no power x_σ^m is a leading mononomial x^{j_κ} of a basis element g_κ. (We have previously used this fact in section 9.4 where zero manifolds appeared in a singular fashion.) An associated normal set basis of $\mathcal{R}[\mathcal{I}]$ satisfies $\mathcal{N} \supset \{ b_\mu \in T^s \ : \ $ no x^{j_κ} is a divisor of $b_\mu \}$. Moreover, the dimensionality of the unbounded parts of the normal set $\mathcal{N}$ indicates the highest dimension of a component of the zero set of $\mathcal{I}$: If the normal set exponents $j = (j_1, \ldots, j_s) \in \mathbb{N}_0^s$ include one or several d-dimensional coordinate subspaces of $\mathbb{N}_0^s$, then $Z[\mathcal{I}]$ contains at least one d-dimensional manifold.

The normal set of a positive-dimensional ideal may or may not also possess *bounded parts*, viz. monomials x^j for which there exists no variable x_σ such that *all* monomials $x^j x_\sigma^k$, $k \in \mathbb{N}_0$, are in $\mathcal{N}$. The number of these "niche" monomials equals the number of dual space basis functionals which constitute evaluations *not along a manifold*. These may be function and derivative evaluations at isolated zeros, or *external* derivative evaluations at multiple d-points on a manifold; cf. section 9.4.4.

All these statements are well-known facts from algebraic geometry, where they are usually formulated in a more abstract and general fashion. In our context, they provide an intuitive background for our computational approach; therefore we illustrate them by a few examples.

Example 11.1: In Exercise 1 of section 9.4 we considered the polynomial system (9.60). For $c \neq 0, 1$, the $\mathtt{tdeg\,(x1,x2,x3)}$ Groebner basis of $P(c)$ has the leading monomials x_1^2, $x_1 x_2$, $x_1 x_3$, x_2^3, $x_2^2 x_3$, which define the normal set $\mathcal{N} = \{ x_3^\kappa, x_2 x_3^\kappa, \kappa \in \mathbb{N}_0, x_1, x_2^2 \}$. The presence of the coordinate subspace $(0, 0, \kappa)$ of $\mathbb{N}_0^3$ in the exponent set of $\mathcal{N}$ indicates the presence of one or more one-dimensional zero manifolds; here they are the two straight lines $(t, 0, t)$, $(t, 1, t +$ $1)$, $t \in \mathbb{C}$. The presence of the two niche monomials x_1, x_2^2 indicates the presence of two isolated dual space basis elements. One of these is the evaluation at the isolated zero $(1, 1, 1)$, the other one the evaluation of the external derivative ∂_{010} at the 2-fold d-point $(1, 0, 1)$.

For $c = 0$, the Groebner basis loses the leading monomial x_2^3. This opens the way for another one-dimensional coordinate subspace $(0, \kappa, 0) \subset \mathbb{N}_0^3$; at the same time, x_2^2 ceases to be a niche monomial. This corresponds to the additional zero manifold $(1, t, 1)$; the previously external x_2-derivative at $(1, 0, 1)$ has now become internal for this manifold.

For $c = 1$, the Groebner basis loses the further leading monomial $x_2^2 x_3$. Now the 2-dimensional coordinate subspace $(0, \kappa_2, \kappa_3)$ is in the exponent set of $\mathcal{N}$ while the niche monomial x_1 still persists. The zero set now consists of the two-dimensional manifold (t_1, t_2, t_1+t_2), $t_i \in \mathbb{C}$, and the isolated zero $(1, 1, 1)$. $\quad\square$

Example 11.2: Take the example (9.58) of section 9.4.3, with a at the critical value $a_0 = \sqrt{3}/2$. The $\mathtt{tdeg\,(x1,x2,x3)}$ Groebner basis of $P(a_0)$ has the leading monomials

$$x_1 x_3^4, \ x_1 x_2^2, \ x_1 x_2 x_3^2, \ x_2^2 x_3^2, \ x_1^2 x_2, \ x_1 x_2^2 .$$

This leaves all three coordinate axes of $\mathbb{N}_0^3$ in the exponent set of $\mathcal{N}$ plus one adjacent parallel for each; this corresponds to the one-dimensional closed-loop manifold M_0 of section 9.4.3. The 4 niche monomials $x_1 x_2$, $x_1 x_2 x_3$, $x_1 x_3^2$, $x_1 x_3^3$ of $\mathcal{N}$ correspond to the 4 isolated zeros of $P(a_0)$; the remaining 12 zeros of $P(a)$ move into simple d-points on M_0 for $a \to a_0$. $\quad\square$

11.1.2 Finite Sections of Infinite Multiplication Matrices

The multiplicative structure of a quotient ring $\mathcal{R}[\mathcal{I}] \subset \mathcal{P}^s$ with monomial basis $\mathcal{N} = \{b_\mu\}$ is specified by the *linear maps* $b_\mu \to x_\sigma b_\mu$, $\sigma = 1(1)s$. For a finite basis $\mathcal{N}$, with its elements $b_\mu(x)$ arranged into a vector $\mathbf{b}(x) = (b_1, \ldots, b_m)^T$, these maps are represented by the *multiplication matrices* $A_\sigma \in \mathbb{C}^{m \times m}$, $\sigma = 1(1)s$; cf. section 2.2.1. For infinite-dimensional quotient rings as we consider them now, it is not so clear how one should represent their multiplicative structure for computational purposes; standard multiplication matrices A_σ would possess infinitely many rows and columns.

At first, we observe that $x_\sigma b_\mu(x) \bmod \mathcal{I}$ is nontrivial only if $x_\sigma b_\mu(x) \notin \mathcal{N}$; even in this case, the representative of $[x_\sigma b_\mu]_\mathcal{I}$ in span $\mathcal{N}$ is a *finite* linear combination of elements in $\mathcal{N}$. Thus, each *row* of an A_σ contains only a *finite* number of nonzero elements. Furthermore, the polynomials $x_\sigma b_\mu(x) \bmod \mathcal{I}$ for b_μ from the *unbounded* parts of $\mathcal{N}$ can be generated in a recursive fashion; thus, the complete information about the multiplicative structure must be contained in a *finite number of rows*.

These observations suggest the following considerations: Let $\mathcal{G}[\mathcal{I}] = \{g_1, \ldots, g_k\}$ be the reduced Groebner basis of a positive-dimensional ideal $\mathcal{I} \subset \mathcal{P}^s$ and write $g_\kappa(x) = x^{j_\kappa} - \mathrm{NF}_\mathcal{I}[x^{j_\kappa}](x)$. Let $\nu_\sigma := \max_\kappa (j_\kappa)_\sigma$ and form the finite subset $\mathcal{N}' \subset \mathcal{N}$ of all monomials x^j with

$$\sigma\text{-component of } j \ \leq \ \nu_\sigma, \quad \sigma = 1(1)s; \tag{11.1}$$

then append to $\mathcal{N}'$ those monomials occurring in $\mathrm{NF}[x^{j_\kappa}]$, $\kappa = 1(1)k$, not yet in $\mathcal{N}'$ and form the closed hull $\hat{\mathcal{N}} \subset \mathcal{N}$ of $\mathcal{N}'$. For Example 11.1, $c \neq 0, 1$, e.g., we obtain initially the set $\mathcal{N}' = \{1, x_3, x_2, x_1, x_2 x_3, x_2^2\}$ but we have to append the monomial x_3^2 occurring in p_1 while the closure adds no further monomials.

Then we consider those monomials in $\hat{\mathcal{N}}$ which have a positive neighbor in $\mathcal{N}$ but not in $\hat{\mathcal{N}}$; their removal from $\hat{\mathcal{N}}$ creates the set $\check{\mathcal{N}} \subset \hat{\mathcal{N}}$. Since $\mathcal{N}$ is infinite, $\check{\mathcal{N}}$ is a proper subset of $\hat{\mathcal{N}}$. In Example 11.1, $c \neq 0, 1$, we remove $x_2 x_3$, x_3^2 and obtain $\check{\mathcal{N}} = \{1, x_3, x_2, x_1, x_2^2\}$.

Now we form the border set $B[\check{\mathcal{N}}]$: it consists of elements from $\hat{\mathcal{N}} \setminus \check{\mathcal{N}}$, of leading monomials x^{j_κ} of the g_κ, and, generally, some multiples of such leading monomials. We check the monomials which occur in the normal forms of these multiples: If some monomial is not in $\hat{\mathcal{N}} \cup \{x^{j_\kappa}\}$, we append it to $\hat{\mathcal{N}}$ but not to $\check{\mathcal{N}}$. This finishes the construction of the finite subsets $\check{\mathcal{N}}$ and $\hat{\mathcal{N}}$ of $\mathcal{N}$. In Example 11.1, $c \neq 0, 1$, $B[\check{\mathcal{N}}] = \{x_3^2, x_2 x_3, x_1 x_3, x_1 x_2, x_1^2, x_2^2 x_3, x_2^3, x_1 x_2^2\}$, with $x_3^2, x_2 x_3 \in \hat{\mathcal{N}} \setminus \check{\mathcal{N}}$ and $x_1 x_3, x_1 x_2, x_1^2, x_2^2 x_3, x_2^3 \in \{x^{j_\kappa}\}$. $\mathrm{NF}[x_1 x_2^2] = x_1 x_3 + (1 - c) x_2^2 + x_1 - (1 - c) x_2 - x_3$, with $x_1 x_3$ a leading monomial and the other monomials in $\hat{\mathcal{N}}$ (even in $\check{\mathcal{N}}$) so that we are finished.

Let $\check{\mathbf{b}}$ and $\hat{\mathbf{b}}$ be the vectors of the monomials in $\check{\mathcal{N}}$ and $\hat{\mathcal{N}}$, resp., in a fixed order, with the monomials in $\hat{\mathcal{N}} \setminus \check{\mathcal{N}}$ behind those of $\check{\mathcal{N}}$. By the above construction, the normal forms of the monomials in $x_\sigma \check{\mathbf{b}}$, $\sigma = 1(1)s$, are in span $\hat{\mathbf{b}}$. Thus we may form matrices $\overline{A}_\sigma \in \mathbb{C}^{\check{n} \times \hat{n}}$ which satisfy

$$x_\sigma \check{\mathbf{b}}(x) \equiv \overline{A}_\sigma \hat{\mathbf{b}} \quad \bmod \mathcal{I}, \quad \sigma = 1(1)s. \tag{11.2}$$

These matrices are the *left-upper submatrices* of the infinite multiplication matrices of $\mathcal{R}[\mathcal{I}]$ for an infinite normal set vector whose first $\hat{n}$ elements are those in $\hat{\mathbf{b}}$.

Definition 11.1. The finite normal set sections $\check{N}$ and $\hat{N}$ defined above will be called *inner* and *outer* normal sets of $\mathcal{R}[\mathcal{I}]$, the matrices $\overline{A}_\sigma$ *finite multiplication matrices* of $\mathcal{R}[\mathcal{I}]$. $\square$

To facilitate formal manipulations, we introduce $\overline{I} := (\, I \,|\, 0\,) \in \mathbb{C}^{\check{n}\times\hat{n}}$, and we note that $\check{\mathbf{b}} = \overline{I}\,\hat{\mathbf{b}}$. Then (11.2) may be written as

$$(\overline{A}_\sigma - x_\sigma\,\overline{I})\,\hat{\mathbf{b}}(x) \equiv 0 \quad \mathrm{mod}\ \mathcal{I}, \quad \sigma = 1(1)s, \tag{11.3}$$

and we have

Theorem 11.1. At each zero $z = (\zeta_1, \ldots, \zeta_s)$ of $\mathcal{I}$ (isolated or on a manifold),

$$(\overline{A}_\sigma - \zeta_\sigma\,\overline{I})\,\hat{\mathbf{b}}(z) = 0, \qquad \sigma = 1(1)s. \tag{11.4}$$

On each d-dimensional zero manifold M of $\mathcal{I}$, with a local parametrization $z = z(t)$, $t \in D \subset \mathbb{C}^d$, there holds

$$(\overline{A}_\sigma - \zeta_\sigma(t)\,\overline{I})\,\hat{\mathbf{b}}(z(t)) \overset{t}{\equiv} 0, \qquad \sigma = 1(1)s. \tag{11.5}$$

Theorem 11.1 suggests that the computational solution of the *rectangular matrix eigenproblem* (11.3) may permit the determination of isolated and manifold solutions of the polynomial system P whose quotient ring $\mathcal{R}[\langle P\rangle]$ has the inner and outer normal sets $\check{N}$ and $\hat{N}$ and the finite multiplication matrices $\overline{A}_\sigma$ for $\check{N}$, $\hat{N}$.

Example 11.3: As in Example 11.1, we consider the polyomial system (9.60), with indeterminate c; inner and outer normal sets for $\mathcal{R}[\langle P(c)\rangle]$ have been determined above Definition 11.1. With the associated normal set vectors $\check{\mathbf{b}} = (1, x_3, x_2, x_1, x_2^2)^T$ and $\hat{\mathbf{b}} = (\,\ldots, x_2 x_3, x_3^2)^T$, we obtain the finite multiplication matrices

$$\overline{A}_1 = \begin{pmatrix} 0 & 0 & 0 & 1 & 0 & 0 & 0 \\ 0 & -1 & 1 & 1 & 0 & -1 & 1 \\ 0 & -1 & c & 1 & -c & 1 & 0 \\ 0 & -1 & 2-c & 1 & c & -2 & 1 \\ 0 & -1 & c-1 & 1 & 1-c & 1 & 0 \end{pmatrix}, \quad \overline{A}_2 = \begin{pmatrix} 0 & 0 & 1 & 0 & 0 & 0 & 0 \\ 0 & 0 & 0 & 0 & 0 & 1 & 0 \\ 0 & 0 & 0 & 0 & 1 & 0 & 0 \\ 0 & -1 & c & 1 & -c & 1 & 0 \\ 0 & 0 & 0 & 0 & 1 & 0 & 0 \end{pmatrix},$$

$$\overline{A}_3 = \begin{pmatrix} 0 & 1 & 0 & 0 & 0 & 0 & 0 \\ 0 & 0 & 0 & 0 & 0 & 0 & 1 \\ 0 & 0 & 0 & 0 & 0 & 1 & 0 \\ 0 & -1 & 1 & 1 & 0 & -1 & 1 \\ 0 & 0 & -1 & 0 & 1 & 1 & 0 \end{pmatrix}.$$

Since all row sums in the $\overline{A}_\sigma$ are 1, it is obvious that the isolated zero $(1,1,1)$ of (9.60) satisfies (11.4). For the zero manifold $\zeta_1(t) = t$, $\zeta_2(t) = 0$, $\zeta_3(t) = t$, we find

$$\overline{A}_1 \cdot \begin{pmatrix} 1 \\ t \\ 0 \\ t \\ 0 \\ 0 \\ t^2 \end{pmatrix} - t \cdot \begin{pmatrix} 1 \\ t \\ 0 \\ t \\ 0 \end{pmatrix} \overset{t}{\equiv} 0$$

and the analogous relations (11.5) for $\sigma = 2, 3$. Similarly, the zero manifold $(t, 1, t+1)$ satisfies

$$\overline{A}_1 \cdot \begin{pmatrix} 1 \\ t+1 \\ 1 \\ t \\ 1 \\ t+1 \\ (t+1)^2 \end{pmatrix} - t \cdot \begin{pmatrix} 1 \\ t+1 \\ 1 \\ t \\ 1 \end{pmatrix} \overset{t}{\equiv} 0$$

and the other relations (11.5), which is not quite as evident. $\square$

Example 11.3 illustrates the meaning of (11.3): Besides a number of trivial relations its left-hand side contains the following polynomials from $\mathcal{I}$:

$$bb_j(x) := x^j - \mathrm{NF}_{\mathcal{I}}[x^j], \quad \forall j : x^j \in B[\check{\mathcal{N}}], \; x^j \notin \hat{\mathcal{N}}, \tag{11.6}$$

where the monomials in $\mathrm{NF}_{\mathcal{I}}[x^j]$ are from $\hat{\mathcal{N}}$.

Proposition 11.2. For $\check{\mathcal{N}}$ and $\hat{\mathcal{N}}$ constructed as described above, (11.6) is a basis of $\mathcal{I}$.
Proof: By our construction of $\check{\mathcal{N}}$ and $\hat{\mathcal{N}}$ from a Groebner basis $\{g_\kappa\}$ of $\mathcal{I}$, the leading monomials of the g_κ are among the x^j of (11.6). Thus, $\{bb_j\} \supset \{g_\kappa\}$. On the other hand, all bb_j are in $\mathcal{I}$. $\square$

Definition 11.2. For a positive-dimensional ideal $\mathcal{I}$, the polynomial set $B_{\check{\mathcal{N}}}[\mathcal{I}] := \{bb_j, \; x^j \in B[\check{\mathcal{N}}], \; x^j \notin \hat{\mathcal{N}}\}$ of (11.6) will be called a *finite border basis* of $\mathcal{I}$. $\square$

11.1.3 Extension of the Central Theorem

Proposition 11.2 establishes immediately that each solution $z \in \mathbb{C}^s$ of (11.4) and $z(t) \in \mathbb{C}^s$, $t \in D \subset \mathbb{C}^d$, of (11.5) is a solution of the underlying polynomial system P. However, (11.4) and (11.5) assume that the components of $\check{\mathbf{b}}(z)$ and $\hat{\mathbf{b}}(z)$ are *internally consistent*, i.e. that the x^j-component of $\hat{\mathbf{b}}(z)$ equals z^j. This is guaranteed by the structure of the finite multiplication matrices $\overline{A}_\sigma$:

Proposition 11.3. In the situation of section 11.1.2, consider vectors $z = (\zeta_\sigma) \in \mathbb{C}^s$ and $\hat{\beta} \in \mathbb{C}^{\hat{n}}$ which satisfy

$$(\overline{A}_\sigma - \zeta_\sigma \overline{I})\,\hat{\beta} = 0, \qquad \sigma = 1(1)s.$$

Then a component β_μ of $\hat{\beta}$ which corresponds to a monomial x^{j_μ} in the outer normal set vector $\hat{\mathbf{b}}(x)$ satisfies $\beta_\mu = z^{j_\mu}$.
Proof: Due to the closedness of $\hat{\mathcal{N}}$, for each component x^{j_μ} of $\hat{\mathbf{b}}(x)$ except 1 there exists at least one component in $\hat{\mathbf{b}}(x)$ which is a negative neighbor $x^{j'} = x^{j_\mu}/x_\sigma$ of x^{j_μ}. Therefore, there is a trivial row (one 1, zeros otherwise) in the corresponding $\overline{A}_\sigma$ which ascertains that the x^{j_μ}-component z^{j_μ} of a solution of (11.3) equals $\zeta_\sigma z^{j'}$. Recursively, this establishes the equality for all components of $\hat{\beta}$. $\square$

Corollary 11.4. In the situation of Proposition 11.3, if $z = z(t)$ and $\hat{\beta} = \hat{\beta}(t)$ for $t \in D \subset \mathbb{C}^d$, then

$$(\overline{A}_\sigma - \zeta_\sigma(t)\,\overline{I})\,\hat{\beta}(t) \;\overset{!}{\equiv}\; 0\,, \qquad \sigma = 1(1)s\,,$$

implies $\beta_\mu(t) = z(t)^{j_\mu}$, $t \in D$.

Together, Theorem 11.1, Proposition 11.3 and Corollary 11.4 yield

Theorem 11.5 (Extended Central Theorem). For a positive-dimensional polynomial system $P \subset \mathcal{P}^s$, let the quotient ring $\mathcal{R}[\langle P \rangle]$ have an outer normal set vector $\hat{\mathbf{b}} \in (\mathcal{T}^s)^{\hat{n}}$ and associated finite multiplication matrices $\overline{A}_\sigma \in \mathbb{C}^{\check{n} \times \hat{n}}$. Then the *joint* point and parametric eigensolutions of the singular matrix eigenproblems

$$(\overline{A}_\sigma - x_\sigma\,\overline{I})\,\hat{\mathbf{b}}(x) \;=\; 0\,, \qquad \sigma = 1(1)s\,, \tag{11.7}$$

represent all isolated zeros and zero manifolds, resp., of P.

Note that we have not addressed the question of multiplicities for isolated zeros or zero manifolds in Theorem 11.5; this is not immediately possible from our considerations so far but requires a further analysis of singular matrix eigenproblems on the one hand and of algebraic varieties on the other hand.

Exercises

1. Take the polynomial system $P(c)$ of (9.60) and consider the cases $c = 0$ and $c = 1$.

   (a) Compute the `tdeg(x1,x2,x3)` Groebner bases and confirm the statements in Example 11.1. Then form inner and outer normal sets for $\langle P \rangle$.

   (b) Form the finite multiplication matrices for the cases $c = 0$ and $c = 1$, respectively. Confirm the assertions (11.4) and (11.5) of Theorem 11.1.

2. Take the system $P(a_0)$ of Example 11.2 and proceed as in (a) above. Form the finite multiplication matrices and confirm (11.4) for the 4 isolated zeros (t, t, t), with $t = \pm \frac{1}{3}\sqrt{-9 \pm 6\sqrt{3}}$.

3. Take the system $P = \begin{cases} p_1(x_1, x_2) &=& x_1^3 + x_1 x_2^2 - x_1^2 - x_2^2 - x_1 - 1\,, \\ p_2(x_1, x_2) &=& x_1^2 x_2 + x_2^3 + x_1^2 + x_2^2 - x_2 - 1\,; \end{cases}$ and proceed like in Exercise 1.

11.2 Singular Matrix Eigenproblems

All the observations in section 11.1 become meaningful only if we are able to find the point and parametric solutions of problems of the form (11.7) for rectangular matrices $\overline{A} \in \mathbb{C}^{\check{n} \times \hat{n}}$, $\check{n} < \hat{n}$. Following Kronecker, such problems are called *singular matrix eigenproblems*. In this section, we indicate how their solution may be algorithmically determined. A more detailed analysis of singular matrix eigenproblems may be found in [11.3].

11.2.1 The Solution Space of a Singular Matrix Eigenproblem

We consider the singular matrix eigenproblem (cf. (11.3))

$$(\overline{A} - \lambda \overline{I})\,z \;=\; 0\,. \tag{11.8}$$

Immediately we note that, for $\overline{A} \in \mathbb{C}^{\check{n} \times \hat{n}}$ with $\check{n} < \hat{n}$, the dimension of the *kernel* of $\overline{A} - \lambda \overline{I}$ is *positive* for all $\lambda \in \mathbb{C}$. Thus, there exist solutions $z(\lambda) \in \mathbb{C}^{\hat{n}}$ with λ as a *parameter*.

Proposition 11.6. Equation (11.8) has $r = \hat{n} - \check{n}$ parametric solutions of the form

$$z_\rho(\lambda) \;=\; \begin{pmatrix} \vdots \\ z_{\rho\mu}(\lambda) \\ \vdots \end{pmatrix} \;=\; \sum_{\kappa=0}^{k_\rho} \lambda^\kappa \, z_{\rho\kappa}, \quad \text{with} \quad \begin{array}{l} z_{\rho\mu}(\lambda) \in \mathcal{P}_{k_\rho}, \; \mu = 1(1)\hat{n}, \\[4pt] z_{\rho\kappa} \in \mathbb{C}^{\hat{n}}, \; \kappa = 0(1)k_\rho. \end{array} \tag{11.9}$$

The general parametric solution of (11.8) is

$$z(\lambda) \;=\; \sum_{\rho=1}^{r} c_\rho(\lambda)\, z_\rho(\lambda), \quad \text{with } \textit{arbitrary} \text{ coefficient functions } c_\rho(\lambda). \tag{11.10}$$

Proof: Substitution of $z(\lambda) = \sum_{\kappa=0}^{k} \lambda^\kappa z_\kappa$ into (11.8) yields the linear system

$$\begin{array}{rcl}
\overline{A}\, z_0 & = & 0, \\
-\overline{I}\, z_0 + \overline{A}\, z_1 & = & 0, \\
-\overline{I}\, z_1 + \overline{A}\, z_2 & = & 0, \\
\ddots \qquad \vdots & & \\
-\overline{I}\, z_{k-1} + \overline{A}\, z_k & = & 0, \\
-\overline{I}\, z_k & = & 0,
\end{array} \tag{11.11}$$

for $(z_0, \ldots, z_k)^T \in \mathbb{C}^{(k+1)\hat{n}}$. Since the matrix of (11.11) has $(k+2)\check{n}$ rows, it must have nontrivial solutions for sufficiently large k.

The r-dimensional kernel of the last equation is spanned by $z_{1k_1} = (.. 0 .. |1, 0, ..)^T, \ldots, z_{rk_r} = (.. 0 .. |0, .., 1)^T$. For each $z_{\rho k_\rho}$, with $k_\rho \geq 0$ not yet fixed, we may solve (11.11) from bottom to top until we obtain a $z_{\rho 0} \in \ker \overline{A}$; the number of steps determines k_ρ.

With (11.10), $(\overline{A} - \lambda \overline{I})\, z(\lambda) = \sum_\rho c_\rho(\lambda)\, (\overline{A} - \lambda \overline{I})\, z_\rho(\lambda) = 0.$ $\square$

While $\operatorname{rk} (\overline{A} - \lambda \overline{I}) \leq \check{n}$ for all $\lambda \in \mathbb{C}$, there may be particular values $\lambda_\nu \in \mathbb{C}$ for which this rank is $< \check{n}$. For these *regular eigenvalues* λ_ν, there exist regular eigenvectors $\dot{z}_\nu \in \mathbb{C}^{\hat{n}}$ (or invariant subspaces if the rank deficiency is greater than 1), just like for quadratic matrices.

Proposition 11.7. Iff $n_0 := \check{n} - \sum_{\rho=1}^{r} k_\rho > 0$, there exist n_0 regular eigenvalues $\dot{\lambda}_\nu$ (not necessarily distinct) for (11.8), with associated eigenvectors for simple eigenvalues or invariant subspaces for multiple eigenvalues.

Proof: (Sketch) According to (11.11), we have

$$\overline{A} \begin{pmatrix} | & & | & & | & & | \\ z_{11} & .. & z_{1k_1} & .. & z_{r1} & .. & z_{rk_r} \\ | & & | & & | & & | \end{pmatrix} = \overline{I} \begin{pmatrix} | & & | & & | & & | \\ z_{10} & .. & z_{1,k_1-1} & .. & z_{r0} & .. & z_{r,k_r-1} \\ | & & | & & | & & | \end{pmatrix}. \tag{11.12}$$

In [11.3], it is shown that the $\sum k_\rho$ vectors $z_{\rho\kappa} \in \mathbb{C}^{\hat{n}}$ and $\overline{I} z_{\rho\kappa} \in \mathbb{C}^{\check{n}}$, resp., are linearly independent. In $\mathbb{C}^{\hat{n}}$, we have the further r vectors $z_{\rho 0}$ which span $\ker \overline{A}$. If $\sum k_\rho = \check{n}$, the $\hat{n}$

vectors $z_{\rho\kappa}$, $\kappa = 0(1)k_\rho$, span the $\mathbb{C}^{\hat{n}}$ and the $\check{n}$ vectors $\overline{I}z_{\rho\kappa}$, $\kappa = 0(1)k_\rho - 1$, span the $\mathbb{C}^{\check{n}}$; thus there is no room for further eigenvectors.

If $\sum k_\rho < \check{n}$, there are n_0 further basis vectors in $\mathbb{C}^{\hat{n}}$ which are mapped into linear combinations of the remaining n_0 basis vectors of $\mathbb{C}^{\check{n}}$ and the $\overline{I}z_{\rho\kappa}$. With (11.12), we can modify the supplementary basis vectors of $\mathbb{C}^{\hat{n}}$ such that their images by $\overline{A}$ are linear combinations of the accordingly modified supplementary basis vectors of $\mathbb{C}^{\check{n}}$ only. Let the matrix $B \in \mathbb{C}^{n_0 \times n_0}$ represent these linear combinations; then the eigenvalues of B are the regular eigenvalues λ_ν of (11.8) and the associated eigenvectors are determined by the eigenvectors of B. If B has multiple eigenvalues, their invariant subspaces determine those of $\overline{A}$. $\quad\square$

After the determination of the parametric eigensolutions of (11.8), the value of n_0 and thus the existence of "point" eigensolutions is known. Note that a regular eigensolution $\dot{z}$ is really an $(r + 1)$-dimensional *subspace*: For λ_ν, any vector $z = \gamma_0 \dot{z}_\nu + \sum_{\rho=1}^{r} \gamma_\rho z_\rho(\lambda_\nu)$ satisfies $(\overline{A} - \lambda_\nu \overline{I}) z = 0$.

$$
\textit{Example 11.4: Take } \overline{A} = \begin{pmatrix}
0 & 1 & 0 & 0 & 0 & 0 & 0 & 0 \\
0 & 0 & 0 & 1 & 0 & 0 & 0 & 0 \\
0 & 0 & 0 & 0 & 1 & 0 & 0 & 0 \\
0 & 0 & 0 & 0 & 0 & 0 & 1 & 0 \\
0 & 0 & 0 & 0 & 0 & 0 & 0 & 1 \\
1 & 1 & 0 & -1 & 0 & -1 & -1 & 0
\end{pmatrix}, \text{ with } \check{n} = 6,\ \hat{n} = 8,\ r = 2.
$$

The general parametric solution of the singular eigenproblem for $\overline{A}$ is (cf. Example 11.5)

$$
z(\lambda) = c_1(\lambda) \begin{pmatrix} 0 \\ 0 \\ 1 \\ 0 \\ \lambda \\ 0 \\ 0 \\ \lambda^2 \end{pmatrix} + c_2(\lambda) \begin{pmatrix} 1 \\ \lambda \\ 0 \\ \lambda^2 \\ 0 \\ 1 - \lambda^2 \\ \lambda^3 \\ 0 \end{pmatrix}, \quad \text{with } k_1 = 2,\ k_2 = 3,\ \text{ so that } n_0 = 1.
$$

A particular regular eigensolution $\dot{z} = (1, -1, 1, 1, -1, 1, -1, 1)^T$ is obtained for $\dot{\lambda} = -1$; it is easily checked that it is not obtainable from $z(\lambda)$ for any choice of λ and the coefficients. Naturally, the addition of the general eigensolution $z(-1)$ with arbitrary c_1, c_2 leaves this property unchanged.

It is also easily checked that

$$
rk\,(\,z_{10}\ z_{11}\ z_{12}\ z_{20}\ z_{21}\ z_{22}\ z_{23}\ \dot{z}\,) = 8 \text{ and } rk\overline{I}\,(\,z_{10}\ z_{11}\ z_{20}\ z_{21}\ z_{22}\ \dot{z}\,) = 6,
$$

as claimed above. $\quad\square$

11.2.2 Algorithmic Determination of Parametric Eigensolutions

We want to determine $r = \hat{n} - \check{n}$ eigensolutions of the form (11.9) which satisfy (11.11). At first, we note that $\overline{I}z = \check{b} \in \mathbb{C}^{\check{n}}$ implies $z = \begin{pmatrix} \check{b} \\ \hat{b} \end{pmatrix}$, with $\hat{b} \in \mathbb{C}^r$ arbitrary. For $b \in \mathbb{C}^{\hat{n}}$, we introduce the notation $b =: \begin{pmatrix} \check{b} \\ \grave{b} \end{pmatrix}$, and accordingly $\overline{A} =: \begin{pmatrix} \check{A} & \grave{A} \end{pmatrix}$ for $\overline{A} \in \mathbb{C}^{\check{n} \times \hat{n}}$. Then,

we have $\overline{A} \begin{pmatrix} \check{b} \\ \grave{b} \end{pmatrix} = \check{A} \check{b} + \grave{A} \grave{b} \in \mathbb{C}^{\check{n}}$. Now, the bottom-up solution of (11.11) yields, with

unknown k and indeterminate $\grave{b}_\kappa \in \mathbb{C}^r$, $z_k = \begin{pmatrix} 0 \\ \grave{b}_k \end{pmatrix}$ and

$$
\begin{aligned}
\overline{A}\, z_k &= \grave{A}\, \grave{b}_k , & z_{k-1} &= \begin{pmatrix} \grave{A}\, \grave{b}_k \\ \grave{b}_{k-1} \end{pmatrix}, \\[2mm]
\overline{A}\, z_{k-1} &= \check{A}\grave{A}\, \grave{b}_k + \grave{A}\, \grave{b}_{k-1} , & z_{k-2} &= \begin{pmatrix} \check{A}\grave{A}\, \grave{b}_k + \grave{A}\, \grave{b}_{k-1} \\ \grave{b}_{k-2} \end{pmatrix}, \\[2mm]
& \quad \dots & & \quad \dots \\[2mm]
\overline{A}\, z_{k-\kappa} &= \check{A}^\kappa \grave{A}\, \grave{b}_k + \check{A}^{\kappa-1}\grave{A}\, \grave{b}_{k-1} + \dots + \grave{A}\, \grave{b}_{k-\kappa} .
\end{aligned}
$$
$$\tag{11.13}$$

Thus $\overline{A}\, z_{k-\kappa} = 0$ requires a nontrivial solution of

$$
\begin{pmatrix} \grave{A} & \check{A}\grave{A} & \dots & \check{A}^\kappa \grave{A} \end{pmatrix} \begin{pmatrix} \grave{b}_{k-\kappa} \\ \vdots \\ \grave{b}_k \end{pmatrix} = 0 \quad \text{or} \quad \mathrm{rk}\begin{pmatrix} \grave{A} & \check{A}\grave{A} & \dots & \check{A}^\kappa \grave{A} \end{pmatrix} < (\kappa + 1)\, r .
$$

As we look for r linearly independent sets of parameters $\grave{b}_k, \dots, \grave{b}_{k-\kappa}$, we form the matrices $\check{A}^{\kappa'}\grave{A}$ until

$$
\mathrm{rk}\begin{pmatrix} \grave{A} & \check{A}\grave{A} & \dots & \check{A}^\kappa \grave{A} \end{pmatrix} \le \kappa\, r ,
$$

which must happen, at the latest, for $\kappa \ge \check{n}/r$. The successive rank determination can easily be achieved by a successive triangular decomposition of the matrix as it grows by r columns in each step.

At each occasion where the rank increases by less than r upon an increase of κ, we may compute immediately the parameter set(s) made possible in order to obtain a $z_\rho(\lambda)$ of minimal degree. For example, if $\mathrm{rk}\,\grave{A} = r - r_0$, there are r_0 parameter vectors $\grave{b}_0^{(\rho)}$ such that $z_\rho(\lambda) = \begin{pmatrix} 0 \\ \grave{b}_0^{(\rho)} \end{pmatrix}$ satisfies $(\overline{A} - \lambda \overline{I})\, z_\rho = 0$ for all λ; these are the (potential) *constant* parametric solutions of (11.8), with $k_\rho = 0$. But in determining the z_ρ in order of increasing k_ρ, we have to keep the vectors $(\grave{b}_{k_\rho}^{(\rho)}, \grave{b}_{k_\rho-1}^{(\rho)}, \dots)^T$ *supplemented by zeros to equal length* linearly independent.

The determination of the z_ρ from the $\grave{b}_\kappa^{(\rho)}$ follows (11.13); with a numbering of the $\grave{b}_\kappa^{(\rho)}$ from 0 to k_ρ, we obtain

$$
z_\rho(\lambda) = \sum_\kappa^{k_\rho} \lambda^\kappa z_{\rho\kappa} , \quad \text{with } z_{\rho\kappa} = \begin{pmatrix} \sum_{\kappa'=0}^{k_\rho-\kappa-1} \check{A}^{\kappa'} \grave{A}\, \grave{b}_{\kappa+\kappa'+1}^{(\rho)} \\ \grave{b}_\kappa^{(\rho)} \end{pmatrix} . \tag{11.14}
$$

The matrices needed in the evaluation have been formed previously in the recursive build-up of the $\check{A}^{\kappa'}\grave{A}$.

Example 11.5: For the matrix $\overline{A}$ of Example 11.4, the matrix $(\grave{A}\ \check{A}\grave{A}\ \ldots)$ becomes

$$\begin{pmatrix} 0 & 0 & 0 & 0 & 1 & 0 & 0 & : \\ 0 & 0 & 1 & 0 & 0 & 0 & 0 & : \\ 0 & 0 & 0 & 1 & 0 & 0 & 0 & : \\ 1 & 0 & 0 & 0 & 0 & 0 & 0 & : \\ 0 & 1 & 0 & 0 & 0 & 0 & 0 & : \\ -1 & 0 & 0 & 0 & 1 & 0 & 0 & : \end{pmatrix}.$$

A first linear dependence appears for $\kappa_1 = 2$ (zero column 6); for $\kappa_2 = 3$, we have a further zero column 7. The associated parameter sets are (scaled) $b^{(1)} = ((0,0), (0,0), (0,1))$ and $b^{(2)} = ((0,0), (0,0), (0,0), (1,0))$. Note that the two final pairs which represent b_k for $k = 2$ and 3, resp., must span the $\mathbb{C}^r$.

When we evaluate (11.14) with the parameter sets $b^{(1)}$ and $b^{(2)}$, we obtain the parametric eigensolutions $z_1(\lambda)$ and $z_2(\lambda)$ specified in Example 11.4. □

11.2.3 Algorithmic Determination of Regular Eigensolutions

By Proposition 11.7, (11.8) has n_0 regular or point eigenvalues $\dot{\lambda}_\nu$, where $n_0 := \check{n} - \sum_\rho k_\rho$. After the determination of the parametric eigensolutions, the k_ρ are known. Reference [11.3] describes an algorithmic procedure which uses the z_ρ to transform (11.8) into a quadratic matrix eigenproblem of dimension n_0 which yields the regular eigenvalues and information for the determination of the associated eigenvectors. This approach has been implemented by Wu and his collaborators.

Independently of the determination of the parametric eigensolutions, one may use the definition of regular eigenvalues $\dot{\lambda}$ above Proposition 11.7: rk $(\overline{A} - \dot{\lambda}\,\overline{I}) < \check{n}$. It is well known that, for a quadratic matrix A, an LU-decomposition of $A - \lambda I$ with indeterminate λ yields the characteristic polynomial of A as the only element in the last row of U. For the rectangular matrix $\overline{A} - \lambda\overline{I}$, the last row of $\overline{U} \in \mathbb{C}^{\check{n} \times \hat{n}}$ has $r + 1$ elements which are polynomials in $\dot{\lambda}$. A regular eigenvalue $\dot{\lambda}$ must be a simultaneous zero of these polynomials.

Algorithmically, this approach is expensive for large, dense problems, but finite multiplication matrices $\overline{A}_\sigma$ are generally sparse and have many trivial rows, with a 1 not in the main diagonal. The LU-decomposition may use these 1's as pivots; this speeds up the decomposition considerably and yields polynomials in the last row of a degree much smaller than $\check{n}$.

A regular eigenvalue $\dot{\lambda}$ of $\overline{A}$ must also be an eigenvalue of the quadratic matrix $\check{A}$; this follows from an LU-decomposition of $\overline{A} - \lambda\,\overline{I}$ with diagonal pivots which yields the characteristic polynomial of $\check{A}$ as polynomial in the last row of $\overline{U}$. Hence, one can also find the eigenvalues λ_ν of $\check{A}$ and check the matrices $\overline{A} - \lambda_\nu\,\overline{I}$ for a rank deficiency.

For a simple regular eigenvalue $\dot{\lambda}$, with rk $(\overline{A} - \dot{\lambda}\,\overline{I}) = \check{n} - 1$, the kernel of $\overline{A} - \dot{\lambda}\,\overline{I}$ has dimension $r + 1$, where r of the spanning vectors are the $z_\rho(\dot{\lambda})$, $\rho = 1(1)r$. If the parametric solutions are known, a supplementary basis vector of ker $(\overline{A} - \dot{\lambda}\,\overline{I})$ can easily be found; otherwise a distinction is not possible.

Example 11.6: For $\overline{A}$ from Example 11.4, an LU-decomposition of $\overline{A} - \dot{\lambda}\,\overline{I}$ which uses the 1's in the 4 trivial rows as initial pivots and a $\dot{\lambda}$-free element in the 5th and last elimination step, generates the polynomials $(1 + \dot{\lambda})^2(1 - \dot{\lambda})$, $(1 + \dot{\lambda})$, 0 in the last row of $\overline{U}$. This yields the only regular eigenvalue $\lambda = -1$ of $\overline{A}$; cf. Example 11.4. Or we find the eigenvalues of $\check{A}$ as $-1, 0, 0, 0, 0, 0$; a rank reduction for $\overline{A} - \lambda\,\overline{I}$ occurs only for $\lambda = -1$.

With $\lambda = -1$, $\overline{A} - \lambda\,\overline{I}$ has the 3-dimensional kernel $(\gamma_1, -\gamma_1, \gamma_2, \gamma_1, -\gamma_2, \gamma_3, -\gamma_1, \gamma_2)^T$. With $\gamma_2 = 1$, $\gamma_1 = \gamma_3 = 0$, we have $z_1(-1)$, with $\gamma_1 = 1$, $\gamma_2 = \gamma_3 = 0$, $z_2(-1)$. Any choice of the γ_ρ which is linearly independent of these two yields a regular eigensolution $\check{z}$ which is not an evaluation at $\check{\lambda}$ of the general parametric eigensolution. For $\gamma_i = 1$, $i = 1(1)3$, we obtain the regular eigensolution $\check{z}$ specified in Example 11.4. $\quad\square$

Exercises

1. Determine the parametric eigensolutions of the finite multiplication matrices of Exercise 11.1-2.

2. Determine the parametric and regular eigensolutions of the finite multiplication matrices of Exercise 11.1-3; cf. Example 11.4.

11.3   Zero Sets from Finite Multiplication Matrices

We consider a positive-dimensional system $P \subset \mathcal{P}^s$, with the inner and outer normal sets $\check{\mathcal{N}}$ and $\hat{\mathcal{N}}$ and the associated finite multiplication matrices $\overline{A}_\sigma$ of the quotient ring $\mathcal{R}[\langle P \rangle]$. How can we use the solutions of the singular matrix eigenproblems (11.8) for the $\overline{A}_\sigma$ to determine the zero set of P?

11.3.1   One-Dimensional Zero Sets

By section 11.2, the singular eigenproblem (11.7) for, say, $\overline{A}_s$ has a parametric solution

$$z(x_s) = \sum_{\rho=1}^{r} c_\rho(x_s)\, z_\rho(x_s), \qquad c_\rho : \mathbb{C} \to \mathbb{C} \text{ arbitrary}, \qquad (11.15)$$

and, possibly, one or several regular eigenvalues $\dot{x}_s$ with eigenmanifolds

$$z = \dot{z} + \sum_{\rho=1}^{r} \dot{c}_\rho\, z_\rho(\dot{x}_s), \qquad \dot{c}_\rho \in \mathbb{C} \text{ arbitrary}. \qquad (11.16)$$

By Theorem 11.5, we must determine the functions c_ρ such that $z(x_s)$ can be interpreted as $\hat{\mathbf{b}}(x_1(x_s), x_2(x_s), \ldots, x_s) = (\hat{b}_\mu(x_1(x_s), x_2(x_s), \ldots, x_s), \mu = 1(1)\hat{n})^T$. Let $\hat{\mathbf{b}}(x) = (1, x_s, .., x_2, x_1, \ldots)^T$; then (11.8) requires that the first component of $z(x_s)$ is 1 and the 2nd component is x_s. The components 3 through $s + 1$ of $z(x_s)$ must be interpreted as $x_{s-1}(x_s), \ldots, x_1(x_s)$. This shows that the use of $\overline{A}_s$ can work only if the zero manifold(s) of P permit(s) a local parametrization by x_s; we assume that this is the case.

     With this interpretation of the first $s+1$ components of the general parametric eigensolution of $(\overline{A}_s - x_s\,\overline{I})\, z(x_s) = 0$, the further x^{j_μ}-components of $z(x_s)$ must equal $x_1(x_s)^{j_{\mu 1}} x_2(x_s)^{j_{\mu 2}} \ldots x_s^{j_{\mu s}}$. And, by (11.8), we must have

$$(\overline{A}_\sigma - x_\sigma(x_s)\,\overline{I})\, z(x_s) = 0, \quad \sigma = 1(1)s - 1. \qquad (11.17)$$

All these requirements are either trivially satisfied or they yield equations for the coefficient functions $c_\rho(x_s)$. By Theorem 11.5, if P has a zero manifold parametrizable by x_s, these equations have a solution; vice versa, each independent solution set $\{c_\rho(x_s)\}$ yields a local representation $M := \{(x_1(x_s), x_2(x_1), \ldots, x_s, x_s \in D \subset \mathbb{C}\}$ of a zero manifold of P. If there is no such zero manifold, there cannot be a solution $\{c_\rho(x_s)\}$ and vice versa.

Analogously, we can determine the coefficients $c_\rho \in \mathbb{C}$ in a regular eigensolution (11.16) to obtain potential isolated solutions of P.

Example 11.7: We take

$$P = \begin{cases} p_1(x_1, x_2) & = & x_1^3 + x_1 x_2^2 - x_1^2 - x_2^2 - x_1 + 1, \\ p_2(x_1, x_2) & = & x_1^2 x_2 + x_2^3 + x_1^2 + x_2^2 - x_2 - 1; \end{cases}$$

cf. Exercise 11.1-3 and Example 11.4 for $\check{N}, \hat{N}$, the $\overline{A}_\sigma$, and the solution of the singular eigenproblem for $\overline{A}_2$. The requirements on the 1st and 2nd component of

$$z(x_2) = c_1(x_2) \begin{pmatrix} 0 \\ 0 \\ 1 \\ 0 \\ x_2 \\ 0 \\ 0 \\ x_2^2 \end{pmatrix} + c_2(x_2) \begin{pmatrix} 1 \\ x_2 \\ 0 \\ x_2^2 \\ 0 \\ 1 - x_2^2 \\ x_2^3 \\ 0 \end{pmatrix}$$

yield $c_2(x_2) \equiv 1$ immediately; from the 3rd component, we obtain $x_1(x_2) = c_1(x_2)$. With these relations, the internal consistency condition for the 6th component x_2^2 requires $1 - x_2^2 = x_1(x_2)^2$, or $x_1(x_2) = \pm\sqrt{1 - x_2^2}$. The remaining consistency conditions are now automatically satisfied. Also, $(\overline{A}_1 - x_1(x_2)\,\overline{I})\,z(x_2)$ vanishes as is easily checked. Thus, we have established $x_1 = \pm\sqrt{1 - x_1^2}$ or $x_1^2 + x_2^2 - 1 = 0$ as a zero manifold of P. Note that $c_1(x_2) = x_1(x_2) = \pm\sqrt{1 - x_2^2}$; thus the coefficient functions c_ρ need not be polynomial.

For the regular eigenmanifold

$$z(\dot{x}_2) = (1 + \dot{c}_2, -1 - \dot{c}_2, 1 + \dot{c}_1, 1 + \dot{c}_2, -1 - \dot{c}_1, 1, -1 - \dot{c}_2, 1 + \dot{c}_1)^T$$

of Example 11.4, the 1st component yields $\dot{c}_2 = 0$ and the 2nd one $x_2 = -1$. The 6th component yields $x_1^2 = 1$ which would permit $x_1 = \pm 1$, with $\dot{c}_1 = 0$ or -2, resp.; but $x_1 = -1$ does not generate an eigenvector of $\overline{A}_1$ for the eigenvalue -1. $\square$

Admittedly, this example is very simple-minded and any computer algebra system will factor $p_1 = (x_1 - 1)(x_1^2 + x_2^2 - 1)$, $p_2 = (x_2 + 1)(x_1^2 + x_2^2 - 1)$. However, the above resolution could have been performed fully automatically, without the intricacies of a factorization algorithm. We display another less transparent example which emphasizes the same aspect.

Example 11.8: We consider the symmetric system `cyclic4` which is known to be singular:

$$P = \begin{cases} p_1(x_1, x_2, x_3, x_4) & = & x_1 + x_2 + x_3 + x_4, \\ p_2(x_1, x_2, x_3, x_4) & = & x_1 x_2 + x_2 x_3 + x_3 x_4 + x_4 x_1, \\ p_3(x_1, x_2, x_3, x_4) & = & x_1 x_2 x_3 + x_2 x_3 x_4 + x_3 x_4 x_1 + x_4 x_1 x_2, \\ p_4(x_1, x_2, x_3, x_4) & = & x_1 x_2 x_3 x_4 - 1. \end{cases}$$

We find a normal set from the `tdeg(x1,x2,x3,x4)` Groebner basis with leading monomials x_1, x_2^2, $x_2x_3^2$, $x_2x_4^4$, $x_3^3x_4^2$, $x_3^2x_4^4$, which implies (cf. (11.1)) $\nu_1 = 1$, $\nu_2 = 2$, $\nu_3 = 3$, $\nu_4 = 4$. The correspondingly restricted normal set has to be augmented by x_4^5 which appears in one of the basis elements. The outer normal set $\hat{\mathcal{N}}$ has 23 elements; the 4 elements which border the infinite sections of $\mathcal{N}$ are removed to form the 19 element inner normal set $\check{\mathcal{N}}$. Thus the finite multiplication matrices are 19 by 23.

We work with $\overline{A}_4$ because it has only 3 nontrivial rows. Since $r = 4$, the general parametric eigensolution of $(\overline{A}_4 - x_4\,\overline{I})\,z(x_4) = 0$ consists of 4 components according to Proposition 11.6. Since the columns no.20 and 21 of $\overline{A}_4$ are zero columns, there are two parametric components $z_\rho(x_4)$ with $k_\rho = 0$ and only one 1 in element 20 or 21, respectively. The other two z_ρ are obtained by the algorithm in section 11.2.2; they have degrees 4 and 7, respectively:

$$z_3(x_4) = (0, 0, 1, 0, 0, x_4, 0, 0, -x_4, 0, x_4^2, 0, 0, -x_4^2, 0, x_4^3, 0, 0, 0, 0, 0, 0, x_4^4)^T$$

$$z_4(x_4) = (x_4^2, x_4^3, 0, -x_4^3, x_4^4, 0, -x_4^4, 1, 0, x_4^5, 0, -x_4^5, x_4, 0, x_4^6, 0, -x_4^6, x_4^2, x_4^3, 0, 0, x_4^7, 0)^T.$$

Since we have ordered $\hat{\mathbf{b}}$ by increasing term order, the first 4 elements of $\hat{\mathbf{b}}(x)$ are 1, x_4, x_3, x_2; note that x_1 is not in the normal set. From the internal consistency of z_4, we obtain $c_4(x_4) = 1/x_4^2$ and $x_2(x_4) = -x_4$. From z_3, we have $x_3(x_4) = c_3$ and from the 8th element (with $b_8 = x_3^2$) of z_4 we obtain $x_3^2 = c_3^2 = c_4 \cdot 1 = 1/x_4^2$. Thus there are two solutions for c_3, viz. $c_3(x_4) = \pm 1/x_4$. From the first row of $\overline{A}_1$ (or directly from p_1), we obtain $x_1(x_4) = -x_3(x_4) = \mp 1/x_4$ and we have generated the well-known two solution manifolds of P

$$(-1/x_4, \ -x_4, \ 1/x_4, \ x_4) \quad \text{and} \quad (1/x_4, \ -x_4, \ -1/x_4, \ x_4), \quad \text{with } x_4 \in \mathbb{C} \setminus \{0\}.$$

By Proposition 11.7, we have $n_0 = 19 - 11 = 8$ so that we have to look for 8 regular eigensolutions. According to section 11.2.3, we may compute the eigenvalues of $\check{A}_3$ to obtain candidates for regular eigenvalues. This yields $1, -1, i, -i, 0$. There are indeed 4 different regular eigenvectors of $\overline{A}_4$ for $\lambda = 0$, but $x_4 = 0$ is the only forbidden value. The other 4 values may be confirmed as regular eigenvalues of $\overline{A}_4$, with a rank deficiency 2 for each; the 8 associated isolated solutions of P coincide with the points on the two manifolds with $x_4 = 1, -1, i, -i$. Thus we have recognized these points on the manifolds as *2-fold d-points*; cf. section 9.4.4. Actually, when we perturb the system P by, say, small constants, the manifolds disappear, but each of the eight 2-fold d-points splits into *two* isolated zeros of $\tilde{P}$ which has a total of 16 zeros. $\square$

The determination of the vectors z_ρ in (11.15) is exclusively a matter of linear algebra and can always be performed. The nonlinear part of the determination of a parametric representation of a positive-dimensional zero manifold of a polynomial system is contained in the determination of the *coefficient functions* $c_\rho(x_s)$ in (11.15). Obviously, the internal consistency conditions for (11.15) and the requirements (11.17) lead to a *polynomial* system in the c_ρ, with x_s as a parameter. Therefore, in sufficiently simple cases, we may be able to find *explicit* solutions $c_1(x_s), \ldots, c_r(x_s)$. If this is not the case, we may try to use a different parameter variable, or to substitute a suitable function of an independent parameter for x_s. It is not clear at this point which difficulties may arise and under what conditions they can or cannot be overcome.

11.3.2   Multi-Dimensional Zero Sets

In the approach of section 11.3.1 for the determination of the coefficient functions c_ρ, $\rho = 1(1)r$, in (11.15), it may happen that the system of conditions on the c_ρ functions has a solution manifold. We may then be able to introduce a further parameter (besides x_s) into the c_ρ so that (11.15) represents a two-dimensional zero manifold of P. That parameter may be directly related to one of the other variables x_σ. In principle, we can also assume higher-dimensional solutions of the c_ρ system and correspondingly many additional parameters in $z(x_s)$. However, in all but simple situations, we must anticipate severe difficulties in finding *explicit* solutions. Therefore, we restrict ourselves to the display of examples.

Example 11.9: We consider the polynomial system $P(c)$ of (9.60) with $c = 1$; cf. Exercise 11.1-1. We have the normal sets $\mathcal{N} = \{1, x_3, x_2, x_1\}$ and $\hat{\mathcal{N}} = \{\ldots, x_3^2, x_2 x_3, x_2^2\}$ and the finite multiplication matrices

$$\overline{A}_2 = \begin{pmatrix} 0 & 0 & 1 & 0 & 0 & 0 & 0 \\ 0 & 0 & 0 & 0 & 0 & 1 & 0 \\ 0 & 0 & 0 & 0 & 0 & 0 & 1 \\ 0 & -1 & 1 & 1 & 0 & 1 & -1 \end{pmatrix}, \quad \overline{A}_3 = \begin{pmatrix} 0 & 1 & 0 & 0 & 0 & 0 & 0 \\ 0 & 0 & 0 & 0 & 1 & 0 & 0 \\ 0 & 0 & 0 & 0 & 0 & 1 & 0 \\ 0 & -1 & 1 & 1 & 1 & -1 & 0 \end{pmatrix},$$

and a slightly less trivial $\overline{A}_1$. The general parametric solution of the singular eigenproblem for $\overline{A}_3$ is

$$z(x_3) = c_1(x_3) \begin{pmatrix} 0 \\ 0 \\ 0 \\ 0 \\ 0 \\ 0 \\ 1 \end{pmatrix} + c_2(x_3) \begin{pmatrix} 0 \\ 0 \\ 1 \\ -1 \\ 0 \\ x_3 \\ 0 \end{pmatrix} + c_3(x_3) \begin{pmatrix} 1 \\ x_3 \\ 0 \\ x_3 \\ x_3^2 \\ 0 \\ 0 \end{pmatrix},$$

with $n_0 = 4 - (0 + 1 + 2) = 1$; thus there must also be one regular solution z_0, which is easily seen as $(1, 1, 1, 1, 1, 1, 1)^T$, or $x_1 = x_2 = x_3 = 1$.

The internal consistency of $z(x_3)$ with $\hat{\mathbf{b}}$ requires $c_3 = 1$, $c_2 = x_2$, $-c_2 + x_3 = x_1$, and, from the 7th component of $z(x_1)$, $c_1 = c_2^2$. With an *arbitrary* c_2 and c_1 from the last equation, the two conditions (11.17) are satisfied. Thus we have $c_2 = x_2$ as a second parameter and $x_1 = x_3 - x_2$, $x_2, x_3 \in \mathbb{C}$, as the parametric representation of the two-dimensional solution manifold; cf. Example 11.1. $\square$

11.3.3   Direct Computation of Two-Dimensional Eigensolutions

When we know the existence of a two-dimensional zero manifold and potential parameters (say x_1, x_2) from a Groebner basis of P, we may attempt to find directly a simultaneous solution $z(x_1, x_2)$ of

$$(\overline{A}_1 - x_1 \overline{I}) z(x_1, x_2) = 0 \quad \text{and} \quad (\overline{A}_2 - x_2 \overline{I}) z(x_1, x_2) = 0. \tag{11.18}$$

In analogy to the approach in Proposition 11.6, we look for particular solutions

$$z_\rho(x_1, x_2) = \sum_{\kappa_1, \kappa_2 = 0}^{\kappa_1 + \kappa_2 = k} x_1^{\kappa_1} x_2^{\kappa_2} z_{\rho \kappa_1 \kappa_2}, \quad \text{with } z_{\rho \kappa_1 \kappa_2} \in \mathbb{C}^{\hat{n}}, \tag{11.19}$$

to form the general parametric solution of (11.18) as

$$z(x_1, x_2) = \sum_{\rho=1}^{r_{12}} c_\rho(x_1, x_2)\, z_\rho(x_1, x_2)\,. \tag{11.20}$$

If (11.18) is to have a nontrivial solution for arbitrary x_1, x_2, the $2\check{n} \times \hat{n}$ matrix

$$\overline{A}_{12}(x_1, x_1) := \begin{pmatrix} \overline{A}_1 - x_1 \overline{I} \\ \overline{A}_2 - x_2 \overline{I} \end{pmatrix}$$

must have a rank below $\hat{n}$ uniformly in x_1, x_2; then $r_{12} = \hat{n} - \mathrm{rk}\,\overline{A}_{12}$. Generally, r_{12} will be smaller than the r for either $\overline{A}_1$ or $\overline{A}_2$; this reduces the number of coefficient functions to be determined. Also, if our assumption about the solution manifold(s) has been correct, the system for the coefficient functions can have isolated solutions only.

In place of the linear system (11.11), we now have the double recursion (we have dropped the subscript ρ from (11.19))

$$\begin{array}{rclcl}
\overline{A}_1\, z_{00} &=& \overline{A}_2\, z_{00} &=& 0\,, \\
\overline{A}_1\, z_{10} - \overline{I}\, z_{00} &=& \overline{A}_2\, z_{10} &=& 0\,, \\
\overline{A}_1\, z_{01} &=& \overline{A}_2\, z_{01} - \overline{I}\, z_{00} &=& 0\,, \\[1.2ex]
\overline{A}_1\, z_{20} - \overline{I}\, z_{10} &=& \overline{A}_2\, z_{20} &=& 0\,, \\
\overline{A}_1\, z_{11} - \overline{I}\, z_{01} &=& \overline{A}_2\, z_{11} - \overline{I}\, z_{10} &=& 0\,, \\
\overline{A}_1\, z_{02} &=& \overline{A}_2\, z_{02} - \overline{I}\, z_{01} &=& 0\,,
\end{array} \tag{11.21}$$

etc.

The recursion ends when all $z_{\kappa, k-\kappa}$, $\kappa = 0(1)k$, are in $\ker \overline{I}$. Again it can be seen that this must happen for some finite k.

Actually, for an algorithmic solution, one can proceed like in section 11.2.2 and start with vectors from $\ker \overline{I}$ for the $z_{\kappa_1 \kappa_2}$, $\kappa_1 + \kappa_2 = k$, with an initially unknown k. However, since we must also satisfy (see (11.21))

$$\overline{A}_1\, z_{\kappa, k-\kappa} = \overline{I}\, z_{\kappa-1, k-\kappa} = \overline{A}_2\, z_{\kappa-1, k-\kappa+1}\,,$$

these starting vectors must satisfy

$$\begin{array}{rclcl}
\overline{A}_1\, z_{0k} & & &=& 0\,, \\
\overline{A}_1\, z_{\kappa, k-\kappa} \;-\; \overline{A}_2\, z_{\kappa-1, k-\kappa+1} &=& 0\,, & \kappa = 1(1)k \\
\;-\; \overline{A}_2\, z_{k0} &=& 0\,.
\end{array} \tag{11.22}$$

The rank of this system for $z_{\kappa, k-\kappa} \in \ker \overline{I}$ also determines the appropriate value of k. We skip further details and turn immediately to an example:

Example 11.10: In the situation of Example 11.9, we know the existence of a 2-dimensional zero set. This is confirmed by the fact that $\mathrm{rk}\,\overline{A}_{12}(x_1, x_2) = 6 < \hat{n} = 7$ uniformly in x_1, x_2; thus, $r_{12} = 1$ and we expect *one* $z_\rho(x_1, x_2)$ only.

When we set our starting vectors $z_{\kappa,k-\kappa} = \begin{pmatrix} 0 \\ I \end{pmatrix} \cdot \alpha_\kappa$, $\alpha_\kappa \in \mathbb{C}^3$, $\kappa = 0(1)k$, and test the homogeneous system (11.22) for nontrivial solutions α_κ, we find that $k = 2$ yields a unique starting set $z_{\kappa,2-\kappa}$, $\kappa = 0(1)2$.

The bottom to top recursion, in analogy to section 11.2.2, yields

$$
z(x_1, x_2) = c \left(
\begin{pmatrix} 1 \\ 0 \\ 0 \\ 0 \\ 0 \\ 0 \\ 0 \end{pmatrix}
+ x_1 \begin{pmatrix} 0 \\ 1 \\ 0 \\ 1 \\ 0 \\ 0 \\ 0 \end{pmatrix}
+ x_2 \begin{pmatrix} 0 \\ 1 \\ 1 \\ 0 \\ 0 \\ 0 \\ 0 \end{pmatrix}
+ x_1^2 \begin{pmatrix} 0 \\ 0 \\ 0 \\ 0 \\ 1 \\ 0 \\ 0 \end{pmatrix}
+ x_1 x_2 \begin{pmatrix} 0 \\ 0 \\ 0 \\ 0 \\ 0 \\ 2 \\ 1 \\ 0 \end{pmatrix}
+ x_2^2 \begin{pmatrix} 0 \\ 0 \\ 0 \\ 1 \\ 1 \\ 1 \\ 1 \end{pmatrix}
\right)
$$

$$
= c \begin{pmatrix} 1 \\ x_1 + x_2 \\ x_2 \\ x_1 \\ (x_1 + x_2)^2 \\ x_2(x_1 + x_2) \\ x_2^2 \end{pmatrix}.
$$

With the obvious $c = 1$, the coefficient determination has become trivial; the internal consistency of $z(x_1, x_2)$ as well as the satisfaction of $(\overline{A}_3 - x_3(x_1, x_2)\,\overline{I})\, z(x_1, x_2)$ is immediately verified. $\square$

Exercises

1. By the approach in section 11.3.1, find a representation of the 1-dimensional zero manifold of $P(a_0)$ in Example 11.2; cf. Exercises 11.1-2 and 11.2-1. Why do you get two separate parameter representations although there is only 1 (real) manifold? Try to visualize the manifold by plots.

2. Perform the detailed computations in Example 11.10.

3. Consider the dense quadratic polynomial $p(x, y, z) = \sum_{|j|\leq 2} \alpha_{j_1 j_2 j_3}\, x^{j_1} y^{j_2} z^{j_3} \in \mathcal{P}^3$.

   (a) For $\alpha_{002} \neq 0$, solve $p = 0$ for z to obtain two local parametric representations $z_\rho(x, y)$, $\rho = 1, 2$, of the quadratic manifold $p = 0$.

   (b) Find this representation by the approaches of sections 11.3.2 and 11.3.3.

   (c) From the expression for $z_\rho(x, y)$, find numerical coefficient sets which lead to special situations. How do these affect the procedures in (b) ?

11.4 A Quasi-0-Dimensional Approach

The following concept is well known in polynomial algebra; cf., e.g., [2.13].

Definition 11.3. Consider a term order on T^s and an ideal $\mathcal{I} \subset \mathcal{P}^s$. A subset X_0 of $\{x_1, \ldots, x_s\}$ is a *maximal independent subset* mod $\mathcal{I}$ with respect to the term order if no monomial over

X_0 is a leading monomial of a polynomial in $\mathcal{I}$ and if there are no such subsets of a larger cardinality. $\square$

Proposition 11.8. The cardinality d of a maximal independent subset mod $\mathcal{I} \subset \mathcal{P}^s$ does not depend on the term order; it equals the dimension of $\mathcal{I}$.

In this section, we will treat the variables in a maximal independent subset mod $\langle P \rangle$ as *parameters* in the 0-dimensional system in the remaining variables.

11.4.1 Quotient Rings with Parameters

We consider a positive-dimensional system $P \subset \mathcal{P}^s$, $s \geq 2$, and assume that the variables have been numbered such that the set $X_0 = \{x_{\bar{s}+1}, \ldots, x_s\} \neq \emptyset$, $1 \leq \bar{s} < s$, constitutes a maximal independent set of $s_0 := s - \bar{s}$ variables. Then we consider P as a polynomial $\bar{P}$ in the variables from $\bar{X} = \{x_1, \ldots, x_{\bar{s}}\}$ and treat the $x_\sigma \in X_0$ as parameters.

By Proposition 11.8, $\langle \bar{P} \rangle \subset \mathcal{P}^{\bar{s}}$ is a 0-dimensional ideal so that a normal set basis $\bar{\mathcal{N}} \subset T^{\bar{s}}$ of the quotient ring $\mathcal{R}[\langle \bar{P} \rangle] = \mathcal{P}^{\bar{s}}/\langle \bar{P} \rangle$ is finite, with $|\bar{\mathcal{N}}| =: \bar{m}$. Let $\bar{\mathbf{b}} \in (T^{\bar{s}})^{\bar{m}}$ be the associated basis vector, with first component 1 and the $x_\sigma \in \bar{X}$ in the following components. The multiplicative structure of $\mathcal{R}[\langle \bar{P} \rangle]$ may be represented, with reference to $\bar{\mathbf{b}}$, by $\bar{m} \times \bar{m}$ multiplication matrices $\bar{A}_\sigma$, $\sigma = 1(1)\bar{s}$, as usual. Naturally, the elements of the nontrivial rows of the $\bar{A}_\sigma$ are polynomials in the $x_\sigma \in X_0$. Thus we are faced with matrix eigenproblems of the form

$$(\bar{A}_\sigma(x_{\bar{s}+1}, \ldots, x_s) - \lambda(x_{\bar{s}+1}, \ldots, x_s)\, I)\, z(x_{\bar{s}+1}, \ldots, x_s) = 0, \quad \sigma = 1(1)\bar{s}. \tag{11.23}$$

If we are able to determine a joint eigenvector $z \in (\mathcal{P}^{s_0})^{\bar{m}}$ of the A_σ, $\sigma = 1(1)\bar{s}$, with a first component 1, then the further components of z must represent a zero $(\zeta_1, \ldots, \zeta_{\bar{s}})$ of $\bar{P}$; cf. section 2.4.

Generally, some or all of the components of z (except the first one, after normalization) depend explicitly on some or all of the $x_\sigma \in X_0$. Thus, the components of z for the $x_\sigma \in \bar{X}$ contain a parametric representation of an s_0-dimensional zero manifold of P in terms of the parameters $x_{\bar{s}}, \ldots, x_s$. This holds also when none of the z-components depends on some particular $x_\sigma \in X_0$; then $(\zeta_1, \ldots, \zeta_{\bar{s}})$ is a zero of $\langle \bar{P} \rangle$ for arbitrary values of that parameter and the zero manifold of P extends parallel to the subspace of that x_σ.

Beyond such vectors which are joint eigenvectors of the $\bar{A}_\sigma$ uniformly in the parameter variables, there may be further eigenvectors of the individual $\bar{A}_\sigma$ which become *joint* eigenvectors only upon the restriction of the s_0 parameter variables $x_\sigma \in X_0$ to a manifold M' of a dimension $m' < s_0$ in the $\mathbb{C}^{s_0}$. The case $m' = 0$, with *fixed values* for each parameter variable and hence also for the $x_\sigma \in \bar{X}$, yields the potential isolated point solutions of the system P. Cases with $m' > 0$ generate m'-dimensional zero manifolds of P. However, in the computation of a Groebner basis for $\bar{P}$, ordinary Groebner basis software will cancel common factors which are polynomials in the indeterminate parameter variables from $\bar{X}_0$ only. Thus, we may lose solutions of P which depend on the vanishing of such factors; cf. the two examples below. We will indicate a way to avoid this in the following section.

All this may work well in simple examples, with integer coefficients and other nice features; it may appear more attractive than the approach via singular matrix eigenproblems. But we must be aware of the fact that an explicit solution of the parametric eigenproblems (11.23) is, generally, beyond reach for polynomial systems P related to some real-world phenomenon. A potential successful utilization of the above approach will often have to employ clever tricks valid only in the presence of some special circumstances. Again we restrict ourselves to the presentation of examples.

Example 11.11: Again, we consider the polynomial system $P(c)$ of (9.60), at first with $c = 2$ (two 1-dimensional zero manifolds) and then with $c = 1$ (2-dimensional zero manifold); cf. previous examples.

The `tdeg(x1,x2,x3)` Groebner basis for $P(2)$ has the leading monomials x_1^2, $x_1 x_2$, x_2^3, $x_1 x_3$, $x_2^2 x_3$. By Definition 11.3, this marks $X_0 = \{x_3\}$ as a maximal independent set of variables, so that $\bar{X} = \{x_1, x_2\}$, $\bar{s} = 2$, $s_0 = 1$. The `tdeg(x1,x2)` Groebner basis for $\bar{P}(x_1, x_2)$, with x_3 as a parameter, has leading monomials x_1, x_2^2; for $\mathbf{b} = (1, x_2)^T$, we obtain

$$\bar{A}_1 = \begin{pmatrix} x_3 & -1 \\ 0 & x_3 - 1 \end{pmatrix}, \quad \bar{A}_2 = \begin{pmatrix} 0 & 1 \\ 0 & 1 \end{pmatrix}.$$

Both $\bar{A}_\sigma$ have the eigenvectors $(1, 0)^T$ and $(1, 1)^T$ which implies $x_2 = 0$ and $x_2 = 1$, resp., along the zero manifolds. The associated parameter representations $x_1 = x_3$ and $x_1 = x_3 - 1$, resp., are obtained from the Groebner basis element $x_1 + x_2 - x_3$.

The isolated zero $(1,1,1)$ of $P(2)$ which is not on either manifold has been lost; this occurs because a common factor $x_3 - 1$ has been cancelled during the Groebner basis computation.

For $P(1)$, the maximal independent set is $X_0 = \{x_2, x_3\}$. When we consider x_1 as the only variable, the ideal $\langle \bar{P}(1) \rangle$ becomes simply $\langle x_1 + x_2 - x_3 \rangle$; this displays the 2-dimensional zero manifold of $P(1)$ but has lost the isolated zero. $\square$

Example 11.12: We consider the system $P(a_0)$ of Example 11.2. The leading monomials of the `tdeg(x1,x2,x3)` Groebner basis admit each set $\{x_\sigma\}$ as X_0. When we use our approach with $X_0 = \{x_3\}$ and form the Groebner basis of $\langle P(a_0) \rangle$ for `tdeg(x1,x2)`, we meet a surprise: The Groebner basis is simply

$$(3 x_3^2 + 1) x_1 + (3 x_3^2 + 1) x_2 + 4 x_3, \quad (3 x_3^2 + 1) x_2^2 + 4 x_3 x_2 + (x_3^2 - 1),$$

with the normal set $\bar{\mathcal{N}} = \{1, x_2\}$. Thus, from the quadratic equation in x_2, we find directly the parameter respresentation

$$x_1 = \frac{-2 x_3 \pm \sqrt{6 x_3^2 - 3 x_3^4 + 1}}{3 x_3^2 + 1}, \quad x_2 = \frac{-2 x_3 \mp \sqrt{6 x_3^2 - 3 x_3^4 + 1}}{3 x_3^2 + 1}, \quad x_3 \in \mathbb{C} \setminus \{\pm \frac{i}{\sqrt{3}}\}.$$

For $x_3 \in \mathbb{R}$, the two branches of the zero manifold meet at $\pm(\frac{1}{3}\sqrt{9 + 6\sqrt{3}}, \frac{1}{3}\sqrt{9 + 6\sqrt{3}}, -\frac{1}{3}\sqrt{9 + 6\sqrt{3}})$ so that there is a closed loop curve in $\mathbb{R}^3$.

The 4 isolated zeros (t, t, t), with $t = \pm\frac{1}{3}\sqrt{-9 \pm 6\sqrt{3}}$, are lost; cf. Exercise 11.1-2. $\square$

11.4.2 A Modified Approach

According to an idea by Wu and collaborators, the loss of lower-dimensional zero set components may be avoided when we stick to the quotient ring $\mathcal{R}[\langle P \rangle]$ but express its multiplicative structure in terms of a reduced normal set. This set $\mathcal{N}_0$ is the *restriction* of the infinite normal set of $\langle P \rangle$ to the $\bar{X}$ subspace and hence finite ($|\mathcal{N}_0| =: m_0$). Let $\mathbf{b}_0(x_1, \ldots, x_{\bar{s}})$ be the associated normal set vector. In the $m_0 \times m_0$ multiplication matrices $A_{0\sigma}$, $\sigma = 1(1)\bar{s}$, with reference to $\mathbf{b}_0$, the $x_\sigma \in X_0$ are treated as indeterminates as previously.

However, these $A_{0\sigma}$ utilize only the information from the Groebner basis elements whose leading monomials do not contain any variables from X_0. Assume that some further basis elements contain a variable $x_\sigma \in X_0$ linearly. Then the representation of multiplication by this x_σ with reference to $\mathbf{b}_0$ will employ the respective Groebner basis elements. Now, we may proceed as described in the previous section, with the essential difference that "joint eigenvectors" refers also to these further multiplication matrices.

When there are Groebner basis elements with leading monomials not linear in the X_0-variables, one may also have to represent multiplication by polynomials which involve all the monomials of X_0-variables which appear as factors in leading monomials of the Groebner basis of $\langle P \rangle$.

Example 11.11, continued: When we use the original Groebner basis of $P(2)$ to express the multiplicative structure in terms of $\mathbf{b}_0 = (1, x_2, x_1, x_2^2)^T$, we obtain

$$
A_{01} = \begin{pmatrix} 0 & 0 & 1 & 0 \\ -x_3 & x_3 + 2 & 1 & -2 \\ (x_3 - 1)\,x_3 & -2\,x_3 & 1 & 2 \\ -x_3 & 1 + x_3 & 1 & -1 \end{pmatrix}, \quad
A_{02} = \begin{pmatrix} 0 & 1 & 0 & 0 \\ 0 & 0 & 0 & 1 \\ x_3 & x_3 + 2 & 1 & -2 \\ 0 & 0 & 0 & 1 \end{pmatrix},
$$

$$
A_{03} = \begin{pmatrix} x_3 & 0 & 0 & 0 \\ 0 & x_3 & 0 & 0 \\ (x_3 - 1)\,x_3 & 1 - x_3 & 1 & 0 \\ 0 & x_3 - 1 & 0 & 1 \end{pmatrix}.
$$

A_{01} has eigenvectors $(1, 0, x_3, 0)^T$, $(1, 1, x_3 - 1, 1)^T$, $(1, 1, 2 - x_3, 1)^T$ and a further one. The first two eigenvectors are immediately verified as eigenvectors of A_{02} and A_{03}; this yields the zero manifolds $(x_3, 0, x_3)$ and $(x_3 - 1, 1, x_3)$. The third eigenvector which does not represent a zero manifold actually agrees also with A_{02} but not with A_{03}; this emphasizes the importance of not dropping the $A_{0\sigma}$ for $\sigma > \bar{s}$. But when we substitute $x_3 = 1$, this eigenvector becomes $(1, 1, 1, 1)^T$ and is consistent with A_{03}; this yields the isolated zero $(1,1,1)$ of P.

For $P(1)$, the multiplicative structure with reference to $\mathbf{b}_0 = (1, x_1)^T$ is represented by

$$
A_{01} = \begin{pmatrix} 0 & 1 \\ x_2 - x_3 & 1 - x_2 + x_3 \end{pmatrix}, \quad
A_{02} = \begin{pmatrix} x_2 & 0 \\ (x_2 - 1)(x_3 - x_2) & 1 \end{pmatrix},
$$

$$
A_{03} = \begin{pmatrix} x_3 & 0 \\ (x_3 - 1)(x_3 - x_2) & 1 \end{pmatrix}.
$$

A_{01} has the eigenvectors $(1, x_3 - x_2)^T$ and $(1, 1)^T$. The first one is also an eigenvector of A_{02} and A_{03} for the zero manifold $x_1 = x_3 - x_2$. The second one agrees with the other two $A_{0\sigma}$ only for $x_2 = x_3 = 1$, i.e. for the isolated zero $(1,1,1)$. $\square$

Exercises

1. The two 2-dimensional zero manifolds of

$$P = \{\, x_1 + 2x_2 + x_3 - (x_4 + x_5)^2,\ x_1 - 2x_2 + x_3 - (x_4 - x_5)^2,\ x_1 x_3 - x_2^2 \,\} \subset \mathcal{P}^5$$

may immediately be parametrized in terms of x_4, x_5. However, the `tdeg(x1,x2,x3,x4,x5)` Groebner basis suggest $X_0 = \{x_1, x_3\}$. Use the approach of section 11.4.1 to find the parametrization of the zero sets in terms of x_1, x_3.

Historical and Bibliographical Notes 11

When a system P of $n < s$ polynomials in $\mathcal{P}^s$ is a complete intersection, its zero set $Z[\langle P \rangle]$ contains at least one manifold of dimension $s - n$; there may or may not be further components of a smaller dimension including isolated zeros. In this situation, it is not a priori clear what would constitute a "better" description of the complete zero set. Supposedly, the only generally desirable "result" is a separate description of the individual components of $Z[\langle P \rangle]$; a special case has been considered in section 7.4. Another case of importance is the identification of potential isolated zeros; cf. section 11.2.3.

When Wu Wenda and his coworkers had realized how an extension of the quotient ring centered approach to positive-dimensional systems was possible ([11.1]), we were quite enthusiastic about the potential consequences; a first report [11.2] from 1993 illustrates that. Later, our expectations became more modest. Also, the fact that the resonance of our research remained small, prevented a meaningful evaluation of the potential values of our approach.

Since this Chapter 11 contains only some glimpses on the subject of positive-dimensional systems which are related to our approach for 0-dimensional systems, we refrain from indicating a historical perspective. Some further results of our research which have not been included here concern the computation of *normal forms* modulo a positive-dimensional $\langle P \rangle$, with respect to suitably selected normal sets. May others continue and expand our work if they find it interesting.

References

[11.1] Y.Zh. Huang, W.D. Wu: A Modified Version of an Algorithm for Solving Multivariate Polynomial Systems, Academia Sinica MM Research Preprint no. 5 (1990), 23–29.

[11.2] H.J. Stetter: Multivariate Polynomial Equations as Matrix Eigenproblems, in: Contributions In Numerical Mathematics, World Sci. Ser. Appl. Anal., 2, World Sci. Publishing, River Edge, NJ, 1993, 355–371.

[11.3] F.R. Gantmacher: The Theory of Matrices, vol. 2, Chelsea Publ. Co., New York, 1989. (Translation of the Russian edition of 1959.)

Index